Wolfgang P. Schleich
Quantum Optics in Phase Space

Wolfgang P. Schleich

Quantum Optics in Phase Space

Berlin · Weinheim · New York · Chichester · Brisbane · Singapore · Toronto

Author:
Prof. Dr. Wolfgang P. Schleich, Abteilung für Quantenphysik, Universität Ulm, Germany
e-mail: Wolfgang.Schleich@physik.uni-ulm.de

1st edition

Library of Congress Card No: applied for

British Library Cataloguing-in-Publication Data: A catalogue record for this book is available from the British Library.

Die Deutsche Bibliothek – CIP Cataloguing-in-Publication-Data
A catalogue record for this publication is available from Die Deutsche Bibliothek

> This book was carefully produced. Nevertheless, authors, editors, and publishers do not warrant the information contained therein to be free of errors. Readers are advised to keep in mind that statements, data, illustrations, procedural details or other items may inadvertently be inaccurate.

© WILEY-VCH Verlag Berlin GmbH, Berlin (Federal Republic of Germany), 2001
ISBN 978-3-527-29435-0
Printed on non-acid paper.

Printed and bound in Great Britain by
CPI Antony Rowe, Chippenham and Eastbourne

WILEY-VCH Verlag Berlin GmbH
Bühringstrasse 10
D-13086 Berlin

"Faith maintained is one of the great gifts bestowed by fellow man"

Dedicated to the two people who always had faith that this book would be completed

Kathy and Michael

Preface

During the winter semester of 1992/93 I taught for the first time the course *Quantum Optics I* at the University of Ulm, which was followed by part II in the summer semester of 1993. When I offered the course a second time the University was kind enough to financially support two diplom students, Erwin Mayr and Daniel Krähmer, who had already taken this class in the previous year to transform my hand-written notes and sketches of drawings into a legible form. Erwin and Daniel have done a tremendous job. Since then I have taught this course many times and collected more and more material which was included into this manuscript by other graduate students of the Abteilung. It has served many generations of students at the University of Ulm as a first introduction to the field of quantum optics.

During one of his many visits to Ulm, Michael Poulson, a close friend from the VCH-Wiley publishing house saw the manuscript on my desk. "I want to publish these notes" was his immediate reaction. Michael had complete faith that this manuscript would eventually be turned into a publishable book. He wanted the material to be expanded to include problems, experiments and an exhaustive list of references. The goal was to convert the existing manuscript of about 150 pages into a book of about 250 pages. His trust in me was so great that he started advertising *Quantum Optics in Phase Space* before we had even signed a contract. I believe the present result satisfies the criteria Michael had put forward with one exception – the number of pages.

At Christmas of 1996 we finally signed a contract and Michael was extremely relieved. I still remember his words "now I have finally succeeded in signing you up for the book". A week later his untimely death during Christmas vacation added a new meaning to this sentence and a purposeful dimension to his faith and expectation; I was determined more than ever to deliver what I had promised.

Eventually Erwin and Daniel graduated and their new professional life did not allow them to devote more time to continue the project. Since that fateful Christmas of 1996, many students have helped me transform my class notes into various sections of the book continuing the work that Erwin and Daniel had begun. Stephan Meneghini took over and for several years he was instrumental in typing the manuscript. But also he graduated during the course of the project. In the final phase of the book his role was taken over by Florian Haug. I am enormously grateful to all of them for their assistance. What started out with 200 pages at Erwin and Daniel's departure eventually expanded and reached its present 700 page size.

Similarly, the field of quantum optics has expanded enormously over the last 10 years. This fact reflects itself in the variety of textbooks that have been published on this topic. It is impossible to represent all branches of this rapidly moving field in one single book. As a consequence many current topics are left out in the present one, such as quantum information or Bose-Einstein condensation. The main theme of the book is quantum phase space and the application of semi-classical concepts, such as WKB techniques to problems of quantum optics. In the present American hype of the "e-mail and the information highway" some people have suggested to call the book "phase-space.com."

Many friends and colleagues have read through various parts of the book and have made useful comments. In this regard I want to mention especially I. Bialynicki-Birula, J.H. Eberly, H.J. Kimble, D. Kobe, R.F. O'Connell, H. Walther, K. Wódkiewicz and E. Wolf. Special thanks go to M. König who has worked very carefully through the whole book and has made numerous constructive remarks. In the final stage all members of the Abteilung have proofread the entire book. Many thanks to G. Alber, M. Bienert, M. Cirone, O. Crasser, A. Delgado, D. Fischer, M. Freyberger, F. Haug, V. Kozlov, H. Mack, W. Merkel, G. Metikas, M. Mussinger, K. Vogel, J. Wichmann and V.P. Yakovlev. K. Vogel was also instrumental in putting the index together. I am grateful to my secretaries B. Casel, R. Knöpfle and U. Thomas who were helpful in collecting the literature.

Various chapters of the book have been tested in two lecture series given at the University of Texas at Austin. The penetrating questions of J.H. Eberly, M. Fink, D. Heinzen, J. Keto, M. Raizen, W.C. Schieve and E.C.G. Sudarshan have helped to sharpen my arguments through the extremely lively discussions during and after the classes. They have enormously helped to improve the presentation of the material. The kind hospitality of and the always friendly atmosphere at the physics department at UT Austin are greatly appreciated.

Many science organisations have supported the research summarized in the present book. In this context I want to mention especially the Deutsche Forschungsgemeinschaft and the Leibniz Program, the European Community, the Heraeus Foundation, the Humboldt Foundation and the University of Ulm. All have graciously financed my students, assistants and visitors. Many thanks to all of them.

The quiet periods in Denton, Texas with my understanding father-in-law, H.C. Phillips who always refers to me as his "blue electron son-in-law" were very conducive to completing this book. Moreover, I greatfully acknowledge the kind hospitality at the physics department at North Texas State University, Denton.

Last, but not least, I want to express my sincere thanks to my teachers. G. Süßmann, whose lectures at the Ludwig-Maximilians-Universität in München woke my interest in theoretical physics and made me change my degree from high school teacher to physicist. Süßmann's broad and deep interest in the whole field of physics and not just a special area has always impressed me and hopefully this book reflects his influence. M. O. Scully and H. Walther have introduced me to the field of quantum optics 20 years ago. I was fortunate enough to closely work with them on various problems of quantum optics and they have strongly influenced my view of the field. Through my collaborations with them I have gained many insights. A different angle of physics came through my years in Texas working with J.A. Wheeler. He taught me that many phenomena in physics become transparent when viewed using WKB techniques combined with the concept of phase space. In this sense the origin of this book stems from my years in Austin, Texas working with John on interference in phase space.

Special thanks go to my publishers Wiley-VCH and, in particular, to the innocent successor of Michael Poulson, Michael Bär, for his patience in awaiting the final outcome of *Quantum Optics in Phase Space*. Indeed they have suffered along with me in my trials of writing a comprehensive textbook on the application of phase space to quantum optics.

Above all I want to thank my parents, who encouraged me to think deeply and who made it possible for me to get the education necessary to pursue my studies. A special thanks goes to my wife Kathy and Michael Poulson, who never gave up their faith in me that this book would ever be finished even when other people close to me have made bets that the book would never (or not) be completed before the year 2050. Michael Poulson once said "I am not worried about the book being finished because Kathy will make sure you get it done for both of us". With these fateful words he was right; may he rest in peace.

Wolfgang P. Schleich
Ulm, November 2000

Contents

1 **What's Quantum Optics?**   1
    1.1   On the Road to Quantum Optics 1
    1.2   Resonance Fluorescence 2
         1.2.1   Elastic Peak: Light as a Wave 2
         1.2.2   Mollow-Three-Peak Spectrum 3
         1.2.3   Anti-Bunching 5
    1.3   Squeezing the Fluctuations 7
         1.3.1   What is a Squeezed State? 7
         1.3.2   Squeezed States in the Optical Parametric Oscillator 9
         1.3.3   Oscillatory Photon Statistics 12
         1.3.4   Interference in Phase Space 13
    1.4   Jaynes-Cummings-Paul Model 14
         1.4.1   Single Two-Level Atom plus a Single Mode 15
         1.4.2   Time Scales 15
    1.5   Cavity QED 16
         1.5.1   An Amazing Maser 16
         1.5.2   Cavity QED in the Optical Domain 19
    1.6   de Broglie Optics 22
         1.6.1   Electron and Neutron Optics 22
         1.6.2   Atom Optics 23
         1.6.3   Atom Optics in Quantized Light Fields 25
    1.7   Quantum Motion in Paul Traps 26
         1.7.1   Analogy to Cavity QED 26
         1.7.2   Quantum Information Processing 26
    1.8   Two-Photon Interferometry and More 28
    1.9   Outline of the Book 29

2 **Ante**   35
    2.1   Position and Momentum Eigenstates 36
         2.1.1   Properties of Eigenstates 36
         2.1.2   Derivative of Wave Function 38
         2.1.3   Fourier Transform Connects x- and p-Space 39
    2.2   Energy Eigenstate 40
         2.2.1   Arbitrary Representation 41
         2.2.2   Position Representation 42
    2.3   Density Operator: A Brief Introduction 44

		2.3.1	A State Vector is not Enough!	44
		2.3.2	Definition and Properties	48
		2.3.3	Trace of Operator	49
		2.3.4	Examples of a Density Operator	51
	2.4		Time Evolution of Quantum States	53
		2.4.1	Motion of a Wave Packet	54
		2.4.2	Time Evolution due to Interaction	55
		2.4.3	Time Dependent Hamiltonian	57
		2.4.4	Time Evolution of Density Operator	61

3 Wigner Function — 67

	3.1		Jump Start of the Wigner Function	68
	3.2		Properties of the Wigner Function	69
		3.2.1	Marginals	69
		3.2.2	Overlap of Quantum States as Overlap in Phase Space	71
		3.2.3	Shape of Wigner Function	72
	3.3		Time Evolution of Wigner Function	74
		3.3.1	von Neumann Equation in Phase Space	74
		3.3.2	Quantum Liouville Equation	75
	3.4		Wigner Function Determined by Phase Space	76
		3.4.1	Definition of Moyal Function	76
		3.4.2	Phase Space Equations for Moyal Functions	77
	3.5		Phase Space Equations for Energy Eigenstates	78
		3.5.1	Power Expansion in Planck's Constant	79
		3.5.2	Model Differential Equation	81
	3.6		Harmonic Oscillator	84
		3.6.1	Wigner Function as Wave Function	85
		3.6.2	Phase Space Enforces Energy Quantization	86
	3.7		Evaluation of Quantum Mechanical Averages	87
		3.7.1	Operator Ordering	88
		3.7.2	Examples of Weyl-Wigner Ordering	90

4 Quantum States in Phase Space — 99

	4.1		Energy Eigenstate	100
		4.1.1	Simple Phase Space Representation	100
		4.1.2	Large-m Limit	101
		4.1.3	Wigner Function	105
	4.2		Coherent State	108
		4.2.1	Definition of a Coherent State	109
		4.2.2	Energy Distribution	110
		4.2.3	Time Evolution	113
	4.3		Squeezed State	119
		4.3.1	Definition of a Squeezed State	121
		4.3.2	Energy Distribution: Exact Treatment	125
		4.3.3	Energy Distribution: Asymptotic Treatment	128
		4.3.4	Limit Towards Squeezed Vacuum	132

		4.3.5	Time Evolution	135
	4.4		Rotated Quadrature States	136
		4.4.1	Wigner Function of Position and Momentum States	137
		4.4.2	Position Wave Function of Rotated Quadrature States	140
		4.4.3	Wigner Function of Rotated Quadrature States	142
	4.5		Quantum State Reconstruction	143
		4.5.1	Tomographic Cuts through Wigner Function	143
		4.5.2	Radon Transformation	144

5 Waves à la WKB — 153

- 5.1 Probability for Classical Motion . . . 153
- 5.2 Probability Amplitudes for Quantum Motion . . . 155
  - 5.2.1 An Educated Guess . . . 156
  - 5.2.2 Range of Validity of WKB Wave Function . . . 158
- 5.3 Energy Quantization . . . 159
  - 5.3.1 Determining the Phase . . . 159
  - 5.3.2 Bohr-Sommerfeld-Kramers Quantization . . . 161
- 5.4 Summary . . . 163
  - 5.4.1 Construction of Primitive WKB Wave Function . . . 163
  - 5.4.2 Uniform Asymptotic Expansion . . . 164

6 WKB and Berry Phase — 171

- 6.1 Berry Phase and Adiabatic Approximation . . . 172
  - 6.1.1 Adiabatic Theorem . . . 172
  - 6.1.2 Analysis of Geometrical Phase . . . 174
  - 6.1.3 Geometrical Phase as a Flux in Hilbert Space . . . 175
- 6.2 WKB Wave Functions from Adiabaticity . . . 176
  - 6.2.1 Energy Eigenvalue Problem as Propagation Problem . . . 177
  - 6.2.2 Dynamical and Geometrical Phase . . . 181
  - 6.2.3 WKB Waves Rederived . . . 183
- 6.3 Non-Adiabatic Berry Phase . . . 185
  - 6.3.1 Derivation of the Aharonov-Anandan Phase . . . 186
  - 6.3.2 Time Evolution in Harmonic Oscillator . . . 187

7 Interference in Phase space — 189

- 7.1 Outline of the Idea . . . 189
- 7.2 Derivation of Area-of-Overlap Formalism . . . 192
  - 7.2.1 Jumps Viewed From Position Space . . . 192
  - 7.2.2 Jumps Viewed From Phase Space . . . 197
- 7.3 Application to Franck-Condon Transitions . . . 200
- 7.4 Generalization . . . 201

8 Applications of Interference in Phase Space — 205

- 8.1 Connection to Interference in Phase Space . . . 205
- 8.2 Energy Eigenstates . . . 206
- 8.3 Coherent State . . . 208
  - 8.3.1 Elementary Approach . . . 209

		8.3.2	Influence of Internal Structure 212

        8.3.2   Influence of Internal Structure 212
  8.4   Squeezed State 213
        8.4.1   Oscillations from Interference in Phase Space 213
        8.4.2   Giant Oscillations 216
        8.4.3   Summary 218
  8.5   The Question of Phase States 221
        8.5.1   Amplitude and Phase in a Classical Oscillator 221
        8.5.2   Definition of a Phase State 223
        8.5.3   Phase Distribution of a Quantum State 227

9 Wave Packet Dynamics   233
  9.1   What are Wave Packets? 233
  9.2   Fractional and Full Revivals 234
  9.3   Natural Time Scales 237
        9.3.1   Hierarchy of Time Scales 237
        9.3.2   Generic Signal 239
  9.4   New Representations of the Signal 241
        9.4.1   The Early Stage of the Evolution 241
        9.4.2   Intermediate Times 244
  9.5   Fractional Revivals Made Simple 246
        9.5.1   Gauss Sums 246
        9.5.2   Shape Function 246

10 Field Quantization   255
  10.1 Wave Equations for the Potentials 256
        10.1.1 Derivation of the Wave Equations 256
        10.1.2 Gauge Invariance of Electrodynamics 257
        10.1.3 Solution of the Wave Equation 260
  10.2 Mode Structure in a Box 262
        10.2.1 Solutions of Helmholtz Equation 262
        10.2.2 Polarization Vectors from Gauge Condition 263
        10.2.3 Discreteness of Modes from Boundaries 264
        10.2.4 Boundary Conditions on the Magnetic Field 264
        10.2.5 Orthonormality of Mode Functions 265
  10.3 The Field as a Set of Harmonic Oscillators 266
        10.3.1 Energy in the Resonator 267
        10.3.2 Quantization of the Radiation Field 269
  10.4 The Casimir Effect 272
        10.4.1 Zero-Point Energy of a Rectangular Resonator 272
        10.4.2 Zero-Point Energy of Free Space 274
        10.4.3 Difference of Two Infinite Energies 275
        10.4.4 Casimir Force: Theory and Experiment 276
  10.5 Operators of the Vector Potential and Fields 278
        10.5.1 Vector Potential 278
        10.5.2 Electric Field Operator 280
        10.5.3 Magnetic Field Operator 281

10.6		Number States of the Radiation Field	281
	10.6.1	Photons and Anti-Photons	282
	10.6.2	Multi-Mode Case	282
	10.6.3	Superposition and Entangled States	283

11 Field States   291

11.1		Properties of the Quantized Electric Field	291
	11.1.1	Photon Number States	292
	11.1.2	Electromagnetic Field Eigenstates	293
11.2		Coherent States Revisited	295
	11.2.1	Eigenvalue Equation	295
	11.2.2	Coherent State as a Displaced Vacuum	297
	11.2.3	Photon Statistics of a Coherent State	298
	11.2.4	Electric Field Distribution of a Coherent State	299
	11.2.5	Over–completeness of Coherent States	301
	11.2.6	Expansion into Coherent States	303
	11.2.7	Electric Field Expectation Values	305
11.3		Schrödinger Cat State	306
	11.3.1	The Original Cat Paradox	306
	11.3.2	Definition of the Field Cat State	307
	11.3.3	Wigner Phase Space Representation	307
	11.3.4	Photon Statistics	310

12 Phase Space Functions   321

12.1		There is more than Wigner Phase Space	321
	12.1.1	Who Needs Phase Space Functions?	321
	12.1.2	Another Description of Phase Space	322
12.2		The Husimi-Kano Q-Function	324
	12.2.1	Definition of Q-Function	324
	12.2.2	Q-Functions of Specific Quantum States	324
12.3		Averages Using Phase Space Functions	330
	12.3.1	Heuristic Argument	330
	12.3.2	Rigorous Treatment	333
12.4		The Glauber-Sudarshan P-Distribution	337
	12.4.1	Definition of P-Distribution	337
	12.4.2	Connection between Q- and P-Function	338
	12.4.3	P-Function from Q-Function	339
	12.4.4	Examples of P-Distributions	341

13 Optical Interferometry   349

13.1		Beam Splitter	350
	13.1.1	Classical Treatment	350
	13.1.2	Symmetric Beam Splitter	352
	13.1.3	Transition to Quantum Mechanics	353
	13.1.4	Transformation of Quantum States	353
	13.1.5	Count Statistics at the Exit Ports	356
13.2		Homodyne Detector	357

		13.2.1	Classical Considerations	357

- 13.2.1 Classical Considerations 357
- 13.2.2 Quantum Treatment 358
- 13.3 Eight-Port Interferometer 361
  - 13.3.1 Quantum State of the Output Modes 361
  - 13.3.2 Photon Count Statistics 363
  - 13.3.3 Simultaneous Measurement and EPR 365
  - 13.3.4 Q-Function Measurement 367
- 13.4 Measured Phase Operators 370
  - 13.4.1 Measurement of Classical Trigonometry 370
  - 13.4.2 Measurement of Quantum Trigonometry 372
  - 13.4.3 Two-Mode Phase Operators 374

14 Atom-Field Interaction 381
- 14.1 How to Construct the Interaction? 382
- 14.2 Vector Potential-Momentum Coupling 382
  - 14.2.1 Gauge Principle Determines Minimal Coupling 383
  - 14.2.2 Interaction of an Atom with a Field 386
- 14.3 Dipole Approximation 389
  - 14.3.1 Expansion of Vector Potential 389
  - 14.3.2 $\vec{A}\cdot\vec{p}$-Interaction 390
  - 14.3.3 Various Forms of the $\vec{A}\cdot\vec{p}$ Interaction 390
  - 14.3.4 Higher Order Corrections 392
- 14.4 Electric Field-Dipole Interaction 393
  - 14.4.1 Dipole Approximation 393
  - 14.4.2 Röntgen Hamiltonians and Others 393
- 14.5 Subsystems, Interaction and Entanglement 395
- 14.6 Equivalence of $\vec{A}\cdot\vec{p}$ and $\vec{r}\cdot\vec{E}$ 396
  - 14.6.1 Classical Transformation of Lagrangian 397
  - 14.6.2 Quantum Mechanical Treatment 399
  - 14.6.3 Matrix elements of $\vec{A}\cdot\vec{p}$ and $\vec{r}\cdot\vec{E}$ 399
- 14.7 Equivalence of Hamiltonians $H^{(1)}$ and $\widetilde{H}^{(1)}$ 400
- 14.8 Simple Model for Atom-Field Interaction 402
  - 14.8.1 Derivation of the Hamiltonian 402
  - 14.8.2 Rotating-Wave Approximation 406

15 Jaynes-Cummings-Paul Model: Dynamics 413
- 15.1 Resonant Jaynes-Cummings-Paul Model 413
  - 15.1.1 Time Evolution Operator Using Operator Algebra . . 414
  - 15.1.2 Interpretation of Time Evolution Operator 416
  - 15.1.3 State Vector of Combined System 418
  - 15.1.4 Dynamics Represented in State Space 418
- 15.2 Role of Detuning 420
  - 15.2.1 Atomic and Field States 420
  - 15.2.2 Rabi Equations 422
- 15.3 Solution of Rabi Equations 423
  - 15.3.1 Laplace Transformation 424

	15.3.2	Inverse Laplace Transformation	425
15.4	Discussion of Solution		426
	15.4.1	General Considerations	427
	15.4.2	Resonant Case	427
	15.4.3	Far Off-Resonant Case	429

16 State Preparation and Entanglement — 435

- 16.1 Measurements on Entangled Systems ... 435
  - 16.1.1 How to Get Probabilities ... 436
  - 16.1.2 State of the Subsystem after a Measurement ... 439
  - 16.1.3 Experimental Setup ... 440
- 16.2 Collapse, Revivals and Fractional Revivals ... 444
  - 16.2.1 Inversion as Tool for Measuring Internal Dynamics ... 444
  - 16.2.2 Experiments on Collapse and Revivals ... 447
- 16.3 Quantum State Preparation ... 451
  - 16.3.1 State Preparation with a Dispersive Interaction ... 451
  - 16.3.2 Generation of Schrödinger Cats ... 454
- 16.4 Quantum State Engineering ... 454
  - 16.4.1 Outline of the Method ... 454
  - 16.4.2 Inverse Problem ... 458
  - 16.4.3 Example: Preparation of a Phase State ... 461

17 Paul Trap — 473

- 17.1 Basics of Trapping Ions ... 474
  - 17.1.1 No Static Trapping in Three Dimensions ... 474
  - 17.1.2 Dynamical Trapping ... 475
- 17.2 Laser Cooling ... 479
- 17.3 Motion of an Ion in a Paul Trap ... 480
  - 17.3.1 Reduction to Classical Problem ... 481
  - 17.3.2 Motion as a Sequence of Squeezing and Rotations ... 483
  - 17.3.3 Dynamics in Wigner Phase Space ... 486
  - 17.3.4 Floquet Solution ... 490
- 17.4 Model Hamiltonian ... 494
  - 17.4.1 Transformation to Interaction Picture ... 495
  - 17.4.2 Lamb-Dicke Regime ... 496
  - 17.4.3 Multi-Phonon Jaynes-Cummings-Paul Model ... 498
- 17.5 Effective Potential Approximation ... 500

18 Damping and Amplification — 507

- 18.1 Damping and Amplification of a Cavity Field ... 508
- 18.2 Density Operator of a Subsystem ... 509
  - 18.2.1 Coarse-Grained Equation of Motion ... 509
  - 18.2.2 Time Independent Hamiltonian ... 511
- 18.3 Reservoir of Two-Level Atoms ... 511
  - 18.3.1 Approximate Treatment ... 512
  - 18.3.2 Density Operator in Number Representation ... 514
  - 18.3.3 Exact Master Equation ... 519

	18.3.4	Summary 522
18.4	One-Atom Maser 522	
	18.4.1	Density Operator Equation 523
	18.4.2	Equation of Motion for the Photon Statistics 524
	18.4.3	Phase Diffusion 529
18.5	Atom–Reservoir Interaction 532	
	18.5.1	Model and Equation of Motion 532
	18.5.2	First Order Contribution 533
	18.5.3	Bloch Equations 535
	18.5.4	Second Order Contribution 537
	18.5.5	Lamb Shift 539
	18.5.6	Weisskopf-Wigner Decay 540

19 Atom Optics in Quantized Light Fields 549

	19.1	Formulation of Problem 549
	19.1.1	Dynamics 549
	19.1.2	Time Evolution of Probability Amplitudes 552
19.2	Reduction to One-Dimensional Scattering 554	
	19.2.1	Slowly Varying Approximation 554
	19.2.2	From Two Dimensions to One 555
	19.2.3	State Vector 556
19.3	Raman-Nath Approximation 557	
	19.3.1	Heuristic Arguments 557
	19.3.2	Probability Amplitudes 558
19.4	Deflection of Atoms 559	
	19.4.1	Measurement Schemes and Scattering Conditions 559
	19.4.2	Kapitza-Dirac Regime 562
	19.4.3	Kapitza-Dirac Scattering with a Mask 568
19.5	Interference in Phase Space 571	
	19.5.1	How to Represent the Quantum State? 572
	19.5.2	Area of Overlap 572
	19.5.3	Expression for Probability Amplitude 573

20 Wigner Functions in Atom Optics 579

	20.1	Model 579
20.2	Equation of Motion for Wigner Functions 581	
20.3	Motion in Phase Space 582	
	20.3.1	Harmonic Approximation 583
	20.3.2	Motion of the Atom in the Cavity 583
	20.3.3	Motion of the Atom outside the Cavity 585
	20.3.4	Snap Shots of the Wigner Function 586
20.4	Quantum Lens 587	
	20.4.1	Distributions of Atoms in Space 587
	20.4.2	Focal Length and Deflection Angle 589
20.5	Photon and Momentum Statistics 590	
20.6	Heuristic Approach 592	

| | 20.6.1 | Focal Length . . . . | 592 |
| | 20.6.2 | Focal Size . . . . | 594 |

A Energy Wave Functions of Harmonic Oscillator   597
    A.1   Polynomial Ansatz 597
    A.2   Asymptotic Behavior 599
        A.2.1   Energy Wave Function as a Contour Integral 600
        A.2.2   Evaluation of the Integral I_m 600
        A.2.3   Asymptotic Limit of f_m 603
        A.2.4   Bohr's Correspondence Principle 603

B Time Dependent Operators   605
    B.1   Caution when Differentiating Operators 605
    B.2   Time Ordering 606
        B.2.1   Product of Two Terms 607
        B.2.2   Product of n Terms 608

C Süßmann Measure   611
    C.1   Why Other Measures Fail 611
    C.2   One Way out of the Problem 612
    C.3   Generalization to Higher Dimensions 613

D Phase Space Equations   615
    D.1   Formulation of the Problem 615
    D.2   Fourier Transform of Matrix Elements 616
    D.3   Kinetic Energy Terms 617
    D.4   Potential Energy Terms 619
    D.5   Summary 620

E Airy Function   621
    E.1   Definition and Differential Equation 621
    E.2   Asymptotic Expansion 622
        E.2.1   Oscillatory Regime 623
        E.2.2   Decaying Regime 624
        E.2.3   Stokes Phenomenon 625

F Radial Equation   629

G Asymptotics of a Poissonian   633

H Toolbox for Integrals   635
    H.1   Method of Stationary Phase 635
        H.1.1   One-Dimensional Integrals 635
        H.1.2   Multi-Dimensional Integrals 637
    H.2   Cornu Spiral 639

I   Area of Overlap — 643
- I.1   Diamond Transformed into a Rectangle 643
- I.2   Area of Diamond 644
- I.3   Area of Overlap as Probability 646

J P-Distributions — 649
- J.1   Thermal State 649
- J.2   Photon Number State 650
- J.3   Squeezed State 651

K   Homodyne Kernel — 655
- K.1   Explicit Evaluation of Kernel 655
- K.2   Strong Local Oscillator Limit 656

L   Beyond the Dipole Approximation — 659
- L.1   First Order Taylor Expansion 659
  - L.1.1   Expansion of the Hamiltonian 659
  - L.1.2   Extension to Operators 661
- L.2   Classical Gauge Transformation 661
  - L.2.1   Lagrangian with Center-of-Mass Motion 662
  - L.2.2   Complete Time Derivative 663
  - L.2.3   Hamiltonian Including Center-of-Mass Motion 663
- L.3   Quantum Mechanical Gauge Transformation 664
  - L.3.1   Gauge Potential 664
  - L.3.2   Schrödinger equation for $\tilde{\Phi}$ 667

M   Effective Hamiltonian — 669

N   Oscillator Reservoir — 671
- N.1   Second Order Contribution 671
  - N.1.1   Evaluation of Double Commutator 671
  - N.1.2   Trace over Reservoir 673
- N.2   Symmetry Relations in Trace 673
  - N.2.1   Complex Conjugates 674
  - N.2.2   Commutator Between Field Operators 674
- N.3   Master Equation 675
- N.4   Explicit Expressions for Γ, β and $\tilde{G}$ 676
- N.5   Integration over Time 677

O   Bessel Functions — 679
- O.1   Definition 679
- O.2   Asymptotic Expansion 680

P   Square Root of δ — 683

Q   Further Reading — 685

Index — 688

1 What's Quantum Optics?

What is quantum optics? This is a rather personal question. A well-known scientist in this field once gave the following authoritative answer: "Whatever I do defines quantum optics!" On a more objective basis one is tempted to define this branch of physics by the pun: "Quantum optics is that branch of optics where the quantum features of light matter."

Which discovery in physics marks the birthday of quantum optics? Many phenomena come to our mind. Is it the discovery of the quantum, the development of QED, or the maser/laser? Or is it none of the above?

In this chapter we answer this question in a back handed way by summarizing some path breaking experiments that define quantum optics. Admittedly this list is not complete and chosen in a rather subjective way. The rapidly moving field of quantum optics demonstrates most clearly that even after 100 years of quantum physics there is still a lot to be learned from Planck's original discovery.

1.1 On the Road to Quantum Optics

More than hundred years ago M. Planck was struggling with the experimental data of black body radiation obtained at the *Physikalisch-Technische Reichsanstalt* in Berlin by H. Rubens and F. Kurlbaum. From todays point of view these experiments look rather academic. However, they were motivated by industrial applications. Indeed, standards had to be developed in order to describe light bulbs. This need triggered one of the most important problems in the physics of the 20*th* century: Classical electromagnetic theory cannot explain the measured black body spectrum. In a desperate but courageous attempt Planck postulated that the oscillators in the walls of the cavity can only absorb and emit radiation in discrete units. This revolutionary idea of discreteness rather than a continuum provided the celebrated radiation formula and was the starting point of quantum mechanics.

Nowadays we associate the quantization with the field rather than with the mechanical oscillators in the wall. However, wave and matrix mechanics were first developed for massive particles and then, later, transferred to the electromagnetic field leading to quantized electrodynamics.

The field of quantum electrodynamics, QED, which deals with the interaction of quantized matter with quantized electromagnetic fields started with P.A.M. Dirac.

He was the first to derive the Einstein A and B coefficients of spontaneous and induced emission. The field of QED culminated on the one hand with the experimental discovery of the level shift in the hydrogen atom by W.E. Lamb and R.C. Retherford and the measurement of the anomalous moment of the electron by H.M. Foley and P. Kusch. On the other hand the theoretical works of S. Tomonaga, J. Schwinger and R. Feynman showed how to avoid the infinities that had plagued the theory since the thirties. The incredible agreement between theory and experiment established nowadays in many QED systems confirms beyond any doubt the quantized nature of light.

The development of the ammonium maser by C.H. Townes, J. Gordon and H. Zeiger, and the laser by T. Maiman following the paper *Optical Masers* by A. Schawlow and Townes has opened the new field of quantum electronics. Motivated by the experiments on the maser and building on his own theoretical work on water-vapor absorption W.E. Lamb developed a theory of the maser during the years 1954–1956. Later he worked out a complete semi-classical theory of laser action. Independently the group of H. Haken in Stuttgart developed their own approach. In the semiclassical treatment of Lamb and Haken the electromagnetic field was described classically and the atom quantum mechanically.

Since then laser theory has come a long way from the early approaches using birth and death equations via the semiclassical theory of the laser to the fully quantized version. The three approaches to the quantum theory of the laser are the Fokker-Planck method, pursued by H. Haken and H. Risken, the noise operator method by M. Lax and W.H. Louisell and the density matrix techniques by M.O. Scully and W.E. Lamb. Earlier the quantum theory of photon counting has been developed by R. Glauber.

Unfortunately, the quantum effects of the laser were scarce. The photon statistics of the laser and the phase diffusion were the only quantum effects that could be measured.

1.2 Resonance Fluorescence

Quantum optics has received an enormous push from the phenomenon of resonance fluorescence. The light emitted from an atom which is driven by a classical monochromatic electromagnetic field shows interesting quantum effects in its spectrum and in the statistics. We now briefly review this corner stone of quantum optics.

1.2.1 Elastic Peak: Light as a Wave

Resonance fluorescence is an old problem that has been discussed for the first time in great detail by W. Heitler in his classic book "The Quantum Theory of Radiation". He pointed out that the emitted radiation has the same frequency as the incident radiation. Thus, the spectrum is a delta-function. In this sense the atom is just a driven dipole and therefore radiates with the frequency of the driving field. This elastic component in the scattered light has been observed experimentally and is shown in Fig. 1.1.

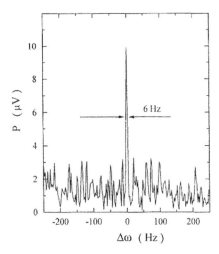

Fig. 1.1: Heterodyne spectrum of the elastic fluorescence component of a single trapped ^{24}Mg$^+$ ion. A narrow peak emerges when the frequency difference between the heterodyne signal and the driving field vanishes. Taken from J.T. Höffges et al., Opt. Comm. **133**, 170 (1997).

For this measurement a single magnesium ion was stored in a modified Paul trap shown in Fig. 1.2. A laser was driving an electronic transition of the ion and the emitted radiation was superimposed with the driving field. The resulting signal, commonly referred to as heterodyne signal, was analyzed with respect to frequency. This spectrum displays a narrow structure at the frequency of the incident radiation as shown in Fig. 1.1. Theoretically, the width of this line should be zero. However, it is determined by the spectrum of the exciting light.

In this context it is interesting to note that this experiment is also a verification of the wave nature of light. Indeed, the elastic peak is so narrow since the emitted wave has a fixed phase relation with the driving field. The emitted light is therefore a wave. This experiment clearly supports the wave rather than the particle concept. However, as we will see in the next section we can slightly rearrange the experiment such that the particle aspect of light emerges. This is one more manifestation of Bohr's principle of complementarity.

1.2.2 Mollow-Three-Peak Spectrum

However, this delta-function peak in the spectrum is only one part of the problem. In the late sixties B.R. Mollow investigated resonance fluorescence using quantum electrodynamics and found that the spectrum depends on the intensity of the incident radiation. For low intensities the Heitler result is valid. However, for larger intensities the spectrum displays a more complicated structure: In addition to this elastic delta-function peak there exist three broad, incoherent contributions centered at the incident frequency and at two side bands. They are shifted by a frequency

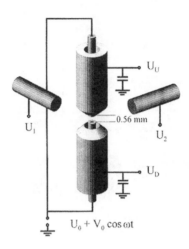

Fig. 1.2: Electrode configuration of the endcap trap. The trap consists of two co-linearly arranged cylinders corresponding to the cap electrodes of the traditional Paul trap. The ring electrode is simulated by two hollow cylinders which are concentric with each of the cylindrical endcaps. Additional electrodes allow for the compensation of stray electric field components. The open structure offers a large detection solid angle and good access for laser beams. Taken from J.T. Höffges *et al.*, Opt. Comm. **133**, 170 (1997).

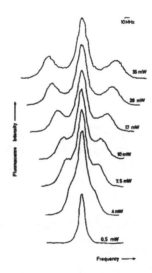

Fig. 1.3: Experimental three-peak Mollow spectrum for increasing laser intensity. We note the emergence of the side peaks. The elastic peak ideally represented by a delta-function located on top of the central peak is not shown. Taken from W. Hartig *et al.*, Z. Physik A **278**, 205 (1976).

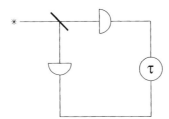

Fig. 1.4: Measurement of second order correlation function. Light from a source passes a beam splitter and falls onto two detectors. We are interested in the distribution of consecutive clicks of the two detectors. The first photon hitting a detector starts the clock, the second photon hitting the other detector stops the clock. As light source we could use a thermal light bulb, a laser or the resonance fluorescence of a single ion driven by a laser field.

determined by the electric field of the incident wave. These incoherent peaks have a different width determined by the natural line width Γ of the atom. Indeed, the central peak has a width of $\Gamma/2$ whereas the sidebands have the width $3\Gamma/4$.

This spectrum has been measured experimentally by the groups of C.R. Stroud, S. Ezekiel and H. Walther in the mid-seventies. In Fig. 1.3 we show the emergence of the three-peak Mollow spectrum for increasing laser intensity.

1.2.3 Anti-Bunching

A new chapter in the book of resonance fluorescence was opened in the mid-seventies when H. Carmichael and D.F. Walls in New Zealand, and H.J. Kimble and L. Mandel in the US independently from each other analyzed the statistics of the light. They found a time delay between two successive photons emitted from the atom. The light is anti-bunched. This behavior is in sharp contrast to thermal light where the photons come in bunches. It is interesting to note that also a laser has a non-vanishing probability for two photons arriving right after each other.

One way of measuring the effect of bunching or anti-bunching is the Hanbury Brown and Twiss arrangement shown in Fig. 1.4. Light falls through a beamsplitter onto two detectors. We can measure the delay time between two consecutive clicks on the two detectors: The first photon triggers the detector in one arm, the second photon fires the second detector in the other arm. Repeating the experiment many times we measure the distribution of delay times.

This experiment performed in the late fifties by H. Hanbury Brown and R.Q. Twiss using sunlight was the starting point of the quantum theory of photon counting pioneered by R. Glauber. For an insight into the importance of this role we refer to the Les Houches lectures of R. Glauber. Glauber's theory of photon counting is based on correlation functions of the electromagnetic field. In this formalism photon anti-bunching expresses itself in the behavior of the second order correlation function $g^{(2)}(\tau)$ as a function of the delay τ and, in particular, at $\tau = 0$.

Figure 1.5 shows the dependence of $g^{(2)}$ on the delay τ for three typical light sources, namely a thermal light source, a laser and resonance fluorescence. We note that for $\tau \to \infty$ all curves approach unity. However, the starting points for all curves,

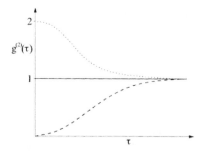

Fig. 1.5: Second order correlation function as a function of delay τ. When the light source in the Hanbury Brown and Twiss experiment is a light bulb the second order correlation function $g^{(2)}(\tau)$ (dotted line) exhibits a dominant maximum at short delay times: It is therefore more probable to find two photons right after each other rather than at larger delays. The light is bunched. When the source is a laser the light obeys Poissonian statistics and $g^{(2)}(\tau)$ is independent of the delay (solid line). However, resonance fluorescence displays quite a different behavior (dashed line): The light is antibunched since the probability that two photons follow right after each other is very small.

that is, $g^{(2)}$ at $\tau = 0$, are different. In case of thermal light it begins at $g^{(2)}(0) = 2$ and approaches unity from above. Hence, it is more likely to find two photons arriving after each other. For a laser the distribution is independent of the delay. However, the light emitted by an atom that is driven by a laser field is quite different. Here, the probability to find a photon just after one has been detected is zero and hence, $g^{(2)}(0) = 0$. Hence, the curve approaches unity from below. In this case we need the full quantum theory of radiation to describe the resonance fluorescence light.

These theoretical predictions have been verified experimentally by the groups of L. Mandel and H. Walther using atomic beams. The development of Paul traps for ions and magneto-optical traps for atoms has opened a new era in these experimental studies of resonance fluorescence. Now it is possible to observe the radiation of a single particle and thus anti-bunched light from a single ion, atom or molecule. In Fig. 1.6 we show the measured second order correlation function for resonance fluorescence of a single magnesium ion. These curves clearly show that it is highly unlikely to have two photons emitted right after each other.

The phenomenon of anti-bunching observed with the help of a single ion is particularly interesting in the context of the heterodyne experiments shown in Fig. 1.1 since in both experiments we are analyzing the same radiation. In the heterodyne measurement of the resonance fluorescence we find a narrow spectrum confirming the wave nature of the emitted light. However, when we perform a correlation experiment of the same light we observe the particle nature. Hence, resonance fluorescence provides us in this way with another striking demonstration of wave-particle duality.

Closely related to the phenomenon of anti-bunching is the effect of sub-Poissonian statistics. R. Glauber has shown that a classical current radiates an electromagnetic field in a coherent state. Its photon statistics, that is the probability to find m photons is then a Poissonian distribution. However, the photon statistics of the radiation emitted by a driven atom is narrower than a Poissonian: It enjoys sub-

1.3 Squeezing the Fluctuations

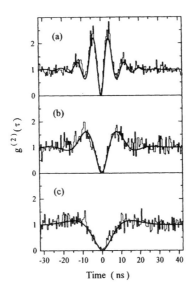

Fig. 1.6: Photon anti-bunching in the resonance fluorescence of a single ^{24}Mg$^+$ ion stored in the endcap trap shown in Fig. 1.2. The second order correlation function $g^{(2)}(\tau)$ displays a striking minimum at zero delay times. We show the curves for three characteristic detunings. The integration time was limited by the storage time of the ion. In the last case (c) the storage time was 220 minutes. Taken from J.T. Höffges et al., Opt. Comm. **133**, 170 (1997).

Poissonian statistics. This effect has been observed in the group of L. Mandel for an atomic beam, and for a single ion in the group of H. Walther.

1.3 Squeezing the Fluctuations

Recently resonance fluorescence brought out another impressive quantum effect of the radiation field. Its fluctuations are squeezed as shown by the experimental curve in Fig. 1.7.

1.3.1 What is a Squeezed State?

In order to uniquely describe the state of a classical, mechanical harmonic oscillator we need both, the amplitude and the phase of the oscillator. Likewise, we need amplitude and phase to describe uniquely the electromagnetic field. In its most elementary version we represent the electromagnetic field by a vector in complex space as shown in Fig. 1.8. Here it is worthwhile to mention that we do not mean the complete electric field vector $\vec{E}$ but only one component of it.

However, when we quantize the field a whole distribution of vectors is needed. This distribution function provides for every point of complex space a weight factor. One could imagine that this weight function represents the probability to find a specific electric field vector. Unfortunately, quantum mechanics does not allow such

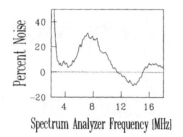

Fig. 1.7: Experimental observation of squeezing in resonance fluorescence. The dotted horizontal line marks the shot-noise limit defined by the vacuum fluctuations of the electromagnetic field. We note that in a small frequency domain above 12 MHz the fluctuations of the resonance fluorescence light go below this limit. Taken from Lu *et al.*, Phys. Rev. Lett. **81**, 3635 (1998).

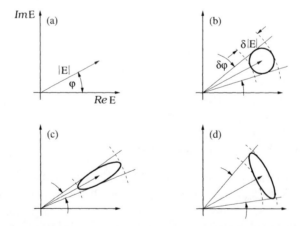

Fig. 1.8: Representation of the electromagnetic field in complex space, that is, phase space as a vector (a). Due to the quantum mechanics of the field the end point of the vector can lie anywhere in a domain of phase space with minimum area $2\pi\hbar$. This uncertainty domain can be a circle (b) which results in a symmetric distribution of the fluctuations. It can also be an ellipse with an asymmetric distribution (c,d). In this case we have squeezed either the fluctuations in phase (c) or amplitude (d). The electromagnetic field is in a squeezed state.

a probability interpretation. Indeed, this complex space is phase space spanned by two conjugate variables and Heisenberg's uncertainty relation prevents us from characterizing a quantum mechanical system by a single point in phase space. Therefore, we cannot attribute a probability to a point in phase space. Nevertheless, quasi-probability distributions exist and are useful in the quantum mechanical description of electromagnetic fields as discussed in great detail throughout the book. However, all of these distributions have some diseases and do not allow a probability interpretation.

In the most elementary case this distribution of field vectors is symmetric in the two variables of complex space with respect to the average field. However, in various interferometric applications it is important to measure the phase of the field with high precision. In this case we are not so interested in the amplitude. It is therefore advantageous to redistribute the quantum fluctuations in an asymmetric way. Since we have to conserve the area in phase space or rather the volume underneath the distribution a reduction of the fluctuations in one variable leads to an increase of the fluctuations in the other variable. This phenomenon has certain analogies to squeezing tooth paste out of its tube. Therefore, the expression *squeezing the fluctuations* has become popular.

Squeezed states have been mentioned in the early papers of R. Glauber and have been studied in great detail by H.P. Yuen, J.N. Hollenhorst and D. Stoler. They have been realized experimentally for the first time by the group of R. Slusher at Bell Labs in 1985. Shortly after this experiment J. Kimble and his group, then at The University of Texas at Austin produced highly squeezed states. By now they are produced routinely.

1.3.2 Squeezed States in the Optical Parametric Oscillator

Figure 1.9 shows a setup for squeezing the vacuum fluctuations. It makes heavily use of the field of nonlinear optics and the optical parametric oscillator (OPO). This is a device which creates light of frequency 2ω from light of frequency ω. This phenomenon is commonly referred to as *second harmonic generation*.

We note that also the inverse process is possible: We can create light of frequency ω from light of frequency 2ω. This process is called *parametric down-conversion*.

A Pico-Introduction into Nonlinear Optics

In order to achieve second-harmonic generation or down-conversion we need a medium such as a crystal for which the polarization P is not only a linear function of the electric field E but also involves quadratic or higher powers of E, that is

$$P = \chi^{(1)} E + \chi^{(2)} E^2 + \ldots . \tag{1.1}$$

Moreover, we recall from classical electrodynamics that the polarization drives the wave equation

$$\Box E \equiv \left[\vec{\nabla}^2 - \frac{1}{c^2} \frac{\partial^2}{\partial t^2} \right] E = -\mu \frac{\partial^2}{\partial t^2} P$$

for the electric field.

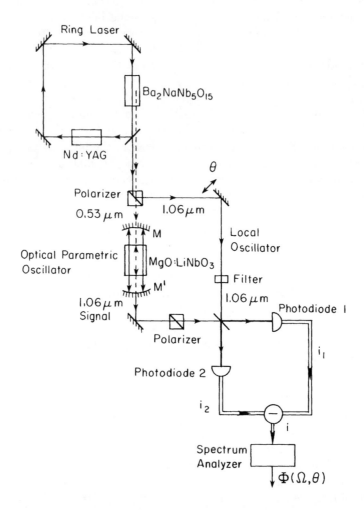

Fig. 1.9: Experimental setup for generating and observing squeezed light. A ring laser (top) with a nonlinear crystal creates light of frequency ω and 2ω. A frequency sensitive beam splitter transmits radiation of the frequency 2ω, but reflects it for ω. The transmitted light of 2ω drives a cavity with a nonlinear crystal (center). As a result light of frequency ω emerges. A beam splitter combines the new light beam and the reference beam, both with frequency ω. At the two exit ports two detectors measure the light intensities and transform them into electric currents i_1 and i_2. A spectrum analyzer (bottom) observes the fluctuations in the difference i of the two currents. Taken from L.A. Wu et al., J. Opt. Soc. Am. B **4**, 1465 (1987).

For the sake of simplicity we now suppress the space dependence and solely concentrate on the time dependence. In order to bring out the essentials we concentrate on the OPO.

With the help of the well-known trigonometric relation

$$\cos^2(\omega t) = \frac{1}{2}(1 + \cos(2\omega t))$$

we note that a nonlinear medium creates out of the field

$$E = E_1 \cos(\omega t)$$

a constant field E_0 and a field

$$E' = E_2 \cos(2\omega t),$$

of frequency 2ω. From light of frequency ω we have therefore generated light of frequency 2ω.

A similar analysis explains the phenomenon of subharmonic generation.

How to Squeeze?

We now return to the question of squeezing in the OPO. In order to create a squeezed state we need an asymmetric treatment of the two phase space variables. The nonlinear interaction of Eq. (1.1) just achieves that. The light created in an OPO, that is in second harmonic generation is squeezed.

In Fig. 1.9 we summarize a famous experiment creating squeezed light. However, here the phenomenon of subharmonic generation was used. Light from a ring laser with frequency 2ω feeds the cavity of the optical parametric oscillator. Due to the nonlinear medium the subharmonic is generated and light of the frequency ω exits the cavity. It is mixed with light of the same frequency which has not passed through the cavity and has been deflected by a frequency dependent beam splitter. A movable mirror adjusts the phase of this field. Since it is a strong field we refer to it as the local oscillator.

Detection of Squeezing

But how to measure the suppression of quantum fluctuations? The crucial instrument for measuring fluctuations is the so-called homodyne detector shown at the bottom of Fig. 1.9. Here, we mix the squeezed light with a strong classical field using a beam splitter. We measure the resulting light intensities expressed in photo electron currents i_1 and i_2 in the two output ports of the beam splitter and subtract them from each other. We record their difference $i_-(t)$ as a function of time. This current fluctuates around a mean value $\langle i_- \rangle$ where the average denotes the time average. The statistics of these fluctuations provides us with the full distribution of the current difference and, in particular, of its second moment V which is a measure of the width of the distribution. This experiment is performed for a fixed phase θ between the two input fields of the beam splitter.

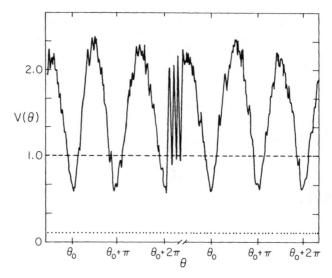

Fig. 1.10: Width V of the photo-current distribution as a function of the phase difference θ between the two input ports of the homodyne detector. In its most elementary phase space representation the vacuum state is a circle and therefore rotationally symmetric. It does not have a preferred phase. Consequently, when we mix the vacuum state with a local oscillator the width V is independent of θ. In contrast, a squeezed state is represented by an ellipse which displays a preferred direction in phase space. Hence, the width V depends on the phase angle. In the domains around $\theta_0 + k\pi$ where $k = 0, 1, 2, \ldots$ the fluctuations fall below the vacuum level. The light is squeezed. In the intermediate domains the fluctuations are larger than the corresponding ones of the vacuum. Taken from L.A. Wu et al., J. Opt. Soc. Am. B **4**, 1465 (1987).

When we block the input port from the parametric oscillator only the vacuum gets mixed with the local oscillator. Since the vacuum is rotationally symmetric the fluctuations V are independent of the relative phase θ as shown by the dashed horizontal line in Fig. 1.10.

However, the squeezed state shows an asymmetry in phase space and therefore the fluctuations are phase sensitive. When we now vary the phase the fluctuations can fall below or reach above the level set by the vacuum fluctuations as indicated by the oscillatory curve in Fig. 1.10.

1.3.3 Oscillatory Photon Statistics

Squeezed states are characterized by an asymmetric distribution of quantum fluctuations. However, this is only one of many remarkable properties of these states. Indeed, they also display a rather unusual distribution of light quanta.

The coherent state, the most elementary approximation of a laser field, exhibits a Poissonian distribution of photons. In contrast, a squeezed vacuum enjoys an oscillatory photon statistics as measured in the experiment by the group of J. Mlynek and shown in Fig. 1.11.

1.3 Squeezing the Fluctuations

There are many explanations for this surprising effect. The most direct explanation uses the concept of interference in phase space.

1.3.4 Interference in Phase Space

This concept rests on a semiclassical interpretation of the quantum mechanical scalar product $\langle\chi|\psi\rangle$ between two quantum states $|\chi\rangle$ and $|\psi\rangle$. It associates an area of overlap in phase space with this complex-valued probability amplitude $\langle\chi|\psi\rangle$. Indeed, we represent the two quantum states $|\chi\rangle$ and $|\psi\rangle$ in phase space. When there is only one overlap then this area represents the corresponding probability, that is the absolute value squared of the scalar product. When there is more than one overlap the areas *interfere*. In this case each area represents an interfering probability amplitude with a given amplitude and phase. The amplitude is the square root of the area whereas the phase is determined by other appropriate domains in phase space.

We summarize the concept of interference in phase space by the formula

$$\langle\chi|\psi\rangle = \sum_j \left(\begin{array}{c} j\text{th area} \\ \text{of overlap in} \\ \text{phase space} \end{array}\right)^{1/2} \exp\left[i\left(\begin{array}{c} j\text{th area} \\ \text{enclosed by} \\ \text{central lines} \end{array}\right)\right] \qquad (1.2)$$

for the scalar product of two quantum states $|\chi\rangle$ and $|\psi\rangle$.

Photon Statistics of Squeezed Vacuum

In order to gain insight into the photon statistics of the squeezed vacuum we evaluate the scalar product $\langle m|\psi_{\text{sq}}\rangle$ between the mth photon number eigenstate $|m\rangle$ and the squeezed vacuum state $|\psi_{\text{sq}}\rangle$. We represent the mth number state by a band in phase space and the squeezed state by a highly elongated ellipse as shown in Fig. 1.12. Consequently, the band creates two symmetrically located areas of overlap A_m and

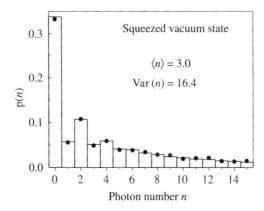

Fig. 1.11: The experimentally reconstructed photon number distribution of a squeezed vacuum state shows distinct odd-even oscillations. Taken from G. Breitenbach et al., Nature **387**, 471 (1997).

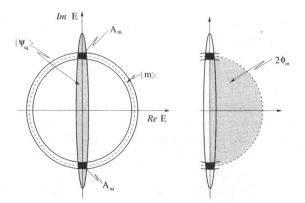

Fig. 1.12: Interference in phase space as an explanation for the oscillatory photon statistics of a highly squeezed vacuum. The circular band representing a photon number state cuts out of the highly elongated ellipse representing a squeezed vacuum two symmetrically located areas of overlap (left). The total probability amplitude to find m photons in a squeezed vacuum is hence given by the interference of these two areas in phase space. The phase difference is the area enclosed by the central lines of the two states of interest (right).

the scalar product

$$\langle m | \psi_{\text{sq}} \rangle = A_m^{1/2} e^{i\phi_m} + A_m^{1/2} e^{-i\phi_m} \quad (1.3)$$

is then the sum of two interfering areas of phase space. The phase difference $2\phi_m$ between the two areas is governed by the area enclosed by the center line of the ellipse and the band as shown in the right of Fig. 1.12.

Young's Double-Slit Experiment in Phase Space

This example brings out most clearly that the physics of *interference in phase space* is very similar to the familiar double-slit experiment. In both cases there are two interfering contributions to the total probability of detecting a specified outcome. In one case, the two contributions come from the two slits. In the other case, they result from the two distinct areas of overlap in phase space. The phase difference in the double-slit experiment is measured by the difference in optical path length from the centers of the two slits to the point of detection. Likewise, the probability amplitudes for the two contributions in Eq. (1.3) have a phase difference. In this sense Young's famous interference experiment is generalized to *interference in phase space*.

1.4 Jaynes-Cummings-Paul Model

How to describe the interaction of an atom with light? On first sight this is quite a formidable task since there are many degrees of freedom involved. There is the atom with its nucleus and many electrons. In the simplest case we can consider a hydrogen atom consisting of a single proton and a single electron. The atom moves as a unit.

The electron itself performs a motion relative to the proton. Under appropriate conditions both motions have to be treated quantum mechanically.

The atom itself represents a dipole $\vec{\wp} \equiv e\vec{r}$ which interacts with an electric field $\vec{E}$ via the interaction Hamiltonian

$$H_{\vec{r} \cdot \vec{E}} \equiv -\vec{\wp} \cdot \vec{E}(\vec{R}, t). \tag{1.4}$$

Here $\vec{r}$ and $\vec{R}$ denote the relative coordinate of the electron and the center-of-mass of the atom, respectively. Therefore, the dipole contains the internal degrees of freedom, that is the atomic levels and $\vec{R}$ describes the center-of-mass motion.

1.4.1 Single Two-Level Atom plus a Single Mode

In the full quantum version of the theory, that is in non-relativistic quantum electrodynamics, the internal coordinate $\vec{r}$, the center-of-mass $\vec{R}$ and the electric field $\vec{E}$ become operators. The simplest situation is when there are only two states of the electron, that is two internal levels involved, and only one mode of the electromagnetic field interacts with these two levels causing transitions. This model has been proposed in the early days of maser theory by E.T. Jaynes and F.W. Cummings, and independently by H. Paul.

Due to the simplicity of this model it can be solved analytically. Despite the simplicity it contains a lot of physics. For many years it has been considered a theorist's toy with no immediate application. However, the method of optical pumping developed by A. Kastler, allows us to make the two-level approximation. Moreover, superconducting microwave cavities and high-quality optical mirrors create single mode cavities with extremely high quality factors of the order of $Q = 3 \cdot 10^{10}$ corresponding to an average lifetime of a photon in the cavity of 0.2 s. Therefore, nowadays the Jaynes-Cummings-Paul model is one of the corner stones of quantum optics.

1.4.2 Time Scales

Thus, the most elementary system of cavity QED is a single two-level atom interacting with a single mode of a cavity. The dynamics of this system is governed by the Schrödinger equation and the time evolution is therefore unitary: There is an exchange of a single quantum of excitation between the atom and the field. This periodic exchange is governed by the so-called single photon Rabi frequency

$$q_0 \equiv \frac{\wp \mathcal{E}_0}{\hbar}$$

determined by the dipole moment $\wp$, Planck's constant $\hbar$ and the electric field $\mathcal{E}_0$. But what electric field?

In the quantized version of electrodynamics the electric field operator is proportional to the so-called vacuum electric field

$$\mathcal{E}_0 \equiv \sqrt{\frac{\hbar \Omega}{\varepsilon_0 V}}. \tag{1.5}$$

Here ε_0, Ω and V denote the electric permittivity, the frequency of the resonator and the volume of the mode of interest, respectively.

We can therefore increase the strength of the interaction by choosing a larger dipole moment and/or by increasing the vacuum electric field by decreasing the mode volume. This is the strategy of microwave or optical resonators.

In a realistic experiment atoms decay spontaneously. This process is associated with a decay rate $\gamma_\perp$. Similarly, the electric field in the resonator leaks out with a rate κ. These decays make the time evolution non-unitary and we need a description in terms of the density operator rather than the state vector. This is the topic of a later section. However, here we only discuss the time scales involved.

Moreover, in most experiments the atoms move through the resonator and only interact with the cavity field for a time T. We conclude by noting that in the experiments on cavity QED discussed in the next section there is the special hierarchy g_0, κ, $\gamma_\perp$ and T^{-1} between the four rates.

1.5 Cavity QED

For decades resonators for electromagnetic radiation had decay times which were short compared to the time scales associated with the internal dynamics of an atom interacting with these cavity fields. The theory of the laser made heavily use of that fact to simplify the resulting equations. However, recently new resonators have been developed in the microwave and in the optical regime. They have very long decay times, that is high quality factors. As a consequence, an atom can absorb, reemit and reabsorb the same photon many times. The atom undergoes many Rabi cycles before the field has decayed in the cavity. This new resonator technology is the basis of the new era of cavity QED.

1.5.1 An Amazing Maser

From the interaction Hamiltonian, Eq. (1.4), we recognize that we can obtain a strong coupling between the atom and the electric field when we increase the dipole moment and/or when we increase the electric field. Since the dipole is essentially determined by the separation of the electron from the nucleus highly excited states of the electron provide large dipole moments. It is therefore advantageous to work with such Rydberg atoms.

Indeed, tuneable lasers allow the controlled preparation of highly excited atoms with main quantum numbers n of the order of 60. Due to the fact that the separation of the electron from the nucleus grows with n^2 the electron in such a Rydberg atom is far away from the nucleus and therefore the atom shows a large dipole moment. Since the coupling of the atom to the field is achieved through the dipole moment Rydberg atoms exhibit an unusually strong coupling.

The transitions between neighboring high quantum numbers lie in the microwave domain. Therefore, the combination of high-quality microwave resonators with Rydberg atom spectroscopy offers unique opportunities. The groups of H. Walther in Garching and S. Haroche in Paris have capitalized on these advantages and have constructed masers based on this Rydberg atom-microwave cavity approach.

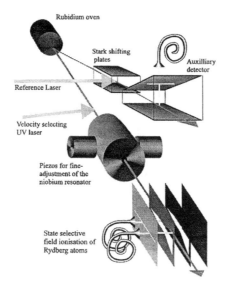

Fig. 1.13: Experimental setup of the Garching one-atom maser. Atoms leaving the rubidium oven (top) are excited into the $63p_{3/2}$ Rydberg state using a UV-laser (left beam line). Since the laser beam is under an angle relative to the atomic beam a specific velocity subclass of atoms is excited. This creates a rather well-defined interaction time for the atoms with the cavity field. After the atoms have passed the microwave cavity they are detected using state selective field ionization. Two piezo translators allow to tune the cavity. A reference beam (right beam line) is used to stabilize the laser frequency to a Stark shifted atomic resonance, allowing the velocity of the atoms to be tuned continuously. Taken from M. Weidinger *et al.*, Phys. Rev. Lett. **82**, 3795 (1999).

One-Atom Maser

These masers are amazing since they mase even when on average there is less than one atom in the cavity. The principle experimental arrangement of the Garching one-atom maser shown in Fig. 1.13 is rather simple: A dilute atomic beam of Rydberg atoms prepared by a laser passes a high-quality microwave resonator. When the field is on resonance with an atomic transition the atom can deposit its excitation in the field. The next atom interacts with this so modified field and can also deposit its excitation. When the decay time of the cavity is long compared to the transit time of the atoms and the characteristic time of the internal dynamics a field can build up in the cavity.

In Fig. 1.14 we show the resonance line of the first one-atom maser. Here the number of excited atoms is measured as a function of the cavity detuning. When the atom is on resonance with the laser field the number of atoms in the excited state drops dramatically because all atoms now deposit their excitation and amplify the field.

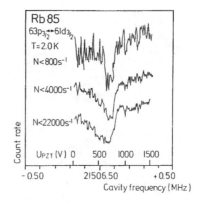

Fig. 1.14: Resonance line of the first one-atom maser: Number of excited atoms versus cavity detuning. Rubidium atoms in the excited state $63p_{3/2}$ traverse the microwave resonator and interact with the cavity field. The number of atoms remaining in this state is measured at the exit of the cavity. When the cavity is tuned through the transition frequency $\Delta\omega = 21506.5\,\text{MHz}$ to the $61d_{3/2}$ level the number of excited atoms sharply decreases and a complicated resonance line forms. When the flux N of atoms is reduced the resonance gets sharper. The microwave black body radiation provided by the cavity corresponds to a temperature of 2 K. Taken from D. Meschede *et al.*, Phys. Rev. Lett. **54**, 551 (1985).

Radiation of the Maser

The radiation of this maser is very different from ordinary laser light. It shows highly non-classical features such as anti-bunching, sub-Poissonian statistics and squeezing. Moreover, this device allows us to study fundamental questions of quantum mechanics such as entanglement of quantum systems, state preparation and decoherence.

There is a fundamental difference between an ordinary laser and the one-atom maser: The high Q-factor of the cavity implies that no light can escape. Therefore, the non-classical radiation of the one-atom maser cannot be coupled out of the resonator. Any method to take the field out would spoil the Q-factor and hence the quantum state of the radiation. This feature implies that we cannot make straightforward measurements on the maser field using ordinary photo detectors. The only tool to probe the radiation in the cavity are the atoms themselves. They play two roles in this game: On one hand they generate the maser field and on the other hand they probe it.

Since the wave length of the maser line lies in the microwave regime the black body radiation created by the temperature of the cavity is a great concern. Indeed, according to Planck the average number $\bar{n}$ of black body photons in a cavity with temperature T and resonance frequency Ω reads

$$\bar{n} = \frac{1}{\exp\left(\frac{\hbar\Omega}{k_B T}\right) - 1}. \tag{1.6}$$

Here k_B denotes the Boltzmann constant.

When we substitute typical numbers into this formula, we find that the average number of photons varies between $\bar{n} = 3.6$ for 4.3 K, $\bar{n} = 1.5$ at 2 K and $\bar{n} = 0.054$ at 0.3 K. Therefore, in order to avoid black body photons the cavity has to be cooled to extremely low temperatures. The first experiments on the one-atom maser have been performed around 2 K. Nowadays temperatures of 0.3 K have been achieved.

1.5.2 Cavity QED in the Optical Domain

In the optical domain essentially two routes have been pursued in the context of cavity QED: *(i)* Resonators with high quality mirrors and *(ii)* glass spheres based on whispering gallery modes.

The key ingredients of optical resonators are the mirrors. They need an enormous reflectivity and were originally engineered to minimize back-reflections in ring laser gyroscopes. Such gyros are optical navigational instruments that are in commercial airplanes and submarines. Therefore, for a long time, the quality of these mirrors was highly classified and they were not commercially available. With the end of the cold war this situation changed.

However, there is also a disadvantage of the optical approach. Since low-lying states are involved the dipole moments for the interaction are not as large as in the microwave domain. Fortunately, this can be compensated by increasing the vacuum electric field. According to Eq. (1.5) this amounts to decreasing the size of the resonator.

Motion of Atoms Through a Cavity: Atom-Cavity Microscope

The approach of optical resonators was pioneered by the group of J. Kimble, presently at Caltech (Pasadena). They showed the so-called vacuum Rabi splitting in the optical domain. Most recently they have used a high-Q optical cavity to measure the motion of atoms along a standing light wave. Individual atoms were dropped in the gravitational field from a magneto-optical trap (MOT) as shown in Fig. 1.15. Such a MOT uses forces imposed on the atom by the interaction between the dipole and the inhomogeneous magnetic field. After the trap has been switched off the atoms fall through the resonator. Since the separation of the two mirrors is extremely small, in the order of 100 μm, most atoms miss the entrance slit. However, a few make it into the cavity. A laser feeds radiation into the cavity and a photo detector measures the transmitted light. Due to the strong coupling of the atoms with the field they change the transmitted intensity and the light at the detector drops while the atom transverses the cavity as shown in Fig. 1.16. In this way we can observe single atoms passing a cavity and interacting with its field.

In these experiments the one-photon Rabi frequency g_0, the cavity field decay rate κ, the atomic dipole decay rate $\gamma_\perp$ and the inverse atomic transit time T^{-1} are

$$\left(g_0, \kappa, \gamma_\perp, T^{-1}\right)/2\pi = (11, 3.5, 2.5, 0.001) \text{ MHz}.$$

Kimble and coworkers even go to the extreme of reconstructing the path of a single atom using the light leaving the cavity. In this sense this device is an atom-cavity microscope. Moreover, they even store the atom in the cavity using the

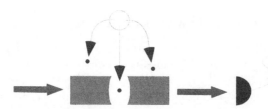

Fig. 1.15: Atoms falling through a high-Q optical cavity. Cesium atoms are stored in a magneto-optical trap (MOT), roughly 7 mm above a Fabry-Perot cavity. When the trap is switched off, some atoms fall through the resonator which is pumped by a laser creating a standing light field. A photo detector measures the transmitted light. Taken from H. Mabuchi et al., Opt. Lett. **21**, 1393 (1996).

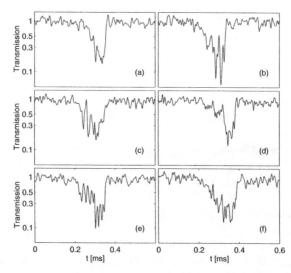

Fig. 1.16: Six examples of transit signals associated with the passage of individual atoms through the optical cavity. When an atom traverses the cavity the index of refraction in the cavity changes and the transmitted light intensity drops. The oscillations in the minimum correspond to the atom moving over various periods of the standing wave. Taken from H. Mabuchi et al., Opt. Lett. **21**, 1393 (1996).

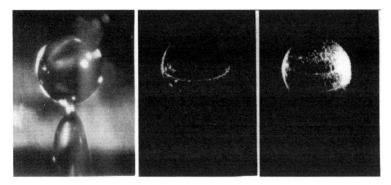

Fig. 1.17: Whispering gallery mode micro-resonator based on a glass sphere. On the left we show this sphere under external illumination whereas the pictures in the middle and on the right display the radiating modes as a narrow or a broad band along the equator, respectively. Taken from M.L. Gorodetsky and V.S. Ilchenko, Opt. Comm. **113**, 133 (1994).

electromagnetic field corresponding to a single photon: Single atoms are bound in orbit by single photons.

We conclude this discussion by emphasizing that a very similar experiment was performed by the group of G. Rempe originally in Konstanz, now in Garching. However, in contrast to the experiment by Kimble the atoms are launched from below and have to climb against gravity. This setup leads to extremely long interaction times provided the cavity is located close to the turning points of the motion.

One-Atom Laser

The group of M. Feld at MIT used a resonator made out of such gyroscope mirrors to replace the microwave cavity of an one-atom maser configuration by a high-Q optical resonator. The resulting device has the advantage that low-lying atomic transitions are sufficient and no cooling of the cavity is necessary. Moreover, one can use ordinary photon detectors to make measurements on the radiation field. For a comparison of this laser to the Garching one-atom maser, the cavity QED experiments in Paris and the optical resonator experiments in Pasadena or Konstanz/Garching see the article by Kimble mentioned in the list of references at the end of this chapter.

Whispering Gallery Modes

The second avenue towards optical resonators with high Q-factors are glass spheres such as the one shown in Fig. 1.17. This approach was pioneered by the group of V. Braginsky in Moscow. The spheres contain modes that propagate on the inside very close to the surface and that can be excited from the outside by a laser field. Quality factors of the order of 10^8 have been reached with this technique in the groups of J. Kimble and S. Haroche.

In analogy to acoustical waves these modes are called whispering gallery modes. This name stems from the fact that in St. Paul's cathedral in London one can hear a whisper propagate along the gallery.

22   1 What's Quantum Optics?

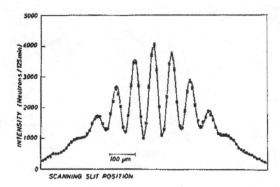

Fig. 1.18: A double-slit diffraction pattern measured with very cold neutrons of wave length 2 nm corresponding to a velocity of 200 ms^{-1}. The solid line represents a first principle prediction from quantum mechanics including all features of the experimental apparatus. Taken from A. Zeilinger *et al.*, Rev. Mod. Phys. **60**, 1067 (1988).

The main disadvantage of these resonators is that they are filled with the glass medium. Therefore, it is not easy to send an atom in a controlled way through the resonator field. However, one can dope the medium with an atom and achieve lasing in this way. Another possibility to make use of these resonators is to let atoms interact with the evanescent field of these whispering gallery modes. Theoretical suggestions to scatter atoms from these fields or even trap atoms have been made.

1.6   de Broglie Optics

One prediction of quantum mechanics is the wave nature of massive particles, that is we can associate a de Broglie wave with their center-of-mass motion. The famous experiment by C.J. Davisson and R.H. Germer in 1926 on scattering of electrons from a Ni-crystal was a sensation at its time. It brought out clearly the wave nature of electrons.

1.6.1   Electron and Neutron Optics

The electron microscope is an important application of de Broglie waves. In the mean time lenses, prisms and interferometers for electrons have been created. Moreover, electron interferometers have been extremely useful for fundamental tests of quantum mechanics.

Similarly we can consider the wave nature of neutrons. Also here the huge field of neutron optics has developed. In Fig. 1.18 we show as an example the double-slit diffraction pattern obtained from neutrons clearly confirming their wave nature. In particular, neutron interferometers have been used extensively to study fundamental questions of quantum mechanics. Here, neutron interferometry has placed a stringent upper boundary on the strength of possible nonlinear contributions in the Schrödinger equation. Moreover, the fact that a 4π rotation is needed to rotate a spin has been

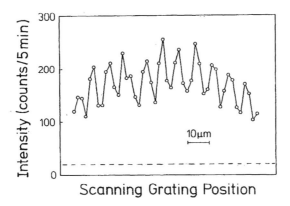

Fig. 1.19: Double-slit interference pattern from an atomic de Broglie wave. An intense atomic beam of helium atoms produced by a supersonic gas expansion diffracts from a double slit produced by a micro-fabricated transmission structure. The resulting atomic density profile in the far field shows interference fringes. The dashed line is the detector background and the line connecting the experimental points is to guide the eye. Taken from O. Carnal *et al.*, Phys. Rev. Lett. **66**, 2689 (1991).

shown using neutron interferometry. Moreover, many concepts of quantum optics have been transferred to this area which has led to the name neutron quantum optics.

1.6.2 Atom Optics

Due to the mechanical effects of light on atoms the laser has become a perfect tool to control the center-of-mass motion of atoms. Here we have to mention especially the field of laser cooling where we can reduce the kinetic energy of the atoms and create temperatures in the micro-Kelvin domain. Moreover, atoms are also very popular for atom optics purposes since they enjoy rich internal degrees of freedom on which the laser can act.

In Fig. 1.19 we show the diffraction pattern of a double-slit experiment of atomic de Broglie waves. Here, the slit was provided by a mechanical mask. However, also a laser beam can be used to provide a diffraction grating. Here again we use the dipole interaction of the atom and the fact that the field depends on the position of the atom. In this way deflection, focusing and interferometry of atomic de Broglie waves has been demonstrated experimentally.

As an example we show in Fig. 1.20 the deflection of a helium beam due to the interaction with a standing light field. Depending on the internal states of the atom we obtain a coherent separation of the beam into two beams as indicated in Fig. 1.21. This is reminiscent of the Stern-Gerlach experiment. Here the separation is caused by the interaction of an inhomogeneous magnetic field with the magnetic dipole of the atom. In the present case it is the interaction of the electric dipole with the inhomogeneous electric field.

This coherent separation also suggests that we can use the interaction of atoms

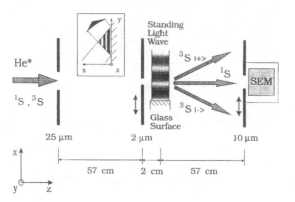

Fig. 1.20: Experimental setup for the optical Stern-Gerlach effect. Helium atoms in the ^{1}S and ^{3}S state traverse a standing light wave with a period larger than the width of the atomic beam. This has been made possible by reflecting the standing wave from a glass plate under an angle. As shown by the insert this creates a standing wave pattern of a period much larger than the optical wave length. The atoms in the ^{3}S state get deflected into two directions corresponding to their internal states and in complete analogy to the familiar Stern-Gerlach effect. The atoms in the ^{1}S state do not feel the laser field and continue without interaction. Taken from T. Sleator et al., Phys. Rev. Lett. **68**, 1996 (1992).

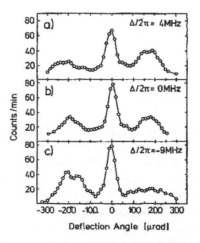

Fig. 1.21: Atomic intensity profile in the optical Stern-Gerlach effect for three characteristic detunings of the laser field. The two side maxima correspond to the two deflection angles of the atoms in the triplet state. The central peak at position zero is due to the undeflected singlet-state atoms. Taken from T. Sleator et al., Phys. Rev. Lett. **68**, 1996 (1992).

1.6 de Broglie Optics

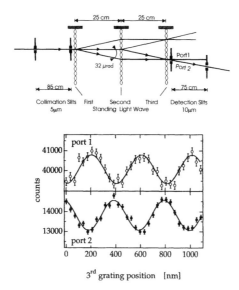

Fig. 1.22: A sequence of three standing laser fields divides and recombines coherently many atomic beams (top). At the exit of the interferometer two such beams are recombined. The count rates in the two exit ports are out of phase as shown at the bottom. Taken from E.M. Rasel *et al.*, Phys. Rev. Lett. **75**, 2633 (1995).

with light fields to obtain coherent beam splitters for atoms. When we combine three such standing laser fields as shown in Fig. 1.22 we can obtain an atom interferometer. The resulting count rates in the two arms of the interferometer are strictly out of phase as depicted at the bottom of Fig. 1.22.

We conclude by noting that in the mean time rotation sensors, that is gyroscopes based on atom interferometry are competing with optical gyroscopes. Moreover, interferometry with even more macroscopic objects has been performed. Now there exist interferometers for large molecules such as fullerenes.

Even more interesting is the case when the de Broglie wavelength of the atoms becomes of the order of their separation. Then the individual atoms loose their identity and their wave functions start to overlap. In the case of bosonic atoms there exists Bose-Einstein condensation and the atoms all fall into the same state of a trap.

1.6.3 Atom Optics in Quantized Light Fields

When we now consider the interaction of an atom with quantized light the roles of matter and light are interchanged. Whereas originally matter was treated as a particle and light as a wave we now bring out the wave nature of matter and the particle nature of light. This has opened a completely new field namely atom optics and quantized light fields. It is the marriage of the two fields atom optics and cavity quantum electrodynamics.

So far no experiment has been performed in this realm. However, K.A.H. van Leeuwen (Eindhoven) has built up a huge atomic beam machine to consider the discrete deflection of an atomic beam due to the discreteness of the photon number. Unfortunately, no results are available yet.

1.7 Quantum Motion in Paul Traps

The techniques of laser cooling have made it possible to reduce the kinetic energy of ions stored in a trap to a level where the center-of-mass motion of the ion has to be treated quantum mechanically. Since to a first approximation the binding potential of a quadrupole trap such as the endcap shown in Fig. 1.2 is quadratic the Hamiltonian of the center-of-mass motion is that of a harmonic oscillator.

1.7.1 Analogy to Cavity QED

Moreover, the ion has internal degrees of freedom. Again we consider for the sake of simplicity a two-level ion. For the case of an ion interacting with a classical standing wave while it is moving in the trap the Hamiltonian of the total system is a generalization of the one of the Jaynes-Cummings-Paul model.

When the extension of the vibratory ground state of the trap is small compared to the period of the light wave, that is in the Lamb-Dicke limit, this Hamiltonian reduces to that of the Jaynes-Cummings-Paul model. In this case this system of the ion stored in the Paul trap and interacting with the classical wave is a mechanical analogue of cavity QED situation. The role of the quanta of excitation of the field is now played by the quanta of vibration, that is photons are replaced by phonons. Again there is a periodic exchange between the excitations in the vibration and in the internal states. This exchange depends on the quantum mechanical state of the vibration.

In Fig. 1.23 we show the internal dynamics for three initial quantum states. In particular we show the probability to find the atom in the ground state as a function of time when it initially started in the ground state for a phonon eigenstate (top), a thermal state (middle) and a squeezed state (bottom). The solid line is the prediction of a generalized Jaynes-Cummings-Paul model.

1.7.2 Quantum Information Processing

The central difference between classical mechanics and quantum mechanics lies in the concept of entanglement. This phrase, in German *Verschränkung*, has been coined by E. Schrödinger in a paper of 1935 summarizing the present status of quantum mechanics. It expresses the fact that after an interaction two quantum mechanical systems cannot be separated, that is their respective quantum states are not product states anymore. This entanglement is alien to the classical world. It is the essential ingredient of the so-called Einstein-Podolsky-Rosen situation.

In the Jaynes-Cummings-Paul model we have the interaction of the internal degrees of freedom with the photon field. In the Paul trap analogy we have the inter-

1.7 Quantum Motion in Paul Traps

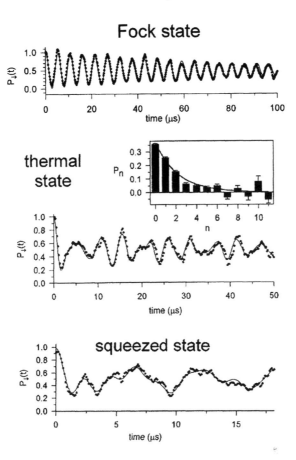

Fig. 1.23: Gallery of quantum states of motion causing characteristic internal dynamics of an ion stored in a Paul trap and interacting with a classical standing light field. When the motion is in a phonon eigenstate the internal dynamics displays damped Rabi oscillations (top). When the motion is initially in a thermal state (middle) the internal dynamics displays rather irregular oscillations which allow to reconstruct the initial exponential phonon distribution shown in the insert. The case of a squeezed state of motion (bottom) again shows rather irregular oscillations. Taken from D.M. Meekhof *et al.*, Phys. Rev. Lett. **76**, 1796 (1996).

action of the internal degrees of freedom with the center-of-mass motion. In both cases this interaction results in an entanglement of the degrees of freedom.

Recently it has been proposed to use quantum degrees of freedom to store information. Indeed, we can use the internal levels of the atom or ion as qubits. Of course, we can also use the superposition of ground and excited state. This is analogous to the so-called optical computer which solely capitalizes on the interference property of classical electromagnetic fields. However, in the so-called quantum computer we now use the entanglement to add a quantum ingredient. This can lead to an exponential speed-up of the computer for certain type of problems.

Indeed, problems in computing are classified according to the number of steps necessary to perform the task. Let us consider the problem of multiplying two numbers with N digits. There certainly exists algorithms that perform this task. When we now increase the number of digits the problem gets more involved. However, independent of the algorithm the increase in steps depends polynomially on N. In contrast, when we want to factor a large number of N digits this increase depends exponentially on N. Recently P. Shor has developed a new algorithm based on quantum entanglement that only needs a polynomial effort. This is of great importance in cryptography since the codes are based on the impossibility of factorizing large numbers.

However, these considerations are based on unitary time evolution. Every real quantum system exhibits dissipation. There is always a small fraction of light leaking out of a resonator, or there is heating in a Paul trap due to the external time dependent driving field. Quantum states are extremely fragile. A loss of a single photon can destroy the superposition. This has led to the development of error correcting codes.

So far single quantum gates have been realized experimentally. The group of D. Wineland (Boulder) has realized as a controlled-not gate based on a single ion moving in a Paul trap and interacting with a classical standing light wave. Here the ion has to be cooled to low temperatures so that the discreteness of the energy eigenstates in the trap becomes important. In the domain of quantum electrodynamics the group of J. Kimble has created a phase gate by sending atoms through a high-Q optical cavity.

However, the quantum computer consisting of many ions in a trap, has not been realized yet. One of the many challenging problems is to cool all the ions to the ground state. Recently two ions oscillating in a specific mode have been cooled to the ground state.

1.8 Two-Photon Interferometry and More

A parametric oscillator creates light of frequency 2ω from light of frequency ω. In the language of light quanta we can imagine that we have started from two quanta with energy $\hbar\omega$ and have created a single quantum of energy $2\hbar\omega$. However, also the inverse process can occur. Light of frequency 2ω propagates through a crystal to emerge as light with frequency ω. Hence, one quantum of energy $2\hbar\omega$ creates two quanta of energy $\hbar\omega$. Since, these two quanta are created due to an interaction of the light with the crystal the quanta are entangled. They are therefore perfect objects to investigate the concepts of entanglement.

Many phenomena of quantum mechanics resulting from entanglement have now been tested using this work horse of entangled photons. Space does not allow us to go into more detail and we only mention the tests of Bell-inequalities, quantum teleportation and quantum dense coding as a few examples. For a more comprehensive discussion we refer to the references at the end of this chapter.

1.9 Outline of the Book

This collection of quantum optics experiments clearly shows that this field has matured and now contains a wealth of phenomena. It is therefore impossible to include all topics in this area in a single book. Moreover, in the mean time many textbooks on quantum optics have appeared. We have therefore selected for the present book those topics which have so far not been treated extensively in other books. Moreover, we focus on quantum optics phenomena that become transparent when viewed from phase space. We have tried to present the material in a self-contained way so that a student with elementary knowledge of quantum mechanics can follow it.

The emphasis of the book is on phase space as the underlying foundation of quantum optics. In this context it is quite amusing to recall that it was the quantization of phase space volume that led M. Planck to the correct radiation formula. We show that many of these phase space ideas are still extremely useful to understand many quantum optics phenomena. In particular, the semiclassical formulation of quantum mechanics á la Wentzel-Kramers-Brillouin (WKB) sometimes referred to as *asymptotology* serves as our guiding principle. In this context semiclassical does not exclude the quantum nature of light. On the contrary in this formalism we assume that there is a macroscopic excitation in the field but fully take into account the interference nature of quantum mechanics.

We illustrate this formalism first for mechanical oscillators. This is justified for two reasons: *(i)* The electromagnetic field is a collection of harmonic oscillators and therefore everything derived for mechanical oscillators can immediately be transfered to field oscillators. *(ii)* Electrons in atoms, the two nuclei of a diatomic molecule or an ion in a Paul trap are mechanical oscillators whose motion have to be treated quantum mechanically.

The book is organized as follows: After a brief review of the essential concepts of quantum mechanics we turn to the visualization of quantum states in phase space using the Wigner function. This representation brings out striking properties of quantum states such as the oscillatory photon statistics of highly squeezed states or the possibility of reconstructing quantum states using tomography. Many of these effects make their appearance in this semiclassical limit. We therefore turn to a brief review of the WKB method and connect it with Berry phase. This leads directly to the concept of interference in phase space and wave packet dynamics.

The Wigner function is only one of an infinite set of phase space distribution functions. These generalized distribution functions follow from appropriately ordered coherent state representations of the density operator. This is of great importance in the field of cavity quantum electrodynamics. We therefore first review the quantization of the radiation field and then turn to the discussion of various quantum states. Again, phase space is the common theme connecting these topics. Multiports, that is an arrangement of various beam splitters and phase shifters allow to measure such phase space distribution functions.

So far, there is no interaction between matter and light. We therefore address the question of how to construct the interaction between the atom and the field and discuss extensively the Jaynes-Cummings-Paul model. Quantum measurements and state preparation based on quantum entanglement are then the next logical topic.

The Paul trap is analogous to cavity QED but represents a further development of the Jaynes-Cummings-Paul model and that for two reasons: *(i)* It is not confined to one-photon transitions. *(ii)* The external potential governing the center-of-mass motion is explicitly time dependent.

The examples above have all been treated with pure states. We now turn to systems in which a density matrix treatment is necessary. We derive a density matrix equation for the damping or amplification of a cavity field. This immediately leads to the density matrix of the one-atom maser. Spontaneous emission of an atom can also be derived from such a density matrix approach. Another system with a necessity for the density matrix approach stems from realm of atom optics. We consider the motion of an atom through a quantized standing wave. Again phase space provides deeper insight into the deflection and focusing of atomic beams from electromagnetic fields.

References

On the Road to Quantum Optics

For a historical overview of laser physics see for example
M. Bertolotti, *From Masers to Lasers*, Adam Hilger, Bristol, 1983

and

W.E. Lamb, W.P. Schleich, M.O. Scully and **C.H. Townes**, *Laser Physics: Quantum Controversy in Action*, Rev. Mod. Phys. **71**, S263–S273 (1999)

For a summary of the various schools of laser physics see
H. Haken, *Laser Theory*, Springer Verlag, Heidelberg and Berlin, 1984. Appeared originally in Encyclopedia of Physics, edited by S. Flügge, Springer, Heidelberg, 1970
W.H. Louisell, *Quantum Statistical Properties of Radiation*, Wiley, New York, 1973
M. Sargent III, M.O. Scully and **W.E. Lamb Jr.** *Laser Theory* Addison-Wesley, Reading, Mass., 1974
M.O. Scully and **M.S. Zubairy**, *Quantum Optics*, Cambridge U.P. New York, 1996

A summary of the state of the field of quantum optics is presented in the proceedings of the following summer schools and conferences:
C. DeWitt, A. Blandin and **C. Cohen-Tannoudji**, *Quantum Optics and Electronics*, Gordon and Breach, New York, 1965
R.J. Glauber, *Quantum Optics: Proceedings of the International School of Physics "Enrico Fermi"*, Academic, New York, 1969

Current Trends in Quantum Optics

An excellent overview can be found in
R. Bonifacio, *Mysteries, Puzzles, and Paradoxes in Quantum Mechanics*, American Institute of Physics, Woodbury, 1999
T.W. Hänsch and **H. Walther**, *Laser spectroscopy and quantum optics*, Rev. Mod. Phys. **71**, S242–S252 (1999)

Resonance Fluorescence

For the early considerations see

W. Heitler, *The Quantum Theory of Radiation*, Oxford University Press, Oxford, 1930

B.R. Mollow, *Power Spectrum of Light Scatterd by Two-Level Systems*, Phys. Rev. **188**, 1969–1975 (1969)

For a nice pedagogical and complete treatment of the theory of resonance fluorescence see

C. Cohen-Tannoudji, in: *Frontiers in Laser Spectroscopy*, ed. by R. Balian, S. Haroche and S. Liberman, North-Holland, Amsterdam, Vol. I, 1977

For the experiments in resonance fluorescence see

F. Schuda, C.R. Stroud Jr. and **M. Hercher**, *Observation of the resonant Stark effect at optical frequencies*, J. Phys. B **7**, L198–L202 (1974)

F.Y. Wu, R.E. Grove and **S. Ezekiel**, *Investigation of the Spectrum of Resonance Fluorescence Induced by a Monochromatic Field*, Phys. Rev. Lett. **35**, 1426–1429 (1975)

W. Hartig, W. Rasmussen, R. Schieder and **H. Walther**, *Study of the Frequency Distribution of the Fluorescent Light Induced by Monochromatic Radiation*, Z. Physik A **278**, 205–210 (1976)

J.T. Höffges, H.W. Baldauf, T. Eichler, S.R. Helmfrid and **H. Walther**, *Heterodyne measurement of the fluorescent radiation of a single trapped ion*, Opt. Comm. **133**, 170–174 (1997)

J. T. Höffges, H. W. Baldauf, W. Lange and **H. Walther**, *Heterodyne measurement of the resonance fluorescence of a single ion*, J. Mod. Opt. **44**, 1999–2010 (1997)

For summaries of the early photon anti-bunching experiments in Rochester and Garching see

L. Mandel and **E. Wolf**, *Optical Coherence and Quantum Optics*, Cambridge U.P., New York, 1995

J.D. Cresser, J. Häger, G. Leuchs, M. Rateike and **H. Walter**, in: *Dissipative Systems in Quantum Optics*, ed. by R. Bonifacio and L. Lugiato, Topics in Current Physics, **27**, Springer, Berlin, 1982

For the experimental observation of squeezing in resonance fluorescence see

Z.H. Lu, S. Bali and **J.E. Thomas**, *Observation of Squeezing in the Phase-Dependent Fluorescence Spectra of Two-Level Atoms*, Phys. Rev. Lett. **81**, 3635–3638 (1998)

Squeezed Light

L.A. Wu, M. Xiao and **H.J. Kimble**, *Squeezed states of light from an optical parametric oscillator*, J. Opt. Soc. Am. B **4**, 1465–1475 (1987)

For a measurement of the oscillatory photon statistics of the squeezed vacuum see

G. Breitenbach, S. Schiller and **J. Mlynek**, *Measurement of the quantum states of squeezed light*, Nature **387**, 471–475 (1997)

One-Atom Maser

D. Meschede, H. Walther and G. Müller, *One-Atom Maser*, Phys. Rev. Lett. **54**, 551–554 (1985)

M. Weidinger, B.T.H. Varcoe, R. Heerlein and H. Walther, *Trapping States in the Micromaser*, Phys. Rev. Lett. **82**, 3795–3798 (1999)

High Finesse Optical Cavities

M.G. Raizen, R.J. Thompson, R.J. Brecha, H.J. Kimble and H.J. Carmichael, *Normal-Mode Splitting and Linewidth Averaging for Two-State Atoms in an Optical Cavity*, Phys. Rev. Lett. **63**, 240–243 (1989)

H.J. Kimble, *Nonclassical light 20 years later: an assessment of the voyage into Hilbert space*, Phil. Trans. R. Soc. Lond. A **355**, 2327–2342 (1997)

H. Mabuchi, Q.A. Turchette, M.S. Chapman and H.J. Kimble, *Real-time detection of individual atoms falling through a high-finesse optical cavity*, Opt. Lett. **21**, 1393–1395 (1996)

P. Münstermann, T. Fischer, P. Maunz, P.W.H. Pinkse and G. Rempe, *Dynamics of Single Atom Motion Observed in a High-Finesse Cavity*, Phys. Rev. Lett. **82**, 3791–3794 (1999)

P.W.H. Pinkse, T. Fischer, P. Maunz and G. Rempe, *Trapping an atom with single photons*, Nature **404**, 365–368 (2000)

C.J. Hood, T.W. Lynn, A.C. Doherty, A.S. Parkins and H.J. Kimble, *The Atom-Cavity Microscope: Single Atoms Bound in Orbit by Single Photons*, Science **287**, 1447–1453 (2000)

For the concept of micro-resonators based on whispering gallery modes see

V.B. Braginsky and V.S. Ilchenko, *Properties of optical dielectric microresonators*, Sov. Phys. Doklady **234**, 306–308 (1987)

V.B. Braginsky, M.L. Gorodetsky and V.S. Ilchenko, *Quality-factor and nonlinear properties of optical whispering gallery modes*, Phys. Lett. A **137**, 393–397 (1989)

M.L. Gorodetsky and V.S. Ilchenko, *High-Q optical whispering gallery microresonators: precession approach for spherical mode analysis and emission patterns with prism couplers*, Opt. Comm. **113**, 133-143 (1994)

de Broglie Optics

For the original paper on scattering electrons off nickel see

C. Davisson and L.H. Germer, *The Scattering of Electrons by a Single Crystal of Nickel*, Nature, **119**, 558–560 (1927)

This together with many other seminal papers, that were originally published in Nature, is reprinted in a special collection of papers to celebrate the 100th anniversary of the American physical society:

C. Davisson and L.H. Germer, in: *A celebration of physics*, ed. by P. Campbell, Nature, London, 1999

For an introduction into electron interferometry see
F. Hasselbach, H. Kiesel and **P. Sonnentag**, *Exploration of the Fundamentals of Quantum Mechanics by Charged Particle Interferometry*, in: Decoherence: Theoretical, Experimental, and Conceptual Problems, ed. by Ph. Blanchard, D. Giulini, E. Joos, C. Kiefer and I.-O. Stamatescu, Springer, Berlin, 2000

For an overview of neutron optics see for example
H. Rauch and **S.A. Werner**, *Neutron Interferometry. Lessons in Experimental Quantum Mechanics*, Oxford University Press, Oxford, 1999
A. Zeilinger, R. Gähler, C.G. Shull, W. Treimer and **W. Hampe**, *Single- and double-slit diffraction of neutrons*, Rev. Mod. Phys. **60**, 1067–1073 (1988)

Pioneering experiments on atom optics are for example
O. Carnal and **J. Mlynek**, *Young's Double-Slit Experiment with Atoms: A Simple Atom Interferometer*, Phys. Rev. Lett. **66**, 2689–2692 (1991)
T. Sleator, T. Pfau, V. Balykin, O. Carnal and **J. Mlynek**, *Experimental Demonstration of the Optical Stern-Gerlach Effect*, Phys. Rev. Lett. **68**, 1996–1999 (1992)
E.M. Rasel, M.K. Oberthaler, H. Batelaan, J. Schmiedmayer and **A. Zeilinger**, *Atom Wave Interferometry with Diffraction Gratings of Light*, Phys. Rev. Lett. **75**, 2633–2637 (1995)

Even macroscopic molecules such as C_{60} show interference in a double-slit experiment
M. Arndt, O. Nairz, J. Vos-Andreae, C. Keller, G. van der Zouw and **A. Zeilinger**, *Wave-particle duality of C_{60} molecules*, Nature **401**, 680–682 (1999)

Quantum Computing

The concept of a quantum computer dates back to
R.P. Feynman, *Simulating Physics with Computers*, Int. J. Theor. Phys. **21**, 467–488 (1982)
D. Deutsch, *Quantum theory, the Church-Turing principle and the universal quantum computer*, Proc. R. Soc. Lond. A **400**, 97–117 (1985)

For the experimental realization of a quantum gate based on an ion in the Paul trap see
C. Monroe, D.M. Meekhof, B.E. King, W.M. Itano and **D.J. Wineland**, *Demonstration of a Fundamental Quantum Logic Gate*, Phys. Rev. Lett **75**, 4714–4717 (1995)

For the quantum phase gate experiment see
Q.A. Turchette, C.J. Hood, W. Lange, H. Mabuchi and **H.J. Kimble**, *Measurement of Conditional Phase Shifts for Quantum Logic*, Phys. Rev. Lett. **75**, 4710–4713 (1995)

Theoretical proposals for quantum computers based on ion traps or cavity QED can be found in
J.I. Cirac and **P. Zoller**, *Quantum Computations with Cold Trapped Ions*, Phys. Rev. Lett. **74**, 4091–4094 (1995)
T. Pellizzari, S.A. Gardiner, J.I. Cirac and **P. Zoller**, *Decoherence, Continuous Observation and Quantum Computing: A Cavity QED Model*, Phys. Rev. Lett. **75**, 3788–3791 (1995)

A summary on quantum information, quantum computation and quantum communication can be found in the special issue
Modern Studies of Basic Quantum Concepts and Phenomena, ed. by E.B. Karlsson and E. Brändas, Physica Scripta, **T76**, 1998
This issue was reprinted as a book by World Scientific, Singapore, 1998
D. Bouwmeester, A.K. Ekert and **A. Zeilinger**, *The Physics of Quantum Information*, Springer, Heidelberg, 2000
For a nice presentation of number theory and classical cryptography see
N. Koblitz, *A Course in Number Theory and Cryptography*, Springer, New York, 1994

Two-Photon Experiments

For a summary of the two-photon experiments to test the foundations of quantum physics see
A. Zeilinger, *Experiment and the Foundations of Quantum Physics*, Rev. Mod. Phys. **71**, S288–S297 (1999)
L. Mandel, *Quantum Effects in One-Photon and Two-Photon Interference*, Rev. Mod. Phys. **71**, S274–S282 (1999)
V.B. Braginsky and **F. Ya. Khalili**, in: *Quantum Measurement*, edited by K. Thorne, Cambridge University Press, Cambridge, 1992

2 Ante

In the card game of poker each player has to put up a stake before the deal to build up the pot. This is summarized by the expression *ante*. In the same sense when we want to discuss quantum optics, and in particular, the influence of quantum mechanics on optics, we have to build up our pot of knowledge on quantum physics. We therefore in this chapter briefly summarize the essential ingrediences of quantum mechanics. In particular, we focus on those concepts which are heavily used in the field of quantum optics.

We divide classical mechanics into two branches: *Kinematics* describes motion without going into the origin of the motion. In contrast *dynamics* asks for the origin of the motion. In the same spirit *quantum kinematics* describes the quantum states and quantum dynamics their time evolution. Quantum kinematics relies on five axioms: *(i)* All information about a quantum system is contained in the state vector, *(ii)* the state vector is a vector in Hilbert space, *(iii)* the absolute value squared of the wave function gives the probability density, *(iv)* the observables are hermitian operators and *(v)* the operators satisfy certain commutation relations. Quantum dynamics follows from the Schrödinger equation or the von Neumann equation.

In this chapter we first discuss quantum kinematics and then turn to the dynamics. Throughout this chapter we focus on a quantum system consisting of a particle of mass M moving in a one-dimensional potential. We denote the position of the particle by x and its momentum by p. As mentioned above in quantum mechanics the position and momentum become hermitian operators $\hat{x}$ and $\hat{p}$. In the first section we discuss the properties of their eigenstates $|x\rangle$ and $|p\rangle$.

Another important operator in quantum mechanics is the Hamiltonian

$$\hat{H} \equiv \frac{\hat{p}^2}{2M} + U(\hat{x}). \tag{2.1}$$

In the non-relativistic case it consists of the operators $\hat{p}^2/(2M)$ and $U(\hat{x})$ of kinetic and potential energy, respectively.

The eigenstates $|m\rangle$ of the Hamiltonian are the energy eigenstates defined by

$$\hat{H}|m\rangle = E_m |m\rangle.$$

The harmonic oscillator with a quadratic potential

$$U(x) \equiv \frac{1}{2} M\Omega^2 x^2$$

is the paradigm of physics. In the field of quantum optics the harmonic oscillator is of particular importance and that for two reasons: *(i)* Recently ions in traps and atoms in standing waves have been cooled to temperatures where their motion has to be treated quantum mechanically. *(ii)* When we quantize the electro-magnetic field each mode is a harmonic oscillator. Also quantized field effects have been observed recently. These applications serve as a motivation to briefly rederive the position distribution of energy eigenstates of a harmonic oscillator.

The state of a quantum system is a vector in Hilbert space. This Hilbert space is enormously big. We need in general a lot of numbers to specify uniquely a quantum state. In many cases this is simply not possible. Then we cannot describe the system by a state but have to resort to a density operator description. In the last section of this chapter we briefly summarize various properties of the density operator.

We then turn to the topic of quantum dynamics. The Hamiltonian determines through the Schrödinger equation

$$i\hbar \frac{d}{dt} |\psi\rangle = \hat{H} |\psi\rangle$$

the time evolution of the quantum state $|\psi\rangle$. We briefly summarize various techniques of solving this equation. In particular we discuss formal solutions of the Schrödinger equation and the von Neumann equation in terms of the Volterra-Schlesinger product integral and multi-commutators.

2.1 Position and Momentum Eigenstates

As mentioned above in quantum mechanics the position and momentum of a particle become hermitian operators $\hat{x}$ and $\hat{p}$. Moreover, a state vector $|\psi\rangle$ describes the particle's motion. The obvious examples for such vectors are the eigenstates $|x\rangle$ and $|p\rangle$ of the position and momentum operators $\hat{x}$ and $\hat{p}$. They are defined by the eigenvalue equations

$$\hat{x} |x\rangle = x |x\rangle$$

and

$$\hat{p} |p\rangle = p |p\rangle .$$

Here x and p are the eigenvalues. Since the operators are hermitian these eigenvalues are real. Moreover, they can take any value, that is the spectrum of the operator is continuous.

2.1.1 Properties of Eigenstates

When we multiply the first equation by $\langle x'|$ and make use of the fact that $\hat{x}$ is hermitian, that is

$$\langle x'| \hat{x} = \langle x'| x',$$

we find

$$(x' - x) \langle x'| x \rangle = 0, \qquad (2.2)$$

2.1 Position and Momentum Eigenstates

which has the solution
$$\langle x'| x \rangle = \delta(x' - x). \tag{2.3}$$

Hence, two position eigenstates $|x'\rangle$ and $|x\rangle$ corresponding to two different eigenvalues x' and x are orthogonal. However, due to the δ-function they are not normalized to unity. The same holds true for the momentum eigenstates.

Hilbert space is defined for all square integrable functions. Obviously, the position and the momentum eigenstates are not members of this space. This rather subtle mathematical problem gives rise to many complications which we do not want to dwell on in the frame work of this book. For more details and solutions we refer to the literature.

Completeness Relation and Expansion of States

The eigenstates $|x\rangle$ of the position operator form a complete set of states. In other words we can expand any quantum state $|\psi\rangle$ into position eigenstates. Here the completeness relation
$$\mathbb{1} = \int_{-\infty}^{\infty} dx \, |x\rangle \langle x|$$
yields the representation
$$|\psi\rangle = \mathbb{1}\,|\psi\rangle = \int_{-\infty}^{\infty} dx \, |x\rangle \langle x|\,|\psi\rangle \equiv \int_{-\infty}^{\infty} dx \, \psi(x) \, |x\rangle \tag{2.4}$$
where we have introduced the wave function
$$\psi(x) \equiv \langle x|\,|\psi\rangle \equiv \langle x|\,\psi\rangle$$
of the state $|\psi\rangle$ in position space.

Likewise, we can represent the state $|\psi\rangle$ in momentum eigenstates by using the completeness relation
$$\mathbb{1} = \int_{-\infty}^{\infty} dp \, |p\rangle \langle p|.$$

This yields the decomposition
$$|\psi\rangle = \mathbb{1}\,|\psi\rangle = \int_{-\infty}^{\infty} dp \, |p\rangle \langle p|\,|\psi\rangle \equiv \int_{-\infty}^{\infty} dp \, \psi(p) \, |p\rangle \tag{2.5}$$
where the wave function in momentum space reads
$$\psi(p) \equiv \langle p|\,|\psi\rangle \equiv \langle p|\,\psi\rangle.$$

Here we have introduced a notation that indicates that the two wave functions $\psi(x)$ and $\psi(p)$ are two representations of the same state $|\psi\rangle$. However, we do not obtain one from the other by simply replacing the arguments. In the next subsections we recall that it is the Fourier transform which relates the two representations.

2.1.2 Derivative of Wave Function

In the preceding section we have seen that the scalar product $\langle x|\psi\rangle$ is the wave function of the state $|\psi\rangle$ in position space. What is the object $\langle x|\hat{p}|\psi\rangle$ in terms of the wave function? Obviously it is the position representation of the quantum state $|\psi'\rangle \equiv \hat{p}|\psi\rangle$ obtained by acting with the momentum operator $\hat{p}$ on the state $|\psi\rangle$. But what is the state $|\psi'\rangle$ and how does it relate to the wave function $\psi(x)$? We do recall, of course, that this object is the derivative of the wave function with respect to position. Indeed, in position space the momentum operator is a derivative with respect to the position. We now rederive this well known fact.

In classical mechanics position and momentum are conjugate variables. In quantum mechanics the corresponding operators satisfy the Heisenberg commutation relation

$$[\hat{x}, \hat{p}] \equiv \hat{x}\hat{p} - \hat{p}\hat{x} = i\hbar. \tag{2.6}$$

This formula ties position space to momentum space. To bring this out more clearly we now calculate the quantity $\langle x|\hat{p}|\psi\rangle$.

When we use the position representation Eq. (2.4) of the state $|\psi\rangle$ we arrive at

$$\langle x|\hat{p}|\psi\rangle = \int_{-\infty}^{\infty} dx'\, \langle x|\hat{p}|x'\rangle\, \langle x'|\psi\rangle. \tag{2.7}$$

We find the quantity $\langle x|\hat{p}|x'\rangle$ by multiplying the Heisenberg commutation relation Eq. (2.6) with $\langle x|$ and $|x'\rangle$, which yields

$$\langle x|\hat{x}\hat{p} - \hat{p}\hat{x}|x'\rangle = i\hbar\delta(x - x').$$

Here we have used the orthogonality relation Eq. (2.3).

We use the eigenvalue equation of the position eigenstates and recall that $\hat{x}$ is a hermitian operator. This allows us to apply the operator $\hat{x}$ onto $\langle x|$. Therefore, this equation simplifies to

$$(x - x')\langle x|\hat{p}|x'\rangle = i\hbar\delta(x - x')$$

which provides

$$\langle x|\hat{p}|x'\rangle = i\hbar\frac{\delta(x - x')}{x - x'}.$$

The right hand side is essentially the first derivative of a δ-function. This comes out most clearly when we substitute this result back into the expression Eq. (2.7) for

$$\langle x|\hat{p}|\psi\rangle = \int_{-\infty}^{\infty} dx'\, i\hbar\frac{\delta(x - x')}{x - x'} \langle x'|\psi\rangle.$$

To perform the integration we expand the wave function $\psi(x') \equiv \langle x'|\psi\rangle$ into a Taylor series around x which yields

$$\psi(x') = \psi(x) - \frac{d\psi}{dx}(x - x') + \frac{1}{2}\frac{d^2\psi}{dx^2}(x - x')^2 + \ldots$$

2.1 Position and Momentum Eigenstates

When we substitute this expansion into the integral we arrive at

$$\langle x|\hat{p}|\psi\rangle = i\hbar\left(\psi(x)\int_{-\infty}^{\infty}dx'\,\frac{\delta(x-x')}{x-x'} - \frac{d\psi}{dx}\int_{-\infty}^{\infty}dx'\delta(x-x')\right.$$
$$\left. + \frac{1}{2}\frac{d^2\psi}{dx^2}\int_{-\infty}^{\infty}dx'\delta(x-x')(x-x') + \ldots\right).$$

The first integral vanishes due to the antisymmetry of the integrand. The third integral also vanishes because the function $x - x'$ vanishes for $x' = x$. Hence, we find

$$\langle x|\hat{p}|\psi\rangle = -i\hbar\frac{d}{dx}\psi(x). \qquad (2.8)$$

We conclude the subsection by noting that the position distribution of the state vector obtained by applying the momentum operator to the state is the derivative of the wave function of the state. This is the well known result

$$\hat{p} = \frac{\hbar}{i}\frac{d}{dx},$$

that is the momentum operator in position representation is the derivative with respect to position.

2.1.3 Fourier Transform Connects x- and p-Space

What is the connection between position and momentum wave functions? To answer this question we multiply the expansion

$$|\psi\rangle = \int dp\,\psi(p)\,|p\rangle$$

of $|\psi\rangle$ in momentum eigenstates, Eq. (2.5), by $\langle x|$ which yields

$$\psi(x) \equiv \langle x|\psi\rangle = \int_{-\infty}^{\infty}dp\,\psi(p)\,\langle x|p\rangle.$$

We are therefore forced to find the scalar product $\langle x|p\rangle$ between a position and a momentum eigenstate. In the notation used above this means that we have to find the wave function $p(x) \equiv \langle x|p\rangle$ of the momentum eigenstate in position space.

We apply the result, Eq. (2.8) of the preceding section to the special case $|\psi\rangle = |p\rangle$ which yields

$$\langle x|\hat{p}|p\rangle = p\,\langle x|p\rangle = p\cdot p(x) = -i\hbar\frac{d}{dx}p(x).$$

The resulting differential equation

$$\frac{dp(x)}{dx} = \frac{i}{\hbar}p\cdot p(x)$$

has the solution

$$p(x) = \langle x|p\rangle = \mathcal{N}e^{ipx/\hbar}. \qquad (2.9)$$

We find the normalization constant $\mathcal{N}$ from the orthogonality relation Eq. (2.3) of the position eigenstates by inserting the completeness relation of the momentum states, that is

$$\delta(x-x') = \int_{-\infty}^{\infty} dp\, \langle x|p\rangle \langle p|x'\rangle = \int_{-\infty}^{\infty} dp\, p(x) \cdot p^*(x').$$

Here the asterisk denotes the complex conjugate.

Indeed, when we substitute the expression Eq. (2.9) for the wave function $p(x)$ into this integral we arrive at

$$\delta(x-x') = |\mathcal{N}|^2 \int_{-\infty}^{\infty} dp\, e^{-i(x-x')p/\hbar}.$$

We recall the integral definition of the δ-function

$$\delta(x) = \frac{1}{2\pi\hbar} \int_{-\infty}^{\infty} dp\, e^{ixp/\hbar}$$

and therefore find

$$\mathcal{N} = \sqrt{\frac{1}{2\pi\hbar}}.$$

In conclusion the scalar product between a position and a momentum eigenstate reads

$$\langle x|p\rangle = (2\pi\hbar)^{-1/2} \exp\left(\frac{i}{\hbar}xp\right). \tag{2.10}$$

This allows us to find the explicit relation

$$\psi(x) = (2\pi\hbar)^{-1/2} \int_{-\infty}^{\infty} dp\, \psi(p) e^{ixp/\hbar}$$

between the position wave function $\psi(x)$ and the momentum wave function $\psi(p)$.

2.2 Energy Eigenstate

We now turn to the eigenstates of the Hamiltonian Eq. (2.1), that is the energy eigenstates $|m\rangle$. Here we focus on the case of the harmonic oscillator. This model plays an important role not only in the context of mechanical oscillators, but also for field oscillators. In Chapter 10 these states appear again and denote the degree to which the field oscillator is excited. In this case m denotes the number of excitations, or loosely speaking, the number of photons. That is the reason why sometimes we also refer to these states as number states. Sometimes we also refer to them as Fock states after the Russian scientist V.A. Fock.

Energy eigenstates $|m\rangle$ of the harmonic oscillator are defined by the eigenvalue equation

$$\hat{H}|m\rangle = E_m|m\rangle, \tag{2.11}$$

where the Hamiltonian reads

$$\hat{H} = \frac{\hat{p}^2}{2M} + \frac{1}{2}M\Omega^2 \hat{x}^2.$$

Here M denotes the mass of the oscillator, Ω its eigen-frequency, and $\hat{x}$ and $\hat{p}$ are the operators describing its position and momentum, respectively. The energy of the mth eigenstate is E_m.

The Dirac notation of a quantum state does not rely on specific representation. In particular, the abstract state vector $|m\rangle$ does not provide any insight into the properties of the state. It allows abstract calculations to obtain answers to questions such as: What is the probability to find the oscillator particle at position x or to find the momentum p. However, to obtain these answers we have to resort to the position representation or the momentum representation. In the present section we now discuss these representations for the energy eigenstates.

2.2.1 Arbitrary Representation

We now consider a hermitian operator $\hat{X}$ which could be for example the position or the momentum operator. The eigenstates $|X\rangle$ of the operator $\hat{X}$ defined by the eigenvalue equation

$$\hat{X}|X\rangle = X|X\rangle$$

with the eigenvalue X are complete. They allow us to represent the unity operator $\mathbb{1}$ by

$$\mathbb{1} = \int dX\, |X\rangle \langle X|.$$

Here we have assumed that the spectrum, that is the distribution of eigenvalues, is continuous. The range of the eigenvalues determines the range of the integration. For the sake of simplicity we have suppressed the limits of the integral. In the case of a discrete spectrum we have to replace the integral by a summation.

The completeness relation allows us to express the energy eigenstate $|m\rangle$ as a continuous superposition of eigenstates $|X\rangle$ of the operator $\hat{X}$. Indeed, we find by multiplying the ket vector $|m\rangle$ by the unity operator

$$|m\rangle = \mathbb{1}|m\rangle = \int dX\, |X\rangle \langle X|m\rangle \equiv \int dX\, u_m(X)|X\rangle.$$

Here the function $u_m(X) = \langle X||m\rangle \equiv \langle X|m\rangle$ is the wave function of the energy eigenstate $|m\rangle$ in X- representation. How can we find this function?

We find the equation determining the wave function $u_m(X)$ of an energy eigenstate of a harmonic oscillator in an arbitrary representation $|X\rangle$ by multiplying the eigenvalue equation Eq. (2.11) from the left with the state $\langle X|$. This yields

$$\langle X|\hat{H}|m\rangle = E_m \langle X|m\rangle = E_m u_m(X).$$

The problem is obviously to evaluate the Hamiltonian acting on $|X\rangle$. We now show this for the special case of $\hat{X}$ being the position operator $\hat{x}$.

2.2.2 Position Representation

In the present section we consider the energy eigenstate in position representation. With the help of the completeness relation

$$\mathbf{1} = \int dx \, |x\rangle \langle x|$$

we find the decomposition

$$|m\rangle = \mathbf{1}|m\rangle = \int dx \, |x\rangle \langle x|m\rangle \equiv \int dx \, u_m(x) \, |x\rangle$$

of the mth energy eigenstate. Here $u_m(x) \equiv \langle x|m\rangle$ is the position wave function of the energy eigenstate $|m\rangle$.

In Appendix A.1 we show that as a consequence of the square integrability

$$\langle m|m\rangle = \int dx \, \langle m|x\rangle \langle x|m\rangle = \int dx \, u_m^*(x) u_m(x) = \int dx \, |u_m(x)|^2 = 1$$

of the eigenfunctions $|m\rangle$ the spectrum of the Hamiltonian is discrete.

Schrödinger Equation in Coordinate Representation

We now derive the position representation of the eigenvalue equation (2.11). Hence, after multiplying the eigenvalue equation Eq. (2.11) by $\langle x|$ we find

$$\langle x| \hat{H} |m\rangle = E_m \langle x|m\rangle \equiv E_m u_m(x)$$

and with the Hamiltonian Eq. (2.1)

$$-\langle x| \hat{p}^2 |m\rangle + 2M(E_m - \tfrac{1}{2} M\Omega^2 x^2) u_m(x) = 0.$$

As mentioned above the problem is to evaluate the first term in this equation since the position eigenstates $|x\rangle$ are not eigenstates of the momentum operator $\hat{p}$. However, we can express this term in terms of the position wave function when we use the completeness relation

$$\mathbf{1} = \int dp \, |p\rangle \langle p|$$

of the momentum eigenstates $|p\rangle$. Indeed, we find

$$\langle x| \hat{p}^2 |m\rangle = \int dp \, \langle x| \hat{p}^2 |p\rangle \langle p|m\rangle = \int dp \, p^2 \, \langle x|p\rangle \langle p|m\rangle.$$

In the next step we make use of the position representation Eq. (2.10)

$$\langle x|p\rangle = \frac{1}{\sqrt{2\pi\hbar}} e^{ipx/\hbar} \tag{2.12}$$

of a momentum eigenstate and write

$$\langle x| \hat{p}^2 |m\rangle = \int dp \, p^2 \frac{1}{\sqrt{2\pi\hbar}} e^{ipx/\hbar} \langle p|m\rangle = \frac{\partial^2}{\partial(\tfrac{i}{\hbar}x)^2} \int dp \, \frac{1}{\sqrt{2\pi\hbar}} e^{ipx/\hbar} \langle p|m\rangle.$$

We identify the plane wave in the integral as the position representation Eq. (2.10) of the momentum eigenstate and obtain

$$\langle x|\hat{p}^2|m\rangle = -\hbar^2 \frac{\partial^2}{\partial x^2} \underbrace{\langle x| \int dp\, |p\rangle\langle p|m\rangle}_{1} = -\hbar^2 \frac{\partial^2}{\partial x^2} u_m(x). \qquad (2.13)$$

Here we have used in the last step the completeness relation of the momentum eigenstates.

With this result the eigenvalue equation (2.11) reads in position representation

$$\frac{d^2 u_m(x)}{dx^2} + \frac{2M}{\hbar^2}\left(E_m - \frac{1}{2}M\Omega^2 x^2\right) u_m(x) = 0. \qquad (2.14)$$

Obviously this result is consistent with the notion of replacing the kinetic energy operator by the second derivative with respect to position multiplied by $-\hbar^2/(2M)$.

Solution of the Schrödinger Equation

The Schrödinger equation in position representation, Eq. (2.14), is the differential equation of the parabolic cylinder function. To bring this out more clearly and to read off the energy eigenvalues we now transform the equation into a more convenient form by introducing scaled variables. For the energy we use $E_m \equiv \hbar\Omega \eta_m$ and as a position variable we introduce $\xi \equiv \kappa x$, where

$$\kappa \equiv \sqrt{M\Omega/\hbar}. \qquad (2.15)$$

In our new variable ξ the Schrödinger equation then reads

$$\frac{d^2 u_m(\xi)}{d\xi^2} + (2\eta_m - \xi^2) u_m(\xi) = 0. \qquad (2.16)$$

In Appendix A.1 we discuss the solutions of this equation and perform an asymptotic expansion of the resulting functions. However, strictly speaking this equation is not quite the standard differential equation

$$\frac{d^2 u(y)}{dy^2} - \left(a + \frac{1}{4}y^2\right) u(y) = 0$$

defining the parabolic cylinder function. The transformation $y \equiv \xi\sqrt{2}$ and the identification $a \equiv -\eta_m$ connects the two equations.

We have to solve the first equation subjected to the following boundary condition: The solutions must vanish when ξ goes to infinity in order to guarantee their normalizability.

In Appendix A.1 we show that normalizable wave functions only occur for $\eta_m = m + 1/2$. In this case the solutions take the form

$$u_m(\xi) = \mathcal{N}_m H_m(\xi)\, e^{-\frac{1}{2}\xi^2} \qquad (2.17a)$$

where the normalization constant

$$\mathcal{N}_m = \left(\frac{1}{\pi}\right)^{1/4} \frac{1}{(2^m \, m!)^{1/2}} \qquad (2.17\text{b})$$

and $H_m(\xi)$ is the mth Hermite polynomial.

We conclude this section by presenting the wave functions

$$u_m(x) = \mathcal{N}_m \, H_m(\kappa x) \, e^{-\frac{1}{2}(\kappa x)^2}$$

in the unscaled variables where the normalization constant is

$$\mathcal{N}_m = \left(\frac{\kappa^2}{\pi}\right)^{1/4} \frac{1}{(2^m \, m!)^{1/2}}.$$

The energy eigenvalues read

$$E_m = \hbar\Omega(m + 1/2).$$

In the case of the harmonic oscillator the position and momentum variables both enter quadratically. Hence, the corresponding wave functions are very similar. In particular, the momentum wave function is also given by Hermite polynomials as discussed in the problem 2.1.

2.3 Density Operator: A Brief Introduction

The state vector contains the full information about a quantum system. However, in many cases we do not know every detail of our system. This could be for example because the system has too many degrees of freedom. In particular, when it is coupled to a reservoir we cannot keep track of the motion of the constituents. An atom that undergoes spontaneous emission or the damping of the radiation in a cavity are such systems coupled to a reservoir. In this case we cannot describe the system by a state vector but need a new concept.

2.3.1 A State Vector is not Enough!

We cannot describe every quantum state by a state vector, that is by a wave function. Why is that? In order to answer this question and to be specific we consider a particle in a one-dimensional harmonic oscillator potential well. However, the treatment is more general.

Classical Probabilities plus Interference

We first discuss the case when the particle is in a single energy eigenstate $|m\rangle$. Here we do not address the question of how to prepare this state but rather focus on a different issue: What is the probability $W(x)\,dx$ to find the particle between the positions x and $x + dx$?

2.3 Density Operator: A Brief Introduction

Since there is still no dynamics we consider an ensemble of identically prepared particles and make a single position measurement on each. Quantum mechanics predicts the resulting probability density $W(x)$: According to the Born interpretation $W(x)$ is the square of the energy wave function $u_m(x)$ in position representation. Here we have made use of the fact, that the energy wave functions of the harmonic oscillator are real. In general these wave functions do not have to be real. In this case the probability density is the absolute value squared of the wave function, that is

$$W_{|m\rangle}(x) = |\langle x|m\rangle|^2 = |u_m(x)|^2.$$

In order to keep the treatment as general as possible we throughout this section treat $u_m(x)$ as a complex quantity.

In this context it is worthwhile mentioning that in general we do not distinguish between probability density and probability. Moreover, we usually refer to $W(x)$ as the probability to find the particle at position x, when we really mean $W(x)\,dx$ and the interval between x and $x + dx$.

Next, we consider a more complicated case when the oscillator is in a superposition

$$|\psi\rangle = \sum_{m=0}^{\infty} \psi_m |m\rangle \tag{2.18}$$

of many energy eigenstates. Here ψ_m are complex-valued expansion coefficients. Again we do not ask how to obtain this superposition but rather ask for the probability to find the particle at the position x. The wave function

$$\psi(x) \equiv \langle x|\psi\rangle = \sum_{m=0}^{\infty} \psi_m u_m(x)$$

in this case is a superposition of energy wave functions. Hence, the probability to find the particle at position x reads

$$W(x) = |\psi(x)|^2 = \sum_{m,n=0}^{\infty} \psi_m^* \psi_n u_m^*(x) u_n(x).$$

We can now decompose the double sum into the terms where $m = n$ and the terms $m \neq n$. In this case we find

$$W(x) = \sum_{m=0}^{\infty} |\psi_m|^2 |u_m(x)|^2 + \sum_{m \neq n} \psi_m^* \psi_n u_m^*(x) u_n(x)$$

$$- \sum_{m=0}^{\infty} P_m W_{|m\rangle}(x) + \sum_{m \neq n} \psi_m^* \psi_n u_m^*(x) u_n(x).$$

The individual terms in the first sum have a distinct meaning: Each term represents the probability to find the particle at the position x given it is in the mth energy eigenstate. The probability to be in the mth energy eigenstate is $P_m \equiv \psi_m^* \psi_m$.

We note, however, that the probability to find the particle at the position x provided it is prepared in this superposition state $|\psi\rangle$, Eq. (2.18), is not simply the sum of these probabilities but also involves the double sum containing the products $\psi_m^* \psi_n u_m^*(x) u_n(x)$ for $m \neq n$. These additional terms are the crucial distinction between classical and quantum mechanics.

Averaged Quantum States?

We now turn to a slightly different question. We assume that the only information we have about the quantum state of the particle is a family of probabilities P_m to be in the mth energy eigenstate. Can we still describe the particle by a superposition state? What are the expansion coefficients ψ_m?

Why not choose them as the square root of the probabilities P_m? This is only a special case of the more general situation

$$\psi_m(\phi_m) = \sqrt{P_m}\, e^{i\phi_m}.$$

But what fixes the phase angles ϕ_m?

Since we do not have any information about them we have to average over all possible realizations, that is, we perform an ensemble average. The most elementary probability distribution for these phases is a constant distribution which produces the average

$$\overline{e^{i\phi_m}} = \int_{-\pi}^{\pi} d\phi_m\, e^{i\phi_m}\, \frac{1}{2\pi} = 0.$$

In this case we have to average the probability

$$W(x) = \sum_{m=0}^{\infty} P_m |u_m(x)|^2 + \sum_{m \neq n} \sqrt{P_m \cdot P_n}\, e^{i(\phi_n - \phi_m)} u_m^*(x) u_n(x)$$

to find the particle at the position x with respect to a constant distribution of phases. Indeed, the averaged probability distribution $\overline{W}(x)$ reads

$$\overline{W}(x) = \sum_{m=0}^{\infty} P_m |u_m(x)|^2.$$

Obviously, using this averaging procedure, we can get the position distribution correctly. But can we represent the quantum state of the oscillator characterized by probabilities and random phases by a state vector

$$|\psi\rangle = \sum_{m=0}^{\infty} \sqrt{P_m}\, e^{i\phi_m} |m\rangle,$$

that is a superposition of energy eigenstates? The answer is a clear no! Indeed, when we take the average of the quantum state we obviously get zero.

Quantum State as an Operator

We gain more insight into the appropriate definition of the state by rewriting the probability to find the particle at the position x. For this purpose we again recall the relation

$$W(x) = |\psi(x)|^2 = |\langle x|\psi\rangle|^2 = \langle x|\psi\rangle \langle\psi|x\rangle = \langle x| [|\psi\rangle \langle\psi|] |x\rangle.$$

2.3 Density Operator: A Brief Introduction

This formulation of the probability distribution suggests to introduce the hermitian operator

$$\hat{\rho} \equiv |\psi\rangle \langle \psi|$$

which yields the compact expression

$$W(x) = \langle x| \hat{\rho} |x\rangle$$

for the probability.

Here we emphasize that the order of the bra and ket vectors $\langle\psi|$ and $|\psi\rangle$ is important: For the order $\langle\psi|\psi\rangle$ we have the scalar product providing us with a number. However, when we have the opposite order we arrive at an operator

$$[|\psi\rangle \langle\psi|] |\phi\rangle = [\langle\psi|\phi\rangle] |\psi\rangle$$

which acts on another state $|\phi\rangle$.

For the superposition state

$$|\psi\rangle = \sum_{m=0}^{\infty} \psi_m |m\rangle$$

this operator reads

$$\hat{\rho} = \sum_{m,n=0}^{\infty} \psi_m \psi_n^* |m\rangle \langle n| = \sum_{m,n=0}^{\infty} \rho_{m,n} |m\rangle \langle n|. \qquad (2.19)$$

We note that in general all kind of combinations m and n of the operators $|m\rangle \langle n|$ emerge. The complex-valued numbers $\rho_{m,n} = \psi_m \psi_n^*$ form a matrix consisting of products made out of the expansion coefficients ψ_n. Since we can use the operator $\hat{\rho}$ to calculate probability densities such as the position probability and others, the operator $\hat{\rho}$ carries the name density operator and the matrix elements $\rho_{m,n}$ form the density matrix in the energy representation.

What is the operator $\hat{\rho}$ when we now average over the phases? To answer this question we substitute the expression

$$\psi_m = \sqrt{P_m} e^{i\phi_m}$$

for the expansion coefficients ψ_m into the definition of $\hat{\rho}$ and find

$$\hat{\rho} = \sum_{m,n=0}^{\infty} \sqrt{P_m \cdot P_n} e^{i(\phi_m - \phi_n)} |m\rangle \langle n|$$

$$= \sum_{m=0}^{\infty} P_m |m\rangle \langle m| + \sum_{m \neq n}^{\infty} \sqrt{P_m \cdot P_n} e^{i(\phi_m - \phi_n)} |m\rangle \langle n|,$$

which when averaged over the various realizations of the phases ϕ_m and ϕ_n yields

$$\overline{\hat{\rho}} = \sum_{m=0}^{\infty} P_m |m\rangle \langle m| = \sum_{m=0}^{\infty} \rho_{m,m} |m\rangle \langle m|.$$

Here, in contrast to the superposition state only the terms $m = n$ in the operators $|m\rangle\langle n|$ appear.

In conclusion, we have two kinds of averages: The first one results from quantum mechanics and the fact that a quantum state can only provide a statistical description. The second average is a classical one. It reflects the fact that we do not have complete information about the system: We do not even know in which quantum state the system is. This gives rise to the averaged density operator $\bar{\hat{\rho}}$. Whereas in the first case we can describe the state by a state vector in the second case we have to resort to the density operator formalism. Sometimes we refer to state vectors as pure states whereas averaged density operators describe mixed states. In the remainder of the book we do not distinguish between $\hat{\rho}$ and $\bar{\hat{\rho}}$ and always write $\hat{\rho}$ even when we deal with a mixed state.

2.3.2 Definition and Properties

In the preceding section we have motivated the density operator as the most general description of a quantum state. We are forced to use the density operator concept whenever we do not have enough information about the system. In the example discussed above we do not know the phases of the probability amplitudes. In the present section we briefly summarize various properties of the density operator.

We consider a quantum system which can assume any state $|\psi_j\rangle$ with $j = 0, 1, \ldots$. When the system is in a mixed state we cannot say that it is in the state $|\psi_1\rangle$, $|\psi_2\rangle$ or any other state $|\psi_j\rangle$. Moreover, we cannot describe its state by a superposition

$$|\psi\rangle = \sum_{j=0}^{\infty} \psi_j |\psi_j\rangle.$$

Since in a mixed state we only know the probability $P_{|\psi_j\rangle}$ with which the state $|\psi_j\rangle$ appears it is the operator

$$\hat{\rho} = \sum_{j=0}^{\infty} P_{|\psi_j\rangle} |\psi_j\rangle\langle\psi_j| = \sum_{j=0}^{\infty} \rho_{j,j} |\psi_j\rangle\langle\psi_j|, \qquad (2.20)$$

which describes the "quantum state" of the system. This operator obviously only involves diagonal elements of the operator $|\psi_j\rangle\langle\psi_k|$ and the density matrix $\rho_{j,k} \equiv P_j \delta_{jk}$ is diagonal in the $|\psi_j\rangle$-representation.

We calculate the expectation value

$$\langle\psi_m|\hat{\rho}|\psi_m\rangle = \sum_{j=0}^{\infty} P_{|\psi_j\rangle} \langle\psi_m|\psi_j\rangle\langle\psi_j|\psi_m\rangle$$

and assume that the system of states $|\psi_m\rangle$ forms an orthonormal set with

$$\langle\psi_m|\psi_j\rangle = \delta_{j,m}.$$

Hence we recognize, that the diagonal element

$$\langle\psi_m|\hat{\rho}|\psi_m\rangle = P_{|\psi_m\rangle}$$

of the density matrix is the probability $P_{|\psi_m\rangle}$ to find the quantum system in the state $|\psi_m\rangle$.

Since the quantities $P_{|\psi_m\rangle}$ are probabilities they have to add up to unity, that is,

$$\operatorname{Tr}\hat{\rho} \equiv \sum_{m=0}^{\infty} \langle\psi_m|\hat{\rho}|\psi_m\rangle = \sum_{m=0}^{\infty} P_{|\psi_m\rangle} = 1.$$

Here we have introduced the trace operation, that is performed a summation over the diagonal elements of the density operator.

2.3.3 Trace of Operator

The trace operation is not only defined for the density operator, but for any operator $\hat{O}$. Indeed, the definition of the trace of the operator $\hat{O}$ reads

$$\operatorname{Tr}\hat{O} \equiv \sum_{m=0}^{\infty} \langle\psi_m|\hat{O}|\psi_m\rangle \tag{2.21}$$

where $|\psi_m\rangle$ is a complete set of states.

We now show that this operation is the sum over the diagonal elements of the operator $\hat{O}$ in the representation $|\psi_m\rangle$. For this purpose we represent the operator $\hat{O}$ in the states $|\psi_m\rangle$. We multiply $\hat{O}$ from the left and from the right by the unity operator represented by the completeness relation

$$\mathbb{1} = \sum_{m=0}^{\infty} |\psi_m\rangle\langle\psi_m| \tag{2.22}$$

and find

$$\hat{O} = \mathbb{1}\hat{O}\mathbb{1} = \sum_{m,n=0}^{\infty} |\psi_m\rangle\langle\psi_m|\hat{O}|\psi_n\rangle\langle\psi_n|.$$

Hence, the operator $\hat{O}$ expressed in the basis $|\psi_m\rangle$ takes the form

$$\hat{O} = \sum_{m,n=0}^{\infty} O_{m,n} |\psi_m\rangle\langle\psi_n|,$$

where

$$O_{m,n} \equiv \langle\psi_m|\hat{O}|\psi_n\rangle$$

are the matrix elements of the operator in the states $|\psi_m\rangle$. Note that the diagonal elements are just $O_{m,m} = \langle\psi_m|\hat{O}|\psi_m\rangle$ and the trace as defined is indeed the sum over the diagonal elements.

Trace is Independent of Representation

We now show that the above definition of the trace of an operator is independent of the complete set of states $|\psi_m\rangle$ used to calculate the diagonal elements of the operator. For this purpose we consider a different complete set of orthonormal states $|\phi_n\rangle$.

When we insert the completeness relation

$$\mathbb{1} = \sum_{n=0}^{\infty} |\phi_n\rangle \langle\phi_n|$$

of the states $|\phi_n\rangle$ into the definition Eq. (2.21) of the trace of an operator $\hat{O}$ we find

$$\operatorname{Tr} \hat{O} = \sum_{l=0}^{\infty} \langle\psi_l| \mathbb{1}\hat{O}\mathbb{1} |\psi_l\rangle = \sum_{l,m,n=0}^{\infty} \langle\psi_l|\phi_m\rangle \langle\phi_m|\hat{O}|\phi_n\rangle \langle\phi_n|\psi_l\rangle ,$$

that is

$$\operatorname{Tr} \hat{O} = \sum_{m,n=0}^{\infty} \langle\phi_m|\hat{O}|\phi_n\rangle \langle\phi_n| \left(\sum_{l=0}^{\infty} |\psi_l\rangle \langle\psi_l|\right) |\phi_m\rangle .$$

With the help of the completeness relation Eq. (2.22) of the states $|\psi_l\rangle$ we arrive at

$$\operatorname{Tr} \hat{O} = \sum_{m,n=0}^{\infty} \langle\phi_m|\hat{O}|\phi_n\rangle \langle\phi_n|\phi_m\rangle ,$$

or

$$\operatorname{Tr} \hat{O} = \sum_{m=0}^{\infty} \langle\phi_m|\hat{O}|\phi_m\rangle ,$$

where we have used the orthonormality relation $\langle\phi_n|\phi_m\rangle = \delta_{m,n}$ of the states $|\phi_j\rangle$. Indeed, the trace is independent of the representation.

Expectation Value is the Trace

The trace operation plays a central role in quantum mechanics since it allows us to calculate expectation values of operators. Such a quantity involves *two* averages: (*i*) We first have to perform the quantum mechanical average, that is the expectation value of the operator $\hat{O}$ in the state $|\psi_m\rangle$, via $\langle\hat{O}\rangle_{|\psi_m\rangle} = \langle\psi_m|\hat{O}|\psi_m\rangle$, and (*ii*) we have to average over the distribution of the states $|\psi_m\rangle$, each appearing with the probability $P_{|\psi_m\rangle}$. Hence we find

$$\langle\hat{O}\rangle = \sum_{m=0}^{\infty} \langle\psi_m|\hat{O}|\psi_m\rangle P_{|\psi_m\rangle}.$$

On the other hand when we now calculate the expression

$$\operatorname{Tr}(\hat{O}\hat{\rho}) = \sum_{l=0}^{\infty} \langle\psi_l|\hat{O}\hat{\rho}|\psi_l\rangle = \sum_{l=0}^{\infty} \langle\psi_l|\hat{O}\left(\sum_{m=0}^{\infty} P_{|\psi_m\rangle} |\psi_m\rangle \langle\psi_m|\right) |\psi_l\rangle$$

$$= \sum_{l,m=0}^{\infty} \langle\psi_l|\hat{O}|\psi_m\rangle P_{|\psi_m\rangle} \langle\psi_m|\psi_l\rangle$$

we arrive at the relation

$$\operatorname{Tr}(\hat{O}\hat{\rho}) = \sum_{m=0}^{\infty} \langle\psi_m|\hat{O}|\psi_m\rangle P_{|\psi_m\rangle} ,$$

which proves the statement

$$\langle\hat{O}\rangle = \operatorname{Tr}(\hat{O}\hat{\rho}).$$

The expectation value of an operator $\hat{O}$ in a quantum system described by a density operator $\hat{\rho}$ is the trace of the product of $\hat{O}$ and $\hat{\rho}$.

An Example

We now illustrate the method of calculating a quantum mechanical expectation value of an operator using the trace formalism. For this purpose we evaluate the expectation value of the density operator itself, that is,

$$\langle \hat{\rho} \rangle = \text{Tr}(\hat{\rho}\hat{\rho}) = \sum_{l=0}^{\infty} \langle \psi_l | \hat{\rho}^2 | \psi_l \rangle.$$

When we insert the definition of the density operator we find

$$\langle \hat{\rho} \rangle = \sum_{l=0}^{\infty} \langle \psi_l | \left(\sum_{m=0}^{\infty} P_{|\psi_m\rangle} |\psi_m\rangle \langle \psi_m| \right) \left(\sum_{n=0}^{\infty} P_{|\psi_n\rangle} |\psi_n\rangle \langle \psi_n| \right) |\psi_l\rangle,$$

or

$$\langle \hat{\rho} \rangle = \sum_{l,m,n=0}^{\infty} P_{|\psi_m\rangle} P_{|\psi_n\rangle} \langle \psi_l | \psi_m \rangle \langle \psi_m | \psi_n \rangle \langle \psi_n | \psi_l \rangle.$$

With the help of the orthonormality relations we arrive at

$$\langle \hat{\rho} \rangle = \sum_{l=0}^{\infty} P_{|\psi_l\rangle}^2. \tag{2.23}$$

The normalization condition

$$\sum_{l=0}^{\infty} P_{|\psi_l\rangle} = 1$$

ensures that the individual probabilities $P_{|\psi_l\rangle}$ are smaller than unity. Hence, we have the inequality

$$P_{|\psi_l\rangle}^2 \leq P_{|\psi_l\rangle}$$

which yields

$$\sum_{l=0}^{\infty} P_{|\psi_l\rangle}^2 \leq \sum_{l=0}^{\infty} P_{|\psi_l\rangle} = 1.$$

The equal sign occurs if and only if there is a single state $|\psi\rangle$ contributing to the sum, that is

$$P_{|\psi_m\rangle} = \delta_{l,m}.$$

In this case the quantum system is described by a single state, and is therefore a pure state, rather than a mixture of states.

We can now return to our original question concerning the expectation value of the density operator Eq. (2.23) and find the inequality

$$\langle \hat{\rho} \rangle \leq 1,$$

where the equal sign occurs for a pure state only.

2.3.4 Examples of a Density Operator

We now illustrate this formalism by comparing and contrasting the density operators of a *thermal state* and a *thermal phase state* of a harmonic oscillator. Throughout this section we use the energy eigenstates $|n\rangle$ of the harmonic oscillator to represent the density operator, that is, $|\psi_n\rangle = |n\rangle$.

Thermal State

In a thermal state the nth energy eigenstate is occupied with a probability $P_n = Ne^{-n\beta}$ where $\beta \equiv \hbar\Omega/(k_B T)$. Here Ω, k_B and T denote the frequency of the oscillator, the Boltzmann constant and the temperature corresponding to the thermal state, respectively. In this case the density operator reads

$$\hat{\rho} = N \sum_{n=0}^{\infty} e^{-n\beta} |n\rangle\langle n|.$$

Here the normalization constant N follows from the condition

$$1 = \text{Tr}\,\hat{\rho} = \sum_{n=0}^{\infty} \langle n|\hat{\rho}|n\rangle = N \sum_{n=0}^{\infty} e^{-n\beta} = \frac{N}{1 - e^{-\beta}},$$

that is,

$$N = 1 - e^{-\beta}.$$

Hence, the density operator takes the explicit form

$$\hat{\rho} = (1 - e^{-\beta}) \sum_{n=0}^{\infty} e^{-n\beta} |n\rangle\langle n|.$$

The temperature T is intimately related to the average energy

$$<\hat{H}> \equiv \text{Tr}(\hat{H}\hat{\rho}) = \sum_{n=0}^{\infty} \langle n|\hat{H}\hat{\rho}|n\rangle = \sum_{n=0}^{\infty} \langle n|\hbar\Omega\left(n + \frac{1}{2}\right)\hat{\rho}|n\rangle$$

in this thermal state. In the last step we have recalled the energy eigenvalue equation.

We therefore have to evaluate the quantity

$$<\hat{H}> = \hbar\Omega \sum_{n=0}^{\infty} \left(n + \frac{1}{2}\right) W_n = \hbar\Omega \left[\left(\sum_{n=0}^{\infty} nW_n\right) + \frac{1}{2}\right] = \hbar\Omega \left(n_{\text{th}} + \frac{1}{2}\right)$$

where

$$W_n \equiv \langle n|\hat{\rho}|n\rangle = Ne^{-\beta n} = (1 - e^{-\beta})e^{-\beta n}$$

and

$$n_{\text{th}} \equiv \sum_{n=0}^{\infty} nW_n$$

are the energy distribution and the average number of thermal quanta in this state, respectively.

When we substitute the energy distribution into the expression for the average number of thermal quanta we can easily perform the summation over n and find

$$n_{\text{th}} = (1 - e^{-\beta}) \sum_{n=0}^{\infty} ne^{-\beta n} = -(1 - e^{-\beta})\frac{\partial}{\partial \beta} \sum_{n=0}^{\infty} e^{-n\beta} = -(1 - e^{-\beta})\frac{\partial}{\partial \beta}\left(\frac{1}{1 - e^{-\beta}}\right),$$

which yields

$$n_{\text{th}} = \frac{e^{-\beta}}{1 - e^{-\beta}} = \frac{1}{\exp\left(\frac{\hbar\Omega}{k_B T}\right) - 1}.$$

Hence, the average number of thermal quanta is governed by the temperature T.
When we use the relation
$$e^\beta = \frac{n_{\text{th}} + 1}{n_{\text{th}}}$$
we arrive at the expression
$$\hat{\rho}_{\text{th}} = \frac{1}{n_{\text{th}} + 1} \sum_{n=0}^{\infty} \left(\frac{n_{\text{th}}}{n_{\text{th}} + 1}\right)^n |n\rangle\langle n| \qquad (2.24)$$
for the density operator of a thermal state of average energy $<\hat{H}> = \hbar\Omega(n_{\text{th}} + 1/2)$.

Thermal Phase State

The difference between a mixed state and a pure state stands out most clearly in a comparison between a thermal state and the thermal phase state
$$|\varphi_0\rangle \equiv \frac{1}{\sqrt{n_{\text{th}} + 1}} \sum_{n=0}^{\infty} \left(\frac{n_{\text{th}}}{n_{\text{th}} + 1}\right)^{n/2} |n\rangle. \qquad (2.25)$$
We note that the state $|\varphi_0\rangle$ does enjoy the same energy distribution
$$W_n = |\langle n|\varphi_0\rangle|^2 = \frac{1}{n_{\text{th}} + 1} \left(\frac{n_{\text{th}}}{n_{\text{th}} + 1}\right)^n$$
as the thermal state. Nevertheless this state is quite different from a thermal state, since it is a quantum mechanical superposition of number states. This manifests itself in the density operator
$$\hat{\rho}_\varphi = |\varphi_0\rangle\langle\varphi_0| = \frac{1}{n_{\text{th}} + 1} \sum_{m,n=0}^{\infty} \left(\frac{n_{\text{th}}}{n_{\text{th}} + 1}\right)^{(m+n)/2} |m\rangle\langle n|.$$

We note that in contrast to the density operator $\hat{\rho}_{\text{th}}$ of a thermal field which only has diagonal elements $\langle n|\hat{\rho}_{\text{th}}|n\rangle$, this operator has also off-diagonal elements. They reflect the fact that the state $|\varphi_0\rangle$ consists of a quantum mechanical superposition of photon number states, that is the interference of individual number states.

2.4 Time Evolution of Quantum States

So far, we have considered quantum kinematics, that is we have discussed the properties of quantum states in the absence of time evolution. We now turn to quantum dynamics and consider the time evolution of states following from the Schrödinger equation
$$i\hbar \frac{d}{dt} |\psi(t)\rangle = \hat{H} |\psi(t)\rangle. \qquad (2.26)$$
Here $\hat{H}$ is the Hamiltonian of the problem at hand.

2.4.1 Motion of a Wave Packet

In the case of a non-relativistic particle moving in a potential U the Hamiltonian, Eq. (2.1),

$$\hat{H} \equiv \frac{\hat{p}^2}{2M} + U(\hat{x})$$

contains the operators of kinetic and potential energy.

Time Evolution Operator

When U is time independent also the Hamiltonian $\hat{H} \equiv \hat{H}_0$ is time independent. Therefore, a formal solution of the Schrödinger equation reads

$$|\psi(t)\rangle = \exp\left[-\frac{i}{\hbar}\hat{H}_0 \cdot (t - t_0)\right] |\psi(t_0)\rangle \equiv \hat{\mathcal{U}}(t, t_0) |\psi(t_0)\rangle . \qquad (2.27)$$

We can easily prove this relation by differentiating this expression with respect to time. For this purpose we recall from Appendix B that in general the differentiation of an exponential operator

$$\hat{\mathcal{U}}(t) \equiv \exp[\hat{B}(t)]$$

is not trivial. Only when the operators $\hat{B}(t)$ and $\hat{B}(t')$ at different times t and t' commute with each other, that is

$$\left[\hat{B}(t), \hat{B}(t')\right] = 0, \qquad (2.28)$$

the differentiation of the exponential operator is identical to a c-number exponential: The exponential reproduces itself and is multiplied by the time derivative of the argument.

Since in the present case

$$\hat{B} \equiv -\frac{i}{\hbar}\hat{H}_0 \cdot t$$

the condition Eq. (2.28) is satisfied.

The unitary time evolution operator $\hat{\mathcal{U}}(t, t_0)$ connects the quantum states $|\psi(t_0)\rangle$ and $|\psi(t)\rangle$ at the two different times $t_0 < t$. Solving the time dependent Schrödinger equation therefore amounts to finding $\hat{\mathcal{U}}$.

Equation (2.27) gives an exact expression for the time evolution operator $\hat{\mathcal{U}}$. We emphasize, however, that this expression is only a formal one. The Hamiltonian involves the square of the momentum operator and through the potential U the position operator. Since these operators do not commute we cannot decompose the exponential in Eq. (2.27) into two separate exponentials containing only momentum or position operators. Therefore, we cannot apply the operators successively, both operators have to be applied together.

Time Evolution in Energy Representation

When the initial state $|\psi(t_0)\rangle$ is an eigenstate $|m\rangle$ of $\hat{H}_0$, that is

$$\hat{H}_0 |m\rangle = E_m |m\rangle ,$$

this expression simplifies to

$$|\psi(t)\rangle = \exp\left[-\frac{i}{\hbar}E_m(t-t_0)\right]|m\rangle,$$

and the initial state only acquires an overall phase. This phase is rather uninteresting since it cancels when we calculate observables such as distribution functions or expectation values.

The situation becomes more interesting when our initial state is not an eigenstate of $\hat{H}_0$. Since the energy eigenstates $|m\rangle$ are complete, we can expand the state

$$|\psi(t_0)\rangle = \sum_{m=0}^{\infty} \psi_m |m\rangle, \qquad (2.29)$$

where $\psi_m \equiv \langle m|\psi(t_0)\rangle$ denote the expansion coefficients. Hence, the initial state $|\psi(t_0)\rangle$ is a superposition of energy eigenstates. We call such an object *wave packet*.

When we substitute the expansion Eq. (2.29) into the formal solution Eq. (2.27) of the Schrödinger equation and use the fact that the states $|m\rangle$ are eigenstates of $\hat{H}_0$ we find

$$|\psi(t)\rangle = \sum_{m=0}^{\infty} \psi_m \exp\left[-\frac{i}{\hbar}E_m(t-t_0)\right]|m\rangle.$$

In this case we do not obtain an overall phase: Each energy eigenstate acquires a different phase and consequently we obtain a superposition of energy eigenstates with different phases. This describes the motion of the initial wave packet.

We return to the problem of wave packet dynamics in more detail in Chapter 4 where we consider coherent and squeezed wave packets in a harmonic potential. Moreover, in Chapter 9 we analyze the motion of a nuclear wave packet in an anharmonic potential such as provided by the electronic states of a diatomic molecule. In the context of the resonant Jaynes-Cummings-Paul model we use in Sec. 15.1 the expression Eq. (2.27) for the time evolution operator to derive the state vector.

2.4.2 Time Evolution due to Interaction

So far, we have considered a closed system which does not interact with the outside world. The time evolution of the system resulted from the fact that the initial state was not an eigenstate of the Hamiltonian. In other words, the initial wave packet did not fit into the potential U. In the present section we consider the time evolution of a quantum system enforced by its interaction with an external force.

One possible quantum system could be a charged particle moving in a potential $U(x)$ provided for example by a Paul trap. We let the ion interact with a spatially homogeneous electric field $E_0(t)$ and describe the interaction by the Hamiltonian

$$\hat{H}_{\text{int}}(t) \equiv -ex \cdot E_0(t).$$

In this case we have the total Hamiltonian

$$\hat{H}(t) = \hat{H}_0 + \hat{H}_{\text{int}}(t) \qquad (2.30)$$

consisting of the Hamiltonians $\hat{H}_0$ and $\hat{H}_{\text{int}}$ of the particle and its interaction. The Schrödinger equation then reads

$$i\hbar \frac{d}{dt} |\psi(t)\rangle = \hat{H}(t) |\psi(t)\rangle = \left[\hat{H}_0 + \hat{H}_{\text{int}}(t)\right] |\psi(t)\rangle. \qquad (2.31)$$

Even when the initial state is an eigenstate of $\hat{H}_0$ it is not necessarily an eigenstate of the total Hamiltonian. Therefore, the time evolution is more complicated.

To gain more insight we first consider the most elementary case of a static field E_0. Then the total Hamiltonian is still time independent and we can use the formal solution Eq. (2.27). However, when the electric field originates for example from a laser field it is time dependent and consequently the interaction Hamiltonian is time dependent. In this case it is not straightforward to solve the time dependent Schrödinger equation.

We are tempted to generalize the formal solution Eq. (2.27) and try the ansatz

$$|\psi(t)\rangle = \exp\left[-\frac{i}{\hbar} \int_{t_0}^{t} dt' \, \hat{H}(t')\right] |\psi(t_0)\rangle. \qquad (2.32)$$

However, as shown in Appendix B this expression is only valid when the Hamiltonians $\hat{H}(t)$ at different times t commute. For arbitrary time dependence the time evolution operator connecting the states $|\psi(t)\rangle$ and $|\psi(t_0)\rangle$ is more complicated. We devote the next sections to deriving formal solutions for this case.

Interaction Picture

We conclude this section by noting that even when the interaction Hamiltonian is not explicitly time dependent we can arrive at a Schrödinger type equation with a time dependent Hamiltonian. When we try the ansatz

$$|\psi(t)\rangle = \exp\left[-\frac{i}{\hbar} \hat{H}_0 \cdot (t - t_0)\right] \left|\psi^{(\text{I})}(t)\right\rangle$$

we find from the Schrödinger equation Eq. (2.31) the transformed equation

$$i\hbar \frac{d}{dt} \left|\psi^{(\text{I})}(t)\right\rangle = \hat{H}_{\text{int}}^{(\text{I})}(t) \left|\psi^{(\text{I})}(t)\right\rangle$$

where we have introduced the Hamiltonian

$$\hat{H}_{\text{int}}^{(\text{I})}(t) \equiv \exp\left[-\frac{i}{\hbar} \hat{H}_0 \cdot (t - t_0)\right] \hat{H}_{\text{int}}(t) \exp\left[-\frac{i}{\hbar} \hat{H}_0 \cdot (t - t_0)\right].$$

Hence, the resulting equation of motion for the transformed state $\left|\psi^{(\text{I})}(t)\right\rangle$ takes the same form as the Schrödinger equation Eq. (2.31). However, we have to replace the total Hamiltonian by the transformed interaction Hamiltonian. We note that due to the time dependence of the transformation this Hamiltonian has become time dependent even when it was time independent before the transformation. Since this formulation emphasizes the interaction it is called the *interaction picture*.

We will use the interaction picture in Sec. 14.8 where we discuss the case of an atom interacting with a quantized field. Here the resulting Hamiltonian is indeed explicitly time dependent.

2.4.3 Time Dependent Hamiltonian

We now discuss various formal solutions of the time dependent Schrödinger equation

$$i\hbar \frac{d}{dt} |\psi(t)\rangle = \hat{H}(t) |\psi(t)\rangle. \tag{2.33}$$

Throughout this section we are not concerned with the origin of the time dependence in the Hamiltonian. It could be the result of a time dependent interaction or a transformation into the interaction picture as discussed before.

We derive various expressions for the time evolution operator $\hat{\mathcal{U}}(t, t_0)$ in the presence of a time dependent Hamiltonian $\hat{H}(t)$. The first derivation rests on infinitesimal transformations. They allow us to write the time evolution operator as an infinite product. We point out an interesting relation to the Volterra-Schlesinger product integral which is a generalization of the Riemann integral. The second approach relies on rewriting the Schrödinger equation as an integral equation and then iterating it. The third formal solution introduces the time ordering operator $\hat{T}$ which plays a prominent role in quantum electrodynamics.

Solution by Infinitesimal Transformations

Our first approach starts from the state vector $|\psi(t_0)\rangle$ at time t_0 and calculates the state vector $|\psi(t + t_0)\rangle$ at a time $t \equiv t_0 + \Delta t$ which is only infinitesimally larger. This allows us to replace the time derivative in the Schrödinger equation by the difference. We therefore find from

$$i\hbar \left. \frac{d|\psi(t)\rangle}{dt} \right|_{t_0} \cong i\hbar \frac{|\psi(t_0 + \Delta t)\rangle - |\psi(t_0)\rangle}{\Delta t} = \hat{H}(t_0) |\psi(t_0)\rangle$$

the state

$$|\psi(t_0 + \Delta t)\rangle = \left[\mathbb{1} - \frac{i}{\hbar} \Delta t \hat{H}(t_0) \right] |\psi(t_0)\rangle \tag{2.34}$$

at the time $t_0 + \Delta t$. Indeed, starting from the state at time t_0 we have made a step into the future by performing an infinitesimal transformation.

We now consider the next step necessary to reach the time $t_0 + 2\Delta t$. From Eq. (2.34) we obtain

$$\begin{aligned} |\psi(t_0 + 2\Delta t)\rangle &= |\psi(t_0 + \Delta t + \Delta t)\rangle \\ &= \left[\mathbb{1} - \frac{i}{\hbar} \Delta t \hat{H}(t_0 + \Delta t) \right] |\psi(t_0 + \Delta t)\rangle \\ &= \left[\mathbb{1} - \frac{i}{\hbar} \Delta t \hat{H}(t_0 + \Delta t) \right] \left[\mathbb{1} - \frac{i}{\hbar} \Delta t \hat{H}(t_0) \right] |\psi(t_0)\rangle. \end{aligned} \tag{2.35}$$

We note that there is an important ordering in this expression: The Hamiltonian $\hat{H}(t_0)$ at an earlier time t_0 acts before the Hamiltonian $\hat{H}(t_0 + \Delta t)$ at a later time $t_0 + \Delta t$. The time evolution of the state progresses by multiplying the infinitesimal time evolution operator

$$\delta \hat{\mathcal{U}}(t) \equiv \mathbb{1} - \frac{i}{\hbar} \Delta t \hat{H}(t)$$

from the left onto the state. The times in the argument of $\delta\hat{\mathcal{U}}(t)$ increase from right to left, with earlier times always being to the right of later times.

In order to achieve the finite time step from time t_0 to t we have to perform many such infinitesimal transformations after each other. For this purpose we first divide the time interval $t - t_0$ into N infinitesimal intervals $\Delta t \equiv (t - t_0)/N$ starting at times $t_\nu \equiv t_0 + \nu \cdot \Delta t$, where $\nu = 0, \ldots, N$. Then we let the number N of intervals approach infinity.

When we generalize the second order iteration Eq. (2.35) we find the expression

$$|\psi(t)\rangle = \hat{\mathcal{U}}(t, t_0) |\psi(t_0)\rangle,$$

where the time evolution operator

$$\hat{\mathcal{U}}(t, t_0) = \lim_{N \to \infty} \delta\hat{\mathcal{U}}(t_{N-1}) \cdot \delta\hat{\mathcal{U}}(t_{N-2}) \ldots \delta\hat{\mathcal{U}}(t_0) = \lim_{N \to \infty} \prod_{\nu=0}^{N-1} \left[\mathbb{1} - \frac{i}{\hbar}\Delta t \hat{H}(t_\nu)\right] \quad (2.36)$$

is an infinite product. Again we emphasize that in this product there is a strict order: Operators at earlier times act before operators at later times.

It is instructive to rewrite the infinite product. For this purpose we note that each factor in the product can be represented by the exponential of a logarithm. Indeed, we have

$$\mathbb{1} - \frac{i}{\hbar}\Delta t \hat{H}(t_\nu) = \exp\left\{\ln\left[\mathbb{1} - \frac{i}{\hbar}\Delta t \hat{H}(t_\nu)\right]\right\}.$$

Hence, the time evolution operator is a product of exponential operators. However, since the operators do not necessarily commute we cannot combine the arguments of the exponentials. This is only possible when the Hamiltonians $\hat{H}(t_\nu)$ at different times t_ν commute. Only in this case we have

$$e^{\hat{A}} \cdot e^{\hat{B}} = e^{\hat{A}+\hat{B}}$$

and hence,

$$\hat{\mathcal{U}}(t, t_0) = \lim_{N \to \infty} \exp\left\{\sum_{\nu=0}^{N-1} \ln\left[\mathbb{1} - \frac{i}{\hbar}\Delta t \hat{H}(t_\nu)\right]\right\}.$$

Since $\Delta t \to 0$ for $N \to \infty$ we can expand the logarithm $\ln(1 + z) \cong z$ and therefore find assuming that the limit exists

$$\hat{\mathcal{U}}(t, t_0) = \lim_{N \to \infty} \exp\left[-\frac{i}{\hbar} \sum_{\nu=0}^{N-1} \Delta t \hat{H}(t_\nu)\right]$$

$$= \exp\left[-\frac{i}{\hbar} \int_0^{t-t_0} dt'' \hat{H}(t_0 + t'')\right] = \exp\left[-\frac{i}{\hbar} \int_{t_0}^{t} dt' \hat{H}(t')\right]$$

in agreement with Eq. (2.32). Here we have made use of the definition

$$\lim_{N \to \infty} \sum_{\nu=0}^{N} \Delta t\, A(t_\nu) \equiv \int_0^t dt'\, A(t')$$

of the Riemann integral. In the last step we have introduced the shifted integration variable $t' \equiv t_0 + t''$

2.4 Time Evolution of Quantum States

Volterra-Schlesinger Product Integral

The representation Eq. (2.36) of the time evolution operator $\hat{U}$ as an infinite product of operators is a generalization of the Riemann integral in two respects: Instead of a sum we have a product and instead of c-numbers we have operators. Following Volterra and Schlesinger we define the so-called product integral via

$$\lim_{N\to\infty} \prod_{\nu=0}^{N-1} \Delta t \hat{A}(t_\nu) \equiv \prod_0^{t-t_0} dt'\, \hat{A}(t_0 + t')$$

and the time evolution operator reads

$$\hat{U}(t, t_0) = \prod_0^{t-t_0} dt' \left[\mathbb{1} - \frac{i}{\hbar}\hat{H}(t_0 + t')\right].$$

Unfortunately this formula is in general not very useful since the product integral is extremely difficult to perform and leads to higher transcendental functions.

Integral Equation

We now turn to yet another representation of the time evolution operator. For this purpose we first transform the Schrödinger equation into the integral equation

$$|\psi(t)\rangle = |\psi(t_0)\rangle - \frac{i}{\hbar} \int_{t_0}^t dt'\, \hat{H}(t') |\psi(t')\rangle. \tag{2.37}$$

Indeed, when we differentiate this expression we immediately obtain the Schrödinger equation Eq. (2.33).

When we now write this equation for $t = t'$, that is

$$|\psi(t')\rangle = |\psi(t_0)\rangle - \frac{i}{\hbar} \int_{t_0}^{t'} dt''\, \hat{H}(t'') |\psi(t'')\rangle$$

and substitute it into Eq. (2.37) we arrive at

$$|\psi(t)\rangle = |\psi(t_0)\rangle - \frac{i}{\hbar} \int_{t_0}^t dt'\, \hat{H}(t') |\psi(t_0)\rangle + \left(-\frac{i}{\hbar}\right)^2 \int_{t_0}^t dt' \int_{t_0}^{t'} dt''\, \hat{H}(t')\hat{H}(t'') |\psi(t'')\rangle.$$

(2.38)

The third term on the right hand side of this equation brings out clearly the time ordering. Since the integration over the variable t'' only extends to t' the Hamiltonian $\hat{H}(t'')$ at an earlier time always stands to the right of the Hamiltonian $\hat{H}(t')$ at a later time. We therefore have automatically taken into account time ordering.

We can now keep iterating till we find the formal solution

$$|\psi(t)\rangle = \hat{U}(t, t_0) |\psi(t_0)\rangle$$

where the time evolution operator $\hat{\mathcal{U}}$ reads

$$\hat{\mathcal{U}}(t, t_0) \equiv \sum_{n=0}^{\infty} \left(\frac{-i}{\hbar}\right)^n \int_{t_0}^{t} dt_n \int_{t_0}^{t_n} dt_{n-1} \ldots \int_{t_0}^{t_2} dt_1 \, \hat{H}(t_n)\hat{H}(t_{n-1})\ldots \hat{H}(t_2)\hat{H}(t_1). \quad (2.39)$$

Here we have introduced the integration variables $t_n, t_{n-1}, \ldots, t_2, t_1$ which fulfill the hierarchy $t_n \geq t_{n-1} \geq \ldots \geq t_1 \geq t_0$. This guarantees time ordering.

Time Ordering Operator

In Sec. 2.4.2 we have tried the ansatz

$$\hat{\mathcal{U}}(t, t_0) \equiv \exp\left[-\frac{i}{\hbar} \int_{t_0}^{t} dt' \hat{H}(t')\right]$$

for the time evolution operator of a time dependent Hamiltonian. Unfortunately, the resulting state is only a solution of the Schrödinger equation provided the Hamiltonians at different times commute. Causality enforces time ordering. We can therefore find the exact time evolution operator by enforcing this time ordering. For this purpose we introduce the time ordering operator $\hat{T}$. This operator guarantees that all operators at earlier times are to the right of operators at later times, that is

$$\hat{T}\left[\hat{A}(t')\,\hat{B}(t'')\right] = \begin{cases} \hat{B}(t'')\,\hat{A}(t') & \text{for } t'' > t' \\ \hat{A}(t')\,\hat{B}(t'') & \text{for } t' > t''. \end{cases}$$

The time evolution operator then reads

$$\hat{\mathcal{U}}(t, t_0) = \hat{T}\left\{\exp\left[-\frac{i}{\hbar}\int_0^t dt' \hat{H}(t')\right]\right\}.$$

This expression is identical to the representation Eq. (2.39) of the time evolution operator. To derive this result we first expand the exponential, that is

$$\hat{T}\left\{\exp\left[-\frac{i}{\hbar}\int_0^t dt' \hat{H}(t')\right]\right\} = \hat{T}\left\{\sum_{n=0}^{\infty}\left(\frac{-i}{\hbar}\right)^n \frac{1}{n!}\left[\int_0^t dt' \hat{H}(t')\right]^n\right\}$$

$$= \sum_{n=0}^{\infty}\left(\frac{-i}{\hbar}\right)^n \frac{1}{n!}\hat{T}\left[\int_0^t dt' \hat{H}(t')\right]^n \quad (2.40)$$

and note the relation

$$\hat{T}\left[\int_{t_0}^{t} dt' \hat{H}(t')\right]^n = n! \int_{t_0}^{t} dt_n \int_{t_0}^{t_n} dt_{n-1} \ldots \int_{t_0}^{t_2} dt_1 \hat{H}(t_n)\hat{H}(t_{n-1})\ldots \hat{H}(t_1)$$

derived in Appendix B. This procedure immediately yields Eq. (2.39).

2.4.4 Time Evolution of Density Operator

We conclude this chapter by extending our considerations of quantum dynamics to density operators. We first rederive the equation of motion for the density operator and then present a formal solution in terms of double-commutators.

Von Neumann Equation

We note that the time dependence of the density operator

$$\hat{\rho}(t) \equiv \sum_{j=0}^{\infty} P_{|\psi_j\rangle} |\psi_j(t)\rangle\langle\psi_j(t)|$$

results from time dependence of the state vector $|\psi_j\rangle$. Hence, when we differentiate the definition Eq. (2.20) of the density operator with respect to time we find

$$\frac{d\hat{\rho}}{dt} = \sum_{j=0}^{\infty} P_{|\psi_j\rangle} \left[\frac{d|\psi_j\rangle}{dt}\langle\psi_j| + |\psi_j\rangle\frac{d\langle\psi_j|}{dt} \right],$$

which with the help of the Schrödinger equation

$$i\hbar \frac{d|\psi_j\rangle}{dt} = \hat{H}|\psi_j\rangle$$

reduces to

$$\frac{d\hat{\rho}}{dt} = -\frac{i}{\hbar}\sum_{j=0}^{\infty} P_{|\psi_j\rangle}\left[\hat{H}|\psi_j\rangle\langle\psi_j| - |\psi_j\rangle\langle\psi_j|\hat{H}\right].$$

Here $\hat{H}$ denotes the Hamiltonian of the system.

We cast this equation in the compact form

$$\frac{d\hat{\rho}}{dt} = -\frac{i}{\hbar}[\hat{H}, \hat{\rho}] \qquad (2.41)$$

of the von Neumann equation.

In complete analogy with the state vector formulation of quantum mechanics discussed in Eq. (2.27) we can present the formally exact expression

$$\hat{\rho}(t) = \exp\left[-\frac{i}{\hbar}\hat{H}_0(t-t_0)\right]\hat{\rho}(t_0)\exp\left[\frac{i}{\hbar}\hat{H}_0(t-t_0)\right] = \hat{\mathcal{U}}(t,t_0)\hat{\rho}(t_0)\hat{\mathcal{U}}^\dagger(t,t_0) \qquad (2.42)$$

for the density operator when the Hamiltonian is time independent, that is $H \equiv H_0$. Indeed, by differentiation with respect to time we can easily prove this result. In Sec. 18.3.3 we use this technique to derive the density matrix equation of the one-atom maser.

Formal Solution

Unfortunately the formal solution Eq. (2.42) is only valid for time independent Hamiltonians. In Sec. 18.5. we investigate the interaction of an atom with a reservoir of harmonic oscillators representing electromagnetic field modes. In this case the Hamiltonian is explicitly time dependent. We therefore need a different tool to find formal solutions.

To this end we formally integrate the von Neumann equation Eq. (2.41) and arrive at

$$\hat{\rho}(t) = \hat{\rho}(t_0) - \frac{i}{\hbar} \int_{t_0}^{t} dt_1 \, [\hat{H}(t_1), \hat{\rho}(t_1)], \qquad (2.43)$$

where $\hat{\rho}(t_0)$ denotes the density operator of the system at time t_0. We now solve this integral equation using perturbation theory and therefore substitute the expression Eq. (2.43) for the time t_1, that is, the formula

$$\hat{\rho}(t_1) = \hat{\rho}(t_0) - \frac{i}{\hbar} \int_{t_0}^{t_1} dt_2 \, [\hat{H}(t_2), \hat{\rho}(t_2)]$$

into the commutator at the right-hand side of Eq. (2.43) and find

$$\hat{\rho}(t) = \hat{\rho}(t_0) - \frac{i}{\hbar} \int_{t_0}^{t} dt_1 \, [\hat{H}(t_1), \hat{\rho}(t_0)]$$

$$+ \left(-\frac{i}{\hbar}\right)^2 \int_{t_0}^{t} dt_1 \int_{t_0}^{t_1} dt_2 \, [\hat{H}(t_1), [\hat{H}(t_2), \hat{\rho}(t_2)]].$$

Hence, the first two terms on the right hand side only involve the density operator $\hat{\rho}(t_0)$ at the time t_0, the last integrals, however, containing the double commutator involve the density operator $\hat{\rho}(t_2)$, that is, at all times. To eliminate this term we continue to substitute Eq. (2.43) into these multi-commutators. We then arrive at

$$\hat{\rho}(t) = \hat{\rho}(t_0) + \sum_{s=1}^{\infty} \left(-\frac{i}{\hbar}\right)^s$$

$$\times \int_{t_0}^{t} dt_1 \int_{t_0}^{t_1} dt_2 \cdots \int_{t_0}^{t_{s-1}} dt_s \, [\hat{H}(t_1), [\hat{H}(t_2), \cdots, [\hat{H}(t_s), \hat{\rho}(t_0)] \cdots]]. \qquad (2.44)$$

We emphasize that this formal solution of the von Neumann equation is exact. In Chapter 18 we use Eq. (refeq:mblastminutechange1) to derive a density matrix equation for a cavity field damped by a reservoir of two-level atoms.

Problems

2.1 Momentum Wave Function

Derive the momentum wave function of an energy eigenstate of a harmonic oscillator.

2.2 Square Integrability

Show that a discrete energy spectrum implies square integrability of the energy wave functions.

Hint: Write Eq. (2.2) for energy eigenstates.

2.3 Matrix Element of Kinetic Energy in Position Space

Derive the matrix element

$$\langle \vec{r} | \frac{\hat{p}^2}{2M} | \vec{r}' \rangle = \frac{1}{2M} \left(\frac{\hbar}{i} \vec{\nabla} \right)^2 \delta(\vec{r} - \vec{r}')$$

by inserting a complete set of momentum states $|\vec{p}\rangle$.

2.4 Propagator of the Free Particle

The Green's function or the propagator $G(x,t|x',t')$ allows us to calculate the wave function $\psi(x,t)$ at time t given the wave function $\psi(x',t')$ at an earlier time t'.

(a) Show that for a time independent Hamiltonian $\hat{H}$ the Green's function

$$G(x,t|x',t') \equiv \langle x | \exp(-i\hat{H}(t-t')/\hbar) | x' \rangle$$

provides the representation

$$\psi(x,t) = \int_{-\infty}^{\infty} dx' \, G(x,t|x',t') \, \psi(x',t').$$

(b) Show that the propagator for a free particle moving according to the Hamiltonian

$$\hat{H}_0 \equiv \frac{\hat{p}^2}{2M}$$

reads

$$G(x,t|x',t') = \sqrt{\frac{M}{2\pi i \hbar (t-t')}} \exp\left(i \frac{M}{2\hbar} \frac{(x-x')^2}{(t-t')} \right).$$

2.5 Properties of the Density Operator

The density operator of a system is given by

$$\hat{\rho} = \sum_j P_{|\psi_j\rangle} |\psi_j\rangle\langle\psi_j|, \quad \text{where} \quad 0 \leq P_{|\psi_j\rangle} \leq 1, \quad \sum_j P_{|\psi_j\rangle} = 1. \quad (2.45)$$

Prove the following statements:

(a) The eigenvalues of an arbitrary density operator are real.

(b) The eigenvalues λ_i of an arbitrary density operator satisfy $0 \leq \lambda_i \leq 1$.

(c) $\sum_i \lambda_i = 1$.

(d) $\sum_i \lambda_i^2 \leq 1$.

Note that the states $|\psi_j\rangle$ in Eq. (2.45) are in general not orthogonal. Therefore, Eq. (2.45) is not a spectral decomposition of $\hat{\rho}$.

2.6 Hermite Polynomials and Thermal States

(a) The Hermite polynomials are defined by

$$H_n(x) \equiv (-1)^n e^{x^2} \left(\frac{d}{dx}\right)^n e^{-x^2}.$$

Prove the relations

$$\sum_{n=0}^{\infty} \frac{H_n(x)}{n!} z^n = e^{2xz - z^2}$$

and

$$H_n(x) = \frac{2^n e^{x^2}}{i^n \sqrt{\pi}} \int_{-\infty}^{+\infty} z^n e^{-z^2 + 2ixz} \, dz.$$

(b) Use the above integral representation of the Hermite polynomial to derive the relation

$$\sum_{n=0}^{\infty} \frac{H_n(x) H_n(y)}{n!} \zeta^n = \frac{\exp\left[x^2 - \frac{(x - 2y\zeta)^2}{1 - 4\zeta^2}\right]}{\sqrt{1 - 4\zeta^2}}$$

valid for $|\zeta| < 1/2$ and $\zeta \in \mathbb{R}$. What is the result for $|\zeta| = 1/2$?

(c) The density operator of a harmonic oscillator of frequency Ω in thermal equilibrium with a heat bath of temperature T reads

$$\hat{\rho} = N \sum_{m=0}^{\infty} e^{-m\beta} |m\rangle\langle m|,$$

where $\beta \equiv \hbar\Omega/(k_B T)$ and $N \equiv 1 - e^{-\beta}$.

Derive the position representation

$$\rho(x, x') \equiv \langle x|\hat{\rho}|x'\rangle = \sqrt{\frac{M\Omega}{\hbar\pi} \tanh\left(\frac{\hbar\Omega}{2k_B T}\right)}$$

$$\times \exp\left\{-\frac{M\Omega}{4\hbar} \tanh\left(\frac{\hbar\Omega}{2k_B T}\right)(x + x')^2 - \frac{M\Omega}{4\hbar} \coth\left(\frac{\hbar\Omega}{2k_B T}\right)(x - x')^2\right\}$$

of the density operator.

References

Introduction to Quantum Mechanics

For a review of the concepts of quantum mechanics we refer to
J. von Neumann, *Mathematical Foundations of Quantum Mechanics*, Springer, Berlin, 1932
P.A.M. Dirac, *The Principles of Quantum Mechanics*, Clarendon Press, Oxford, 1958
L.D. Landau and **E.M. Lifshitz**, *Quantum Mechanics*, Pergamon Press, New York, 1958
D. Bohm, *Quantum Theory*, Prentice Hall, Englewood Cliffs, 1951
R.P. Feynman, R.B. Leighton and **M. Sands**, *The Feynman Lectures on Physics, Vol. 3, Quantum Mechanics*, Addison-Wesley, Reading, Mass., 1965
L.I. Schiff, *Quantum Mechanics*, McGraw-Hill, New York, 1968
C. Cohen-Tannoudji, B. Diu and **F. Laloë**, *Quantum Mechanics Vol. I and II*, Wiley, New York, 1977
A. Bohm, *Quantum Mechanics*, Springer, Heidelberg, 1993
I. Bialynicki-Birula, M. Cieplak and **J. Kaminski**, *Theory of Quanta*, Oxford University Press, 1992

The expression *quantum kinematics* was coined by
H. Weyl, *Gruppentheorie und Quantenmechanik*, Hirzel, Leipzig, 1928
and discussed in a nice pedagogical way by
J. Schwinger, *Hermann Weyl and Quantum Kinematics*, in: *Exact Sciences and their Philosophical Foundations*, Vorträge des internationalen Hermann-Weyl-Kongresses, Kiel, 1985, edited by W. Deppert, K. Hübner, A. Oberschelp and V. Weidemann, Verlag P. Lang, Frankfurt, 1988

Time Evolution Operator in Quantum Electrodynamics

For the representation of the time evolution operator as an infinite product see
S. Tomonaga, *On a Relativistically Invariant Formulation of the Quantum Theory of Wave Fields*, Prog. Theoret. Phys. **1** 27–39 (1946)
The time ordering operator has been introduced by
F.J. Dyson, *The Radiation Theories of Tomonaga, Schwinger, and Feynman*, Phys. Rev. **75**, 486–502 (1949)
These historic papers are reprinted in
J. Schwinger, *Selected Papers on Quantum Electrodynamics*, Dover Publications, New York, 1958

The Volterra-Schlesinger Product Integral

The relation between a set of linear differential equations and the product integral see
L. Schlesinger, *Vorlesungen über lineare Differentialgleichungen*, B.G. Teubner, Leipzig, 1908

L. **Schlesinger**, *Einführung in die Theorie der gewöhnlichen Differentialgleichungen auf funktionentheoretischer Grundlage*, De Gruyter, Berlin, 1922

For a summary of the properties of the product integral see

L. **Schlesinger**, *Neue Grundlagen für einen Infinitesimalkalkul der Matrizen*, Math. Z. **33**, 33–61 (1931)

G. **Rasch**, *Zur Theorie und Anwendung des Produktintegrals*, J. rein. u. angew. Math. **171**, 65–119 (1934)

For the connection of the Volterra-Schlesinger product integral and the infinite product representation of the time evolution operator in quantum electrodynamics see

H. **Salecker**, *Quantenelektrodynamische Selbstenergie und exakte Lösungen der Schrödinger-Gleichung II*, Z. Naturforschg. **5a**, 480–492 (1950)

Historical Papers on Density Operator

The concept of the density operator was introduced independently by

J. **von Neumann**, *Wahrscheinlichkeitstheoretischer Aufbau der Quantenmechanik*, Göttinger Nachr. 245–272 (1927)

J. **von Neumann**, *Thermodynamik quantenmechanischer Gesamtheiten*, Göttinger Nachr. 273–291 (1927)

L.D. **Landau**, *Das Dämpfungsproblem in der Wellenmechanik*, Z. Physik **45**, 430–441 (1927)

3 Wigner Function

Quantum mechanics describes a microscopic system in terms of a state vector $|\psi\rangle$ or a density operator $\hat{\rho}$. These are rather abstract objects and it is hard to read off their properties. However, there exists a representation of quantum mechanics which brings out directly the properties of a quantum state. This representation lives in phase space and rests on the concept of the Wigner function. In the present chapter we review the basic properties of this phase space formulation of quantum mechanics.

Phase space is also the appropriate stage to consider the connection between classical and quantum mechanics. Due to the statistical interpretation pioneered by M. Born we cannot compare the predictions of quantum mechanics with the corresponding classical ones made for a single particle. With the help of modern quantum optics, in particular, devices such as the Paul trap or the one-atom maser, we can now perform experiments on single quantum particles. A single measurement provides only a single data point. Quantum mechanics is a statistical theory and therefore cannot predict the outcome of this single measurement. An exception is of course an outcome whose probability vanishes exactly. This event can never occur. When we repeat the measurements many times we obtain a histogram in agreement with the quantum mechanical prediction.

Therefore, it is more appropriate to compare quantum mechanics for a single particle with classical mechanics of an ensemble of particles. The framework well suited for this purpose is statistical mechanics, and in particular, the Liouville equation for the phase space distribution of an ensemble of classical particles. M. Born in his Nobel Prize lecture advocated such a comparison between the two theories and made a detailed study using an elementary example of quantum mechanics: The particle in the box. In this book we show that phase space brings out most clearly the differences and similarities between classical and quantum mechanics. For this purpose we study quantum mechanical phase space distribution functions. In the present chapter we focus on the Wigner function.

We consider the quantum mechanical motion of a particle of mass M. For the sake of simplicity we restrict ourselves to a one-dimensional motion described by the position and momentum operators $\hat{x}$ and $\hat{p}$, respectively. Due to the commutation relation $[\hat{x}, \hat{p}] = i\hbar$ between $\hat{x}$ and $\hat{p}$ it is not possible to define a genuine phase space distribution. However, we can define an object which depends on the eigenvalues x and p, but this distribution has some diseases. In particular, it can take on negative values. We shall later show that the central concept of *interference of probability amplitudes* reflects itself in these negative parts of the Wigner function.

3.1 Jump Start of the Wigner Function

The Wigner phase space distribution of a quantum state described by a density operator $\hat{\rho}$ reads

$$W(x,p) \equiv \frac{1}{2\pi\hbar} \int_{-\infty}^{\infty} d\xi \, \exp\left(-\frac{i}{\hbar}p\xi\right) \left\langle x + \frac{1}{2}\xi \middle| \hat{\rho} \middle| x - \frac{1}{2}\xi \right\rangle.$$

This quantity appeared in a paper by E.P. Wigner in 1932 in the context of quantum mechanical corrections to thermodynamic equilibrium. However, no motivation for this expression was given. In the present section we provide such a jump shot at the Wigner function.

Our goal is to describe the motion of a particle from position x' to x''. Heisenberg's matrix mechanics serves as our guide for this problem. Matrix mechanics was motivated by the transitions in an atom from a level n' to the level n''. Heisenberg recognized that it is not the motion in the level n' nor in n'' which is important but the quantum jump from n' to n''. In this spirit we shall consider a quantum jump from x' to x'', that is a quantum jump over the distance $\xi \equiv x'' - x'$. Moreover, Heisenberg noted the strength of the transition is given by an appropriate operator matrix element. For a dipole transition it is the dipole moment $\hat{\mu}$ represented by $\langle n'' | \hat{\mu} | n' \rangle$.

In the case of the particle moving in one dimension the appropriate operator describing the state of the particle is the density operator $\hat{\rho}$. Hence this quantum jump analogy suggests the matrix element $\langle x'' | \hat{\rho} | x' \rangle$ describing the quantum jump from the position eigenstate $|x'\rangle$ to the position eigenstate $|x''\rangle$. We can define a center of the jump via $x \equiv (x' + x'')/2$ and introduce instead of the two coordinates x' and x'' the center x and the distance ξ of the quantum jump via

$$x' = x - \frac{1}{2}\xi \quad \text{and} \quad x'' = x + \frac{1}{2}\xi.$$

Our quantum jump then reads $\langle x + \frac{1}{2}\xi | \hat{\rho} | x - \frac{1}{2}\xi \rangle$. We certainly associate the momentum p of the particle with the jump from x' to x'', that is with ξ. Since the momentum distribution follows by Fourier transform from the position distribution we perform a Fourier transform with respect to the quantum jump, that is

$$W(x,p) = \frac{1}{2\pi\hbar} \int_{-\infty}^{\infty} d\xi \, \exp\left(-\frac{i}{\hbar}p\xi\right) \left\langle x + \frac{1}{2}\xi \middle| \hat{\rho} \middle| x - \frac{1}{2}\xi \right\rangle. \tag{3.1}$$

Here we have included a normalization factor $1/(2\pi\hbar)$ which ensures the property

$$\int_{-\infty}^{\infty} dx \int_{-\infty}^{\infty} dp \, W(x,p) = 1$$

as discussed in the next section.

The Wigner function is therefore a Fourier transform

$$W(x,p) = \frac{1}{2\pi\hbar} \int_{-\infty}^{\infty} d\xi \, \exp\left(-\frac{i}{\hbar}p\xi\right) \tilde{\rho}(x,\xi)$$

of the density operator $\rho(x'', x') \equiv \langle x''|\rho|x'\rangle$ in position representation expressed in the variables $x \equiv (x'+x'')/2$ and $\xi \equiv x''-x'$ corresponding to the center of the jump and the jump distance, respectively, that is

$$\tilde{\rho}(x,\xi) = \rho(x'',x') = \rho(x+\frac{1}{2}\xi, x-\frac{1}{2}\xi) = \left\langle x+\frac{1}{2}\xi \middle| \hat{\rho} \middle| x-\frac{1}{2}\xi \right\rangle.$$

Indeed, the matrix element $\langle x''|\rho|x'\rangle$ depends on two position variables. We Fourier transform one of them and as a consequence we still have two variables: The Fourier variable of the jump, which we call p, and the center of the jump, which we call x. Both quantities are c-numbers and not operators. Therefore, the Wigner function W depends on two classical variables x and p. However, it is not quite clear yet, that these variables correspond to position and moment which span the phase space in which the Wigner function lives. We derive this feature in the next section.

We conclude by mentioning that in the case of a pure state $|\psi\rangle$, that is for $\hat{\rho} = |\psi\rangle\langle\psi|$ this expression reduces to

$$W(x,p) = \frac{1}{2\pi\hbar} \int_{-\infty}^{\infty} d\xi \, \exp\left(-\frac{i}{\hbar}p\xi\right) \psi^*\left(x-\tfrac{1}{2}\xi\right) \psi\left(x+\tfrac{1}{2}\xi\right) \quad (3.2)$$

where $\psi(x) \equiv \langle x|\psi\rangle$ is the position representation of the state $|\psi\rangle$. Hence the Wigner function is the Fourier transform of the shifted position wave functions of the state $|\psi\rangle$.

3.2 Properties of the Wigner Function

So far we have motivated the introduction of the Wigner function by the need for a pictorial representation of the abstract notion of a quantum state. However, as we will see later, the Wigner function can achieve a lot more: It allows us to calculate quantum mechanical expectation values using concepts of classical statistical mechanics. In this respect, however, the Wigner function is not unique. Indeed, as discussed in Chapter 12, there is an infinite amount of phase space distribution functions that achieve this goal. Nevertheless, the Wigner function is unique in the sense that it has simple properties. The discussion of these features is the topic of the present section.

3.2.1 Marginals

We now integrate the Wigner function either over the variable p or the variable x and show that we find the probability distribution of position or momentum, respectively. We start our analysis with integration over p.

Position Distribution

When we integrate both sides of Eq. (3.1) over p and interchange the integrations over ξ and p we find

$$\int_{-\infty}^{\infty} dp\, W(x,p) = \int_{-\infty}^{\infty} d\xi\, \langle x + \tfrac{1}{2}\xi|\, \hat{\rho}\, |x - \tfrac{1}{2}\xi\rangle\, \frac{1}{2\pi\hbar} \int_{-\infty}^{\infty} dp\, \exp\left(-\frac{i}{\hbar} p\xi\right)$$

which with the help of the relation

$$\frac{1}{2\pi\hbar} \int_{-\infty}^{\infty} dp\, \exp\left(-\frac{i}{\hbar} p\xi\right) = \delta(\xi) \qquad (3.3)$$

reduces to

$$\int_{-\infty}^{\infty} dp\, W(x,p) = \int_{-\infty}^{\infty} d\xi\, \langle x + \tfrac{1}{2}\xi|\, \hat{\rho}\, |x - \tfrac{1}{2}\xi\rangle\, \delta(\xi),$$

that is

$$\int_{-\infty}^{\infty} dp\, W(x,p) = \langle x|\, \hat{\rho}\, |x\rangle \equiv W(x).$$

Hence the Wigner function has the property that when we integrate over the momentum variable we find the probability distribution $W(x)$ for the position.

Momentum Distribution

Analogously we get the momentum distribution by integration over position x. Indeed, by introducing the new integration variables $x' = x - \tfrac{1}{2}\xi$ and $x'' = x + \tfrac{1}{2}\xi$, that is the original positions, we find

$$\int_{-\infty}^{\infty} dx\, W(x,p) = \frac{1}{2\pi\hbar} \int_{-\infty}^{\infty} dx \int_{-\infty}^{\infty} d\xi\, \exp\left(-\frac{i}{\hbar} p\xi\right) \langle x + \tfrac{1}{2}\xi|\, \hat{\rho}\, |x - \tfrac{1}{2}\xi\rangle$$

$$= \frac{1}{2\pi\hbar} \int_{-\infty}^{\infty} dx' \int_{-\infty}^{\infty} dx''\, \exp\left(-\frac{i}{\hbar} p(x'' - x')\right) \langle x''|\, \hat{\rho}\, |x'\rangle.$$

In order to connect this expression with the momentum distribution $W(p) \equiv \langle p|\, \hat{\rho}\, |p\rangle$ we insert two complete sets

$$\mathbb{1} = \int dx'\, |x'\rangle\langle x'| \qquad \text{and} \qquad \mathbb{1} = \int dx''\, |x''\rangle\langle x''|$$

of position eigenstates into the definition of the momentum distribution

$$W(p) = \langle p|\, \hat{\rho}\, |p\rangle = \int_{-\infty}^{\infty} dx' \int_{-\infty}^{\infty} dx''\, \langle p|x''\rangle\langle x''|\, \hat{\rho}\, |x'\rangle\langle x'|p\rangle$$

and recall the momentum representation of the position eigenstate $|x''\rangle$

$$\langle p|x''\rangle = \frac{1}{\sqrt{2\pi\hbar}} \exp\left(-\frac{i}{\hbar} px''\right) \qquad (3.4)$$

derived in Sec. 2.1.3. We therefore recognize the formula

$$W(p) = \frac{1}{2\pi\hbar} \int_{-\infty}^{\infty} dx' \int_{-\infty}^{\infty} dx'' \exp\left(-\frac{i}{\hbar}p(x''-x')\right) \langle x''|\hat{\rho}|x'\rangle.$$

The Wigner function integrated over position provides the momentum distribution.

3.2.2 Overlap of Quantum States as Overlap in Phase Space

There exists another interesting property of the Wigner function, namely the trace product rule

$$\mathrm{Tr}(\hat{\rho}_1 \hat{\rho}_2) = 2\pi\hbar \int_{-\infty}^{\infty} dx \int_{-\infty}^{\infty} dp\, W_{\hat{\rho}_1}(x,p) W_{\hat{\rho}_2}(x,p). \tag{3.5}$$

Here

$$W_{\hat{\rho}_j}(x,p) \equiv \frac{1}{2\pi\hbar} \int_{-\infty}^{\infty} d\xi \exp\left(-\frac{i}{\hbar}p\xi\right) \left\langle x+\frac{1}{2}\xi\left|\hat{\rho}_j\right|x-\frac{1}{2}\xi\right\rangle \tag{3.6}$$

denote the Wigner functions of the density operators $\hat{\rho}_j$ with $j = 1$ and 2.

The trace of the product of two density operators is determined by the product of the two corresponding Wigner functions integrated over phase space. This trace product rule plays an important role in quantum physics. It allows us to gain deeper insight into the shape of quantum states as we show in the next section.

In the case of pure states $\hat{\rho}_j = |\psi_j\rangle\langle\psi_j|$ we find

$$\mathrm{Tr}(\hat{\rho}_1 \hat{\rho}_2) = \mathrm{Tr}(|\psi_1\rangle\langle\psi_1|\psi_2\rangle\langle\psi_2|) = |\langle\psi_1|\psi_2\rangle|^2$$

and the trace product rule, Eq. (3.5), reduces to

$$|\langle\psi_1|\psi_2\rangle|^2 = 2\pi\hbar \int_{-\infty}^{\infty} dx \int_{-\infty}^{\infty} dp\, W_{|\psi_1\rangle}(x,p) W_{|\psi_2\rangle}(x,p). \tag{3.7}$$

Hence the modulus square of the scalar product between the quantum states $|\psi_1\rangle$ and $|\psi_2\rangle$ is the product of the Wigner function $W_{|\psi_1\rangle}$ and $W_{|\psi_2\rangle}$ of the two states integrated over the phase space.

We now turn to the derivation of Eq. (3.5). For this purpose we substitute the definitions Eq. (3.6) into the right side of Eq. (3.5) and interchange the order of integrations which yields

$$\mathcal{R} \equiv 2\pi\hbar \int_{-\infty}^{\infty} dx \int_{-\infty}^{\infty} dp\, W_{\hat{\rho}_1}(x,p) W_{\hat{\rho}_2}(x,p)$$

$$= \int_{-\infty}^{\infty} dx \int_{-\infty}^{\infty} d\xi_1 \int_{-\infty}^{\infty} d\xi_2 \frac{1}{2\pi\hbar} \int_{-\infty}^{\infty} dp\, \exp\left[-\frac{i}{\hbar}p(\xi_1+\xi_2)\right]$$

$$\times \left\langle x+\frac{1}{2}\xi_1\left|\hat{\rho}_1\right|x-\frac{1}{2}\xi_1\right\rangle \left\langle x+\frac{1}{2}\xi_2\left|\hat{\rho}_2\right|x-\frac{1}{2}\xi_2\right\rangle.$$

When we recall the integral representation (3.3) of the delta function we arrive at

$$\mathcal{R} = \int_{-\infty}^{\infty} dx \int_{-\infty}^{\infty} d\xi \left\langle x + \frac{1}{2}\xi \middle| \hat{\rho}_1 \middle| x - \frac{1}{2}\xi \right\rangle \left\langle x - \frac{1}{2}\xi \middle| \hat{\rho}_2 \middle| x + \frac{1}{2}\xi \right\rangle$$

where we have integrated over ξ_2 using the delta function with $\xi_2 = -\xi_1 \equiv -\xi$.

We introduce the integration variables $x'' \equiv x + \xi/2$ and $x' \equiv x - \xi/2$ and find

$$\mathcal{R} = \int_{-\infty}^{\infty} dx' \int_{-\infty}^{\infty} dx'' \left\langle x'' \middle| \hat{\rho}_1 \middle| x' \right\rangle \left\langle x' \middle| \hat{\rho}_2 \middle| x'' \right\rangle,$$

which with the completeness relation

$$\int_{-\infty}^{\infty} dx' \, |x'\rangle \langle x'| \equiv \mathbb{1}$$

reduces to

$$\mathcal{R} = \int_{-\infty}^{\infty} dx'' \left\langle x'' \middle| \hat{\rho}_1 \hat{\rho}_2 \middle| x'' \right\rangle = \mathrm{Tr}(\hat{\rho}_1 \hat{\rho}_2).$$

This expression is indeed the left hand side of Eq. (3.5).

3.2.3 Shape of Wigner Function

From the general definition of the Wigner function Eq. (3.1) and the trace product rule, Eq. (3.5), we can read off various properties of the shape of a Wigner function. We cannot squeeze a state to a phase space domain smaller than $2\pi\hbar$. Moreover, the Wigner function of a normalizable state cannot take on arbitrary large values and most importantly it can become negative.

Size of a Quantum State

When we consider two identical density operators $\hat{\rho}_1 = \hat{\rho}_2 = \hat{\rho}$ and recall that

$$\mathrm{Tr}\,\hat{\rho}^2 \leq 1$$

where the equal sign holds only for a pure state, we find from Eq. (3.5)

$$2\pi\hbar \int_{-\infty}^{\infty} dx \int_{-\infty}^{\infty} dp \, W_{\hat{\rho}}^2(x,p) \leq 1, \tag{3.8}$$

or

$$2\pi\hbar \leq \frac{1}{\int_{-\infty}^{\infty} dx \int_{-\infty}^{\infty} dp \, W_{\hat{\rho}}^2(x,p)}.$$

In Appendix C we show that the right hand side is a measure for the area in phase space occupied by the quantum state. Hence, this relation expresses the familiar fact that a pure quantum state takes up an area of $2\pi\hbar$ in phase space whereas the corresponding area occupied by a mixed state is larger.

Upper Bound of Wigner Function

The Wigner function cannot take on arbitrarily large values. There exists an upper bound $1/(\pi\hbar)$ for the function. To prove this fact we recall the definition

$$W(x,p) = \frac{1}{2\pi\hbar} \int_{-\infty}^{\infty} d\xi \, \exp\left(-\frac{i}{\hbar}p\xi\right) \psi^*\left(x - \tfrac{1}{2}\xi\right) \psi\left(x + \tfrac{1}{2}\xi\right)$$

of the Wigner function for a pure state. We then interpret this integral as the scalar product

$$W(x,p) = \frac{1}{\pi\hbar} \int_{-\infty}^{\infty} d\xi \, \phi_1^*(\xi) \phi_2(\xi) = \frac{1}{\pi\hbar} \langle \phi_1 | \phi_2 \rangle$$

between the two wave functions

$$\phi_1(\xi) \equiv \frac{1}{\sqrt{2}} \exp\left(\frac{i}{\hbar}p\xi\right) \psi\left(x - \tfrac{1}{2}\xi\right) \quad \text{and} \quad \phi_2(\xi) \equiv \frac{1}{\sqrt{2}} \psi\left(x + \tfrac{1}{2}\xi\right)$$

corresponding to the states $|\phi_1\rangle$ and $|\phi_2\rangle$. Here we have included a factor $\sqrt{2}$ in the definition of the wave functions ϕ_1 and ϕ_2 as to ensure the normalization

$$\langle \phi_j | \phi_j \rangle \equiv \int_{-\infty}^{\infty} d\xi \, |\phi_j(\xi)|^2 = \int_{-\infty}^{\infty} d\left(\tfrac{\xi}{2}\right) \left|\psi\left(x - \tfrac{1}{2}\xi\right)\right|^2 = 1.$$

In the last step we have assumed that the wave function $\psi = \psi(x)$ is normalized.

These definitions allow us to estimate the Wigner function

$$|W(x,p)| = \frac{1}{\pi\hbar} |\langle \phi_1 | \phi_2 \rangle|$$

with the help of the Cauchy-Schwarz inequality

$$|\langle \phi_1 | \phi_2 \rangle|^2 \leq \langle \phi_1 | \phi_1 \rangle \cdot \langle \phi_2 | \phi_2 \rangle = 1 \cdot 1$$

valid for two normalized states $|\phi_1\rangle$ and $|\phi_2\rangle$.

Hence, we have established the inequality

$$|W(x,p)| \leq \frac{1}{\pi\hbar}.$$

The Wigner function of a pure normalizable state cannot take on values larger than $1/(\pi\hbar)$.

Wigner Functions Can Take on Negative Values

We again return to the remarkable property, Eq. (3.5), for the trace of the product of two density operators. For the case of two operators $\hat{\rho}_1$ and $\hat{\rho}_2$ such that

$$\text{Tr}(\hat{\rho}_1 \hat{\rho}_2) = 0,$$

this relation implies

$$\int_{-\infty}^{\infty} dx \int_{-\infty}^{\infty} dp\, W_{\hat{\rho}_1}(x,p) W_{\hat{\rho}_2}(x,p) = 0, \tag{3.9}$$

that is the product of the two Wigner functions integrated over the whole phase space has to vanish.

This condition enforces that the Wigner function $W_{\hat{\rho}_1}$ or/and $W_{\hat{\rho}_2}$ must take on negative values. In particular, we show in Chapter 4 that the Wigner function of an energy eigenstate of a harmonic oscillator can take on negative values. This surprising feature makes it impossible to interpret the Wigner function as a true probability distribution. Nevertheless, the Wigner function is useful to calculate quantum mechanical expectation values.

The condition Eq. (3.9) does not exclude the possibility that there exist Wigner functions which are positive everywhere. We recall that this equation holds only for two orthogonal states. For example in Chapter 4 we discuss the Wigner functions of coherent and squeezed wave packets. They are Gaussians and therefore positive everywhere. This is closely related to the Hudson-Piquet theorem which states that the only non-negative Wigner function is a Gaussian distribution.

3.3 Time Evolution of Wigner Function

The Wigner function is an important tool to transform the operator equation of motion for the density operator into a c-number equation. However, this equation is rather complicated and brings out the nonlocal nature of the Wigner function. In the present section we derive this equation of motion. In particular, we consider the quantum mechanical motion of a particle in a potential $U(x)$.

3.3.1 von Neumann Equation in Phase Space

We derive the equation of motion for the Wigner function by taking matrix elements of the left and the right hand side of the von Neumann equation with the bra and ket vectors $\langle x + \tfrac{1}{2}\xi|$ and $|x - \tfrac{1}{2}\xi\rangle$. Indeed, when we start from

$$\frac{\partial \hat{\rho}}{\partial t} = -\frac{i}{\hbar}[\hat{H}, \hat{\rho}]$$

we arrive at

$$\frac{\partial}{\partial t}\langle x + \tfrac{1}{2}\xi|\hat{\rho}|x - \tfrac{1}{2}\xi\rangle = \langle x + \tfrac{1}{2}\xi|\frac{\partial \hat{\rho}}{\partial t}|x - \tfrac{1}{2}\xi\rangle = -\frac{i}{\hbar}\langle x + \tfrac{1}{2}\xi|[\hat{H}, \hat{\rho}]|x - \tfrac{1}{2}\xi\rangle.$$

We multiply both sides by $(2\pi\hbar)^{-1}\exp(-ip\xi/\hbar)$ and integrate over the variable ξ. This yields on the left hand side the time derivative of the Wigner function whereas the right hand side is more complicated. Indeed, we find

$$\frac{\partial W}{\partial t} = -\frac{i}{\hbar}\int_{-\infty}^{\infty} d\xi \frac{\exp(-ip\xi/\hbar)}{2\pi\hbar}\langle x + \tfrac{1}{2}\xi|\left[\frac{\hat{p}^2}{2M} + U(\hat{x})\right]\hat{\rho} - \hat{\rho}\left[\frac{\hat{p}^2}{2M} + U(\hat{x})\right]|x - \tfrac{1}{2}\xi\rangle,$$

where we have used the Hamiltonian

$$\hat{H} = \frac{\hat{p}^2}{2M} + U(\hat{x}).$$

Therefore, the equation of motion for the Wigner function reads

$$\frac{\partial W}{\partial t} = \mathcal{T} + \mathcal{U},$$

where the quantity

$$\mathcal{T} \equiv -\frac{i}{\hbar}\frac{1}{2M}\frac{1}{2\pi\hbar}\int_{-\infty}^{\infty} d\xi\, e^{-ip\xi/\hbar}\langle x + \tfrac{1}{2}\xi|\hat{p}^2\hat{\rho} - \hat{\rho}\hat{p}^2|x - \tfrac{1}{2}\xi\rangle$$

contains the kinetic part of the Hamiltonian and

$$\mathcal{U} \equiv -\frac{i}{\hbar}\frac{1}{2\pi\hbar}\int_{-\infty}^{\infty} d\xi\, e^{-ip\xi/\hbar}\langle x + \tfrac{1}{2}\xi|U(\hat{x})\hat{\rho} - \hat{\rho}U(\hat{x})|x - \tfrac{1}{2}\xi\rangle$$

accounts for the potential energy.

In Appendix D we evaluate the terms $\mathcal{T}$ and $\mathcal{U}$ separately, and express them as derivatives of the Wigner function. We find the relations

$$\mathcal{U} = \sum_{l=0}^{\infty} \frac{(i\hbar/2)^{2l}}{(2l+1)!} \frac{d^{2l+1}U(x)}{dx^{2l+1}} \frac{\partial^{2l+1}}{\partial p^{2l+1}} W(x,p;t) \tag{3.10}$$

and

$$\mathcal{T} = -\frac{p}{M}\frac{\partial}{\partial x} W(x,p;t). \tag{3.11}$$

Hence depending on the form of the potential $U(x)$ the potential energy term $\mathcal{U}$ brings in higher derivatives of the Wigner function with respect to the momentum. In contrast, the kinetic part $\mathcal{T}$ of the Wigner function involves only a first derivative with respect to position.

3.3.2 Quantum Liouville Equation

We are now in a position to express the equation of motion for the Wigner function. When we combine the potential and kinetic term Eqs. (3.10) and (3.11) we arrive at

$$\left(\frac{\partial}{\partial t} + \frac{p}{M}\frac{\partial}{\partial x} - \frac{dU(x)}{dx}\frac{\partial}{\partial p}\right)W(x,p;t) = \sum_{l=1}^{\infty}\frac{(-1)^l(\hbar/2)^{2l}}{(2l+1)!}\frac{d^{2l+1}U(x)}{dx^{2l+1}}\frac{\partial^{2l+1}}{\partial p^{2l+1}}W(x,p;t). \tag{3.12}$$

It is instructive to consider the classical limit of this equation. For this purpose we formally set $\hbar = 0$. Provided the derivatives of the Wigner function on the right hand side of this equation do not become singular, the right hand side vanishes. We therefore recover the Liouville equation

$$\left(\frac{\partial}{\partial t} + \frac{p}{M}\frac{\partial}{\partial x} - \frac{dU(x)}{dx}\frac{\partial}{\partial p}\right)W(x,p;t) = 0$$

of classical statistical mechanics. In this sense the terms on the right hand side of Eq. (3.12) constitute the quantum mechanical corrections to the classical Liouville equation.

Is it possible that the time evolution of quantum states is still governed by classical mechanics even for $\hbar \neq 0$? Indeed, the classical Liouville equation is identical to the quantum mechanical equation of motion of the Wigner function, provided we evolve the state in a potential that contains only terms up to second order in position. In this case each point in Wigner function phase space moves according to the classical equations of motion. The quantum mechanical properties of a system are then hidden in the initial condition. Whereas in classical mechanics any normalizable, non-negative distribution function is allowed, this is no longer true in quantum mechanics. The laws of quantum mechanics govern the class of functions which may represent the quantum state of a system.

3.4 Wigner Function Determined by Phase Space

The definitions Eqs. (3.1) and (3.2) of the Wigner function are in terms of the density operator or a bilinear form of the wave function. The general strategy to obtain the Wigner function is then to start from the quantum state expressed in $\hat{\rho}$ or $|\psi\rangle$, evaluate them at the shifted positions $x \pm \xi/2$ and perform the Fourier transform of the jump ξ. This procedure suggests that the road to the Wigner function always has to go through Schrödinger quantum mechanics since we have to find the state.

However, there exists the possibility to directly calculate the Wigner function from phase space by solving *two* coupled partial differential equations. In fact, these equations determine a larger class of phase space functions known as Moyal functions.

3.4.1 Definition of Moyal Function

Before we discuss these phase space differential equations we first motivate the significance of the Moyal functions. For this purpose we start from the time dependent Wigner function represented in the form

$$W(x,p;t) = \frac{1}{2\pi\hbar} \int_{-\infty}^{\infty} d\xi \, e^{-ip\xi/\hbar} \, \langle x + \tfrac{1}{2}\xi | \hat{\rho}(t) | x - \tfrac{1}{2}\xi \rangle.$$

We rewrite this equation using the relation

$$\hat{\rho}(t) = e^{-i\hat{H}t/\hbar} \, \hat{\rho}(0) \, e^{i\hat{H}t/\hbar}$$

derived in Sec. 2.4.4 and find

$$W(x,p;t) = \frac{1}{2\pi\hbar} \int_{-\infty}^{\infty} d\xi \, e^{-ip\xi/\hbar} \, \langle x + \tfrac{1}{2}\xi | e^{-i\hat{H}t/\hbar} \, \hat{\rho}(0) \, e^{i\hat{H}t/\hbar} | x - \tfrac{1}{2}\xi \rangle.$$

Next we insert two completeness relations

$$\sum\!\!\!\!\!\!\int dE \, |E\rangle\langle E| = \mathbf{1}$$

in energy representation, where the symbol $\Sigma\hspace{-0.9em}\int$ denotes summation plus integration over all energy eigenstates of the discrete and continuous spectrum. When we make use of the energy eigenvalue equation

$$\hat{H}|E\rangle = E|E\rangle \tag{3.13}$$

we arrive at

$$W(x,p;t) = \Sigma\hspace{-0.9em}\int dE' \Sigma\hspace{-0.9em}\int dE'' e^{-i(E''-E')t/\hbar} \langle E''|\hat{\rho}(0)|E'\rangle W_{|E''\rangle\langle E'|}(x,p), \tag{3.14}$$

where we have defined the Moyal function

$$W_{|E''\rangle\langle E'|}(x,p) \equiv \frac{1}{2\pi\hbar} \int_{-\infty}^{\infty} d\xi\, e^{-ip\xi/\hbar} \langle x+\tfrac{1}{2}\xi|E''\rangle\langle E'|x-\tfrac{1}{2}\xi\rangle. \tag{3.15}$$

We note that this function is a generalization of the Wigner function. Indeed, the diagonal Moyal function $W_{|E\rangle\langle E|}(x,p) \equiv W_{|E\rangle}(x,p)$ is the usual Wigner function of the energy eigenstate $|E\rangle$.

Equation (3.14) brings out the importance of the Moyal function: The time independent Moyal functions determine the time evolution of the Wigner function with the initial density operator in energy representation as weight factors.

3.4.2 Phase Space Equations for Moyal Functions

We now return to the question of how to find the Wigner function from phase space without going through Schrödinger quantum mechanics. For this purpose we derive partial differential equations for the Moyal functions.

Our starting point is the anti-commutator

$$\frac{1}{2}\{|E''\rangle\langle E'|, \hat{H}\} \equiv \frac{1}{2}\left(|E''\rangle\langle E'|\hat{H} + \hat{H}|E''\rangle\langle E'|\right) = \frac{E'+E''}{2}|E''\rangle\langle E'|$$

and the commutator

$$\frac{i}{\hbar}\left[|E''\rangle\langle E'|, \hat{H}\right] \equiv \frac{i}{\hbar}\left[|E''\rangle\langle E'|\hat{H} - \hat{H}|E''\rangle\langle E'|\right] = \frac{i}{\hbar}(E'-E'')|E''\rangle\langle E'|$$

between the Hamiltonian $\hat{H}$ and the operator $|E''\rangle\langle E'|$.

Here we have made use of the energy eigenvalue equation Eq. (3.13) and have multiplied by factors which will turn out to be useful later.

When we now multiply both sides of the two equations from the right by $|x-\tfrac{1}{2}\xi\rangle$ and from the left by $(2\pi\hbar)^{-1} e^{-ip\xi/\hbar} \langle x+\tfrac{1}{2}\xi|$ and integrate over ξ, we find

$$\frac{1}{2\pi\hbar}\int_{-\infty}^{\infty} d\xi\, e^{-ip\xi/\hbar} \langle x+\tfrac{1}{2}\xi|\frac{1}{2}\{|E''\rangle\langle E'|, \hat{H}\}|x-\tfrac{1}{2}\xi\rangle = \frac{E'+E''}{2} W_{|E''\rangle\langle E'|}$$

and

$$\frac{1}{2\pi\hbar}\int_{-\infty}^{\infty} d\xi\, e^{-ip\xi/\hbar} \langle x+\tfrac{1}{2}\xi|\frac{i}{\hbar}[|E''\rangle\langle E'|, \hat{H}]|x-\tfrac{1}{2}\xi\rangle = \frac{i}{\hbar}(E'-E'') W_{|E''\rangle\langle E'|}.$$

3 Wigner Function

Hence, the right hand sides of both equations involve already the Moyal function. To express the left hand side in terms of derivatives of the Moyal function we substitute the Hamiltonian

$$\hat{H} = \frac{\hat{p}^2}{2M} + U(\hat{x})$$

into the expression and follow the lines of the derivation of the quantum Liouville equation. For the details of this derivation we refer to Appendix D.

We arrive at the result

$$\left[\frac{p^2}{2M} + U - \frac{\hbar^2}{8M} \frac{\partial^2}{\partial x^2} + \sum_{l=1}^{\infty} \frac{(-1)^l (\hbar/2)^{2l}}{(2l)!} \frac{d^{2l}U}{dx^{2l}} \frac{\partial^{2l}}{\partial p^{2l}}\right] W_{|E''\rangle\langle E'|} = \frac{E' + E''}{2} W_{|E''\rangle\langle E'|}$$

(3.16a)

and

$$\left[\frac{p}{M} \frac{\partial}{\partial x} - \frac{dU}{dx} \frac{\partial}{\partial p} - \sum_{l=1}^{\infty} \frac{(-1)^l (\hbar/2)^{2l}}{(2l+1)!} \frac{d^{2l+1}U}{dx^{2l+1}} \frac{\partial^{2l+1}}{\partial p^{2l+1}}\right] W_{|E''\rangle\langle E'|} = \frac{i}{\hbar} (E'' - E') W_{|E''\rangle\langle E'|}.$$

(3.16b)

These equations determine the functional dependence of the Moyal function $W_{|E''\rangle\langle E'|}$ on their arguments position x and momentum p.

We note that the even derivatives of the potential only enter into the first equation whereas the odd derivatives appear only in the second equation. For an arbitrary potential the equations are of infinite order. One might wonder if it helps that the series is a power series in Planck's constant. This expansion suggests to only include the lowest order terms. However, this procedure gives wrong results since Planck's constant appears always in front of the highest derivative. Such differential equations are notoriously difficult to analyze as we briefly discuss in the next section.

3.5 Phase Space Equations for Energy Eigenstates

To gain some insight into these equations we now turn to the case of an energy eigenstate. Here $E' = E''$ and hence, the two equations reduce to the partial differential equations

$$\left[\frac{p^2}{2M} + U - \frac{\hbar^2}{8M} \frac{\partial^2}{\partial x^2} + \sum_{l=1}^{\infty} \frac{(-1)^l (\hbar/2)^{2l}}{(2l)!} \frac{d^{2l}U}{dx^{2l}} \frac{\partial^{2l}}{\partial p^{2l}}\right] W_{|E\rangle}(x,p) = E\, W_{|E\rangle}(x,p)$$

(3.17)

and

$$\left[\frac{p}{M} \frac{\partial}{\partial x} - \frac{dU}{dx} \frac{\partial}{\partial p} - \sum_{l=1}^{\infty} \frac{(-1)^l (\hbar/2)^{2l}}{(2l+1)!} \frac{d^{2l+1}U}{dx^{2l+1}} \frac{\partial^{2l+1}}{\partial p^{2l+1}}\right] W_{|E\rangle}(x,p) = 0 \qquad (3.18)$$

for the Wigner function $W_{|E\rangle}$ of the energy eigenstate $|E\rangle$. We note that the second equation is the stationary form of the quantum Liouville equation Eq. (3.12). The first equation, however, is an eigenvalue equation for the energy E. We illustrate this fact in the next section using the example of an energy eigenstate of a harmonic oscillator. However, in the present section we concentrate on the influence of the power expansion of $\hbar$ on the Wigner function.

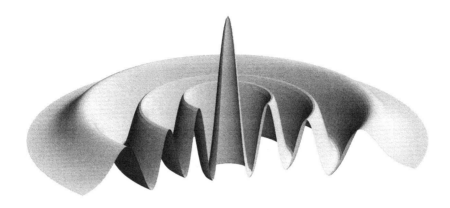

Fig. 3.1: Wigner function of the 6th energy eigenstate of a harmonic oscillator.

3.5.1 Power Expansion in Planck's Constant

In Fig. 3.1 and on the cover page of this book we show the Wigner function of the sixth energy eigenstate of a harmonic oscillator. We recognize a dominant maximum along the classical trajectory

$$E_{qm} = \frac{p_{qm}^2(x)}{2M} + U(x) \tag{3.19}$$

determined by the quantum mechanical energy E_{qm}.

On first sight this phrase might sound like an oxymoron. We mean the phase space trajectory as determined by the conservation of energy. However, for the purpose of the present discussion we have chosen the energy E_{qm} to be determined by Schrödinger quantum mechanics.

The phase space domain circumnavigated by the classical orbit displays characteristic oscillations. Moreover, there is a striking maximum at the origin of phase space. Outside of the trajectory the Wigner function decays as to ensure the normalization condition.

This behavior is not restricted to the energy eigenstates of the harmonic oscillator. In Fig. 3.2 we show the Wigner function of an energy eigenstate of a Morse oscillator. Again, in the phase space domain circumnavigated by the classical trajectory dramatic oscillations and cusps arise. In contrast to the case of the harmonic oscillator here the classical trajectory does not appear that pronounced. Moreover, ripples appear in the domain outside of the classical orbit.

We now show how these properties arise from the phase space equations Eqs. (3.17) and (3.18). In particular, we focus on the dependence of the solutions of these equations on Planck's constant.

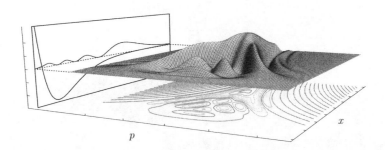

Fig. 3.2: Probability density (background) and Wigner function (foreground) of an energy eigenstate of a Morse oscillator. The Wigner function displays complicated structures in the phase space region circumvented by the classical trajectory corresponding to the quantum energy. The phase space domain outside of the trajectory shows ripples. Taken from M. Hug *et al.*, Phys. Rev. A **57**, 3206 (1998).

Classical Limit

We begin the discussion of these two equations by analyzing the classical case where $\hbar \equiv 0$. Here the two equations for the Wigner function reduce to

$$\left[\frac{p^2}{2M} + U(x) - E\right] W_{|E\rangle}(x, p) = 0$$

and

$$\left[\frac{p}{M}\frac{\partial}{\partial x} - \frac{dU}{dx}\frac{\partial}{\partial p}\right] W_{|E\rangle}(x, p) = 0.$$

In this derivation we have assumed that the derivatives of the Wigner function do not become singular, when $\hbar \equiv 0$.

The first equation immediately yields

$$W_{|E\rangle}(x, p) = \delta\left[\frac{p^2}{2M} + U(x) - E\right]$$

which is a delta function along the classical trajectory of Energy E. We note that this solution satisfies the second equation. It is also worthwhile mentioning that there is no constraint on the energy: In the limit of $\hbar = 0$ the eigenvalue equation for the Wigner function looses its power to determine the energy.

Semiclassical Limit

When we take into account the next term containing $\hbar$ in the expansion to obtain the correction to the above δ-function result we find a solution that is dramatically different. The hard δ-function of classical physics gets softened by quantum mechanics and becomes an Airy function.

3.5 Phase Space Equations for Energy Eigenstates

In order to bring out the consequences of a non-vanishing Planck's constant we now slightly rewrite Eq. (3.17). Due to the complexity of these equations in the case of an arbitrary potential, we cannot present an exact analytic solution. We therefore in the present section motivate the behavior of the solution of the two coupled phase space equations. We emphasize that this treatment is not exact but tries to give an idea.

We substitute the eigenvalue E_{qm} in form of the phase space trajectory Eq. (3.19) into Eq. (3.17) which yields

$$\left\{ -\frac{p_{qm}^2(x) - p^2}{2M} - \frac{\hbar^2}{8M}\frac{\partial^2}{\partial x^2} - \frac{\hbar^2}{8}\frac{d^2U}{dx^2}\frac{\partial^2}{\partial p^2} + \frac{\hbar^4}{16\cdot 24}\frac{d^4U}{dx^4}\frac{\partial^4}{\partial p^4} \right.$$
$$\left. + \sum_{l=3}^{\infty} \frac{(i\hbar/2)^{2l}}{2l!}\frac{d^{2l}U}{dx^{2l}}\frac{\partial^{2l}}{\partial p^{2l}} \right\} W_{|E_{qm}\rangle}(x,p) = 0.$$

Close to the classical trajectory we can approximate the difference of the squares of the momenta by the quantity

$$p_{qm}^2 - p^2 = (p_{qm} - p)(p_{qm} + p) \cong 2p_{qm}(p_{qm} - p).$$

Hence, we have reduced the dependence of p from a quadratic to a linear function. Moreover, for the time being we neglect the second derivative with respect to x. We thus have an equation of the form

$$\left\{ -\frac{p_{qm}}{M}(p_{qm} - p) - \frac{\hbar^2}{8}\frac{d^2U}{dx^2}\frac{\partial^2}{\partial p^2} + \frac{\hbar^4}{16\cdot 24}\frac{d^4U}{dx^4}\frac{\partial^4}{\partial p^4} \right.$$
$$\left. + \sum_{l=3}^{\infty} \frac{(i\hbar/2)^{2l}}{2l!}\frac{d^{2l}U}{dx^{2l}}\frac{\partial^{2l}}{\partial p^{2l}} \right\} W_{|E_{qm}\rangle}(x,p) = 0. \quad (3.20)$$

When we concentrate on the dependence of the Wigner function on the momentum for a fixed position the above equation takes the form

$$\left(-\alpha(p_{qm} - p) + \hbar^2\beta\frac{d^2}{dp^2} + \hbar^4\gamma\frac{d^4}{dp^4} + \ldots \right) W_{|E_{qm}\rangle}(x,p) = 0$$

of an ordinary differential equation of infinite order with various constants $\alpha, \beta, \gamma, \ldots$ determined by Eq. (3.20).

3.5.2 Model Differential Equation

In the preceding section we have shown that the differential equations determining the Wigner function are power expansions in terms of Planck's constant. The highest derivative brings in the highest power of $\hbar$. Such equations have subtle limits as $\hbar$ approaches 0. In order to bring out these subtleties we now consider the model differential equation

$$\left(-y + \varepsilon^2\frac{d^2}{dy^2} + \varepsilon^4\frac{d^4}{dy^4} \right) W(y) = 0 \quad (3.21)$$

for the function $W = W(y; \varepsilon)$. Here the parameter ε plays the role of Planck's constant and we focus on the behavior of the solution $W(y; \varepsilon)$ as ε approaches zero.

Solution by Fourier Transform

In order to solve the complete differential equation including all derivatives with respect to the variable y we introduce the Fourier transform

$$\widetilde{W}(k) \equiv \int_{-\infty}^{\infty} dy \, e^{iky} \, W(y)$$

of W and its inverse

$$W(y) \equiv \frac{1}{2\pi} \int_{-\infty}^{\infty} dk \, e^{-iky} \, \widetilde{W}(k). \tag{3.22}$$

Here we have assumed that the function W decays fast enough so that the Fourier transform exists.

When we substitute the representation Eq. (3.22) of W in terms of $\widetilde{W}$ into the differential equation Eq. (3.21) we arrive at

$$\frac{1}{2\pi} \int_{-\infty}^{\infty} dk \left[y + \varepsilon^2 k^2 - \varepsilon^4 k^4 \right] e^{-iky} \, \widetilde{W}(k) = 0. \tag{3.23}$$

We can express the term y as a derivative of the exponential with respect to k, that is

$$y \, e^{-iky} = \frac{1}{(-i)} \frac{d}{dk} \left(e^{-iky} \right),$$

which after integrating by parts yields

$$\int_{-\infty}^{\infty} dk \, y \, e^{-iky} \, \widetilde{W}(k) = \frac{1}{(-i)} \int_{-\infty}^{\infty} dk \left(\frac{d}{dk} e^{-iky} \right) \widetilde{W}(k)$$

$$= \frac{1}{(-i)} \left\{ e^{-iky} \, \widetilde{W}(k) \bigg|_{-\infty}^{\infty} - \int_{-\infty}^{\infty} dk \, e^{-iky} \frac{d\widetilde{W}(k)}{dk} \right\}.$$

In order to secure the existence of the the Fourier transform the boundary terms have to vanish and we find the relation

$$\int_{-\infty}^{\infty} dy \, e^{-iky} \, \widetilde{W}(k) = \int_{-\infty}^{\infty} dk \, e^{-iky} \frac{1}{i} \frac{d\widetilde{W}(k)}{dk}.$$

Thus Eq. (3.23) takes the form

$$\frac{1}{2\pi} \int_{-\infty}^{\infty} dk \left[\frac{1}{i} \frac{d\widetilde{W}}{dk} + \left(\varepsilon^2 k^2 - \varepsilon^4 k^4 \right) \widetilde{W}(k) \right] = 0,$$

which corresponds to the ordinary differential equation

$$\frac{d\widetilde{W}(k)}{dk} = -i \left(\varepsilon^2 k^2 - \varepsilon^4 k^4 \right) \widetilde{W}(k)$$

with the solution

$$\widetilde{W}(k) = \exp\left[-i\left(\frac{\varepsilon^2}{3}k^3 - \frac{\varepsilon^4}{5}k^5\right)\right].$$

Here we have chosen for the sake of simplicity the initial condition $\widetilde{W}(k=0) = 1$.

When we substitute this expression back into the Fourier ansatz Eq. (3.22) we arrive at

$$W_4 \equiv W(y;\varepsilon) = \frac{1}{2\pi}\int_{-\infty}^{\infty} dk \, \exp\left\{-i\left[-\frac{\varepsilon^4}{5}k^5 + \frac{\varepsilon^2}{3}k^3 + yk\right]\right\}. \qquad (3.24)$$

This expression involves powers of ε up to the fourth order in ε as indicated by the subscript 4 of W.

Such an integral appears in the theory of diffraction of light and carries the name swallow tail integral. It has been studied in great detail in the context of uniform asymptotic expansion of integrals.

From Swallow Tail via Airy to δ-function

The solution Eq. (3.24) shines light on the power expansion of the differential equation in terms of ε. Indeed, the same powers appear in the argument of the exponential.

For $\varepsilon \equiv 0$ we find the integral representation

$$W_0 \equiv W(y;\varepsilon = 0) = \delta(y) \equiv \frac{1}{2\pi}\int_{-\infty}^{\infty} dk \, e^{-iky}$$

of the delta function in complete accordance with the solution of the equation

$$y\,W(y) = 0$$

following from Eq. (3.21) in the limit of $\varepsilon = 0$.

However, retaining only the term ε^2 we arrive at the function

$$W_2 \equiv W(y;\varepsilon) = \frac{1}{2\pi}\int_{-\infty}^{\infty} dk \, \exp\left[-i\left(\frac{\varepsilon^2}{3}k^3 + yk\right)\right],$$

which with the help of the substitution $\varepsilon^2 k^3 = \xi^3$, that is $\varepsilon^{2/3} k = \xi$, takes the form

$$W(y) = \varepsilon^{-2/3}\,\mathrm{Ai}\left(\varepsilon^{-2/3} y\right). \qquad (3.25)$$

Here we have introduced the Airy function

$$\mathrm{Ai}(y) \equiv \int_{-\infty}^{\infty} d\xi \, \exp\left[i\left(\frac{1}{3}\xi^3 + y\xi\right)\right] \qquad (3.26)$$

familiar from diffraction theory.

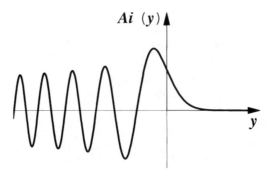

Fig. 3.3: Airy function of real argument y. The function is bound with a dominant maximum close to the origin. For positive arguments the Airy function decays exponentially whereas for negative arguments it displays characteristic oscillations.

In Appendix E we summarize the most important properties of the Airy function. For the present discussion however, it suffices to recall the behavior shown in Fig. 3.3: For positive arguments the Airy function is positive and decays exponentially. For negative arguments it displays oscillations and decays slowly. There is a dominant maximum in the neighborhood of $y = 0$ where the second derivative of the Airy function vanishes. This fact results from the differential equation

$$\left(\frac{d^2}{dy^2} - y\right) \text{Ai}(y) = 0.$$

Moreover, we find immediately from the integral representation Eq. (3.26) of the Airy function the normalization property

$$\int_{-\infty}^{\infty} dy \, \text{Ai}(y) = 1.$$

The fact that ε is non-vanishing has drastically changed the solution W: Whereas for $\varepsilon \equiv 0$ we had a delta function located at $y = 0$ we now have an oscillatory function whose dominant maximum lies in the neighborhood of $y = 0$.

As ε approaches zero, the prefactor $\varepsilon^{-2/3}$ in W multiplying the Airy function increases the amplitude of the oscillations as well as of the dominant maximum. Moreover, due to the appearance of the factor $\varepsilon^{-2/3}$ in the argument of the Airy function the period of the oscillations decreases. At the same time the dominant maximum moves closer to $y = 0$. Hence the function $W(y;\varepsilon)$ given by Eq. (3.25) approaches a delta function in the limit of $\varepsilon \to 0$.

3.6 Harmonic Oscillator

Many textbooks on quantum mechanics state that the harmonic oscillator is a classical system. This claim is motivated by the fact that for a quadratic potential the equation of motion for the Wigner function reduces to the classical Liouville equation as discussed in the preceding section. However, energy eigenstates depend on

Planck's constant and are therefore quantum mechanical objects. They follow from the first of the two equations discussed before.

3.6.1 Wigner Function as Wave Function

To illustrate this we now solve the two coupled equations for the case of an energy eigenstate of the harmonic oscillator. We show that in this example the energy eigenvalue equation in phase space for the one-dimensional harmonic oscillator reduces to the Schrödinger equation of a two-dimensional harmonic oscillator.

For the quadratic potential

$$U(x) = \frac{1}{2} M\Omega^2 x^2$$

the two equations (3.17) and (3.18) reduce to

$$\left[\frac{p^2}{2M} + \frac{1}{2} M\Omega^2 x^2 - \frac{\hbar^2}{8M} \frac{\partial^2}{\partial x^2} - \frac{\hbar^2 M\Omega^2}{8} \frac{\partial^2}{\partial p^2} \right] W_{|E\rangle}(x,p) = E\, W_{|E\rangle}(x,p) \quad (3.27\text{a})$$

and

$$\left[\frac{p}{M} \frac{\partial}{\partial x} - M\Omega^2 x \frac{\partial}{\partial p} \right] W_{|E\rangle}(x,p) = 0. \quad (3.27\text{b})$$

When we recall $\kappa \equiv (M\Omega/\hbar)^{1/2}$ and introduce the dimensionless position $\xi \equiv \kappa x$, momentum $\zeta \equiv p/(\hbar\kappa)$ and energy $\eta \equiv E/(\hbar\Omega)$ the two equations take the form

$$\frac{1}{4} \left[\frac{\partial^2}{\partial \zeta^2} + \frac{\partial^2}{\partial \xi^2} \right] W_{|\eta\rangle}(\xi, \zeta) + \left[2\eta - (\zeta^2 + \xi^2) \right] W_{|\eta\rangle}(\xi, \zeta) = 0$$

and

$$\left(\zeta \frac{\partial}{\partial \xi} - \xi \frac{\partial}{\partial \zeta} \right) W_{|\eta\rangle}(\xi, \zeta) = 0.$$

The first equation involves the Laplacian with respect to the two phase space variables ξ and ζ. Moreover, it brings in both variables in a quadratic way. Hence, this eigenvalue equation for the energy eigenstate of a one-dimensional harmonic oscillator is completely analogous to the Schrödinger equation for the energy eigenstate of a two-dimensional harmonic oscillator. Consequently, we can find the Wigner function by expanding into products of harmonic oscillator wave functions containing Hermite polynomials.

However, the second phase space equation enforces a special symmetry on these solutions. Indeed, the Wigner function can only depend on the sum of the squares of the two phase space variables, that is it can only depend on the energy. In order to show this we make the ansatz

$$W_{|E\rangle}(\xi, \zeta) = W(y)$$

where

$$y(\xi, \zeta) \equiv \xi^2 + \zeta^2. \quad (3.28)$$

86   3 Wigner Function

With the help of the relations

$$\frac{\partial W_{|\eta\rangle}}{\partial \xi} = \frac{dW}{dy} 2\xi \quad \text{and} \quad \frac{\partial W_{|\eta\rangle}}{\partial \zeta} = \frac{dW}{dy} 2\zeta \qquad (3.29)$$

we arrive at

$$\left(\zeta \cdot \frac{\partial}{\partial \xi} - \xi \cdot \frac{\partial}{\partial \zeta}\right) W_{|\eta\rangle}(\xi, \zeta) = (\zeta \cdot 2\xi - \xi \cdot 2\zeta)\frac{dW(y)}{dy} = 0. \qquad (3.30)$$

We now turn to the eigenvalue equation for the Wigner function which due to the symmetry imposed by the classical Liouville equation turns into a ordinary differential equation. With the help of the relations

$$\frac{\partial^2 W_{|\eta\rangle}}{\partial \xi^2} = \frac{d^2 W(y)}{dy^2} 4\xi^2 + 2\frac{dW(y)}{dy} \quad \text{and} \quad \frac{\partial^2 W_{|\eta\rangle}}{\partial \zeta^2} = \frac{d^2 W(y)}{dy^2} 4\zeta^2 + 2\frac{dW(y)}{dy}$$

following from the chain rule Eq. (3.29) and the definition Eq. (3.28) of y we find the differential equation

$$\left(y\frac{d^2}{dy^2} + \frac{d}{dy} + 2\eta - y\right) W(y) = 0 \qquad (3.31)$$

of second order discussed in more detail in Appendix F.

3.6.2 Phase Space Enforces Energy Quantization

In Sec. 2.2.2 we have determined the position representation of an energy eigenstate of a harmonic oscillator. The appropriate boundary conditions – decay of the wave function in the classically forbidden regime – select from all possible solutions of the corresponding Schrödinger equation the energy wave function. Therefore, the boundary conditions in position space enforce the discreteness and the values of the eigenenergies.

Similarly, we find the Wigner function of this energy eigenstate by solving a Schrödinger type differential equation, namely, Eq. (3.27a). This equation is a partial differential equation in phase space. Consequently the boundary conditions in phase space determine the eigenenergies.

Since the differential equation Eq. (3.31) is of second order it has two independent solutions. The fact that the Wigner function is normalized puts a constraint on the behavior of these solutions: They must decay as $y \to \infty$.

In Appendix F we solve this radial equation subjected to this boundary condition and show that the function

$$W(y) = \mathcal{N}_m e^{-y} L_m(2y)$$

consisting of the mth Laguerre polynomial represents a normalizable solution of the ordinary differential equation Eq. (3.31) provided

$$\eta = \left(m + \frac{1}{2}\right).$$

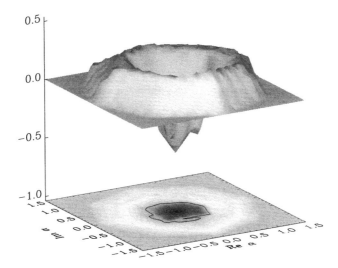

Fig. 3.4: Experimentally reconstructed Wigner function of an approximate $|n=1\rangle$ number state. The black contour represents $W_{|1\rangle} = 0$. The negative values around the origin highlight the non-classical character of this state. Taken from Leibfried et al., Phys. Rev. Lett. **77**, 4281 (1996).

This is the energy quantization following from phase space quantization.

In the phase space variables x and p the Wigner function of the mth energy eigenstate reads

$$W_{|m\rangle}(x,p) = \frac{(-1)^m}{\pi\hbar} \exp\left\{-\left[\left(\frac{p}{\hbar\kappa}\right)^2 + (\kappa x)^2\right]\right\} L_m\left\{2\left[\left(\frac{p}{\hbar\kappa}\right)^2 + (\kappa x)^2\right]\right\}. \quad (3.32)$$

In the next chapter we discuss the properties of this function in more detail. Here we only mention that it has been measured experimentally for an ion moving in the harmonic oscillator potential provided by a Paul trap as shown in Fig. 3.4.

3.7 Evaluation of Quantum Mechanical Averages

The Wigner function provides a vivid representation of the quantum state at hand. However, the quantum state is only one side of the coin. The other is the hermitian operator corresponding to this or that observable. Only with the two concepts together we can make contact with the experiment. Indeed, the quantum state is needed to evaluate expectation values of operators.

In classical statistical physics we evaluate averages of functions $A(x,p)$ which depend on the phase space variables x and p with the help of a classical distribution $W_{\text{cl}}(x,p)$ via the relation

$$\langle A_{\text{cl}}(x,p)\rangle = \int_{-\infty}^{\infty} dx \int_{-\infty}^{\infty} dp \, A_{\text{cl}}(x,p) \, W_{\text{cl}}(x,p). \quad (3.33)$$

3 Wigner Function

Does a similar method exist in quantum mechanics? In the quantum mechanical case the role of the classical phase space distribution is taken up by the Wigner function. It is therefore suggestive to calculate averages of a quantum mechanical operator $\hat{A}$ in a way

$$\langle \hat{A}(\hat{x},\hat{p}) \rangle = \int_{-\infty}^{\infty} dx \int_{-\infty}^{\infty} dp\, A(x,p) W(x,p)$$

similar to Eq. (3.33). Here $A(x,p)$ is the c-number representation of the operator $\hat{A}(\hat{x},\hat{p})$ such that the phase space integration with respect to the Wigner function of the state yields the correct quantum mechanical result. In the present section we show that there exists a well-defined procedure which allows us to calculate these averages in complete accordance with quantum mechanics.

3.7.1 Operator Ordering

Needless to say it is not trivial to find this c-number representation of a given operator. The following most elementary example brings out clearly the problems associated with this concept.

Examples

We consider the operator product

$$\hat{A} \equiv \hat{x} \cdot \hat{p}$$

consisting of the position and the momentum operators $\hat{x}$ and $\hat{p}$, respectively. For the present discussion it is not relevant that $\hat{A}$ is not hermitian.

We now try to find a classical c-number representation. One possible approach is to replace all operators $\hat{x}$ and $\hat{p}$ by c-numbers x and p. However the results of this naive approach depend on the stage at which we have taken this classical limit. In particular, when we start from the operator $\hat{A}$ and replace $\hat{x}$ and $\hat{p}$ we arrive

$$\hat{A} \to A(x,p) \equiv x \cdot p. \tag{3.34}$$

On the other hand we can rewrite the operator $\hat{A}$ in an exact way by applying first the commutation relation

$$[\hat{p},\hat{x}] = \frac{\hbar}{i},$$

which yields

$$\hat{A} = \hat{p} \cdot \hat{x} - [\hat{p},\hat{x}] = \hat{p} \cdot \hat{x} - \frac{\hbar}{i}.$$

When we now replace operators by c-numbers we arrive at

$$\bar{A} = p \cdot x - \frac{\hbar}{i}.$$

Obviously this expression is different from the one defined in Eq. (3.34) by the commutator.

Similarly we can cast the operator $\hat{A}$ in the form

$$\hat{A} = \frac{1}{2}(\hat{x} \cdot \hat{p} + \hat{p} \cdot \hat{x}) - \frac{1}{2}[\hat{p}, \hat{x}] = \frac{1}{2}(\hat{x} \cdot \hat{p} + \hat{p} \cdot \hat{x}) - \frac{\hbar}{2i}.$$

Here the position and the momentum operators appear in a symmetric way. In this case the naive c-number representation reads

$$A_S \equiv x \cdot p - \frac{\hbar}{2i}.$$

This elementary example clearly shows that there exist many classical limits of the same operators depending on how we have chosen to order the noncommuting operators before we have replaced them by c-numbers. This problem is known in the literature as the ordering problem. Consequently the averaging procedure using the Wigner function can only be in agreement with the quantum mechanical result provided the correct ordering has been chosen. In the next section we show that indeed the Wigner function allows to calculate expectation values of symmetrically ordered operators.

Weyl-Wigner Ordering

In the definition of the Wigner function

$$W(x, p) = \frac{1}{2\pi\hbar} \int_{-\infty}^{\infty} d\xi \, \exp\left(-\frac{i}{\hbar} p\xi\right) \langle x + \tfrac{1}{2}\xi | \hat{\rho} | x - \tfrac{1}{2}\xi \rangle$$

we represent the density operator $\hat{\rho}$ in terms of a classical function $W(x,p)$ that lives in phase space. We can generalize this concept of replacing quantum mechanical operators by classical c-numbers to arbitrary operators $\hat{A}$ by defining the so-called Weyl-Wigner correspondence

$$A(x, p) \equiv \int_{-\infty}^{\infty} d\xi \, \exp\left(-\frac{i}{\hbar} p\xi\right) \langle x + \tfrac{1}{2}\xi | \hat{A} | x - \tfrac{1}{2}\xi \rangle. \tag{3.35}$$

In this way we have defined a classical phase space representation of the operator $\hat{A}$ which allows us to evaluate quantum mechanical expectation values such as $\langle \hat{A} \rangle$ in terms of the Wigner phase space distribution function $W(x,p)$.

Indeed, when we recall the definition

$$\langle \hat{A} \rangle = \text{Tr}(\hat{A}\hat{\rho})$$

of the expectation value of the operator $\hat{A}$ in terms of the density operator we find with the help of the trace product formula Eq. (3.5) the relation

$$\langle \hat{A} \rangle = \int_{-\infty}^{\infty} dx \int_{-\infty}^{\infty} dp \, A(x, p) \, W(x, p).$$

3.7.2 Examples of Weyl-Wigner Ordering

We now illustrate the concept of Weyl-Wigner ordering using two examples: The operator product $\hat{A} = \hat{x} \cdot \hat{p}$ shows that the Wigner function allows us to calculate symmetrically ordered operators in a quasi-classical way. The Weyl-Wigner transform of the operator product $\hat{H}\hat{\rho}$ consisting of the Hamiltonian and the density operator enables us to make the connection with the phase space equations Eqs. (3.17) and (3.18).

Symmetric Ordering

In the case of the operator product $\hat{A} = \hat{x} \cdot \hat{p}$ we find from the definition Eq. (3.35) of the Weyl-Wigner ordering

$$A(x,p) = \int_{-\infty}^{\infty} d\xi \, \exp\left(-\frac{i}{\hbar}p\xi\right) \left\langle x + \tfrac{1}{2}\xi \right| \hat{x} \cdot \hat{p} \left| x - \tfrac{1}{2}\xi \right\rangle,$$

or

$$A(x,p) = \int d\xi \, \exp\left(-\frac{i}{\hbar}p\xi\right) \left(x + \tfrac{1}{2}\xi\right) \left\langle x + \tfrac{1}{2}\xi \right| \hat{p} \left| x - \tfrac{1}{2}\xi \right\rangle.$$

When we insert a complete set of momentum states we find

$$A(x,p) = \int d\xi \int dp' \, \exp\left(-\frac{i}{\hbar}p\xi\right) \left(x + \tfrac{1}{2}\xi\right) p' \left\langle x + \tfrac{1}{2}\xi \middle| p' \right\rangle \left\langle p' \middle| x - \tfrac{1}{2}\xi \right\rangle,$$

or

$$A(x,p) = \frac{1}{2\pi\hbar} \int d\xi \int dp' \, \exp\left[-\frac{i}{\hbar}(p - p')\xi\right] p' \left(x + \tfrac{1}{2}\xi\right),$$

where in the last step we have made use of the relation

$$\langle x | p \rangle = \frac{1}{\sqrt{2\pi\hbar}} \exp\left(\frac{i}{\hbar} x \cdot p\right).$$

We cast the integrals in the form

$$A(x,p) = \left[x + \frac{1}{2}\frac{\partial}{\partial(-ip/\hbar)}\right] \int_{-\infty}^{\infty} dp' \, p' \frac{1}{2\pi\hbar} \int_{-\infty}^{\infty} d\xi \, \exp\left[-\frac{i}{\hbar}(p - p')\xi\right]$$

and perform them with the help of the delta function at $p = p'$ leading to

$$A(x,p) = x \cdot p - \frac{\hbar}{2i}\frac{\partial}{\partial p} p = x \cdot p - \frac{\hbar}{2i}.$$

Hence, for this example the Weyl-Wigner ordering corresponds to the symmetric ordering.

We conclude by emphasizing that in general the Weyl-Wigner correspondence corresponds to symmetric ordering.

3.7 Evaluation of Quantum Mechanical Averages

Schrödinger Equation in Phase Space

We now illustrate this technique of representing quantum mechanical operators by c-numbers for the case of the time independent Schrödinger equation. In particular, we show that we arrive at the two coupled equations in phase space Eqs. (3.17) and (3.18) determining the Wigner function of an energy eigenstate.

We start from the eigenvalue equation

$$\hat{H}\,|E\rangle = E\,|E\rangle$$

of an energy eigenstate $|E\rangle$ in the form

$$\hat{H}\,(|E\rangle\langle E|) = E\,(|E\rangle\langle E|)$$

of an operator equation. Hence, on the left hand side we have the product of two operators whereas on the right hand side we have a single operator multiplied by a c-number. In the Weyl-Wigner correspondence the right hand side of this equation turns into the Wigner function $W_{|E\rangle}$ of the energy eigenstate multiplied by the eigenvalue. The left hand side is more complicated, since it involves the product of two operators.

We recall from Problem 3.4 that for a product $\hat{A}\hat{B}$ of two operators $\hat{A}$ and $\hat{B}$ the Weyl-Wigner correspondence reads

$$(AB)(x,p) = A\left(x - \frac{\hbar}{2i}\frac{\partial}{\partial p}, p + \frac{\hbar}{2i}\frac{\partial}{\partial x}\right)B(x,p).$$

According to this prescription we have to replace the position and momentum variables in A by operators containing position and momentum and their derivatives, that is

$$x \to x - \frac{\hbar}{2i}\frac{\partial}{\partial p} \quad \text{and} \quad p \to p + \frac{\hbar}{2i}\frac{\partial}{\partial x}.$$

These operators sometimes carry the name *Bopp operators* since they are associated with F. Bopp who held the famous Sommerfeld chair at the University of Munich.

With the identification $\hat{A} \equiv \hat{H}$ and $\hat{B} \equiv |E\rangle\langle E| \equiv \hat{\rho}$ we find the formula

$$(H\rho)(x,p) = H\left(x - \frac{\hbar}{2i}\frac{\partial}{\partial p}, p + \frac{\hbar}{2i}\frac{\partial}{\partial x}\right)W_{|E\rangle}(x,p).$$

Hence, the Weyl-Wigner representation of the time independent Schrödinger equation

$$\hat{H}\hat{\rho} = E\,\hat{\rho}$$

reads

$$H\left(x - \frac{\hbar}{2i}\frac{\partial}{\partial p}, p + \frac{\hbar}{2i}\frac{\partial}{\partial x}\right)W_{|E\rangle}(x,p) = E\,W_{|E\rangle}(x,p).$$

For the Hamiltonian

$$\hat{H} = \frac{\hat{p}^2}{2M} + U(\hat{x})$$

of a particle in the potential U the above expression for the Wigner function of the energy eigenstate reads

$$\left[\frac{1}{2M}\left(p+\frac{\hbar}{2i}\frac{\partial}{\partial x}\right)^2 + U\left(x - \frac{\hbar}{2i}\frac{\partial}{\partial p}\right)\right] W_{|E\rangle}(x,p) = E\, W_{|E\rangle}(x,p).$$

When we take the real and imaginary part of this equation we arrive at

$$\left\{\frac{p^2}{2M} - \frac{\hbar^2}{8M}\frac{\partial^2}{\partial x^2} + \frac{1}{2}\left[U\left(x - \frac{\hbar}{2i}\frac{\partial}{\partial p}\right) + U\left(x + \frac{\hbar}{2i}\frac{\partial}{\partial p}\right)\right]\right\} W_{|E\rangle}(x,p) = E\, W_{|E\rangle}(x,p)$$

and

$$\left\{\frac{p}{M}\frac{\partial}{\partial x} - \frac{1}{i\hbar}\left[U\left(x - \frac{\hbar}{2i}\frac{\partial}{\partial p}\right) - U\left(x + \frac{\hbar}{2i}\frac{\partial}{\partial p}\right)\right]\right\} W_{|E\rangle}(x,p) = 0.$$

The Taylor expansion of the potential U finally yields Eqs. (3.17) and (3.18).

The first equation plays the role of the Schrödinger energy eigenvalue equation. Since it is a partial differential equation in phase space it depends on two variables. Moreover, the potential energy enters in a rather complicated way bringing in a combination of the position variable as well as a momentum derivative. The second equation is the stationary part of the Liouville equation. It is remarkable that both equations contain either the sum or the difference of the potential evaluated at the two operators.

Problems

3.1 Time Evolution of the Wigner Function in Quadratic Potentials

Consider a potential of the form

$$U(x) = \frac{\alpha}{2}x^2 + \beta x + \gamma.$$

Prove that the time evolution of the Wigner function is given by

$$W(x,p,t) = W_0\left(x_0(x,p,t), p_0(x,p,t)\right).$$

Here, $W_0(x,p)$ is the Wigner function at time $t = 0$, and $(x_0(x,p,t), p_0(x,p,t))$ is the phase space point at which a *classical* particle would have to start at time $t = 0$ in order to reach the point (x, p) at time t.

3.2 Wigner Function of a Thermal State

(a) Calculate the Wigner function $W(x,p)$ corresponding to the density operator of a harmonic oscillator in thermal equilibrium with a heat bath of temperature. Use the result of Problem 2.6 for the position representation $\rho(x, x') = \langle x|\hat{\rho}|x'\rangle$ of the density operator.

(b) Investigate the behavior of the Wigner function $W(x,p)$ for high and low temperatures and interpret your result.

3.3 Moyal Function for a Harmonic Oscillator

Show that the Moyal function for the harmonic oscillator reads

$$W_{|m\rangle\langle n|}(\eta) = \frac{2(-1)^n}{\pi\hbar}\sqrt{\frac{n!}{m!}}(2\eta^*)^{m-n}\exp\left(-2|\eta|^2\right)L_n^{m-n}\left(4|\eta|^2\right)$$

where we have introduced the abbreviation

$$2\left[\left(\frac{p}{\hbar\kappa}\right)^2 + (\kappa x)^2\right] = \frac{4}{\hbar\Omega}\left(\frac{p^2}{2M} + \frac{1}{2}M\Omega^2 x^2\right) = 4\eta(x,p).$$

This expression is valid for $m \leq n$. How does this relation look like for $m > n$?

Hint: See Groenewold (1946) or Buzek and Knight (1995).

3.4 Weyl-Wigner-Ordering

Show that for a product $\hat{A}\hat{B}$ of two operators $\hat{A}$ and $\hat{B}$ the Weyl-Wigner correspondence reads

$$(AB)(x,p) = A(x - \frac{\hbar}{2i}\frac{\partial}{\partial p}, p + \frac{\hbar}{2i}\frac{\partial}{\partial x})B(x,p).$$

3.5 Classical Limit

According to Eqs. (3.32) and (2.15) the Wigner function of the mth energy eigenstate reads

$$W_{|m\rangle}(x,p) = \frac{(-1)^m}{\pi\hbar}\exp\left[-\frac{2}{\hbar\Omega}\left(\frac{p^2}{2M} + \frac{1}{2}M\Omega^2 x^2\right)\right]L_m\left[\frac{4}{\hbar\Omega}\left(\frac{p^2}{2M} + \frac{1}{2}M\Omega^2 x^2\right)\right].$$

Consider the classical limit $\hbar \to 0$. How does the δ function along the classical energy E arise?

Hint: Obviously it is not meaningful to just let $\hbar$ go to 0 and keep m fixed. In order to see the classical limit we need to keep the energy $E = (m+1/2)\hbar\Omega$ of the state fixed as $\hbar$ approaches 0. Hence, $m/\hbar$ has to stay constant or in other words m has to go to infinity. Use the asymptotics of Laguerre polynomials.

3.6 Wigner Function and Tunneling

Start from the two coupled phase space equations Eqs. (3.17) and (3.17) and derive the equations determining the Wigner function of an energy eigenstate of an inverted harmonic oscillator of potential

$$U(x) = -\frac{1}{2}M\Omega^2 x^2.$$

Show that for energies far below zero and far above zero the Wigner function is approximately an Airy function along the classical trajectory of energy E. For zero energy the Wigner function is the zeroth order Bessel function. Discuss the phenomenon of tunneling in the language of Wigner functions.

Hint: See Balazs and Voros (1990).

References

Classical-Quantum Transition

For a discussion of the comparison of quantum mechanics of a single particle and classical statistical mechanics of an ensemble of particles see

M. Born, *Continuity, Determinism, and Reality*, Dän. Math.-Phys. Kl. **30**, 3–26 (1955)

M. Born, *Vorhersagbarkeit in der klassischen Mechanik*, Z. Physik **153**, 372–388 (1958)

M. Born, *Statistical Interpretation of Quantum Mechanics*, Science **122**, 675–679 (1955)

Original Papers on Wigner Functions

The concept of this phase space distribution function originated from

E.P. Wigner, *On the quantum correction for thermodynamic equilibrium*, Phys. Rev. **40**, 749–759 (1932)

There is an interesting footnote in this paper in which the author states that this function has been proposed earlier for another purpose with L. Szilard. However, no such paper has been published. In the context of scattering theory P.A.M. Dirac and W. Heisenberg have used expressions similar to the Wigner function. Dirac even worked out the equation of motion.

P.A.M. Dirac, *Note on Exchange Phenomena in the Thomas Atom*, Proc. Camb. Phil. Soc. **26**, 376–385 (1930)

W. Heisenberg, *Über die inkohärente Streuung von Röntgenstrahlen*, Physik. Zeitschr. **32**, 737–740 (1931)

The Moyal function was introduced in

J.E. Moyal, *Quantum mechanics as a statistical theory*, Proc. Camb. Phil. Soc. **45**, 99–124 (1947)

Early ideas on the Moyal function and, in particular, the Moyal function for harmonic oscillators can be found in

H.J. Groenewold, *On the principles of elementary quantum mechanics*, Physica **12**, 405–460 (1946)

Reviews of Wigner Functions

For a review of the properties of Wigner functions see, for example,

V.I. Tatarskii, *The Wigner representation of quantum mechanics*, Sov. Phys. Usp. **26**, 311–327 (1983)

N.L. Balazs and **B.K. Jennings**, *Wigner's functions and other distribution functions in Mock phase spaces*, Phys. Rep. **104**, 347–391 (1984)

M. Hillery, R.F. O'Connell, M.O. Scully and **E.P. Wigner**, *Distribution functions in physics: fundamentals*, Phys. Rep. **106**, 121–167 (1984)

B.-G. Englert, *On the operator bases underlying Wigner's, Kirkwood's and Glauber's phase space functions*, J. Phys. A **22**, 625–640 (1989)

I. Bialynicki-Birula, M. Cieplak and J. Kaminski, *Theory of Quanta*, Oxford University Press, 1992

A.M. Ozorio de Almeida, *The Weyl representation in classical and quantum mechanics*, Phys. Rep. **295**, 265–342 (1998)

E. Scheibe, *Die Reduktion physikalischer Theorien*, Band II, Springer, Heidelberg, 1999

The heuristic approach to the Wigner function of Sec. 3.1 can be found in

G. Süßmann and W.P. Schleich, *A jump shot at the Wigner distribution*, Physics Today, **44** (10), 146–148 (1991)

Scattering theory using the Wigner function can be found in

P. Carruthers and F. Zachariasen, *Quantum collision theory with phase space distributions*, Rev. Mod. Phys. **55**, 245–285 (1983)

For an application of the Wigner function to time-frequency analysis see

L. Cohen, *Time-Frequency Analysis*, Prentice Hall, Englewood Cliffs (1995)

Wigner Functions for Electric Field

Since there exists a close analogy between the wave function and the classical electric field Wigner functions in classical optics have enjoyed great theoretical and experimental interest. See for example

H.O. Bartelt, K.-H. Brenner and A.W. Lohmann, *The Wigner distribution function and its optical production*, Opt. Commun. **32**, 32–38 (1980)

D. Dragoman, *The Wigner distribution function in Optics and Optoelectronics*, Progress in Optics, edited by E. Wolf, Volume XXXVII, North Holland, Amsterdam, 1997

D. Dragoman, M. Dragoman, J. Bähr and K.-H. Brenner, *Phase-space measurements of micro-optical objects*, Appl. Opt. **38**, 5019–5023 (1999)

The Wigner function is also intimately connected to the definition of the time dependent spectrum put forward by

J.H. Eberly and K. Wódkiewicz, *The time-dependent physical spectrum of light*, J. Opt. Soc. Am. **67**, 1252–1261 (1977)

and shown in

K.-H. Brenner and K. Wódkiewicz, *The time-dependent physical spectrum of light and the Wigner distribution function*, Opt. Comm. **43**, 103–106 (1982)

Asymptotology of Wigner Functions

The asymptotic limit of Wigner functions has been studied in great detail by

M.V. Berry, *Semi-classical mechanics in phase space: a study of Wigner's function*, Phil. Trans. R. Soc. London **287 A**, 237–271 (1977)

M.V. Berry, *Quantum scars of classical closed orbits in phase space*, Proc. R. Soc. London **423 A**, 219–231 (1989)

M.V. Berry, *Fringes decorating anticaustics in ergodic wave functions*, Proc. R. Soc. London **424**, 279–288 (1989)

Interpretation of Wigner Functions

For the quantum mechanical scalar product of two quantum states as the overlap between two Wigner functions in phase space see

R.F. O'Connell and **E.P. Wigner**, *Quantum-mechanical distribution functions: conditions for uniqueness*, Phys. Lett. **83A**, 145–148 (1981)

For an interpretation of the quantum mechanical measurement resulting in the phase space overlap of two Wigner functions see

R.F. O'Connell and **A.K. Rajagopal**, *New interpretation of the scalar product in Hilbert space*, Phys. Rev. Lett. **48**, 525–526 (1982)

R.F. O'Connell and **D.F. Walls**, *Operational approach to phase-space measurements in quantum mechanics*, Nature (London) **312**, 257–258 (1984)

K. Wódkiewicz, *Operational approach to phase-space measurements in quantum mechanics*, Phys. Rev. Lett. **52**, 1064–1067 (1984)

S. Stenholm, *The Wigner Function: I. The Physical Interpretation*, Euro. J. Phys. **1**, 244–248 (1980)

V. Buzek and **P.L. Knight**, *Quantum Interference, Superposition States of Light and Nonclassical Effects*, Prog. Opt. **34**, 1–158 (1995)

For the Weyl-Wigner correspondence and its implications on the quantum theory of measurement see

F. Bopp, *Statistische Mechanik bei Störung des Zustandes eines physikalischen Systems durch die Beobachtung*, in: *W. Heisenberg und die Physik unserer Zeit*, ed. by F. Bopp, Vieweg, Braunschweig, 1961

Dynamics from Wigner Functions

M. Grønager and **N. E. Henriksen**, *Quantum dynamics via a time propagator in Wigner's phase space*, J. Chem. Phys. **102**, 5387–5395 (1995)

H. Feldmeier and **J. Schnack**, *Fermionic Molecular Dynamics*, Prog. Part. Nucl. Phys. **39**, 393–442 (1997)

Wigner Functions Determined from Phase Space

The two equations for the Moyal functions have been derived and discussed by

D.B. Fairlie, *The formulation of quantum mechanics in terms of phase space functions*, Proc. Camb. Phil. Soc. **60**, 581–586 (1964)

W. Kundt, *Classical statistics as a limiting case of quantum statistics*, Z. Naturforsch. A **22**, 1333–1336 (1967)

L. Wang and **R.F. O'Connell**, *Quantum mechanics without wave function*, Found. Phys. **18**, 1023–1033, (1988)

J.P. Dahl, in *Energy Storage and Redistribution in Molecules*, ed. J. Hinze, Plenum Press, New York, 1983

J.P. Dahl, in *Semiclassical Descriptions of Atomic and Nuclear Collisions*, eds. J. Bang and J. de Boer, Elsevier, Amsterdam, 1985

For a discussion of the equations in the case of an energy eigenstate of an anharmonic oscillator see

M. Hug, C. Menke and **W.P. Schleich**, *How to calculate the Wigner function from phase space*, J. Phys. A **31**, L217–L224 (1998)

M. Hug, C. Menke and **W.P. Schleich**, *Modified spectral method in phase space: Calculation of the Wigner function. I. Fundamentals*, Phys. Rev. A **57**, 3188–3205 (1998)

M. Hug, C. Menke and **W.P. Schleich**, *Modified spectral method in phase space: Calculation of the Wigner function. II. Generalizations*, Phys. Rev. A **57**, 3206–3224 (1998)

For a compact formulation of the Moyal equations using the star product see

T. Curtright, D. Fairlie and **C. Zachos**, *Features of time-independent Wigner functions*, Phys. Rev. D **58**, 025002-1–14 (1998)

C. Zachos and **T. Curtright**, *Phase-Space Quantization of Field Theory*, Prog. Theor. Phys. Suppl. **135**, 244–258 (1999)

For the Wigner function of the Morse oscillator see

J.P. Dahl and **M. Springborg**, *The Morse oscillator in position space, momentum space and phase space*, J. Chem. Phys. **88**, 4535–4547 (1988)

Wigner Function and Negative Probability

For a discussion of the Hudson-Piquet theorem we refer to

R.L. Hudson, *When is the Wigner quasi-probability density non-negative?*, Rep. Math. Phys. **6**, 249–252 (1974)

C. Piquet *Fonctions de type positif associées à deux opérateurs hermitiens*, C. R. Acad. Sc. Paris **279 A**, 107–109 (1974)

The concept of negative probability was pushed by

M.S. Bartlett, *Negative Probability*, Proc. Cambridge Philos. Soc. **41**, 71–73 (1945)

R.P. Feynman, in *Negative Probabilities in Quantum Mechanics*, ed. by B. Hiley and F. Peat, Routledge, London, 1987

See also

M.O. Scully, H. Walther and **W.P. Schleich**, *Feynman's approach to negative probability in quantum mechanics*, Phys. Rev. A **49**, 1562–1566 (1994)

The negative parts of the Wigner function of the first excited state of a harmonic oscillator has been measured by

D. Leibfried, D.M. Meekhof, B.E. King, C. Monroe, W.M. Itano and **D.J. Wineland**, *Experimental Determination of the Motional Quantum State of a Trapped Atom*, Phys. Rev. Lett. **77** 4281–4285 (1996)

For a discussion of the phenomenon of tunneling in the language of the Wigner function and negative probabilities see

N.L. Balazs and **A. Voros**, *Wigner's Function and Tunneling*, Annals of Physics **199**, 123–140 (1990)

Quantum mechanical phase space distribution functions do not necessarily have to take on negative values. There exist distributions which are always positive and still provide the correct, quantum mechanical marginals. However, they are not bilinear in the wave function anymore, see
L. Cohen, *Positive and negative joint quantum distributions*, in: *Frontiers of Nonequilibrium Statistical Physics*, edited by G.T. Moore and M.O. Scully, Plenum Press, New York, 1986

Area in Phase Space

For an estimate of the area in phase space occupied by a Wigner function see
G. Süssmann, *Uncertainty Relation: From Inequality to Equality*, Z. Naturforsch. **52a**, 49–52 (1997)

4 Quantum States in Phase Space

The harmonic oscillator plays a prominent role in physics, and in particular, in quantum physics. Since it can be solved exactly, it is the favorite toy of theoreticians. However, it also serves as a model for real systems. For example, in Chapter 10 we show that a single mode of the electromagnetic field in a resonator is a harmonic oscillator. Moreover, laser cooling has allowed us to observe the quantum motion of a single ion stored in a Paul trap. Since this Paul trap provides an approximate quadratic potential for the ion, this system is a realistic mechanical harmonic oscillator.

In this chapter we study quantum states of a simple harmonic oscillator system described by the Hamiltonian

$$\hat{H} \equiv \frac{\hat{p}^2}{2M} + \frac{1}{2} M\Omega^2 \hat{x}^2, \tag{4.1}$$

where M denotes the mass of the oscillator and Ω is its eigen frequency. The operators $\hat{x}$ and $\hat{p}$ describe its position and momentum, respectively.

We concentrate on energy eigenstates, coherent and squeezed states and rotated quadrature states. In particular, we discuss the energy distribution of these states. For the case of a field oscillator this corresponds to the photon statistics of an electromagnetic field. In the case of the vibratory motion the energy distribution corresponds to occupation probability in the individual phonon modes. We show that the energy distribution of a coherent state is a Poissonian, whereas the corresponding distribution of a highly squeezed state displays characteristic oscillations. We derive simple analytical expressions for these distributions in the limit of large quantum numbers. In this way we meet the first examples of the main theme of the book: Complex phenomena become simple in the appropriate asymptotic limit. Following M.V. Berry we refer to this field *asymptotology*.

Another focus of this chapter is the time dependence of the position and momentum distributions of the above mentioned states. We find these distributions from the time evolution of the corresponding Wigner functions. This feature allows us to reconstruct the Wigner function of quantum states from the quadrature distribution functions.

4.1 Energy Eigenstate

The main theme of the book is asymptotology. We consider the wave functions of a quantum system in the asymptotic limit of large quantum numbers. Here we present the first example for such an asymptotic discussion by considering the energy wave function of a harmonic oscillator in the limit of $m \gg 1$. We show that the wave function at a given position consists of a coherent superposition of a right going and a left going wave with a fixed phase difference. These quantities have simple geometrical meanings in phase space. However, we emphasize that this property is not limited to the wave function of the harmonic oscillator but can be applied to the wave functions of any arbitrary potential as we discuss in Chapter 5.

4.1.1 Simple Phase Space Representation

In Sec. 2.2 we have already discussed the energy eigenstates of a harmonic oscillator. In particular, we have derived the wave function

$$u_m(x) = \mathcal{N}_m H_m(\kappa x) e^{-\frac{1}{2}(\kappa x)^2} \tag{4.2}$$

of an energy eigenstate in position space. The normalization constant is

$$\mathcal{N}_m \equiv \left(\frac{\kappa^2}{\pi}\right)^{1/4} \frac{1}{(2^m \, m!)^{1/2}} \tag{4.3}$$

where $\kappa \equiv \sqrt{M\Omega/\hbar}$.

How to represent an energy eigenstate in phase space? In classical mechanics the motion of a particle of well-defined energy E is described by an ellipse in phase space spanned by the variables position x and momentum p. This follows from the conservation of energy in the form

$$\frac{p^2}{2M} + \frac{x^2}{2/(M\Omega^2)} = E.$$

Hence, for a quantum particle in an energy eigenstate $|m\rangle$ of energy

$$E_m = \hbar\Omega\left(m + \tfrac{1}{2}\right)$$

we find the phase space trajectory

$$\frac{p^2}{2M} + \frac{x^2}{2/(M\Omega^2)} = \hbar\Omega\left(m + \tfrac{1}{2}\right).$$

It is convenient to introduce the dimensionless position κx and the dimensionless momentum $p/(\hbar\kappa)$ and consider the phase space spanned by these variables. Indeed, now the phase space trajectory following from the conservation of energy

$$\left(\frac{p}{\hbar\kappa}\right)^2 + (\kappa x)^2 = 2\left(m + \tfrac{1}{2}\right)$$

4.1 Energy Eigenstate

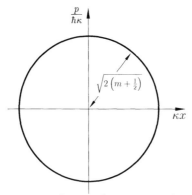

Fig. 4.1: Elementary phase space picture of an energy eigenstate. We consider phase space spanned by the dimensionless position κx and the dimensionless momentum $p/(\hbar\kappa)$. In these variables the phase space trajectory corresponding to the eigenstate of energy $E_m = \hbar\Omega(m+1/2)$ is a circle of radius $\sqrt{2(m+1/2)}$ traversed in clockwise direction.

is a circle of radius $\sqrt{2(m+1/2)}$ as shown in Fig. 4.1.

Why not associate this circular phase space orbit as the most elementary representation of an energy eigenstate? We now show that in the limit of large quantum numbers the energy wave function $u_m(x)$ can indeed be represented as a line integral in such a phase space.

4.1.2 Large-m Limit

For this purpose we first cast the wave function Eq. (4.2) into a different form. We note that the position dependence occurs in the Gaussian and in the Hermite polynomials. The polynomials depend on the quantum number m. For a discussion of the large-m limit it is therefore useful to rewrite the Hermite polynomials.

Energy Wave Function as a Contour Integral

We recall the representation

$$H_m(x) = -2^m \, m! \, \frac{1}{2\pi i} \oint dz \, z^{-(m+1)} \, e^{xz - z^2/4}$$

of the Hermite polynomial H_m in terms of a Cauchy integral in complex z-space. Here the contour circumnavigates the origin $z = 0$ in the clockwise direction.

We verify this formula by substituting the generating function

$$e^{xz - z^2/4} = \sum_{n=0}^{\infty} \frac{H_n(x)}{n!} \left(\frac{z}{2}\right)^n$$

of the Hermite polynomials into the integral on the right hand side of the equation. We therefore find

$$-\sum_{n=0}^{\infty} H_n(x) 2^{m-n} \frac{m!}{n!} \frac{1}{2\pi i} \oint dz \, z^{-(m-n+1)} = H_m(x)$$

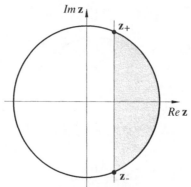

Fig. 4.2: Energy wave function of a harmonic oscillator from a contour integral in complex space. We can obtain the wave function at position x by a line integral in complex z-space. The path circumvents the origin in the clockwise direction. In the limit of large quantum numbers the main contribution to this line integral arises from two points $z_\pm$ only. They have the same real part determined by κx, but their imaginary parts $\pm p_m(x)/(\hbar\kappa)$ determined by the classical momentum $p_m(x)$ differ by a sign. Hence, the two points are located symmetrically with respect to the real axis. For different positions x these points $z_\pm$ lie on a circle of radius $\sqrt{2(m+1/2)}$. The phase difference between them is the shaded area enclosed by the circle and the vertical line at κx.

where in the last step we have used the relation

$$\oint dz\, z^{-(k+1)} = -ir^{-k}\int_0^{2\pi} d\varphi\, e^{-ik\varphi} = 2\pi i \delta_{k,0}$$

and the polar representation $z = re^{i\varphi}$ with $dz = iz d\varphi$.

When we substitute this expression into Eq. (4.2) we arrive at

$$u_m(x) = -\left(\frac{\kappa^2}{\pi}\right)^{1/4}(2^m m!)^{1/2}\frac{1}{2\pi i}\oint dz\, z^{-(m+1)} e^{\kappa x z - z^2/4}\, e^{-(\kappa x)^2/2}. \quad (4.4)$$

Hence, we have expressed the energy wave function by a contour line in complex z-space.

Asymptotics

We now relate this complex space to classical phase space. For this purpose we show in Appendix A that the dominant contributions to the line integral in Eq. (4.4) result from the two symmetrically located points

$$z_\pm = \kappa x \pm i[2(m+\tfrac{1}{2}) - \kappa^2 x^2]^{1/2} \equiv \kappa x \pm i p_m(x)/(\hbar\kappa),$$

shown in Fig. 4.2.

Note, that the real and the imaginary part of these points are determined by the position x and momentum

$$p_m(x) \equiv \sqrt{2M\left(E_m - \tfrac{1}{2}M\Omega^2 x^2\right)} \quad (4.5)$$

4.1 Energy Eigenstate

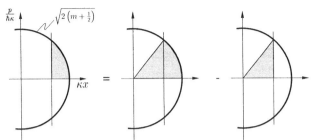

Fig. 4.3: Shaded area as the difference between the segment and the triangle defined by the real axis, the vertical line at κx and the circle of radius $\sqrt{2(m+1/2)}$.

of the mechanical oscillator. Moreover, these points lie on a circle of radius $\sqrt{2(m+1/2)}$ since

$$|z_\pm|^2 = (\kappa x)^2 + \left(\frac{p_m(x)}{\hbar\kappa}\right)^2 = 2(m+1/2).$$

This result is consistent with the naive phase space picture of an energy eigenstate of energy $E_m = \hbar\Omega(m+1/2)$ as a circular phase space trajectory of radius $\sqrt{2(m+1/2)}$.

In the large-m limit the energy wave function takes the form

$$u_m(x) \cong \sqrt{\mathcal{A}_m}\, e^{i\phi_m} + \sqrt{\mathcal{A}_m}\, e^{-i\phi_m} \qquad (4.6a)$$

where the amplitudes

$$\mathcal{A}_m(x) \equiv \frac{1}{2\pi} \frac{\kappa^2 \hbar}{p_m(x)} \qquad (4.6b)$$

and the phases

$$\phi_m(x) \equiv S_m(x) - \frac{\pi}{4} \qquad (4.6c)$$

with

$$S_m(x) = (m+\tfrac{1}{2}) \arctan\left\{\left[2(m+\tfrac{1}{2}) - \kappa^2 x^2\right]^{1/2} (\kappa x)^{-1}\right\} - \tfrac{1}{2}\kappa x \left[2(m+\tfrac{1}{2}) - \kappa^2 x^2\right]^{1/2}$$

are position dependent.

Hence, the energy wave function in the large-m limit consists of a superposition of a right going and a left going wave with a fixed phase relation. The phase difference $2\phi_m$ between the two counter propagating waves is determined by the constant $\pi/4$ and the phase $S_m(x)$. The latter quantity seems on first sight to be rather complicated. However, it has a simple geometrical interpretation in phase space: As shown in Fig. 4.3 it is the area of the phase space enclosed by the vertical line at κx and a circle of radius $\xi_m \equiv \sqrt{2(m+1/2)}$, that is

$$S_m(x) \equiv \int_{\kappa x}^{\xi_m} d\xi\, \tilde{p}_m(\xi)$$

where
$$\tilde{p}_m(\xi) \equiv \sqrt{2(m+1/2) - \xi^2}$$
is a dimensionless momentum of a harmonic oscillator of dimensionless coordinate ξ.

When we substitute the energy $E_m = \hbar\Omega(m+1/2)$ into the momentum p_m Eq. (4.5) and recall the definition of $\kappa = \sqrt{M\Omega/\hbar}$ we find the connection
$$p_m(x) = \hbar\kappa\sqrt{2(m+1/2) - (\kappa x)^2} \equiv \hbar\kappa\tilde{p}_m(\kappa x)$$
between the dimensionless momentum $\tilde{p}_m(\xi)$ and the regular momentum $p_m(x)$. Hence, the phase $S_m(x)$ takes the form
$$S_m(x) = \frac{1}{\hbar}\int_x^{\xi_m/\kappa} dx'\, p_m(x').$$

Since the energy wave function is a standing wave the amplitudes $\mathcal{A}_m$ of the right and the left going wave have to be identical. Note that this amplitude involves the square root of the classical momentum in the denominator. At the classical turning point $x_m \equiv \xi_m/\kappa$ the momentum vanishes and hence, this amplitude and the approximation of the wave function become singular. Therefore, this approximation is only valid appropriately away from the turning points.

In Chapter 5 we will discuss this problem of the turning point in more detail. In particular, we derive the solution of the Schrödinger equation in the neighborhood of the turning point in terms of the Airy function. This allows us to express the wave function using a uniform asymptotic expansion. We will return to this question later.

Summary

The asymptotic analysis performed in Appendix A and summarized in this section shows that the energy wave function
$$u_m(x) = \left(\frac{\kappa^2}{\pi}\right)^{1/4} H_m(\kappa x)\, e^{-\frac{1}{2}(\kappa x)^2}$$
of a harmonic oscillator displays in the large-m limit the form
$$u_m(x) \cong \left(\frac{2}{\pi}\right)^{1/2}\left(\frac{\hbar\kappa^2}{p_m(x)}\right)^{1/2} \cos\left[\frac{1}{\hbar}\int_x^{\xi_m/\kappa} dx'\, p_m(x') - \pi/4\right]. \quad (4.7)$$

In Chapter 5 we show that the approximate expressions Eqs. (4.6) for u_m is identical to the standard WKB wave function.

Sometimes it is convenient to express this wave function in terms of its dimensionless position variable $\xi \equiv \kappa x$. In this case the position wave function
$$u_m(\xi) \equiv \pi^{-1/4}\frac{1}{(2^m m!)^{1/2}} H_m(\xi)\, e^{-\frac{1}{2}\xi^2} \quad (4.8)$$

takes the form

$$u_m(\xi) \cong \left(\frac{2}{\pi}\right)^{1/2} [\tilde{p}_m(\xi)]^{-1/2} \cos\left[\int_\xi^{\xi_m} d\xi' \, \tilde{p}_m(\xi') - \pi/4\right]. \qquad (4.9)$$

For the sake of simplicity we have not introduced two different symbols for the energy wave function in the variables x versus $\xi \equiv \kappa x$ although the functions are slightly different.

4.1.3 Wigner Function

In Chapter 3 we have introduced the concept of the Wigner function as a possible extension of a classical phase space distribution function to quantum phase space. We have already derived the Wigner function of an energy eigenstate of a harmonic oscillator. The derivation in Chapter 3 relied on the partial differential equation in phase space. In the present section we start from the definition of the Wigner function as a Fourier integral of shifted wave functions and perform the relevant integration.

Evaluation of Wigner Integral

We substitute the position representation

$$u_m(x) = \left(\frac{\kappa^2}{\pi}\right)^{1/4} H_m(\kappa x) \, e^{-\frac{1}{2}(\kappa x)^2}$$

of an energy eigenstate in position space into the definition

$$W(x,p) = \frac{1}{2\pi\hbar} \int_{-\infty}^{\infty} dy \, \exp\left(-\tfrac{i}{\hbar} py\right) \psi^*\!\left(x - \tfrac{1}{2}y\right) \psi\!\left(x + \tfrac{1}{2}y\right) \qquad (4.10)$$

of the Wigner function (3.2) and arrive at

$$W_{|m\rangle}(x,p) = \frac{1}{2\pi\hbar} \frac{1}{\sqrt{\pi}} \frac{e^{-\kappa^2 x^2}}{2^m \, m!} \int_{-\infty}^{\infty} d\xi \, \exp\left(-i\frac{p}{\hbar\kappa}\xi - \tfrac{1}{4}\xi^2\right) H_m(\kappa x + \tfrac{1}{2}\xi) \, H_m(\kappa x - \tfrac{1}{2}\xi).$$

We complete the square in the exponent

$$-\left(\frac{\xi}{2}\right)^2 - 2\frac{\xi}{2} i \frac{p}{\hbar\kappa} - \left(i\frac{p}{\hbar\kappa}\right)^2 - \left(\frac{p}{\hbar\kappa}\right)^2 = -\left(\frac{\xi}{2} + i\frac{p}{\hbar\kappa}\right)^2 - \left(\frac{p}{\hbar\kappa}\right)^2,$$

and introduce the new integration variable

$$\zeta \equiv \frac{\xi}{2} + i\frac{p}{\hbar\kappa},$$

4 Quantum States in Phase Space

which yields

$$W_{|m\rangle}(x,p) = \frac{1}{\pi\hbar} \exp\left[-(\kappa x)^2 - \left(\frac{p}{\hbar\kappa}\right)^2\right] (2^m m!)^{-1}$$

$$\times \frac{1}{\sqrt{\pi}} \int_{-\infty}^{\infty} d\zeta \, e^{-\zeta^2} H_m(\kappa x + \zeta - i\tfrac{p}{\hbar\kappa}) H_m(\kappa x - \zeta + i\tfrac{p}{\hbar\kappa}).$$

With the symmetry relation $H_m(-\zeta) = (-1)^m H_m(\zeta)$ for the Hermite polynomials we have

$$W_{|m\rangle}(x,p) = \frac{(-1)^m}{\pi\hbar} \exp\left[-(\kappa x)^2 - \left(\frac{p}{\hbar\kappa}\right)^2\right] (2^m m!)^{-1}$$

$$\times \frac{1}{\sqrt{\pi}} \int_{-\infty}^{\infty} d\zeta \, e^{-\zeta^2} H_m(\zeta - i\tfrac{p}{\hbar\kappa} + \kappa x) H_m(\zeta - i\tfrac{p}{\hbar\kappa} - \kappa x),$$

which now allows us to apply the integral relation

$$(2^m m!)^{-1} \frac{1}{\sqrt{\pi}} \int_{-\infty}^{\infty} d\zeta \, e^{-\zeta^2} H_m(\zeta + \zeta_1) H_m(\zeta + \zeta_2) = L_m(-2\zeta_1 \zeta_2)$$

where L_m denotes the mth Laguerre polynomial.

We note the formula

$$\left(-i\frac{p}{\hbar\kappa} + \kappa x\right)\left(-i\frac{p}{\hbar\kappa} - \kappa x\right) = -\left(\frac{p}{\hbar\kappa}\right)^2 - (\kappa x)^2$$

to find the expression

$$W_{|m\rangle}(x,p) = \frac{(-1)^m}{\pi\hbar} \exp\left\{-\left[\left(\frac{p}{\hbar\kappa}\right)^2 + (\kappa x)^2\right]\right\} L_m\left\{2\left[\left(\frac{p}{\hbar\kappa}\right)^2 + (\kappa x)^2\right]\right\}$$

for the Wigner function of the mth energy eigenstate.

The argument of the Laguerre polynomial is the energy

$$2\left[\left(\frac{p}{\hbar\kappa}\right)^2 + (\kappa x)^2\right] = \frac{4}{\hbar\Omega}\left(\frac{p^2}{2M} + \frac{1}{2}M\Omega^2 x^2\right) \equiv 4\eta(x,p)$$

of the oscillator – kinetic energy plus potential energy – scaled in units of the energy separation $\hbar\Omega$ of the individual oscillators. The factor 4 arises in order to find the correct kinetic and potential energy.

Shape of the Wigner Function

Expressed in terms of the dimensionless energy the Wigner function of the mth energy eigenstate of the harmonic oscillator takes the compact form

$$W_{|m\rangle}(x,p) = \frac{(-1)^m}{\pi\hbar} e^{-2\eta(x,p)} L_m[4\eta(x,p)].$$

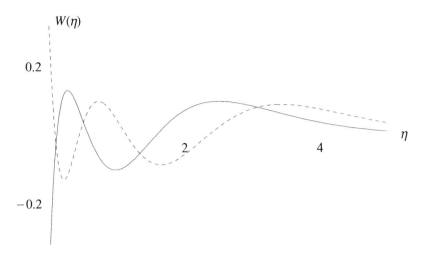

Fig. 4.4: Wigner function $W_{|m\rangle}(\eta)$ of an energy eigenstate $|m\rangle$ as a function of the dimensionless energy η for $m = 3$ (solid line) and $m = 4$ (dashed line).

Since only the dimensionless energy η enters this expression the Wigner function is constant along the contours in phase space of constant energy, that is along ellipses in phase space. The dependence of W_m on the energy, however, is rather interesting. Since the mth Laguerre polynomial $L_m(\zeta)$ is a polynomial of mth degree it enjoys m zeros as a function of ζ. Hence, the function oscillates between positive and negative values as shown in Fig. 4.4: The Wigner function of the mth energy eigenstate consists of wave fronts and wave troughs.

Here it is important to note that $L_m(0) = 1$ and hence

$$W_{|m\rangle}(\eta = 0) = \frac{(-1)^m}{\pi \hbar}.$$

Therefore, the Wigner function of the mth energy eigenstate at the origin of phase space changes its sign depending on the quantum number being even or odd as indicated in Fig. 4.4.

The change between positive and negative values of the Wigner function at the origin from one m to another has an important consequence: The last wave front is always a wave crest, that is here the Wigner function is always positive. This last crest is located in the neighborhood of the quantized phase space trajectory.

In Fig. 4.5 we show the Wigner function together with the corresponding distributions for position and momentum obtained by integrating the Wigner function along the momentum and position axes, respectively. The Wigner function is oscillating in the region enclosed by the classical phase space trajectory and therefore the value of the integral strongly depends along which path we integrate. This causes the oscillations in the probability distributions for position and momentum.

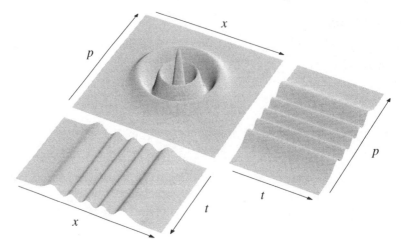

Fig. 4.5: Wigner function of the number state $|m=4\rangle$ and the corresponding position and momentum distributions. Since the number states are the eigenstates of the harmonic oscillator the Wigner function as well as the position and momentum distributions are stationary. These distributions are obtained by integrating the Wigner function along the momentum and position axes, respectively. Due to the oscillations in the Wigner function the marginal distributions are also oscillatory.

4.2 Coherent State

Coherent states are of central importance to quantum mechanics and, in particular, to quantum optics. They are the states of a harmonic oscillator system which mimic in the best possible way the classical motion of a particle in a quadratic potential. They were discovered for the mechanical oscillator by E. Schrödinger in order to avoid the unwanted features of the spreading of wave packets. In the context of field oscillators, that is quantized electromagnetic fields, they were investigated in great detail by R. Glauber, J. Klauder and E.C.G. Sudarshan. Due to their central role in quantum optics we devote a lot of space to the discussion of their properties. In the present section we only give a brief introduction using a mechanical oscillator analogue. We derive the complete formalism of coherent states in Sec. 11.2.

In this section we first define the coherent state as the state resulting from a sudden displacement of the quadratic potential. We then discuss the energy distribution of coherent states. This distribution is governed by the overlap integral of a coherent state and of an energy eigenstate. We calculate this overlap integral in two ways: first by using the exact wave functions for these states and second by using a rather crude approximation which, however, brings out most clearly the underlying physics. We then proceed with a discussion of the time evolution of a coherent state in which we establish the connection with the motion of a classical particle in a harmonic potential.

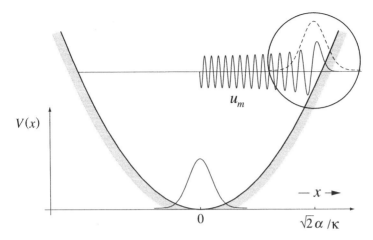

Fig. 4.6: The sudden displacement of a harmonic oscillator potential $U(x) = M\Omega^2 x^2/2$ by an amount $x_0 = \sqrt{2}\,\alpha/\kappa$, and its simultaneous lowering corresponding to an energy $M\Omega^2 x_0^2/2 = \alpha^2 \hbar \Omega$, creates out of its Gaussian ground state $u_0 = u_0(x)$ (shown by the solid curve at the potential minimum) a *coherent* state $\psi_{\text{coh}} = \psi_{\text{coh}}(x)$ located at $x = \sqrt{2}\,\alpha/\kappa$. This state, whose wave function is depicted here by the dashed curve, is not an eigenstate of the oscillator potential. The energy distribution of the coherent state, that is, the probability W_m of finding the mth energy eigenstate $|m\rangle$ in $|\psi_{\text{coh}}\rangle$ is governed by the overlap $w_m(|\psi_{\text{coh}}\rangle)$ between u_m and ψ_{coh}, as depicted inside the circle. For simplicity we have chosen the specific quantum number $m = 55$ and show $u_{m=55}$ for only positive x-values. The displacement parameter has the value $\alpha = 7$.

4.2.1 Definition of a Coherent State

We consider a mechanical oscillator such as a ball rolling without friction in a quadratic potential. We start from the ground state of the oscillator with wave function

$$u_0(x) \equiv \left(\frac{\kappa^2}{\pi}\right)^{1/4} \exp(-\frac{1}{2}\kappa^2 x^2).$$

We obtain a coherent state by suddenly displacing this ground state. In the mechanical model we achieve this displacement by suddenly displacing the origin of the harmonic oscillator by an amount $x_0 = \sqrt{2}\,\alpha/\kappa$, and by simultaneously lowering the potential energy by $M\Omega x_0^2/2 = \alpha^2 \hbar \Omega$ as illustrated in Fig. 4.6.

In the present context *sudden* implies that the displacement must occur on a time scale shorter than any time scale involved in the problem. The particle should not have time to react to the displacement. Hence, the wave function does not change and has now to be expressed in the variables of the new displaced potential which yields

$$\psi_{\text{coh}}(x) \equiv \left(\frac{\kappa^2}{\pi}\right)^{1/4} \exp\left[-\frac{1}{2}(\kappa x - \sqrt{2}\,\alpha)^2\right]. \qquad (4.11)$$

The opposite case is the slow and therefore adiabatic displacement of the potential. Here the particle stays always in the ground state.

4.2.2 Energy Distribution

The wave function ψ_{coh} is not an eigenfunction of the harmonic oscillator potential, that is, the coherent state is not a stationary state and thus undergoes a time development as discussed in the next section. Consequently, the coherent state shows a spread in energy. But how large is this spread? This is the question addressed in the present section.

The energy distribution of a coherent state is not the classically expected delta function located at the classical value

$$\overline{m} = (1/2)(\text{spring constant})(\text{displacement})^2 = (1/2) \cdot M\Omega^2 \cdot (\sqrt{2}\,\alpha/\kappa)^2 = \alpha^2 \hbar\Omega,$$

but rather follows from standard quantum mechanics as

$$W_m \equiv |w_m(|\psi_{\mathrm{coh}}\rangle)|^2 = |\langle m|\psi_{\mathrm{coh}}\rangle|^2, \tag{4.12a}$$

where

$$w_m(|\psi_{\mathrm{coh}}\rangle) \equiv \int_{-\infty}^{\infty} dx\, u_m(x)\, \psi_{\mathrm{coh}}(x) \tag{4.12b}$$

is the overlap integral between the wave functions of the coherent state, Eq. (4.11), and the mth energy eigenstate, given by Eq. (4.2).

Exact Treatment

We now evaluate the overlap $w_m(|\psi_{\mathrm{coh}}\rangle)$ between the mth energy wave function u_m and the coherent state wave function ψ_{coh} given by the integral

$$w_m(|\psi_{\mathrm{coh}}\rangle) = [2^m\, m!]^{-1/2} \frac{1}{\sqrt{\pi}} \int_{-\infty}^{\infty} d\xi\, H_m(\xi) \exp\left(-\frac{1}{2}\xi^2\right) \exp\left[-\frac{1}{2}(\xi - \sqrt{2}\,\alpha)^2\right]$$

in an exact way. Here we have introduced the dimensionless integration variable $\xi \equiv \kappa x$.

The relation

$$\frac{1}{2}\xi^2 + \frac{1}{2}(\xi - \sqrt{2}\,\alpha)^2 = \left(\xi - \frac{1}{\sqrt{2}}\alpha\right)^2 + \frac{1}{2}\alpha^2$$

allows us to combine the two quadratic contributions in the exponential functions and with the help of the integral formula

$$\int_{-\infty}^{\infty} dy\, H_m(y) \exp\left[-(y-y_0)^2\right] = \pi^{1/2} (2y_0)^m,$$

we can perform the integration. Indeed, the probability amplitude $w_m(|\psi_{\mathrm{coh}}\rangle)$ of finding the mth energy eigenstate in the coherent state $|\psi_{\mathrm{coh}}\rangle$ reads

$$w_m(|\psi_{\mathrm{coh}}\rangle) = \frac{\alpha^m}{\sqrt{m!}} e^{-\alpha^2/2}. \tag{4.13}$$

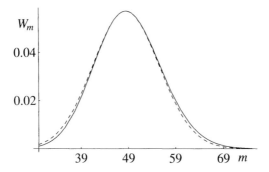

Fig. 4.7: The probability W_m of finding the mth energy eigenstate $|m\rangle$ in a coherent state is given by a Poisson distribution (solid line, Eq. (4.14)). This exact distribution and its asymptotic limit for large displacements (broken line, Eq. (4.15)) are almost indistinguishable in the neighborhood of the maximum, $m \cong \alpha^2 - 1/2$. We have chosen a displacement $\alpha = 7$. Every classical oscillator initially at rest and then subject to such a displacement will undergo the same, sudden increase in energy, an increase which in our units is $\alpha^2 = 49$. Curves, it should be recognized, are not curves, because m is never other than an integer.

The energy spread of a coherent state is therefore a Poissonian distribution

$$W_m(|\psi_{\text{coh}}\rangle) = \frac{\alpha^{2m}}{m!} e^{-\alpha^2}. \tag{4.14}$$

This Poissonian distribution and, in particular, the discreteness of the quantum numbers m has been measured for the case of an electromagnetic field in a cavity and for the vibratory motion of a single ion stored in a Paul trap. For these results we refer to Fig. 16.8 and Fig. 16.9

We conclude this subsection by discussing this result in the limit of large displacement, that is for $\alpha \gg 1$. According to Appendix G the Poissonian distribution, Eq. (4.14), allows the asymptotic expansion

$$W_m(|\psi_{\text{coh}}\rangle) \cong \frac{1}{\sqrt{2\pi}\,\alpha} \exp\left[-\left(\frac{m+\frac{1}{2}-\alpha^2}{\sqrt{2}\,\alpha}\right)^2\right] \tag{4.15}$$

depicted in Fig. 4.7 by a broken line. Hence, the Poissonian distribution reduces to a Gaussian distribution centered at $\alpha^2 - 1/2$. Likewise the probability amplitude

$$w_m(|\psi_{\text{coh}}\rangle) \cong \left(\sqrt{2\pi}\alpha\right)^{-1/2} \exp\left[-\frac{1}{2}\left(\frac{m+\frac{1}{2}-\alpha^2}{\sqrt{2}\,\alpha}\right)^2\right]$$

of finding the mth energy eigenstate is a Gaussian.

Asymptotic Treatment

In the preceding section we have derived the Poissonian energy distribution of a coherent state by evaluating the overlap integral in an exact way. Moreover, we have

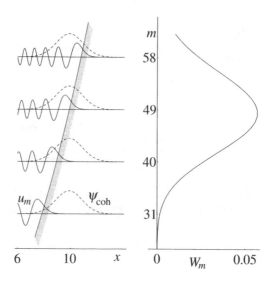

Fig. 4.8: The probability $W_m = W_m(|\psi_{\text{coh}}\rangle)$ of finding the mth energy eigenstate $|m\rangle$ in $|\psi_{\text{coh}}\rangle$, shown on the right hand side of this figure, is governed by the overlap $w_m(|\psi_{\text{coh}}\rangle)$ in position space between the two wave functions u_m and ψ_{coh}, displayed for particular choices of m on the left hand side of the picture. For quantum numbers $m \cong \kappa^2 x_0^2/2 = \alpha^2$, the right wall of the harmonic oscillator $U(x) = M\Omega^2 x^2/2$ of Fig. 4.6 is essentially a straight line. Hence, the energy wave function in the neighborhood of the classical turning point x_m is an Airy function which has its dominant maximum shortly before the potential wall. The width of the coherent state, shown here by the dashed curves, is large compared to the wavelength of the energy eigenwave u_m. Therefore, the oscillatory part of u_m averages out when integrated together with ψ_{coh} over x. Only the last maximum of positive value contributes to the integral $w_m(|\psi_{\text{coh}}\rangle)$. This maximum is narrow compared to ψ_{coh} and hence acts as a delta function located at $x = x_m$. The overlap integral $w_m(|\psi_{\text{coh}}\rangle)$ reduces to the "reading out" of the value of $\psi_{\text{coh}}(x) = (\kappa^2/\pi)^{1/4} \exp[-(\kappa x - \sqrt{2}\alpha)^2/2]$ at the turning point x_m of the mth energy eigenstate, that is, $w_m(|\psi_{\text{coh}}\rangle) \cong (\kappa^2/\pi)^{1/4} \exp[-(\kappa x_m - \sqrt{2}\alpha)^2/2]$. This procedure provides in a direct way the Gaussian approximation of the exact Poissonian distribution of a coherent state, shown on the right hand side of the figure.

found the Gaussian asymptotic limit of this distribution. Deeper insight into this result Eq. (4.15) springs from a treatment of the overlap integral Eq. (4.12b) using the large-m limit of the wave functions u_m of the harmonic oscillator.

From Fig. 4.8 we note that the wave function of the coherent state is broad compared to the wavelength of the energy wave function u_m, Eq. (4.2). Hence the oscillatory part of u_m averages out when we evaluate the integral Eq. (4.12b). The main contribution arises from the last bump of u_m in the neighborhood of the turning point $x_m \equiv \sqrt{2(m+1/2)}/\kappa$, a bump which acts essentially as a delta function. Hence, for the present discussion we can approximate u_m by

$$u_m(x) \cong N\delta(x - x_m)$$

where N is a constant which we determine from the normalization of the energy distribution.

In this approximation the overlap integral Eq. (4.12b) reduces to

$$w_m\left(|\psi_{\text{coh}}\rangle\right) \cong \psi_{\text{coh}}(x_m) = N\pi^{-1/4}\exp\left[-\frac{1}{2}(\kappa x_m - \sqrt{2}\,\alpha)^2\right].$$

Hence, the probability amplitude $w_m\left(|\psi_{\text{coh}}\rangle\right)$ follows in its dependence on m the wave function of the coherent state ψ_{coh} in the x-variable. Therefore, the probability W_m to find the mth energy eigenstate shows a single maximum at $m \cong \alpha^2$ as indicated on the right hand side of Fig. 4.8.

We conclude this discussion by rewriting the probability amplitude W_m in a more compact form. This allows us to make contact with the expression derived in the preceding section.

When we note that

$$\kappa x_m - \sqrt{2}\,\alpha = \sqrt{2(m+\tfrac{1}{2})} - \sqrt{2}\,\alpha = \frac{m+\tfrac{1}{2}-\alpha^2}{\tfrac{1}{2}\left(\sqrt{2(m+\tfrac{1}{2})}+\sqrt{2}\,\alpha\right)} \cong \frac{m+\tfrac{1}{2}-\alpha^2}{\sqrt{2}\,\alpha}$$

we find

$$w_m \cong N\pi^{-1/4}\exp\left[-\frac{1}{2}\left(\frac{m+\tfrac{1}{2}-\alpha^2}{\sqrt{2}\,\alpha}\right)^2\right].$$

After normalizing this result we find the expression Eq. (4.15).

4.2.3 Time Evolution

We have created a coherent state from the ground state of the oscillator by suddenly displacing the potential. The so created Gaussian wave packet Eq. (4.11) – having the potential energy $M\Omega x_0^2/2 = \alpha^2\hbar\Omega$ – bounces back and forth between the classical turning points of the vibratory motion corresponding to this energy. It has just the "right" width to keep its shape while oscillating as we show now. We pursue three different approaches: (i) we calculate in an exact way the time evolution of a state vector (ii) we follow the time evolution in the position distribution or (iii) we use the time evolution of the Wigner function.

State Vector Treatment

In the present section we discuss the properties of a coherent state which moves in a harmonic potential. For this purpose we solve the Schrödinger equation

$$i\hbar\frac{d}{dt}|\psi(t)\rangle = \hat{H}|\psi(t)\rangle$$

for the case where the system at time $t=0$ is in the coherent state $|\psi_{\text{coh}}\rangle$.

The formal solution of this problem reads

$$|\psi(t)\rangle = \exp\left(-\frac{i}{\hbar}\hat{H}t\right)|\psi_{\text{coh}}\rangle.$$

We can evaluate the action of the Hamiltonian on the initial coherent state most conveniently when we use the expansion

$$|\psi_{\text{coh}}\rangle = \sum_{m=0}^{\infty} |m\rangle\langle m|\psi_{\text{coh}}\rangle = \sum_{m=0}^{\infty} w_m\left(|\psi_{\text{coh}}\rangle\right) |m\rangle \qquad (4.16\text{a})$$

of the initial coherent state into energy eigenstates $|m\rangle$. According to Eq. (4.13) the expansion coefficients read

$$w_m = \frac{\alpha^m}{\sqrt{m!}} e^{-\alpha^2/2}. \qquad (4.16\text{b})$$

When we recall the energy eigenvalue equation

$$\hat{H}|m\rangle = \hbar\Omega(m + \frac{1}{2})|m\rangle$$

the time evolution of the coherent state reads

$$|\psi(t)\rangle = \sum_{m=0}^{\infty} w_m \, e^{-i\Omega t(m+1/2)} |m\rangle.$$

We substitute w_m using Eq. (4.16b) into this expression and arrive at the sum

$$|\psi(t)\rangle = e^{-i\Omega t/2} \sum_{m=0}^{\infty} \frac{(\alpha \, e^{-i\Omega t})^m}{\sqrt{m!}} e^{-\alpha^2/2} |m\rangle. \qquad (4.17)$$

Here we have combined the time dependent phase factors with the probability amplitudes w_m of the initial coherent state of amplitude α. We note that the time dependence has changed the amplitude α into $\alpha e^{-i\Omega t}$. It essentially appears in the probability amplitude as if we now have a coherent state of complex displacement. This suggests that at time t the state

$$|\psi(t)\rangle = e^{-i\Omega t/2} \left|\psi_{\text{coh}} \, e^{-i\Omega t}\right\rangle$$

of the oscillator is again a coherent state with the same modulus of the amplitude but a different phase. In Chapter 11 we present a more general definition of a coherent state where the displacements can have real and imaginary parts. Also the Wigner function treatment of the next section shows that the initial Wigner function just moves on a circle in phase space without changing its shape. This indicates again that the coherent state remains a coherent state.

We also note that after one period $T \equiv 2\pi/\Omega$ of the oscillation the state vector has picked up a phase $\Omega \cdot T \cdot 1/2 = \pi$. This phase is a consequence of the zero point energy of the harmonic oscillator. We can relate it to the so-called Berry phase, which we will discuss in detail in Chapter 6. In the present section we do not want to get into the subtleties of how to measure this phase. Here we only note that since this is an overall phase it cannot be detected by observing a single oscillator only.

We conclude this section by emphasizing that under time development in a harmonic oscillator a coherent state remains a coherent state.

Position Distribution

In the preceding section we have discussed the time evolution of a coherent state using the state vector. In the present section we use this treatment to discuss the time dependence of the wave function

$$\psi_{\text{coh}}(x,t) \equiv \langle x|\psi_{\text{coh}}(t)\rangle \tag{4.18}$$

in position space. Here we follow closely the original paper by E. Schrödinger from 1926.

When we substitute the expansion Eq. (4.17) of the state vector into energy eigenstates into the above expression for the wave function $\psi_{\text{coh}}(x,t)$ of the coherent state and use the position representation $u_m(x)$, Eq. (4.2), of the mth energy eigenstate we find

$$\psi_{\text{coh}}(x,t) = \left(\frac{\kappa^2}{\pi}\right)^{1/4} \exp\left[-\frac{1}{2}\left(i\Omega t + \alpha^2 + \kappa^2 x^2\right)\right] \sum_{m=0}^{\infty} \frac{H_m(\kappa x)}{m!} \left(\frac{\alpha e^{-i\Omega t}}{\sqrt{2}}\right)^m$$

We evaluate the sum on the right hand side of this equation with the help of the generating function

$$\sum_{m=0}^{\infty} \frac{H_m(x)}{m!} z^m = e^{2xz - z^2}$$

for the Hermite polynomials and arrive at

$$\psi_{\text{coh}}(x,t) = \left(\frac{\kappa^2}{\pi}\right)^{1/4} \exp\left\{-\frac{1}{2}\left[i\Omega t + \alpha^2 + \kappa^2 x^2 + \alpha^2 e^{-2i\Omega t} - 2\sqrt{2}\,\kappa x\,\alpha e^{-i\Omega t}\right]\right\}. \tag{4.19}$$

In order to gain more insight into this rather complicated expression we decompose the exponent into real and imaginary part. With the help of the relations

$$\alpha^2\left(1 + e^{-2i\Omega t}\right) = 2\alpha^2 \cos^2(\Omega t) - i\alpha^2 \sin(2\Omega t) =$$
$$2\left[\text{Re}\left(\alpha e^{-i\Omega t}\right)\right]^2 - i\,2\,\text{Re}\left(\alpha e^{-i\Omega t}\right)\text{Im}\left(\alpha e^{-i\Omega t}\right)$$

we finally get

$$\psi_{\text{coh}}(x,t) = \left(\frac{\kappa^2}{\pi}\right)^{1/4} \exp\left\{-\frac{1}{2}\left[\kappa x - \sqrt{2}\,\text{Re}\left(\alpha e^{-i\Omega t}\right)\right]^2\right\} \exp\left\{i\hbar\kappa\sqrt{2}\,\text{Im}\left(\alpha e^{-i\Omega t}\right) x/\hbar\right\}$$
$$\times \exp\left\{-i\left[\Omega t/2 + \text{Re}\left(\alpha e^{-i\Omega t}\right)\text{Im}\left(\alpha e^{-i\Omega t}\right)\right]\right\}. \tag{4.20}$$

Here we have combined the imaginary part of the phase factor in a way to bring out most clearly the physics. Moreover, we have always kept the real-valued displacement parameter α together with the time dependent exponential $e^{-i\Omega t}$. In principle we could have taken α out of these real and imaginary parts in Eq. (4.20). However, as mentioned in the previous section the displacement parameter can also take on complex values. This case is therefore already included in our treatment.

The time dependent wave function of the coherent state consists of the product of three terms. The first contribution is a Gaussian in position which has a maximum at the time dependent position

$$\bar{x}(t) \equiv \frac{\sqrt{2}}{\kappa} \operatorname{Re}\left(\alpha e^{-i\Omega t}\right).$$

The second term is a plane wave in position with a time dependent momentum

$$\bar{p}(t) \equiv \hbar\kappa\sqrt{2}\operatorname{Im}\left(\alpha e^{-i\Omega t}\right)$$

The third term is an overall phase which is independent of position but depends on time. It consists of the sum of the phase

$$\phi_{\mathrm{zp}}(t) = \Omega t/2$$

due to the zero point energy of the oscillator and the area

$$A_\Delta \equiv \frac{1}{2}\bar{x}(t)\cdot\bar{p}(t)/\hbar = \operatorname{Re}\left(\alpha e^{-i\Omega t}\right)\cdot\operatorname{Im}\left(\alpha e^{-i\Omega t}\right)$$

of the rectangular triangle defined by $\bar{x}$ and $\bar{p}$. In this notation the wave function of the coherent state reads

$$\psi_{\mathrm{coh}}(x,t) = \left(\frac{\kappa^2}{\pi}\right)^{1/4} \exp\left(-\frac{\kappa^2}{2}[x-\bar{x}(t)]^2\right)$$

$$\times \exp\left[i\frac{\bar{p}(t)x}{\hbar}\right]\exp\{-i[\phi_{\mathrm{zp}}(t)+A_\Delta(t)]\}. \tag{4.21}$$

In Fig. 4.9 we discuss the dynamical behavior of this wave function in detail and focus on the influence of the individual terms. In particular, we show the wave function in its real and imaginary part as a function of position. In this context the wave function is a line that winds through three-dimensional space defined by position and real and imaginary parts of the wave function.

We conclude this section by discussing the probability distribution in position space. From the above expression of the wave function we find immediately

$$|\psi_{\mathrm{coh}}(x,t)|^2 = \frac{\kappa}{\sqrt{\pi}}\exp\left\{-\kappa^2[x-\bar{x}(t)]^2\right\}. \tag{4.22}$$

This expression represents a Gaussian wave packet which does not change its shape. In particular, its width $\Delta x = \kappa^{-1} = \sqrt{\hbar/(M\Omega)}$ defined by the point where the Gaussian has decayed to $1/e$ is *not* time dependent. The center of the wave packet lies at $\bar{x}(t)$ and follows the classical equations of motion.

In momentum representation we obtain a similar result as we show explicitly in the next section using the Wigner function. In particular we show that the corresponding width $\Delta p = \sqrt{\hbar M\Omega}$ of the momentum distribution is time independent.

Moreover, the product of the widths in position and momentum is $\hbar$. This means that the state is a state of minimum uncertainty.

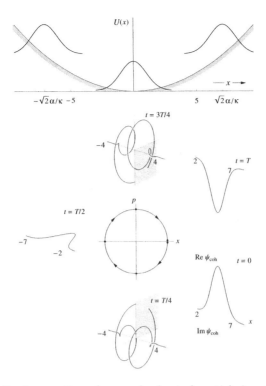

Fig. 4.9: During vibratory motion of a pseudo-classical particle in a harmonic oscillator potential, $U(x) = M\Omega x^2/2$, represented by a coherent state – a ground state displaced by an amount $x_0 = \sqrt{2}\,\alpha/\kappa$ – the corresponding position probability $W_x(t) = |\psi_{\text{coh}}(x,t)|^2$ keeps its Gaussian envelope (top). The wave function $\psi_{\text{coh}}(x,t)$, however, creates a complicated trajectory in complex space while the particle traverses its circular phase space orbit. We exemplify this behavior for special times by the insets below. At time $t = 0$ the real-valued Gaussian wave function is centered at $x = \sqrt{2}\,\alpha/\kappa = 3\sqrt{2}/\kappa$. When the wave packet runs down the potential hill, transforming its potential energy into kinetic energy, the Gaussian envelope of $\psi_{\text{coh}}(x,t)$ tilts over towards the negative imaginary axis. Its ends curl up, creating a Gaussian *cork screw* as indicated for the time $t = T/4$. Climbing up the west hill side the twisted wave curve unscrews but continues to tilt as a result of the zero point energy induced phase term $-i\Omega t/2$. At half an oscillation period, that is at $t = T/2$, when the particle rests at the left turning point $x = -\sqrt{2}\,\alpha/\kappa$ the wave function is a Gaussian again, however, now purely imaginary. As the particle reverses its motion and accelerates down the hill $\psi_{\text{coh}}(x,t)$ rolls itself up into the by-now familiar cork screw, depicted for $t = 3T/4$. Its central maximum pushes at an angle $-3\pi/4$ through the complex plane. Note, however, that as a result of the positive momentum the helicity of the cork screw has changed its sign. After one complete revolution, that is at time $t = T$, the twisted thread has recovered and turned into a purely real Gaussian again, however, now of purely negative values. Hence after one period $t = T$ of the motion the wave function of a coherent state, that is, its position probability amplitude $\psi_{\text{coh}}(x,t)$ has not returned to its original form but has picked up a *dynamical* phase $-\pi$ as a result of the zero point energy of the harmonic oscillator.

Wigner Function Treatment

From the time dependent wave function, Eq. (4.21), one can calculate the time dependent Wigner function, Eq. (4.10), of the coherent state using the standard integral definition of a Wigner function.

However, in the present section we pursue a different approach: We first find the Wigner function of the coherent state at time $t = 0$ and then obtain the Wigner function at later times by solving the Liouville equation for the Wigner function discussed in Chapter 3. When we integrate this result over position or momentum we find the time dependent momentum or position distribution.

We start by calculating the Wigner function at time $t = 0$. When we substitute the wave function Eq. (4.11) into the definition of the Wigner function we arrive – in the case of real α – at

$$W(x,p) = \frac{1}{\pi\hbar} \exp\left[-(\kappa x - \sqrt{2}\alpha)^2 - [p/(\hbar\kappa)]^2\right]. \tag{4.23}$$

From the equation of motion for the Wigner function, Eq. (3.12), we know that in the case of the harmonic oscillator the Wigner function evolves according to the classical Liouville equation. Therefore, the time evolution of the Wigner function in the quadratic potential can be written as

$$W(x,p;t) = W\Big(x_0(x,p;t), p_0(x,p;t); 0\Big). \tag{4.24}$$

where the expressions

$$x_0(x,p;t) = \cos(\Omega t)\, x - \sin(\Omega t)\, \frac{p}{\hbar\kappa}\frac{1}{\kappa} \tag{4.25a}$$

$$p_0(x,p;t) = \sin(\Omega t)\, \kappa x \hbar\kappa + \cos(\Omega t)\, p \tag{4.25b}$$

connect the phase space point (x,p) at time t with the initial point (x_0, p_0) at time $t = 0$ according to the classical equations of motion. This relation brings out the fact that momentum and position in phase space are on the same footing and are interchangeable.

When we take the Wigner function (4.23) as initial condition we therefore find

$$W(x,p;t) = \frac{1}{\pi\hbar} \exp\left[-\Big(\cos(\Omega t)\,\kappa x - \sin(\Omega t)\,\frac{p}{\hbar\kappa} - \sqrt{2}\alpha\Big)^2 \right.$$
$$\left. - \Big(\sin(\Omega t)\,\kappa x + \cos(\Omega t)\,\frac{p}{\hbar\kappa}\Big)^2\right]. \tag{4.26}$$

Hence, the Wigner function undergoes a rotation in phase space along the classical trajectory. Moreover, its width does not change as a function of time. In the next section we consider a squeezed state where, indeed, the width changes.

We conclude this section by calculating the momentum distribution by integrating the time dependent Wigner function Eq. (4.26) over the variable x. After performing a Gauss integral we arrive at

$$|\psi(p,t)|^2 = \frac{1}{\hbar\kappa\sqrt{\pi}} \exp\left[-\Big(\frac{p}{\hbar\kappa} + \sqrt{2}\alpha\,\sin(\Omega t)\Big)^2\right]. \tag{4.27}$$

4.3 Squeezed State   119

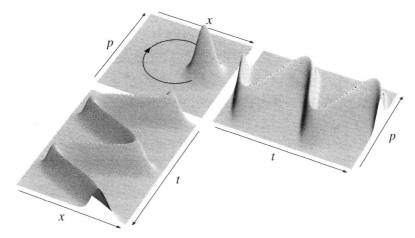

Fig. 4.10: Time evolution of a coherent state. The Wigner function is a symmetric Gaussian bell and moves in phase space along a circle. We only show the initial Wigner function and indicate its rotation by a circle. As it moves the marginal distributions in the form of Gaussians perform harmonic oscillation with constant and equal widths. Here we have chosen the displacement parameter $\alpha = 2$.

Indeed, the momentum distribution is also a Gaussian centered at $\bar{p}(t)$ with a time independent width.

In Fig. 4.10 we show the Wigner function (4.26) together with the corresponding distributions for position and momentum, Eqs. (4.22) and (4.27), as they evolve in time. We depict the Wigner function for time $t = 0$ only, whereas the marginal distributions we show for all times. One can see that the wave packets in position and momentum space oscillate back and forth in a harmonic way and that they keep their shape during the motion.

We emphasize that this motion of the wave packet has been observed in the case of the electromagnetic field. The top figure in the middle column of Fig. 4.11 shows the electromagnetic field distribution as a function of time. It is in complete agreement with the time dependence of the position distribution of a coherent state shown in Fig. 4.10. A similar curve has also been measured for the vibratory motion of a diatomic molecule.

4.3 Squeezed State

In the preceding section we have created a coherent state of a mechanical, harmonic oscillator from its ground state by suddenly lowering and displacing the quadratic potential. We have seen that the so created wave packet oscillates back and forth between the classical turning points and keeps its shape. The width of the wave packet is identical to ground state wave packet of the oscillator.

Already in the early days of quantum mechanics E.H. Kennard considered the time evolution of wave packets that at time $t = 0$ were either broader or narrower than the ground state wave packet. In contrast to coherent states such wave packets

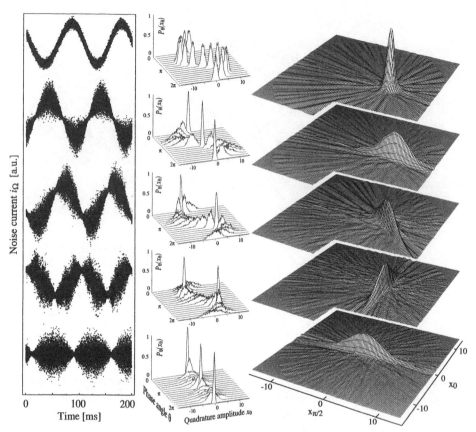

Fig. 4.11: Noise traces (left), quadrature distributions $P_\theta(x_\theta) \equiv W(X_\theta)$, and reconstructed Wigner functions (right) of generated quantum states. From the top: Coherent state, phase-squeezed state in the $\phi = 48°$-quadrature, amplitude-squeezed state, squeezed vacuum state. The noise traces as a function of time show the oscillation of the electric fields in a 4π interval for the upper four states, whereas for the squeezed vacuum (belonging to a different set of measurements) a 3π interval is shown. The quadrature distributions (center) can be interpreted as the time evolution of wave packets (position probability densities) during one oscillation period. For the reconstruction of the quantum states a π interval suffices. Taken from G. Breitenbach *et al.*, Nature **387**, 471 (1997).

oscillate in their width as they move back and forth in the oscillator potential. These states have taken on a prominent role in the recent years in quantum optics. In this field they are referred to as *squeezed states*. This name stems from the fact that they are narrower or broader than the ground state wave packet. These states have been studied extensively theoretically and have become important in the context of molecular physics and Paul traps. In particular, in the context of electromagnetic fields squeezed states of light have been investigated theoretically as well as experimentally. In 1985 Bell Labs created the first squeezed state of light.

In the present section we give a brief introduction into the physics of squeezed states. We first define the state in terms of a mechanical oscillator such as a pen-

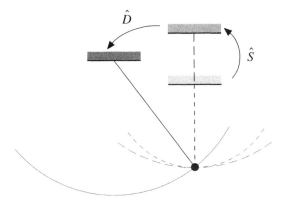

Fig. 4.12: Mechanical model for the generation of a squeezed state of a harmonic oscillator represented here by a pendulum where we restrict ourselves to small angles. We first apply the squeeze operator $\hat{S}$ and then the displacement operator $\hat{D}$. Squeezing is performed by lifting the point of suspension and simultaneously lengthening the thread. This effectively changes the frequency of the oscillator. The displacement is performed by suddenly moving the point of suspension along a circle with its center at the mass point.

dulum. We discuss the energy distribution and show that a highly squeezed state displays an oscillatory energy distribution. Moreover, we illustrate the time evolution of a squeezed state using the Wigner function.

4.3.1 Definition of a Squeezed State

We start our discussion by defining a squeezed state using the mechanical model, illustrated in Fig. 4.12. As in the case of the coherent state we start from the ground state of the oscillator that is from the state of the pendulum hanging down from the ceiling. We suddenly lift the point of suspension and simultaneously lengthen the thread. This effectively changes the frequency of the oscillator. In the present section we show that this corresponds to effectively squeezing the wave packet.

We then again suddenly displace the potential which corresponds to suddenly moving the point of suspension on a circle keeping the length of the thread fixed. Hence, we obtain a squeezed state when, in addition to those changes necessary to generate the coherent state, we also suddenly change the frequency of the oscillator from Ω to Ω/s, where $s > 0$.

We note, however, that the order in which we displace and squeeze is important. The state we obtain from the ground state after squeezing and displacing it, is different from the state we obtain by displacing and then squeezing. Both states are squeezed states. However, they differ by their displacement parameter as we show in this section.

Mathematical Formulation

We now formulate this mechanical definition of a squeezed state in mathematical terms. For this purpose we start again from the ground state wave function

$$u_0(x) = \left(\frac{\kappa^2}{\pi}\right)^{1/4} \exp\left(-\frac{1}{2}\kappa^2 x^2\right) \tag{4.28}$$

of a harmonic oscillator with mass M and frequency Ω, and $\kappa \equiv \sqrt{M\Omega/\hbar}$.

A sudden change of the oscillator frequency from Ω to Ω/s leaves the ground state unchanged but, of course, the value of κ changes to

$$\kappa' = \sqrt{\frac{M\Omega/s}{\hbar}} = \frac{1}{\sqrt{s}}\kappa.$$

We now express the original ground state wave function, Eq. (4.28), in terms of the new potential, that is we now express the wave function using this new value κ'. In this new coordinate system the wave function reads

$$\psi_0(x) = \left(\frac{s\kappa'^2}{\pi}\right)^{1/4} \exp\left(-\frac{s}{2}\kappa'^2 x^2\right).$$

Hence, only for $s = 1$ this state is also the ground state of the new potential. In this case we have not changed the length of the pendulum. However, for $s \neq 1$ this state is referred as the squeezed ground state or squeezed vacuum. The name results from the change of the width of the distribution as we discuss now. For the sake of simplicity we suppress in the remainder of this chapter the prime at the parameter κ and define the squeezed ground state by the wave function

$$\psi_0(x) = \left(\frac{s\kappa^2}{\pi}\right)^{1/4} \exp\left(-\frac{s}{2}\kappa^2 x^2\right). \tag{4.29}$$

We introduce a squeeze operator $\hat{S}(s)$ which is defined by its action

$$\hat{S}(s)\,\psi(x) = s^{1/4}\,\psi(\sqrt{s}\,x). \tag{4.30}$$

This notation also brings out a different point of view on the squeezed state: We could say that we keep the potential unchanged but we now actively squeeze the ground state. The qualitative result will be the same.

Distributions of Fluctuations

We now turn to the discussion of the widths of the probability distributions of a squeezed state. This treatment brings out most clearly the origin of the name squeezed state. We start with the position probability distribution

$$W(x) = |\psi_0(x)|^2 = \left(\frac{s\kappa^2}{\pi}\right)^{1/2} \exp\left[-s\kappa^2 x^2\right] \tag{4.31}$$

4.3 Squeezed State

of the squeezed state.

We note that for $s > 1$ the width

$$\Delta x = \frac{1}{\sqrt{s}}\frac{1}{\kappa}$$

is smaller than that of the ground state probability distribution

$$W(x) = |u_0(x)|^2 = \left(\frac{\kappa^2}{\pi}\right)^{1/2}\exp\left(-\kappa^2 x^2\right)$$

of the oscillator. Hence, the fluctuations in position of $\psi_0(x)$ are squeezed relative to those of the ground state.

This squeezing is at the expense of the fluctuations in the conjugate variable p. Indeed, the momentum distribution

$$W(p) = |\psi_0(p)|^2 = \frac{1}{\hbar\kappa}\frac{1}{\sqrt{s\pi}}\exp\left[-\frac{1}{s}\left(\frac{p}{\hbar\kappa}\right)^2\right] \tag{4.32}$$

of the squeezed ground state enjoys a width

$$\Delta p = \sqrt{s}\,\hbar\kappa.$$

Hence, for $s > 1$ the width Δp is correspondingly larger as to maintain the minimum uncertainty relation

$$\Delta x \cdot \Delta p = \frac{1}{\sqrt{s}}\frac{1}{\kappa}\cdot\sqrt{s}\hbar\kappa = \hbar.$$

For $0 < s < 1$ we find squeezing of the fluctuations in momentum at the expense of the fluctuations in the position. For $s = 1$ we recover the coherent state which corresponds to a symmetric distribution of the fluctuations on x and p.

In Fig. 4.13 we show the experimentally measured widths of the squeezed vacuum state of the electromagnetic field in the optical parametric oscillator experiment described in more detail in Sec. 1.3.2. The data follow clearly the hyperbola predicted by the minimum uncertainty relation.

Hence, for $s \neq 1$ the fluctuations in one of the two conjugate variables is reduced below the corresponding ground state fluctuations. This has provided the name squeezed state. However, R. Glauber who has given in an early paper one of the first suggestions to create a squeezed state has argued that this name is misleading. It is not the state but the fluctuations that are squeezed. Therefore, a more appropriate name would be *sub-fluctuant* or *super-fluctuant states*. Unfortunately these names, although describing the physics correctly, have not caught on.

Nevertheless, we can argue that indeed the state is squeezed. That comes out most clearly in the Wigner function of a squeezed state. When we substitute the wave function $\psi_0(x)$ of a squeezed ground state into the definition of the Wigner function and perform the Gaussian integrals we arrive at

$$W(x,p) = \frac{1}{\pi\hbar}\exp\left[-s(\kappa x)^2 - \frac{1}{s}\left(\frac{p}{\hbar\kappa}\right)^2\right].$$

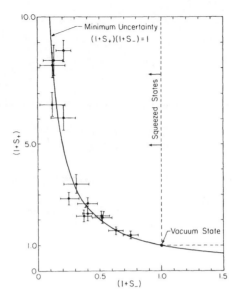

Fig. 4.13: Experimentally measured minimum uncertainty relation of a squeezed vacuum state of the electromagnetic field. All states minimizing the uncertainty relation lie on a hyperbola in the space defined by the fluctuations of the two conjugate variables. Dots represent experimental values whereas the solid line is the hyperbola predicted by theory. Taken from L.A. Wu et al., J. Opt. Soc. Am. B **4**, 1465 (1987).

Hence, for $s \neq 1$ the Gaussian is elongated in one direction and squeezed in the other. In that sense we can argue that the state in form of the Wigner function is squeezed.

The contour lines of the Wigner function are ellipses and the major and the minor axis are governed by the squeezing parameter s. In the case of strong squeezing the contour lines take the form of a cigar. Due to the Gaussian weight factor this has led to the name *Gaussian cigar* for the squeezed state Wigner function.

We emphasize that apart from the squeezing parameter there is an additional parameter entering the definition of a squeezed state. So far we have only considered the case when the minor and the major axis of the ellipse are aligned with the x- and p-axis. However, they can take an angle. This will become important when we consider the time evolution of a squeezed state in Sec. 4.3.5.

Generalized Squeezed States

So far, we have only discussed a squeezed ground state. When we now in addition displace the squeezed ground state we obtain the general squeezed state with the wave function

$$\psi_{\rm sq}(x) \equiv \hat{D}(\alpha)\, \hat{S}(s)\, u_0(x) = \left(\frac{s\kappa^2}{\pi}\right)^{1/4} \exp\left[-\frac{s}{2}\left(\kappa x - \sqrt{2}\,\alpha\right)^2\right], \qquad (4.33)$$

where α is a real number. The parameter $s > 0$ describes the width of the wave packet, as discussed above.

4.3 Squeezed State

In this definition of the generalized squeezed states we have first squeezed the ground state and then displaced it. We now investigate what happens if we interchange the squeezing operator $\hat{S}(s)$ and the displacement operator $\hat{D}(\alpha)$.

For this purpose we consider the state

$$\tilde{\psi}_{\text{sq}}(x) \equiv \hat{S}(s)\,\hat{D}(\alpha)\,u_0(x) = \hat{S}(s)\left(\frac{\kappa^2}{\pi}\right)\exp\left[-\frac{1}{2}(\kappa x - \sqrt{2}\,\alpha)^2\right].$$

When we make use of the relation (4.30) defining the action of the squeezing operator on a wave function we arrive at

$$\tilde{\psi}_{\text{sq}}(x) = \left(\frac{s\kappa^2}{\pi}\right)^{1/4}\exp\left[-\frac{1}{2}(\sqrt{s}\,\kappa x - \sqrt{2}\,\alpha)^2\right]$$

$$= \left(\frac{s\kappa^2}{\pi}\right)^{1/4}\exp\left[-\frac{s}{2}\left(\kappa x - \sqrt{2}\,\frac{\alpha}{\sqrt{s}}\right)^2\right] = \hat{D}(\alpha/\sqrt{s})\,\hat{S}(s)\,u_0(x).$$

Hence, the squeezed state we obtain when we first displace and then squeeze the ground state differs by a factor of $1/\sqrt{s}$ in the effective displacement from the squeezed state obtained from the reverse order of operations.

4.3.2 Energy Distribution: Exact Treatment

The squeezed state as defined in the preceding section is not an energy eigenstate of the new potential. Hence, the state shows a spread in energy. Its energy distribution

$$W_m \equiv |w_m(|\psi_{\text{sq}}\rangle)|^2 \equiv |\langle m|\psi_{\text{sq}}\rangle|^2 \tag{4.34a}$$

is given by the overlap

$$w_m(|\psi_{\text{sq}}\rangle) \equiv \int_{-\infty}^{\infty} dx\, u_m(x)\,\psi_{\text{sq}}(x) \tag{4.34b}$$

between the wave functions of the energy eigenstate and the squeezed state. In the present and the next sections we evaluate this integral in an exact and in an approximate way.

Aligned Squeezed State

When we substitute the expressions Eqs. (4.2) and (4.33) for the energy wave function and the squeezed state into the overlap integral Eq.(4.34b) we arrive at

$$w_m(|\psi_{\text{sq}}\rangle) = s^{1/4}\,[2^m\,m!]^{-1/2}\,\frac{1}{\sqrt{\pi}}\int_{-\infty}^{\infty} d\xi\, H_m(\xi)\exp\left(-\frac{1}{2}\xi^2\right)\exp\left[-\frac{s}{2}(\xi - \sqrt{2}\,\alpha)^2\right],$$

where we have introduced the dimensionless integration variable $\xi \equiv \kappa x$.

With the help of the relations

$$\frac{1}{2}\left[\xi^2 + s(\xi - \sqrt{2}\,\alpha)^2\right] = \left\{[(s+1)/2]^{1/2}\,\xi - s(s+1)^{-1/2}\alpha\right\}^2 + \frac{s}{s+1}\alpha^2$$

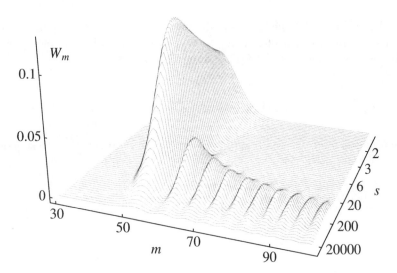

Fig. 4.14: Probabilities W_m of finding the mth energy eigenstate $|m\rangle$ in a squeezed state for different choices of the squeeze parameter s. All curves are plotted for the same value $\alpha = 7$ of the displacement parameter. The rearmost curve (no squeeze at all, $s = 1$) shows the ideal Poisson distribution associated with a coherent state. Curves that are further forward display oscillations in the probability distribution of excitation. As the squeezing becomes extreme ($s \to \infty$) there are more and more of these oscillations at ever higher values of m and the percentage of probability under any one peak goes to zero (foremost curve, a mere line, $W_m \cong 0$).

and

$$\int_{-\infty}^{\infty} dy\, \exp\!\left[-(y-y_0)^2\right] H_m(\lambda y) = \pi^{1/2}(1-\lambda^2)^{m/2}\, H_m\!\left[\frac{\lambda}{(1-\lambda^2)^{1/2}}\, y_0\right]$$

we can evaluate this integral. After minor algebra we arrive at the probability amplitude

$$w_m = \left(\frac{2\sqrt{s}}{s+1}\right)^{1/2}\left(\frac{s-1}{s+1}\right)^{m/2}(2^m m!)^{-1/2} H_m\!\left[\frac{s\sqrt{2}\,\alpha}{(s^2-1)^{1/2}}\right]\exp\!\left(-\frac{s\alpha^2}{s+1}\right)$$

to find the mth energy eigenstate in the squeezed state.

From this expression we obtain the energy distribution

$$W_m = \frac{2\sqrt{s}}{s+1}\left(\frac{s-1}{s+1}\right)^{m}(2^m m!)^{-1} H_m^2\!\left[\frac{s\sqrt{2}\,\alpha}{(s^2-1)^{1/2}}\right]\exp\!\left(-\frac{2s\alpha^2}{s+1}\right). \tag{4.35}$$

In Fig. 4.14 we display this probability W_m as a function of the squeeze parameter s for a fixed displacement $\alpha = 7$. We emphasize that in this figure and subsequent ones, curves are not really curves, because m is always an integer.

We note that for $s = 1$, that is for a coherent state the argument of the Hermite polynomial goes to infinity while at the same time the prefactor $(s-1)^m$ approaches zero. When we recall the asymptotic behavior

$$H_m(y) \sim (2y)^m$$

of the Hermite polynomials for $y \gg 1$ the energy distribution Eq. (4.35) simplifies to the familiar Poisson distribution

$$W_m = \frac{\alpha^{2m}}{m!} e^{-\alpha^2}.$$

This Poissonian character of W_m is shown in Fig. 4.14 by the rearmost curve.

When we increase the squeeze, that is increase s, the distribution gets narrower than the Poissonian distribution, that is, sub-Poissonian. However, for strong squeezing $s \to \infty$ the distribution starts to oscillate as shown in the foreground of Fig. 4.14. Moreover, the first maximum – the dominant contribution in the absence of oscillations – decreases rapidly as more and more maxima arise.

The rapid variations obscured by the Hermite polynomials in Eq. (4.35) become more visible when we apply in Chapter 8 the concept of interference in phase space developed in Chapter 7. However, the squeezed ground state, that is, $\alpha = 0$, already provides some insight. Since $H_{2m+1}(0) = 0$ and $H_{2m}(0) \neq 0$, it follows from Eq. (4.35) that

$$W_{2m+1} = 0 \quad \text{while} \quad W_{2m} \neq 0.$$

Thus we have period-two-oscillations. We emphasize that these oscillations have been measured experimentally using *quantum state tomography* as shown in Fig. 1.11.

Rotated Squeezed States

So far the major and the minor axis of a squeezed state cigar have been aligned with the coordinate system of phase space. We now briefly discuss the energy distribution of a rotated squeezed state.

In Problem 4.2 we show that in this case the energy distribution reads

$$W_m^{(\text{exact})} = \frac{2\sqrt{s}}{s+1} \left(\frac{s-1}{s+1}\right)^m (2^m m!)^{-1} \left| H_m \left(\frac{\sqrt{2}\alpha}{\sqrt{s^2-1}} (s\cos\varphi + i\sin\varphi) \right) \right|^2$$
$$\times \exp\left[-\frac{2\alpha^2}{s+1} (s\cos^2\varphi + \sin^2\varphi) \right] \quad (4.36)$$

where φ denotes the angle to the momentum axis.

In Fig. 4.15 we compare the energy distribution of a nonrotated and a rotated squeezed state of identical displacement and squeezing parameters. We note that the oscillations in the the nonrotated case have a single but large period. In contrast the corresponding curve for the rotated state displays two effects: (i) There is a rapid oscillation, and (ii) a slow modulation on top of it. These effects are hidden in these rather complicated expressions of the Hermite polynomials. They stand up most clearly in the asymptotic expansions discussed in the next subsection.

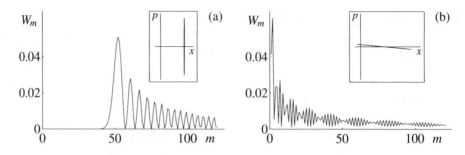

Fig. 4.15: Energy distribution of a highly squeezed state. In (a) the Gaussian cigar is parallel to the p-axis, whereas in (b) it is rotated by 85° with respect to the p-axis as shown by the insets. The parameters are $s = 201$ and $\alpha = 7$.

4.3.3 Energy Distribution: Asymptotic Treatment

We now want to gain insight into the dependence of the energy distribution $W_m(|\psi_{\text{sq}}\rangle)$ of a squeezed state Eq. (4.35) on the parameters characterizing this state. For this purpose we make use of the special form of the energy wave function in the large-m limit discussed in Sec. 4.1.2.

Naive Picture

In Fig. 4.16 we show the overlap $w_m(|\psi_{\text{sq}}\rangle)$ between the wave functions u_m and ψ_{sq} of the energy eigenstate and the squeezed state for specific quantum numbers m. Here we consider the situation of a state highly squeezed in the x-variable, that is, for $s \gg 1$.

When we compare Fig. 4.16 to the corresponding picture for a coherent state, Fig. 4.8, we find that now the wave function ψ_{sq} is *narrow* compared to the wavelength of the energy wave function u_m. Hence, ψ_{sq} acts essentially as a delta function located at $x = \sqrt{2}\,\alpha/\kappa$, that is

$$\psi_{\text{sq}}(x) \cong N\delta(x - \sqrt{2}\,\alpha/\kappa), \tag{4.37}$$

where N is a constant which we have to determine at the end of the calculation from the normalization of the energy distribution.

In this approximation the squeezed state wave function distills out the value of u_m at $x = \sqrt{2}\,\alpha/\kappa$, that is,

$$w_m(|\psi_{\text{sq}}\rangle) \cong N u_m(x = \sqrt{2}\,\alpha/\kappa). \tag{4.38}$$

From Fig. 4.16 and Eq. (4.38) we note that for increasing values of m the probability amplitude $w_m(|\psi_{\text{sq}}\rangle)$ of a highly squeezed state follows the energy wave function u_m at $x = \sqrt{2}\,\alpha/\kappa$ in its dependence on m. For m-values smaller than α^2, we are in the classically forbidden regime of the energy wave function. This results in a small probability of finding the mth energy eigenstate as shown on the right hand side of Fig. 4.16.

For $m \cong \alpha^2$ the turning point hump rests on top of the squeezed state function giving rise to a maximum in the overlap $w_m(|\psi_{\text{sq}}\rangle)$ and thus also in $W_m(|\psi_{\text{sq}}\rangle)$. For

4.3 Squeezed State   129

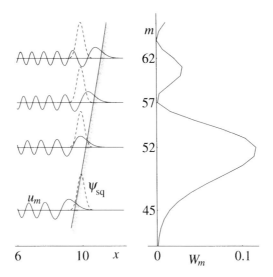

Fig. 4.16: The energy distribution of a highly squeezed state $W_m(|\psi_{\rm sq}\rangle)$, shown in its oscillatory behavior on the right hand side of the figure, results from the overlap $w_m(|\psi_{\rm sq}\rangle)$ between the mth energy wave function u_m and the squeezed state wave $\psi_{\rm sq}$, depicted on the left hand side in the neighborhood of $x = \sqrt{2}\,\alpha/\kappa$ by solid and by dashed curves respectively. In contrast to the coherent state discussion of Figs. 4.6 and 4.8 the Gaussian $\psi_{\rm sq}$ is narrow compared to the wavelength of u_m. This Gaussian acts essentially as a delta function located at $x = \sqrt{2}\,\alpha/\kappa$ and maps out the oscillations of u_m in the coordinate x, namely $u_m(x) \sim \cos(S_m(x) - \pi/4)$, onto oscillations in the energy distribution W_m, that is, $W_m(|\psi_{\rm sq}\rangle) \sim \cos\left[S_m(x = \sqrt{2}\,\alpha/\kappa) - \pi/4\right]$. The oscillatory energy distribution of a highly squeezed state is a result of consecutive wave fronts of u_m moving through the "narrow slit" provided by the squeezed state. For definiteness we have chosen for the values of the displacement and squeezing parameters $\alpha = 7$ and $s = 21$.

quantum numbers m appropriately larger than α^2 the squeezed state wave function probes *the oscillatory regime of* u_m leading to an *oscillatory behavior* of $W_m(|\psi_{\rm sq}\rangle)$. This domain in $W_m(|\psi_{\rm sq}\rangle)$ we now investigate in more detail.

Overlap Integral as a Taylor Expansion

We now return to the overlap integral

$$w_m(|\psi_{\rm sq}\rangle) = \left(\frac{s}{\pi}\right)^{1/4} \int_{-\infty}^{\infty} d\xi\, u_m(\xi)\, \exp\left[-\frac{s}{2}(\xi - \sqrt{2}\,\alpha)^2\right] \tag{4.39}$$

between the position wave functions $u_m(\xi)$ and $\psi_{\rm sq}(\xi)$ of the energy eigenstate and the squeezed state. In order to simplify the notation we have introduced the dimensionless coordinate $\xi \equiv \kappa x$ in which the energy wave function takes the form given by Eq. (4.8).

A more complete treatment of the overlap integral Eq. (4.39) does not invoke the delta function approximation of $\psi_{\rm sq}$, Eq. (4.37), but expands the energy wave

function
$$u_m(\xi) = \sum_{k=0}^{\infty} \frac{1}{k!} \left.\frac{d^k u_m(\xi)}{d\xi^k}\right|_{\xi=\sqrt{2}\alpha} (\xi - \sqrt{2}\alpha)^k$$

into a Taylor series around $\xi = \sqrt{2}\alpha$.

When we substitute this expansion together with the wave function Eq. (4.33) of the squeezed state into the overlap integral Eq. (4.34b) we arrive at

$$w_m(|\psi_{\text{sq}}\rangle) = \left(\frac{s}{\pi}\right)^{1/4} \sum_{k=0}^{\infty} \frac{1}{k!} \left.\frac{d^k u_m(\xi)}{d\xi^k}\right|_{\xi=\sqrt{2}\alpha} \int_{-\infty}^{\infty} dy\, y^k \exp\left(-\frac{s}{2}y^2\right).$$

Due to the anti-symmetry of the integrand the integral vanishes for odd powers of y, that is

$$\int_{-\infty}^{\infty} dy\, y^{2k+1} \exp\left(-\frac{s}{2}y^2\right) = 0.$$

Hence, only even terms contribute to the sum. With the help of the integral relation

$$\int_{-\infty}^{\infty} dy\, y^{2k} \exp\left(-\frac{s}{2}y^2\right) = \pi^{1/2} \left(\frac{2}{s}\right)^{k+1/2} 2^{-2k} \frac{(2k)!}{k!}$$

we therefore arrive at

$$w_m(|\psi_{\text{sq}}\rangle) = 2^{3/4} \pi^{1/2} \left(\frac{1}{2\pi s}\right)^{1/4} \sum_{k=0}^{\infty} \frac{(2s)^{-k}}{k!} \left.\frac{d^{2k} u_m(\xi)}{d\xi^{2k}}\right|_{\xi=\sqrt{2}\alpha}. \tag{4.40}$$

Even Derivatives of Energy Wave Function

So far the calculation is exact. However, in this form the expression Eq. (4.40) is not very useful. We have to find the even derivatives of the energy wave function and then perform the summation. One possibility to achieve this goal is to start by differentiating the Schrödinger equation

$$\frac{d^2 u_m(\xi)}{d\xi^2} = -[\tilde{p}_m(\xi)]^2 u_m(\xi)$$

with the dimensionless momentum

$$\tilde{p}_m(\xi) \equiv \sqrt{2(m+1/2) - \xi^2} \tag{4.41}$$

with respect to ξ. Unfortunately, due to the product rule this becomes a rather complicated procedure.

However, in the present discussion we are only interested in an approximate analytical result. We therefore now focus on the oscillatory domain of $u_m(\xi)$, that is, on the regime $|\xi| < \xi_m \equiv \sqrt{2(m+1/2)}$. Since $\xi = \sqrt{2}\alpha$, this also corresponds to $m > \alpha^2$, that is the oscillatory domain of $W_m(|\psi_{\text{sq}}\rangle)$. Here the momentum $\tilde{p}_m(\xi)$ is slowly varying and we can neglect the variation of $\tilde{p}_m(\xi)$ compared to the variation of $u_m(\xi)$. Hence, we arrive at the approximate relation

$$\frac{d^{2k} u_m}{d\xi^{2k}} \cong (-1)^k \, [\tilde{p}_m(\xi)]^{2k} \, u_m(\xi). \tag{4.42}$$

We obtain the same result when we start from the large-m limit Eq. (4.9)

$$u_m(\xi) \cong \left(\frac{2}{\pi}\right)^{1/2} [\tilde{p}_m(\xi)]^{-1/2} \cos\left[\int_\xi^{\xi_m} d\xi' \, \tilde{p}_m(\xi') - \pi/4\right]$$

of the energy wave function. In performing the differentiation of u_m, Eq. (4.7), we again neglect the slow variation of $[\tilde{p}_m(\xi)]^{-1/2}$ compared to the variation of the cosine and arrive at Eq. (4.42).

Approximate Analytical Energy Distribution

With the help of the approximate expression Eq. (4.42) for the even derivatives of the energy wave function and the definition Eq. (4.41) for the dimensionless momentum the formula Eq. (4.40) for the squeezed state probability amplitude reduces to

$$w_m(|\psi_{\text{sq}}\rangle) = 2 \left(\frac{1}{2\pi s}\right)^{1/4} \sum_{k=0}^{\infty} \frac{\left[-(m+\tfrac{1}{2} - \alpha^2)/s\right]^k}{k!} \left(m + \frac{1}{2} - \alpha^2\right)^{-1/4} \cos\phi_m.$$

The remaining sum is an exponential and we therefore arrive at the compact formula

$$w_m(|\psi_{\text{sq}}\rangle) = 2\, \mathcal{A}_m^{1/2} \cos\phi_m \tag{4.43a}$$

for the probability amplitude of finding the mth energy eigenstate in a highly squeezed state.

Here we have defined

$$\mathcal{A}_m \equiv \frac{1}{\sqrt{2\pi s}} \frac{\exp\left[-2(m+\tfrac{1}{2} - \alpha^2)/s\right]}{\sqrt{m + \tfrac{1}{2} - \alpha^2}} \tag{4.43b}$$

and

$$\phi_m \equiv \int_{\sqrt{2}\alpha}^{\xi_m} d\xi \, \tilde{p}_m(\xi) - \frac{\pi}{4}$$

$$= (m + \tfrac{1}{2}) \arctan\left[\frac{\sqrt{m + \tfrac{1}{2} - \alpha^2}}{\alpha}\right] - \alpha\sqrt{m + \tfrac{1}{2} - \alpha^2} - \frac{\pi}{4}. \tag{4.43c}$$

Thus the energy distribution in the limit of strong squeezing takes the form

$$W_m = |w_m(|\psi_{\text{sq}}\rangle)|^2 \equiv |\langle m|\psi_{\text{sq}}\rangle|^2 = 4\,\mathcal{A}_m \cos^2\phi_m, \tag{4.44}$$

where the amplitude $\mathcal{A}_m$ and the phase ϕ_m are defined in Eqs. (4.43).

We conclude this subsection by comparing in Fig. 4.17(a) this approximate result for W_m to the exact one, Eq. (4.35). Here we show the relative deviation $\Delta W_m \equiv |W_m - W_m^{(\text{exact})}|/W_m^{(\text{exact})}$ as a function of m. We emphasize the excellent agreement. The deviation is larger for m-values where $W_m^{(\text{exact})}$ is almost vanishing, since then we divide by a small number.

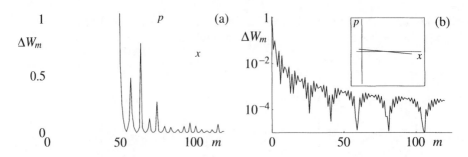

Fig. 4.17: Relative errors $\Delta W_m \equiv |W_m - W_m^{(\text{exact})}|/W_m^{(\text{exact})}$ of asymptotic energy distributions for (a) the nonrotated squeezed state, Eq. (4.44), and (b) the rotated squeezed state, Eq. (4.48), shown in the insets. In the nonrotated case the energy distribution displayed in Fig. 4.15(a) has almost vanishing values for $m \lesssim \alpha^2$. To focus on the oscillatory behavior of W_m we have plotted ΔW_m in (a) only for the region $m \gtrsim \alpha^2$. Since in the rotated case W_m exhibits oscillations starting from the vacuum state $m = 0$, we show ΔW_m for the total m-region in (b). To bring out the deviations most clearly we have used a logarithmic plot. We have chosen $s = 201$, $\alpha = 7$. The angle φ of the rotation in (b) is $85°$.

4.3.4 Limit Towards Squeezed Vacuum

We now consider the energy distribution Eq. (4.44) in the limit of $m \gg \alpha^2$. With the help of the asymptotic expression

$$\arctan \beta = \frac{\pi}{2} - \frac{1}{\beta} + \ldots \tag{4.45}$$

valid for $\beta \gg 1$, the phase ϕ_m, Eq. (4.43c), reduces to

$$\phi_m \cong m\frac{\pi}{2} - 2\alpha\sqrt{m + 1/2} \tag{4.46}$$

and hence the energy distribution, Eq. (4.44), reads

$$W_m \cong 2\,\mathcal{A}_m \left[1 + (-1)^m \cos\left(4\alpha\sqrt{m + 1/2}\right)\right]. \tag{4.47}$$

The term $(-1)^m$ causes rapid oscillations as we go from even to odd values of m. These odd-even oscillations are modulated by the slowly varying envelope from the term $\cos(4\alpha\sqrt{m + 1/2})$ as shown in Fig. 4.18.

In the case of a squeezed vacuum, $\alpha = 0$, Eq. (4.47), simplifies to

$$W_m \cong 2\,\mathcal{A}_m[1 + (-1)^m]$$

and we are left with rapid odd-even oscillations, only. The period of the slow modulation is then infinite.

Rotated Squeezed States

We now turn to the case of a rotated squeezed state. In the preceding section we have already given the exact result and have seen from Fig. 4.15 that two types

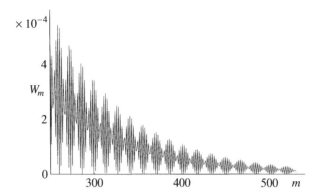

Fig. 4.18: The energy distribution, Eq. (4.35), of a nonrotated squeezed state displays in its exponential tail, that is for $m \gg \alpha^2$, rapid odd-even oscillations with a slowly varying amplitude. The parameters are $s = 201$ and $\alpha = 2$.

of oscillations in the energy distribution arise. These oscillations come out most clearly in the asymptotic expression. However, it is not straightforward to find this asymptotics. The concept of interference in phase space discussed in Chapter 7 provides immediately this expression. For the details of this derivation we refer to this chapter. Here we only quote the result and discuss it.

The energy distribution of a highly squeezed state rotated by an angle φ with respect to the momentum axis reads

$$W_m = 2A_m[\cosh \kappa_m + \cos(2\Phi_m)] \tag{4.48}$$

where

$$A_m \equiv \frac{1}{\sqrt{2\pi s}} \frac{\exp\{-2[m + \tfrac{1}{2} - \alpha^2 \cos(2\varphi)]/s\}}{(m + \tfrac{1}{2} - \alpha^2 \cos^2 \varphi)^{1/2}} \tag{4.49}$$

and

$$\kappa_m \equiv \frac{4}{s} \alpha \sin \varphi \left(m + \frac{1}{2} - \alpha^2 \cos^2 \varphi\right)^{1/2} \tag{4.50}$$

with

$$\Phi_m \equiv \phi_m(\sqrt{2}\,\alpha \cos \varphi)$$
$$- \left(m + \frac{1}{2}\right) \arctan \left[\frac{\left(m + \tfrac{1}{2} - \alpha^2 \cos^2 \varphi\right)^{1/2}}{\alpha \cos \varphi}\right]$$
$$- \alpha \cos \varphi \left(m + \frac{1}{2} - \alpha^2 \cos^2 \varphi\right)^{1/2} - \frac{\pi}{4}.$$

In Fig. 4.17(b) we compare the so-calculated energy distribution of the rotated squeezed state, Eq. (4.48), to the exact expression Eq. (4.36) and find excellent agreement.

We now turn to the discussion of the giant oscillations using the above asymptotic expressions. We first recognize that the oscillations make their appearance for angles

φ close to $\pi/2$. We therefore introduce the angle $\theta \equiv \pi/2 - \varphi$. In the limit of $\theta \ll 1$, or more precisely, in the limit of

$$\frac{\left(m + \frac{1}{2} - \alpha^2 \sin^2 \theta\right)^{1/2}}{\alpha \sin \theta} \cong \frac{\sqrt{m + \frac{1}{2}}}{\alpha \theta}\left[1 - \frac{1}{2}\frac{(\alpha\theta)^2}{m + \frac{1}{2}}\right] \cong \frac{\sqrt{m + \frac{1}{2}}}{\alpha \theta} \gg 1,$$

we find with the help of the asymptotic expansion of $\arctan \beta$, Eq. (4.45), for the phase Φ_m, Eq. (8.25), the expression

$$\Phi_m \cong m\frac{\pi}{2} - 2\alpha\theta\sqrt{m + \frac{1}{2}}. \tag{4.51}$$

When we compare this expression to the phase of the nonrotated squeezed state in the same limit, Eq. (4.46), we note that for $m \gg \alpha^2$. the phase Φ_m displays the same m-dependence. We hence expect to find a similar behavior of W_m for the rotated squeezed state. Indeed, Eq. (4.51) brings Eq. (4.48) to

$$W_m = 2A_m\left[\cosh \kappa_m + (-1)^m \cos\left(4\alpha\theta\sqrt{m + 1/2}\right)\right], \tag{4.52}$$

that is, a form very similar to Eq. (4.47).

Again the term $(-1)^m$ results in the rapid odd-even oscillations with an amplitude $\cos(4\alpha\theta\sqrt{m + 1/2})$ which is responsible for the giant oscillations. They disappear at m-values where the cosine vanishes, that is when

$$4\alpha\theta\sqrt{m_n + 1/2} = (2n + 1)\frac{\pi}{2}.$$

Therefore, the envelope has nodes at

$$m_n = \left[\frac{(2n + 1)\pi}{8\alpha\theta}\right]^2 - \frac{1}{2}. \tag{4.53}$$

The difference Δm_n between two successive nodes, that is the length of the nth giant oscillation, reads

$$\Delta m_n \equiv m_{n+1} - m_n = \frac{1}{8}\left(\frac{\pi}{\alpha\theta}\right)^2(n + 1). \tag{4.54}$$

Hence, for increasing φ, that is for decreasing θ, the length Δm_n increases and becomes infinite for $\theta = 0$. For a given displacement α the optimal angle θ_{opt} follows from the requirement that only a few giant oscillations are present in the regime of m-values displayed in a given figure. For the case of Fig. 4.15(b) this implies that $\Delta m_{n=1}$ has to be of the order 10. Indeed, we find from Eq. (4.54)

$$\Delta m_{n=1} = \frac{1}{4}\frac{\pi^2}{\alpha^2 \theta_{\text{opt}}^2} \cong 10$$

that is

$$\theta_{\text{opt}} = \frac{\pi}{2\sqrt{10}\,\alpha} \cong 4°. \tag{4.55}$$

We note that this condition is independent of the squeezing parameter s. However, this does not imply that the appearance of the giant oscillations is independent of s. According to Eq. (4.52) these oscillations only manifest themselves when the first contribution in Eq. (4.52), $\cosh \kappa_m$, is of the order of the second term, that is when $\cosh \kappa_m \cong 1$. This yields the condition

$$\kappa_m \cong \frac{4}{s} \alpha \left(1 - \frac{1}{2}\theta^2\right) \sqrt{m + \frac{1}{2}} \ll 1$$

for the onset of the giant oscillations.

Outlook

We conclude this section by emphasizing that the energy distribution of a highly squeezed state displays oscillations as a function of quantum number m. These oscillations come out most clearly in the appropriate asymptotic expansion of the energy distribution resulting in a very compact formula. The physical meaning of this formula will become clear when we discuss the concept of interference in phase space.

4.3.5 Time Evolution

In this section we discuss some aspects of the time evolution of squeezed states. The most intuitive way of doing this is to employ the Wigner function. Again we calculate the Wigner function at time $t = 0$ by substituting the wave function of the squeezed state, Eq. (4.33), into the definition of the Wigner function, Eq. (4.10). After performing the integration we arrive at

$$W(x, p) = \frac{1}{\pi \hbar} \exp\left[-s(\kappa x - \sqrt{2}\alpha)^2 - \frac{1}{s}\left(\frac{p}{\hbar \kappa}\right)^2\right]. \tag{4.56}$$

Again we find the time evolution of this Wigner function by replacing the initial positions x_0 and momenta p_0 by the classical trajectories

$$x_0(x, p, t) = \cos(\Omega t)\, x - \sin(\Omega t)\, \frac{p}{\hbar \kappa}\frac{1}{\kappa}$$

$$p_0(x, p, t) = \sin(\Omega t)\, \kappa x \hbar \kappa + \cos(\Omega t)\, p$$

leading from x_0 and p_0 to x and p. Hence, the Wigner function at time t reads

$$W(x, p, t) = \frac{1}{\pi \hbar} \exp\left[-s\left(\cos(\Omega t)\, \kappa x - \sin(\Omega t)\, \frac{p}{\hbar \kappa} - \sqrt{2}\alpha\right)^2 \right.$$
$$\left. - \frac{1}{s}\left(\sin(\Omega t)\, \kappa x + \cos(\Omega t)\, \frac{p}{\hbar \kappa}\right)^2\right]. \tag{4.57}$$

Again we obtain the probability distributions for position and momentum as a function of time by integrating this expression over p and x. In particular, we find for the position distribution

$$|\psi_{\text{sq}}(x, t)|^2 = \frac{\kappa}{\sqrt{\pi s_x(t)}} \exp\left[-\frac{1}{s_x(t)}\left(\kappa x - \sqrt{2}\alpha \cos(\omega t)\right)^2\right],$$

136   4 Quantum States in Phase Space

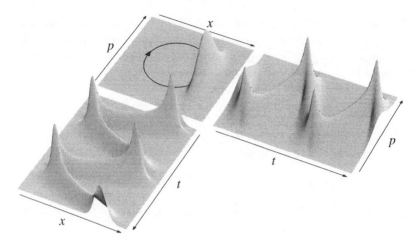

Fig. 4.19: Time evolution of a squeezed state with $\alpha = 2$ and squeezing parameter $s = 4$. The Wigner function is an asymmetric "squeezed" Gaussian and moves in phase space along a circle. We only show the initial Wigner function and indicate its rotation by a circle. As it moves, the marginal distributions in the form of Gaussians perform harmonic oscillation. In contrast to the case of a coherent state the widths are now oscillating in time: A large width in momentum implies a small width in position and vice versa. We have a breathing wave packet.

where
$$s_x(t) = s\, \sin^2(\Omega t) + \frac{1}{s} \cos^2(\Omega t),$$
and for the momentum distribution
$$|\psi_{\text{sq}}(p,t)|^2 = \frac{1}{\hbar\kappa\sqrt{\pi s_p(t)}} \exp\left[-\frac{1}{s_p(t)}\left(\frac{p}{\hbar\kappa} + \sqrt{2}\,\alpha\, \sin(\Omega t)\right)^2\right],$$
where
$$s_p(t) = s\, \cos^2(\Omega t) + \frac{1}{s} \sin^2(\Omega t).$$

We emphasize that only for $s = 1$, that is, for a coherent state the widths of these Gaussians determined by s_x and s_p are time independent. This comes out most clearly in Fig. 4.19 where we depict the Wigner function and the corresponding marginal distributions as they evolve in time. In contrast to the corresponding figure of a coherent state, Fig. 4.10, these curves show huge maxima at the turning points indicating strong time dependent oscillations of the squeezed wave packet.

4.4 Rotated Quadrature States

In Sec. 4.5 we address the question of how to measure the Wigner function. The essential ingredient of this scheme is the rotated dimensionless quadrature operator

$$\hat{X}_\theta \equiv \cos\theta\, (\kappa\hat{x}) + \sin\theta \left(\frac{\hat{p}}{\hbar\kappa}\right). \qquad (4.58)$$

We note that for $\theta = 0$ the operator $\hat{X}_{\theta=0}$ is up to the scale factor κ identical to the position operator $\hat{x}$, whereas for $\theta = \pi/2$ it is up to a scale factor $1/(\hbar\kappa)$ identical to the momentum operator $\hat{p}$.

Since the quadrature operator $\hat{X}_\theta$ satisfies the periodicity condition

$$\hat{X}_{\theta+\pi} = -\hat{X}_\theta$$

we can restrict the regime of θ to an interval of π. As we show later it is convenient to choose the interval $-\pi/2 \leq \theta \leq \pi/2$.

The rotated quadrature operator is a linear combination of the position and the momentum operator. The angle θ fixes the respective weights of the two operators. In this section we show that this angle amounts to a rotation of the coordinate system in phase space. This comes out most clearly in the Wigner function of the eigenstates $|X_\theta\rangle$.

Since the position and the momentum operators have a continuous spectrum the rotated quadrature operator $\hat{X}_\theta$ has also a continuous spectrum. Moreover, since $\hat{x}$ and $\hat{p}$ are hermitian, the operator $\hat{X}_\theta$ is hermitian, too. This guarantees that the eigenvalues X_θ are real and the rotated quadrature operator is an observable. We show in Chapter 13 that in the case of an electromagnetic field oscillator the homodyne detector corresponds to a measurement of this operator. For every angle θ there is a continuous family of eigenstates $|X_\theta\rangle$ with eigenvalues X_θ. In addition, the states depend on the angle θ.

4.4.1 Wigner Function of Position and Momentum States

In the present section we calculate the Wigner function of a position or momentum eigenstate $|x\rangle$ or $|p\rangle$. Since these states are only normalized with respect to δ-functions the Wigner functions of these states only exist in the sense of a distribution. In particular, they violate the rule that the square of the Wigner function integrated over all phase space is determined by $2\pi\hbar$.

Position Eigenstates

We first consider the case of a position eigenstate $|x_0\rangle$ and recall the definition, Eq. (3.1), of the Wigner function

$$W_{|x_0\rangle}(x,p) = \frac{1}{2\pi\hbar} \int_{-\infty}^{\infty} dy\, e^{-ipy/\hbar} \langle x + \tfrac{1}{2}y|x_0\rangle\langle x_0|x - \tfrac{1}{2}y\rangle.$$

With the help of the orthonormality relation

$$\langle x|y\rangle = \delta(x-y)$$

we arrive at

$$W_{|x_0\rangle}(x,p) = \frac{1}{2\pi\hbar} \int_{-\infty}^{\infty} dy\, e^{-ipy/\hbar}\, \delta(x + \tfrac{1}{2}y - x_0)\, \delta(x_0 - x + \tfrac{1}{2}y).$$

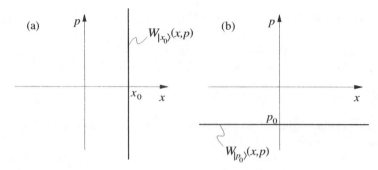

Fig. 4.20: The Wigner functions of (a) a position eigenstate $|x_0\rangle$ and (b) a momentum eigenstate $|p_0\rangle$ are infinitely thin but infinitely tall walls which run parallel to either the momentum or position axis.

We can use one of the δ-functions to perform the integral which yields

$$W_{|x_0\rangle}(x,p) = \frac{1}{\pi\hbar} e^{-ip2(x-x_0)/\hbar} \delta(2(x-x_0)).$$

When we recall the relations

$$\delta(ax) = \frac{1}{|a|}\delta(x)$$

and

$$f(x)\,\delta(x-y) = f(y)\,\delta(x-y)$$

we find

$$W_{|x_0\rangle}(x,p) = \frac{1}{2\pi\hbar}\delta(x-x_0). \tag{4.59}$$

Hence the Wigner function of a position eigenstate $|x_0\rangle$ is a delta function at $x = x_0$. We note that the momentum variable does not enter at all. Therefore, the Wigner function is an infinitely thin but infinitely tall wall whose base line runs parallel to the momentum axis as indicated in Fig. 4.20(a). This Wigner function representation of a position eigenstate confirms the naive picture of such a state: The position is well-defined but no knowledge exists about momentum.

Momentum Eigenstate

Analogously we find for the Wigner function

$$W_{|p_0\rangle}(x,p) = \frac{1}{2\pi\hbar}\int_{-\infty}^{\infty} dy\, e^{-ipy/\hbar}\,\langle x+\tfrac{1}{2}y|p_0\rangle\langle p_0|x-\tfrac{1}{2}y\rangle$$

of a momentum eigenstate $|p_0\rangle$ the expression

$$W_{|p_0\rangle}(x,p) = \frac{1}{2\pi\hbar}\frac{1}{2\pi\hbar}\int_{-\infty}^{\infty} dy\, e^{-i(p-p_0)y/\hbar},$$

where in the last step we have used the formula, Eq. (2.10),

$$\langle x|p\rangle = (2\pi\hbar)^{-1/2} \exp\left(\frac{i}{\hbar}xp\right)$$

for the scalar product between a momentum state and a position state.

When we recall the integral definition of the delta function we find that the Wigner function

$$W_{|p_0\rangle}(x,p) = \frac{1}{2\pi\hbar}\delta(p - p_0)$$

of a momentum eigenstate is a delta function at p_0. Again, it is an infinitely tall, but infinitely thin wall along p_0 as shown in Fig. 4.20(b).

Marginals of Wigner Functions

In Sec. 3.2.2 we have shown that the modulus square of the scalar product between two quantum states $|\psi\rangle$ and $|\phi\rangle$ is the product of the Wigner functions $W_{|\psi\rangle}$ and $W_{|\phi\rangle}$ of the two states integrated over phase space, that is

$$|\langle\phi|\psi\rangle|^2 = 2\pi\hbar \int_{-\infty}^{\infty} dx \int_{-\infty}^{\infty} dp\, W_{|\phi\rangle}(x,p)\, W_{|\psi\rangle}(x,p). \tag{4.60}$$

We note that the marginal property

$$W_{|\psi\rangle}(x_0) \equiv |\langle x_0|\psi\rangle|^2 = \int_{-\infty}^{\infty} dp\, W_{|\psi\rangle}(x_0, p) \tag{4.61}$$

of the Wigner function is a special case of this formula. Indeed, when we make use of the expression Eq. (4.59) for the Wigner function of a position eigenstate, we find from

$$W_{|\psi\rangle}(x_0) = 2\pi\hbar \int_{-\infty}^{\infty} dx \int_{-\infty}^{\infty} dp\, W_{|x_0\rangle}(x,p)\, W_{|\psi\rangle}(x,p)$$

the result

$$W_{|\psi\rangle}(x_0) = \int_{-\infty}^{\infty} dx \int_{-\infty}^{\infty} dp\, \delta(x - x_0)\, W_{|\psi\rangle}(x,p) = \int_{-\infty}^{\infty} dp\, W_{|\psi\rangle}(x_0, p).$$

This relation allows us to understand geometrically the remarkable property, Eq. (4.61), of the Wigner function: The probability distribution results from the integration over the conjugate variable. We envision the Wigner function as a landscape with hills and valleys extending over x-p phase space as illustrated for example in Fig. 3.1 by the Wigner function of the 6th energy eigenstate of a harmonic oscillator. The probability to find the particle at the position x_0 is thus the volume, or the weighted area cut out of the Wigner function by a thin plate located at x_0 aligned parallel to the momentum axis.

4.4.2 Position Wave Function of Rotated Quadrature States

Our goal is to derive the Wigner function $W_{|X_\theta\rangle}$ of the eigenstate $|X_\theta\rangle$. For this purpose we need the wave function $\mathcal{X}(x; X_\theta) \equiv \langle x | X_\theta \rangle$ in position representation. This expression must obviously depend on the position variable x and the eigenvalue X_θ specifying the state for a fixed angle θ. We denote this wave function by $\mathcal{X}(x; X_\theta)$.

Eigenvalue Equation in Position Space

There exist many possibilities to derive a formula for $\mathcal{X}$. Here we only pursue the one which starts from the eigenvalue equation

$$\hat{X}_\theta |X_\theta\rangle = X_\theta |X_\theta\rangle$$

for the rotated quadrature eigenstates.

We multiply both sides by $\langle x|$ and recall the definition of the quadrature operator which yields

$$\langle x| \cos\theta \, (\kappa \hat{x}) + \sin\theta \left(\frac{\hat{p}}{\hbar\kappa}\right) |X_\theta\rangle = X_\theta \langle x | X_\theta \rangle.$$

The x-representation

$$\hat{p} \equiv \frac{\hbar}{i} \frac{d}{dx}$$

of the momentum operator finally yields

$$\left(\cos\theta\, \kappa x + \frac{1}{i\kappa} \sin\theta \frac{d}{dx}\right) \mathcal{X}(x; X_\theta) = X_\theta\, \mathcal{X}(x; X_\theta),$$

that is

$$\frac{d\mathcal{X}(x; X_\theta)}{dx} = -i\kappa \frac{\cos\theta\, \kappa x - X_\theta}{\sin\theta} \mathcal{X}(x; X_\theta).$$

Since this is an ordinary differential equation of first order for the wave function of the rotated quadrature operator we immediately find the solution

$$\mathcal{X}(x; X_\theta) = \mathcal{N}(\theta, X_\theta) \exp\left\{-i \frac{[\cos\theta\, \kappa x - X_\theta]^2}{2\sin\theta \cos\theta}\right\}. \tag{4.62}$$

The position wave function of a rotated quadrature state is a Gaussian. In contrast to a coherent state it is purely imaginary. Therefore, the wave function can not be normalized to one. In complete analogy to position or momentum eigenstates we normalize the quadrature eigenstates with respect to δ-functions. This partially determines the factor $\mathcal{N}$.

The eigenvalue equation is a differential equation of first order in the position x. Hence the function $\mathcal{N}$ can still depend on the angle θ and on the eigenvalue X_θ. Therefore, the eigenvalue equation defines families of quadrature eigenstates which differ by their choice of $\mathcal{N}$.

Evaluation of Normalization Constant

Since the rotated quadrature states are eigenstates of a hermitian operator they satisfy the completeness relation

$$\int dX_\theta \, |X_\theta\rangle \langle X_\theta| = \mathbb{1}.$$

We emphasize that this relation holds true for any angle θ. The integration is over the complete range of eigenvalues X_θ.

We find $\mathcal{N}$ by projecting with $\langle x|$ and $|y\rangle$ onto the completeness relation and noting the orthogonality relation which yields

$$\int dX_\theta \, \langle x | X_\theta \rangle \langle X_\theta | y \rangle = \int dX_\theta \, \mathcal{X}(x; X_\theta) \mathcal{X}^*(y; X_\theta) = \delta(x - y).$$

With the help of the expression Eq. (4.62) for the position wave function $\mathcal{X}(x; X_\theta)$ of the rotated quadrature state we therefore find

$$\delta(x-y) = \int_{-\infty}^{\infty} dX_\theta \, |\mathcal{N}(\theta, X_\theta)|^2 \exp\left\{ \frac{-i}{2\sin\theta\cos\theta} \left[(\cos\theta \, \kappa x - X_\theta)^2 - (\cos\theta \, \kappa y - X_\theta)^2 \right] \right\}.$$

The terms quadratic in X_θ in the exponent cancel each other and the cross terms lead to the integral

$$\delta(x-y) = \exp\left[-\frac{i\cos\theta}{2\sin\theta} \kappa^2 (x^2 - y^2) \right] \int_{-\infty}^{\infty} dX_\theta \, |\mathcal{N}(\theta, X_\theta)|^2 \exp\left[i(x-y) \frac{\kappa}{\sin\theta} X_\theta \right].$$

When we introduce the new integration variable $k \equiv \kappa X_\theta / \sin\theta$ the equation determining $\mathcal{N}$ takes the form

$$\delta(x-y) = \exp\left[-\frac{i\cos\theta}{2\sin\theta} \kappa^2 (x^2 - y^2) \right] \frac{\sin\theta}{\kappa} \int_{-\infty}^{\infty} dk \, \left| \mathcal{N}\left(\theta, \frac{\sin\theta}{\kappa} k\right) \right|^2 \exp[ik(x-y)].$$

So far, the dependence of $\mathcal{N}$ on X_θ, and thus on k is not determined yet. However, at this point we realize that the integral over k is a δ-function in the variable $x - y$ provided $|\mathcal{N}|^2$ is independent of k and hence independent of X_θ. In the remainder of the book we therefore choose a family of quadrature states X_θ such that $|\mathcal{N}|^2$ depends on θ only.

Under this condition we find

$$\delta(x-y) = \exp\left[-\frac{i\cos\theta}{2\sin\theta} \kappa^2 (x^2 - y^2) \right] 2\pi \frac{\sin\theta}{\kappa} |\mathcal{N}(\theta)|^2 \delta(x-y),$$

or

$$\delta(x-y) = 2\pi \frac{\sin\theta}{\kappa} |\mathcal{N}(\theta)|^2 \delta(x-y).$$

Here we have made use of the relation

$$f(x)\,\delta(x-y) = f(y)\,\delta(x-y).$$

Hence, we find for the normalization condition

$$|\mathcal{N}(\theta)|^2 = \frac{\kappa}{2\pi \sin\theta}.$$

We conclude by summarizing the main result of this section. The wave function $\mathcal{X}(x; X_\theta)$ of a rotated quadrature eigenstate $|X_\theta\rangle$ in position space reads

$$\mathcal{X}(x; X_\theta) = \left(\frac{\kappa}{2\pi \sin\theta}\right)^{1/2} \exp\left\{-i\frac{[X_\theta - \cos\theta\, \kappa x]^2}{2\sin\theta \cos\theta}\right\}.$$

We emphasize that the angle θ is restricted to the regime of $-\pi/2 \le \theta \le \pi/2$. This domain has the advantage that the normalization constant $\mathcal{N}$ is always purely real.

4.4.3 Wigner Function of Rotated Quadrature States

We are now in a position to calculate the Wigner function

$$W_{|X_\theta\rangle}(x,p) = \frac{1}{2\pi\hbar} \int_{-\infty}^{\infty} dy\, e^{-ipy/\hbar} \langle x + \tfrac{1}{2}y|X_\theta\rangle\langle X_\theta|x - \tfrac{1}{2}y\rangle$$

of the eigenstate $|X_\theta\rangle$. With the help of the expression Eq. (4.62) for the wave function $\mathcal{X}(x; X_\theta)$ of the quadrature state this integral

$$W_{|X_\theta\rangle}(x,p) = \frac{1}{2\pi\hbar} \int_{-\infty}^{\infty} dy\, e^{-ipy/\hbar} \mathcal{X}(x + \tfrac{1}{2}y; X_\theta)\mathcal{X}^*(x - \tfrac{1}{2}y; X_\theta)$$

takes the form

$$W_{|X_\theta\rangle}(x,p) = \frac{1}{2\pi\hbar}|\mathcal{N}(\theta)|^2 \int_{-\infty}^{\infty} dy\, e^{-ipy/\hbar}$$

$$\times \exp\left\{\frac{-i}{2\sin\theta\cos\theta}\left[(X_\theta - \cos\theta\,\kappa(x + \tfrac{1}{2}y))^2 - (X_\theta - \cos\theta\,\kappa(x - \tfrac{1}{2}y))^2\right]\right\}.$$

When we simplify the exponent in the integral and use the expression for the normalization constant we find

$$W_{|X_\theta\rangle}(x,p) = \frac{1}{2\pi\hbar}\frac{\kappa}{2\pi\sin\theta} \int_{-\infty}^{\infty} dy\, \exp\left[i\left(X_\theta - \cos\theta\,\kappa x - \sin\theta\,\frac{p}{\hbar\kappa}\right)\frac{\kappa}{\sin\theta}y\right],$$

that is

$$W_{|X_\theta\rangle}(x,p) = \frac{1}{2\pi\hbar}\delta\left[X_\theta - \left(\cos\theta\,\kappa x + \sin\theta\,\frac{p}{\hbar\kappa}\right)\right]. \tag{4.63}$$

Hence the Wigner function of the rotated quadrature state is a delta function along the line

$$X_\theta = \cos\theta\,\kappa x + \sin\theta\,\frac{p}{\hbar\kappa}.$$

4.5 Quantum State Reconstruction

In Chapter 3 we have introduced the Wigner function as a representation of a quantum state. We recall that any scheme making measurements on a quantum system can only provide probability distributions. Since the quantum state contains all information available about a quantum system we can definitely calculate all probability distributions starting from this state. We now ask the inverse question: Is it possible to use a set of probability distributions to reconstruct the quantum state?

This question dates back to the early days of quantum mechanics and, in particular, to the handbook article by W. Pauli. He was wondering if it is possible to find the wave function in amplitude *and* phase from the probability distributions in position and momentum. Pauli did not answer the question. However, simple counterexamples show that in general this is not possible. As discussed in Problem 4.3 we need more distributions than these two.

In the field of classical optics this problem is summarized by the name of *phase retrieval*. In this context we want to determine the amplitude *and* phase of an electromagnetic field by making appropriate intensity measurements. Again we need many intensity distributions to reconstruct the electromagnetic field.

The essential ingredient of this reconstruction scheme is the rotated, dimensionless quadrature operator X_θ discussed in the preceding section.

4.5.1 Tomographic Cuts through Wigner Function

Can we reconstruct the quantum state as represented for example by the Wigner function using various probability distributions? Obviously, the position and the momentum distributions were not enough. What about the quadrature distributions

$$W(X_\theta) \equiv \langle X_\theta | \hat{\rho} | X_\theta \rangle = \mathrm{Tr}\,\{|X_\theta\rangle \langle X_\theta| \hat{\rho}\}$$

for various angles θ?

We now show that indeed the knowledge of these distributions for all angles $-\pi/2 \leq \theta \leq \pi/2$ is necessary to reconstruct the Wigner function. For this purpose we relate the quadrature distributions $W(X_\theta)$ of the quantum state to its Wigner function. According to the trace product rule, Eq. (3.5), we can represent the quadrature distribution $W(X_\theta)$ by the phase space integral

$$W(X_\theta) = 2\pi\hbar \int_{-\infty}^{\infty} dx \int_{-\infty}^{\infty} dp\, W_{|X_\theta\rangle}(x,p)\, W_{\hat{\rho}}(x,p)$$

of the product of the two corresponding Wigner functions. With the help of the expression Eq. (4.63) for the Wigner function $W_{|X_\theta\rangle}(x,p)$ of the rotated quadrature eigenstate this formula reads

$$W(X_\theta) = \int_{-\infty}^{\infty} dx \int_{-\infty}^{\infty} dp\, \delta\!\left[X_\theta - \left(\cos\theta\, \kappa x + \sin\theta\, \frac{p}{\hbar\kappa}\right)\right] W_{\hat{\rho}}(x,p). \qquad (4.64)$$

We obtain the probability distribution $W(X_\theta)$ as the overlap between the Wigner function $W_{\hat{\rho}}$ of the state of interest and an infinitely thin phase space strip represented

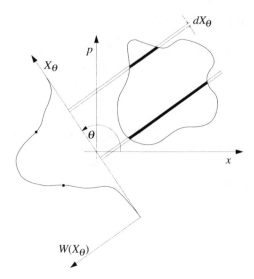

Fig. 4.21: The Wigner function $W_{\hat{\rho}}(x,p)$ represented by one of its contour lines. Every overlap between the infinitely thin phase-space strips at X_θ leads to a value $W(X_\theta)$. All these strip-overlaps for a fixed phase space angle θ constitute the complete probability distribution $W(X_\theta)$.

by the δ-function. Hence every strip given by the eigenvalue X_θ defines a line along which we have to integrate the Wigner function. This integration procedure results in the distribution $W(X_\theta)$ for a fixed phase space angle θ as shown schematically in Fig. 4.21. Therefore, we can obtain all distributions $W(X_\theta)$ from the Wigner function. This is not surprising since $W_{\hat{\rho}}$ contains all the information about the quantum state. But can we invert the procedure?

4.5.2 Radon Transformation

In Chapter 13 we show for the case of a light field that a homodyne detector measures the quadrature distributions $W(X_\theta)$. Suppose we have measured these distributions for all angles $-\pi/2 \leq \theta \leq \pi/2$. Is it then possible to reconstruct the complete Wigner function based on this information? The answer is affirmative and rests on the Radon transformation.

Inversion Scheme

In order to invert the relation Eq. (4.64) and express the Wigner function in terms of quadrature distributions we write the δ-function in Eq. (4.64) as a Fourier integral and find

$$W(X_\theta) = \frac{1}{2\pi} \int_{-\infty}^{\infty} dt\, e^{-itX_\theta} \int_{-\infty}^{\infty} dx \int_{-\infty}^{\infty} dp\, \exp\left[i\kappa x(t\cos\theta) + i\frac{p}{\hbar\kappa}(t\sin\theta)\right] W_{\hat{\rho}}(x,p).$$

4.5 Quantum State Reconstruction

We multiply this equation by $e^{it'X_\theta}$ and integrate over X_θ which yields

$$\int_{-\infty}^{\infty} dX_\theta \, e^{it'X_\theta} W(X_\theta) = \int_{-\infty}^{\infty} dx \int_{-\infty}^{\infty} dp \, \exp\left(i\kappa x \, \zeta + i\frac{p}{\hbar\kappa}\eta\right) W_{\hat{\rho}}(x,p). \tag{4.65}$$

Here we have introduced the abbreviations $\zeta \equiv t' \cos\theta$ and $\eta \equiv t' \sin\theta$.

The expression on the right hand side of this equation is a two-dimensional Fourier transformation which we now invert to find the Wigner function $W_{\hat{\rho}}$. At first sight this inversion seems to be straightforward. However, the range of the integration variables t' and θ is such that we do not cover the complete (ζ, η)-space. Indeed, t' is defined on the whole axis, that is $-\infty < t' < \infty$. In contrast θ leaves out half of the plane since $-\pi/2 \leq \theta \leq \pi/2$.

We multiply both sides of Eq. (4.65) by the factor $|t'| \exp(-i\kappa x'\zeta - i\frac{p'}{\hbar\kappa}\eta)$ and integrate over the complete range of t' and θ. We therefore arrive at

$$\int_{-\infty}^{\infty} dt' |t'| \int_{-\pi/2}^{\pi/2} d\theta \, \exp(-i\kappa x'\zeta - i\frac{p'}{\hbar\kappa}\eta) \int_{-\infty}^{\infty} dX_\theta \, e^{it'X_\theta} W(X_\theta) =$$

$$\int_{-\infty}^{\infty} dx \int_{-\infty}^{\infty} dp \, W_{\hat{\rho}}(x,p) \int_{-\infty}^{\infty} dt' |t'| \int_{-\pi/2}^{\pi/2} d\theta \, \exp\left(i\kappa(x-x')\zeta + i\frac{p-p'}{\hbar\kappa}\eta\right)$$

where on the right hand side of this equation we have interchanged the order of integration.

Integral Relation

We now focus on the integral

$$\mathcal{I} \equiv \int_{-\infty}^{\infty} dt' |t'| \int_{-\pi/2}^{\pi/2} d\theta \, \exp\left(i\kappa(x-x')\zeta + i\frac{p-p'}{\hbar\kappa}\eta\right)$$

consisting of the two integrals over t' and θ and show that they are the integral representations of two delta functions in position and momentum. For this purpose we first decompose the domain of the integration variable t' into the positive and negative values

$$\mathcal{I} = \left(\int_0^{\infty} dt' |t'| \int_{-\pi/2}^{\pi/2} d\theta + \int_{-\infty}^{0} dt' |t'| \int_{-\pi/2}^{\pi/2} d\theta\right) \exp\left(i\kappa(x-x') t' \cos\theta + i\frac{p-p'}{\hbar\kappa} t' \sin\theta\right)$$

where in the last step we have recalled the abbreviations $\zeta \equiv t' \cos\theta$ and $\eta \equiv t' \sin\theta$. When we perform the substitutions $t'' \equiv -t'$ and $\theta \equiv \theta' + \pi$ in the second integral and recall the relations $t' \cos\theta = -t'' \cos(\theta' + \pi) = t'' \cos\theta'$ and $t' \sin\theta = -t'' \sin(\theta' + \pi) = t'' \sin\theta'$ we arrive at

$$\mathcal{I} = \left(\int_0^{\infty} dt' |t'| \int_{-\pi/2}^{\pi/2} d\theta + \int_0^{\infty} dt' |t'| \int_{\pi/2}^{3\pi/2} d\theta\right) \exp\left(i\kappa(x-x') t' \cos\theta + i\frac{p-p'}{\hbar\kappa} t' \sin\theta\right)$$

$$= \int_0^\infty dt' \, |t'| \int_{-\pi/2}^{3\pi/2} d\theta \, \exp\left(i\kappa(x-x')t'\cos\theta + i\frac{p-p'}{\hbar\kappa}t'\sin\theta\right).$$

Indeed, we now cover the whole (ζ, η)-plane.

When we express the polar coordinates in cartesian coordinates the integral reads

$$\mathcal{I} = \int_{-\infty}^\infty d\zeta \, \exp[i\kappa(x-x')\zeta] \int_{-\infty}^\infty d\eta \, \exp\left(-i\frac{p-p'}{\hbar\kappa}\eta\right) = (2\pi)^2 \hbar \, \delta(x-x')\delta(p-p'),$$

which is just the product of two δ-functions.

Measurement of Wigner Function

With this result we invert Eq. (4.65) and find

$$W_{\hat{\rho}}(x,p) = \frac{1}{4\pi^2\hbar} \int_{-\infty}^\infty dt \, |t| \int_{-\pi/2}^{\pi/2} d\theta \int_{-\infty}^\infty dX_\theta \, \exp\left[it\left(X_\theta - \kappa x\cos\theta - \frac{p}{\hbar\kappa}\sin\theta\right)\right] W(X_\theta),$$

(4.66)

where we have replaced all the primed variables by their unprimed counterparts.

Equation (4.66) is the answer to our question whether it is possible to reconstruct the Wigner function from the knowledge of all probability distributions $W(X_\theta)$. Indeed, we obtain $W_{\hat{\rho}}$ from the transformation Eq. (4.66) of an *ensemble* of distributions $W(X_\theta)$ defined by all the phases between $\theta = -\pi/2$ and $\theta = \pi/2$.

In Chapter 13 we show that a homodyne detector measures the required ensemble of distributions $W(X_\theta)$: We fix $\theta_1 = 0$ by the phase of the local oscillator and determine the distribution $W(X_{\theta_1})$, as it was already described. Then we choose a second phase $\theta_2 = 2\pi/N$ and again measure the resulting distribution. We continue until we reach $\theta_N = N\pi/N = \pi$. This procedure gives us a sequence of distributions $\{W(X_{\theta_1}), ..., W(X_{\theta_N})\}$. The resulting discrete ensemble is a good approximation to our true ensemble which is in fact continuous.

We then use this discretized version of our ensemble, perform the transformation Eq. (4.66) numerically and regain $W_{\hat{\rho}}$. It is in this sense that we may speak of a measurement of the Wigner function $W_{\hat{\rho}}$. This measurement technique is sometimes called homodyne-tomography, since $W_{\hat{\rho}}$ is reconstructed from all distributions $W(X_\theta)$ which are projections of the original Wigner function on rotated planes. This resembles very much the well-known tomography techniques used to produce three-dimensional computer graphics of the human body.

Problems

4.1 Energy Distribution of Squeezed Vacuum

In Sec. 4.3.2 we have calculated the energy distribution of a squeezed state, Eq. (4.35). In the case of squeezed vacuum this distribution reads

$$W_m = \frac{2\sqrt{s}}{s+1}\left(\frac{s-1}{s+1}\right)^m (2^m m!)^{-1} H_m^2(0).$$

Show with the relations for the Hermite polynomial proved in Problem 2.6a that

$$H_m(0) = \begin{cases} 0 & \text{for } m \text{ odd} \\ (-2)^{m/2} \prod_{\ell=1}^{m/2} (2\ell - 1) & \text{for } m \text{ even} \end{cases}.$$

What is a physical explanation for the fact that the energy distribution does not change when s is replaced by $1/s$? Calculate the mean energy.

Hint: See Vogel and Schleich (1991).

4.2 Energy Distribution of a Rotated Squeezed State

Derive the energy distribution of a rotated squeezed state.

Hint: See Schleich and Wheeler (1987b).

4.3 State Reconstruction from Probability Distributions?

Do the two probability distributions in position and momentum of a quantum state uniquely determine the state?

Hint: This is an old problem that even baffled W. Pauli (1933) and H. Reichenbach (1948). For a discussion see A. Vogt (1978) and A. Orlowski and H. Paul (1995).

4.4 Squeezed Fock States

So far we have discussed squeezed states which originate from the vacuum state. However, it is also possible to apply the squeezing operator $\hat{S}$ on an energy eigenstate $|n\rangle$. Discuss the properties of these states and, in particular, their time evolution in a harmonic oscillator. This time evolution was observed experimentally using an ensemble of cold Cs-atoms moving in a far detuned standing electromagnetic wave as shown in Fig. 4.22.

4.5 Superposition State

Discuss the time evolution of the superposition of the ground state and the first excited state of a harmonic oscillator. The corresponding experiment is shown in Fig. 4.23.

References

Relief of Special Functions

For the behavior of various special functions and, in particular, of Hermite and Laguerre polynomials depicted by three-dimensional plots see
E. Jahnke and **F. Emde**, *Tables of Functions*, Dover, New York, 1945

Even in today's world of computers with *Mathematica* and *Maple* this book with its three-dimensional reliefs of special functions is still impressive.

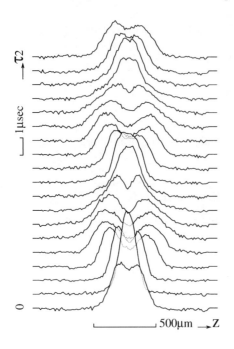

Fig. 4.22: Measured wave packet dynamics of a ensemble of cold Cs-atoms in a far-detuned standing electromagnetic wave. When the thermal energy of the atoms is small they move close to the minima of the standing wave. Therefore, we can approximate the potential by a harmonic oscillator. Time evolution of a squeezed energy eigenstate $|n=1\rangle$ depicted by the corresponding position distributions at various times. Taken from M. Morinaga et al., Phys. Rev. Lett. **83**, 4037 (1999).

Elementary Wave Packets

For the original work on wave packets of the mechanical oscillator see
E. Schrödinger, *Der stetige Übergang von der Mikro- zur Makromechanik*, Naturwissenschaften **14**, 664–666 (1926)
E.H. Kennard, *Zur Quantenmechanik einfacher Bewegungstypen*, Z. Physik **44**, 326–352 (1927)

For the physics of displaced energy eigenstates see
I.R. Senitzky, *Quantum Effects in the Interaction between Electrons and High-Frequency Fields*, Phys. Rev. **95**, 904–913 (1954)

For the experimental realization of atomic wave packets in a standing electromagnetic wave see
M. Morinaga, I. Bouchoule, J.-C. Karam and **C. Salomon**, *Manipulation of Motional Quantum States of Neutral Atoms*, Phys. Rev. Lett. **83**, 4037–4040 (1999)

Squeezed States

For a discussion of squeezed states of light see
D. Stoler, *Equivalence Classes of Minimum Uncertainty Packets*, Phys. Rev. D **1**, 3217–3219 (1970)
H.P. Yuen, *Two-Photon Coherent States of the Radiation Field*, Phys. Rev. A **13**, 2226–2243 (1976)

The physics of squeezed states is summarized in
D.F. Walls, *Squeezed States of Light*, Nature (London) **306**, 141–146 (1983)
R.W. Henry and **S.C. Glotzer**, *A Squeezed States Primer*, Am. J. Phys. **56**, 318–328 (1988)

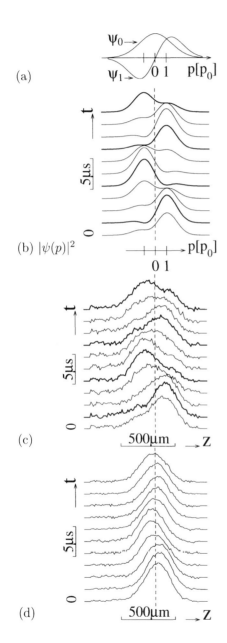

Fig. 4.23: Comparison between theory (a) and experiment (b) about the time evolution of a superposition state consisting of the ground state and the first excited state of a harmonic oscillator. The two corresponding wave functions are shown on the top. For the experiment we use an ensemble of cold Cs-atoms moving in a far detuned standing wave. On the bottom (d) we display the observed time evolution for a coherent state of a harmonic oscillator. Taken from M. Morinaga *et al.*, Phys. Rev. Lett. **83**, 4037 (1999).

M. Nieto, *What are squeezed states really like?*, in: *Frontiers in Nonequilibrium Statistical Physics*, edited by G.T. Moore and M.O. Scully, Plenum Press, New York, 1986

Many of the pioneering papers in the field of squeezed states have been reprinted in the book

P. Meystre and **D.F. Walls**, *Nonclassical Effects in Quantum Optics*, American Institute of Physics, New York, 1991

For the state of the art of squeezed states in quantum optics see the special issues edited by

H.J. Kimble and **D.F. Walls**, J. Opt. Soc. Am. B **4**, (10) (1987)

P. Knight and **R. Loudon**, J. Mod. Opt. **34** (6) (1984)

E. Giacobino and **C. Fabre**, Appl. Phys. B **55** (3) (1992)

For the experimental verification of the squeezed vacuum state as a state of minimum uncertainty see

L.A. Wu, **M. Xiao** and **H.J. Kimble**, *Squeezed states of light from an optical parametric oscillator*, J. Opt. Soc. Am. B **4**, 1465–1475 (1987)

For the discussion of the energy distribution of coherent and squeezed states in the appropriate limit see

W.P. Schleich and **J.A. Wheeler**, *Oscillations in the Photon Distribution of Squeezed States and Interference in Phase Space*, Nature (London) **326**, 574–577 (1987a)

W.P. Schleich and **J.A. Wheeler**, *Oscillations in Photon Distribution of Squeezed States*, J. Opt. Soc. Am. B **4, 1715–1722 (1987b)**

K. Vogel and **W.P. Schleich**, in: *Fundamental Systems in Quantum Optics*, eds. J. Dalibard, J.M. Raimond and J. Zinn-Justin, Elsevier, Amsterdam, 1991

Quantum State Reconstruction

For a discussion of the Pauli problem as defined in

W. Pauli, *Die allgemeinen Prinzipien der Wellenmechanik*, Handbuch der Physik, Vol. 24, edited by H. Geiger and K. Scheel, Springer Verlag, Berlin, 1933

see

H. Reichenbach, *Philosophic Foundations of Quantum Mechanics*, University of California Press, Berkeley, 1948

A. Vogt, *Position and Momentum Distributions do not Determine the Quantum Mechanical State*, in: *Mathematical Foundations of Quantum Theory*, ed. by A.R. Marlow, Academic Press, New York, 1978

A. Orlowski and **H. Paul**, *Phase retrieval in quantum mechanics*, Phys. Rev. A **50**, 921–924 (1995)

For a review of quantum state measurement see

M. Freyberger, P. Bardroff, C. Leichtle, C. Schrade and **W.P. Schleich**, *The art of measuring quantum states*, Physics World **10 (11), 41–45 (1997)**

U. Leonhardt, *Measuring the Quantum State of Light*, Cambridge University Press, Cambridge, 1997

W.P. Schleich and **M. Raymer**, *Quantum State Preparation and Measurement*, J. Mod. Opt. **44** (11/12) (1997)

D. Leibfried, **T. Pfau** and **C. Monroe**, *Shadows and Mirrors: Reconstructing Quantum States of Atom Motion*, Physics Today **51** (4), 22–28 (1998)

D.G. Welsch, **W. Vogel** and **T. Opatrny**, *Progress in Optics*, ed. by E. Wolf, North-Holland, Amsterdam, 1998

Quantum State Tomography

The tomography experiment was first performed by a group at the University of Oregon, see

D.T. Smithey, M. Beck, M.G. Raymer and **A. Faridani**, *Measurement of the Wigner distribution and the density matrix of a light mode using optical homodyne tomography: application to squeezed states and the vacuum*, Phys. Rev. Lett. **70**, 1244–1247 (1993)

For a whole gallery of measured Wigner functions see

G. Breitenbach, S. Schiller and **J. Mlynek**, *Measurement of the quantum states of squeezed light*, Nature **387**, 471–475 (1997)

For proposals of measuring the Wigner function see

A. Royer, *Measurement of the Wigner function*, Phys. Rev. Lett. **55**, 2745–2748 (1985)

K. Vogel and **H. Risken**, *Determination of quasiprobability distributions in terms of probability distributions for the rotated quadrature phase*, Phys. Rev. A **40**, 2847–2849 (1989)

Closely related to this is the work by

W. Vogel and **W.P. Schleich**, *Phase distribution of a quantum state without using phase states*, Phys. Rev. A **44**, 7642–7646 (1991)

The method of quantum state tomography relies on the inverse Radon transform. For the original paper on the Radon transform see

J. Radon, *Über die Bestimmung von Funktionen durch ihre Integralwerte längs gewisser Mannigfaltigkeiten*, Ber. Verh. Sächs. Akad. Wiss. Leipzig, Math.-Nat. Kl. **69**, 262–277 (1917)

This paper is also reprinted in

J. Radon, in *75 Years of Radon Transform*, ed. by S. Gindikin and P. Michor, International Press, Boston, 1994

5 Waves à la WKB

How can we describe the motion of a non-relativistic particle in a binding potential? We may choose classical mechanics as our framework and talk about phase space trajectories; that is one possibility. To find the energy wave functions of this oscillator represents the quantum mechanical counterpart of this classical approach. Unfortunately, an analytical treatment of the corresponding Schrödinger equation is limited to a very few special potentials such as the harmonic oscillator, the Morse potential, and a few more. For the most part we have to resort to numerical solutions. However, the analytical as well as the numerical solutions often hide striking and remarkable properties of the problems at hand. Such hidden features only come to light in the semiclassical limit of quantum mechanics – the topic of the present section.

This approach is associated with the names of G. Wentzel, H. Kramers and L. Brillouin who in 1926 just after the invention of wave mechanics pioneered the field of approximate solutions of the time independent Schrödinger equation. In the present chapter we briefly summarize this approach and note that it has proven to be extremely useful in the field of quantum optics and quantum chaos.

5.1 Probability for Classical Motion

Consider a particle with mass M and energy E moving in the one-dimensional potential $U = U(x)$ shown in Fig. 5.1(a). We describe its motion by the coordinate x and the momentum p. As outlined in Chapter 3 the appropriate comparison between classical and quantum mechanics relies on statistical mechanics and the use of an ensemble of classical particles. Therefore, when we refer to a single classical particle we in this chapter mean an ensemble of classical particles.

Conservation of energy, expressed here as,

$$E = \frac{p^2(x)}{2M} + U(x) \tag{5.1}$$

provides immediately the classical x-p oscillator phase space trajectory

$$p^{(\text{cl})}(x; E) \equiv \pm p(x; E) \equiv \pm\sqrt{2M[E - U(x)]} \tag{5.2}$$

illustrated in Fig. 5.1(b).

154   5 Waves à la WKB

When we have no knowledge of the initial position $x_0 \equiv x|_{t=0}$ of the particle nor of its momentum $p_0 \equiv p|_{t=0}$, but only of its energy E, we associate with this motion in phase space a classical probability $W^{(\text{cl})}(x; E)$ to find the particle at the position x. This probability is governed by the reciprocal of the classical momentum $p(x; E)$, Eq. (5.2). In other words, $W^{(\text{cl})}(x; E)$ is large where the particle's velocity is small, but it is small when p assumes large values. More precisely we define

$$W^{(\text{cl})}(x; E) = \frac{1}{2} \mathcal{N}^2 \frac{1}{p(x; E)}. \tag{5.3}$$

At the turning points of the motion, denoted by ϑ and ξ, the momentum p,

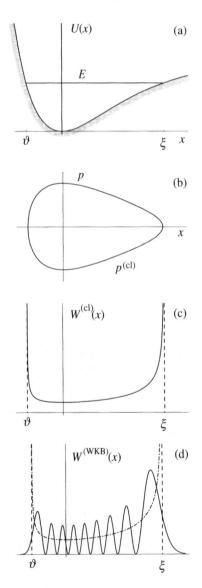

Fig. 5.1: A particle of mass M and energy E vibrating in the potential $U = U(x)$ between the turning points ϑ and ξ shown in (a) traverses a closed circuit in oscillator phase space constructed out of coordinate x and momentum p. This phase space trajectory depicted in (b) is given by the classical momentum $p^{(\text{cl})}(x) = \pm p(x; E)$. The classical probability of finding the particle at position x, denoted $W^{(\text{cl})}(x; E)$, governed by the inverse of p and shown in (c), exhibits only a single minimum, namely at the minimum of U. Here the velocity of the particle is at its maximal value. The probability diverges at the turning points ϑ and ξ of the classical motion where the momentum vanishes. The area underneath the curve is normalized to unity. The semiclassical probability $W^{(\text{WKB})}(x)$, that is the solid curve of (d), displays many maxima and minima as a result of *interference* between right- and left-going waves, that is, as a consequence of the standing wave corresponding to the energy eigenfunction. The quantity $W^{(\text{WKB})}(x)$ does not diverge at the turning points, but exhibits Airy function-type maxima in the neighborhood of ϑ and ξ. Moreover, it reaches into the classically forbidden regime. The envelope of $W^{(\text{WKB})}(x)$ is *twice* the classical probability curve, depicted in (d) by the dashed curve, that is $2W^{(\text{cl})}(x)$, to maintain the normalization $\int dx\, W^{(\text{WKB})}(x) = 1$, in the presence of the interference-induced oscillations. Classical probability governed by the inverse of the momentum and interference between waves constitute the essential ingredients of semiclassical quantum theory.

Eq. (5.2), vanishes and hence $W(x; E)$, Eq. (5.3), is infinite as shown in Fig. 5.1(c). Nevertheless, $W^{(\text{cl})}(x; E)$ is still normalizable, that is

$$\int_{\vartheta}^{\xi} dx\, W^{(\text{cl})}(x; E) = 1. \tag{5.4}$$

When we substitute the definition Eq. (5.3) of the classical probability $W^{(\text{cl})}$ into the normalization condition Eq. (5.4) and note that $p(x) = M dx/dt$, we can determine the normalization constant $\mathcal{N}$ by

$$1 = \frac{1}{2}\mathcal{N}^2 \int_{\vartheta}^{\xi} \frac{dx}{p(x; E)} = \frac{1}{2}\frac{\mathcal{N}^2}{M}\int_0^{T/2} dt = \frac{1}{4}\frac{\mathcal{N}^2}{M}\oint dt = \frac{1}{4}\frac{\mathcal{N}^2}{M} T$$

that is,

$$\mathcal{N} = 2\left(\frac{M}{T}\right)^{1/2}, \tag{5.5}$$

where $T \equiv \oint dt$ denotes the period of the orbit corresponding to the energy E. To identify the reciprocal of the classical momentum as a measure of the position probability – that is the message of this subsection.

5.2 Probability Amplitudes for Quantum Motion

We achieve the transition from classical mechanics to quantum mechanics by multiplying the expression Eq. (5.1) for the conservation of energy, by the energy wave function $u(x)$, and by replacing the momentum p by the differential operator $(\hbar/i)\times d/dx$. This prescription yields

$$\frac{d^2 u}{dx^2} + \frac{2M}{\hbar^2}[E - U(x)]u = 0. \tag{5.6a}$$

With the help of the definition Eq. (5.2) of the classical momentum p the Schrödinger equation reads

$$u''(x) + \left[\frac{p(x)}{\hbar}\right]^2 u(x) = 0. \tag{5.6b}$$

Here, and in the remainder of this chapter, a prime denotes a differentiation with respect to the position coordinate x.

We now derive an approximate solution of the above Schrödinger equation which is valid between the two turning points. Here we do not follow the standard derivation discussed in many textbooks on quantum mechanics. This approach expands the phase of the wave function in powers of Planck's constant. We rather try to find the solution by combining the classical probability concept discussed in the preceding section with the idea of interference of probability amplitudes.

5.2.1 An Educated Guess

To gain insight into an approximate solution of the time independent Schrödinger equation we first consider the special case of a constant potential $U(x) = U_0$. Here the classical momentum Eq. (5.2) is constant, that is

$$p(x;E) = \sqrt{2M(E-U_0)} \equiv p_0.$$

Therefore, we can integrate the Schrödinger equation Eq. (5.6b) in an exact way and find

$$u(x) = N \cos\left[\frac{1}{\hbar}\int_x^\xi d\tilde{x}\, p_0 - \alpha\right]$$

where N and α are constants.

Wave Ansatz

Guided by this example, we now approach the case of a non-constant classical momentum p by proposing the *ansatz*

$$u^{(\text{wave})}(x) \equiv \cos\left[\frac{1}{\hbar}\int_x^\xi d\tilde{x}\, p(\tilde{x};E) - \alpha\right] \tag{5.7}$$

which satisfies the differential equation

$$\left[u^{(\text{wave})}\right]'' + \left\{\left(\frac{p}{\hbar}\right)^2 - \left(\frac{p}{\hbar}\right)' \tan\left[\frac{1}{\hbar}\int_x^\xi d\tilde{x}\, p(\tilde{x};E) - \alpha\right]\right\} u^{(\text{wave})} = 0,$$

an equation similar to the Schrödinger equation (5.6).

Hence, the *ansatz* of Eq. (5.7) is close to the real solution if we can neglect the second contribution in the bracket. Nevertheless, our *ansatz* exhibits a more serious deficiency than just not being able to solve the Schrödinger equation Eq. (5.6) in an exact way. It does not satisfy the correspondence principle: Indeed, in the limit of large quantum numbers, that is large energies, the results of quantum mechanics *must* approach those of classical mechanics. In particular, the quantum mechanical probability of finding the particle of energy E at the position x, given by

$$W(x) \equiv [u(x)]^2, \tag{5.8}$$

must transmute itself into Eq. (5.3) in this limit.

Particle Ansatz

The correspondence principle and the comparison between the classical and quantum mechanical probabilities, Eqs. (5.3) and (5.8) respectively, suggest an *ansatz* quite different from the wave *ansatz*, Eq. (5.7). This is the conjecture

$$u^{(\text{cl})}(x) = \frac{\mathcal{N}}{\sqrt{p(x;E)}} = \mathcal{N}e^{-\frac{1}{2}\ln p}. \tag{5.9}$$

5.2 Probability Amplitudes for Quantum Motion

Here the constant $\mathcal{N}$ is defined as in Eq. (5.5).

Unfortunately, $u^{(\text{cl})}$ does not satisfy the Schrödinger equation (5.6) either but leads instead to the rather complicated equation

$$\left[u^{(\text{cl})}\right]'' + \left\{\frac{1}{2}(\ln p)'' - \left[\frac{1}{2}(\ln p)'\right]^2\right\} u^{(\text{cl})} = 0.$$

This discussion shows most clearly that we are not allowed to describe the motion of the particle via a wave solution, à la Eq. (5.7). Nor are we entitled to apply the particle ansatz Eq. (5.9) motivated by the classical probability concept.

Quantum Mechanical Particle: Both Wave and Particle

The function which describes this feature best is the *product* of the wave *ansatz*, Eq. (5.7), and of the particle *ansatz*, Eq. (5.9). Indeed, the function

$$u^{(\text{WKB})}(x) \equiv u^{(\text{cl})}(x)\, u^{(\text{wave})}(x) = \frac{\mathcal{N}}{\sqrt{p(x)}} \cos\left[\frac{1}{\hbar}\int_x^\xi d\tilde{x}\, p(\tilde{x}) - \alpha\right] \quad (5.10)$$

satisfies the Schrödinger-type equation

$$\left[u^{(\text{WKB})}\right]'' + \left\{\left(\frac{p}{\hbar}\right)^2 - \left(\frac{1}{\sqrt{p}}\right)'' \sqrt{p}\right\} u^{(\text{WKB})} = 0, \quad (5.11)$$

as well as the correspondence principle.

To test this we calculate the probability of finding the particle at the position x, namely

$$W^{(\text{WKB})}(x) \equiv \left[u^{(\text{cl})}(x)\right]^2 \left[u^{(\text{wave})}(x)\right]^2 \quad (5.12)$$

which follows from Eqs. (5.8) and (5.10). We first note that also this ansatz does not lead at once to the classical result in the correspondence limit. Indeed no matter how large the energy the probability always displays oscillations as a function of position as shown in Fig. 5.1(d). However, when we now average this probability over the rapid oscillations of $u^{(\text{wave})}$, we arrive at the averaged probability function

$$\overline{W}^{(\text{WKB})}(x) \equiv \left[u^{(\text{cl})}(x)\right]^2 \left\langle\left[u^{(\text{wave})}(x)\right]^2\right\rangle_x$$

$$= \frac{\mathcal{N}^2}{p(x)} \frac{1}{\xi - \vartheta} \int_\vartheta^\xi dx\, \frac{1}{2}\left\{1 + \cos\left[2\int_x^\xi dx\, p(x) - 2\alpha\right]\right\} \quad (5.13)$$

which then reduces to

$$\overline{W}^{(\text{WKB})}(x) \cong \frac{1}{2}\frac{\mathcal{N}^2}{p(x)} = W^{(\text{cl})}(x).$$

In the last step we have made use of the definition Eq. (5.3) of the classical probability.

Due to the oscillations of $u^{(\text{WKB})}$ resulting from $u^{(\text{wave})}$, the probability $W^{(\text{WKB})}(x)$, Eq. (5.12), never approaches the classical probability Eq. (5.3). It is only the *averaged* probability $\overline{W}^{(\text{WKB})}(x)$ which satisfies the correspondence principle. In this sense the classical limit of quantum mechanics is very different from the limit of special relativity leading to Galilean relativity in the limit of small velocities. In the latter case it is enough to let a parameter such as the ratio of the velocity of the particle and the light velocity go to zero and the equations of special relativity reduce to the Newtonian equations. In quantum mechanics no such uniform limit exists. We need an additional averaging procedure.

5.2.2 Range of Validity of WKB Wave Function

Our educated guess suggests that

$$u^{(\text{WKB})}(x) = \frac{\mathcal{N}}{\sqrt{p(x)}} \cos\left[\frac{1}{\hbar}\int_x^\xi d\tilde{x}\, p(\tilde{x}) - \alpha\right]$$

is an approximate solution of the one-dimensional Schrödinger equation for an energy eigenstate. But how good is this approximation and what are the limits of applications?

To answer this question we now compare the Schrödinger equation Eq. (5.6b) to the equation for $u^{(\text{WKB})}$, Eq. (5.11), and find that the expression $u^{(\text{WKB})}$ is a solution of the Schrödinger equation, provided that

$$p^2 \gg \sqrt{p}\left|\left(\frac{1}{\sqrt{p}}\right)''\right|\hbar^2. \tag{5.14}$$

When we differentiate the classical momentum, Eq. (5.2), we find

$$\left(\frac{1}{\sqrt{p}}\right)' = \frac{M}{2}\left(\frac{1}{\sqrt{p}}\right)^5 U'$$

and hence

$$\left(\frac{1}{\sqrt{p}}\right)'' = \frac{5M^2}{4}\left(\frac{1}{\sqrt{p}}\right)^9 U'^2 + \frac{M}{2}\left(\frac{1}{\sqrt{p}}\right)^5 U''.$$

We substitute these expressions into Eq. (5.14), which yields the validity condition

$$\frac{\hbar^2}{2M}\left|\frac{5}{16}\frac{[U'(x)]^2}{[E-U(x)]^3} + \frac{1}{4}\frac{U''(x)}{[E-U(x)]^2}\right| \ll 1 \tag{5.15}$$

of the WKB solution Eq. (5.10).

Clearly the condition is violated in the neighborhood of the turning points ϑ and ξ, where $U(\vartheta) = U(\xi) = E$. Hence, the WKB solution, Eq. (5.10), can only be valid in the domain between the two turning points, that is, in the region where $\vartheta \ll x \ll \xi$.

Moreover, according to Eq. (5.15), the potential U – or rather the rate of change of U – puts constraints on the applicability of the WKB technique. Only slowly varying potentials are allowed. Potentials with cusps, that is, with sharp corners, lie outside of this approach and have to be treated by a different method – a WKB technique which can "grapple" with a corner. For a more detailed discussion we refer to the literature at the end of this chapter. However, we emphasize that in general the WKB wave functions are in excellent agreement with the exact wave functions.

5.3 Energy Quantization

We now return to the construction of the WKB solution which with the normalization Eq. (5.5) reads

$$u^{(\text{WKB})}(x; E) = 2 \left(\frac{M}{T\, p(x; E)} \right)^{1/2} \cos\left[\frac{1}{\hbar} \int_x^\xi d\tilde{x}\, p(\tilde{x}; E) - \alpha \right]. \tag{5.16}$$

In order for the solution Eq. (5.16) to represent an energy eigenfunction, we have to fix the two remaining free constants: the energy E and the phase α. We deduce these quantities by approximately solving the Schrödinger equation in the neighborhood of the turning points ϑ and ξ. When we find the asymptotic continuation of this approximate solution into the oscillatory domain of $u^{(\text{WKB})}$, that is the domain where $\vartheta < x < \xi$, this function reduces to the expression of Eq. (5.16) – provided $\alpha = \pi/4$. We now show this explicitly.

5.3.1 Determining the Phase

In order to find the phase α we now expand the potential $U(x)$ into a Taylor series

$$U(x) = U(\xi) + \left.\frac{dU}{dx}\right|_{x=\xi} (x - \xi) + \ldots \cong E + U' \cdot (x - \xi)$$

around the right turning point ξ including the linear term only. Hence, in the neighborhood of ξ the Schrödinger equation (5.6) of an energy wave function reads

$$u''(x) - k^3 (x - \xi)\, u(x) = 0 \tag{5.17}$$

where we have introduced the quantity

$$k \equiv \left(\frac{2MU'}{\hbar^2} \right)^{1/3}.$$

We emphasize that k has the units of one over length and therefore plays the role of a wave number.

This allows us to cast the Schrödinger equation in dimensionless variables $y \equiv kx$, that is,

$$\frac{d^2 u}{d(kx)^2} - k(x - \xi)\, u = 0 \tag{5.18}$$

This equation shows that the behavior of an energy wave function at the turning point follows from a Schrödinger equation in a linear potential. The specific form of the potential only enters through its first derivative at the turning point.

Schrödinger Equation in a Linear Potential

We therefore have to solve the time independent Schrödinger equation in a linear potential. G. Breit was the first who expressed the solution in terms of the Airy function

$$\text{Ai}(y) \equiv \frac{1}{2\pi} \int_{-\infty}^{\infty} dt \, \exp\left[i\left(\frac{1}{3}t^3 + yt\right)\right]. \tag{5.19}$$

In Appendix E we discuss various properties of the Airy function. In particular, we show that it satisfies the differential equation

$$\frac{d^2 \text{Ai}(y)}{dy^2} - y \, \text{Ai}(y) = 0. \tag{5.20}$$

A comparison between Eq. (5.18) and Eq. (5.20) suggests the solution

$$u(x) = N \, \text{Ai}\left[k(x - \xi)\right], \tag{5.21}$$

where N is a constant.

By differentiating this expression with respect to position and by making use of the differential equation Eq. (5.20) of the Airy function it is easy to verify that the so-defined $u(x)$ satisfies the approximate Schrödinger equation Eq. (5.17) corresponding to the linear potential.

Matching the two Solutions

So far we have obtained an exact solution of the Schrödinger equation in the neighborhood of the turning point ξ. In order to determine the phase α we now have to match that solution with the oscillatory wave function Eq. (5.16) valid appropriately away from the turning points.

For this purpose we recall from Appendix E the appropriate asymptotic expansion of the Airy function. From the solution Eq. (5.21) we recognize that a positive argument of the Airy function corresponds to the classically forbidden domain of the potential whereas a negative argument describes the classically allowed regime

Since we want to match the WKB wave in the allowed regime, Eq. (5.16), with the Airy function expression at the turning point we use the asymptotic expansion

$$\text{Ai}(z) \cong \frac{1}{\sqrt{\pi}} |z|^{-1/4} \cos\left(\frac{2}{3}|z|^{3/2} - \frac{\pi}{4}\right) \tag{5.22}$$

valid for negative arguments z where $z \equiv -k(\xi - x)$.

Therefore, the expression for the wave function, Eq. (5.21), close to the turning point takes the form

$$u(x) \cong N \frac{1}{\sqrt{\pi}} [k(\xi - x)]^{-1/4} \cos\left[\frac{2}{3}[k(\xi - x)]^{3/2} - \frac{\pi}{4}\right]. \tag{5.23}$$

We now compare this result to the WKB-wave function

$$u^{(\text{WKB})}(x) \cong 2 \left(\frac{M}{Tp(x)}\right)^{1/2} \cos\left[\frac{1}{\hbar} \int_x^\xi d\tilde{x}\, p(\tilde{x}) - \alpha\right]$$

5.3 Energy Quantization

given by Eq. (5.16). In order to obtain a correct comparison between the two wave functions we have to linearize the potential in the WKB wave function. Indeed, we have derived the Airy function result by linearizing the potential.

For this purpose we consider the approximate momentum

$$p(x) \equiv \sqrt{2M[E - U(x)]} = \sqrt{2MU'(x)}\sqrt{\xi - x} = \hbar k^{3/2}\sqrt{\xi - x} \qquad (5.24)$$

associated with a linear potential and calculate the phase

$$S(x) \equiv \frac{1}{\hbar}\int_x^\xi d\tilde{x}\, p(\tilde{x}) = \frac{1}{\hbar}\int_x^\xi d\tilde{x}\, \sqrt{2M[E - U(\tilde{x})]}$$

which takes the form

$$S(x) \cong k^{3/2}\int_x^\xi d\tilde{x}\, \sqrt{\xi - \tilde{x}} = \frac{2}{3}[k(\xi - x)]^{3/2}. \qquad (5.25)$$

Hence, the WKB wave function reads

$$u^{(\mathrm{WKB})} \cong 2\left(\frac{M}{T\hbar k}\right)^{1/2}[k(\xi - x)]^{-1/4}\cos\left[\frac{2}{3}[k(\xi - x)]^{3/2} - \alpha\right]. \qquad (5.26)$$

A comparison between the WKB wave function Eq. (5.26) and the wave function Eq. (5.23) extended into the oscillatory regime starting from the turning point therefore yields

$$\alpha = \frac{\pi}{4}$$

and

$$N \equiv \sqrt{\pi}\, 2\left(\frac{M}{T\hbar k}\right)^{1/2}.$$

Accordingly, we know that at the turning points the WKB wave has a phase of $-\pi/4$ as illustrated in Fig. 5.2. This is the simplest form of the so called Maslov index which arises when we perform a WKB analysis in higher dimensional systems. For a more detailed discussion we refer to the literature mentioned in the references.

Moreover, this analysis also allows us to find the complete wave function

$$u(x) \cong \sqrt{\pi}\, 2\left(\frac{M}{T\hbar k}\right)^{1/2}\mathrm{Ai}\left[-k(\xi - x)\right]. \qquad (5.27)$$

in the neighborhood of the turning point.

5.3.2 Bohr-Sommerfeld-Kramers Quantization

In the preceding subsection we have found the phase of the WKB wave function by matching it with the Airy function expanded from the right turning point ξ. Of course we can also apply the same procedure to the left turning point ϑ. This provides another expansion into the oscillatory regime of the energy wave function. Obviously the expansions from the left and from the right have to lead to identical

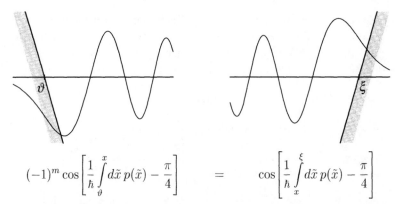

Fig. 5.2: The phase of a WKB wave function at the turning points ϑ and ξ of the classical motion is $-\pi/4$, that is, $u_m^{(\text{WKB})}(x=\vartheta,\xi) \propto \cos(-\pi/4)$. Hence, the oscillatory part of $u_m^{(\text{WKB})}$ reads $\cos(\int_\vartheta^x dx\, p(x)/\hbar - \pi/4)$ or $\cos(\int_x^\xi dx\, p(x)/\hbar - \pi/4)$. To guarantee the uniqueness of the WKB wave function $u_m^{(\text{WKB})}$ at any point x within the classically allowed region, the value of $u_m^{(\text{WKB})}(x)$ obtained by expanding from the right turning point ξ must be identical to the one found from an expansion starting from the left turning point ϑ. Odd quantum numbers imply an odd number of zeros. Hence, the wave function in the neighborhood of ϑ has a different sign than at ξ. We have to incorporate an additional phase change of $(-1)^m = \cos(m\pi)$ in the uniqueness condition which leads to the energy quantization $\oint dx\, p(x) = 2\pi\hbar(m+1/2)$.

results at any point in between. This is the condition that enforces the quantization of energies.

Odd quantum numbers imply an odd number of zeros. Hence, the wave function of odd quantum numbers has a different sign in the neighborhood of ϑ than at ξ. This leads to the uniqueness condition

$$(-1)^m \cos\left[\frac{1}{\hbar}\int_\vartheta^x d\tilde{x}\, p(\tilde{x}) - \frac{\pi}{4}\right] = \cos\left[\frac{1}{\hbar}\int_x^\xi d\tilde{x}\, p(\tilde{x}) - \frac{\pi}{4}\right].$$

When we combine the prefactor $(-1)^m$ into the phase of the wave

$$\cos\left[\frac{1}{\hbar}\int_\vartheta^x d\tilde{x}\, p(\tilde{x}) - m\pi - \frac{\pi}{4}\right] - \cos\left[\frac{1}{\hbar}\int_x^\xi d\tilde{x}\, p(\tilde{x}) - \frac{\pi}{4}\right] = 0$$

and make use of the relation

$$\cos\alpha - \cos\beta = 2\sin\frac{\beta+\alpha}{2}\sin\frac{\beta-\alpha}{2}$$

we arrive at

$$2\sin\left\{\frac{1}{2}\left[\frac{1}{\hbar}\int_\vartheta^\xi dx\, p(x) - m\pi - \frac{\pi}{2}\right]\right\}\sin\left\{\frac{1}{2}\left[\frac{1}{\hbar}\int_\vartheta^x d\tilde{x}\, p(\tilde{x}) - \frac{1}{\hbar}\int_x^\xi d\tilde{x}\, p(\tilde{x}) - m\pi\right]\right\} = 0$$

Since this equation has to be valid for arbitrary x the first sine must vanish which yields

$$\frac{1}{2}\left[\frac{1}{\hbar}\int_{\vartheta}^{\xi} dx\, p(x) - m\pi - \frac{\pi}{2}\right] = 0.$$

This is the Bohr-Sommerfeld-Kramers quantization condition

$$2\int_{\vartheta}^{\xi} dx\, p(x) \equiv \oint dx\, p(x) \equiv J = 2\pi\hbar\left(m+\frac{1}{2}\right).$$

We note that the action integral J is quantized in terms of half integers of $2\pi\hbar$. This quantization in terms of half integers rather than integers results from the phase shift $\alpha = \pi/4$ due to the behavior of the Airy function approximation of the wave function. This becomes clear when we trace back the origin of the factor $1/2$. It was H. Kramers who derived this result in 1926.

This quantization of action implies quantization of energy. Indeed, when we recall that the momentum $p(x)$ contains the energy E as a free parameter we realize that only specific energies E_m can meet the Bohr-Sommerfeld-Kramers condition

$$\oint dx\, p_m(x) \equiv \oint dx\, p(x; E_m) \equiv \oint dx\, \sqrt{2M[E_m - U(x)]} = 2\pi\hbar\left(m+\frac{1}{2}\right).$$

We conclude this section by noting that the WKB method of finding energy wave functions cannot be applied to more dimensional non-separable systems. Examples include the hydrogen atom in a strong constant magnetic field or billiard problems. These are topics of quantum chaos studies. Here, semiclassical techniques turn out to be extremely useful. However, this goes beyond the scope of this book.

5.4 Summary

We conclude this discussion of the WKB wave function by summarizing the important results. In particular, we show the various steps that lead to the WKB wave function of the mth energy eigenstate and then briefly discuss the uniform asymptotic expansion of the wave function.

5.4.1 Construction of Primitive WKB Wave Function

In order to construct the WKB wave function we start from the conservation of energy

$$E = \frac{p^2(x)}{2M} + U(x)$$

of a particle of mass M moving in the binding potential $U(x)$. We solve for the classical momentum

$$p(x; E) \equiv \sqrt{2M[E - U(x)]} \tag{5.28}$$

and substitute this expression into the Bohr-Sommerfeld-Kramers quantization condition

$$2\int_{\vartheta}^{\xi} dx\, p(x; E) = \oint dx\, p(x; E) = 2\pi\hbar\left(m + \frac{1}{2}\right).$$

and solve for the energy E. This procedure provides discrete energies E_m and through the definition of the momentum, Eq. 5.28, we find the quantized momentum

$$p_m(x) \equiv p(x; E = E_m) = \sqrt{2M[E_m - U(x)]}$$

of the particle in the mth eigenstate with energy E_m.

The turning points ϑ_m and ξ_m of the mth orbit follow from the condition $E_m = U(\vartheta_m) = U(\xi_m)$. This allows us to calculate the period

$$T_m \equiv M \oint \frac{dx}{p_m(x)} = 2M \int_{\vartheta_m}^{\xi_m} \frac{dx}{p_m(x)}$$

of the mth energy eigenstate. Moreover, we can find the phase

$$\varphi_m(x) \equiv S_m(x) - \frac{\pi}{4} \equiv \frac{1}{\hbar}\int_x^{\xi_m} d\tilde{x}\, p_m(\tilde{x}) - \frac{\pi}{4}, \tag{5.29}$$

of the WKB wave at the position x.

Hence, the WKB wave function of the mth energy eigenstate reads

$$u_m^{(\text{WKB})}(x) \equiv 2\left(\frac{M}{T_m\, p_m(x)}\right)^{1/2} \cos\left[\frac{1}{\hbar}\int_x^{\xi_m} d\tilde{x}\, p_m(\tilde{x}) - \frac{\pi}{4}\right]. \tag{5.30}$$

Note that this expression is only valid inside the oscillatory regime of the wave function and appropriately away from the turning points. At the turning point the momentum vanishes, and hence, the WKB wave function becomes infinite. Moreover, we have set the value of the wave function to be zero in the classically forbidden regime. Since this approximation of the exact wave function is a rather primitive one this form of the WKB wave function is called *primitive WKB wave function*.

5.4.2 Uniform Asymptotic Expansion

We can overcome the problem of the singularity of the primitive WKB wave function at the turning point by making use of the Airy function solution Eq. (5.27). Moreover, we can use the asymptotic expression of the Airy function for positive arguments derived in Appendix E to find a simple expression for the wave function in the forbidden regime as discussed in Problem 5.1.

In this way we can represent the wave function in various regimes of position space by simple expressions. However, these expressions are only valid in these particular domains. The so-called uniform asymptotic expansion provides an expression that is uniformly valid through the whole regime. Here we do not derive this expression but rather motivate it.

Oscillatory Regime

For positions left of the right turning point ξ_m, that is for $x \leq \xi_m$ this solution reads

$$u_m(x) = 2\left(\frac{\pi M}{T_m p_m(x)}\right)^{1/2} \left[\tfrac{3}{2} S_m(x)\right]^{1/6} Ai\left(-\left[\tfrac{3}{2} S_m(x)\right]^{2/3}\right). \tag{5.31}$$

Indeed, by substituting the asymptotic expression of the Airy function in the oscillatory regime we can verify that this formula reduces to the primitive WKB wave function. Moreover, for positions close to the turning point we can use the asymptotic expressions Eqs. (5.24) and (5.25) for the momentum $p_m(x)$ and the phase $S_m(x)$ to arrive at the the wave function Eq. (5.27) valid close to the turning point. Hence, this formula combines both limiting regimes in one expression.

Classically Forbidden Regime

For positions in the forbidden regime, that is for $x \geq \xi_m$, the potential $U(x)$ is larger than the energy E. Therefore we have to replace the momentum $p_m(x)$ by

$$\overline{p}_m(x) \equiv \sqrt{2M[U(x) - E]}$$

and the phase $S_m(x)$ by

$$\overline{S}_m(x) \equiv \frac{1}{\hbar} \int_{\xi_m}^{x} d\tilde{x}\, \overline{p}_m(\tilde{x}).$$

The expression for the wave function in the forbidden regime therefore reads

$$u_m(x) = 2\left(\frac{\pi M}{T_m \overline{p}_m(x)}\right)^{1/2} \left[\tfrac{3}{2} \overline{S}_m(x)\right]^{1/6} Ai\left(\left[\tfrac{3}{2} \overline{S}_m(x)\right]^{2/3}\right). \tag{5.32}$$

Again we can convince ourselves that this formula is correct by substituting the asymptotic expansion of the Airy function for large positive arguments into this expression and arriving at the primitive WKB result valid in the classically forbidden regime. Moreover, at the turning point this result reduces to the one obtained from the one valid left of the turning point.

Problems

5.1 Primitive WKB Wave Function in the Forbidden Regime

In order to derive the primitive WKB wave function in the classically forbidden regime where $U(x) > E$ cast the Schrödinger equation Eq. (5.6a) in the form

$$u''(x) - \left[\frac{\overline{p}(x)}{\hbar}\right]^2 u(x) = 0.$$

where $\overline{p} \equiv \sqrt{2M(U(x) - E)}$. Apply the ideas of Sec. 5.2 to find an approximate solution which is decaying away from the turning point. By matching this solution to the one valid at the turning point, Eq. (5.21) find the primitive WKB wave function in the classically forbidden regime.

5.2 Uniform WKB Wave Function

Derive the expressions Eqs. (5.31) and (5.32) for the uniform WKB wave function by starting from the Schrödinger equation Eq. (5.6a) and introducing the phase $S(x)$ as a new variable. Compare the resulting differential equation to the one of the Airy function.

5.3 WKB for the Hydrogen Atom: Langer Transformation

The radial part of the hydrogenic wave functions obeys the equation

$$-\frac{\hbar^2}{2M}\frac{d^2 u}{dr^2} - \frac{e^2}{r}u + \frac{\hbar^2}{2M}\frac{\ell(\ell+1)}{r^2}u = Eu$$

with $\ell = 0, 1, 2, \ldots$ This equation can be interpreted as the one-dimensional Schrödinger equation for a particle in the effective potential

$$U_{\text{eff}}(r) = -\frac{e^2}{r} + \frac{\hbar^2}{2M}\frac{\ell(\ell+1)}{r^2}.$$

(a) What are the energy eigenvalues obtained from the Bohr-Sommerfeld-Kramers quantization rule?
Hint: Introduce the scaled variables

$$\rho = \frac{Me^2}{\hbar^2} r \quad \text{and} \quad \eta = \frac{\hbar^2}{Me^4} E.$$

(b) What happens for $\ell = 0$?
(c) Perform the transformation

$$\rho = e^\xi \qquad u = e^{\xi/2} v$$

and interpret the resulting equation again as the one-dimensional Schrödinger equation for a particle in an effective potential.

(d) What are the energy eigenvalues obtained in this case from the Bohr-Sommerfeld-Kramers quantization rule?

Hint: See for example Langer (1937) and Yost et al. (1939).

5.4 Morse Oscillator

We consider the Morse potential

$$U(x) \equiv D(e^{-2ax} - 2e^{-ax})$$

with $D > 0$, $a > 0$.

(a) Sketch the potential. For which energies would the motion of a classical particle in the potential $U(x)$ be bound?

(b) Find the period of a bound classical particle with energy E in the potential $U(x)$.

(c) What are energy eigenvalues calculated from the Bohr-Sommerfeld-Kramers quantization rule

$$\oint p\,dx = 2\pi\hbar(m+1/2)?$$

(d) Calculate the WKB wave functions.

(e) Sketch the trajectories $p^2/2M + U(x) = E_m$. What is the geometrical interpretation of the phase in the WKB wave functions?

Hint: The resulting integrals reduce to standard integrals with the help of the substitution $u = e^{\pm ax}$.

The Schrödinger equation for the potential $U(x)$ can also be solved exactly see, for example, Landau and Lifshitz: *Quantum Mechanics* (1977). The energy eigenvalues calculated in WKB approximation coincide with the exact values.

5.5 WKB Wave Function of a Harmonic Oscillator

Follow the steps provided in Sec. 5.4 to derive the WKB approximation for the energy wave functions $u_m(x)$ of the harmonic oscillator potential $U(x) \equiv \frac{1}{2}M\Omega^2 x^2$. Compare this result to the asymptotic expression Eq. (4.6) derived in Sec. 4.1.2 by applying the method of steepest descent to the integral representation of the Hermite polynomial.

References

Original WKB Papers

The original papers on the WKB approximation are stimulating to read. See for example

G. Wentzel, *Eine Verallgemeinerung der Quantenbedingungen für die Zwecke der Wellenmechanik*, Z. Phys. **38**, 518–529 (1926)

H.A. Kramers, *Wellenmechanik und halbzahlige Quantisierung*, Z. Phys. **39**, 828–840 (1926)

L. Brillouin, *La méchanique ondulatoire de Schrödinger; une méthode générale de résolution par approximations successives*, C.R. Acad. Sci. Paris **183**, 24–26 (1926)

For the problem of a linear potential see

G. Breit, *The Propagation of Schrödinger Waves in a Uniform Field of Force*, Phys. Rev. **32**, 273–276 (1928)

Introductory articles and books on WKB include

P. Debye, *Wellenmechanik and Korrespondenzprinzip*, Physik. Zeitschr. **28**, 170–174 (1927)

W. Pauli, *Die allgemeinen Prinzipien der Wellenmechanik*, Handbuch der Physik, Vol. 24, edited by H. Geiger and K. Scheel, Springer Verlag, Berlin, 1933

H.A. Kramers, *Quantentheorie des Elektrons und der Stahlung*, Vol. 2, Hand- und Jahrbuch der Chemischen Physik, Eucken-Wolf, Leipzig, 1938

For the appropriate transformation of the variables in the hydrogen atom see for example

R.E. Langer, *On the Connection Formulas and the Solutions of the Wave Equation*, Phys. Rev. **51**, 669–676 (1937)

F.L. Yost, J.A. Wheeler and G. Breit, *Coulomb Wave Functions in Repulsive Fields*, Phys. Rev. **49**, 174–189 (1939)

J. Hainz and H. Grabert, *Centrifugal terms in the WKB approximation and semiclassical quantization of hydrogen*, Phys. Rev. A **60**, 1698–1701 (1999)

Advanced WKB

For more advanced literature we refer to

N. Fröman and P.O. Fröman, *JWKB Approximation: Contributions to the Theory*, North-Holland, Amsterdam, 1958

J. Heading, *An Introduction to Phase Integral Methods*, Methuen, London, 1962

W. Wasow, *Asymptotic Expansions for Ordinary Differential Equations*, Wiley, New York, 1965

M.V. Berry and K.E. Mount, *Semiclassical Approximations in Wave Mechanics*, Rep. Prog. Phys. **35**, 315–397 (1972)

J.A. Wheeler, in: *Studies in Mathematical Physics*, ed. by E.H. Lieb, B. Simon and A.S. Wightman, Princeton University Press, Princeton, 1976

N.G. de Bruijn, *Asymptotic Methods in Analysis*, North-Holland, Amsterdam, 1985

M.V. Berry, *Semi-classical mechanics in phase space: a study of Wigner's function*, Phil. Trans. R. Soc. London **287 A**, 237–271 (1977)

M.V. Berry, *Quantum scars of classical closed orbits in phase space*, Proc. R. Soc. London **423 A**, 219–231 (1989)

M.V. Berry, *Fringes decorating anticaustics in ergodic wave functions*, Proc. R. Soc. London **424**, 279–288 (1989)

For a treatment of potentials with sharp corners using WKB see

J. Bestle, W.P. Schleich and J.A. Wheeler, *Anti-Stealth: WKB grapples with a corner*, Appl. Phys. B **60**, 289–299 (1995)

Uniform Asymptotic Expansion

For a summary of the uniform asymptotic expansion see

C. Chester, B. Friedmann and F. Ursell, *An extension of the method of steepest descents*, Proc. Camb. Philos. Soc. **53**, 599–611 (1957)

N. Bleistein and R.A. Handelsmann, *Asymptotic Expansions of Integrals*, Holt, Rinehard and Winston, New York, 1975

WKB and Einstein-Brillouin-Keller (EBK) Quantization

A nice introduction into higher dimensional WKB methods and EBK quantization can be found in
I.C. Percival, *Semiclassical theory of bound states*, Adv. Chem. Phys. **36**, 1–61 (1977)

For applications of WKB techniques in higher dimensions and, in particular, in chaotic systems such as the hydrogen atom in a magnetic field or a stadium problem see
H. Friedrich, *Theoretical atomic physics*, Springer-Verlag, Heidelberg, 1998
V.P. Maslov and **M.V. Fedoriuk**, *Semi-Classical Approximation in Quantum Mechanics*, Reidel, Boston, 1981

6 WKB and Berry Phase

In Chapter 4 we have generated coherent and squeezed states of a harmonic oscillator such as a pendulum by *suddenly* displacing the point of suspension or by changing its length. In this way we have gone from the ground state of the pendulum to states which are not energy eigenstates. What happens when we consider the other extreme, that is, when we start from the ground state and now *slowly* change the length of the pendulum?

In this context slowly means slow compared to all time scales of the system, that is we are considering adiabatic changes. Hence, for every moment of time there exist instantaneous energy eigenstates $|m(t)\rangle$. Paul Ehrenfest in the early days of quantum mechanics discovered that under adiabatic changes the pendulum stays in such an instantaneous eigenstate. However, it acquires a phase. This phase consists of two parts: *(i)* A dynamical phase which originates from the fact that a stationary state undergoes unitary time evolution and *(ii)* a geometrical phase which is related to the topology of parameter space. The latter is the so-called Berry phase.

The energy wave function of a binding potential follows from the time independent Schrödinger equation. When the potential is slowly changing in position we can approximate the wave function by the WKB wave. This behavior is reminiscent of the above discussion of adiabatic changes. Indeed, there is a close connection between the WKB wave function and the Berry phase. We have already seen that the WKB wave function contains a phase which changes continuously over many multiples of 2π as we are moving from one turning point to the other. However, at the turning point the phase jumps by $-\pi/2$.

In the present chapter we show that these two phases in the WKB wave function can indeed be interpreted as a dynamical and a topological phase. For this purpose in Sec. 6.1 we give a brief introduction to the concept of the Berry phase and then rederive in Sec. 6.2 the WKB wave function in a language which brings out most clearly the analogy to the Berry phase.

Moreover, the requirement of an adiabatic change is not necessary. Following Y. Aharonov and J. Anandan we can also define a Berry phase for non-adiabatic changes. We conclude this chapter by outlining this idea in Sec. 6.3. In particular, we make the connection with the time evolution of a quantum state in a harmonic oscillator discussed in Chapter 4.

6.1 Berry Phase and Adiabatic Approximation

In the present section we consider the time evolution of a quantum state when the Hamiltonian $\hat{H}$ of the system depends on a time dependent external parameter $\vec{R}(t)$, that is $\hat{H} = \hat{H}[\vec{R}(t)]$. In particular, we focus on the case when $\vec{R}(t)$ describes a closed loop in parameter space such that $\vec{R}(t=0) = \vec{R}(t=T)$. Here T denotes the period of this cycle. We calculate the phase acquired during this cyclic evolution.

For this purpose we first discuss the adiabatic approximation. We find that under adiabatic time evolution a quantum system initially prepared in an energy eigenstate remains in an instantaneous energy eigenstate but acquires a phase.

6.1.1 Adiabatic Theorem

The time evolution of any quantum state $|\Psi(t)\rangle$ follows from the Schrödinger equation

$$i\hbar \frac{d}{dt} |\Psi(t)\rangle = \hat{H}[\vec{R}(t)] |\Psi(t)\rangle. \tag{6.1}$$

An instantaneous energy eigenstate $|n[\vec{R}(t)]\rangle$ of energy $E_n[\vec{R}(t)]$ to the Hamiltonian $\hat{H}[\vec{R}(t)]$ is defined by

$$\hat{H}[\vec{R}(t)] \left|n[\vec{R}(t)]\right\rangle = E_n[\vec{R}(t)] \left|n[\vec{R}(t)]\right\rangle. \tag{6.2}$$

We now consider the time evolution of an arbitrary state $|\Psi(t=0)\rangle$ and expand it into instantaneous energy eigenstates according to the relation

$$|\Psi(t=0)\rangle = \sum_n \Psi_n(t=0) |n[\vec{R}(t=0)]\rangle. \tag{6.3}$$

In order to solve the time dependent Schrödinger equation Eq. (6.1) we make the ansatz

$$|\Psi(t)\rangle = \sum_n \Psi_n(t) \exp[-i\phi_n^{(d)}(t)] |n[\vec{R}(t)]\rangle. \tag{6.4}$$

Here we have included the dynamical phase

$$\phi_n^{(d)}(t) \equiv \frac{1}{\hbar} \int_0^t dt' E_n[\vec{R}(t')] \tag{6.5}$$

which is the obvious generalization of the usual phase factor $E_n t/\hbar$ to a time dependent system.

We find the differential equation for the expansion coefficients $\Psi_n(t)$ by substituting the ansatz Eq. (6.4) into the Schrödinger equation Eq. (6.1) which yields

$$i\hbar \frac{d}{dt} |\Psi(t)\rangle = i\hbar \sum_n \left\{ \left[\left(\frac{d}{dt} \Psi_n \right) + \Psi_n \left(-i \frac{d}{dt} \phi_n^{(d)} \right) \right] \exp\left[-i\phi_n^{(d)}\right] |n\rangle \right. \\ \left. + \Psi_n \exp\left[-i\phi_n^{(d)}\right] \left(\frac{d}{dt} |n\rangle \right) \right\}.$$

Here we have suppressed for the sake of simplicity the argument $\vec{R}(t)$ of the instantaneous energy eigenstate $|n\rangle$. We emphasize, however, that these states depend

6.1 Berry Phase and Adiabatic Approximation

implicitly on time through $\vec{R}(t)$. This time dependence manifests itself in the above equation in the last term, that is in the derivative of the state $|n\rangle$ with respect to time.

When we recall from the definition of the dynamical phase, Eq. (6.5), the relation

$$\frac{d}{dt}\phi_n^{(d)}(t) = \frac{1}{\hbar}E_n[\vec{R}(t)]$$

the expression resulting from the left hand side of the Schrödinger equation takes the form

$$i\hbar\frac{d}{dt}|\Psi(t)\rangle = i\hbar\sum_n \exp\left[-i\phi_n^{(d)}\right]\left[\left(\frac{d}{dt}\Psi_n\right)|n\rangle + \Psi_n\left(\frac{d}{dt}|n\rangle\right)\right]$$
$$+ \sum_n E_n\Psi_n \exp\left[-i\phi_n^{(d)}\right]|n\rangle. \tag{6.6a}$$

The eigenvalue equation Eq. (6.2) for $|n[\vec{R}(t)]\rangle$ allows us to evaluate the right hand side of the Schrödinger equation for the ansatz Eq. (6.4) of the state vector $|\Psi(t)\rangle$ which yields

$$\hat{H}|\Psi(t)\rangle = \sum_n E_n\Psi_n \exp\left[-i\phi_n^{(d)}\right]|n\rangle. \tag{6.6b}$$

When we compare the right hand side of Eqs. (6.6a) we arrive at

$$\sum_n \exp\left[-i\phi_n^{(d)}\right]\left[\left(\frac{d}{dt}\Psi_n\right)|n\rangle + \Psi_n\left(\frac{d}{dt}|n\rangle\right)\right] = 0. \tag{6.7}$$

We now project onto the energy eigenstate $|m[\vec{R}(t)]\rangle$ and assume the orthonormality condition

$$\langle m|n\rangle = \delta_{m,n}.$$

This leads to the differential equation

$$\frac{d}{dt}\Psi_m = -\langle m|\frac{d}{dt}|m\rangle\Psi_m - \sum_{m\neq n}\exp\left[-i\left(\phi_n^{(d)} - \phi_m^{(d)}\right)\right]\langle m|\frac{d}{dt}|n\rangle\Psi_n. \tag{6.8}$$

Here we have excluded the term Ψ_m from the sum.

We recognize that the matrix element

$$I_{m,n}(t) \equiv \langle m|\frac{d}{dt}|n\rangle$$

couples the nth energy eigenstate with the mth state. Even if we start out initially in a single energy eigenstate the unitary time evolution provided by the Schrödinger equation populates other energy eigenstates. Three factors determine the transition probability from the nth level to the mth level: *(i)* The matrix element $I_{m,n}$, *(ii)* the difference $\phi_n^{(d)}(t) - \phi_m^{(d)}(t)$ of the dynamical phases, *(iii)* and the initial probability amplitude Ψ_n of the nth level.

We first discuss the role of the matrix element $I_{m,n}$ and try to get an estimate for its size. For this purpose we bring out the explicit time dependence through the time dependent parameter $\vec{R}(t)$. We replace differentiation with respect to time by differentiation with respect to the parameter $\vec{R}$ times the rate of change of $\vec{R}$. We hence find

$$I_{m,n} = \left\langle m[\vec{R}(t)] \left| \frac{d}{dt} \right| n[\vec{R}(t)] \right\rangle = \left\langle m[\vec{R}(t)] \left| \vec{\nabla}_{\vec{R}} \right| n[\vec{R}(t)] \right\rangle \frac{d}{dt}\vec{R}(t) \quad (6.9)$$

which indicates that the size of $I_{m,n}$ is governed by the rate of change of $\vec{R}$. Provided this rate is small compared to the time scales involved in the problem the coupling due to the matrix element $I_{m,n}$ is small.

We also note that for $n \neq m$ the individual terms in the sum contain phase factors which vary as a function of time. These oscillations also reduce the effective coupling of the mth state to the remaining states.

The adiabatic theorem following from Eq. (6.8) therefore states that a system initially in an energy eigenstate stays in it provided the change in $\vec{R}(t)$ is adiabatic. In this case there are no transitions to other instantaneous energy eigenstates. Hence, under time evolution the state $|m[\vec{R}(t=0)]\rangle$ remains an energy eigenstate $|m[\vec{R}(t)]\rangle$, but acquires a phase, as we now show.

6.1.2 Analysis of Geometrical Phase

In order to find this phase we consider the time evolution of an energy eigenstate $|m[\vec{R}(t=0)]\rangle$ under an adiabatic evolution of $\vec{R}(t)$. This adiabaticity condition allows us to neglect in Eq. (6.8) the coupling to the other energy eigenstates $|n[\vec{R}(t)]\rangle$. We therefore arrive at

$$\frac{d}{dt}\Psi_m(t) \cong -\left\langle m[\vec{R}] \left| \vec{\nabla}_{\vec{R}} \right| m[\vec{R}] \right\rangle \cdot \frac{d\vec{R}}{dt} \Psi_m(t) \quad (6.10)$$

where we have used the representation Eq. (6.9) of the matrix element $I_{m,m}$.

We now show that the term $\langle m[\vec{R}]|\vec{\nabla}_{\vec{R}}|m[\vec{R}]\rangle$ is purely imaginary. Indeed, we find by differentiating the normalization condition $1 \equiv \langle m[\vec{R}]|m[\vec{R}]\rangle$ the relation

$$0 = \vec{\nabla}_{\vec{R}} \left(\langle m[\vec{R}] | m[\vec{R}] \rangle \right) = \langle \vec{\nabla}_{\vec{R}} m[\vec{R}] | m[\vec{R}] \rangle + \langle m[\vec{R}] | \vec{\nabla}_{\vec{R}} | m[\vec{R}] \rangle ,$$

or

$$0 = \langle m[\vec{R}] | \vec{\nabla}_{\vec{R}} | m[\vec{R}] \rangle^* + \langle m[\vec{R}] | \vec{\nabla}_{\vec{R}} | m[\vec{R}] \rangle = 2\,\text{Re}\left\{ \langle m[\vec{R}] | \vec{\nabla}_{\vec{R}} | m[\vec{R}] \rangle \right\},$$

that is, the real part vanishes.

Hence, indeed, the probability amplitude acquires a phase determined by the differential equation

$$\frac{d}{dt}\Psi_m(t) = -i\,\text{Im}\left[\langle m[\vec{R}] | \vec{\nabla}_{\vec{R}} | m[\vec{R}] \rangle \right] \cdot \frac{d\vec{R}}{dt} \Psi_m(t)$$

which yields

$$\Psi_m(t) = \exp\left[-i\phi_m^{(g)}(t) \right] \Psi_m(t=0).$$

Here we have introduced the geometrical phase

$$\phi_m^{(g)}(t) \equiv \int_{\vec{R}(t=0)}^{\vec{R}(t)} d\vec{R}'\, \text{Im}\left\{\langle m[\vec{R}']| \vec{\nabla}_{\vec{R}'} |m[\vec{R}']\rangle\right\}.$$

The name geometrical or topological phase originates from the fact that the integral brings in the geometry of Hilbert space. Indeed, it is the scalar product between the state vector $|m[\vec{R}]\rangle$ and $\vec{\nabla}_{\vec{R}}|m[\vec{R}]\rangle$ that determines the phase. To be more precise it is the geometry measured along the path traversed by $\vec{R}(t)$.

We conclude this section by discussing the adiabatic evolution of the quantum state $|m[\vec{R}(t=0)]\rangle$. In this case the initial condition for the expansion coefficients $\Psi_n(t=0)$ in Eq. (6.3) reads

$$\Psi_n(t=0) = \delta_{n,m}$$

and the state

$$|\Psi(t)\rangle = \exp\left[-i\phi_m^{(g)}(t)\right] \exp\left[-i\phi_m^{(d)}(t)\right] |m[\vec{R}(t)]\rangle \qquad (6.11)$$

acquires apart from the dynamical phase $\phi_m^{(d)}(t)$ the geometrical phase $\phi_m^{(g)}(t)$.

6.1.3 Geometrical Phase as a Flux in Hilbert Space

We conclude this section by considering the case of a cyclic evolution with a closed path $\mathcal{C}$ in parameter space. Here T denotes the period of the cycle.

According to Eq. (6.11) the state vector after one cycle is not identical to the initial vector. Apart from the dynamical phase

$$\phi_m^{(d)}(T) \equiv \frac{1}{\hbar} \int_0^T dt\, E_m[\vec{R}(t)]$$

it has acquired the geometrical phase

$$\phi_m^{(g)} \equiv \oint_{\mathcal{C}} d\vec{R}\, \text{Im}\left\{\langle m[\vec{R}]| \vec{\nabla}_{\vec{R}} |m[\vec{R}]\rangle\right\}. \qquad (6.12)$$

This expression brings in elements of differential geometry. Indeed, it is reminiscent of the concept of measuring the curvature of a surface by parallel transport of a tangent vector. Here we propagate a vector that is tangent to a curved surface along a path on this surface as shown in Fig. 6.1. When we consider a closed path the vector after one cycle is not identical to the initial vector: The two vectors have a non-vanishing angle between them. This angle is a measure for the curvature of the surface.

In the case of the geometrical phase we are measuring the curvature of parameter space. To bring this out more clearly we can apply Stokes' theorem to transform the line integral along $\mathcal{C}$ into a surface integral. The geometric phase

$$\phi_m^{(g)} \equiv \int_{\mathcal{F}} d\vec{f} \cdot \vec{\mathcal{B}}_m[\vec{R}]$$

is then the flux of the field

$$\vec{\mathcal{B}}_m[\vec{R}] \equiv \vec{\nabla}_{\vec{R}} \times \text{Im}\left\{\langle m[\vec{R}]| \vec{\nabla}_{\vec{R}} |m[\vec{R}]\rangle\right\}$$

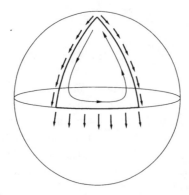

Fig. 6.1: Parallel transport on a sphere as a measure of curvature. We consider a vector that is always tangentially on the surface of the sphere and propagate it along a closed path on the sphere. When we return to the starting point the initial and the propagated vector are not identical but enclose an angle. This angle measures the curvature of the sphere.

through the area $\mathcal{F}$ with normal vector $\vec{f}$ enclosed by the path $\mathcal{C}$.

We recognize that the geometrical phase $\phi_m^{(g)}$ is similar to the Aharonov-Bohm phase. Here electrons scatter from a vector potential created by a long and thin solenoid. This device has a constant magnetic field inside but no magnetic field outside. Hence, the contour lines of the vector potential are circles around the solenoid. The wave function of an electron circumventing the solenoid on the left hand side experiences a phase shift that is different from the phase shift experienced on the right hand side. The total Aharonov-Bohm phase shift determining the interference pattern in the far field is then the line integral

$$\phi^{(g)} \sim \int_{\mathcal{C}} d\vec{R} \cdot \vec{A}(\vec{R}) = \int_{\mathcal{F}} d\vec{f} \cdot \left[\vec{\nabla}_{\vec{R}} \times \vec{A}(\vec{R})\right] = \int_{\mathcal{F}} d\vec{f} \cdot \vec{B}(\vec{R})$$

of the vector potential along the closed path $\mathcal{C}$ of the electrons that is the flux of the magnetic field $\vec{B}$ through the area $\mathcal{F}$ defined by the path $\mathcal{C}$. This analogy brings out most clearly that the geometrical phase measures a flux in Hilbert space.

We conclude this section by mentioning that there are many more subtleties of the geometric phase. Space does not allow us to go deeper into this exciting topic. However, it is worthwhile mentioning that this geometrical phase has been observed for the case of the polarization of light going through twisted fiber wave guides. Moreover, most recently it has been claimed that one can derive Fermi-Dirac and Bose-Einstein statistics from it. However, this is beyond the scope of the present book and we refer to the literature.

6.2 WKB Wave Functions from Adiabaticity

In Sec. 5.2.1 we have motivated the form of an energy wave function such as the one shown in Fig. 6.2 corresponding to an arbitrary but slowly varying binding potential $U(x)$. In the present section we derive this result by bringing the time

6.2 WKB Wave Functions from Adiabaticity

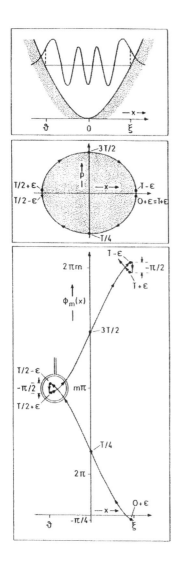

Fig. 6.2: Phase of a WKB energy wave function in its dependence on position. The mth energy wave function in a binding potential undergoes oscillations and displays m nodes (top). The particle corresponding to this energy performs a closed orbit in phase space (middle). The area enclosed is $2\pi\hbar(m + 1/2)$. The phase of the WKB wave function increases as we move from the right turning point ξ to the left turning point ϑ and back (bottom). The phase accumulated on the trip from one side to the other is $\pi/4 + m\pi + \pi/4 = (m + 1/2)\pi$. The two contributions of $\pi/4$ arise from the fact that the phase of the WKB wave at the turning point is $-\pi/4$. In contrast to this monotonic increase the phase undergoes a jump by $-\pi/2$ at the turning points, that is when the particle crosses the x-axis in phase space. The total phase accumulated in one complete cycle is thus $2(m + 1/2)\pi - 2\pi/2 = 2\pi m$ and the wave function is identical to the initial one.

independent Schrödinger equation for the wave function $u(x)$ into the form of a Schrödinger equation for a two–state system. This approach allows us to connect the WKB-approximation to the concept of the geometrical phase.

6.2.1 Energy Eigenvalue Problem as Propagation Problem

We start from the time independent Schrödinger equation

$$u''(x) + k^2(x)u(x) = 0 \qquad (6.13)$$

with
$$p(x) \equiv \hbar k(x) \equiv \sqrt{2M[E - U(x)]}$$

and introduce formally the two-component vector

$$|\Psi(x)\rangle \equiv \begin{pmatrix} u(x) \\ u'(x) \end{pmatrix}. \tag{6.14}$$

Throughout the section the prime denotes differentiation with respect to x.

We find the Schrödinger type equation

$$i\frac{d}{dx}|\Psi(x)\rangle = \mathcal{H}(x)|\Psi(x)\rangle \tag{6.15}$$

for the propagation of the vector $|\Psi\rangle$. Here we have used the word *propagation* rather than *evolution* to emphasize the fact that now position plays the role of time in the Schrödinger equation.

However, in contrast to the Hamiltonian of the familiar Schrödinger equation the Hamiltonian-like quantity

$$\mathcal{H}(x) \equiv \begin{pmatrix} 0 & i \\ -ik^2(x) & 0 \end{pmatrix} \tag{6.16}$$

is not hermitian. Moreover, $\mathcal{H}$ only contains off-diagonal elements. It therefore leads to a system of two coupled equations. Since this coupling is position dependent, it is not trivial to solve this propagation equation in an exact way. However, we can still search for approximate solutions.

For this purpose we first find the instantaneous eigenstates of $\mathcal{H}$. Since $\mathcal{H}$ is not hermitian, we have left and right eigenstates.

Instantaneous Eigenvalues and Eigenstates

The eigenvalues
$$\lambda_\pm(x) \equiv \pm k(x) \tag{6.17}$$

of $\mathcal{H}(x)$ are real and position dependent. They give rise to the instantaneous eigenstates

$$|u_\pm(x)\rangle = \begin{pmatrix} 1 \\ \mp ik(x) \end{pmatrix} \tag{6.18}$$

which satisfy the eigenvalue equation

$$\mathcal{H}(x)|u_\pm(x)\rangle = \lambda_\pm(x)|u_\pm(x)\rangle. \tag{6.19}$$

This equation is completely analogous to the equation Eq. (6.2) for the instantaneous energy eigenstates $|m[\vec{R}(t)]\rangle$ with $\vec{R}(t)$ being replaced by x.

We note that the states $|u_\pm\rangle$ are not orthogonal on each other. However, they are orthogonal on the states

$$\langle v_\pm(x)| \equiv (\mp ik(x), 1) \tag{6.20}$$

6.2 WKB Wave Functions from Adiabaticity

which are the left sided eigenstates of $\mathcal{H}$, that is

$$\langle v_\pm | \mathcal{H} = \lambda_\pm \langle v_\pm |.$$

Indeed, we have

$$\langle v_+ | u_- \rangle = \langle v_- | u_+ \rangle = 0, \tag{6.21}$$

following from the definitions, Eqs. (6.18) and (6.20) of the eigenvectors.

Moreover, since $\mathcal{H}$ is not hermitian the scalar product between the left and right eigenstate of the same eigenvalue cannot be normalized and we find

$$\langle v_\pm | u_\pm \rangle = \mp 2ik. \tag{6.22}$$

Table: Key Elements in Berry Phase versus WKB

	Berry phase	WKB
evolution equation	$i\hbar \frac{\partial}{\partial t} |\Psi(t)\rangle = \hat{H}[\vec{R}(t)] |\Psi(t)\rangle$	$i\frac{d}{dx} |\Psi(x)\rangle = \mathcal{H}(x) |\Psi(x)\rangle$
Hamiltonian	hermitian	$\mathcal{H}(x) \equiv \begin{pmatrix} 0 & i \\ -ik^2(x) & 0 \end{pmatrix}$
state	$|m[\vec{R}(t)]\rangle$	$|\Psi(x)\rangle \equiv \begin{pmatrix} u(x) \\ u'(x) \end{pmatrix}$
eigenvalue equation	$\hat{H}[\vec{R}(t)] |m[\vec{R}(t)]\rangle =$ $E_m[\vec{R}(t)] |m[\vec{R}(t)]\rangle$	$\mathcal{H}(x) |u_\pm(x)\rangle =$ $\lambda_\pm(x) |u_\pm(x)\rangle$
evolution	$|\Psi(t)\rangle =$ $\exp\left[-i\phi_m^{(g)} - i\phi_m^{(d)}\right] |m[\vec{R}(t)]\rangle$	$|\Psi(x)\rangle =$ $\exp\left[-\phi_\pm^{(g)} - i\phi_\pm^{(d)}\right] |u_\pm(x)\rangle$
dynamical phase $\phi_m^{(d)}$	$\frac{1}{\hbar} \int_0^t dt' E_m[\vec{R}(t')]$	$\int_0^x d\tilde{x}\, \lambda_\pm(\tilde{x})$
geometrical phase $\phi_m^{(g)}$	$\int_C d\vec{R}\, \text{Im}\left\{\langle m[\vec{R}] | \vec{\nabla}_{\vec{R}} | m[\vec{R}]\rangle\right\}$	$\oint dx\, \frac{\langle v_\pm | d/dx | u_\pm \rangle}{\langle v_\pm | u_\pm \rangle}$

Ansatz

We now expand the vector $|\Psi\rangle$ into the eigenstates $|u_\pm\rangle$. This approach is analogous to the expansion Eq. (6.3) of the state vector into instantaneous energy eigenstates. Again, we first separate out the dynamical phases

$$\phi_\pm^{(d)}(x) \equiv \int_{x_0}^{x} d\tilde{x}\, \lambda_\pm(\tilde{x}) \tag{6.23}$$

acquired by these eigenstates when they start at the position x_0

We therefore start from the ansatz

$$|\Psi(x)\rangle = \Psi_+(x) \exp\left[-i\phi_+^{(d)}(x)\right] |u_+(x)\rangle + \Psi_-(x) \exp\left[-i\phi_-^{(d)}(x)\right] |u_-(x)\rangle \quad (6.24)$$

which when substituted into the propagation equation, Eq. (6.15), yields the equation

$$\exp\left[-i\phi_+^{(d)}\right]\left[\left(\frac{d}{dx}\Psi_+\right)|u_+\rangle + \Psi_+\left(\frac{d}{dx}|u_+\rangle\right)\right] +$$

$$\exp\left[-i\phi_-^{(d)}\right]\left[\left(\frac{d}{dx}\Psi_-\right)|u_-\rangle + \Psi_-\left(\frac{d}{dx}|u_-\rangle\right)\right] = 0.$$

Here we have used the eigenvalue equation Eq. (6.19) for $|u_\pm\rangle$.

When we now recall the orthogonality relations, Eqs. (6.21)

$$\langle v_+ | u_-\rangle = 0 \quad \text{and} \quad \langle v_- | u_+\rangle = 0$$

we arrive at the two coupled equations

$$\frac{d}{dx}\Psi_+(x) = -\frac{\langle v_+|\frac{d}{dx}|u_+\rangle}{\langle v_+|u_+\rangle}\Psi_+(x) - \exp\left[-i\left(\phi_-^{(d)} - \phi_+^{(d)}\right)\right]\frac{\langle v_+|\frac{d}{dx}|u_-\rangle}{\langle v_+|u_+\rangle}\Psi_-(x) \quad (6.25a)$$

and

$$\frac{d}{dx}\Psi_-(x) = -\frac{\langle v_-|\frac{d}{dx}|u_-\rangle}{\langle v_-|u_-\rangle}\Psi_-(x) - \exp\left[-i\left(\phi_+^{(d)} - \phi_-^{(d)}\right)\right]\frac{\langle v_-|\frac{d}{dx}|u_+\rangle}{\langle v_-|u_-\rangle}\Psi_+(x). \quad (6.25b)$$

We recognize that this set of equations is completely analogous to the set of equations Eqs. (6.8) in the discussion of the adiabatic theorem. The role of time there is now played by position. The main difference is that the states are not normalized now, which gives rise to the terms $\langle v_\pm | u_\pm\rangle$ in the denominators. Since there are only two states $|u_\pm\rangle$ involved, the sum in Eq. (6.8) reduces to a single term coupling the probability amplitude Ψ_+ to Ψ_- and vice versa. For a detailed comparison between the original Berry phase and the WKB problem we refer to the table.

Explicit Equations for $\Psi_\pm$

We now make Eqs. (6.25) more explicit. For this purpose we note from the definitions Eq. (6.18) and (6.20) of the eigenstates the relations

$$\langle v_\pm|\frac{d}{dx}|u_\pm\rangle = \mp ik'(x) \quad (6.26a)$$

and

$$\langle v_\pm|\frac{d}{dx}|u_\mp\rangle = \pm ik'(x). \quad (6.26b)$$

With the help of Eqs. (6.22) and (6.26) the expression Eq. (6.23) for the dynamical phases the two coupled equations Eqs. (6.25) reduce to

$$\frac{d}{dx}\Psi_+(x) = -\frac{1}{2}\frac{k'(x)}{k(x)}\Psi_+(x) + \exp\left[2i\int_{x_0}^x d\tilde{x}\, k(\tilde{x})\right]\frac{1}{2}\frac{k'(x)}{k(x)}\Psi_-(x) \quad (6.27a)$$

and

$$\frac{d}{dx}\Psi_-(x) = -\frac{1}{2}\frac{k'(x)}{k(x)}\Psi_-(x) + \exp\left[-2i\int_{x_0}^{x} d\tilde{x}\, k(\tilde{x})\right]\frac{1}{2}\frac{k'(x)}{k(x)}\Psi_+(x). \quad (6.27b)$$

So far the treatment is exact. Indeed, the two coupled differential equations of first order are completely equivalent to the time independent Schrödinger equation Eq. (6.13) of second order.

6.2.2 Dynamical and Geometrical Phase

In the discussion of the adiabatic theorem in Sec. 6.1.1 we have considered the time evolution of a single energy eigenstate under an adiabatic change described by a parameter $\vec{R}(t)$. The adiabaticity has allowed us to neglect the coupling to the other energy eigenstates and we have found a geometrical phase. In the present section we pursue an analogous approach for the set of equations Eqs. (6.27).

We note that the coupling of the amplitude Ψ_+ to the amplitude Ψ_- occurs via the ratio k'/k and a phase factor. This phase factor varies twice as fast as the phase factors in the wave function itself. In lowest order we can therefore neglect the coupling and find from the approximate differential equation

$$\frac{d}{dx}\Psi_\pm(x) \cong -\frac{\langle v_\pm|\frac{d}{dx}|u_\pm\rangle}{\langle v_\pm|u_\pm\rangle}\Psi_\pm(x) = -\frac{1}{2}\frac{k'(x)}{k(x)}\Psi_\pm(x) \quad (6.28)$$

the approximate solution

$$\Psi_\pm(x) \cong \exp\left[-\frac{1}{2}\ln\frac{k(x)}{k(x_0)}\right]\Psi_\pm(x_0). \quad (6.29)$$

Here x_0 denotes the initial position. We emphasize that the approximate propagation equation Eq. (6.28) is completely analogous to the corresponding time evolution equation Eq. (6.10)

Thus the approximate solution for the state reads

$$|\Psi_\pm(x)\rangle \cong \exp\left[-\phi_\pm^{(g)}(x)\right]\exp\left[-i\phi_\pm^{(d)}(x)\right]\Psi_\pm(x_0)|u_\pm\rangle, \quad (6.30)$$

where we have introduced

$$\phi_\pm^{(g)}(x) \equiv \int_{x_0}^{x} dx\, \frac{\langle v_+|\frac{d}{dx}|u_+\rangle}{\langle v_\pm|u_\pm\rangle} = \int_{x_0}^{x} dx'\, \frac{1}{2}\frac{k'(x)}{k(x)} = \frac{1}{2}\ln\frac{k(x)}{k(x_0)}. \quad (6.31)$$

The vectors $|u_\pm\rangle$ acquire two types of phases as we adiabatically change the position. The phase

$$\phi_\pm^{(d)}(x) = \pm\int_{x_0}^{x} d\tilde{x}\, k(\tilde{x})$$

which results from the change of the instantaneous state vectors $|u_\pm\rangle$. This is analogous to the dynamical phase. The phase $\phi_\pm^{(g)}$ is the analogue of the Berry phase Eq. (6.12)

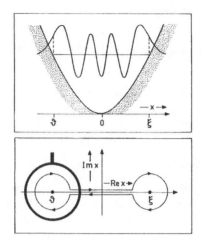

Fig. 6.3: Phase jump of the WKB wave function in the neighborhood of a turning point explained by a path in complex space. Usually the energy wave function of a binding potential depends on the real-valued position x (top). However, in order to bring out the phase change at the turning points we slightly deform the path connecting the two turning points: Whenever we get close to a turning point we circumvent it by moving on a circle in the complex plane. As a result the straight paths connecting the two circles are slightly off the real axis. The resulting behavior of the wave function at the left turning point is shown in Fig. 6.4.

However, there is a subtle difference. The wave number k is real. Hence in contrast to the Berry phase Eq. (6.12) which is purely imaginary it seems that now the geometrical phase Eq. (6.31) is real. Thus it represents a damping associated with the classical part of the wave function. We recognize this fact by representing the phase factor by

$$\exp\left(-\phi_{\pm}^{(g)}\right) = \exp\left[-\frac{1}{2}\ln\frac{k(x)}{k(x_0)}\right] = \sqrt{\frac{k(x_0)}{k(x)}} \sim \frac{1}{\sqrt{k(x)}}.$$

However, at the turning point the wave number vanishes and the phase becomes infinite. When we circumvent this singularity by going into complex space as suggested by Fig. 6.3 the phase accumulated is $\pi/2$. Indeed, we find this value by first expanding the wave number

$$k(x) \equiv \sqrt{\frac{2M}{\hbar^2}\left[E - U(x)\right]} \cong \sqrt{\frac{2M}{\hbar^2}|U'(\vartheta)|}\,(x - \vartheta)^{\frac{1}{2}}$$

around the left turning point ϑ.

Moreover, the derivative of the wave number reads

$$k'(x) = \sqrt{\frac{2M}{\hbar^2}|U'(\vartheta)|}\,\frac{1}{2}(x - \vartheta)^{-\frac{1}{2}}.$$

Hence the geometrical phase accumulated while circumventing the turning point ϑ in complex space by moving along a circular path $\mathcal{C}$ parameterized by $z = \vartheta - \rho e^{-i\theta}$ of radius ρ is given by

$$\phi_\pm^{(g)}(\theta) = \frac{1}{2}\int_\mathcal{C} dz\, \frac{k'(z)}{k(z)} = \frac{1}{4}\int_\mathcal{C} dz\, \frac{1}{(z-\vartheta)} = -\frac{i}{4}\int_0^\theta d\tilde\theta = -i\frac{\theta}{4}.$$

Due to the path in complex space $\phi_\pm^{(g)}$ has become purely imaginary.

Moreover, for a complete circle around the turning point we find

$$\phi_\pm^{(g)}(2\pi) = -i\frac{\pi}{2}.$$

Hence the eigenstates $|u_\pm\rangle$ pick up a real phase shift of $-\pi/2$ as we go around the turning point ϑ.

6.2.3 WKB Waves Rederived

The Schrödinger equation Eq. (6.15) is a propagation equation for a quasi-state vector $|\Psi\rangle$ whose components consist of the energy wave function $u(x)$ and its first derivative $u'(x)$. According Eq. (6.30), an eigenvector $|u_\pm\rangle$, when propagated along the x-axis from one turning point to the other and back, accumulates two phases: The first is of dynamical origin, the second one originates from geometry. It arises from the singularity of the wave number at the turning point.

From Quasi-State Vectors to WKB waves

These features are very reminiscent of the standard WKB wave function

$$u_m^{(WKB)}(x) = \mathcal{N}_m \frac{1}{\sqrt{k_m(x)}} \cos\varphi_m(x)$$

with

$$\varphi_m(x) \equiv S_m(x) - \frac{\pi}{4} = \int_x^\xi dx\, k(x) - \frac{\pi}{4}$$

discussed in Chapter 5.

Indeed, we can immediately identify the dynamical phase $\phi_\pm^{(d)}$ with the phase $S_m(x)$. Moreover, the classical prefactor $k_m^{-\frac{1}{2}}(x)$ is the real part of the geometrical phase, whereas the phase $-\pi/4$ seems to be closely related to the imaginary part. However, this is not quite clear yet since the WKB-phase is $-\pi/4$, whereas the total geometrical phase accumulated going around the turning point is $-\pi/2$. Obviously both phases are closely related since they both originate from the turning point: The WKB-phase emerges from the fact that the wave function reaches into the classically forbidden regime. The Berry phase originates from avoiding the turning point and escaping into complex space.

184 6 WKB and Berry Phase

Moreover, it is not quite clear yet how the quasi-state vectors $|u_\pm\rangle$ are connected to the WKB-wave function. In order to answer these questions we now use Eq. (6.18) to decompose the quasi-state vector $|\Psi\rangle$ in its components

$$|\Psi(x)\rangle = \Psi_\pm(x) \exp\left[-i\phi_\pm^{(d)}(x)\right] |u_\pm(x)\rangle = \Psi_\pm(x) \exp\left[-i\phi_\pm^{(d)}(x)\right] \begin{pmatrix} 1 \\ \mp ik(x) \end{pmatrix}$$

and find making use of the definition Eq. (6.14) of $|\Psi\rangle$ the relations

$$u_\pm(x) = \Psi_\pm(x) \exp\left[-i\phi_\pm^{(d)}(x)\right]$$

and

$$u'_\pm(x) = \mp ik(x) \Psi_\pm(x) \exp\left[-i\phi_\pm^{(d)}(x)\right] = \mp ik(x) u_\pm(x).$$

Hence in the adiabatic approximation the derivative of the wave functions $u_\pm$ reproduces $u_\pm$ and brings in the factor $\mp ik(x)$. This term originates from the differential of the dynamical phase. Therefore, in this approximation we do not differentiate the amplitudes $\Psi_\pm$.

When we now recall the approximate solution Eq. (6.29) we find the expression

$$u_\pm^{(\mathrm{WKB})}(x) \cong \frac{1}{\sqrt{k(x)}} \exp\left[\mp i \int_{x_0}^{x} d\tilde{x}\, k(\tilde{x})\right] \tag{6.32}$$

for the WKB-wave function and its first derivative

$$\frac{d}{dx} u_\pm^{(\mathrm{WKB})} \cong \mp ik(x) \left[\frac{1}{\sqrt{k(x)}} \exp\left[\mp i \int_{x_0}^{x} d\tilde{x}\, k(\tilde{x})\right]\right] = \mp ik(x)\, u_\pm^{(\mathrm{WKB})}. \tag{6.33}$$

Hence, we have indeed found the form of the WKB wave function discussed in Sec. 5.2.1.

A Fresh Look at WKB Waves

We now make the connection between the phases acquired by the quasi–state vectors $|u_\pm\rangle$ propagated along the path shown in Fig. 6.3 and the phases φ_m and $-\pi/4$ in the WKB wave.

For this purpose we follow the change of the wave function as we are moving from the right turning point ξ to the left turning point ϑ. On top of Fig. 6.2 we show the wave function of the mth energy eigenstate in a binding potential. It is caught between the two turning points of the classical motion and displays m nodes. Within the WKB-approximation its energy is determined in such a way that the area enclosed by the classical phase space trajectory is equal to $2\pi\hbar(m+1/2)$ as indicated by the picture in the middle.

As we are moving from the right turning point to the left turning point the phase $\varphi_m(x)$ increases monotonously starting from $-\pi/4$. We are going through m nodes and hence the phase S_m reaches $m\pi$. In order to get to the left turning point we need another phase of $\pi/4$. Hence, the total phase accumulated on our march from

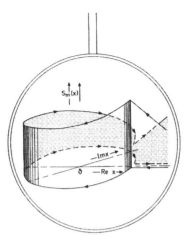

Fig. 6.4: Behavior of the WKB wave function in the neighborhood of the left turning point ϑ. As we approach the turning point from the real axis the phase of the wave function increases. We avoid the turning point by following a circle in the complex plane. As a consequence the phase continuously decreases. After a $2\pi-$ rotation we have obtained a phase difference of $-\pi/2$. As we continue our journey to the right turning point along the real axis the phase increases again. A projection of this circular staircase onto a plane, that is the reduction to a real-valued x-axis, creates a discontinuous jump of $-\pi/2$.

the right to the left is $\pi/4 + m\pi + \pi/4 = (m + 1/2)\pi$ as shown at the bottom of Fig. 6.2.

The same is true on our march from the left to the right. However, there is a subtlety. There is a phase jump of $\pi/2$ at the two turning points. This guarantees that the wave function is unique after one round trip in phase space. But where does this phase jump come from?

We now know that the answer rests on the geometrical phase. Indeed, we can avoid the turning point by escaping into the complex plane. To be more specific we approach the turning point by following along a line parallel to the real axis with a small negative imaginary part. Once we get close to the turning point we go on a circle around it and lose a phase $\pi/2$ on our way down the staircase shown in Fig. 6.4.

6.3 Non-Adiabatic Berry Phase

In the derivation of the Berry phase we have assumed adiabatic changes. However, this restriction is not essential as we discuss in the present section. This concept of a non-adiabatic Berry phase also shines new light on the time evolution of a wave packet in a harmonic oscillator.

6.3.1 Derivation of the Aharonov-Anandan Phase

We now focus on a quantum state $|\psi\rangle$ which due to the Schrödinger time evolution

$$i\hbar \frac{d}{dt}|\psi(t)\rangle = \hat{H}(t)|\psi(t)\rangle$$

has a special property: At a time T the state is identical to the original state up to a phase ϕ, that is

$$|\psi(T)\rangle = e^{i\phi}|\psi(0)\rangle. \tag{6.34}$$

Here the Hamiltonian $\hat{H}$ can have an arbitrary time dependence.

We now introduce the state

$$\left|\tilde{\psi}(t)\right\rangle \equiv e^{-i\varphi(t)}|\psi(t)\rangle \tag{6.35}$$

and construct the time dependent phase φ in such a way as to remove this phase ϕ and achieve

$$\left|\tilde{\psi}(T)\right\rangle = \left|\tilde{\psi}(0)\right\rangle. \tag{6.36}$$

We now show that in this case ϕ consists of a dynamical and a geometrical part.

For this purpose we differentiate the state $|\tilde{\psi}\rangle$ with respect to time and make use of the Schrödinger equation for the time evolution of the state $|\psi\rangle$ which yields

$$\frac{d}{dt}\left|\tilde{\psi}(t)\right\rangle = -i\dot{\varphi}(t)\left|\tilde{\psi}(t)\right\rangle - \frac{i}{\hbar}\exp[-i\varphi(t)]\hat{H}|\psi\rangle.$$

We multiply this equation by $\langle\tilde{\psi}|$ and arrive at

$$\dot{\varphi} = i\left\langle\tilde{\psi}\left|\frac{d}{dt}\right|\tilde{\psi}\right\rangle - \frac{1}{\hbar}\langle\psi|\hat{H}|\psi\rangle.$$

Here we have used the connection formula Eq. (6.35) between the state $|\psi\rangle$ and the state $|\tilde{\psi}\rangle$ to eliminate the phase φ in the last term.

After integration over a time T this equation provides the phase difference

$$\varphi(T) - \varphi(0) = i\int_0^T dt'\left\langle\tilde{\psi}\left|\frac{d}{dt'}\right|\tilde{\psi}\right\rangle - \frac{1}{\hbar}\int_0^T dt'\,\langle\psi|\hat{H}|\psi\rangle. \tag{6.37}$$

We now use this phase

$$\varphi(T) - \varphi(0) \equiv \phi \tag{6.38}$$

to show that it is possible to achieve the relation

$$\left|\tilde{\psi}(T)\right\rangle = \left|\tilde{\psi}(0)\right\rangle.$$

Indeed, from Eqs. (6.34) and (6.35) we find

$$\left|\tilde{\psi}(T)\right\rangle = e^{-i\varphi(T)}|\psi(T)\rangle = e^{-i\varphi(T)}e^{i\phi}|\psi(0)\rangle$$

which with the help of the condition Eq. (6.38) reads

$$\left|\tilde{\psi}(T)\right\rangle = e^{-i\varphi(0)}|\psi(0)\rangle = \left|\tilde{\psi}(0)\right\rangle.$$

In the last step we have used again the connection formula Eq. (6.35).

Hence, the quantum state $|\tilde{\psi}\rangle$ after the time T is identical to the state at time $t = 0$. However, according to Eq. (6.34), the state $|\psi\rangle$ has acquired the phase ϕ which consists of two parts associated with time evolution and geometry as shown by Eq. (6.37).

6.3.2 Time Evolution in Harmonic Oscillator

We now illustrate the concept of the Aharonov-Anandan phase by applying it to the time evolution of a quantum state in a quadratic potential. We identify the geometrical and dynamical phase.

In Sec. 4.2.3 we have found that the time evolution of any quantum state $|\psi_0\rangle$ in a harmonic oscillator of frequency Ω is given by

$$|\psi(t)\rangle = \sum_m \psi_m \exp\left[-i\left(m+\tfrac{1}{2}\right)\Omega t\right]|m\rangle \tag{6.39}$$

where $\psi_m \equiv \langle m|\psi_0\rangle$ are the coefficients for the expansion of the initial state $|\psi_0\rangle$ into energy eigenstates $|m\rangle$ of the harmonic oscillator.

After a period $T = 2\pi/\Omega$ of the oscillator the state reads

$$|\psi(T)\rangle = e^{-i\pi}|\psi(0)\rangle,$$

and up to a phase π is therefore identical to the initial state. Usually we disregard this overall phase factor resulting from the zero-point energy. However, we now show that this term represents the most elementary example of an Aharonov-Anandan phase.

For this purpose we define the state

$$\left|\tilde{\psi}(t)\right\rangle \equiv e^{i\Omega t/2}|\psi(t)\rangle \tag{6.40}$$

which is periodic and thus satisfies the requirement

$$|\tilde{\psi}(T)\rangle = |\tilde{\psi}(0)\rangle,$$

put forward in Eq. (6.36). We now calculate the phase

$$\phi \equiv i\int_0^T dt' \left\langle \tilde{\psi}\left|\frac{d}{dt'}\right|\tilde{\psi}\right\rangle - \frac{1}{\hbar}\int_0^T dt' \langle \psi|\hat{H}|\psi\rangle.$$

using the expressions Eqs. (6.40) and (6.39) for the states $|\tilde{\psi}\rangle$ and $|\psi\rangle$ and arrive at

$$\phi = 2\pi \sum_m m|\psi_m|^2 - 2\pi \sum_m \left(m+\tfrac{1}{2}\right)|\psi_m|^2 = 2\pi\overline{m} - 2\pi\left(\overline{m}+\tfrac{1}{2}\right) = -\pi,$$

where $\overline{m} \equiv \sum_m m|\psi_m|^2$.

Hence, the geometric phase is $2\pi\overline{m}$, whereas the dynamical phase reads $2\pi(\overline{m}+1/2)$.

References

Introduction and Further Developments of Berry Phase

The concept of the geometrical phase is developed in
M.V. Berry, *Quantal phase factors accompanying adiabatic changes*, Proc. Roy. Soc. Lond. A **392**, 45–57 (1984)

For an elementary approach see
M.V. Berry, *The geometric phase*, Sci. Am. **259**(6), 46–52 (1988)
B.R. Holstein, *The adiabatic theorem and Berry phase*, Am. J. Phys. **57**, 1079–1084 (1989)
A. Shaphere and **F. Wilczek**, *Geometric Phases in Physics*, World Scientific, Singapore, 1989

For a generalization of the Berry phase to non-adiabatic changes see
Y. Aharonov and **J. Anandan**, *Phase Change During a Cyclic Quantum Evolution*, Phys. Rev. Lett. **58**, 1593–1596 (1987)

For the connection between the Berry phase and quantum statistics see
M.V. Berry and **J.M. Robbins**, *Quantum indistinguishability*, Proc. Roy. Soc. Lond. **A453**, 1771–1790 (1997)

Parallel Transport

For an explanation of the concept of parallel transport in differential geometry and applied to general relativity see
H.C. Ohanian, *Gravitation and Spacetime*, W.W. Norton, New York, 1976
C.W. Misner, K.S. Thorne and **J.A. Wheeler**, *Gravitation*, W.H. Freeman, New York, 1973

Berry Phase and WKB

For the connection between the WKB wave functions and the Berry phase see
R.G. Littlejohn, *Cyclic Evolution in Quantum Mechanics and the Phases of Bohr-Sommerfeld and Maslov*, Phys. Rev. Lett. **61**, 2159–2162 (1988)
M.V. Berry, *Quantum adiabatic anholonomy*, in: *Anomalies, Phases, Defects*, edited by U. Bregda, G. Garmo and G. Morandi, Naples, Bibliopolis, 1990
M.G. Benedict and **W. Schleich**, *On the Correspondence of Semiclassical and Quantum Phases in Cyclic Evolutions*, Found. Phys. **23**, 389–397 (1993)

Experiments

For experiments related to the Berry phase see
T. Bitter and **D. Dubbers**, *Manifestation of Berry's Topological Phase in Neutron Spin Rotation*, Phys. Rev. Lett. **59**, 251–254 (1997)
D.J. Richardson, A.I. Kilvington, K. Green and **S.K. Lamoreaux**, *Demonstration of Berry's Phase using Stored Ultracold Neutrons*, Phys. Rev. Lett. **61**, 2030–2033 (1988)
R.Y. Chiao, *Optical Manifestations of Berry's Topological Phases: Aharonov-Bohm-like Effects for the Photon*, Proc. 3rd Int. Symp. Foundations of Quantum Mechanics, ed. Shun-ichi Kobayashi *et al.*, The Physical Society of Japan, Tokyo, 1990, p. 80–92

7 Interference in Phase space

A quantum mechanical system, initially prepared in a state $|n\rangle$ and exposed to a sudden change of conditions, undergoes a jump from this level to the state $|m\rangle$ with a probability amplitude

$$w_{m,n} = \langle m|n\rangle = \int_{-\infty}^{\infty} dx\, \langle m|x\rangle \langle x|n\rangle = \int_{-\infty}^{\infty} dx\, u_m(x)\, v_n(x). \tag{7.1}$$

Here $u_m(x)$ and $v_n(x)$ denote the wave functions of the corresponding states in position representation.

The same overlap integral governs the probability amplitude for the radiative transition of a molecule from one vibronic state to another when the dipole moment's variation with internuclear separation x is small over the range of x-values that contribute importantly to the matrix element. We do not ask here how to derive this well-known standard result, but how to discover whether $w_{m,n}$ is big or small, and what makes it so.

We gain insight into this question by applying the concept of interference in phase space: The scalar product is determined by the interference of the areas of overlap between the quantum states of interest. The derivation of this concept is the topic of the present chapter.

We first explain the central ingredients of this approach and then present the mathematical derivation. This makes heavy use of the WKB wave functions and the method of stationary phase discussed in Appendix H. An application of this concept to Franck-Condon transitions in diatomic molecules and a generalization to arbitrary states concludes this chapter.

7.1 Outline of the Idea

For this purpose we first have to find a simple representation of an energy eigenstate and then a simple geometrical phase space algorithm that allows us to evaluate the scalar product. This algorithm is the overlap of phase space areas representing the individual quantum states. But how can we represent these states?

7 Interference in Phase space

In the preceding chapters we have found the Wigner representation of quantum mechanics and in particular of an arbitrary quantum state. This is one possibility. Another even simpler method relies on the WKB analysis of an energy eigenstate performed in the last chapter. It suggests a representation in phase space by a single trajectory as indicated in Fig. 7.1 by the dashed lines. This Kramers trajectory for the mth energy eigenstate encompasses the area $2\pi\hbar(m + 1/2)$ in phase space, as suggested by the Bohr-Sommerfeld-Kramers quantization condition. In contrast we can also view an energy eigenstate as a band in phase space as shown in Fig. 7.1 by the grey band. Indeed, in his original paper on the radiation formula of phase space Max Planck emphasizes that a quantum state is a band in phase space with its boundaries determined by the quantization conditions phase space area $2\pi\hbar m$ on its inner boundary and $2\pi\hbar(m+1)$ on its outer. Hence, the band contains the phase space area $2\pi\hbar$. Since the boundaries of the band are defined by the old Planck-Bohr-Sommerfeld quantization conditions of integer multiples of Planck's constant rather than half integer as proposed by Kramers we call these bands Planck-Bohr-Sommerfeld bands. The Kramers trajectory defined by half integer actions runs midway through the band as shown in Fig. 7.1. The totality of all conceivable final states, that is, the entire collection of Planck-Bohr-Sommerfeld bands, fills out the totality of the phase space. It is this interpretation of a quantum state that allows us to develop a simple algorithm for the evaluation of the scalar product.

Now that we have understood how to represent the individual quantum states we can turn to the question of the scalar product. For this purpose we note from Fig. 7.1 that through this all-encompassing collection of "race tracks" cuts the band of quite different lineage for the initial state of quantum number n. It too has area $2\pi\hbar$.

Under sudden change of the conditions of motion, what is the probability that the system, initially in state n, will transit to this, that, or the other final state m? In answering this question, already in the absence of a closer examination, a simple consideration serves as a guide. We expect the probability for a transition to any final state m to be related to the area or areas of intersection between it and the band for the initial state. No overlap – no transition!

Moreover, the areas cut out of the initial-state band n by all candidate final-state bands add up to the total area $2\pi\hbar$ of that initial band. Furthermore, the probabilities for transition from the initial state n to all final states m add up to unity. Therefore, it is tempting to identify transition probability with $1/(2\pi\hbar)$ times the area of overlap. It would be hard to imagine a simpler algorithm to calculate transition probabilities, nor one that would in a more obvious fashion uphold the sum rule.

Stated so, however, this path to the reckoning of transition probability is too simplistic to be right, and for a simple reason: As shown in Fig. 7.1 the area of overlap typically consists not of one piece but of two pieces, and occasionally more, usually disposed (except in the presence of a magnetic field) symmetrically above and below the coordinate axis. With each of these areas of intersection, quantum mechanics associates not merely a probability but a probability *amplitude*. Young's double-slit experiment provides the relevant guide: *Probabilities* do not add – probability *amplitudes* do. We have no escape but to conclude that we deal here with *interference*

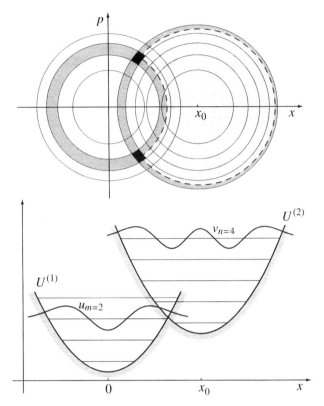

Fig. 7.1: Tools for figuring the transition probability – illustrating the ties between classical theory and quantum theory. Below: The curves for potential energy as a function of the position coordinate x before the transition (above) and after the transition (slightly lower). The horizontal lines indicate allowable energy levels in these potentials before and after this transition. The wave functions u_m and v_n shown here belong to the initial and final states between which we wish to evaluate the transition probability via an integral of the form

$$\int u_m(x) \times \text{(slowly varying function of } x) \times v_n(x)\,dx.$$

Top: Bands in phase space (dark shaded) associated with initial and final state, and the area of overlap between them (blackened). In the case of more than one such overlap (in the present example two) the two, diamond-shaped, dark areas differ in their "momentum": In one dark zone both oscillators move to the "right", whereas in the other to the "left". Thus the total probability amplitude is the sum of contributions $\mathcal{A}_{m,n}^{1/2}\exp(\pm i\varphi_{m,n})$ from the two zones. Here the phase $\varphi_{m,n}$ is fixed by the light shaded area caught between the center lines of the two states. No simpler illustration presents itself for *interference in phase space*.

in phase space.

Hence, we do not figure probability of transition by adding the reduced area of intersection, $\mathcal{A}_{m,n}^{(\diamond)}/2\pi\hbar$, above the coordinate axis, to the identical reduced area of intersection, $\mathcal{A}_{m,n}^{(\diamond)}/2\pi\hbar$, below the axis. Instead, we add a probability amplitude $(\mathcal{A}_{m,n}^{(\diamond)}/2\pi\hbar)^{1/2} \exp(i\varphi_{m,n})$ for the one domain of intersection to a probability amplitude $(\mathcal{A}_{m,n}^{(\diamond)}/2\pi\hbar)^{1/2} \exp(-i\varphi_{m,n})$ associated with the other domain and square the sum to obtain the jump probability,

$$W_{m,n} = \left| \left[\frac{\mathcal{A}_{m,n}^{(\diamond)}}{2\pi\hbar}\right]^{1/2} e^{i\varphi_{m,n}} + \left[\frac{\mathcal{A}_{m,n}^{(\diamond)}}{2\pi\hbar}\right]^{1/2} e^{-i\varphi_{m,n}} \right|^2 = 4 \frac{\mathcal{A}_{m,n}^{(\diamond)}}{2\pi\hbar} \cos^2 \varphi_{m,n}.$$

Here the phase $\varphi_{m,n}$ of the contribution of the upper domain of intersection is governed by the light shaded area in phase space shown in Fig. 7.1 and embraced between the horizontal axis and the Kramers center lines of the two bands of interest. Similarly the phase follows for the lower domain.

7.2 Derivation of Area-of-Overlap Formalism

In the present section we derive the concept of area-of-overlap as a determiner of transition probability outlined in the preceding section. For this purpose we first evaluate the scalar product between two energy eigenstates using their WKB approximations and the method of stationary phase. We then identify the individual terms in the resulting expression as areas in phase space.

7.2.1 Jumps Viewed From Position Space

Approximate the wave functions u_m and v_n by their WKB expressions and then evaluate the integral Eq. (7.1) with the help of stationary phase; that is the approach suggested by Landau many years ago in the context of energy transfer in collisions, pursued successfully in many different physical problems and reviewed in this section with the emphasis on an interpretation in phase space.

Energy Wave Functions in WKB

In the semiclassical limit, that is for large quantum numbers m, we approximate the wave functions u_m in the region between the two classical turning points ϑ_m and ξ_m, ($\vartheta_m < \xi_m$, see Fig. 7.2a) by the WKB wave functions

$$u_m^{(\text{WKB})}(x) = \frac{N_m}{\sqrt{p_m(x)}} \cos[S_m(x) - \pi/4] \tag{7.2a}$$

where

$$p_m(x) \equiv \sqrt{2M[E_m - U^{(1)}(x)]} \tag{7.2b}$$

and

$$S_m(x) \equiv \frac{1}{\hbar} \int_x^{\xi_m} dx\, p_m(x). \tag{7.2c}$$

7.2 Derivation of Area-of-Overlap Formalism

The energy E_m of the mth eigenstate of the potential $U^{(1)} = U^{(1)}(x)$, shown in Fig. 7.2(a), is determined by the Bohr-Sommerfeld-Kramers quantization condition

$$J_m \equiv \oint dx\, p_m(x) = 2\pi\hbar \left(m + \frac{1}{2}\right). \tag{7.2d}$$

According to Eq. (5.5), the normalization constants $\mathcal{N}_m$ are given by

$$\mathcal{N}_m \equiv 2\left(\frac{M}{T_m}\right)^{1/2}, \tag{7.2e}$$

where the period of the mth orbit in $U^{(1)}$ is

$$T_m \equiv 2M \int_{\vartheta_m}^{\xi_m} dx\, p_m^{-1} = \oint dt. \tag{7.2f}$$

Analogously we approximate the wave functions $v_n = v_n(x)$ for x-values appropriately between the classical turning points ζ_n and χ_n ($\zeta_n < \chi_n$, see Fig. 7.2(a)) by

$$v_n^{(\text{WKB})}(x) = \frac{\mathcal{N}_n}{\sqrt{p_n(x)}} \cos[S_n(x) - \pi/4] \tag{7.3a}$$

with

$$p_n(x) \equiv \sqrt{2M[E_n - U^{(2)}(x)]} \tag{7.3b}$$

and

$$S_n(x) \equiv \frac{1}{\hbar} \int_x^{\chi_n} dx\, p_n(x). \tag{7.3c}$$

The energy E_n, corresponding to motion in the potential $U^{(2)} = U^{(2)}(x)$, follows from the Bohr-Sommerfeld-Kramers quantization condition

$$J_n \equiv \oint dx\, p_n(x) = 2\pi\hbar \left(n + \frac{1}{2}\right). \tag{7.3d}$$

The normalization constants $\mathcal{N}_n$ are given by

$$\mathcal{N}_n \equiv 2\left(\frac{M}{T_n}\right)^{1/2}, \tag{7.3e}$$

where

$$T_n = 2M \int_{\zeta_n}^{\chi_n} dx\, p_n^{-1} \tag{7.3f}$$

is the period of the nth orbit in $U^{(2)}$.

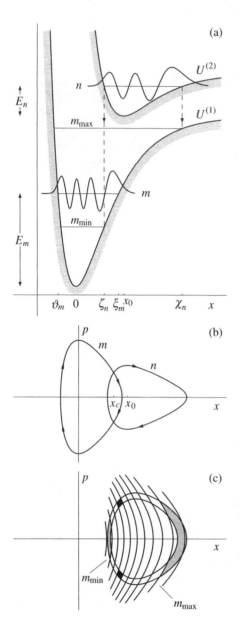

Fig. 7.2: Area-of-overlap method to evaluate jump probability, briefly recapitulated. (a) Potential of the force under which a system oscillates, depicted as a function of coordinate x, before ($U^{(2)}$, with minimum at x_0) and after ($U^{(1)}$, with minimum at $x = 0$) the sudden change in the conditions of motion (capital example: vibration of a diatomic molecule as effected by an electronic transition). Also shown in (a) are the energy level and wave function for the original vibration state n and for a sample one m of the many final states that compete for attention. (b) Phase space trajectories m and n of candidate final state m and given initial state n, as determined from the Kramers quantization rule: The areas enclosed are equal to $2\pi\hbar(m + 1/2)$ and $2\pi\hbar(n + 1/2)$, respectively. (c) Initial state and candidate final state depicted as Planck-Bohr-Sommerfeld bands of area $2\pi\hbar$. The two symmetrically located diamond-shaped *zones* of intersection between bands n and m serve as the two "slits" of the "phase space-double-slit experiment," the outcome of which governs the transition probability $W_{m,n}$. The sum of transition probabilities from n to all final states m, when accurately calculated, adds exactly to $2\pi\hbar$, as expected from the area of the nth initial-band. When the band for the candidate final state makes a tangential intersection with the band for the given initial state, as happens in the neighborhood of the turning points $x = \zeta_n$ and $x = \chi_n$, it has an unusually large overlap with the initial-state band (shaded areas) and, according to the area-of-overlap-concept, yield a large transition probability. For m-values smaller than $m_{\min}$ or larger than $m_{\max}$ there is no overlap. We exclude such cases of no overlap from the present discussion.

7.2 Derivation of Area-of-Overlap Formalism

Points of Stationary Phase

In order to evaluate the probability amplitude $w_{m,n}$ we substitute the WKB expressions Eqs. (7.2a) and (7.3a) for the two energy wave functions $u_m(x)$ and $v_n(x)$. We hence arrive at

$$w_{m,n} = \int_{-\infty}^{\infty} dx \, u_m(x) \, v_n(x) \cong \int_{\zeta_n}^{\xi_m} dx \, u_m^{(\text{WKB})}(x) \, v_n^{(\text{WKB})}(x)$$

where the integration in the last integral is confined to the region common to both wave functions.

When we now use the explicit expressions Eqs. (7.2) and (7.3) with their respective normalization constants we find

$$w_{m,n} \cong \frac{M}{(T_m T_n)^{1/2}} \int_{\zeta_n}^{\xi_m} dx \, [p_m(x) \, p_n(x)]^{-1/2} \{\exp[i S_{m,n}(x)] + \exp[-i S_{m,n}(x)]\}, \quad (7.4)$$

where

$$S_{m,n}(x) \equiv S_m(x) - S_n(x) = \frac{1}{\hbar} \int_x^{\xi_n} dx \, p_m(x) - \frac{1}{\hbar} \int_x^{\chi_n} dx \, p_n(x). \quad (7.5)$$

Here we have made use of the expressions Eqs. (7.2c) and (7.3c) for the phases of the WKB wave functions. Moreover, we have neglected the rapidly oscillating contributions $\exp\{\pm i[S_m(x) + S_n(x)]\}$.

The envelope $[p_m(x) p_n(x)]^{-1/2}$ is slowly varying on the scale of the oscillations resulting from $\exp[i S_{m,n}]$. We therefore employ the method of stationary phase, that is, we expand $S_{m,n}$ into a Taylor series

$$S_{m,n}(x) = S_{m,n}(x_c) + \left.\frac{dS_{m,n}}{dx}\right|_{x=x_c} (x - x_c) + \frac{1}{2} \left.\frac{d^2 S_{m,n}}{dx^2}\right|_{x=x_c} (x - x_c)^2 + \ldots \quad (7.6)$$

around the stationary point x_c, given by

$$\left.\frac{dS_{m,n}}{dx}\right|_{x=x_c} = 0.$$

From the explicit form of $S_{m,n}$, Eq. (7.5), we find the condition

$$p_m(x_c) = p_n(x_c). \quad (7.7)$$

Therefore, the main contribution to the integral (7.4) arises from those points x_c at which the momenta p_m and p_n are equal, that is, from the points of crossing of the two phase space orbits

$$E_m = \frac{p_m^2}{2M} + U^{(1)}(x) \quad \text{and} \quad E_n = \frac{p_n^2}{2M} + U^{(2)}(x) \quad (7.8)$$

as shown in Fig. 7.2(b). The number of such crossings x_c depends on the shape of the potentials $U^{(1)}$ and $U^{(2)}$. In the remainder of the chapter, however, we limit ourselves to potentials giving rise to a *single* crossing point x_c only.

Evaluation of Scalar Product

The phase difference $S_{m,n}$, Eq. (7.5), expressed by the Taylor series, Eq. (7.6), thus reads

$$S_{m,n}(x) \cong S_{m,n}(x_c) + \frac{1}{2\hbar}\left[\frac{dp_n(x)}{dx} - \frac{dp_m(x)}{dx}\right]_{x=x_c}(x-x_c)^2.$$

When we substitute this relation into Eq. (7.4) and evaluate the slowly varying term $[p_m(x)\,p_n(x)]^{-1/2}$ at x_c, we arrive at

$$w_{m,n} \cong M\,[T_m\,p_m(x_c)T_n\,p_n(x_c)]^{-1/2}\bigg\{\exp[iS_{m,n}(x_c)]$$

$$\times \int dx\,\exp\left\{\frac{i}{2\hbar}\left[\frac{dp_n}{dx} - \frac{dp_m}{dx}\right]_{x=x_c}(x-x_c)^2\right\} + \text{c. c.}\bigg\}, \qquad (7.9a)$$

where, according to Eq. (7.5),

$$S_{m,n}(x_c) \equiv \frac{1}{\hbar}\int_{x_c}^{\xi_m} dx\,p_m(x) - \frac{1}{\hbar}\int_{x_c}^{\chi_n} dx\,p_n(x). \qquad (7.9b)$$

denotes the phase difference at the crossing point x_c.

Probability Amplitudes Made Simple

In the remaining integral, Eq. (7.9a), determining the scalar product we have not specified the limits of integration yet. In Eq. (7.4) this integration has been restricted to the region common to both wave functions, that is, $\zeta_n < x < \xi_m$. According to Eqs. (7.7) and (7.9a), however, the main contribution to the integral arises from the neighborhood δx of x_c, that is, from the vicinity of the crossing of the two Kramers phase space orbits, Eq. (7.8). But what determines this neighborhood?

Definitely one factor governing the contributing neighborhood is the quantity

$$k \equiv \frac{1}{(\pi\hbar)^{1/2}}\left|\frac{dp_n(x_c)}{dx} - \frac{dp_m(x_c)}{dx}\right|^{1/2}$$

appearing in the integral Eq. (7.9a). It is a useful measure of the difference in slope of the two Kramers orbits in phase space, Eq. (7.8). With a big angle of crossing the waves quickly get out of step, leading to a short region of constructive interference. The converse is true for a small angle of crossing. Therefore, the magnitude of the probability amplitude $w_{m,n}$ is governed by this angle.

The other important quantity determining $w_{m,n}$ is the neighborhood of x_c which provides different values for the remaining integral. This dependence converges slowly and is discussed in detail with the help of the Cornu spiral in Problem 7.4. We, therefore, extend the integration limits to infinity and recall the integral relation

$$\int_{-\infty}^{\infty} dx\,e^{ikx^2} = \sqrt{\frac{\pi}{|k|}}\,e^{i\,\text{sign}k\,\frac{\pi}{4}}.$$

7.2 Derivation of Area-of-Overlap Formalism

In this case the total probability amplitude, Eq. (7.1), is the sum

$$w_{m,n} \cong \mathcal{A}_{m,n}^{1/2} \exp(i\varphi_{m,n}) + \mathcal{A}_{m,n}^{1/2} \exp(-i\varphi_{m,n}) \qquad (7.10\text{a})$$

of two probability amplitudes. Both have the same absolute value

$$\mathcal{A}_{m,n} \equiv 2\pi\hbar M^2 \left[T_m\, p_m(x_c)\, T_n\, p_n(x_c)\right]^{-1} \left|\frac{dp_n(x_c)}{dx} - \frac{dp_m(x_c)}{dx}\right|^{-1}. \qquad (7.10\text{b})$$

However, they have different phases $\pm\varphi_{m,n}$ given by

$$\varphi_{m,n} \equiv \frac{1}{\hbar}\int_{x_c}^{\xi_m} dx\, p_m(x) - \frac{1}{\hbar}\int_{x_c}^{\chi_n} dx\, p_n(x) + \text{sign}\left[\frac{dp_n(x_c)}{dx} - \frac{dp_m(x_c)}{dx}\right]\frac{\pi}{4}. \qquad (7.10\text{c})$$

Three conclusions stand out from this analysis: *(i)* The quantum mechanical scalar product, Eq. (7.1), between two semiclassical states described by WKB wave functions $u_m^{(\text{WKB})}$ and $v_n^{(\text{WKB})}$, Eqs. (7.2) and (7.3), is the sum of (in this case) two complex-valued *probability amplitudes*, of magnitude $\mathcal{A}_{m,n}^{1/2}$, given by Eq. (7.10b) and having a phase difference $2\varphi_{m,n}$ governed by Eq. (7.10c). *(ii)* These probability amplitudes are the result of the crossing of the Kramers trajectories, Eq. (7.8), corresponding to the two quantum states. *(iii)* The magnitude of each amplitude is partially determined by the angle of the railway switch.

The analysis presented in this section breaks down for quantum states m which touch the elliptical n-orbit of Fig. 7.2(c) tangentially. This situation corresponds to Franck-Condon transitions around $m_{\min}$ and $m_{\max}$ of Fig. 7.2(a), at the turning points of the classical motion. Here the WKB waves, Eqs. (7.2) and (7.3), turn singular and a uniform asymptotic expansion of u_m and v_n is necessary. Since the goal of the present chapter is to bring out the conceptual points of the phase space interpretation of semiclassical techniques, such as the area-of-overlap approach of Sec. 7.2.2 we restrict ourselves to quantum states m which lie between $m_{\min}$ and $m_{\max}$ of Fig. 7.2(a), that is, to m-bands which cut two symmetrically located diamonds out of the n-band. For these m-values, this approach yields good agreement with the exact treatment.

7.2.2 Jumps Viewed From Phase Space

In the preceding section we have shown that the scalar product between two energy eigenstates of two different potentials is a sum of probability amplitudes. In the present section we prove that their absolute values squared $\mathcal{A}_{m,n}$ and the phases $\varphi_{m,n}$ are areas in phase space. For this purpose we first represent the energy eigenstates as bands in phase space and then calculate their overlap in the Bohr correspondence limit.

States as Bands in Phase Space

According to Eq. (7.8), we can associate with the mth energy eigenstate in potential $U^{(1)}$ a *single* phase space orbit

$$E_m = \frac{p_m^2}{2M} + U^{(1)}(x) \qquad (7.11)$$

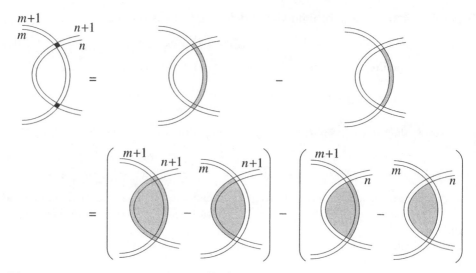

Fig. 7.3: The total area of overlap, $A_{m,n}^{(\text{PBS})}$, between the initial Planck-Bohr-Sommerfeld band n and the candidate final one m, that is, the sum of the two symmetrically located diamond-shaped zones, can be expressed as the double differences of the shaded areas embraced by the corresponding phase space trajectories.

as shown in Fig. 7.2(b). Here the energy E_m follows from the Bohr-Sommerfeld-Kramers quantization condition.

A version better suited for the present purpose, however, associates with the mth energy eigenstate not a single trajectory but a whole *band* in phase space. According to Planck, each state takes up an area $2\pi\hbar$. The simplest definition of the inner edge of this band is thus given by the phase space orbit

$$E_m^{(\text{in})} = \frac{1}{2M}\left[p_m^{(\text{in})}\right]^2 + U^{(1)}(x) \tag{7.12a}$$

where the energy $E_m^{(\text{in})}$ is defined by

$$\oint dx\, p_m^{(\text{in})} = 2\pi\hbar m. \tag{7.12b}$$

The outer edge is thus determined by the orbit

$$E_m^{(\text{out})} = \frac{1}{2M}\left[p_m^{(\text{out})}\right]^2 + U^{(1)}(x) \tag{7.13a}$$

with $E_m^{(\text{out})}$ given by

$$\oint dx\, p_m^{(\text{out})} = 2\pi\hbar(m+1). \tag{7.13b}$$

The Kramers trajectory, Eq. (7.11), "runs" in the middle of the mth band of area $2\pi\hbar$, defined by Eqs. (7.12) and (7.13).

7.2 Derivation of Area-of-Overlap Formalism

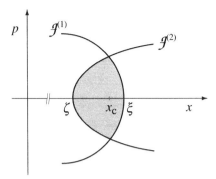

Fig. 7.4: Area $a = a(\mathcal{J}^{(1)}, \mathcal{J}^{(2)})$ in phase space, caught between the two orbits corresponding to the reduced actions $\mathcal{J}^{(1)}$ and $\mathcal{J}^{(2)}$.

Area of Overlap as Double Differences

When we associate with the m state of Fig. 7.2(a) such a band and likewise with the nth state of $U^{(2)}$, the two bands intersect each other in the neighborhood of the crossing point $(x_c, \pm p_m(x_c))$ in phase space, forming two diamond-shaped zones of total area $A_{m,n}^{(\mathrm{PBS})} = 2\mathcal{A}_{m,n}^{(\diamond)}$ as shown in Fig. 7.2(c). No better algorithm for evaluating the overlap integral $w_{m,n}$, Eq. (7.1), of the two wave functions u_m and v_n in *position space* offers itself than the area of overlap $A_{m,n}^{(\mathrm{PBS})}$ in *phase space* of the corresponding Planck-Bohr-Sommerfeld bands, Eqs. (7.12) and (7.13). We now evaluate this area of overlap $A_{m,n}^{(\mathrm{PBS})}$.

According to Fig. 7.3 we can express the quantity $A_{m,n}^{(\mathrm{PBS})}$ in terms of the double differences

$$A_{m,n}^{(\mathrm{PBS})} = a(m+1, n+1) - a(m, n+1) - [a(m+1, n) - a(m, n)] \qquad (7.14)$$

of the area

$$a = a(\mathcal{J}^{(1)}, \mathcal{J}^{(2)})$$
$$= 2\left\{ \int_{\zeta(\mathcal{J}^{(2)})}^{x_c(\mathcal{J}^{(1)}, \mathcal{J}^{(2)})} dx\, p(x; \mathcal{J}^{(2)}) + \int_{x_c(\mathcal{J}^{(1)}, \mathcal{J}^{(2)})}^{\xi(\mathcal{J}^{(1)})} dx\, p(x; \mathcal{J}^{(1)}) \right\} \qquad (7.15)$$

of Fig. 7.4 embraced by the two phase space orbits

$$E^{(j)}(\mathcal{J}^{(j)}) = \frac{1}{2M} p^2(x; \mathcal{J}^{(j)}) + U^{(j)}(x) \qquad (7.16)$$

($j = 1, 2$), corresponding to the two reduced actions

$$\mathcal{J}^{(j)} \equiv \frac{1}{2\pi\hbar} \oint dx\, p(x; \mathcal{J}^{(j)}). \qquad (7.17)$$

The point of intersection x_c of the two orbits, Eq. (7.16), determined by the condition

$$p(x_c; \mathcal{J}^{(1)}) = p(x_c; \mathcal{J}^{(2)}) \qquad (7.18)$$

depends on $\mathcal{J}^{(1)}$ and $\mathcal{J}^{(2)}$. The classical turning points $\zeta = \zeta(\mathcal{J}^{(2)})$ and $\xi = \xi(\mathcal{J}^{(1)})$ follow from the condition

$$p(x = \xi; \mathcal{J}^{(1)}) = p(x = \zeta; \mathcal{J}^{(2)}) = 0. \tag{7.19}$$

and only depend on one of the two reduced actions.

Area of Overlap in the Correspondence Limit

According to Bohr's correspondence principle we can replace in the limit of large quantum numbers m and n the differences by partial differentials, that is

$$a(m+1, n+1) - a(m, n+1) \cong \left.\frac{\partial a(\mathcal{J}^{(1)}, n+1)}{\partial \mathcal{J}^{(1)}}\right|_{\mathcal{J}^{(1)} = m + \frac{1}{2}}$$

and similarly for the other action variable. Equation (7.14) thus reduces to

$$A_{m,n}^{(\mathrm{PBS})} \cong \left.\frac{\partial^2 a(\mathcal{J}^{(1)}, \mathcal{J}^{(2)})}{\partial \mathcal{J}^{(1)} \partial \mathcal{J}^{(2)}}\right|_{\substack{\mathcal{J}^{(1)} = m + \frac{1}{2} \\ \mathcal{J}^{(2)} = n + \frac{1}{2}}}. \tag{7.20}$$

Therefore, in this limit we can calculate the area of overlap, $A_{m,n}^{(\mathrm{PBS})}$, between the states m and n by partially differentiating the area $a = a(\mathcal{J}^{(1)}; \mathcal{J}^{(2)})$, Eq. (7.15), with respect to the reduced actions $\mathcal{J}^{(j)}$. This calculation we perform in Appendix I and find

$$A_{m,n}^{(\mathrm{PBS})} = 2(2\pi\hbar)^2 M^2 \left[T_m\, p_m(x_c)\, T_n\, p_n(x_c)\, |p'_m(x_c) - p'_n(x_c)|\right]^{-1}. \tag{7.21}$$

When we compare this result to the magnitude of one of the contributing probability amplitudes, Eq. (7.10), we arrive at

$$\mathcal{A}_{m,n} \cong \frac{1}{2}\left[\frac{A_{m,n}^{(\mathrm{PBS})}}{2\pi\hbar}\right] = \frac{A_{m,n}^{(\diamond)}}{2\pi\hbar}. \tag{7.22}$$

Thus we can associate the magnitude of the probability amplitude, $\mathcal{A}_{m,n}^{1/2}$, with the square root of the area of *one* of the two symmetrically located diamond-shaped zones divided by $\sqrt{2\pi\hbar}$. The weight factor $2\pi\hbar$ stems from the fact that the probabilities have to add up to unity, whereas the area of the Planck-Bohr-Sommerfeld bands is given by $2\pi\hbar$.

7.3 Application to Franck-Condon Transitions

The area of overlap between two Planck-Bohr-Sommerfeld bands determines the magnitude of the probability amplitude. But why? The mathematics shown above proves it! But can we present a more intuitive argument in favor of this?

The answer is *yes*. Represent the nth energy eigenstate of the potential $U^{(2)}$ shown in Fig. 7.2(a) – the initially occupied quantum state – by the nth elliptical Planck-Bohr-Sommerfeld band of Fig. 7.2(c). Go a step further and visualize this band as a constant flow of particles bounded by the edges of the band, all moving along their respective phase space trajectories. This is similar to trains running along their tracks. In Fig. 7.2(c) we also depict the initially empty energy bands of the potential $U^{(1)}$.

Now induce a Frank-Condon transition from the nth vibratory level of the potential $U^{(2)}$ to any of the levels of $U^{(1)}$ by *suddenly* changing $U^{(2)}$ to $U^{(1)}$. This alteration of the potential causes the m bands to intersect the nth elliptical band and provides the means for the railway switch to redirect the particles from their course on the "n line" to the corresponding "m line".

But how many particles can we find on any particular m line? Obviously all particles of the n band, caught at the instant of the potential change between the edges of the m band, will be re-routed into the new course given by the trajectories representing that particular band. Hence, the number of particles in the mth band is determined by the black, diamond-shaped area of overlap of the two domains.

Since in this case we find two such diamonds, the number of particles in that band is the sum of particles in each diamond, that is, we have to add the two areas of overlap. Nowhere clearer than at this point do we recognize the difference between classical and quantum physics. The orbiting particles represent classical *probabilities* which we add.

In contrast, quantum mechanics deals in terms of probability *amplitudes;* hence the black diamonds represent *interfering* probability amplitudes, and their phase difference is governed by Eq. (7.9b). This quantity also allows a simple geometrical interpretation in phase space. According to Eq. (7.9b) and Eq. (7.10b) (and apart from the sign shift of $\mathrm{sign}[p'_n(x_c) - p'_m(x_c)]\frac{\pi}{4}$) it is the area in phase space caught between the two Kramers trajectories Eq. (7.8), shown in Fig. 7.1.

7.4 Generalization

We can summarize and at the same time generalize the main result of this chapter by the formula

$$\langle \chi | \psi \rangle = \sum_j \left(\frac{j\text{th area of overlap}}{2\pi\hbar} \right)^{1/2} \exp\left[\frac{i}{\hbar} \left(\begin{array}{c} j\text{th area} \\ \text{enclosed by} \\ \text{central lines} \end{array} \right) \right] \qquad (7.23)$$

for the scalar product of two quantum states $|\chi\rangle$ and $|\psi\rangle$. According to this prescription we depict the two quantum states of interest in phase space. We evaluate their areas of overlap and divide each area by $2\pi\hbar$. When we take the square root we arrive at the corresponding probability amplitude of the quantum mechanical scalar product. When there is more than one distinct overlap we have to sum probability amplitudes. The phase difference between the contribution is again governed by an area in phase space. This time it is the area enclosed by the central lines divided by $\hbar$.

The physics of the overlap algorithm, as just examined and summarized in Fig. 7.1, is very similar to the familiar double-slit experiment. In both cases there are two interfering contributions to the total probability of detecting a specified outcome. In one case, the two contributions come from the two slits. In the other case, they result from the two distinct areas of overlap in phase space. The phase difference in the double-slit experiment is measured by the difference in optical path length from the centers of the two slits to the point of detection. Likewise, the probability amplitudes for the two contributions in Eq. (7.10a) have a phase difference. It is governed by the area in oscillator phase space caught between initial and final orbits expressed by Eqs. (7.9b) and (7.10c). In this sense Young's famous interference experiment is generalized to *interference in phase space*.

Problems

7.1 Area of Overlap from Geometry

Calculate the diamond shaped area of overlap between two energy eigenstates shown in Fig. 7.3 and given by Eq. (7.21) just from geometry in phase space.

Hint: Approximate in the region of overlap the phase space trajectories defining the diamond by straight lines as shown in Fig. I.1. Calculate the area by expressing it in terms of the angle enclosed between the two quantum states and their heights. See also the discussion in Appendix I.

7.2 Interference in Phase Space from Wigner Function

How does the concept of interference in phase space discussed in this chapter translate itself into the language of Wigner functions? Start from the relation Eq. (3.7) expressing the scalar product of two quantum states in terms of their Wigner functions. How does quantum interference arise in this approach?

Hint: See Dowling *et al.* (1991).

7.3 The Mulliken Principle

The Mulllican principle postulates the conservation of kinetic energy of the nuclei during a Franck-Condon transition. How is this related to the area of overlap approach?

Hint: See Mulliken (1971).

7.4 Transition Probability Amplitudes and Cornu Spiral

Use the cornu spiral discussed in Appendix H and shown in Fig. H.1 to discuss geometrically the influence of the neighborhood δx of x_c on the value of each of the probability amplitudes. This shows that the approach to the steady-state values Eqs. (7.10) is rather slow. However, this slowness is a happy feature of the "method of interference in phase space." It tells us that it matters little for the final transition probability exactly how the latter loops spiral inward, it matters little how the wave functions vary – if they vary slowly – outside of the zone of constructive interference, that is, of the caustics of the bands in phase space.

Hint: Cast Eq. (7.9a) into the form

$$w_{m,n}(\delta x) = \mathcal{A}_{m,n}^{1/2} \exp[iS_{m,n}(x_c)] \sqrt{2} F(k\delta x) + \text{c. c.}$$

and use the properties of the Cornu spiral

$$F(y) \equiv \int_0^y dt \, \exp(i\pi t^2/2)$$

discussed in Appendix H.

References

Molecular Physics

For a summary of molecular physics see
G. Herzberg, *Molecular Spectra and Molecular Structure*, Vol. I, Spectra of Diatomic Molecules, Nostrand, Princeton, 1965

This book is the standard introduction to molecular physics. It discusses in great detail the Born-Oppenheimer approximation and the Frank-Condon principle.

See also the recent edition of
L. Bergmann and **C. Schäfer**, *Constituents of Matter: Atoms, Molecules, Nuclei and Particles*, W. de Gruyter, Berlin, 1997

For the original paper on the Born-Oppenheimer approximation see
M. Born and **R. Oppenheimer**, *Zur Quantentheorie der Molekeln*, Ann. Phys. (Leipzig) **84**, 457–484 (1927)

For the history of the Frank-Condon principle see the retirement speech of Edward Condon from the presidency of the Americal Physical Society
E.U. Condon, *The Franck-Condon Principle and Related Topics*, Am. J. Phys. **15**, 365–374 (1947)

The scalar product between the energy eigenstates of two displaced anharmonic oscillators using WKB wave functions and the method of stationary phase has been evaluated for the first time in
L.D. Landau, *Zur Theorie der Energieübertragung bei Stößen*, Phys. Z. Sowjet. **1**, 88–98 (1932)

reprinted in
D. Ter Haar, Collected Papers of L.D. Landau, Pergamon, Oxford, 1965
L.D. Landau and **E.M. Lifshitz**, *Quantum Mechanics – Nonrelativistic Theory*, (Pergamon, Oxford, 1958)

Interference in Phase Space

The concept of interference in phase space is spelled out in
J.A. Wheeler, *Franck-Condon effect and squeezed-state physics as double-source interference phenomena*, Lett. Math. Phys. **10**, 201–206 (1985)

and, in particular, in
J.P. Dowling, W.P. Schleich and **J.A. Wheeler**, *Interference in Phase Space*, Ann. Phys. (Leipzig) **48**, 423–502 (1991)

For a summary of this concept see
V. Buzek and **P.L. Knight**, *Quantum Interference, Superposition States of Light and Nonclassical Effects*, Prog. Opt. **34**, 1–158 (1995)

The concept of interference in phase space has been generalized by
G.S. Agarwal, *Interference in Complementary Spaces*, Found. Phys. **25**, 219–228 (1995)

Phase Space Interpretation of WKB

For closely related work evaluating matrix elements using WKB see
A. Jabłoński, *Über das Entstehen der breiten Absorptions- und Fluoreszenzbanden in Farbstofflösungen*, Z. Phys. **73**, 460–469 (1932)
M.S. Child, *Molecular Collision Theory*, Academic Press, London, 1974
E.J. Heller, *Phase Space Interpretation of Semiclassical Theory*, J. Chem. Phys. **67**, 3339–3351 (1977)
E.J. Heller, *Quantum Corrections to Classical Photodissociation Models*, J. Chem. Phys. **68**, 2066–2075 (1978)
H.-W. Lee and **M.O. Scully**, *Wigner Phase-Space Description of a Morse Oscillator*, J. Chem. Phys. **77**, 4604–4610 (1982)
W.H. Miller, *The Classical S-Matrix in Molecular Collisions*, Adv. Chem. Phys., ed. by I. Prigogine and S.A. Rice, Vol. XXX, Wiley, New York, 1975
R.S. Mulliken, *Role of Kinetic Energy in the Franck-Condon Principle*, J. Chem. Phys. **55**, 309–314 (1971)

8 Applications of Interference in Phase Space

In Chapter 4 we have introduced the coherent and squeezed states of the harmonic oscillator and discussed some of their properties. In particular, we have calculated their energy distribution W_m from the overlap integral between the corresponding wave functions.

In this chapter we revisit this problem and make use of the concept of interference in phase space developed in the last chapter: We evaluate the energy distribution by calculating areas of overlap in phase space. For this purpose we have to find the appropriate phase space representations of the two quantum states of interest, that is the energy eigenstate and the coherent or squeezed state. Then we calculate their overlap. In contrast to the previous chapters we now use dimensionless phase space variables. This facilitates the evaluation of the areas of overlap. Moreover the phase space of a single mode of the electromagnetic field has the same dimensionless phase space variables. We conclude this chapter by briefly discussing the problem of phase states in quantum mechanics. Here the concept of interference in phase space turns out to be particularly useful since it provides deeper insight into the definition of phase states.

8.1 Connection to Interference in Phase Space

In Chapter 7 we have given a geometrical representation of the quantum mechanical scalar product between two quantum states as interfering areas in phase space. We now apply these ideas to analyze the energy distribution

$$W_m \equiv |\langle m | \psi \rangle|^2 = |w_m(|\psi\rangle)|^2$$

of a coherent state or a squeezed state. Indeed, this probability follows from the scalar product of an energy eigenstate $|m\rangle$ and the coherent or squeezed state $|\psi\rangle$.

The fact, that we can apply the concept of interference in phase space stands out most clearly when we recall that the overlap integral

$$w_m(|\psi\rangle) \equiv \int_{-\infty}^{\infty} dx\, u_m(x)\, \psi(x)$$

is of the form Eq. (7.1).

Here $u_m(x)$, Eq. (4.2), is the wave function of the mth energy eigenstate of the harmonic oscillator. The wave functions $\psi(x) \equiv \psi_{\text{coh}}(x)$ or $\psi(x) \equiv \psi_{\text{sq}}(x)$ of the coherent or squeezed state, Eqs. (4.11) or (4.33), respectively play the role of $v_n(x)$.

We realize that this is rather problematic since in our derivation of the formalism of interference in phase space we have made use of the WKB-representation of both wave functions $u_m(x)$ and $v_n(x)$. However, in the present example $v_n(x)$ cannot be represented by a WKB-wave function. Nevertheless, a slightly improved formalism provides excellent results.

In the treatment of Chapter 7 we have allowed for arbitrary binding potentials $U^{(1)}$ and $U^{(2)}$. In the present chapter we focus on the special case of a harmonic oscillator potential

$$U^{(1)}(x) \equiv \frac{1}{2} M\Omega^2 x^2.$$

This allows us to introduce the dimensionless phase space variables

$$\tilde{x} \equiv \kappa x \quad \text{and} \quad \tilde{p} \equiv \frac{p}{\hbar \kappa}$$

where

$$\kappa \equiv \sqrt{M\Omega/\hbar}.$$

The elliptical phase space trajectory of the mth energy eigenstate determined by the equation

$$E_m = \frac{1}{2M} p_m^2 + \frac{1}{2} M\Omega^2 x^2 = \hbar\Omega\left(m + \frac{1}{2}\right)$$

then takes the form of a circle

$$\tilde{\eta}_m = m + \frac{1}{2} = \frac{1}{2}\tilde{p}_m^2 + \frac{1}{2}\tilde{x}^2. \tag{8.1}$$

of radius $\sqrt{2(m + \frac{1}{2})}$. Here we have introduced the dimensionless energy

$$\tilde{\eta} \equiv \frac{E}{\hbar\Omega}.$$

For the sake of simplicity we suppress throughout this chapter the tilde on the dimensionless variables.

8.2 Energy Eigenstates

In Chapter 7 we have represented the energy eigenstates of a binding potential as bands in phase space. In the present section we discuss this representation for the case of energy eigenstates of a harmonic oscillator.

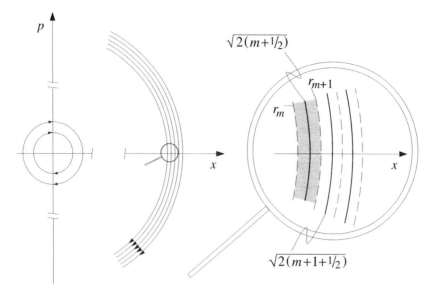

Fig. 8.1: The trajectory corresponding to the mth energy eigenstate of a harmonic oscillator in an appropriately scaled phase space is a circle with radius $\sqrt{2(m+1/2)}$. With the state we associate a band of area 2π, defined by its inner radius $r_m \equiv \sqrt{2m}$ and outer radius $r_{m+1} \equiv \sqrt{2(m+1)}$.

In the dimensionless phase space variables x and p the bands defined in Chapter 7 follow from the two phase space trajectories

$$m = \frac{1}{2}\left[p_m^{(\text{in})}\right]^2 + \frac{1}{2}x^2$$

and

$$m+1 = \frac{1}{2}\left[p_m^{(\text{out})}\right]^2 + \frac{1}{2}x^2$$

setting the inner and outer edges of the circular Planck-Bohr-Sommerfeld band.

Therefore, we associate with the mth number state an occupied band of inner radius $r_m \equiv \sqrt{2m}$ and outer radius $r_{m+1} \equiv \sqrt{2(m+1)}$ as shown in Fig. 8.1. This representation of the mth energy eigenstate by a band is similar to the last positive crest of the corresponding Wigner function discussed in Sec. 4.1.3.

The band takes up an area 2π in phase space. Indeed, this area is given by the difference of the areas of the circles with radius r_{m+1} and r_m, that is,

$$\left(\begin{array}{c}\text{area of}\\ m\text{th band}\end{array}\right) = \pi r_{m+1}^2 - \pi r_m^2 = 2\pi.$$

Hence, the area is independent of the quantum number m.

More insight into this elementary representation of an energy eigenstate springs from the above formula for the area when we rewrite it in the form

$$\left(\begin{array}{c}\text{area of}\\ m\text{th band}\end{array}\right) = \pi\left(r_{m+1}+r_m\right)\left(r_{m+1}-r_m\right) = \pi 2\bar{r}_m \Delta_m = 2\pi. \qquad (8.2)$$

208   8 Applications of Interference in Phase Space

Here we have introduced the averaged radius

$$\bar{r}_m \equiv \frac{1}{2}\left(\sqrt{2(m+1)} + \sqrt{2m}\right)$$

and the radial width

$$\Delta_m \equiv \sqrt{2(m+1)} - \sqrt{2m}.$$

of the band.

The normalization condition, Eq. (8.2), immediately yields the relation

$$\Delta_m = \frac{1}{\bar{r}_m}. \tag{8.3}$$

Since for increasing quantum number the averaged radius $\bar{r}_m$ increases as a square root the width decreases inversely with the square root of m. Hence, the rings get thinner and thinner as m grows.

We conclude this section by briefly discussing the limit of large quantum numbers. In Bohr's semiclassical limit, that is when $m \gg 1$, we can replace in the width Δ_m the difference between neighboring quantum states by the derivative evaluated at the half integer quantum number, that is

$$\Delta_m = r_{m+1} - r_m \cong \left.\frac{\partial r_{m'}}{\partial m'}\right|_{m'=m+1/2} = \left.\frac{\partial \sqrt{2m'}}{\partial m'}\right|_{m'=m+1/2} = \frac{1}{\sqrt{2(m+1/2)}}. \tag{8.4}$$

Indeed, the radial width of the band decreases with increasing quantum number m.

When we compare this expression for Δ_m to Eq.(8.3) we find the approximate relation

$$\bar{r}_m \cong \sqrt{2(m+1/2)} \tag{8.5}$$

for the averaged radius of the mth band. This is the radius of the phase space trajectory corresponding to the half integer action $m + 1/2$.

8.3 Coherent State

In Sec. 4.2 we have shown that the coherent state is a displaced ground state of a harmonic oscillator. Due to this displacement the coherent state is not an energy eigenstate of this oscillator. It displays a Poissonian energy distribution

$$W_m(|\psi_{\text{coh}}\rangle) = \frac{\alpha^{2m}}{m!} e^{-\alpha^2} \cong \frac{1}{\sqrt{2\pi}\alpha} \exp\left[-\left(\frac{m + \frac{1}{2} - \alpha^2}{\sqrt{2}\alpha}\right)^2\right] \tag{8.6}$$

where α denotes the displacement parameter.

What is a *natural* algorithm for estimating the probabilities W_m of Eq. (8.6)? Certainly the above mentioned quantum mechanical scalar product is an excellent *mathematical* tool for this purpose. However, it does not provide any insight into the physics. We now apply the concept of interference in phase space to gain deeper understanding of this result.

For this purpose we have to represent the two states of the scalar product, that is the coherent state $|\psi_{\text{coh}}\rangle$ and the energy eigenstate $|m\rangle$ in phase space. We have already found that the energy eigenstate is a circular band. We now turn to the appropriate representation of the coherent state.

8.3.1 Elementary Approach

A coherent state is a displaced ground state. In our simplified representation of an energy eigenstate as a band in phase the ground state is a circular disk of radius $\sqrt{2}$ centered around the origin. Hence, when we displace this disk to $x_0 = \sqrt{2}\,\alpha$ on the positive x axis as indicated in Fig. 8.2 we find the outer boundary

$$2 = p_{\text{coh}}^2 + (x - \sqrt{2}\,\alpha)^2 \tag{8.7}$$

of this circular representation of the coherent state.

Energy Distribution According to Simple Overlap

The concept of interference in phase space associates the probability W_m to find the mth energy eigenstate in the coherent state with the area of overlap A_m in phase space between the two states.

What does this algorithm yield for W_m? For values of m either smaller or larger than two critical values, that is, for

$$\sqrt{2(m+1/2)} \lesssim \sqrt{2}\,\alpha - \sqrt{2} \quad \text{and} \quad \sqrt{2}\,\alpha + \sqrt{2} \lesssim \sqrt{2(m+1/2)},$$

the bands and the circle do not overlap at all and thus

$$W_m = 0.$$

Bands corresponding to m-values between these end points have a non-vanishing area in common with the circle. The maximum overlap arises for the band that cuts through the center of the circle. This corresponds to the quantum number

$$\sqrt{2}\,\alpha = \sqrt{2(m+1/2)},$$

that is $\alpha^2 = m + 1/2$, in complete agreement with the Gaussian limit Eq. (8.6) of the Poissonian distribution. For neighboring bands the overlaps decrease.

There is one more feature that comes out most clearly in this phase space visualization of the energy distribution: The Poisson statistics is asymmetric with respect to the maximum. We have neglected this property in the Gaussian limit Eq. (8.6). However, it stands out in the overlap formalism: Since the area of each band is constant but the radius increases as the quantum number increases the width Δm of each band decreases. Consequently, the left half of the circle representing the coherent state can accept fewer states than the right half. This leads naturally to an asymmetry in the energy distribution.

Estimate for the Width

We can get an estimate for the width of the distribution by counting the number of bands that cut through the circle. In order to facilitate a comparison with the Gaussian limit of the Poissonian it is advantageous to define the half width Δm of the energy distribution. Our phase space algorithm suggests that this quantity is given by the number of bands that fit into a half circle, that is

$$\Delta m = \frac{\sqrt{2}}{\Delta_m} = \sqrt{2}\,\bar{r}_m \cong 2\sqrt{m+1/2}$$

where we have recalled from Eq. (8.3) the connection between the width Δ_m of the mth band and its averaged radius. Moreover, we have used the asymptotic expansion Eq. (8.5) of $\bar{r}_m$.

We note that only bands with quantum numbers $m \cong \alpha^2$ create a non-vanishing overlap. Hence, we arrive at the expression

$$\Delta m \cong 2\alpha$$

for the half width of the energy distribution.

This result is in contrast to the Gaussian limit of the Poisson distribution which predicts $\sqrt{2}\alpha$. This is a clear indication that our formalism for evaluating probabilities contains the essentials but is not quite correct. This comes out most clearly when we now evaluate the overlap A_m analytically.

Quantitative Analysis

We now show that our formalism predicts a square root rather than a Gaussian dependence on the quantum number m for the area of overlap A_m between the mth Planck-Bohr-Sommerfeld band and the circle, Eq. (8.7).

We define the area of overlap of the two states by

$$A_m \equiv \underset{\substack{\text{overlap } m\text{th band}\\\text{coherent state}}}{\int dx \int dp} \; \frac{1}{2\pi}$$

where we have introduced the factor $1/2\pi$ to maintain the normalization condition

$$\sum_m A_m = \frac{1}{2\pi} \sum_m \underset{\substack{\text{overlap } m\text{th band}\\\text{coherent state}}}{\int dx \int dp} = \frac{1}{2\pi} \cdot \begin{pmatrix} \text{area of coherent} \\ \text{state} = 2\pi \end{pmatrix} = 1.$$

In the preceeding chapter we have found that the probabilities emerge when we divide the area of overlap by $2\pi\hbar$. Since, in our units $\hbar = 1$ this normalization factor is 2π.

We can approximate the area of overlap represented by the mth ring segment by a rectangle of width Δ_m and height

$$2p_{\text{coh}}(x_c) = 2\sqrt{2 - \left(x_c - \sqrt{2}\alpha\right)^2}$$

evaluated at the center $x_c \equiv \bar{r}_m$ of the band. The area of this rectangle reads

$$A_m \cong \frac{1}{2\pi}\Delta_m\, 2p_{\text{coh}}(x_c) = \frac{1}{\pi}\Delta_m\sqrt{2 - \left(x_c - \sqrt{2}\alpha\right)^2}.$$

8.3 Coherent State 211

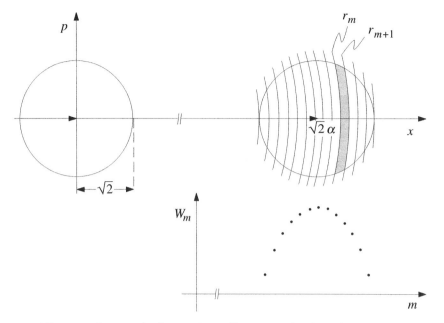

Fig. 8.2: The ground state of a harmonic oscillator, visualized in phase space as a circle of radius $\sqrt{2}$ and displaced from the origin by an amount $\sqrt{2}\,\alpha$ models a coherent state. The area of overlap between the mth band (representing the mth number state) and the circle, the simplest algorithm for determining the energy distribution W_m of a coherent state, yields a "semi-ellipse shaped" distribution, Eq. (8.10), reproducing the qualitative features of the Poisson distribution, as shown in the lower right. When instead we weight each point of phase space according to the circular Gaussian W_{coh}, Eq. (8.11), we get a probability, W_m, in m-values almost indistinguishable from the correct Poissonian result. Indeed, we find Eq. (8.6), depicted in Fig. 4.7 by the broken line.

Moreover, we note from Fig. 8.2 that the main contributions to the overlap arise when the bands cut through the circle. Roughly speaking, in this case the quantum numbers m are of the order of α^2. Since according to Eq. (8.4) the quantum number m appears in Δ_m only in the denominator the dependence on m is very weak and we can therefore replace $m + 1/2$ by α^2 which yields

$$\Delta_m \cong \frac{1}{\sqrt{2}\alpha},$$

and hence

$$A_m \cong \frac{1}{\pi\sqrt{2}\alpha}\sqrt{2 - \left(x_c - \sqrt{2}\,\alpha\right)^2}. \tag{8.8}$$

When we recall that in the large-m limit we have the relation

$$x_c - \sqrt{2}\,\alpha \cong \sqrt{2\left(m + \tfrac{1}{2}\right)} - \sqrt{2}\,\alpha \cong \frac{m + \tfrac{1}{2} - \alpha^2}{\sqrt{2}\,\alpha} \tag{8.9}$$

Eq. (8.8) simplifies to

$$A_m \cong \frac{1}{\pi \, 2\alpha^2} \sqrt{4\alpha^2 - \left(m + \tfrac{1}{2} - \alpha^2\right)^2}. \tag{8.10}$$

The tentative probability distribution $W_m = A_m$ is thus peaked at $m = \alpha^2 - \tfrac{1}{2}$ in agreement with Eq. (8.6). However, it has a half width in m values of 2α in disagreement to the prediction $\sqrt{2}\alpha$. Therefore, this elementary approach is providing some insight into the energy distribution of a coherent state but does not give the full truth yet.

8.3.2 Influence of Internal Structure

The problem of the incorrect prediction of the half width is intimately related to the other problem we have with the result Eq. (8.10) of the elementary approach: It predicts a square root dependence on m rather than a Gaussian. This is due to the fact that so far each point inside the disk carries equal weight. This results in the square root dependence of W_m on m. A better treatment recognizes the smooth spread in x-values

$$|\psi_{\text{coh}}(x)|^2 = \frac{1}{\sqrt{\pi}} \exp[-(x - \sqrt{2}\alpha)^2]$$

and the corresponding spread in momentum

$$|\psi_{\text{coh}}(p)|^2 = \frac{1}{\sqrt{\pi}} \exp\left(-p^2\right).$$

Thus each point in phase space gets weighted according to the Wigner distribution

$$W_{\text{coh}}(x, p) = \frac{1}{\pi} \exp\left[-\left(x - \sqrt{2}\alpha\right)^2 - p^2\right]. \tag{8.11}$$

The circular spot shown in Fig. 8.2 crudely depicts the exponential fall-off of the distribution – an effect analogous to the apodization, for a smooth graduation in transmissivity, often used to improve a lens.

An improved version of the area-of-overlap concept evaluates the weighted overlap

$$\mathcal{A}_m \equiv \int dx \int_{m\text{th band}} dp \, W_{\text{coh}}(x, p)$$

between the mth band and the Gaussian bell Eq. (8.11). Therefore, $\mathcal{A}_m$ represents the volume cut out of the Gaussian bell by the band.

In contrast to the elementary approach we now do not have to introduce a normalization condition since the Wigner function is already normalized. Indeed, we find

$$\sum_m \mathcal{A}_m = \sum_m \int dx \int_{m\text{th band}} dp \, W_{\text{coh}} = \int_{-\infty}^{\infty} dx \int_{-\infty}^{\infty} dp \, W_{\text{coh}} = 1.$$

In the limit $m \gg 1$ we find for $\mathcal{A}_m$, approximated by the mth weighted rectangle,

$$\mathcal{A}_m \cong \Delta_m \exp\left[-\left(\bar{r}_m - \sqrt{2}\alpha\right)^2\right] \frac{1}{\pi} \int_{-\infty}^{\infty} dp \, \exp\left(-p^2\right)$$

which with the help of Eqs. (8.4) and (8.9) reduces to

$$\mathcal{A}_m \cong \frac{1}{\sqrt{2\pi}\,\alpha} \exp\left[-\left(\frac{m+\frac{1}{2}-\alpha^2}{\sqrt{2}\,\alpha}\right)^2\right] = W_m,$$

a result identical to Eq. (8.6).

Hence, in the semiclassical limit the energy distribution W_m of a coherent state is the overlap in phase space between the Gaussian bell of the coherent state and the corresponding Planck-Bohr-Sommerfeld band of the mth number state.

8.4 Squeezed State

We now turn to the case of a squeezed state. In particular, we want to understand the deeper origin of the oscillations in the energy distribution of a highly squeezed state. Here we pursue an approach similar to the one for the coherent state of the last section.

8.4.1 Oscillations from Interference in Phase Space

We represent the squeezed state $|\psi_{\text{sq}}\rangle$ by the phase space distribution

$$W_{\text{sq}}(x,p) = \frac{1}{\pi} \exp[-s(x-\sqrt{2}\,\alpha)^2] \exp\left[-\frac{1}{s}p^2\right] \tag{8.12}$$

derived in Eq. (4.56) and expressed throughout this section in dimensionless phase space variables.

Motivated by the success of the area of overlap approach in the case of the coherent state we now define the weighted overlap

$$\mathcal{A}_m \equiv \int dx \int_{m\text{th band}} dp\, W_{\text{sq}}(x,p),$$

that is, the volume which the mth band cuts out of the Gaussian cigar, Eq. (8.12). What does this algorithm predict for the probability?

For quantum numbers m smaller than α^2 there is no overlap between the mth band and the cigar. Hence, the probability W_m is negligible, in agreement with the exact curve.

For $m \cong \alpha^2$ the bands cut the cigar tangentially creating an unusually large overlap. This gives rise to a dominant maximum in the energy distribution.

However, for $m > \alpha^2$ we find in the limit of $s \gg 1$, that is, for a highly squeezed state, two symmetrically located diamond-shaped areas of overlap as shown in Fig. 8.3(a). Each diamond of area $\mathcal{A}_m$ corresponds to one probability amplitude and hence according to the concept of interference in phase space, Eq. (7.23), we find

$$W_m = |\langle m|\psi_{\text{sq}}\rangle|^2 = \left|\sqrt{\mathcal{A}_m}e^{i\phi_m} + \sqrt{\mathcal{A}_m}e^{-i\phi_m}\right|^2 = 4\mathcal{A}_m \cos^2\phi_m. \tag{8.13}$$

This analysis clearly indicates that our formalism can explain qualitatively the important features of the energy distribution of a squeezed state. We now show that it also is in quantitative agreement.

8 Applications of Interference in Phase Space

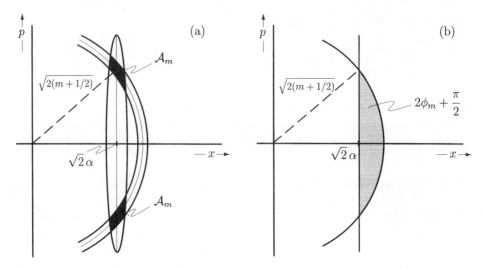

Fig. 8.3: Energy distribution of a highly squeezed state, obtained from interference in phase space. The probability to find the mth energy eigenstate, follows from the weighted area of overlap in phase space between the number state $|m\rangle$ represented by the mth Planck-Bohr-Sommerfeld band and the highly squeezed state depicted here by the Gaussian cigar's elliptical contour line of exponential decay. For m approximately larger than α^2 there exist two such overlaps as shown in (a). These overlaps correspond to two interfering complex-valued probability amplitudes. The probability $\mathcal{A}_m$ associated with a single diamond is determined by the diamond-shaped overlap. The phase difference $2\phi_m$ between the amplitudes is fixed by the shaded domain caught by the central lines of the two states as depicted in (b).

Area of Overlap

For this purpose we now evaluate explicitly the volume cut out of the Gaussian cigar by the mth energy band when there are more than one overlap. This weighted area of overlap reads

$$\mathcal{A}_m = \int_{-\infty}^{\infty} dx \sqrt{\frac{s}{\pi}} \exp[-s(x - \sqrt{2}\alpha)^2] \frac{1}{\sqrt{s\pi}} \int_{\sqrt{2m-x^2}}^{\sqrt{2(m+1)-x^2}} dp \exp\left(-\frac{1}{s}p^2\right).$$

In the limit of $s \gg 1$ we can perform the integration over x by replacing the position distribution by a δ-function at $x = \sqrt{2}\alpha$. This yields

$$\mathcal{A}_m \cong \frac{1}{\sqrt{s\pi}} \int_{\sqrt{2(m-\alpha^2)}}^{\sqrt{2(m+1-\alpha^2)}} dp \exp\left(-\frac{1}{s}p^2\right). \tag{8.14}$$

We note that for $s \gg 1$ the integrand $\exp(-p^2/s)$ is slowly varying over the integration region of extension

$$\delta p_m \equiv \sqrt{2(m+1-\alpha^2)} - \sqrt{2(m-\alpha^2)} = \frac{2}{\sqrt{2(m+1-\alpha^2)} + \sqrt{2(m-\alpha^2)}}. \quad (8.15)$$

Hence, we can approximate the integral over momentum, Eq. (8.14), by taking the value of the function $\exp(-p^2/s)$ at the center

$$\bar{p}_m \equiv \frac{1}{2}\left(\sqrt{2(m+1-\alpha^2)} + \sqrt{2(m-\alpha^2)}\right) \cong \sqrt{2(m+\tfrac{1}{2}-\alpha^2)} \quad (8.16)$$

of the integration regime, times the width δp_m, that is

$$\mathcal{A}_m \cong \delta p_m \frac{1}{\sqrt{s\pi}} \exp\left(-\frac{1}{s}\bar{p}_m^2\right). \quad (8.17)$$

With the help of Eqs. (8.15) and (8.16) we find

$$\delta p_m = \frac{1}{\bar{p}_m} \cong \frac{1}{\sqrt{2(m+\tfrac{1}{2}-\alpha^2)}}$$

and Eq. (8.17) reads

$$\mathcal{A}_m = \frac{1}{\sqrt{2\pi s}} \frac{\exp\left[-2\left(m+\tfrac{1}{2}-\alpha^2\right)/s\right]}{\sqrt{m+\tfrac{1}{2}-\alpha^2}}. \quad (8.18)$$

We note that this expression is identical to the one we have found in Sec. 4.3.2 by performing the appropriate asymptotic limit of the exact energy distribution.

Phase of the Oscillations

The phase ϕ_m associated with each overlap is the area of the shaded segment in Fig. 8.3b. It is the domain which is enclosed by the central lines of the squeezed state $(x = \sqrt{2}\,\alpha)$, of the band $(p_m^2 + x^2 = 2(m+1/2))$ and the x-axis. It reads

$$\phi_m = \binom{\text{area of}}{\text{segment}} - \frac{\pi}{4} = \binom{\text{area of}}{\text{sector}} - \binom{\text{area of}}{\text{triangle}} - \frac{\pi}{4}$$

Here we have included a phase shift of $\pi/4$ resulting from the phase of the energy eigenfunction $u_m(x)$ at the turning point $\zeta_m - \sqrt{2(m+1/2)}$ of the motion corresponding to the energy $m + 1/2$.

When we recall the relations

$$\binom{\text{area of}}{\text{sector}} = \frac{1}{2}\left(\sqrt{2\left(m+\tfrac{1}{2}\right)}\right)^2 \varphi = \left(m+\tfrac{1}{2}\right)\arctan\frac{\sqrt{m+\tfrac{1}{2}-\alpha^2}}{\alpha}$$

and

$$\binom{\text{area of}}{\text{triangle}} = \frac{1}{2}(\sqrt{2}\alpha)\sqrt{2\left(m+\tfrac{1}{2}\right) - 2\alpha^2} = \alpha\sqrt{m+\tfrac{1}{2}-\alpha^2}$$

8 Applications of Interference in Phase Space

we find

$$\phi_m = \left(m + \frac{1}{2}\right) \arctan\left[\frac{1}{\alpha}\sqrt{m + \frac{1}{2} - \alpha^2}\right] - \alpha\sqrt{m + \frac{1}{2} - \alpha^2} - \frac{\pi}{4}. \tag{8.19}$$

We note that this expression for the phase is identical with the result Eq. (4.43cc) of Sec. 4.3.2.

8.4.2 Giant Oscillations

We now turn to the case of a squeezed state rotated by the angle φ relative to the momentum axis. As we have shown in Sec. 4.3.2 the energy distribution of the harmonic oscillator in the squeezed state can display rapid oscillations with a slowly varying envelope as shown in Fig. 4.15(b). In the present section we use the concept of interference in phase space to gain more insight into these giant oscillations and to derive an explicit asymptotic expression for the energy distribution.

We note that in complete analogy to the case $\varphi = 0$ discussed in the preceding section the mth band cuts out of the rotated Gaussian cigar depicted in Fig. 8.4 two interfering diamond-shaped areas of overlap. Hence, Eq. (7.23) yields the expression

$$W_m = \left|\sqrt{\mathcal{A}_m^{(+)}}e^{i\Phi_m} + \sqrt{\mathcal{A}_m^{(-)}}e^{-i\Phi_m}\right|^2 \tag{8.20}$$

for the probability to find the energy $\hbar\omega(m + 1/2)$.

We now evaluate the weighted phase space areas $\mathcal{A}_m^{(\pm)}$ and $\pm\Phi_m$. For the purpose it is convenient to rotate the phase space by the angle φ indicated in Fig. 8.5. In the rotated x'-p'-coordinate system the Gaussian cigar reads

$$W_{\text{sq}}(x', p') = \frac{1}{\pi}\exp\left[-s\left(x' - \sqrt{2}\,\alpha\cos\varphi\right)^2 - \frac{1}{s}(p' + \sqrt{2}\,\alpha\sin\varphi)^2\right]$$

where according to Fig. 8.5 its center is at $x'_c = \sqrt{2}\,\alpha\cos\varphi$ and $p'_c = -\sqrt{2}\,\alpha\sin\varphi$.

Evaluation of Amplitudes

Analogous to the preceding section we calculate the weighted areas of overlap

$$\mathcal{A}_m^{(\pm)} \equiv \iint_{m\text{th band}} dx'dp'\, W_{\text{sq}}(x', p'),$$

shown in Fig. 8.4, by replacing the position distribution by a δ-function at x'_c. We approximate the remaining integrals over the Gaussian momentum distribution

$$\frac{1}{\sqrt{\pi s}}\exp[-(p' + \sqrt{2}\,\alpha\sin\varphi)^2/s]$$

by their values at the centers

$$\pm\bar{p}'_m = \pm\sqrt{2(m + 1/2 - \alpha^2\cos^2\varphi)}. \tag{8.21}$$

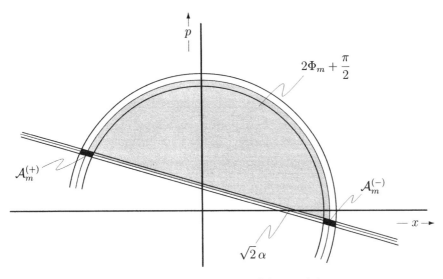

Fig. 8.4: Weighted areas of overlap in phase space $\mathcal{A}_m^{(+)}$ and $\mathcal{A}_m^{(-)}$ between the mth number state band and a rotated, highly squeezed cigar. In contrast to Fig. 8.3a the two black diamonds differ in their weight factors: In the area $\mathcal{A}_m^{(-)}$ the mth Planck-Bohr-Sommerfeld band cuts the cigar close to its center and hence enjoys a large weight, whereas the overlap $\mathcal{A}_m^{(+)}$ is further away from the cigar's center giving rise to a correspondingly smaller contribution. The phase difference $2\Phi_m$ is again fixed by the shaded area caught by the central lines of the two states in complete accordance with Fig. 8.3b.

of the diamonds, times their width $\Delta p_m \cong [2(m+1/2-\alpha^2 \cos^2\varphi)]^{-1/2}$ in momentum. The expressions for Δp_m and $\pm \bar{p}'_m$ follow directly from Fig. 8.5 and yield for

$$\mathcal{A}_m^{(\pm)} \cong \Delta p_m \frac{1}{\sqrt{\pi s}} \exp\left[-\frac{1}{s}\left(\pm \bar{p}'_m + \sqrt{2}\alpha \sin\varphi\right)^2\right]$$

the final result

$$\mathcal{A}_m^{(\pm)} = A_m \exp(\mp \kappa_m) \tag{8.22}$$

where

$$A_m \equiv \frac{1}{\sqrt{2\pi s}} \frac{\exp\{-2[m+\tfrac{1}{2} - \alpha^2 \cos(2\varphi)]/s\}}{(m+\tfrac{1}{2} - \alpha^2 \cos^2\varphi)^{1/2}} \tag{8.23}$$

and

$$\kappa_m \equiv \frac{4}{s}\alpha \sin\varphi \left(m + \frac{1}{2} - \alpha^2 \cos^2\varphi\right)^{1/2}. \tag{8.24}$$

We note that for $\varphi \neq 0$ the two areas of overlap differ in their values, whereas for $\varphi = 0$ we find $\mathcal{A}_m^{(+)} = \mathcal{A}_m^{(-)} = A_m$.

Evaluation of Phases

We now turn to the calculation of the phase Φ_m of the interfering areas. According to Eq. (7.23) it is given by the area enclosed by the central lines of the states involved,

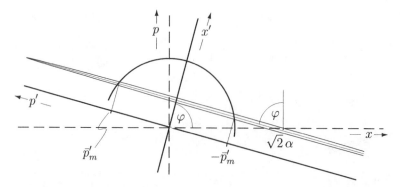

Fig. 8.5: The rotated Gaussian cigar of a highly squeezed state seen from the x'-p' coordinate system, whose p'-axis is parallel to the Gaussian cigar. The center of the Gaussian cigar lies at $x'_c \equiv \sqrt{2}\,\alpha\cos\varphi$ and $p'_c \equiv -\sqrt{2}\,\alpha\sin\varphi$. The central line of the Gaussian cigar is displaced from the p'-axis by $\sqrt{2}\,\alpha\cos\varphi$ (compared to $\sqrt{2}\,\alpha$ in Fig. 8.3b). Hence, the central line of the number state crosses the central line of the Gaussian cigar at the phase space points $x'_c = \sqrt{2}\,\alpha\cos\varphi$ and $\pm\bar{p}'_m \equiv \pm\sqrt{2(m+1/2-\alpha^2\cos^2\varphi)}$.

that is by the segment of Fig. 8.4. In the rotated coordinate system this area is identical to that of the preceding section with $\sqrt{2}\,\alpha$ being replaced by $\sqrt{2}\,\alpha\cos\varphi$ as demonstrated in Fig. 8.5. Hence, we find immediately from Eq. (8.19)

$$\Phi_m \equiv \phi_m(\sqrt{2}\,\alpha\cos\varphi)$$
$$= \left(m+\frac{1}{2}\right)\arctan\left[\frac{\left(m+\frac{1}{2}-\alpha^2\cos^2\varphi\right)^{1/2}}{\alpha\cos\varphi}\right]$$
$$- \alpha\cos\varphi\left(m+\frac{1}{2}-\alpha^2\cos^2\varphi\right)^{1/2} - \frac{\pi}{4}. \qquad (8.25)$$

We are now in a position to give an explicit asymptotic expression for the energy distribution of a rotated squeezed state. Indeed, with the help of Eq. (8.22) the energy distribution, Eq. (8.20), predicted by the concept of interference in phase space now reads

$$W_m = 2A_m[\cosh\kappa_m + \cos(2\Phi_m)]$$

where A_m, κ_m and Φ_m are given by Eqs. (8.23), (8.24) and (8.25), respectively.

We now recall that in Fig. 4.17 we test this asymptotic result against the exact one, Eq. (4.36) and find excellent agreement.

8.4.3 Summary

Interfering areas of overlap in phase space explain the oscillations in the energy distribution of a squeezed wave packet of a mechanical oscillator. When the oscillator is appropriately squeezed in the position variable the probability W_m to find the energy $\hbar\omega(m+1/2)$ displays oscillations of rather large period.

For $m \gg \alpha^2$ these oscillations take on quite a different form: they consist of odd-even oscillations with a slowly varying amplitude. Both types of oscillations are the consequence of the interference between two symmetrically located diamond-shaped areas of overlap between the mth energy band and the squeezed Gaussian cigar. They follow from the energy distribution

$$W_m = \left| \sqrt{\mathcal{A}_m}\, e^{i\phi_m} + \sqrt{\mathcal{A}_m}\, e^{-i\phi_m} \right|^2.$$

We estimate the weighted area $\mathcal{A}_m$ of a single diamond via

$$\mathcal{A}_m = \begin{pmatrix} \text{width of} \\ \text{diamond in} \\ \text{position } x \end{pmatrix} \times \begin{pmatrix} \text{probability} \\ \text{for position} \\ \text{at center of} \\ \text{diamond} \end{pmatrix} \times \begin{pmatrix} \text{width of} \\ \text{diamond in} \\ \text{momentum } p \end{pmatrix} \times \begin{pmatrix} \text{probability} \\ \text{for momentum} \\ \text{at center of} \\ \text{diamond} \end{pmatrix}$$

$$= \sqrt{\frac{\pi}{s}} \times \sqrt{\frac{s}{\pi}} \times [2(m+1/2-\alpha^2)]^{-1/2} \times \frac{1}{\sqrt{\pi s}} \exp[-2(m+1/2-\alpha^2)/s].$$

The phase $\phi_m + \pi/4$ of these probability-amplitude-representing areas is the domain enclosed by the central lines of the squeezed state, the mth band and the x-axis. According to Fig. 8.6(a) it reads

$$\phi_m = \begin{pmatrix} \text{area of} \\ \text{segment} \end{pmatrix} - \frac{\pi}{4} = \begin{pmatrix} \text{area of} \\ \text{sector} \end{pmatrix} - \begin{pmatrix} \text{area of} \\ \text{triangle} \end{pmatrix} - \frac{\pi}{4}$$

$$= \left(m + \frac{1}{2}\right) \arctan\left[\frac{1}{\alpha}\left(m + \frac{1}{2} - \alpha^2\right)^{1/2}\right] - \alpha\left[m + \frac{1}{2} - \alpha^2\right]^{1/2} - \frac{\pi}{4}.$$

We note that for $m \gtrsim \alpha^2$ the phase ϕ_m, that is the area of the segment, increases only slightly when we go from m to $m+1$. Consequently, the oscillations have a rather large period.

However, quite a different behavior sets in when we consider the case of $m \gg \alpha^2$. Following Fig. 8.6(b) we can approximate this phase determining segment via

$$\phi_m \cong \begin{pmatrix} \text{area enclosed by} \\ m\text{th centerline in} \\ \text{positive phase} \\ \text{space quadrant} \end{pmatrix} - \begin{pmatrix} \text{area of} \\ \text{rectangle} \end{pmatrix} - \frac{\pi}{4}$$

$$= \frac{1}{4} 2\pi \left(m + \frac{1}{2}\right) - \sqrt{2}\,\alpha\sqrt{2(m+\frac{1}{2})} - \frac{\pi}{4} = m\frac{\pi}{2} - 2\alpha\sqrt{m + \frac{1}{2}}. \tag{8.26}$$

Hence, ϕ_m consists of a rapidly and a slowly varying function of m. Consequently, W_m displays rapid odd-even oscillations whose amplitudes are modulated as a consequence of the slowly varying contribution. Since $m \gg \alpha^2$ this behavior takes place in the exponential tail of $\mathcal{A}_m$ and we can only bring it to light by considering the limit of extreme squeezing, such that the height $\sqrt{s}$ of the cigar is much larger than the displacement $\sqrt{2}\,\alpha$.

A similar phenomenon appears when the Gaussian cigar encloses an angle φ with the momentum axis: Starting from a critical angle φ rapid odd-even oscillations modulated by a slowly varying envelope appear. They result again from the interference

220   8 Applications of Interference in Phase Space

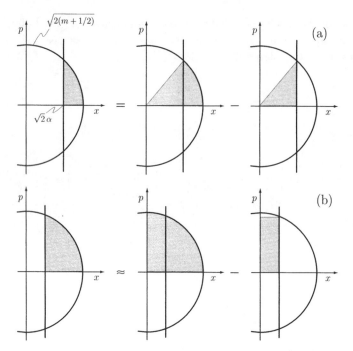

Fig. 8.6: The phase-determining area of the shaded segment is the difference in the areas of the sector and the triangle shown in (a). In the limit of $m \gg \alpha^2$ depicted in (b) we can approximate the area of the phase space segment by the difference in the areas of the positive quadrant of the circle and the rectangle.

of two diamond-shaped phase space overlaps, $\mathcal{A}_m^{(+)}$ and $\mathcal{A}_m^{(-)}$, that is

$$W_m = \left| \sqrt{\mathcal{A}_m^{(+)}} e^{i\Phi_m} + \sqrt{\mathcal{A}_m^{(-)}} e^{-i\Phi_m} \right|^2.$$

The phase $2\Phi_m + \pi/2$ is again the area enclosed by the central lines. We can easily relate Φ_m to the phase ϕ_m of the nonrotated squeezed state when we rotate the coordinate system by the angle φ. In these coordinates the displacement parameter reads $\sqrt{2}\alpha \cos\varphi = \sqrt{2}\alpha \cos(\pi/2 - \theta) = \sqrt{2}\alpha \sin\theta$ and the phase $\Phi_m = \phi_m(\sqrt{2}\alpha \cos\varphi)$ for $\theta \ll 1$ is identical to Eq. (8.26) with α being replaced by $\alpha\theta$. However, in contrast to the energy distribution of the nonrotated squeezed state, here W_m does not have zeros since the two interfering areas of overlap $\mathcal{A}_m^{(\pm)}$ are different. Indeed, we can estimate the areas $\mathcal{A}_m^{(\pm)}$ via

$$\mathcal{A}_m^{(\pm)} = \begin{pmatrix} \text{width of} \\ \text{diamond in} \\ \text{position } x' \end{pmatrix} \times \begin{pmatrix} \text{probability} \\ \text{for position} \\ \text{at center of} \\ \text{diamond} \end{pmatrix} \times \begin{pmatrix} \text{width of} \\ \text{diamond in} \\ \text{momentum } p' \end{pmatrix} \times \begin{pmatrix} \text{probability} \\ \text{for momentum} \\ \text{at center of} \\ \text{diamond} \end{pmatrix}$$

$$= \sqrt{\frac{\pi}{s}} \times \sqrt{\frac{s}{\pi}} \times [2(m+1/2 - \alpha^2 \cos^2\varphi)]^{-1/2}$$

$$\times \frac{1}{\sqrt{\pi s}} \exp\left\{-\left[\pm\frac{1}{s}\sqrt{2(m+1/2-\alpha^2\cos^2\varphi)}+\sqrt{2}\,\alpha\sin\varphi\right]^2\right\}$$

which agrees with Eqs. (8.22)–(8.24).

In this section we have shown how the concept of interference in phase space works for several examples. It proved to be a powerful tool to extract the essential features of scalar products between quantum states.

8.5 The Question of Phase States

The "old" quantum mechanics of Planck, Bohr and Sommerfeld – the so-called *Atommechanik* – relies on action-angle variables. In contrast, the "new" quantum mechanics of Heisenberg and Schrödinger is based on position and momentum variables. It is amazing to note that already a few months after the seminal papers by Heisenberg and Schrödinger, Fritz London asked the question how to reformulate the new quantum mechanics in terms of action-angle variables. He already found out that there was no hermitian operator corresponding to the classical phase variable. In a second paper he then proposed an operator $\widehat{\exp(i\varphi)}$ for the exponential of the phase.

When Dirac quantized the electromagnetic field he also followed the path of action-angle variables. He used a phase operator and ignored the problems pointed out by London. Since that time the problem of the phase operator has occupied generations of physicists. Here we only want to mention the most recent approach by Barnett and Pegg. For a more detailed review we refer to the literature at the end of this chapter.

In various chapters of the book we will return to this problem. However, in the present section we use the *area of overlap principle* to establish a relation between the phase and the energy representation of a quantum state $|\psi\rangle$.

8.5.1 Amplitude and Phase in a Classical Oscillator

In order to bring out the complications associated with the amplitude and phase representation of a quantum oscillator we first consider the analogous problem for a classical harmonic oscillator.

The time evolution of the oscillator is determined by the Newton equation of motion

$$\frac{d^2}{dt^2}x(t)+x(t)=0$$

where t is a dimensionless time scaled in units of the frequency of the oscillator.

Since this is a second order equation we need two quantities to uniquely determine the future of the oscillator: the position $x_0 \equiv x(t_0)$ and the momentum $p_0 \equiv \dot{x}(t_0)$ at a given time t_0. The classical state of the oscillator is characterized by position and momentum and is hence a point in a two-dimensional space spanned by position and momentum. The coordinates of this phase space point are (x_0, p_0).

The Newton equation tells us how this initial point moves in phase space. Indeed, we find for the time dependent position and momentum expressed here as a two dimensional vector the solution

$$\begin{pmatrix} x(t) \\ p(t) \end{pmatrix} = \begin{pmatrix} x_0 \cos t + p_0 \sin t \\ -x_0 \sin t + p_0 \cos t \end{pmatrix},$$

where for the sake of simplicity we have set $t_0 \equiv 0$.

This motion in phase space, parameterized by time, gives rise to a circular phase space trajectory

$$x^2(t) + p^2(t) = x_0^2 + p_0^2 = const$$

which the oscillator traverses with a constant velocity. This comes out most clearly with the help of the trigonometric relations

$$\cos(\alpha - \beta) = \cos(\alpha)\cos(\beta) + \sin(\alpha)\sin(\beta)$$
$$\sin(\alpha - \beta) = \sin(\alpha)\cos(\beta) - \cos(\alpha)\sin(\beta)$$

which casts the vector in the form

$$\begin{pmatrix} x(t) \\ p(t) \end{pmatrix} = \sqrt{x_0^2 + p_0^2} \begin{pmatrix} \cos(\varphi_0 - t) \\ \sin(\varphi_0 - t) \end{pmatrix}. \tag{8.27}$$

By comparison we note that the quantity φ_0 is given by the relations

$$\cos \varphi_0 = \frac{x_0}{\sqrt{x_0^2 + p_0^2}} \quad \text{and} \quad \sin \varphi_0 = \frac{p_0}{\sqrt{x_0^2 + p_0^2}}. \tag{8.28}$$

Indeed, φ_0 is the angle defined by the initial phase space coordinates (x_0, p_0).

Equation (8.27) clearly indicates that the oscillator moves along a circle in phase space as a function of time. The radius is the square root of twice the energy and the angular phase space velocity is unity which, in unscaled variables, is the frequency of the oscillator.

This representation also suggests a different parameterization of the circular phase space trajectory. Instead of using cartesian coordinates polar coordinates are well suited for this motion. We recall that in these dimensionless units the action J is the enclosed area $\pi(x_0^2 + p_0^2)$ and the angle $\varphi(t) \equiv \varphi_0 - t$ is measured relative to the initial phase φ_0. We note that $\varphi(t)$ decreases linearly with time starting from φ_0. This indicates that the motion through phase space is in clockwise direction. Hence, we find

$$\begin{pmatrix} x(t) \\ p(t) \end{pmatrix} = \sqrt{\frac{J}{\pi}} \begin{pmatrix} \cos \varphi(t) \\ \sin \varphi(t) \end{pmatrix}.$$

The action J and the phase φ determine the state of the oscillator in the same way as x and p.

We conclude this section by briefly discussing the transition to quantum mechanics where position and momentum are hermitian, noncommuting operators. Therefore, the concept of a phase becomes problematic. Indeed, the trigonometric functions $\cos \varphi$ and $\sin \varphi$ defined in Eq. (8.28) involve ratios of position and momentum and therefore the ratio of noncommuting operators. This gives rise to an operator ordering problem. We return to this question in Sec. 13.4 where we introduce a second field which allows to overcome this operator ordering problem.

8.5 The Question of Phase States

Another problem associated with the quantum version of a phase is the different spectrum of the conjugate variables. In classical physics the action J and the phase φ are conjugate variables. We therefore expect that their quantum counterparts are also conjugate variables. This however, cannot be true in a strict sense, since the corresponding operators have a very different spectrum. The action variable in the harmonic oscillator is discrete and takes on positive half integer values. In contrast, the phase operator is expected to have a continuous spectrum since the phase variable is continuous. To resort to a discrete set of phase angles is one possible way to resolve this problem of a hermitian phase operator. Many more alternatives offer themselves. Here we do not elaborate on these approaches but rather refer to the literature.

8.5.2 Definition of a Phase State

The goal of the present subsection is to motivate the definition

$$|\varphi\rangle \equiv \frac{1}{\sqrt{2\pi}} \sum_{m=0}^{\infty} \exp\left(i(m+\tfrac{1}{2})\varphi\right) |m\rangle$$

of a phase state $|\varphi\rangle$ without knowing the explicit form of the corresponding phase operator. Our approach puts to use the concept of interference in phase space and relies on our intuition about the appropriate phase space representation of a state of well defined phase.

Energy Representation

We make use of the completeness relation

$$\sum_{m=0}^{\infty} |m\rangle \langle m| = \mathbb{1}$$

of the energy eigenstates $|m\rangle$ to represent the phase state $|\varphi\rangle$ in terms of $|m\rangle$, that is

$$|\varphi\rangle = \sum_{m=0}^{\infty} \langle m|\varphi\rangle |m\rangle \equiv \sum_{m=0}^{\infty} w_m\left[|\varphi\rangle\right] |m\rangle. \qquad (8.29)$$

Hence, we are left with the problem of determining the complex-valued energy probability amplitude

$$w_m[|\varphi\rangle] \equiv \langle m|\varphi\rangle$$

of a phase state.

When we recall that

$$\langle \varphi|m\rangle = \langle m|\varphi\rangle^*$$

we recognize that this quantity is the complex conjugate of the phase probability amplitude

$$w_\varphi[|m\rangle] \equiv \langle \varphi|m\rangle \qquad (8.30)$$

of a number state.

We find the probability amplitude $\langle \varphi|m\rangle$ in a most convenient way by making use of the concept of interference in phase space. For this purpose we have to find

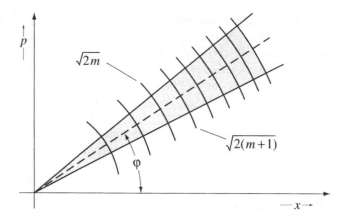

Fig. 8.7: In its most elementary version a phase state $|\varphi\rangle$ may be represented in x-p oscillator phase space as a wedge shaped area emerging from the origin and aligned under an angle φ as indicated by the grey phase space slice. When we decompose this phase state into number states represented in x-p phase space by circular Planck-Bohr-Sommerfeld bands of inner radius $(2m)^{1/2}$ and outer radius $[2(m+1)]^{1/2}$ the expansion coefficients $|\langle\varphi|m\rangle|^2$ follow from the area-of-overlap principle as the area of cross-over between the mth band and the wedge. For increasing m the wedge diverges whereas the width $\Delta_m \cong [2(m+1)]^{1/2} - (2m)^{1/2} \cong [2(m+1/2)]^{-1/2}$ of the Planck-Bohr-Sommerfeld annuli decreases as to keep the area of overlap constant and unity. As a result the expansion of $|\varphi\rangle$ into number states is not absolutely convergent.

a representation of the phase state in phase space. Here we have to rely on our intuition for the phase space representation of such a state of well defined phase: As shown in Fig. 8.7 a phase state $|\varphi\rangle$ may be represented as a wedge emerging from the origin under an angle φ.

In this geometrical picture the expansion Eq. (8.29) of the phase state into energy eigenstates corresponds to a representation of the wedge by a sequence of adjacent ring segments cut out of the wedge by the Planck-Bohr-Sommerfeld bands. We emphasize that each ring segment corresponds to a probability amplitude and therefore has a phase. Consequently, the wedge is built out of a sequence of interfering segments.

Phase Amplitude of Energy Eigenstates

We now calculate the phase probability amplitude $\langle\varphi|m\rangle$ of an energy eigenstate using the concept of interference in phase space. This formalism relates the square of the absolute value of the probability amplitude, $|\langle\varphi|m\rangle|^2$, to the area of overlap in x-p oscillator phase space between the mth circular Planck-Bohr-Sommerfeld band of inner radius $r_m \equiv (2m)^{1/2}$ and outer radius $r_{m+1} \equiv [2(m+1)]^{1/2}$ representing the mth number state and the diverging phase space ray at angle φ shown in Fig. 8.8(a).

8.5 The Question of Phase States 225

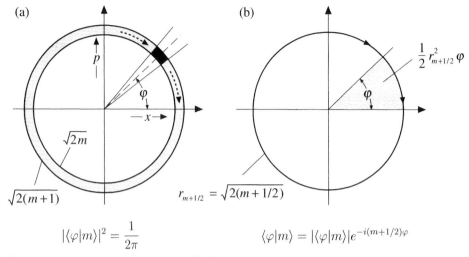

$$|\langle\varphi|m\rangle|^2 = \frac{1}{2\pi} \qquad\qquad \langle\varphi|m\rangle = |\langle\varphi|m\rangle|e^{-i(m+1/2)\varphi}$$

Fig. 8.8: The phase distribution $W_\varphi[|m\rangle]$ of a number state $|m\rangle$ represented in its most elementary version as a circular Planck-Bohr-Sommerfeld band of inner radius $r_m = (2m)^{1/2}$ and outer radius $r_{m+1} = [2(m+1)]^{1/2}$ is given by the area of overlap between this annulus and the diverging beam representing the phase eigenstate, normalized to the 2π-area of the band, that is, $W_\varphi[|m\rangle] \cong 1/(2\pi)$ as shown in (a). The concept of interference in phase space associates with the phase of the complex-valued phase amplitude, $w_\varphi[|m\rangle]$, the area, $r^2_{m+1/2}\varphi/2 = (m+1/2)\varphi$, caught between the center line of the mth band, that is between the circular Kramers trajectory $x^2/2 + p^2/2 = m+1/2$, and the center line of the diverging beam as shown in (b). Due to the zero-point energy correction $1/2$ the phase probability amplitude w_φ is not 2π- but only 4π-periodic.

This area reads

$$|\langle\varphi|m\rangle|^2 \cong \frac{1}{2\pi}\int_{r_m}^{r_{m+1}} dr\, r \int_{-\pi}^{\pi} d\varphi'\, \delta(\varphi-\varphi') = \frac{1}{2\pi}\frac{1}{2}\left(r^2_{m+1}-r^2_m\right) = \frac{1}{2\pi}. \qquad (8.31)$$

Here we have included the factor $(2\pi)^{-1}$ as to achieve the normalization condition

$$\int_{-\pi}^{\pi} d\varphi\, |\langle\varphi|m\rangle|^2 = 1$$

since the area of the Planck-Bohr-Sommerfeld band is 2π.

Thus the phase distribution

$$W_\varphi[|m\rangle] = |\langle\varphi|m\rangle|^2 = \frac{1}{2\pi} \qquad (8.32)$$

of a number state $|m\rangle$ is independent of φ. Hence, the phase in an energy eigenstate is completely undetermined.

The concept of interference in phase space also provides the phase of the probability amplitude $\langle\varphi|m\rangle$. According to Eq. (7.23) it is the area

$$\frac{1}{2}r^2_{m+1/2}\varphi = (m+1/2)\,\varphi, \qquad (8.33)$$

caught between the circular center line of the Planck-Bohr-Sommerfeld band at radius $r_{m+1/2} = [2(m+1/2)]^{1/2}$ and that of the phase space ray shown in Fig. 8.8(b).

With the help of the expression Eq. (8.32) for the phase distribution of an energy eigenstate we thus find the phase probability amplitude

$$\langle \varphi | m \rangle \cong W_\varphi[|m\rangle]^{1/2} \exp[-i(m+1/2)\,\varphi] = \frac{1}{\sqrt{2\pi}} \exp[-i(m+1/2)\,\varphi] \qquad (8.34)$$

of a energy eigenstate.

Phase States in a Truncated Hilbert Space

This result suggests the definition of a phase state

$$|\varphi\rangle \equiv \sum_{m=0}^{\infty} \langle m|\varphi\rangle\, |m\rangle = \frac{1}{\sqrt{2\pi}} \sum_{m=0}^{\infty} \exp\left(+i(m+\tfrac{1}{2})\varphi\right) |m\rangle \qquad (8.35)$$

as a superposition of energy eigenstates.

Unfortunately, this definition has a problem. Indeed, as shown geometrically in Fig. 8.7, each energy eigenstate $|m\rangle$ contributes to the infinite sum in Eq. (8.35) an amount $|\langle m|\varphi\rangle| = (2\pi)^{-1/2}$, that is an amount independent of m. Hence, the expression for $|\varphi\rangle$, Eq. (8.35), is not converging.

When we truncate the sum we arrive at

$$|\varphi\rangle_r = \lim_{r\to\infty} [2\pi(r+1)]^{-1/2} \sum_{m=0}^{r} \exp[i(m+1/2)\varphi]\, |m\rangle. \qquad (8.36)$$

These states have been used by Barnett and Pegg to define a hermitian phase operator. We emphasize, however, that this limit is not a trivial limit. Indeed, Hilbert space of the harmonic oscillator relies on an infinite amount of energy eigenstates. When we now restrict our state space to a finite number of energy eigenstates these states are not complete any more in the ordinary Hilbert space. We therefore live in a different space. Nevertheless, we can use this finite set of energy eigenstates to construct in that space quantum states similar to coherent states or squeezed states or even position and momentum eigenstates. As we increase the dimension of this space we approach the ordinary Hilbert space of the harmonic oscillator.

Unfortunately, space does not allow us to go deeper into these fascinating spaces built out of a finite number of energy eigenstates. We therefore refer to the literature for a more detailed discussion on this topic.

Time Evolution of Phase States

In contrast to the usual convention we have included a phase shift, $\varphi/2$, in the definition Eq. (8.35) of the phase states. This shift makes its appearance through Eqs. (8.33) and (8.34) and results from the zero-point energy of the oscillator. As a consequence, the phase states, Eq. (8.36), are not 2π- but rather 4π-periodic.

This zero-point energy produced phase is irrelevant for calculating moments of φ or constructing a phase operator. It should, however, reveal itself in interference experiments. Moreover, it is convenient when we now consider the time evolution of a phase state.

Indeed, we conclude this motivation by discussing the time evolution

$$|\psi(t)\rangle = \exp\left(-\frac{i}{\hbar}\hat{H}t\right)|\varphi_0\rangle$$

of a phase state $|\psi(t=0)\rangle \equiv |\varphi_0\rangle$ under the Hamiltonian

$$\hat{H} = \hbar\Omega(\hat{m}+1/2).$$

corresponding to a harmonic oscillator.

With the help of the definition Eq. (8.35) of the phase state and the eigenvalue equation for the energy eigenstates we find

$$|\psi(t)\rangle = \frac{1}{\sqrt{2\pi}} \sum_{m=0}^{\infty} \exp\left(i(m+\tfrac{1}{2})\varphi_0\right) \exp\left(-\frac{i}{\hbar}\hat{H}t\right)|m\rangle$$

$$= \frac{1}{\sqrt{2\pi}} \sum_{m=0}^{\infty} \exp\left[i(m+\tfrac{1}{2})(\varphi_0 - \Omega t)\right]|m\rangle$$

and hence

$$|\psi(t)\rangle = |\varphi(t)\rangle = |\varphi_0 - \Omega t\rangle.$$

Under time evolution a phase state remains a phase state. The phase angle decreases linearly in time as expected from the classical considerations of Sec. 8.5.1.

8.5.3 Phase Distribution of a Quantum State

We are now in a position to define the phase probability density of a quantum state $|\psi\rangle$. With the help of the so defined phase states $|\varphi\rangle$ we define this density by the scalar product

$$W(\varphi) \equiv |\langle\varphi|\psi\rangle|^2.$$

Similarly, for a density operator $\hat{\rho}$ the phase probability reads

$$W(\varphi) \equiv \langle\varphi|\hat{\rho}|\varphi\rangle.$$

When we substitute the energy representation

$$|\psi\rangle = \sum_{m=0}^{\infty} \psi_m |m\rangle$$

of a pure state into the above expression for $W(\varphi)$ we find

$$W(\varphi) = \left|\sum_{m=0}^{\infty} \psi_m \langle\varphi|m\rangle\right|^2 = \frac{1}{2\pi}\left|\sum_{m=0}^{\infty} \psi_m \exp(-im\varphi)\right|^2.$$

Here we have made use of Eq. (8.34) for the phase probability amplitude of an energy eigenstate. The constant phase shift $\varphi/2$ cancels due to the absolute value.

Hence, the phase probability density is the absolute value squared of an infinite sum involving the energy probability amplitudes and phase factors.

We conclude this section by relating the phase probability amplitude

$$w_\varphi[|\psi\rangle] \equiv \langle\varphi|\psi\rangle = \sum_{m=0}^{\infty} \langle\varphi|m\rangle\langle m|\psi\rangle,$$

to the photon number probability amplitude

$$w_m[|\psi\rangle] \equiv \langle m|\psi\rangle.$$

With the help of Eq. (8.34) we find that

$$w_\varphi[|\psi\rangle] = \frac{1}{\sqrt{2\pi}} \sum_{m=0}^{\infty} w_m[|\psi\rangle] \exp[-i(m+1/2)\varphi], \tag{8.37}$$

is the discrete Fourier sum of the photon number probability amplitude $w_m[|\psi\rangle]$.

This result is analogous to the familiar fact that the probability amplitude $\psi(p)$ in p is the Fourier transform of the probability amplitude $\psi(x)$ in x. However, because the harmonic oscillator states have positive energy, $m \geq 0$, only the half-sided Fourier sum appears in Eq. (8.37). This fact gives rise to a serious problem when we want to construct the phase operator.

Problems

8.1 WKB-Wave Functions from Interference in Phase Space

Derive the WKB-expression $u_m^{(WKB)}$ Eq. (5.30) of an energy eigenstate from the concept of interference in phase space.

Hint: See Schleich et al. (1989).

8.2 Photon Statistics Using Wigner Functions

Derive the photon statistics of a coherent or a squeezed state using the Wigner function approach.

Hint: See Schleich et al. (1988).

8.3 Poisson Summation Formula

Derive the Poisson Summation Formula

$$\sum_{n=a}^{b} f_n = \sum_{l=-\infty}^{\infty} \int_a^b f(n)\, e^{2\pi i l n} dn + \frac{1}{2}\Big(f(a) + f(b)\Big)$$

valid for any discrete function f_n. Note that $f(n)$ is the continuous extension such that $f(n) = f_n$ at all integers in the range of the summation.

Hint: Interchange the order of integration and summation.

8.4 Properties of Phase States

In Eq. (8.35) we have defined the phase states

$$|\varphi\rangle = \frac{1}{\sqrt{2\pi}} \sum_{m=0}^{\infty} e^{i(m+1/2)\varphi} |m\rangle.$$

(a) Use the Poisson summation formula from Problem 8.3 to derive the formula

$$\langle \varphi | \varphi' \rangle = \frac{1}{2}\delta(\varphi - \varphi') - \mathcal{P}\frac{i}{4\pi \sin[(\varphi - \varphi')/2]}.$$

for the scalar product of two phase states $|\varphi\rangle$ and $|\varphi'\rangle$. Here $\mathcal{P}$ denotes the Cauchy principle part.

This relation shows that phase states corresponding to different angles are *not* orthonormal. This is a consequence of the fact that the sum in the definition (8.35) of the phase state starts from 0 and not from $-\infty$.

(b) Use the WKB representation Eq. (5.30) of the energy wave function $u_m(x)$ and the method of stationary phase to derive the asymptotic expression

$$w_\varphi(x) \equiv \langle x | \varphi \rangle \cong \frac{1}{2\sqrt{\pi}} \frac{\sqrt{x}}{\cos \varphi} \exp\left[\frac{i}{2} x^2 \tan \varphi\right]$$

for the position probability amplitude $w_\varphi(x) \equiv \langle x | \varphi \rangle$ of a phase state.

(c) Use the above expression for the position probability amplitude of a phase state to derive the approximate expression

$$W_\varphi(x, p) = \frac{1}{2\pi} \left(\frac{x}{\cos \varphi}\right)^2 \frac{J_1[2x(p - x \tan \varphi)]}{2x(p - x \tan \varphi)}$$

for the Wigner function of a phase state.

(d) Show that for a squeezed state with position wave function

$$\psi_{\text{sq}}(x) \equiv \langle x | \psi_{\text{sq}} \rangle \equiv \left(\frac{s\kappa^2}{\pi}\right)^{1/4} \exp\left[-\frac{s}{2}\left(\kappa x - \sqrt{2}\alpha\right)^2\right]$$

the phase distribution in the case of a real displacement parameter $\alpha > 0$ and squeezing parameter $s \gg 1$ reads

$$P(\varphi) = \begin{cases} \sqrt{\frac{2}{\pi s}} \frac{\alpha}{\cos^2 \varphi} \exp\left[-\frac{2}{s}\alpha^2 \tan^2 \varphi\right] & \text{for } -\frac{\pi}{2} < \varphi < \frac{\pi}{2} \\ 0 & \text{for } -\pi <, \varphi <, -\frac{\pi}{2} \\ & \text{or } \frac{\pi}{2} < \varphi < \pi. \end{cases}$$

Hint: In the limit of $s \gg 1$ we can use the relation

$$\langle m | \psi_{\text{sq}} \rangle \cong \begin{cases} 0 & \text{for } m < \alpha^2 - \frac{1}{2} \\ \mathcal{A}_m^{1/2} e^{i\phi_m} + \mathcal{A}_m^{1/2} e^{-i\phi_m} & \text{for } m > \alpha^2 - \frac{1}{2}, \end{cases}$$

where $\mathcal{A}_m$ and ϕ_m are defined in Eqs. (8.18) and (8.19). We approximate the sum by an integral and evaluate it with the method of stationary phase. For more details see Schleich et al., (1989).

(e) Show that the phase distribution of a squeezed state exhibits a bifurcation as a function of the displacement or squeezing parameter. This bifurcation has even been observed in the experiments by G. Breitenbach *et al.* (1997).

References

Interference in Phase Space and Photon Statistics

For the application of the concept of interference in phase space to the energy distribution or photon statistics of coherent and squeezed states see

W.P. Schleich, H. Walther and **J.A. Wheeler**, *Area in Phase Space as a Determiner of Transition Probability: Bohr-Sommerfeld Bands, Wigner Ripples and Fresnel Zones*, Found. Phys. **18**, 953–968 (1988)

W.P. Schleich and **J.A. Wheeler**, *Oscillations in Photon Distribution of Squeezed States and Interference in Phase Space*, Nature (London) **326**, 574–577 (1987)

W.P. Schleich and **J.A. Wheeler**, *Oscillations in Photon Distribution of Squeezed States*, J. Opt. Soc. Am. B **4**, 1715–1722 (1987)

W.P. Schleich, D.F. Walls and **J.A. Wheeler**, *Area of Overlap and Interference in Phase Space Versus Wigner Pseudo-Probabilities*, Phys. Rev. A **38**, 1177–1186 (1988)

W.P. Schleich, J.P. Dowling, R.J. Horowicz and **S. Varro**, *Asymptotology in Quantum Optics*, in: *New Frontiers in Quantum Electrodynamics and Quantum Optics*, ed. by A.O. Barut, Plenum Press, New York, 1989, p. 31–61

D. Krähmer, E. Mayr, K. Vogel and **W.P. Schleich**, *Meet a Squeezed State and Interfere in Phase Space*, in Current Trends in Optics, ed. by J.C. Dainty, Academic Press, London, 1994

K. Vogel and **W.P. Schleich**, in: *Fundamental Systems in Quantum Optics*, eds. J. Dalibard, J.M. Raimond and J. Zinn-Justin, Elsevier, Amsterdam, 1991

Phase Operator

The problem of the hermitian phase operator dates back to the papers by

P.A.M. Dirac, *The Quantum Theory of the Emission and Absorption of Radiation*, Proc. R. Soc. London, Ser. **114 A**, 243–269 (1927)

F. London, *Über die Jacobischen Transformationen der Quantenmechanik*, Z. Phys. **37**, 915–925 (1926)

F. London, *Winkelvariable und kanonische Transformationen in der Undulationsmechanik*, Z. Phys. **40**, 193–210 (1927)

For a recent proposal of a hermitian phase operator see

D.T. Pegg and **S.M. Barnett**, *Unitary Phase Operator in Quantum Mechanics*, Europhys. Lett. **6**, 483–487 (1988)

D.T. Pegg and **S.M. Barnett**, *Phase Properties of the Quantized Single-mode Electromagnetic Field*, Phys. Rev. **A39**, 1665–1675 (1989)

and the reviews by

S.M. Barnett and **D.T. Pegg**, *On the Hermitian Optical Phase Operator*, J. Mod. Opt. **36**, 7–19 (1989)

D.T. Pegg, S.M. Barnett and **J.A. Vaccarro**, *Phase in Quantum Electrodynamics*, in: *Quantum Optics V*, edited by D.F. Walls and J. Harvey, Springer Verlag, Heidelberg, 1989

The Hermitian phase operator put forward in this work is based on the Loudon states defined in

R. Loudon, *The Quantum Theory of Light*, 1st ed., Oxford University Press, Oxford, 1973

For a review of the problem of quantum phase and phase dependent measurements see

W.P. Schleich and **S.M. Barnett**, *Quantum Phase and Phase Dependent Measurements*, Physica Scripta **T48** (1993)

H. Paul, *Phase of a Microscopic Electromagnetic Field and its Measurement*, Fortschr. Phys. **22**, 657–689 (1974)

P. Carruthers and **M.M. Nieto**, *Phase and Angle Variables in Quantum Mechanics*, Rev. Mod. Phys. **40**, 411–440 (1968)

R. Lynch, *The Quantum Phase Problem: A Critical Review*, Phys. Rep. **256**, 367–436 (1995)

Interference in Phase Space and Phase States

For an approach towards the problem of phase states using the concept of interference in phase space see

W.P. Schleich, R.J. Horowicz and **S. Varro**, *A Bifurcation in Squeezed State Physics: But How? or Area-of-Overlap in Phase Space as a Guide to the Phase Distribution and the Action-Angle Wigner Distribution of a Squeezed State*, Quantum Optics V, edited by D.F. Walls and J. Harvey (Springer Verlag, Heidelberg, 1989)

W.P. Schleich, R.J. Horowicz and **S. Varro**, *A Bifurcation in the Phase Probability of a Highly Squeezed State*, Phys. Rev. A **40**, 7405–7408 (1989)

For an experimental verification of this phase distribution see

G. Breitenbach and **S. Schiller**, *Homodyne tomography of classical and nonclassical light*, J. Mod. Opt. **44**, 2207–2225 (1997)

Quantum Mechanics in Finite Hilbert Space

For a discussion of various quantum states in a state space of the harmonic oscillator with finite dimensions see

V. Buzek, A.D. Wilson-Gordon, P.L. Knight and **W.K. Lai**, *Coherent States in a finite dimensional basis: their phase properties and relation to coherent states of light*, Phys. Rev. A **45**, 8079–8094 (1992)

P. Figurny, A. Orlowski and **K. Wódkiewicz**, *Squeezed fluctuations of truncated photon operators*, Phys. Rev. A **47**, 5151–5157 (1993)

B. Kneer, *Diskretisierung des elektromagnetischen Feldes in der Quantenoptik*, Diplomarbeit, Universität Ulm 1995

9 Wave Packet Dynamics

In the preceding chapters we have discussed semi-classical techniques in quantum mechanics and have presented various applications. In particular, we have shown that the oscillations in the energy distributions of squeezed states originally hidden behind an opaque curtain of mathematics only emerge when we put to use a WKB analysis. In the present chapter we demonstrate another application of semi-classical tools: We analyze the dynamics of a wave packet.

9.1 What are Wave Packets?

Wave packets consist of a large number of simultaneously excited quantum levels. They appear in the context of atomic and molecular physics, cavity QED and atom optics. Indeed, when we expand, for example, the quantum state $|\psi\rangle$ of the motion of an electron in an atom or of a nucleus in a diatomic molecule into energy eigenstates $|n\rangle$ with eigenenergies $E_n \equiv \hbar\omega_n$ the time evolution of the system reads

$$|\psi(t)\rangle = \sum_n \psi_n e^{-i\omega_n t} |n\rangle$$

where ψ_n denote the expansion coefficients.

A typical signal observed from such a time dependent wave packet is the autocorrelation function

$$\mathcal{C}(t) \equiv |\langle \psi(t) | \psi(0) \rangle| = \left| \sum_n |\psi_n|^2 e^{i\omega_n t} \right|,$$

which measures the overlap between the time-evolved state $|\psi(t)\rangle$ and the original state $|\psi(t=0)\rangle$.

Another example is discussed in Chapter 16 in the context of the Jaynes-Cummings-Paul model. Here, a single mode of the electromagnetic field in a cavity interacts with a single two-level atom. The wave packet consists of a superposition of energy eigenstates of this mode, that is of photon number states. In this sense it is a "photon wave packet".

A typical signal in this context is the atomic inversion

$$\mathcal{I}(t) \equiv -\operatorname{Re} \sum_{n=0}^{\infty} W_n e^{i\sqrt{n}2gt}.$$

The weight factors W_n are the photon statistics of the quantum field and g is the vacuum Rabi frequency.

Despite of the different physical nature of these systems and the studied signals, there is a surprising similarity in the overall structure of the temporal behavior of these signals. It is almost independent of the details of the weights W_n and the absolute sizes of the frequencies $\omega(n)$. Therefore, throughout this chapter we do not specify the physical meaning of these quantities, but only assume rather general properties such as smoothness, normalizability etc. In particular, we study the time evolution of a transient signal of the form

$$\mathcal{S}(t) \equiv \sum_n W_n\, e^{i\omega_n t}, \qquad (9.1)$$

where ω_n and W_n denote the frequency and the weight of the harmonic with the sequential number n.

We note that the autocorrelation function of an atomic or molecular wave packet involves such a signal $\mathcal{S}$. Here the weightfactor $W_n \equiv |\psi_n|^2$ represents the occupation probability of the nth level and the frequency $\omega_n \equiv E_n/\hbar$ follows from the energy spectrum of the atom or the molecule. Similarly the Jaynes-Cummings-Paul model involves a sum of the same form. In this case the frequency spectrum $\omega_n \equiv \sqrt{n}2g$ follows a square root dependence in the summation index as discussed in Chapter 16.

Starting from Eq. (9.1), we present an analytical approach towards the typical features of such transient signals. We derive the quasi-periodical behavior, de-phasing, fractional revivals, and full revivals. All these physical phenomena are a result of quantum beats which represent interference effects between *many* contributing terms in Eq. (9.1). However, due to this very reason it is hard to recognize the fine structure of the signal from the representation of $\mathcal{S}$ given in Eq. (9.1). We therefore use techniques of semiclassical quantum mechanics to derive closed-form expressions in distinguished time intervals of interest which bring out the typical features of the signal $\mathcal{S}$.

9.2 Fractional and Full Revivals

Ultrashort laser pulses have opened a new and fascinating research area – the physics of atomic and molecular wave packets. Short pulses not only allow the excitation of a coherent superposition of many quantum states, but they also provide a tool to monitor the subsequent dynamics.

In Figs. 9.1–9.3 we present three typical examples for transient signals from such coherently prepared systems. Figures 9.1 and 9.2 show the time-resolved emission and the autocorrelation function of an electronic Rydberg wave packet created by a short laser pulse. Figure 9.3 presents the calculated autocorrelation function $\mathcal{C}(t)$ for a vibrational wave packet propagating in the excited potential surface $A^1\Sigma_u^+$ of a sodium dimer.

Although the physical nature of the two systems as well as the displayed observables are rather different, the graphs show some surprising similarities. Initially, they exhibit a sequence of regular peaks. The period T_1 of this pattern corresponds to the

9.2 Fractional and Full Revivals 235

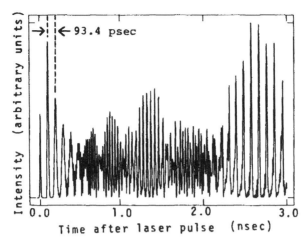

Fig. 9.1: Dynamics of an electronic wave packet monitored by the time-resolved intensity of spontaneous emission. The wave packet was created by a short laser pulse resonant to a manifold of closely-lying Rydberg states in hydrogen centered about the principal quantum number $\bar{n} = 85$. Here the initial beat pattern with period $T_1 = 93.4$ ps repeats itself after approximately $t \simeq 2.6$ ns. Taken from J. Parker and C.R. Stroud Jr., Phys. Rev. Lett. **56**, 716 (1986).

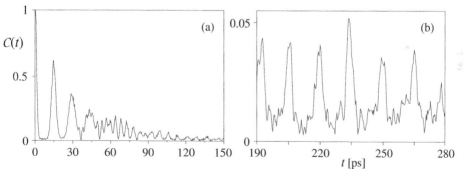

Fig. 9.2: Experimental data of the autocorrelation function $\mathcal{C}(t) = |\langle\psi(t)|\psi(0)\rangle|$ of an atomic wave packet. From (a) we recognize that in the early stage $\mathcal{C}(t)$ is almost periodic with a period $T_1 \cong 15.3$ ps corresponding to the typical energy separation between neighboring eigenstates. However, for larger times this periodicity disappears and new features emerge: At fractions of another characteristic time $T_2 \gg T_1$ the system is again periodic, a phenomenon referred to as fractional revivals. The period is now a fraction of T_1. In the immediate vicinity of the time $T_2 \cong 474$ ps the signal even restores almost completely its initial shape giving rise to full revivals. Moreover, the period T_1 occurs near the time point $T_2/2 \cong 237$ ps as shown in (b), but in this region the signal pattern is shifted by $T_1/2$ with respect to the initial one. These fractional revivals show an asymmetric shape with a fast decay on one side and a slow oscillatory fall down on the other. Taken from J. Wals *et al.*, Physica Scr. **T 58**, 62 (1995).

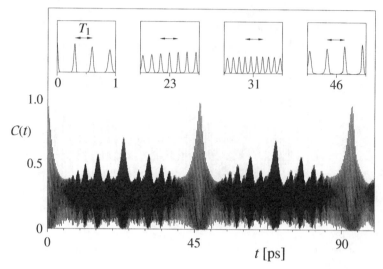

Fig. 9.3: Dynamics of a vibrational wave packet propagating in the excited electronic potential $A^1\Sigma_u^+$ in a sodium dimer. We show the autocorrelation function $\mathcal{C}(t) \equiv |\langle \psi(0)|\psi(t)\rangle|$, where the initial state $|\psi(0)\rangle$ is a replica of the ground state of the lower potential $X^1\Sigma_g^+$. This state may be created by a short laser pulse via a vertical electronic transition, and consists of several vibrational states of the potential $A^1\Sigma_u^+$. For this system the initial periodicity of $T_1 = 300$ fs shown in the inset in the left upper corner of the figure repeats itself after approximately 46 ps as shown by the inset in the right upper corner. The other insets which magnify the behavior of $\mathcal{C}(t)$ over a time duration of 1 ps around $t = 23$ ps and $t = 31$ ps reveal periods different from T_1. To bring this out most clearly we indicate the period T_1 by the arrow.

typical energy separation between neighboring excited levels. After some periods this behavior gradually disappears. However, the initial pattern recurs after a time T_2, which is much longer than T_1. For this reason this time scale T_2 is usually referred to as the *revival time*. For the electronic Rydberg wave packet in Fig. 9.1, T_2 is given by 5.2 ns, whereas for the nuclear wave packet of Fig. 9.3 one finds $T_2 = 94$ ps.

Moreover, these graphs also show, that at *fractions* of this revival time again a periodic structure called *fractional revivals* emerges, however now with a period which is a *fraction* of T_1. This feature comes out most clearly in Fig. 9.3. Full revivals and fractional revivals were observed in a number of experiments in atomic and molecular systems. For a detailed discussion of the experiments we refer to the literature at the end of this chapter.

We conclude by noting that the phenomena of collapse and periodic revivals were only predicted in 1980. J.H. Eberly and his colleagues investigated the time evolution of the atomic inversion predicted by the Jaynes-Cummings-Paul model. We discuss this model and the inversion in more detail in Sec. 16.2. Eberly *et al.* gave the first accurate expressions for the intermediate- and long-time behavior of this cavity QED model, and provided extensive numerical evidence in support of their analytic formulas. Moreover, they coined the term "revival" for this phenomenon. Surprisingly, fractional revivals were not noticed until about a decade later.

9.3 Natural Time Scales

In the present section we rewrite the signal $S(t)$, Eq. (9.1), as to bring out the different time scales in its evolution. Throughout the chapter we assume that the normalized distribution of weight factors W_n has a dominant maximum at the integer $\bar{n} \gg 1$ and a width Δn such that $\bar{n} \gg \Delta n \gg 1$. In this large n regime, that is the semiclassical regime, the frequencies w_n of a physical system depend smoothly on the index n and we use a continuous extension $w(n)$. We note, however, that here we have already assumed, that the underlying physical system is an integrable, that is non-chaotic, system. Otherwise the energy spectrum shows a very complicated behavior with level repulsion and other intricacies. In the present chapter we do not want to get into these problems and assume that $w(n)$ is smooth.

9.3.1 Hierarchy of Time Scales

These properties allow us to expand $w(n)$ in a Taylor series

$$w(n) = w(\bar{n}) + \frac{dw(n)}{dn}\bigg|_{n=\bar{n}} (n-\bar{n}) + \frac{1}{2}\frac{d^2w(n)}{dn^2}\bigg|_{n=\bar{n}} (n-\bar{n})^2$$
$$+ \frac{1}{6}\frac{d^3w(n)}{dn^3}\bigg|_{n=\bar{n}} (n-\bar{n})^3 + \ldots \qquad (9.2)$$

around $\bar{n}$ which we write as

$$w(n) = w(\bar{n}) + \sigma_1\frac{2\pi}{T_1}(n-\bar{n}) + \sigma_2\frac{2\pi}{T_2}(n-\bar{n})^2 + \sigma_3\frac{2\pi}{T_3}(n-\bar{n})^3 + \ldots.$$

Here we have defined

$$\frac{2\pi}{T_j} \equiv \frac{1}{j!}\left|w^{(j)}(\bar{n})\right| = \frac{1}{j!}\left|\frac{d^jw}{dn^j}\right|_{n=\bar{n}}$$

and $\sigma_j = \pm 1$ accounts for the sign of the jth derivative $w^{(j)}(\bar{n})$.

Note, that the value of σ_1 can always be assumed to be $+1$, since the energy of a bound state of a quantum system increases with the quantum number n. However, the signs of σ_j for $j > 1$ depend on the system of interest. For example, in the case of Rydberg atoms we have eigenfrequencies $w(n) \propto -1/n^2$. This creates $\sigma_2 = -1$ and $\sigma_3 = +1$. In the Jaynes-Cummings-Paul model with Rabi frequencies $w(n) \propto \sqrt{n}$ we have $\sigma_2 = -1$ and $\sigma_3 = +1$.

When we insert Eq. (9.2) into Eq. (9.1) we find

$$S(t) = \exp\left[i\,w(\bar{n})t\right] \mathcal{S}(t), \qquad (9.3)$$

where

$$\mathcal{S}(t) \equiv \sum_{m=-\infty}^{\infty} W_{\bar{n}+m} \exp\left[2\pi i\left(\frac{t}{T_1}m + \sigma_2\frac{t}{T_2}m^2 + \sigma_3\frac{t}{T_3}m^3 + \ldots\right)\right]. \qquad (9.4)$$

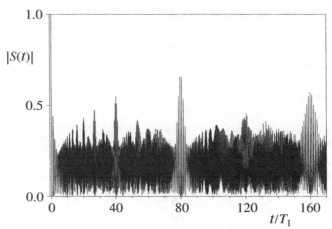

Fig. 9.4: Generic signal represented here by the time dependence of the sum $|S(t)|$, Eq. (9.4), for the case of a Gaussian distribution W_n with variance $\Delta n = 8$. We have chosen the parameters $T_2 = 160 \cdot T_1$ and $T_3 = 1000 \cdot T_2$ with $\sigma_1 = \sigma_2 = \sigma_3 = 1$ and have set all higher time scales in Eq. (9.2) equal to infinity. The sum reveals a fairly complicated temporal behavior.

Here we have introduced the summation index $m \equiv n - \bar{n}$. It is the sum $S(t)$, Eq. (9.4) which we analyze in the remainder of this chapter.

We gain deeper insight into the expansion Eq. (9.2) of ω with respect to n and into the time scales T_j when we recall from Chapter 5 that in the semiclassical limit the action J is proportional to the quantum number n of the bound state. With $J = n\hbar$ and $E = \hbar\omega = H$, where H denotes the Hamiltonian, we therefore arrive at

$$\frac{\partial \omega}{\partial n} = \frac{\partial(\hbar\omega)}{\partial(\hbar n)} = \frac{\partial H}{\partial J}. \tag{9.5}$$

This relation allows us to express the derivatives $\omega^{(j)}$ of the frequency ω with respect to n in terms of derivatives of the Hamiltonian and Planck's constant $\hbar$. Indeed, we find using $J = n\hbar$ the relation

$$\frac{\partial^j \omega}{\partial n^j} = \frac{\partial^{j-1}}{\partial n^{j-1}} \frac{\partial H}{\partial J} = \hbar^{j-1} \frac{\partial^j H}{\partial J^j}. \tag{9.6}$$

We emphasize that for a nonlinear oscillator the Hamiltonian is a nonlinear function of the action. Hence, the individual derivations bring in different powers of Planck's constant. Hence, the expansion (9.2) of ω corresponds to an expansion in powers of $\hbar$. Since the times T_j are the inverse of the derivatives $\omega^{(j)}$, they are proportional to the inverse powers $\hbar^{-j+1}$ and therefore satisfy the hierarchy

$$T_1 \ll T_2 \ll T_3 \ll \ldots. \tag{9.7}$$

Hence, the individual time scales separate. This allows us to analyze the temporal behavior of the signal $S(t)$ Eq. (9.4) in the individual regimes.

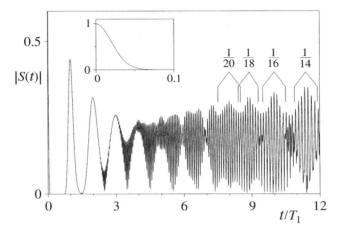

Fig. 9.5: Generic signal of Fig. 9.4 in its early stage, that is for times up to $t = 12 \cdot T_1$. After a rapid decay magnified in the inset we note symmetric peaks with period T_1 which broaden and decay in height. At around $t \cong 3 \cdot T_1$ rapid oscillations set in and a complicated beat structure develops. Note that as soon as this pattern emerges, we find fractional revivals of various orders. In the vicinity of $t = 1/20 \, T_2 = 8 \cdot T_1$ we find 10 peaks within a period of T_1, whereas in the vicinities of $t = 1/18 \, T_2 = 8.9 \cdot T_1$, $t = 1/16 \, T_2 = 10 \cdot T_1$, and $t = 1/14 \, T_2 = 11.4 \cdot T_1$ we find 9, 8, and 7 peaks within a period of T_1, respectively.

9.3.2 Generic Signal

To illustrate the typical temporal behavior of the sum Eq. (9.4) we use the specific example of a Gaussian distribution

$$W_n = \frac{1}{\sqrt{2\pi \Delta n^2}} \exp\left[-\frac{(n-\bar{n})^2}{2\Delta n^2}\right] \qquad (9.8)$$

with the variance Δn. In fact, this example is the most discussed case in the literature, since in pump-probe experiments the exciting laser pulse has often a Gaussian shape which yields a Gaussian or near Gaussian weight function W_n. However, we emphasize that the typical features of transient signals do not change significantly for non-Gaussian weights, as long as W_n is centered about some dominant maximum with a width $\Delta n \gg 1$. We therefore refer to the resulting signal as the *generic* signal.

Note that due to the shift $m \equiv n - \bar{n}$ in the summation over m, Eq. (9.4), the parameter $\bar{n}$ only enters in the times T_j. In the present example we do not specify the functional dependence of $\omega(n)$ on n but choose $T_2 = 160 \cdot T_1$ and $T_3 = 1000 \cdot T_2$. Moreover, we set all higher time scales in the expansion in Eq. (9.2) equal to infinity. Hence in this way $\bar{n}$ does not enter explicitly. In addition we have taken $\sigma_2 = \sigma_3 = 1$.

In Fig. 9.4 we show the overall structure of the sum $|S(t)|$ for the Gaussian Eq. (9.8) with variance $\Delta n = 8$ over a long time interval. Here and in all of the following figures time is scaled in units of T_1. This graph reveals a complicated temporal dependence of $|S(t)|$ similar to the quantities displayed in Figs. 9.1 and 9.3.

Figures 9.5 and 9.6 magnify specific time intervals of Fig. 9.4 in order to resolve

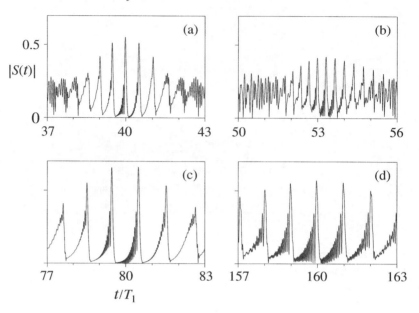

Fig. 9.6: Generic signal of Fig. 9.4 at intermediate times. Here we magnify the behavior of $|S(t)|$ for time intervals of length $6 \cdot T_1$ around distinguished times. The cases (a) and (b) show fractional revivals in the vicinity of $t = 1/4\, T_2 = 40 \cdot T_1$ and $t = 1/3\, T_2 = 53.33 \cdot T_1$, respectively. Note, that the period of the fractional revivals in (a) is given by $T_1/2$ and in (b) by $T_1/3$. A complicated beat pattern arises as soon as neighboring peaks start to overlap considerably as exemplified by the behavior of $|S(t)|$ at the edges of (a) and (b). Moreover, we recognize that the shape of the peaks becomes asymmetric, and oscillations appear on the left hand side of the maximum. The cases (c) and (d) focus on fractional revivals in the vicinity of $t = 1/2\, T_2 = 80 \cdot T_1$, and on the full revivals in the vicinity of $t = 1/1\, T_2 = 160 \cdot T_1$. The period of the fractional revivals in (c) as well as of the full revivals in (d) is given by T_1. However, the fractional revivals are shifted by half of the period T_1. Moreover, the shape of the fractional revivals and the full revivals are highly asymmetric, that is they show a slow oscillatory onset to the left of their maximum, and a rapid decay to the right. In all four cases the height of the peaks is controlled by a slowly varying envelope. The scales of the vertical axes are identical in all four examples.

the fine structure of the signal. Figure 9.5 presents the early stage of the evolution: After a rapid decay shown in the inset we find in the beginning a periodic sequence of symmetric peaks separated by a period of T_1. However, in the course of time the peaks become broader until they overlap and form a complicated beat pattern. In our example, this happens approximately after 3 periods. Later on, as soon as the beats set in, we find fractional revivals of various order which follow each other very closely.

In Fig. 9.6 we present magnified sections of Fig. 9.4 in the vicinity of the times $t = 1/4\, T_2 = 40 \cdot T_1$ (a), $t = 1/3\, T_2 = 53.33 \cdot T_1$ (b), $t = 1/2\, T_2 = 80 \cdot T_1$ (c), and $t = T_2 = 160 \cdot T_1$ (d), respectively. We recognize the following characteristic features: The sum S again involves periodic sequences of peaks, however now the period between two neighboring peaks is given by $T_1/2$, $T_1/3$, T_1, as well as T_1, respectively. Figures 9.6(a)–(c) show fractional revivals whereas Fig. 9.6(d) depicts a full revival. The larger the time $t = 1/r\, T_2$ (here $r = 4, 3, 2, 1$), the larger the difference of the fine structure compared to the symmetric peaks in the initial stage of the evolution. Especially, for larger times the fractional revivals become more and more asymmetric: They show an abrupt break-off on the right domain to their center, whereas on the left domain they decay much slower. Moreover, they show oscillations on top of this slow decay. Note, that similar structures shown in Fig. 9.2(b) were found experimentally by Wals et al. (1994, 1995) in the case of a Rydberg wave packet in rubidium. We further notice, that the heights of the fractional revivals seem to be modulated by a slowly varying amplitude with its center at the time point $t = 1/r\, T_2$: As fractional revivals are located further away from this center, their dominant maximum decreases, the width broadens and the small oscillations on the left to their center smear out.

9.4 New Representations of the Signal

The behavior of the generic signal $S(t)$ displayed in Figs. 9.4, 9.5 and 9.6 is not obvious from the *form* of S in Eq. (9.4). In the next two sections we therefore cast the sum into a form which brings out in a clear way the period of the peaks and the fine details of their shape. In Sec. 9.4.1 we start with the analysis of the initial evolution and in Sec. 9.4.2 we analyze the fine structure of fractional revivals and full revivals.

9.4.1 The Early Stage of the Evolution

In this section we present a method convenient for analyzing the signal at the initial stage of the evolution by using the Poisson summation formula. With the help of this transformation we obtain a new representation of the sum $\mathcal{S}$ which brings out most clearly its features in this time regime.

The phase factor of each term in Eq. (9.4) consists of the product of the factors $\exp(2\pi\, i\, m\, t/T_1)$, $\exp(2\pi\, i\, \sigma_2 m^2\, t/T_2)$, $\exp(2\pi\, i\, \sigma_3 m^3\, t/T_3)$ etc. The relative importance of these factors depends strongly on the specific time under consideration. In the early stage of the evolution, that is for times t of the order of T_1, the main contribution to the phase comes from the first factor $\exp(2\pi\, i\, m\, t/T_1)$. Hence, all the terms in the sum $\mathcal{S}(t)$ are in phase for times t which are multiples of the period T_1. Therefore, we expect a sequence of spikes in the signal located near the time points $t_l = l \cdot T_1$, where $l = 1, 2, \ldots$. However, as time increases, the second factor $\exp(2\pi\, i\, \sigma_2 m^2\, t/T_2)$ becomes also important. Its contribution leads to a growing dephasing of neighboring terms in the sum Eq. (9.4) at the time points t_l which results in a broadening of the spikes.

A Convenient Form of the Signal

To bring this out most clearly we rewrite the sum $S(t)$ with the help of the Poisson summation formula

$$\sum_{m=-\infty}^{\infty} f_m = \sum_{l=-\infty}^{\infty} \int_{-\infty}^{\infty} dm\, f(m)\, \exp(-2\pi i l m), \qquad (9.9)$$

discussed in Problem 8.3. Here $f(m)$ is a continuous extension of the discrete function f_m such that $f(m) = f_m$ for all m.

This yields

$$S(t) = \sum_{l=-\infty}^{\infty} \int_{-\infty}^{\infty} dm\, W(\bar{n}+m)$$
$$\times \exp\left\{ 2\pi i \left[\left(\frac{t}{T_1} - l\right) m + \sigma_2 \frac{t}{T_2} m^2 + \sigma_3 \frac{t}{T_3} m^3 + \ldots \right] \right\} \qquad (9.10)$$

where $W(\bar{n}+m)$ denotes the continuous version of the discrete distribution $W_{\bar{n}+m}$. We note that there are many continuous extensions of the discrete weights W_n. For the example of the Gaussian Eq. (9.8) we choose the extension

$$W(x) = \frac{1}{\sqrt{2\pi \Delta n^2}} \exp\left[-\frac{(x-\bar{n})^2}{2\Delta n^2}\right], \qquad (9.11)$$

but emphasize that the treatment presented is valid for an arbitrary weight distribution W_n.

Structures with Period T_1

To bring this out most clearly we now consider times much smaller than $T_j/(\Delta n)^j$, where $j \geq 3$, and keep the first two terms in the exponent of Eq. (9.10) only. In this case, the integral for the Gaussian distribution Eq. (9.11) is of the form

$$\int_{-\infty}^{\infty} dx\, \exp(-ax^2 + bx) = \sqrt{\frac{\pi}{a}} \exp\left(\frac{b^2}{4a}\right), \qquad (9.12)$$

with

$$a \equiv \frac{1}{2\Delta n^2} - 2\pi i \sigma_2 \frac{t}{T_2} \qquad (9.13)$$

and

$$b \equiv 2\pi i \left(\frac{t}{T_1} - l\right). \qquad (9.14)$$

When we make use of this relation the sum S reads

$$S(t) \cong \sum_{l=-\infty}^{\infty} \frac{1}{\sqrt{1 - i\sigma_2 4\pi \Delta n^2 t/T_2}} \exp\left[-\frac{2\pi^2 \Delta n^2}{1 - i\sigma_2 4\pi \Delta n^2 t/T_2}\left(\frac{t}{T_1} - l\right)^2\right].$$

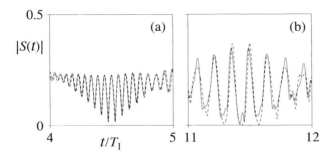

Fig. 9.7: Comparison between exact numerical evaluation of the generic signal $|S(t)|$, Eq. (9.4), (solid line) and the approximate expression Eq. (9.15) (dashed line) in the early stage of the evolution. Whereas for times shown in (a) the two curves are almost indistinguishable, they show deviations for larger times displayed in (b). In both cases the approximation works well for times at the center of each frame and gets worse towards the edges. The scales on the vertical axes in (a) and (b) are identical.

We separate the real and imaginary part in the exponent and arrive at

$$S(t) \cong \sum_{l=-\infty}^{\infty} \frac{1}{\sqrt{1 - i\sigma_2 4\pi \Delta n^2 t/T_2}} \exp\left[-\frac{(t - lT_1)^2}{2\,\sigma_r^2(t)}\right] \exp\left[-i\,\sigma_2 \frac{(t - lT_1)^2}{2\,\sigma_i^2(t)}\right], \quad (9.15)$$

where the widths

$$\sigma_r^2(t) \equiv \left[\frac{1}{4\pi^2 \Delta n^2} + 4\Delta n^2 \left(\frac{t}{T_2}\right)^2\right] T_1^2 \quad (9.16)$$

and

$$\sigma_i^2(t) \equiv \left[\frac{1}{16\pi^3 \Delta n^2 t/T_2} + \frac{1}{\pi}\Delta n^2 \frac{t}{T_2}\right] T_1^2 \quad (9.17)$$

of the real and the imaginary Gaussian increase as a function of time.

Equation (9.15) brings out the physical meaning of the transformation Eq. (9.9). Indeed, the Poisson summation formula allows us to represent a discrete superposition of many harmonics such as the sum S as a sequence of time dependent signals numbered by the index l and arriving one after another: The coherent signal $S(t)$ is now a sequence of complex Gaussians centered at the time points $t_l = l \cdot T_1$. Two consecutive terms of the sum Eq. (9.15) separate in time, when their temporal separation $t_l - t_{l-1} = T_1$ is larger than their width $\delta t_l = 2\sqrt{2}\,\sigma_r(t_l)$, that is, if $T_1 > \delta t_l$. Hence, the application of the Poisson summation formula leads to a significant simplification when the width of each signal in time is shorter than the separation between two signals.

In Fig. 9.7 we compare by a dashed line the approximation Eq. (9.15) to the exact curve shown by the solid line. Whereas initially in (a) a difference between the exact sum and expression Eq. (9.15) can hardly be recognized, the approximation which neglected the cubic term in the exponent, becomes worse for longer times as shown in (b).

9.4.2 Intermediate Times

We now turn to larger times for which the phase difference between two consecutive terms in the sums, Eq. (9.4), is not close to an integer multiple of 2π. In this time regime the representations Eqs. (9.10) and (9.15) are inconvenient since the interference between neighboring signals is important. Indeed, it is the interference of the complex Gaussians in Eq. (9.15) which eventually leads to the formation of fractional revivals, as shown in Figs. 9.5 and 9.7 for $t > 3T_1$. However, neither the period nor the shape of the fractional revivals can be seen from the form of the sum S in Eq. (9.15). Thus, in this time regime the representation Eq. (9.15) of the sum Eq. (9.4) is no more useful. But can we cast the sum S in a form which brings out the period and the shape of the fractional revivals and full revivals?

The answer is yes. The key idea of our approach is a decomposition of the sum Eq. (9.4) into a number of sub-sums, each of which contains only terms whose phases are close to each other. We achieve this by combining each rth term of the original sum Eq. (9.4) to one sub-sum. The particular choice of r depends on the time interval under consideration.

Shift of Origin of Time

Indeed, consider the behavior of S in the neighborhood of the time $t = q/r\, T_2$ of a fractional revival. Here q/r are mutually prime integers. It is advantageous to shift the origin of time into the region of $q/r\, T_2$ and choose it to be an integer multiple l of T_1, that is

$$t \equiv l\, T_1 + \Delta t \equiv \frac{q}{r} T_2 + \epsilon_{q/r} T_1 + \Delta t. \tag{9.18}$$

Here the absolute value of the remainder $\epsilon_{q/r} T_1 \equiv l\, T_1 - \frac{q}{r} T_2$ is less than or equal to half of the period T_1, that is $|\epsilon_{q/r}| \leq 1/2$. This choice allows us to bring the sum S, Eq. (9.4) into the form

$$S(\Delta t) \equiv S(t = \frac{q}{r} T_2 + \epsilon_{q/r} T_1 + \Delta t) = \sum_{m=-\infty}^{\infty} \gamma_m^{(r)}\, \widetilde{W}_m(\Delta t), \tag{9.19}$$

where

$$\gamma_m^{(r)} \equiv \exp\left(2\pi\, i\, \sigma_2 \frac{q}{r} m^2\right) \tag{9.20}$$

and

$$\widetilde{W}_m(\Delta t) \equiv W_{\bar{n}+m} \exp\left\{2\pi i \left[\frac{\Delta t}{T_1} m + \sigma_2 \left(\epsilon_{q/r} + \frac{\Delta t}{T_1}\right) \frac{T_1}{T_2} m^2 \right.\right.$$
$$\left.\left. + \sigma_3 \left(l + \frac{\Delta t}{T_1}\right) \frac{T_1}{T_3} m^3 + \ldots \right]\right\}. \tag{9.21}$$

Here we have used the relation $\exp(2\pi\, i\, mt/T_1) = \exp(2\pi\, i\, ml) \times \exp(2\pi\, i\, m\Delta t/T_1) = \exp(2\pi\, i\, m\Delta t/T_1)$. We note that this representation of the sum S depends on the choice of the origin of time and thus on the fraction q/r. Hence for every different time region under consideration we adopt a different representation of the sum S.

9.4 New Representations of the Signal   245

Decomposition into Sub-Sums and Resummation

We proceed by noting that the function $\gamma_m^{(r)}$, Eq. (9.20), is periodic in m with period r, that is

$$\gamma_{m+r}^{(r)} = \exp\left[2\pi i \sigma_2 q/r\, (m+r)^2\right] = \gamma_m^{(r)}.$$

This periodicity depends only on the denominator r of the fraction q/r. In order to make use of this periodicity we rearrange the summation with the help of the relation

$$\sum_{m=-\infty}^{\infty} a_m = \sum_{p=0}^{r-1} \sum_{k=-\infty}^{\infty} a_{p+kr}, \qquad (9.22)$$

that is we first sum the terms a_m at all multiples kr of this period r and then sum these sub-sums over one period. Since $\gamma_{p+kr}^{(r)} = \gamma_p^{(r)}$, we find

$$S(\Delta t) = \sum_{p=0}^{r-1} \gamma_p^{(r)} \sum_{k=-\infty}^{\infty} \widetilde{W}_{p+kr}(\Delta t). \qquad (9.23)$$

For the sub-sum over the index k we now face the same situation as for the entire sum S, Eq. (9.4), at the initial stage of the temporal evolution. We can therefore apply the Poisson summation formula, Eq. (9.9), to the sub-sums over k which yields

$$S(\Delta t) = \sum_{p=0}^{r-1} \gamma_p^{(r)} \sum_{m=-\infty}^{\infty} \int_{-\infty}^{\infty} dk\, \widetilde{W}(p+kr, \Delta t)\, \exp(-2\pi i\, km),$$

where $\widetilde{W}(x, \Delta t)$ is the continuous version of $\widetilde{W}_m(\Delta t)$, Eq. (9.21). As discussed in Sec. 9.4.1 the Poisson summation formula allows us to represent each time dependent sub-sum as a sequence of time dependent signals numbered by the index m.

When we introduce the new integration variable $x \equiv p + kr$, the integral over x is independent of p, that is

$$S(\Delta t) = \frac{1}{r} \sum_{p=0}^{r-1} \gamma_p^{(r)} \sum_{m=-\infty}^{\infty} \exp\left(2\pi i \frac{p}{r} m\right) \int_{-\infty}^{\infty} dx\, \widetilde{W}(x, \Delta t)\, \exp\left(-2\pi i \frac{m}{r} x\right).$$

Final Result

We interchange the two summations and write the sum S in the form

$$S(t = \frac{q}{r} T_2 + \epsilon_{q/r}\, T_1 + \Delta t) = \sum_{m=-\infty}^{\infty} \mathcal{W}_m^{(r)}\, I_m^{(r)}(\Delta t), \qquad (9.24)$$

where

$$\mathcal{W}_m^{(r)} \equiv \frac{1}{r} \sum_{p=0}^{r-1} \exp\left[2\pi i \left(\sigma_2 p^2 \frac{q}{r} + p\frac{m}{r}\right)\right] \qquad (9.25)$$

are time independent coefficients and the factors

$$I_m^{(r)}(\Delta t) = \int_{-\infty}^{\infty} dx\, W(\bar{n}+x) \exp\left\{2\pi i\left[\left(\frac{\Delta t}{T_1} - \frac{m}{r}\right)x + \sigma_2\left(\epsilon_{q/r} + \frac{\Delta t}{T_1}\right)\frac{T_1}{T_2} x^2\right.\right.$$

$$\left.\left.+ \sigma_3\left(l + \frac{\Delta t}{T_1}\right)\frac{T_1}{T_3} x^3 + \ldots\right]\right\} \qquad (9.26)$$

represent the time dependent signals.

Hence, we have cast the infinite sum Eq. (9.4) into another infinite sum Eq. (9.24). The transformation of the sum, made possible by the shift of the origin of time, Eq. (9.18), together with the decomposition Eq. (9.22) of the sum into sub-sums and the Poisson summation formula, Eq. (9.9), is exact. But what is the advantage of this on first sight more complicated representation of S? It reveals in the most obvious way the fractional revivals as we show in the next section.

9.5 Fractional Revivals Made Simple

In the new representation Eq. (9.24) of the signal $S(t)$, each term contributing to the sum consists of the product of the weight function $\mathcal{W}_m^{(r)}$ and the shape function $I_m^{(r)}(\Delta t)$. We now discuss these constituents in more detail and show that each term $I_m^{(r)}(\Delta t)$ is a fractional revival. However, in complete accordance with Sec. 9.4.1 this statement is only correct and hence the representation Eq. (9.24) is only useful when the temporal width of the signal $I_m^{(r)}(\Delta t)$ is smaller than the separation between two neighboring signals.

9.5.1 Gauss Sums

The factor $\mathcal{W}_m^{(r)}$ is independent of the distribution $W(n)$ and the time Δt. Thus it acts in the sum Eq. (9.24) as a weight. It is a well-known quantity in the context of fractional revivals. According to Eq. (9.25), $\mathcal{W}_m^{(r)}$ is a sum of r complex numbers each with the modulus $1/r$. Hence the modulus of $\mathcal{W}_m^{(r)}$ ranges from zero to unity. The individual terms in the finite sum interfere and this interference depends on the parameters m and r via the phase angle $2\pi m/r$. Moreover, an additional r dependence enters via the phase $2\pi q/r$. Hence the value of $\mathcal{W}_m^{(r)}$ is on first sight a complicated function of m, r and q. However, a detailed analysis reveals the following simple features: (i) For r even, $\mathcal{W}_m^{(r)}$ vanishes for every second value of m, whereas for r odd $\mathcal{W}_m^{(r)}$ is nonzero for every value of m, and (ii) the modulus $|\mathcal{W}_m^{(r)}|$ of each nonzero weighting factor is independent of q; in particular, one finds $|\mathcal{W}_m^{(r)}| = 1/\sqrt{r}$ for r odd and $|\mathcal{W}_m^{(r)}| = \sqrt{2/r}$ for r even. For a more detailed discussion of the properties of Gauss sums we refer to the Problem 9.1.

9.5.2 Shape Function

Now we turn to the time dependent term $I_m^{(r)}(\Delta t)$ which contains the information about the location, the duration and the detailed shape of the signal. To be specific, we employ our previous example of a Gaussian distribution for the function $W(n)$. Moreover, in order to keep the mathematics minimal we consider times such we can neglect the cubic term x^3 and higher contributions. For a discussion of this question see Problem 9.4

For many physical systems this is an excellent approximation. Consider for example the vibrational wave packet propagating in the excited potential surface $A^1\Sigma_u^+$ of

9.5 Fractional Revivals Made Simple

the sodium dimer displayed in Fig. 9.3. Indeed, the time scale T_3 does not affect significantly the shape of the signal for times of the order T_2, since here T_3 is very large compared to T_2. Two prominent examples, where this is *exactly* true for *all* times, are the Morse potential and a box with infinitely steep and infinitely high walls. For these systems, the third derivative and all higher derivatives of the eigenfrequencies $\omega(n)$ with respect to the quantum number n vanish.

In this case we can perform the integral following the calculations of the early stage of the time evolution. We find

$$I_m^{(r)}(\Delta t) = \mathcal{N}(\Delta t) \exp\left[-\frac{\left(\Delta t - \frac{m}{r}T_1\right)^2}{2\,\sigma_r^2(\Delta t)}\right] \exp\left[-i\,\sigma_2 \frac{\left(\Delta t - \frac{m}{r}T_1\right)^2}{2\,\sigma_i^2(\Delta t)}\right],$$

where the complex amplitude

$$\mathcal{N}(\Delta t) \equiv \frac{1}{\sqrt{1 - i\,\sigma_2 4\pi \Delta n^2 (\epsilon_{q/r}T_1 + \Delta t)/T_2}}$$

decreases, but the widths

$$\sigma_r^2(\Delta t) \equiv \left[\frac{1}{4\pi^2 \Delta n^2} + 4\Delta n^2 \left(\frac{\epsilon_{q/r}T_1 + \Delta t}{T_2}\right)^2\right] T_1^2$$

and

$$\sigma_i^2(\Delta t) \equiv \left[\frac{1}{16\pi^3 \Delta n^2 (\epsilon_{q/r}T_1 + \Delta t)/T_2} + \frac{1}{\pi}\Delta n^2 \frac{\epsilon_{q/r}T_1 + \Delta t}{T_2}\right] T_1^2$$

of the real and the imaginary Gaussian increase as a function of time.

Hence, the shape function consists of the product of a complex-valued square root, a real and imaginary Gaussian. The Gaussians only take on non-vanishing values in the neighborhood of integer multiples of T_1/r.

According to Eq. (9.24) the signal is an infinite sum of the shape functions $I_m^{(r)}$. The latter are Gaussians located at times $m\,T_1/r$. When the separation T_1/r between two neighboring Gaussians is larger than their width σ_r they do not overlap. In this case the sum over m that is over the individual Gaussians separates into a time sequence of Gaussians just as in the case of the early time evolution. The only difference is that now the smallest period is the fraction T_1/r.

Hence, the signal in the neighborhood of a fractional revival, that is at a time $t = \frac{q}{r}T_2 + \epsilon_{q/r}T_1 + \Delta t$ consists of a sequence of Gaussians separated by a period of T_1/r provided r is odd or $2T_1/r$ for r even. When the Gaussians do not overlap the mth term in the summation Eq. (9.24) represents the mth fractional revival. Since we have neglected the cubic contribution to the shape function these fractional revivals have the same shape as in the initial stage of the evolution. For a more detailed discussion of the influence of the third order term creating the oscillations on top of the slow decay to the left of the center of each fractional revival shown in Fig. 9.6 we refer to Problem 9.4

When neighboring non-vanishing terms $I_m^{(r)}(\Delta t)$ and $I_{m'}^{(r)}(\Delta t)$ in Eq. (9.24) overlap interferences between these terms arise. Then the phases of the complex Gaussian and the square root start to play an important role. Consequently, the sum S exhibits a more complicated pattern.

Problems

9.1 Properties of Gauss Sums

In Eq. (9.25) we have defined the Gauss sums

$$\mathcal{W}_m^{(r)} \equiv \frac{1}{r} \sum_{p=0}^{r-1} \exp\left[\frac{2\pi i}{r}\left(\sigma_2 p^2 q + pm\right)\right],$$

where q and r do not have a common factor and $\sigma_2 = \pm 1$. Show:

(a) If r is odd then all $\mathcal{W}_m^{(r)}$ are nonzero and $|\mathcal{W}_m^{(r)}| = \sqrt{1/r}$.

(b) If r and $rq/2$ are even then $\mathcal{W}_m^{(r)} = 0$ for odd m and $|\mathcal{W}_m^{(r)}| = \sqrt{2/r}$ for even m.

(c) If r is even and $rq/2$ odd then $\mathcal{W}_m^{(r)} = 0$ for even m and $|\mathcal{W}_m^{(r)}| = \sqrt{2/r}$ for odd m.

Hint: Calculate first

$$\Lambda_\ell^{(r)} = \sum_{m=0}^{r-1} |\mathcal{W}_m^{(r)}|^2 \, e^{\frac{2\pi i}{r} m\ell}.$$

From this expression find $|\mathcal{W}_m^{(r)}|^2$ by making use of the relation

$$|\mathcal{W}_m^{(r)}|^2 = \frac{1}{r} \sum_{\ell=0}^{r-1} \Lambda_\ell^{(r)} \, e^{-\frac{2\pi i}{r} m\ell}.$$

It turns out that almost all $\Lambda_\ell^{(r)}$ vanish.

9.2 Curlicues

In semiclassical quantum mechanics the two limits of time approaching infinity and Planck's constant going to zero do not commute. This fact comes out most clearly in the sums

$$S_N(\tau) \equiv \sum_{n=1}^{N} e^{i\pi n^2 \tau}.$$

We can use the techniques presented in this chapter to cast the sum S_N in a form which allows to gain insight into the functional dependence on the parameters τ and N. Unfortunately, this is rather difficult. For a comprehensive treatment of S_N we refer to the work of M.V. Berry (1988). Here we suggest to evaluate that sum numerically.

(a) Discuss the behavior of this sum in complex space for fixed N as a function of τ. Due to their special form these structures have been called by M.V. Berry curlicues. How are curlicues structures related to the Cornu spiral discussed in Appendix H?

(b) How do these structures depend on N?

According to the Oxford English dictionary a curlicue is a "fantastic curl or twist".

9.3 Time Scales: Orders of Magnitude

In the expansion Eq. (9.2) we have defined the times

$$T_j \equiv \frac{2\pi}{j!\,|\omega^{(j)}(\bar{n})|},$$

where $\omega^{(j)}(\bar{n})$ is the jth derivative of the frequency $E(n)/\hbar$ with respect to the quantum number n taken at the mean value $\bar{n}$. Calculate the values of T_1, T_2, and T_3 for the following three systems:

(a) Resonant Jaynes-Cummings-Paul model where $\omega(n) \equiv \lambda\sqrt{n+1}$ with $\lambda/2\pi \cong 25$ kHz and $\bar{n} \cong 10$

(b) Hydrogen atom with $\bar{n} \cong 50$

(c) Vibration of a diatomic molecule where

$$\omega(n) \cong \omega_0\,(n - \alpha n^2)$$

with $2\pi/\omega_0 \cong 300$ fs, $\alpha \cong 1/100$, and $\bar{n} \cong 10$.

9.4 Wave Packet Dynamics and Complex–Valued Airy Function

Discuss the influence of the third order term m^3 in the signal $S(t)$, Eq. (9.4). For this purpose evaluate the integral $I_m^{(r)}(\Delta t)$ analytically for the case of a Gaussian weight factor and retain the cubic term.

As shown in Leichtle et al. (1996) this yields the modulation of the left side of the Gaussian fractional revivals depicted in Figs. 9.6. The detailed structure of the fractional revivals comes out from the result

$$I_m^{(r)}(\Delta t) = \exp[i\,\Phi_m(\Delta t)]\; G(\Delta t)\; F_m(\Delta t)\; \mathrm{Ai}\,[z_m(\Delta t)]$$

for the integral Eq. (9.26).

Here the functions $G(\Delta t)$ and $F_m(\Delta t)$ are defined by

$$G(\Delta t) \equiv A\,\exp\left[-\lambda\left(\epsilon_{q/r} + \frac{\Delta t}{T_1}\right)^2\right]$$

and

$$F_m(\Delta t) \equiv \exp\left[\mu\left(\frac{\Delta t}{T_1} - \frac{m}{r}\right)\right]$$

and $\mathrm{Ai}(z)$ denotes the Airy function. The quantities Φ_m, A, λ, and μ are real whereas z_m is complex. For the explicit expressions of these quantities see Leichtle et al. (1996).

9.5 Wave Packets of a Particle in a Box

A quantum mechanical particle of mass M caught in a box of length L displays a quadratic energy spectrum. Hence, the time dependent wave function $\psi(x,t)$ reads

$$\psi(x,t) \equiv \sum_{m=1}^{\infty} \psi_m u_m(x) \exp\left(-2\pi i m^2 \frac{t}{T_2}\right)$$

where ψ_m denotes the expansion coefficient of the wave function

$$\phi(x) \equiv \psi(x, t=0)$$

at $t=0$ into the energy eigenfunctions

$$u_m(x) \equiv \sqrt{\frac{2}{L}} \sin\left(m\pi \frac{x}{L}\right)$$

and the revival time T_2 follows from

$$T_2 \equiv \frac{4ML^2}{\pi\hbar}.$$

Use the formalism developed in this chapter to derive the representation

$$\psi\left(x, t = \frac{q}{r}T_2 + \Delta t\right) = \sum_{l=-\infty}^{\infty} W_l^{(r)} \phi\left(x - \frac{l}{r}2L, \Delta t\right)$$

$$- \sum_{l=-\infty}^{\infty} W_l^{(r)} \phi\left(-x + \frac{l}{r}2L, \Delta t\right) \quad (9.27)$$

where

$$\phi(x,t) \equiv \int_0^L dx' \phi(x') G_{\text{free}}(x,t|x',0)$$

is the initial wave function $\phi(x)$ propagated with the help of the Green's function

$$G_{\text{free}}(x,t|x',0) \equiv \sqrt{\frac{M}{2\pi i \hbar t}} \exp\left[i\frac{M}{2\hbar t}(x-x')^2\right]$$

of the free particle.

According to Eq. (9.27) the wave function at fractions of the revival time consists of a superposition of the initial wave packet located at fractions of the length of the box.

Hint: See Stifter et al. (1997).

9.6 Talbot Effect

A periodic structure such as a mechanical grating creates from an incident plane light or de Broglie wave a periodic array of wave packets. The intensity pattern of atoms or light after the grating follows from the superposition of wave packets propagated according to the Green's function of the free particle. These patterns show a remarkable regularity. In particular, there exists a distance from the grating where the original array of wave packets repeats itself. Moreover, at certain fractions of that distance the original array has a smaller period. This phenomenon is called *Talbot effect* and has been observed experimentally for atoms and light.

(a) Propagate a periodic array of wave functions according to the Green's function of the free particle.

(b) Use the Poisson summation formula to map the periodicity in space into the time evolution of a particle with a quadratic energy spectrum.

(c) Use Eq. (9.27) to explain the Talbot effect.

References

Introduction to Wave Packets

Over the last few years numerous papers on the subject of wave packets in atomic and molecular physics have appeared. For a review of electronic wave packets in Rydberg atoms, see

G. Alber and **P. Zoller**, *Laser excitation of electronic wave packets in Rydberg atoms*, Phys. Rep. **199**, 231–280 (1990)

For a review of vibrational wave packets in molecular physics see

M. Gruebele and **A.H. Zewail**, *Ultrafast reaction dynamics*, Phys. Today **43**, 24–33 (1990)

B. Garraway and **K.-A. Suominen**, *Wave-packet dynamics: new physics and chemistry in femto-time*, Rep. Prog. Phys. **58**, 365–419 (1995)

For a recent summary of the state of the art of wave packet dynamics, see

J.A. Yeazell and **T. Uzer**, *The Physics and Chemistry of Wave Packets*, Wiley, New York, 2000

Electronic Wave Packets

For the experimental observation of fractional revivals and full revivals of atomic wave packets, see e.g.

J.A. Yeazell, M. Mallalieu and **C.R. Stroud**, *Observation of the collapse and revival of a Rydberg electronic wave packet*, Phys. Rev. Lett. **64**, 2007–2010 (1990)

J.A. Yeazell and **C.R. Stroud Jr.**, *Observation of fractional revivals in the evolution of a Rydberg atomic wave packet*, Phys. Rev. A **43**, 5153–5156 (1991)

D.R. Meacher, P.E. Meyler, I.G. Hughes and **P. Ewart**, *Observation of the collapse and fractional revival of a Rydberg wavepacket in atomic rubidium*, J. Phys. B **24**, L63–69 (1991)

L. Marmet, H. Held, G. Raithel, J.A. Yeazell and **H. Walther**, *Observation of quasi-Landau wave packets*, Phys. Rev. Lett. **72**, 3779–3782 (1994)

J. Wals, H.H. Fielding, J.F. Christian, L.C. Snoek, W.J. Van der Zande and **H.B. van Linden van den Heuvell**, *Observation of Rydberg wave packet dynamics in a Coulombic and magnetic field*, Phys. Rev. Lett. **72**, 3783–3786 (1994)

J. Wals, H.H. Fielding and H.B. Van Linden van den Heuvell, *The Role of the Quantum Defect and of High-Order Dispersion in Rydberg Wave Packets*, Physica Scr. **T 58**, 62–68 (1995)

Molecular Wave Packets

For molecular wave packets, see

T. Baumert, V. Engel, C. Rottgermann, W.T. Strunz and G. Gerber, *Femtosecond pump-probe study of the spreading and recurrence of a vibrational wave packet in Na_2*, Chem. Phys. Lett. **191**, 639–644 (1992)

Th. Walther, H. Bitto and J.R. Huber, *High-resolution quantum beat spectroscopy in the electronic ground state of a polyatomic molecule by IR-UV-pump-probe method*, Chem. Phys. Lett. **209**, 455–458 (1993)

I. Fischer, D.M. Villeneuve, M.J.J. Vrakking and A. Stolow, *Femtosecond wave-packet dynamics studied by time-resolved zero-kinetic energy photoelectron spectroscopy*, Chem. Phys. **102**, 5566–5569 (1995)

M.J.J. Vrakking, D.M. Villeneuve and A. Stolow, *Observation of fractional revivals of a molecular wave packet*, Phys. Rev. A **54**, R37–40 (1996)

Fractional and Ordinary Revivals

For the theoretical discussion of the revivals in the Jaynes-Cummings-Paul model see

J.H. Eberly, N.B. Narozhny and J.J. Sanchez-Mondragon, *Periodic Spontaneous Collapse and Revival in a Simple Quantum Model*, Phys. Rev. Lett. **44**, 1323–1326 (1980)

General aspects of the dynamics of wave packets in atomic, molecular and quantum optical systems are discussed in

J. Parker and C.R. Stroud Jr., *Coherence and Decay of Rydberg Wave Packets*, Phys. Rev. Lett. **56**, 716–719 (1986)

I.Sh. Averbukh and N.F. Perel'man, *The dynamics of wave packets of highly-excited states of atoms and molecules*, Usp. Fiz. Nauk **161**, 41–81 (1991) [Sov. Phys. Usp. **34**, 572–591 (1991)]

The phenomenon of fractional revivals was proposed theoretically by

I.Sh. Averbukh and N.F. Perel'man, *Fractional revivals: Universality in the long-term evolution of quantum wave packets beyond the correspondence principle dynamics*, Phys. Lett. A **139**, 449–453 (1989)

I.Sh. Averbukh and N.F. Perel'man, *Fractional regenerations of wave packets in the course of long-term evolution of highly excited quantum systems*, Sov. Phys. JETP **69**, 464–469 (1989)

M. Nauenberg, C. Stroud and J. Yeazell, *The classical limit of an atom*, Sci. Am. **270**, No. 6, 24–31 (1994)

M. Nauenberg, *The Transition from Quantum to Classical Mechanics in Atomic Physics*, Comments At. Mol. Phys. **25**, 151–157 (1990)

Theory of Fractional Revivals

The formalism for understanding the phenomenon of revivals and fractional revivals outlined in the present section follows closely the papers by

M. Fleischhauer and **W.P. Schleich**, *Revivals made simple: Poisson summation formula as a key to the revivals in the Jaynes-Cummings model*, Phys. Rev. A **47**, 4258–4269 (1993)

C. Leichtle, I.Sh. Averbukh and **W.P. Schleich**, *Generic structure of multilevel quantum beats*, Phys. Rev. Lett. **77**, 3999–4002 (1996)

C. Leichtle, I. Sh. Averbukh and **W.P. Schleich**, *Multilevel quantum beats: an analytical approach*, Phys. Rev. A **54**, 5299–5312 (1996)

The idea of partially resumming phases comes from the work of

J.T. Winthrop and **C.R. Worthington**, *Theory of Fresnel Images. I. Plane Periodic Objects in Monochromatic Light*, J. Opt. Soc. Am. **55**, 373–381 (1965)

Talbot Effect

In the context of classical optics, a similar phenomenon is known as the Talbot effect: Here, it represents the self-imaging of a grating, which is illuminated by plane waves, in the near field. This effect was first observed by

H.F. Talbot, *Facts relating to Optical Science*, Philos. Mag. **9**, 401–407 (1836)

and explained much later by

L. Rayleigh, *On Copying Diffraction-gratings, and on some Phenomena connected therewith*, Philos. Mag. **11**, 196–205 (1881)

A. Schuster, *A Simple Explanation of Talbot's Bands*, Philos. Mag. **7**, 1–8 (1904)

The Talbot effect was also demonstrated for atomic waves, see

J.F. Clauser and **S. Li**, *Talbot-vonLau atom interferometry with cold slow potassium*, Phys. Rev. A **49**, R2213–R2216 (1994)

M.S. Chapman, C.R. Ekstrom, T.D. Hammond, J. Schmiedmayer, B.E. Tannian, S. Wehinger and **D.E. Pritchard**, *Near-field imaging of atom diffraction gratings: The atomic Talbot effect*, Phys. Rev. A **51**, R14–17 (1995)

For a detailed mathematical treatment of the Talbot effect see

J.F. Clauser and **M.W. Reinsch**, *New Theoretical and Experimental Results in Fresnel Optics with Applications to Matter-Wave and X-Ray Interferometry*, Appl. Phys. B **54**, 380–395 (1992)

and the review by

B. Rohwedder, *Atom Optical Elements Based on Near-field Grating Sequences*, Fortschr. Phys. **47**, 883–911 (1999)

For the experimental verification of the detailed carpet structure of the Talbot effect see

S. Nowak, Ch. Kurtsiefer, C. David and **T. Pfau**, *Higher oder Talbot fringes for atomic matter waves*, Opt. Lett. **22**, 1430–1432 (1997)

The technique introduced in this chapter has also been used to treat the integer, fractional and fractal Talbot effect, see

M.V. Berry, *Quantum Fractals in Boxes*, J. Phys. A **29**, 6617–6629 (1996)

M.V. Berry and **S. Klein**, *Integer, fractional and fractal Talbot effects*, J. Mod. Optics **43**, 2139–2164 (1996)

The same technique has also been used to explain the dynamics of a wave packet in a infinitely deep potential well, see

P. Stifter, W.P. Schleich and **W.E. Lamb**, *The particle in the box revisited*, in: *Proceedings of the conference on "Quantum Optics and Laser Physics"* edited by L. Jin and Y.S. Zhu, Springer, Heidelberg, 1997

Curlicues

For a mathematical description of curlicues along the lines discussed in the present chapter see

M.V. Berry, *Random renormalization in the semiclassical long-time limit of a precessing spin*, Physica D **33**, 26–33 (1988)

M.V. Berry and **J. Goldberg**, *Renormalisation of curlicues*, Nonlinearity **1**, 1–26 (1988)

10 Field Quantization

Over the last years the technology for producing highly reflecting mirrors has made an enormous progress leading to resonators in the optical domain with decay times in the range of microseconds. In the microwave domain superconducting cavities have reached quality factors of the order of 10^{10}. An atom traversing such a cavity can exchange its excitation with the cavity field many times before the field decays. This feature is instrumental for novel light sources such as a laser or maser that operates with one or less than one atom on average. Clearly the light of such devices shows many quantum effects. Moreover, the electromagnetic field strength of the vacuum in such resonators can be of the order of 80 V/m – quite a macroscopic field for the vacuum.

In later chapters we discuss such devices in more detail. We therefore in the present chapter lay the foundations for these studies. We quantize the electromagnetic field in a cavity and show that the field consists of an infinite set of harmonic oscillators. Each oscillator we quantize in the canonical way.

Here we follow the original prescription given by M. Born, W. Heisenberg and P. Jordan. In the famous *Drei-Männer-Arbeit* they quantized the electromagnetic field by expressing the vector potential in terms of its quadrature components. In contrast, P.A.M. Dirac quantized the electromagnetic field by introducing action and angle variables. However, at the same time as Dirac wrote his paper F. London, then at the Universität Stuttgart, showed that no such angle operators exist. Therefore, the approach of Dirac has led to a long standing discussion in quantum optics on the question of the appropriate definition of an hermitian phase operator, as discussed in more detail in Secs. 8.5 and 13.4.

Closely related to the quantization of the radiation field is the concept of the photon. The word *photon* was brought into physics by the chemist G.N. Lewis who originally meant something completely different from what A. Einstein had thought as the light quantum. This makes the question about the wave function for the photon a highly controversial but interesting topic. We will return to this discussion in the next chapter.

10.1 Wave Equations for the Potentials

In the present section we start from Maxwell's equations in free space and derive the wave equations for the vector potential $\vec{A}$ and the scalar potential Φ. We briefly discuss the gauge invariance of electrodynamics. This will become important in Sec. 14.2.1 where we discuss the problem of creating interactions between matter and light. Since we quantize the free radiation field, that is in the absence of charges and currents, we choose the Coulomb gauge which allows us to work solely with the vector potential. We separate variables and obtain the Helmholtz equation for the spatial part $\vec{u}(\vec{r})$ of $\vec{A}(\vec{r}, t)$. The behavior of the electric and magnetic field at the walls of the resonator determines the boundary conditions for $\vec{u}(\vec{r})$.

10.1.1 Derivation of the Wave Equations

We start from Maxwell's equations consisting of the homogeneous equations

$$\vec{\nabla} \cdot \vec{B} = 0 \qquad \vec{\nabla} \times \vec{E} = -\frac{\partial \vec{B}}{\partial t} \qquad (10.1a)$$

and the inhomogeneous equations

$$\vec{\nabla} \cdot \vec{D} = \rho \qquad \vec{\nabla} \times \vec{H} = \frac{\partial \vec{D}}{\partial t} + \vec{j} \quad . \qquad (10.1b)$$

Here ρ and $\vec{j}$ denote the charge and the current and the constitutive relations

$$\vec{D} = \varepsilon_0 \vec{E} \quad \text{and} \quad \vec{B} = \mu_0 \vec{H} \qquad (10.2)$$

relate the magnetic flux density $\vec{B}$ to the magnetic field $\vec{H}$ and the electric displacement current $\vec{D}$ to the electric field $\vec{E}$. The electric permittivity ε_0 and the magnetic permeability μ_0 of free space define the vacuum speed of light

$$c^2 = 1/(\varepsilon_0 \mu_0) \, . \qquad (10.3)$$

We can derive the electric field and the magnetic flux density from the vector potential $\vec{A}$ and the scalar potential Φ by defining the relations

$$\vec{E} \equiv -\vec{\nabla}\Phi - \frac{\partial \vec{A}}{\partial t} \qquad (10.4)$$

and

$$\vec{B} \equiv \vec{\nabla} \times \vec{A}. \qquad (10.5)$$

This ansatz guarantees that the fields $\vec{E}$ and $\vec{B}$ satisfy the homogeneous Maxwell equations (10.1a). Indeed, we find

$$\vec{\nabla} \cdot \vec{B} = \vec{\nabla} \cdot \left(\vec{\nabla} \times \vec{A} \right) = 0$$

and

$$\vec{\nabla} \times \vec{E} = \vec{\nabla} \times \left(-\vec{\nabla}\Phi - \frac{\partial \vec{A}}{\partial t} \right) = -\frac{\partial}{\partial t} \left(\vec{\nabla} \times \vec{A} \right) = -\frac{\partial \vec{B}}{\partial t}.$$

10.1 Wave Equations for the Potentials

We now turn to the inhomogeneous Maxwell equations (10.1b). The constitutive relations, Eqs. (10.2), immediately yield

$$\vec{\nabla} \cdot \vec{E} = \frac{\rho}{\varepsilon_0} \tag{10.6}$$

and

$$\vec{\nabla} \times \vec{B} = \frac{1}{c^2} \frac{\partial \vec{E}}{\partial t} + \mu_0 \vec{j} \tag{10.7}$$

where we have used Eq. (10.3).

When we substitute the representations Eqs. (10.4) and (10.5) of the fields in terms of the potentials into Eq. (10.7) we arrive at

$$\vec{\nabla} \times \left(\vec{\nabla} \times \vec{A} \right) = \frac{1}{c^2} \frac{\partial}{\partial t} \left[-\vec{\nabla}\Phi - \frac{\partial \vec{A}}{\partial t} \right] + \mu_0 \vec{j} = -\frac{1}{c^2} \frac{\partial}{\partial t} \left(\vec{\nabla}\Phi \right) - \frac{1}{c^2} \frac{\partial^2 \vec{A}}{\partial t^2} + \mu_0 \vec{j}. \tag{10.8}$$

We make use of the relation

$$\vec{\nabla} \times \left(\vec{\nabla} \times \vec{A} \right) = \vec{\nabla} \left(\vec{\nabla} \cdot \vec{A} \right) - \vec{\nabla}^2 \vec{A}$$

to rewrite Eq. (10.8) as

$$\vec{\nabla}^2 \vec{A} - \frac{1}{c^2} \frac{\partial^2 \vec{A}}{\partial t^2} = \vec{\nabla} \left[\frac{1}{c^2} \frac{\partial \Phi}{\partial t} + \vec{\nabla} \cdot \vec{A} \right] - \mu_0 \vec{j}. \tag{10.9}$$

We now derive the wave equation for the scalar potential Φ. For this purpose we start from Eq. (10.6) and find with the help of Eq. (10.4)

$$\vec{\nabla} \cdot \vec{E} = -\vec{\nabla} \cdot \left(\vec{\nabla}\Phi \right) - \frac{\partial}{\partial t} \left(\vec{\nabla} \cdot \vec{A} \right) = \frac{\rho}{\varepsilon_0}. \tag{10.10}$$

We note that the left hand side of Eq. (10.9) is the wave operator for the vector potential $\vec{A}$ which is driven by the current $\vec{j}$ and the scalar potential Φ. Similarly the equation for Φ is driven by the charge and the vector potential. Therefore the two equations are coupled. We can eliminate the coupling by choosing the appropriate gauge condition as discussed in the next section.

10.1.2 Gauge Invariance of Electrodynamics

In electrodynamics we have the freedom of a gauge. In the preceeding section we have solved the homogeneous Maxwell's equations by introducing the vector potential $\vec{A}$ and the scalar potential Φ. Indeed, since the magnetic flux density follows from the curl of the vector potential $\vec{A}$, that is

$$\vec{B} \equiv \vec{\nabla} \times \vec{A}.$$

the vector potential is only determined up to a gradient of a scalar field $\Lambda = \Lambda(\vec{r}, t)$. Indeed, the transformed vector potential

$$\vec{A}' \equiv \vec{A} + \vec{\nabla}\Lambda \tag{10.11}$$

creates the same magnetic flux.

Similarly, since the electric field $\vec{E}$ follows from

$$\vec{E} \equiv -\vec{\nabla}\Phi - \frac{\partial \vec{A}}{\partial t}$$

we can include the time derivative $\dot{\Lambda}$ in the scalar potential Φ. Indeed, the so-transformed scalar potential

$$\Phi' \equiv \Phi - \frac{\partial \Lambda}{\partial t} \tag{10.12}$$

creates the same electric field

$$\vec{E}' \equiv -\vec{\nabla}\Phi' - \frac{\partial \vec{A}'}{\partial t} = \vec{E}.$$

Different vector and scalar potentials can lead to the same magnetic and electric field. When we work with potentials we have therefore an additional degree of freedom: We can choose the potentials in a way as to simplify the calculation.

Indeed, we now select potentials such that the two wave equations decouple. This condition restricts the class of all possible potentials. There exist many gauges which are convenient for the specific problem at hand. In the case of quantum electrodynamics or quantum optics the Lorentz gauge and the Coulomb gauge are used frequently. However, the momentum gauge and the time gauge are used in particle physics. For an introduction to these various gauges we refer to the book by Itzykson and Zuber.

Wave Equations in the Lorentz Gauge

The Lorentz gauge connects the vector and scalar potential by the condition

$$\frac{1}{c^2}\frac{\partial \Phi}{\partial t} + \vec{\nabla} \cdot \vec{A} = 0. \tag{10.13}$$

Since it involves the first derivative of the scalar potential with respect to time and the divergence of the vector potential it is Lorentz invariant. This comes out most clearly when we introduce the spacetime coordinates $x_\mu \equiv (ct, x, y, z)$ and the four-vector potential $A_\mu \equiv (\Phi/c, A_x, A_y, A_z)$ with which the Lorentz gauge condition reads

$$\frac{\partial A_\mu}{\partial x_\mu} = 0.$$

Here we have used the Einstein summation convention and have summed over the indices $\mu = 0, 1, 2, 3$.

In the Lorentz gauge the two wave equations decouple and read

$$\vec{\nabla}^2 \vec{A} - \frac{1}{c^2}\frac{\partial^2 \vec{A}}{\partial t^2} = -\mu_0 \vec{j} \tag{10.14}$$

and

$$\vec{\nabla}^2 \Phi - \frac{1}{c^2}\frac{\partial^2 \Phi}{\partial t^2} = -\frac{\rho}{\varepsilon_0}. \tag{10.15}$$

We conclude this section by noting that many potentials satisfy the Lorentz gauge and at the same time provide identical electromagnetical fields. They are all connected by a gauge transformation Λ which satisfies the wave equation

$$\vec{\nabla}^2 \Lambda - \frac{1}{c^2}\frac{\partial^2 \Lambda}{\partial t^2} = 0. \tag{10.16}$$

Indeed, we can derive this equation by substituting the potentials Eqs. (10.11) and (10.12) in the form

$$\vec{A} = \vec{A}' - \vec{\nabla}\Lambda$$

and

$$\Phi = \Phi' + \frac{\partial \Lambda}{\partial t}$$

into the Lorentz condition which yields

$$\frac{1}{c^2}\frac{\partial \Phi'}{\partial t} + \vec{\nabla}\cdot\vec{A}' + \frac{1}{c^2}\frac{\partial^2 \Lambda}{\partial t^2} - \vec{\nabla}^2 \Lambda = 0.$$

Since the transformed potentials have also to satisfy the Lorentz gauge the gauge transformation Λ has to obey the wave equation Eq. (10.16).

Wave Equations in Coulomb Gauge

The Coulomb gauge is defined by the constraint

$$\vec{\nabla}\cdot\vec{A} = 0. \tag{10.17}$$

This condition is not relativistically invariant, since it only involves the derivatives with respect to position. Moreover, in Coulomb gauge the wave equations Eqs. (10.9) and (10.10) reduce to

$$\vec{\nabla}^2 \vec{A} - \frac{1}{c^2}\frac{\partial^2 \vec{A}}{\partial t^2} = \vec{\nabla}\left(\frac{1}{c^2}\frac{\partial \Phi}{\partial t}\right) - \mu_0 \vec{j}$$

and

$$-\vec{\nabla}\cdot\left(\vec{\nabla}\Phi\right) = \frac{\rho}{\varepsilon_0}$$

that is, they do not decouple: The wave equation for the vector potential still contains the scalar potential Φ. However, the equation for Φ has reduced to the Poisson equation.

In the next sections we quantize the radiation field in the absence of charges and currents, that is for $\rho \equiv 0$ and $\vec{j} \equiv 0$. In this case the Poisson equation immediately yields $\Phi \equiv 0$ and the wave equation for the vector potential simplifies to

$$\vec{\nabla}^2 \vec{A} - \frac{1}{c^2}\frac{\partial^2 \vec{A}}{\partial t^2} = 0. \tag{10.18}$$

We will recognize later that the Coulomb gauge reflects the fact that the oscillation direction, that is the polarization of $\vec{A}$ is orthogonal to the wave vector $\vec{k}$.

Hence, the oscillation of the electric field is transverse. This feature is the reason why sometimes the Coulomb gauge is referred to as *transverse gauge*.

Quantization of the electromagnetic field in Coulomb gauge is rather straightforward, since there are only transverse photons. In contrast, the Lorentz gauge involves the longitudinal direction as well as the scalar potential and thus brings in longitudinal as well as scalar photons. In this case the quantization procedure is more complicated and we have to follow the procedure of Gupta and Bleuler.

10.1.3 Solution of the Wave Equation

In the absence of charges and currents we are now left with the problem of solving the wave equation Eq. (10.18) for the vector potential $\vec{A}$ in a cavity with the appropriate boundary conditions at the walls of the resonator. Once we have the solution $\vec{A}(\vec{r},t) = (A_x(\vec{r},t), A_y(\vec{r},t), A_z(\vec{r},t))$, that is the vector potential $\vec{A}$ in all its components A_x, A_y, A_z at all positions $\vec{r}$ and times t, we can calculate via the relations Eqs. (10.4) and (10.5) expressing the fields in terms of the potentials the fields $\vec{E}$ and $\vec{B}$ at any position and at any time.

Separation of Variables

We now solve the wave equation Eq. (10.18) for perfectly conducting walls of the resonator. In this case we try a separation ansatz

$$\vec{A}(\vec{r},t) = \Upsilon\, q(t)\, \vec{v}(\vec{r}), \tag{10.19}$$

where $q(t)$ is a function of time only, and $\vec{v}(\vec{r})$ denotes a vector which depends solely on position $\vec{r}$. Later in the calculation we choose the constant Υ in the most convenient way.

From Eq. (10.18) we find immediately

$$q(t)\vec{\nabla}^2 \vec{v}(\vec{r}) - \frac{1}{c^2}\ddot{q}(t)\vec{v}(\vec{r}) = 0. \tag{10.20}$$

When we now consider each component v_x, v_y, v_z of $\vec{v}$ we arrive at the relation

$$\frac{\vec{\nabla}^2 v_j(\vec{r})}{v_j(\vec{r})} = \frac{1}{c^2}\frac{\ddot{q}(t)}{q(t)} \tag{10.21}$$

for each component $j = x, y, z$.

Since the right-hand side depends only on time and the left-hand side only on the position $\vec{r}$ within the resonator, both sides are independent of $\vec{r}$ or t and hence are constant. The left-hand side contains a second derivative with respect to position. Therefore the constant of separation has the units of (length)$^{-2}$. We call it $-\vec{k}^2 \equiv -k^2$, where $\vec{k}$ is the wave vector determined by the boundary condition. The reason for the minus sign is the fact that the eigenvalues of the Laplacian are negative, as shown in the Problem 10.1.

10.1 Wave Equations for the Potentials

We note that the right-hand side does not depend on the component v_j. Hence, the separation constant cannot depend on j which yields the Helmholtz equation

$$\vec{\nabla}^2 \vec{v}(\vec{r}) + \vec{k}^2 \vec{v}(\vec{r}) = 0 \tag{10.22}$$

for the spatial part of the vector potential and an oscillator equation

$$\ddot{q}(t) + \Omega^2 q(t) = 0, \tag{10.23}$$

for the time dependent part. Here $\Omega \equiv c|\vec{k}| \equiv ck$ is the frequency determined by the boundary conditions through the wave vector $\vec{k}$.

Boundary Conditions

The boundary conditions on $\vec{A}$ follow from the requirements that the tangential component of $\vec{E}$ and that the normal component of $\vec{B}$ vanish. In Coulomb gauge we find

$$\vec{e}_{\|}(\vec{r}) \cdot \vec{E}(\vec{r}, t)\Big|_{\text{bou.}} = -\vec{e}_{\|} \cdot \frac{\partial \vec{A}}{\partial t}\Big|_{\text{bou.}} = -\Upsilon \, \dot{q}(t) \, \vec{e}_{\|}(\vec{r}) \cdot \vec{v}(\vec{r})\Big|_{\text{bou.}} = 0,$$

that is

$$\vec{e}_{\|}(\vec{r}) \cdot \vec{v}(\vec{r})\Big|_{\text{bou.}} = 0. \tag{10.24}$$

Here we have introduced the unit vector $\vec{e}_{\|}(\vec{r})$ tangential to the boundary. Since the boundary can have a complicated shape this vector depends on the position $\vec{r}$ on the boundary.

Moreover, when we introduce the unit vector $\vec{e}_{\perp}$ normal to the boundary we have the condition

$$\vec{e}_{\perp}(\vec{r}) \cdot \vec{B}(\vec{r}, t)\Big|_{\text{bou.}} = \vec{e}_{\perp} \cdot \left[\vec{\nabla} \times \vec{A}\right]\Big|_{\text{bou.}} = \Upsilon \, q(t) \, \vec{e}_{\perp}(\vec{r}) \cdot \left[\vec{\nabla} \times \vec{v}(\vec{r})\right]\Big|_{\text{bou.}} = 0,$$

that is

$$\vec{e}_{\perp}(\vec{r}) \cdot \left[\vec{\nabla} \times \vec{v}(\vec{r})\right]\Big|_{\text{bou.}} = 0. \tag{10.25}$$

On first sight the two conditions on the electric and magnetic field Eqs. (10.24) and (10.25) seem to be independent. However, in Problem 10.2 we show that the requirement on the magnetic field components are satisfied provided the electric field constraints are fulfilled.

The Coulomb gauge itself reads

$$\vec{\nabla} \cdot \vec{v}(\vec{r}) = 0. \tag{10.26}$$

Whereas Eqs. (10.24) and (10.25) only have to hold true on the boundaries, that is on the conducting walls of the resonator, the condition Eq. (10.26) has to be satisfied at every point.

We conclude this section by noting that the conditions Eqs. (10.24) and (10.25) lead to a "discreteness" of the possible wave vectors and hence lead to a discrete set of mode functions $\vec{v}_\ell(\vec{r})$. Here ℓ is a set of integer numbers. However, this discreteness has nothing to do with the quantization of the radiation field. It solely arises from the boundary condition imposed by the form of the resonator on the spatial part of the vector potential. In contrast, the quantization of the radiation field and the concept of the photon is related to the time dependent part of the vector potential, that is to the oscillator equation Eq. (10.23). This is the topic of the next sections.

262   10 Field Quantization

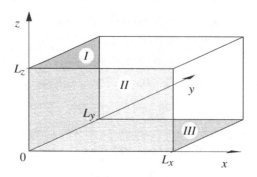

Fig. 10.1: Resonator in the shape of a box. The planes where either $x=0$, $y=0$ or $z=0$ are denoted by I, II or III, respectively.

10.2 Mode Structure in a Box

In the preceding section we have derived the Helmholtz equation for the mode function $\vec{v}(\vec{r})$ and have formulated the boundary conditions imposed on it by the behavior of the electric and magnetic field at the walls of a resonator of arbitrary shape. In the present section we solve these equations for the case of a box-shaped resonator. In particular, we show how the discreteness of the mode structure arises from the boundary conditions.

We emphasize that such a simple model for a resonator is not completely unrealistic. Indeed, the resonator of the one-atom maser is a closed cylindrical box. However, to illustrate the essential ideas and keep the mathematics simple we here consider the rectangular box. For solutions of the Helmholtz equations for more complicated resonator configurations such as open resonators with confocal mirrors used extensively in the cavity QED experiments discussed in Chapter 1 we refer to the literature.

10.2.1 Solutions of Helmholtz Equation

We now consider the rectangular resonator with edges of lengths L_x, L_y and L_z shown in Fig. 10.1. We express the wave vector

$$\vec{k} = k_x \vec{e}_x + k_y \vec{e}_y + k_z \vec{e}_z \tag{10.27}$$

and the position vector

$$\vec{r} = x\vec{e}_x + y\vec{e}_y + z\vec{e}_z \tag{10.28}$$

in cartesian coordinates, that is using unit vectors along the three spatial directions.

We first consider the condition (10.24) on $\vec{v}$ corresponding to the vanishing of the tangential electric field. For the side I of the box this condition reads

$$v_y(x=0,y,z) = v_z(x=0,y,z) = 0. \tag{10.29}$$

Analogously we find for the side II

$$v_x(x,y=0,z) = v_z(x,y=0,z) = 0 \tag{10.30}$$

and for III
$$v_x(x, y, z = 0) = v_y(x, y, z = 0) = 0. \tag{10.31}$$

Hence we are lead to the ansatz
$$\begin{aligned} v_x(x, y, z) &= \mathcal{N}\, n_x \cos(k_x x) \sin(k_y y) \sin(k_z z) \\ v_y(x, y, z) &= \mathcal{N}\, n_y \sin(k_x x) \cos(k_y y) \sin(k_z z) \\ v_z(x, y, z) &= \mathcal{N}\, n_z \sin(k_x x) \sin(k_y y) \cos(k_z z), \end{aligned} \tag{10.32}$$

where we have introduced the unit vector $\vec{e} \equiv (n_x, n_y, n_z)$ which we will determine later as to satisfy the Coulomb gauge condition. Moreover, we have included a normalization factor $\mathcal{N}$.

Note that we have used sine and cosine functions as a basis set for our ansatz. In this way each component v_x, v_y and v_z satisfies the Helmholtz equation. For example, we find
$$\Delta v_x(\vec{r}) + \vec{k}^2 v_x(\vec{r}) = \left(-k_x^2 - k_y^2 - k_z^2\right) v_x(\vec{r}) + \vec{k}^2 v_x(\vec{r}) = 0.$$

Hence we have a choice between linear combinations of sine and cosine functions. However, since we have to satisfy the conditions (10.29)–(10.31) representing a vanishing tangential component of the electric field, we have to choose sine functions for the y and z dependence of v_x as suggested by the condition $v_x(x, y = 0, z) = v_x(x, y, z = 0) = 0$, Eqs. (10.30) and (10.31).

10.2.2 Polarization Vectors from Gauge Condition

Moreover, we have introduced in this ansatz a pure cosine dependence for the x-dependence of v_x and analogously for the y-dependence of v_y or the z-dependence of v_z. In this way we guarantee that the Coulomb gauge condition $\vec{\nabla} \cdot \vec{v}(\vec{r}) = 0$ takes a simple form, namely
$$\vec{\nabla} \cdot \vec{v}(\vec{r}) = \frac{\partial v_x}{\partial x} + \frac{\partial v_y}{\partial y} + \frac{\partial v_z}{\partial z} \tag{10.33}$$
$$= -(n_x k_x + n_y k_y + n_z k_z) \sin(k_x x) \sin(k_y y) \sin(k_z z) \stackrel{!}{=} 0. \tag{10.34}$$

We first assume that none of the components of the $\vec{k}$ vector vanishes. Since the gauge condition Eq. (10.34) must hold true for any position $\vec{r}$ within the resonator, that is for any choice of $\vec{r} = (x, y, z)$, we find
$$n_x k_x + n_y k_y + n_z k_z = \vec{e} \cdot \vec{k} = 0.$$

Hence the unit vector $\vec{e}$ has to be orthogonal to the propagation direction, that is the Coulomb gauge condition reflects the transversality of the wave.

Since there are two linearly independent orthogonal directions on the propagation direction, there are two linearly independent polarization vectors $\vec{e}_1$ and $\vec{e}_2$ for each $\vec{k}$-vector, that is for each set of components k_x, k_y and k_z.

However, there is one exception: If one of the components of $\vec{k}$ vanishes, the gauge condition is satisfied automatically as shown in Eq. (10.34). In this case there is only one such polarization vector. Indeed, when we assume for simplicity $k_y = 0$ we find from Eq. (10.32) the mode function

$$\vec{v}(r) = \vec{e}_y \mathcal{N} \sin(k_x x) \sin(k_z z). \tag{10.35}$$

Hence, in general there are two polarization directions. However, when one of the wave numbers vanishes there is only one. This property will have important consequences in the context of the Casimir effect discussed in Sec. 10.4.

10.2.3 Discreteness of Modes from Boundaries

Now we are left with the problem of finding the wave vector $\vec{k}$. This quantity is fixed by the vanishing of the tangential component of the electric field at the sides I', II' and III' across of the sides I, II and III. Here these conditions read

$$v_y(x = L_x, y, z) = v_z(x = L_x, y, z) = 0$$
$$v_x(x, y = L_y, z) = v_z(x, y = L_y, z) = 0$$
$$v_x(x, y, z = L_z) = v_y(x, y, z = L_z) = 0.$$

These three constraints together with the ansatz Eq. (10.32) immediately lead to the result

$$\begin{aligned} k_x L_x &= l_x \pi \\ k_y L_y &= l_y \pi \\ k_z L_z &= l_z \pi, \end{aligned} \tag{10.36}$$

where l_x, l_y, and l_z are integers. Hence, the components $k_x = l_x \pi / L_x$, $k_y = l_y \pi / L_y$, and $k_z = l_z \pi / L_z$ of the wave vector can only assume integer multiples of π/L_x, π/L_y, and π/L_z. The walls of the resonator, or more precisely the resonator build out of the sides I and I' lead to a discreteness of the x-component of the wave vector. Similarly, the walls II and II', and the walls III and III' enforce the discreteness of the y- and z-component, respectively. We emphasize again that this "quantization" of $\vec{k}$-vectors has nothing to do with the quantization of the electric field which we will discuss in the following section and which is associated with the amplitude $q(t)$.

10.2.4 Boundary Conditions on the Magnetic Field

In the subsection on the boundary conditions we have shown that the requirements for the normal component of $\vec{B}$ are satisfied automatically provided the boundary conditions on the tangential component of $\vec{E}$ are obeyed. In the present section we show for the case of the rectangular cavity that this is indeed correct.

We first focus on the sides I and I' of the resonator. Here the normal vector $\vec{e}_\perp$ is either $\vec{e}_\perp = -\vec{e}_x$ for I, or $\vec{e}_\perp = \vec{e}_x$ for I' and therefore

$$\left(\vec{\nabla} \times \vec{v}(\vec{r})\right)_x \bigg|_{\substack{x=0, \\ x=L_x}} \text{or} = \left(\frac{\partial v_z}{\partial y} - \frac{\partial v_y}{\partial z}\right) \bigg|_{\substack{x=0, \\ x=L_x}} \text{or} = 0.$$

Since $v_z(x = 0, y, z) = v_z(x = L_x, y, z) = v_y(x = 0, y, z) = v_y(x = L_x, y, z) = 0$ this condition is automatically satisfied.

Analogously we find for the sides II and II'

$$\left(\vec{\nabla}\times\vec{v}(\vec{r})\right)_y\bigg|_{\substack{y=0,\\y=L_y}}\text{, or } = \left(\frac{\partial v_x}{\partial z}-\frac{\partial v_z}{\partial x}\right)\bigg|_{\substack{y=0,\\y=L_y}}\text{, or } = 0$$

and finally for the sides III and III'

$$\left(\vec{\nabla}\times\vec{v}(\vec{r})\right)_z\bigg|_{\substack{z=0,\\z=L_z}}\text{, or } = \left(\frac{\partial v_y}{\partial x}-\frac{\partial v_x}{\partial y}\right)\bigg|_{\substack{z=0,\\z=L_z}}\text{, or } = 0.$$

Hence, the boundary conditions on the normal components of the magnetic flux are satisfied automatically.

10.2.5 Orthonormality of Mode Functions

According to the ansatz Eq. (10.32) the spatial part $\vec{v}(\vec{r})$ of the vector potential is given by trigonometric functions. The components k_x, k_y, and k_z of the wave vector are integer multiples l_x, l_y, and l_z of π/L_x, π/L_y, and π/L_z, respectively. This triple of integers we denote by l.

The Coulomb gauge condition $\vec{\nabla}\cdot\vec{A}=0$ enforces for each wave vector two polarization vectors $\vec{e}_1$ and $\vec{e}_2$. An exception is the case when one wave vector component vanishes. Since the polarization vectors depend on this triple we denote the vectors $\vec{e}_{l,1}$ and $\vec{e}_{l,2}$ by $\vec{e}_\ell$. Here the script index ℓ represents a set of four numbers, namely the polarization index (1 or 2) associated with a given set l_x, l_y, and l_z.

We note that two such spatial distributions $\vec{v}_\ell$ or $\vec{v}_{\ell'}$ – two mode functions – satisfy the condition

$$\int d^3r\,\vec{v}_\ell(\vec{r})\cdot\vec{v}_{\ell'}(\vec{r}) = \delta_{\ell,\ell'} \tag{10.37}$$

of orthonormality. Here we integrate over the whole space.

In order to prove the orthonormality relation Eq. (10.37) for the case of the box modes we first note that two mode functions of different polarizations are trivially orthogonal. Hence, it suffices to now consider two mode functions of identical polarization directions which differ in at least one wave number k_x, k_y, or k_z. We now evaluate the integral

$$\mathcal{J} \equiv \int_V d^3r\,\vec{v}_l(\vec{r})\,\vec{v}_{l'}(\vec{r}) = \sum_{i=x,y,z}\int_V d^3r\,(\vec{v}_l(\vec{r}))_i\,(\vec{v}_{l'}(\vec{r}))_i$$

where the integration goes over the volume $V \equiv L_x \cdot L_y \cdot L_z$ of the whole box.

When we recall the form Eq. (10.32) of the mode function we find

$$\mathcal{J} = \mathcal{N}\mathcal{N}'\left\{n_x n'_x\int_0^{L_x}dx\,\cos(k_x x)\cos(k'_x x)\int_0^{L_y}dy\,\sin(k_y y)\sin(k'_y y)\int_0^{L_z}dz\,\sin(k_z z)\sin(k'_z z)\right.$$

$$+n_y n'_y\int_0^{L_x}dx\,\sin(k_x x)\sin(k'_x x)\int_0^{L_y}dy\,\cos(k_y y)\cos(k'_y y)\int_0^{L_z}dz\,\sin(k_z z)\sin(k'_z z)$$

$$\left.+n_z n'_z\int_0^{L_x}dx\,\sin(k_x x)\sin(k'_x x)\int_0^{L_y}dy\,\sin(k_y y)\sin(k'_y y)\int_0^{L_z}dz\,\cos(k_z z)\cos(k'_z z)\right\}.$$

With the help of the integral relations

$$\int_0^{L_i} dx_i \cos\left(\pi l_i \frac{x_i}{L_i}\right) \cos\left(\pi l'_i \frac{x_i}{L_i}\right) = \delta_{l_i, l'_i} \frac{L_i}{2}$$

and

$$\int_0^{L_i} dx_i \sin\left(\pi l_i \frac{x_i}{L_i}\right) \sin\left(\pi l'_i \frac{x_i}{L_i}\right) = \delta_{l_i, l'_i} \frac{L_i}{2}$$

we then arrive at

$$\mathcal{J} = \mathcal{N}^2 \sum_{i=x,y,z} n_i^2 \frac{L_x L_y L_z}{8} \delta_{l_x, l'_x} \delta_{l_y, l'_y} \delta_{l_z, l'_z} = \mathcal{N}^2 \frac{V}{8} \delta_{l, l'} \qquad (10.38)$$

Here we have used the fact that $n_x^2 + n_y^2 + n_z^2 = 1$.

When we compare this result to the normalization condition Eq. (10.37) we find the normalization constant

$$\mathcal{N} \equiv \sqrt{\frac{8}{V}}.$$

We first note that for the example of the box the normalization constant is independent of the mode index ℓ. However, this simplification only holds true for the box. Moreover, we recognize that $\mathcal{N}$ is determined by a fraction of the total volume V. This suggests to introduce the effective mode volume

$$V_\ell \equiv \frac{V}{8}.$$

This definition becomes clear when we recall that for the box each mode function is a sine or a cosine function. In the evaluation of the mode volume we integrate their square which results in a factor $1/2$ of the integration regime. We have three space directions which yields the factor $1/2 \cdot 1/2 \cdot 1/2 = 1/8$.

We conclude this section by noting that for a resonator of a different shape such as a cylindrical, elliptical or confocal resonator, the mode functions assume a different form. However, due to the boundary condition Eq. (10.24) the wave numbers take on discrete values. Moreover, the corresponding mode functions satisfy the orthonormality relation Eq. (10.37).

10.3 The Field as a Set of Harmonic Oscillators

The mode functions $\vec{v}_\ell(\vec{r})$ of the resonator determine the spatial dependence of the vector potential $\vec{A}$ of the electromagnetic field in a resonator. For the case of a rectangular cavity we have shown that the mode functions are orthonormal and complete. This statement is even true for a cavity of rather arbitrary shape. We can therefore expand the vector potential into the mode functions of the specific resonator in question.

10.3 The Field as a Set of Harmonic Oscillators

In the present section we make use of this expansion and calculate the energy

$$\mathcal{H} \equiv \int d^3r \left[\frac{1}{2} \varepsilon_0 \vec{E}^2(\vec{r}, t) + \frac{1}{2} \mu_0 \vec{H}^2(\vec{r}, t) \right] \tag{10.39}$$

of the electromagnetic field in the resonator. Here the integration extends over all space.

In this calculation we do not assume a specific form of the resonator. Our considerations are valid for any resonator that allows such a mode decomposition. We show that the total energy in the resonator is a sum of harmonic oscillator energies corresponding to the individual modes. We quantize these field oscillators in the way familiar from mechanical oscillators. This procedure leads to the quantization of the radiation field. We emphasize again that this quantization appears in the time dependent part of the vector potential.

10.3.1 Energy in the Resonator

So far our analysis has concentrated on the vector potential. However, in order to calculate the energy of the electromagnetic field in the resonator we need the electric and magnetic fields. The solutions Eq. (10.4) and (10.5) of the homogeneous Maxwell equations provide us with the connection between the fields and the potential. Moreover, we have to modify the single mode ansatz Eq. (10.19) and take into account all the modes in the resonator.

Dimensionless Mode Function

In the preceding section we have found the normalization constant $\mathcal{N}$ of the mode functions of a box shaped resonator. This constant is determined by the effective mode volume V_ℓ.

The concept of an effective mode volume is also useful for resonators of arbitrary shape. In this case we again start from the mode function with a normalization constant $\mathcal{N}$ and determine $\mathcal{N}$ from the condition

$$\int d^3r \, \vec{v}_\ell^2(\vec{r}) = 1. \tag{10.40}$$

This relation connects $\mathcal{N}$ with the volume V of the resonator and brings in an algebraic factor due to the integration of the square of oscillatory mode functions. In the case of the box it was a factor 8. This connection allows us to introduce an effective mode volume V_ℓ which combines the volume of the resonator and the algebraic factor.

It is now convenient to separate this mode volume V_ℓ from the mode functions, that is we introduce dimensionless mode functions $\vec{u}_\ell$ via the definition

$$\vec{v}_\ell(\vec{r}) \equiv \frac{1}{\sqrt{V_\ell}} \vec{u}_\ell(\vec{r}).$$

The orthonormality condition Eq. (10.40) then reads

$$\frac{1}{\sqrt{V_\ell V_{\ell'}}} \int d^3r \, \vec{u}_\ell(\vec{r}) \cdot \vec{u}_{\ell'}(\vec{r}) = \delta_{\ell,\ell'}. \tag{10.41}$$

Multi-Mode Expansion: Electric and Magnetic Fields

The resonator supports many modes. We therefore expand the vector potential

$$\vec{A}(\vec{r},t) = \sum_\ell \frac{1}{\sqrt{\varepsilon_0 V_\ell}} q_\ell(t) \vec{u}_\ell(\vec{r}) \tag{10.42}$$

defined in Eq. (10.19) as a discrete sum over the mode functions $\vec{u}_\ell(\vec{r})$. Motivated by the orthonormality relation Eq. (10.41) we have separated the mode volume V_ℓ from the mode function. Moreover, it is convenient to include the electric permittivity ε_0. Hence, the mode volume and the permittivity determine the constant Υ in the separation ansatz Eq. (10.19). So far no condition restricts the mode amplitudes $q_\ell(t)$.

We now calculate the electric field $\vec{E}$ and the magnetic field $\vec{H}$ from the expression Eq. (10.42) for the vector potential using the equations Eqs. (10.4) and (10.5). In Coulomb gauge the electric field then reads

$$\vec{E}(\vec{r},t) = -\frac{\partial \vec{A}}{\partial t} = -\sum_\ell \frac{1}{\sqrt{\varepsilon_0 V_\ell}} \dot{q}_\ell(t) \vec{u}_\ell(\vec{r}), \tag{10.43}$$

and the magnetic field is

$$\vec{H}(\vec{r},t) = \frac{1}{\mu_0} \vec{\nabla} \times \vec{A} = \sum_\ell \frac{1}{\mu_0 \sqrt{\varepsilon_0 V_\ell}} q_\ell(t) \vec{\nabla} \times \vec{u}_\ell(\vec{r}). \tag{10.44}$$

We discuss these relations in more detail in Sec. 10.5 after we have quantized the fields.

Integration over Space

When we substitute Eqs. (10.43) and (10.44) into the integral for the total energy of the field in the resonator we find

$$\mathcal{H} = \frac{1}{2} \sum_{\ell,\ell'} \dot{q}_\ell \dot{q}_{\ell'} \frac{1}{\sqrt{V_\ell V_{\ell'}}} \int d^3r\, \vec{u}_\ell \cdot \vec{u}_{\ell'}$$

$$+ \frac{c^2}{2} \sum_{\ell,\ell'} q_\ell q_{\ell'} \frac{1}{\sqrt{V_\ell V_{\ell'}}} \int d^3r \left(\vec{\nabla} \times \vec{u}_\ell\right) \cdot \left(\vec{\nabla} \times \vec{u}_{\ell'}\right). \tag{10.45}$$

We first express the product of the two curls of the mode functions in the second integral by

$$\left(\vec{\nabla} \times \vec{u}_\ell\right) \cdot \left(\vec{\nabla} \times \vec{u}_{\ell'}\right) = \vec{\nabla} \cdot \left[\vec{u}_{\ell'} \times \left(\vec{\nabla} \times \vec{u}_\ell\right)\right] + \vec{u}_{\ell'} \cdot \left[\vec{\nabla} \times \left(\vec{\nabla} \times \vec{u}_\ell\right)\right].$$

We prove this relation by working out the right hand side of this equation with the help of the vector relation

$$\vec{\nabla} \cdot \left(\vec{f} \times \vec{g}\right) = \vec{g} \cdot \left(\vec{\nabla} \times \vec{f}\right) - \vec{f} \cdot \left(\vec{\nabla} \times \vec{g}\right)$$

and the special choice $\vec{g} \equiv \vec{\nabla} \times \vec{u}_\ell$ and $\vec{f} \equiv \vec{u}_{\ell'}$. Indeed, this yields

$$\vec{\nabla} \cdot \left[\vec{u}_{\ell'} \times \left(\vec{\nabla} \times \vec{u}_\ell\right)\right] = \left(\vec{\nabla} \times \vec{u}_\ell\right) \cdot \left(\vec{\nabla} \times \vec{u}_{\ell'}\right) - \vec{u}_{\ell'} \cdot \left[\vec{\nabla} \times \left(\vec{\nabla} \times \vec{u}_\ell\right)\right],$$

10.3 The Field as a Set of Harmonic Oscillators

that is the above equation.

Moreover, we note that the orthonormality condition (10.41) reduces the first double sum to a single sum. Hence, the energy in the resonator, Eq. (10.45), reads

$$\mathcal{H} = \frac{1}{2} \sum_\ell \dot{q}_\ell^2$$
$$+ \frac{c^2}{2} \sum_{\ell,\ell'} q_\ell q_{\ell'} \frac{1}{\sqrt{V_\ell V_{\ell'}}} \left[\int d^3r \, \vec{\nabla} \cdot \left[\vec{u}_{\ell'} \times \left(\vec{\nabla} \times \vec{u}_\ell \right) \right] + \int d^3r \, \vec{u}_{\ell'} \cdot \left[\vec{\nabla} \times \left(\vec{\nabla} \times \vec{u}_\ell \right) \right] \right].$$

We use the Gauss theorem to convert the first integral over the volume into one over the surface, that is

$$\int d^3r \, \vec{\nabla} \cdot \left[\vec{u}_{\ell'} \times \left(\vec{\nabla} \times \vec{u}_\ell \right) \right] = \int_{\text{surface}} d\vec{S} \cdot \left[\vec{u}_{\ell'} \times \left(\vec{\nabla} \times \vec{u}_\ell \right) \right].$$

Here $d\vec{S}(\vec{r})$ denotes a vector which at the point $\vec{r}$ on the surface of the resonator is orthogonal to the surface.

This surface integral vanishes. Indeed, $\vec{u}_{\ell'}$ is proportional to the spatial part of the electric field and $\vec{\nabla} \times \vec{u}_\ell$ is proportional to the spatial part of the magnetic field. Hence, the integrand contains the vector product of electric and magnetic field. Since the integration extends over the boundary of the resonator the electric and magnetic field have to be evaluated on the boundary. According to the boundary conditions discussed in Sec. 10.1.3 the electric field is orthogonal to the surface whereas the magnetic field is parallel to it. This leads via the vector product to a vector that lies in the surface and is therefore orthogonal to $d\vec{S}(\vec{r})$.

We can easily evaluate the remaining volume integral making use of the Coulomb gauge condition $\vec{\nabla} \cdot \vec{u}_\ell \equiv 0$ and the Helmholtz equation Eq. (10.22) which yields

$$\vec{\nabla} \times \left(\vec{\nabla} \times \vec{u}_\ell \right) = \vec{\nabla} \left(\vec{\nabla} \cdot \vec{u}_\ell \right) - \vec{\nabla}^2 \vec{u}_\ell = -\vec{\nabla}^2 \vec{u}_\ell(\vec{r}) = k_\ell^2 \vec{u}_\ell(\vec{r}) = \left(\frac{\Omega_\ell}{c} \right)^2 \vec{u}_\ell.$$

With the help of the orthonormality relation Eq. (10.41) of the mode functions we finally arrive at

$$\mathcal{H} = \sum_\ell \mathcal{H}_\ell \equiv \sum_\ell \left[\frac{1}{2} \dot{q}_\ell^2 + \frac{1}{2} \Omega_\ell^2 q_\ell^2 \right].$$

Hence, the energy of the electromagnetic field in an arbitrary resonator is the sum of the energies $\mathcal{H}_\ell$ of harmonic oscillators associated with each mode labeled by the index ℓ.

10.3.2 Quantization of the Radiation Field

The energy of the radiation field in a cavity is the sum of the energies in the individual modes. The energy of each mode is that of a harmonic oscillator with a generalized coordinate q_ℓ. The question is now, what fixes the amplitude q_ℓ?

In classical physics the amplitude is not fixed at all and can take on any value. We now show that the quantization of the electromagnetic field prevents q_ℓ from assuming arbitrary values but restricts it to certain regimes depending on the field state. In this section we pursue the quantization prescription in more detail. To bring out the quantum mechanics most clearly we first briefly review the classical harmonic oscillator.

A Classical Harmonic Oscillator: A Pico Review

We start from the Hamiltonian

$$\mathcal{H}_\ell \equiv \frac{1}{2}\dot{q}_\ell^2 + \frac{1}{2}\Omega_\ell^2 q_\ell^2 \tag{10.46}$$

of the ℓth oscillator of unit mass corresponding to the ℓth mode. The two conjugate variables q_ℓ and $p_\ell \equiv \dot{q}_\ell$ obey the Hamilton equations

$$\dot{q}_\ell = \frac{\partial \mathcal{H}_\ell}{\partial p_\ell} = p_\ell \quad \text{and} \quad \dot{p}_\ell = -\frac{\partial \mathcal{H}_\ell}{\partial q_\ell} = -\Omega_\ell^2 q_\ell$$

which indeed lead to the equation of motion

$$\ddot{q}_\ell + \Omega_\ell^2 q_\ell = 0$$

for the amplitudes derived in Eq. (10.23) by the separation ansatz for the wave equation. Hence, $p_\ell \equiv \dot{q}_\ell$ and q_ℓ are conjugate variables.

Since the magnetic field is proportional to q_ℓ and the electric field is proportional to $\dot{q}_\ell = p_\ell$ the two fields behave as the two conjugate variables position and momentum of a mechanical harmonic oscillator. Hence it is convenient to quantize these field oscillators in the same way as the mechanical oscillator by introducing first the complex-valued amplitudes a_ℓ and a_ℓ^* and then turning them into the annihilation and creation operators $\hat{a}_\ell$ and $\hat{a}_\ell^\dagger$ of the mode ℓ.

We start by introducing the amplitudes

$$a_\ell \equiv \frac{1}{\sqrt{2\hbar\Omega_\ell}}(\Omega_\ell q_\ell + ip_\ell) \quad \text{and} \quad a_\ell^* \equiv \frac{1}{\sqrt{2\hbar\Omega_\ell}}(\Omega_\ell q_\ell - ip_\ell).$$

In terms of these amplitudes the conjugate variables q_ℓ and p_ℓ read

$$q_\ell = \sqrt{\frac{\hbar}{2\Omega_\ell}}(a_\ell + a_\ell^*) \quad \text{and} \quad p_\ell = \frac{1}{i}\sqrt{\frac{\hbar\Omega_\ell}{2}}(a_\ell - a_\ell^*). \tag{10.47}$$

We substitute these representations of q_ℓ and p_ℓ into the expression Eq. (10.46) for the Hamiltonian of the ℓth mode and find

$$\mathcal{H}_\ell = -\frac{1}{2}\frac{\hbar\Omega_\ell}{2}\left(a_\ell^2 + (a_\ell^*)^2 - a_\ell a_\ell^* - a_\ell^* a_\ell\right) + \frac{1}{2}\Omega_\ell^2 \frac{\hbar}{2\Omega_\ell}\left(a_\ell^2 + (a_\ell^*)^2 + a_\ell a_\ell^* + a_\ell^* a_\ell\right),$$

or

$$\mathcal{H}_\ell = \frac{1}{2}\hbar\Omega_\ell\left(a_\ell^* a_\ell + a_\ell a_\ell^*\right). \tag{10.48}$$

Here we have kept track of the order in which the complex-valued amplitudes a_ℓ and a_ℓ^* enter the Hamiltonian. This is important when we now replace these amplitudes by operators.

10.3 The Field as a Set of Harmonic Oscillators

Quantization of the Oscillator

We quantize the electromagnetic field by quantizing each mode oscillator. For this purpose we postulate the commutation relation

$$[\hat{q}_\ell, \hat{p}_{\ell'}] = i\hbar\delta_{\ell,\ell'} \tag{10.49}$$

for the operators $\hat{q}_\ell$ and $\hat{p}_{\ell'}$ of the generalized coordinate of the mode ℓ and the generalized momentum of the mode ℓ'. Note that for $\ell' \neq \ell$ we assume

$$[\hat{q}_\ell, \hat{p}_{\ell'}] = 0,$$

since then the operators belong to different modes, that is to two different oscillators and thus commute.

In terms of the annihilation operators

$$\hat{a}_\ell \equiv \frac{1}{\sqrt{2\hbar\Omega_\ell}} (\Omega_\ell \hat{q}_\ell + i\hat{p}_\ell)$$

and the creation operators

$$\hat{a}_\ell^\dagger \equiv \frac{1}{\sqrt{2\hbar\Omega_\ell}} (\Omega_\ell \hat{q}_\ell - i\hat{p}_\ell)$$

the coordinate and momentum operators take the form

$$\hat{q}_\ell = \sqrt{\frac{\hbar}{2\Omega_\ell}} (\hat{a}_\ell + \hat{a}_\ell^\dagger) \quad \text{and} \quad \hat{p}_\ell = \frac{1}{i}\sqrt{\frac{\hbar\Omega_\ell}{2}} (\hat{a}_\ell - \hat{a}_\ell^\dagger). \tag{10.50}$$

Hence, the commutation relation Eq. (10.49) for coordinate and momentum reads

$$i\hbar\delta_{\ell,\ell'} = [\hat{q}_\ell, \hat{p}_{\ell'}] = \frac{\hbar}{2i}[\hat{a}_\ell + \hat{a}_\ell^\dagger, \hat{a}_{\ell'} - \hat{a}_{\ell'}^\dagger] = \frac{\hbar}{2i}\left(-[\hat{a}_\ell, \hat{a}_{\ell'}^\dagger] + [\hat{a}_\ell^\dagger, \hat{a}_{\ell'}]\right) = i\hbar[\hat{a}_\ell, \hat{a}_{\ell'}^\dagger],$$

that is

$$[\hat{a}_\ell, \hat{a}_{\ell'}^\dagger] = \delta_{\ell,\ell'}. \tag{10.51}$$

Here we have used

$$[\hat{a}_\ell, \hat{a}_{\ell'}] = [\hat{a}_\ell^\dagger, \hat{a}_{\ell'}^\dagger] = 0.$$

Hamiltonian of the Radiation Field

We conclude this section by using the commutation relation Eq. (10.51), to express the operator version

$$\hat{\mathcal{H}}_\ell = \frac{1}{2}\hbar\Omega_\ell (\hat{a}_\ell^\dagger \hat{a}_\ell + \hat{a}_\ell \hat{a}_\ell^\dagger)$$

of the classical Hamiltonian $\mathcal{H}_\ell$, Eq. (10.48), of the ℓth oscillator as

$$\hat{\mathcal{H}}_\ell = \hbar\Omega_\ell \left(\hat{a}_\ell^\dagger \hat{a}_\ell + \frac{1}{2}\right), \tag{10.52}$$

272   10 Field Quantization

and therefore the Hamiltonian of the electromagnetic field in the resonator reads

$$\hat{\mathcal{H}} = \sum_\ell \hbar \Omega_\ell \hat{a}_\ell^\dagger \hat{a}_\ell + \mathcal{H}_0. \tag{10.53}$$

Here the second term

$$\mathcal{H}_0 \equiv \sum_\ell \frac{1}{2} \hbar \Omega_\ell$$

in the Hamiltonian is a consequence of the commutation relation. It does not contain any operators and is a sum of all the zero-point energies of the individual mode oscillators. Since there are infinitely many modes, and we sum over all modes this term is infinite. However, we can renormalize the energy and subtract this infinite term. In the Schrödinger equation this amounts to an overall phase that is infinite.

One might therefore think that this term has no observable consequences. However, in the next section we show that due to this term two conducting plates can attract each other even when there is no other radiation field present except the vacuum.

10.4 The Casimir Effect

In the present section we evaluate the contribution

$$\mathcal{H}_0 \equiv \sum_\ell \frac{1}{2} \hbar \Omega_\ell$$

originating from the zero-point energies of the individual mode harmonic oscillators.

Obviously this term is infinite. Nevertheless, there are experimental situations when these zero-point energies can manifest themselves in real effects. For example, when we compare the zero-point energies of two resonators of different size then the difference between the two infinite zero-point energies leads to a finite difference, and especially to a force as we show now.

10.4.1 Zero-Point Energy of a Rectangular Resonator

We can illustrate this phenomenon with the help of two parallel plates of area L^2 which are separated by a distance a as shown in Fig. 10.2. We first derive an explicit expression for the infinite zero-point energy.

The frequency Ω_ℓ of the mode ℓ reads

$$\Omega_\ell = c|\vec{k}_\ell| = c\left(k_x^2 + k_y^2 + k_z^2\right)^{1/2} = c\left[\frac{l_x^2 \pi^2}{a^2} + \left(l_y^2 + l_z^2\right)\frac{\pi^2}{L^2}\right]^{1/2}$$

where we have used the quantization conditions Eq. (10.36). Moreover, for each frequency there exists in general two polarization directions. An exception is when one wave number vanishes. In this case there is only one polarization direction. This is very important for the evaluation of the Casimir effect.

10.4 The Casimir Effect

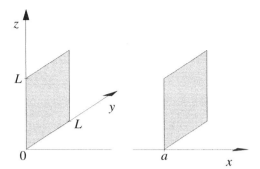

Fig. 10.2: Two parallel plates of area L^2 and separated by a distance a attract each other due to a difference of the infinite zero-point energies in the resonator and outside of the resonator.

Hence we find for the zero-point energy

$$\mathcal{H}_0^{\text{box}} = \sum_\ell \frac{1}{2}\hbar\Omega_\ell = \frac{\hbar c}{2}\sum_{\text{pol}}\sum_{l_x,l_y,l_z=0}^\infty \left[l_x^2\frac{\pi^2}{a^2} + (l_y^2 + l_z^2)\frac{\pi^2}{L^2}\right]^{1/2} \tag{10.54}$$

where the first sum denotes the sum over the polarization directions as indicated above.

We now assume that the plates are large compared with there separation, that is $a \ll L$. In this case the modes numbered by l_y and l_z get denser and approach a continuum. We can therefore replace summation by integration and find

$$\mathcal{H}_0^{\text{box}} = \frac{\hbar c}{2}\sum_{\text{pol}}\sum_{l_x=0}^\infty \int_0^\infty dl_y \int_0^\infty dl_z \left[l_x^2\left(\frac{\pi}{a}\right)^2 + \left(l_y\frac{\pi}{L}\right)^2 + \left(l_z\frac{\pi}{L}\right)^2\right]^{1/2}$$

$$= \frac{\hbar c}{2}\frac{\pi}{a}\sum_{\text{pol}}\sum_{l_x=0}^\infty \int_0^\infty dl_y \int_0^\infty dl_z \left[l_x^2 + \left(l_y\frac{a}{L}\right)^2 + \left(l_z\frac{a}{L}\right)^2\right]^{1/2},$$

$$\mathcal{H}_0^{\text{box}} = \frac{\hbar c \pi^2}{2a^3}\frac{L^2}{\pi}\sum_{\text{pol}}\sum_{l_x=0}^\infty \int_0^\infty d\xi \int_0^\infty d\zeta \left[l_x^2 + \xi^2 + \zeta^2\right]^{1/2},$$

where we have introduced $\xi \equiv l_y a/L$ and $\zeta \equiv l_z a/L$.

When we introduce the polar coordinates $\xi = \sqrt{u}\cos\varphi$ and $\zeta = \sqrt{u}\sin\varphi$ we find with the help of the relation

$$\int_0^\infty d\xi \int_0^\infty d\zeta = \int_0^\infty d(\sqrt{u})\sqrt{u} \int_0^{\pi/2} d\varphi = \int_0^\infty du \frac{1}{2\sqrt{u}}\sqrt{u}\int_0^{\pi/2} d\varphi = \frac{1}{2}\int_0^\infty du \int_0^{\pi/2} d\varphi \tag{10.55}$$

the expression

$$\mathcal{H}_0^{\text{box}} = \frac{\hbar c \pi^2}{8a^3}L^2\sum_{\text{pol}}\sum_{l=0}^\infty \int_0^\infty du \sqrt{l^2 + u}.$$

Here in order to simplify the notation we have set $l \equiv l_x$.

When we now separate the term $l = 0$ and recall that here in contrast to the remainder there is only one polarization direction, we find

$$\mathcal{H}_0^{\text{box}} = \frac{\hbar c \pi^2}{8a^3} L^2 \left[\int_0^\infty du \, \sqrt{u} + 2 \sum_{l=1}^\infty \int_0^\infty du \, \sqrt{l^2 + u} \right]. \qquad (10.56)$$

We note that the zero-point energy of the vacuum-radiation field consists of two parts: (i) a prefactor which depends on the geometry of the setup and involves apart from the constants $\hbar$, c and π the area L^2 of the plates and the separation a and (ii) integrals which are highly divergent. We realize that these integrals do not depend on any geometrical or physical quantity. This infinite expression cannot be evaluated in a sensible way. However, we now re-express the zero-point energy $\mathcal{H}_0^{\text{vac}}$ of the vacuum in the absence of a resonator. When we subtract these two infinite energies from each other we arrive at a finite result.

10.4.2 Zero-Point Energy of Free Space

We now calculate the zero-point energy in the absence of the resonator by not only L, but also the separation a of the two plates approaching infinity. In this case all three discrete summations in Eq. (10.54) are replaced by integrations and we have to evaluate the expression

$$\mathcal{H}_0^{\text{vac}} = \frac{\hbar c}{2} \sum_{\text{pol}} \int_0^\infty dl_x \int_0^\infty dl_y \int_0^\infty dl_z \left[\left(l_x \frac{\pi}{a}\right)^2 + \left(l_y \frac{\pi}{L}\right)^2 + \left(l_z \frac{\pi}{L}\right)^2 \right]^{1/2}$$

which after introducing the new integration variables $\xi \equiv l_y a/L$ and $\zeta \equiv l_z a/L$ reads

$$\mathcal{H}_0^{\text{vac}} = \frac{\hbar c \pi^2}{2a^3} \frac{L^2}{\pi} \sum_{\text{pol}} \int_0^\infty dl_x \int_0^\infty d\xi \int_0^\infty d\zeta \left[l_x^2 + \xi^2 + \zeta^2 \right]^{1/2}.$$

When we recall the transformation Eq. (10.55) to polar coordinates we find

$$\mathcal{H}_0^{\text{vac}} = \frac{\hbar c \pi^2}{8a^3} L^2 2 \int_0^\infty dl \int_0^\infty du \, \sqrt{l^2 + u}, \qquad (10.57)$$

where the factor 2 arises from the sum over the polarizations.

This expression contains the same prefactor consisting of the constants $\hbar$, c and π and the geometrical quantities a and L as the expression Eq. (10.56) for the zero-point energy $\mathcal{H}_0^{\text{box}}$ of the resonator. Moreover, the integration over the variable u corresponding to the modes in y- and z-direction appears in both expressions. However, since there are no plates in the calculation of $\mathcal{H}_0^{\text{vac}}$ there is no discrete summation but an integration.

Note that in Eq. (10.57) we have evaluated the sum over the two polarization directions which yields a factor 2. In contrast to the zero-point energy of the rectangular box with a discrete summation over the modes in x-direction, we now do not have to separate the mode, where one wave number vanishes. Since in free space this mode is a single point in an integral, it cannot make a contribution.

10.4.3 Difference of Two Infinite Energies

We have found that the zero-point energies of both the rectangular resonator and free space are infinite. However, we now have to subtract the infinite contribution of free space from the infinite energy of the rectangular resonator in order to determine the finite Casimir energy. The energy per area is thus

$$\mathcal{V} \equiv \frac{1}{L^2}\left[\mathcal{H}_0^{\text{box}} - \mathcal{H}_0^{\text{vac}}\right] = \frac{\hbar c \pi^2}{4a^3}\tilde{I}. \tag{10.58}$$

Here we have introduced the quantity

$$\tilde{I} \equiv \frac{1}{2}\int_0^\infty du\,\sqrt{u} + \sum_{l=1}^\infty \int_0^\infty du\,\sqrt{l^2+u} - \int_0^\infty dl\int_0^\infty du\,\sqrt{l^2+u}$$

which is the difference

$$\tilde{I} = \frac{1}{2}I(0) + \sum_{l=1}^\infty I(l) - \int_0^\infty dl\, I(l)$$

between the discrete sum and the integral of the terms

$$I(l) \equiv \int_0^\infty du\,\sqrt{l^2+u}.$$

Obviously the integrals $I(l)$ are divergent. Therefore, on first sight the quantity $\tilde{I}$ should also be divergent. However, since it is the difference between the sum and the integral of these divergent quantities a finite result emerges.

We calculate this result using the Euler-MacLaurin formula

$$I = \frac{1}{2}I(0) + \sum_{l=1}^\infty I(l) - \int_0^\infty dl\, I(l) = -\frac{1}{2!}B_2\frac{dI(0)}{dl} - \frac{1}{4!}B_4\frac{d^3I(0)}{dl^3} - \cdots,$$

where the Bernoulli numbers B_j are defined via the generating function

$$\frac{y}{e^y - 1} = \sum_{j=0}^\infty B_j\frac{y^j}{j!}$$

with $B_2 = 1/6$ and $B_4 = -1/30$ as discussed in Problem 10.3.

We evaluate the derivatives by first expressing the integrals in the form

$$I(l) = \int_{l^2}^\infty dv\,\sqrt{v},$$

where in the last step we have introduced the integration variable $v \equiv l^2 + u$. We then find

$$\frac{dI(l)}{dl} = -2l^2$$

and
$$\frac{d^3 I(l)}{dl^3} = -4.$$

Since
$$\left.\frac{d^j I(l)}{dl^j}\right|_{l=0} = 0$$

for $j \geq 4$ the Euler-MacLaurin series terminates after the third derivative.
Moreover, since $dI(0)/dl = 0$ and $B_4 = -1/30$ we find

$$\tilde{I} = -\frac{1}{180}. \tag{10.59}$$

It is the difference between the discrete mode structure given by the two plates and the continuum in the absence of the plates that leads to the finite result Eq. (10.59).

10.4.4 Casimir Force: Theory and Experiment

We are now in a position to calculate the potential energy of two parallel plates due to the zero-point energy.

Force Between Two Parallel Plates

We obtain the energy per unit surface by substituting the result $\tilde{I} = -1/180$, Eq. (10.59), into Eq. (10.58) and find

$$\mathcal{V}(a) = -\frac{1}{720}\frac{\hbar c \pi^2}{a^3},$$

and thus the force per unit area

$$F(a) \equiv -\frac{d\mathcal{V}}{da}$$

reads

$$F(a) = -\frac{\pi^2}{240}\frac{\hbar c}{a^4}. \tag{10.60}$$

This is a quite remarkable result: The difference in the infinite zero-point energies in the presence and absence of condenser plates has led to a finite attractive force between the plates. This force increases with a decreasing separation, namely as the fourth power. Moreover, we recognize that it is a force of pure quantum mechanical origin since it is proportional to Planck's constant $\hbar$. In addition, the velocity of the light enters via the frequency of light and the mode structure. It is interesting to note that a/c is the time the light needs to traverse the distance between the two plates and hence to "probe" the presence of the plates.

Spherical Shell

The Casimir effect has been calculated for many geometrical configurations. It does not always give rise to an attractive force. For example, for a spherical conducting shell we find a repulsive force. This phenomenon has an interesting history.

Motivated by his success in calculating the attractive force between two conducting plates H.B.G. Casimir in 1953 proposed an intriguing model for a charged particle. In its rest frame the particle is regarded as a conducting spherical shell carrying a homogeneous surface charge of total magnitude e. The corresponding electrostatic energy tends to expand the sphere. On the other hand the presence of the conducting boundary condition alters the zero-point energy of the universe. In complete analogy to the arrangement of the parallel plates Casimir argued that the attractive zero-point energy might tend to collapse the sphere. A stable configuration, that is a balance of the two forces arises only for a fixed ratio $e^2/(\hbar c)$. He therefore called this model of a charged particle a *mouse trap* for $e^2/(\hbar c)$.

Unfortunately, this model has a dramatic defect: in 1968 T. Boyer showed that the zero-point energy of a conducting spherical shell of radius a reads

$$\mathcal{V}(a) \cong +0.09 \frac{\hbar c}{2a},$$

and is thus repulsive and not attractive. Hence in contrast to the case of the parallel plates which attract each other due to the vacuum fluctuations the shell expands.

Experimental Verification

For a long time there has been essentially only one experiment trying to verify the Casimir effect. This experiment dating back to 1958 showed an attractive force which as the author M.J. Sparnaay points out is not inconsistent with the formula Eq. (10.60). Unfortunately, this experiment is inconclusive since it displayed effectively 100% uncertainty.

A closely related effect has recently been observed by the group of E. Hinds. They have measured the attraction of an atom by a conducting plate and have found good agreement with theory.

However, the most spectacular experiment on the Casimir effect has been performed recently by S.K. Lamoreaux. Avoiding the experimental difficulties associated with maintaining two plates parallel at the required accuracy he has investigated the Casimir force for the case when the conductors are in the form of a flat plate and a sphere. In this situation the corresponding force reads

$$F_{\mathrm{SF}}(a) = -2\pi R \left(\frac{1}{3} \frac{\pi^2}{240} \frac{\hbar c}{a^3} \right)$$

where R is the radius of curvature of the spherical surface.

One of the main differences to the Casimir force Eq. (10.60) between two parallel plates is that the force F_{SF} is independent of the plate area. Moreover, the power law of the separation a exhibits the inverse cubic rather the fourth power.

In Fig. 10.3 we show a comparison between the measured and the expected Casimir force. The agreement with theory is rather impressive.

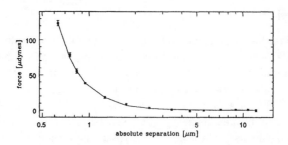

Fig. 10.3: Experimental verification of the Casimir force between a plate and a sphere as a function of the separation. Dots mark experimental points, the solid line is the expected Casimir force. Taken from S.K. Lamoreaux, Phys. Rev. Lett. **78**, 5 (1997).

10.5 Operators of the Vector Potential and Fields

In Sec. 10.3.2 we have quantized the radiation field by introducing commutation relations between the conjugate variables q_ℓ and p_ℓ of the time dependent part of the mode functions. Appropriate combinations of these operators define the annihilation and creation operators $\hat{a}_\ell$ and $\hat{a}_\ell^\dagger$ of the ℓth mode. In the present section we summarize the operators for the vector potential, the electric and magnetic field and discuss their time dependence.

10.5.1 Vector Potential

We first express the vector potential $\vec{A}$, Eq. (10.42), in terms of the creation and annihilation operators using Eq. (10.47), and find

$$\hat{\vec{A}}(\vec{r},t) = \sum_\ell \sqrt{\frac{\hbar}{2\varepsilon_0 V_\ell \Omega_\ell}}\, \vec{u}_\ell(\vec{r}) \left(\hat{a}_\ell(t) + \hat{a}_\ell^\dagger(t)\right). \tag{10.61}$$

Time Dependence of Creation and Annihilation Operators

The time dependence of the vector potential enters through the time dependence of the operators $\hat{a}_\ell$ and $\hat{a}_\ell^\dagger$. Indeed, when we recall the Heisenberg equation of motion

$$\dot{\hat{a}}_\ell = \frac{i}{\hbar}[\hat{\mathcal{H}}, \hat{a}_\ell]$$

for the annihilation operator of the ℓth mode we arrive with the help of the commutation relation Eq. (10.51) at

$$\dot{\hat{a}}_\ell = \frac{i}{\hbar}\sum_{\ell'}\hbar\Omega_{\ell'}\,[\hat{a}_{\ell'}^\dagger \hat{a}_{\ell'}, \hat{a}_\ell] = \frac{i}{\hbar}\sum_{\ell'}\hbar\Omega_{\ell'}\,[\hat{a}_{\ell'}^\dagger, \hat{a}_\ell]\hat{a}_{\ell'} = -\frac{i}{\hbar}\sum_{\ell'}\hbar\Omega_{\ell'}\,\delta_{\ell',\ell}\,\hat{a}_{\ell'},$$

which yields

$$\dot{\hat{a}}_\ell = -i\Omega_\ell\,\hat{a}_\ell. \tag{10.62}$$

Since the solution is

$$\hat{a}_\ell(t) = \hat{a}_\ell(0)e^{-i\Omega_\ell t} \tag{10.63}$$

10.5 Operators of the Vector Potential and Fields

the annihilation operator has a negative phase factor.

When we substitute this solution for $\hat{a}_\ell$ together with the corresponding expression

$$\hat{a}_\ell^\dagger(t) = \hat{a}_\ell^\dagger(0) e^{i\Omega_\ell t}$$

for $\hat{a}_\ell^\dagger$ into the formula Eq. (10.59) for the operator $\hat{\vec{A}}$ of the vector potential we find

$$\hat{\vec{A}}(\vec{r},t) = \sum_\ell \mathcal{A}_\ell \, \vec{u}_\ell(\vec{r}) \frac{1}{\sqrt{2}} \left(\hat{a}_\ell(0) e^{-i\Omega_\ell t} + \hat{a}_\ell^\dagger(0) e^{+i\Omega_\ell t} \right).$$

Here we have introduced the abbreviation

$$\mathcal{A}_\ell \equiv \sqrt{\frac{\hbar}{\varepsilon_0 V_\ell \Omega_\ell}}.$$

The quantity $\mathcal{A}_\ell$ has the units of a vector potential. It involves the mode volume V_ℓ and the mode frequency Ω_ℓ. We recall that the spatial part of the vector potential sits in the dimensionless mode function $\vec{u}_\ell(\vec{r})$ and the operator nature in the annihilation and creation operators $\hat{a}_\ell$ $\hat{a}_\ell^\dagger$. Hence, $\mathcal{A}_\ell$ is the unit of quantization.

Different Representations

We note that the mode operators enter in a way reminiscent of the position operator of a mechanical oscillator. This fact suggests to introduce the dimensionless position operator

$$\hat{x}_\ell(t) \equiv \frac{1}{\sqrt{2}} \left(\hat{a}_\ell(t) + \hat{a}_\ell^\dagger(t) \right) \tag{10.64}$$

of the lth mode. We therefore find

$$\hat{\vec{A}}(\vec{r},t) = \sum_\ell \mathcal{A}_\ell \, \vec{u}_\ell(\vec{r}) \hat{x}_\ell(t),$$

that is, the vector potential is a sum of generalized position operators of the individual modes.

We conclude this section by presenting yet another representation of the vector potential operator

$$\hat{\vec{A}}(\vec{r},t) = \hat{\vec{A}}^{(+)}(\vec{r},t) + \hat{\vec{A}}^{(-)}(\vec{r},t).$$

The positive frequency part

$$\hat{\vec{A}}^{(+)}(\vec{r},t) \equiv \frac{1}{\sqrt{2}} \sum_\ell \mathcal{A}_\ell \, \vec{u}_\ell(\vec{r}) \hat{a}_\ell(0) \, e^{-i\Omega_\ell t}$$

contains annihilation operators only, whereas the negative frequency part

$$\hat{\vec{A}}^{(-)}(\vec{r},t) \equiv \frac{1}{\sqrt{2}} \sum_\ell \mathcal{A}_\ell \, \vec{u}_\ell(\vec{r}) \hat{a}_\ell^\dagger(0) \, e^{+i\Omega_\ell t}$$

contains only the creation operators.

The positive frequency component of the vector potential is the one with the phase $-i\Omega_\ell t$. As R. Glauber always puts it: *"It is the positive frequency component because there is a minus sign in the exponent."*

10.5.2 Electric Field Operator

We now turn to the electric field operator. Equation (10.43) allows us to express the electric field in terms of the time derivative of the vector potential. We start from the mode expansion Eq. (10.61) of the vector potential and use Eq. (10.62) for the derivatives of the annihilation and creation operator to find

$$\hat{\vec{E}}(\vec{r},t) = i\sum_\ell \sqrt{\frac{\hbar\Omega_\ell}{2\varepsilon_0 V_\ell}}\, \vec{u}_\ell(\vec{r})\left(\hat{a}_\ell(t) - \hat{a}_\ell^\dagger(t)\right),$$

or

$$\hat{\vec{E}}(\vec{r},t) = \sum_\ell \mathcal{E}_\ell \vec{u}_\ell(\vec{r})\frac{i}{\sqrt{2}}\left(\hat{a}_\ell(0)e^{-i\Omega_\ell t} - \hat{a}_\ell^\dagger(0)e^{i\Omega_\ell t}\right). \tag{10.65}$$

Here we have introduced the abbreviation

$$\mathcal{E}_\ell \equiv \sqrt{\frac{\hbar\Omega_\ell}{\varepsilon_0 V_\ell}} = \Omega_\ell \mathcal{A}_\ell. \tag{10.66}$$

This quantity has the dimension of an electric field. Sometimes it is referred to as the electric field per photon, or sometimes vacuum electric field. The origin of this, somewhat misleading name will become clear when we consider a single mode of the resonator. Similarly to the vector potential the quantity $\mathcal{E}_\ell$ is the unit of quantization of the electric field.

Different Representations

The form Eq. (10.65) of the electric field operator suggests to introduce the dimensionless momentum operator

$$\hat{\pi}_\ell(t) \equiv \frac{1}{\sqrt{2}\,i}\left(\hat{a}_\ell(t) - \hat{a}_\ell^\dagger(t)\right) \tag{10.67}$$

of the ℓth mode. The resulting form

$$\hat{\vec{E}}(\vec{r},t) = -\sum_\ell \mathcal{E}_\ell\, \vec{u}_\ell(\vec{r})\,\hat{\pi}_\ell(t) \tag{10.68}$$

of the electric field operator $\hat{\vec{E}}$ shows that $\hat{\vec{E}}$ is conjugate to the vector potential $\hat{\vec{A}}$ which involves the position operator $\hat{x}_\ell(t)$.

In complete analogy to the vector potential we can also decompose the electric field operator, Eq. (10.65)

$$\hat{\vec{E}}(\vec{r},t) = \hat{\vec{E}}^{(+)}(\vec{r},t) + \hat{\vec{E}}^{(-)}(\vec{r},t)$$

into its positive frequency component

$$\hat{\vec{E}}^{(+)}(\vec{r},t) \equiv \frac{i}{\sqrt{2}}\sum_\ell \mathcal{E}_\ell\, \vec{u}_\ell(\vec{r})\hat{a}_\ell(0)\,e^{-i\Omega_\ell t}$$

and a negative frequency part

$$\hat{\vec{E}}^{(-)}(\vec{r},t) \equiv \frac{-i}{\sqrt{2}}\sum_\ell \mathcal{E}_\ell\, \vec{u}_\ell(\vec{r})\hat{a}_\ell^\dagger(0)e^{+i\Omega_\ell t}.$$

Again the two terms contain only annihilation or creation operators.

10.5.3 Magnetic Field Operator

We conclude this section by presenting the operator

$$\hat{\vec{H}}(\vec{r},t) = \frac{1}{\mu_0}\hat{\vec{B}}(\vec{r},t) = \frac{1}{\mu_0}\sum_\ell \sqrt{\frac{\hbar}{2\varepsilon_0 V_\ell \Omega_\ell}} \left(\vec{\nabla} \times \vec{u}_\ell(\vec{r})\right)\left(\hat{a}_\ell(t) + \hat{a}_\ell^\dagger(t)\right) \qquad (10.69)$$

for the magnetic field $\hat{\vec{H}}$ which in terms of the dimensionless position operator $\hat{x}_\ell$ reads

$$\hat{\vec{H}}(\vec{r},t) = \sum_\ell \frac{1}{\mu_0} \mathcal{A}_\ell \left(\vec{\nabla} \times \vec{u}_\ell(\vec{r})\right)\hat{x}_\ell(t).$$

In contrast to the operators for the vector potential and the electric field we cannot introduce in a straightforward way a unit of magnetic field quantization since the curl of the mode function appears which has to be included in such a definition.

Since the electric and magnetic field operators involve the conjugate variables, that is the generalized momentum and the generalized position operator the electric and magnetic field do not commute with each other. This fact has important consequences for the question of measuring simultaneously electric and magnetic field components. The problem of general commutation relations between electromagnetic field operators and their measurability has been studied in great detail by N. Bohr and L. Rosenfeld. For more details we refer to the literature at the end of this chapter.

10.6 Number States of the Radiation Field

We have quantized the radiation field by first decomposing it into an infinite set of modes and then quantizing the individual harmonic oscillator associated with each mode. In this way we have promoted the vector potential as well as the electric and the magnetic fields from c-numbers to operators. Hence, we cannot describe the electromagnetic field by a complex- or real-valued function, but have to use the concept of a quantum state. All information about the field is now contained in the quantum state. We emphasize again that this information is only related to the time dependent amplitude operator $\hat{q}_\ell$ of the vector potential and not to the spatial part contained in the mode function.

Which states should we use as our fundamental basis set? Should we use the eigenstates of the electric field operator, or of the magnetic field or, of the total energy? For many problems the eigenstates of the Hamiltonian of the radiation field are the most convenient states. They are related to the concept of the photon.

We first briefly define the properties of photon number states and then discuss the expectation value of the energy of the Hamiltonian. We briefly turn to a superposition of number states and entangled states. Throughout this section we do not give a prescription for preparing such number states of the radiation field. This is a highly nontrivial problem and will be discussed in more detail in Chapter 16.

10.6.1 Photons and Anti-Photons

From Eq. (10.52) we note that the Hamiltonian, that is the energy operator of the lth mode in the absence of the zero-point energy reads

$$\hat{\mathcal{H}}_\ell \equiv \hbar\Omega_\ell\,\hat{n}_\ell,$$

where we have introduced the number operator

$$\hat{n}_\ell \equiv \hat{a}_\ell^\dagger \hat{a}_\ell. \tag{10.70}$$

The eigenstates $|n_\ell\rangle$ of the number operator of the lth mode satisfy the eigenvalue equation

$$\hat{n}_\ell|n_\ell\rangle = n_\ell|n_\ell\rangle, \tag{10.71}$$

where n_ℓ is the eigenvalue.

This eigenvalue equation is analogous to the one of the energy eigenstates of a mechanical harmonic oscillator discussed in Sec. 2.2. We recall that the oscillator can take on integer multiples of the energy $\hbar\Omega_\ell$. Therefore, the number operator can take on the eigenvalues $n_\ell = 0, 1, 2, \ldots$. This has an important consequence for the electromagnetic field when the ℓth mode is in the number state $|n_\ell\rangle$. Indeed, in this case this mode has a well-defined excitation n_ℓ giving rise to the energy $n_\ell \cdot \hbar\Omega_\ell$.

These discrete excitations of the mode oscillator and hence the radiation field are commonly referred to as *photons*. This name dates back to 1926 when the physical chemist G.N. Lewis introduced the word photon in a letter to the editor of *Nature*. In this letter Lewis speculated that the transmission of radiation from one atom to another was carried by a new particle which he coined photon. Lewis considered this to be a real particle but specifically denied that it was the light quantum of M. Planck, A. Einstein and N. Bohr. For a more detailed discussion of the history of the photon and why it should be used with care we refer to the work by W.E. Lamb, especially his article entitled "Anti-Photon".

We conclude this section by emphasizing that it is quite misleading to associate with these discrete excitations a particle notion. The quantum part of the light sits only in the time dependent part of the vector potential and not in the spatial part. Therefore, it is not straightforward to define a wave function in position space for a photon. Various suggestions have been made but this question is still under debate.

10.6.2 Multi-Mode Case

So far, we have only discussed a single mode. However, the radiation field in the resonator consists of infinitely many modes. Hence the state vector $|\Psi\rangle$ of the radiation field has to include an infinite number of states. For the sake of simplicity we first consider the state

$$|\Psi\rangle \equiv |\{n_\ell\}\rangle \equiv |n_1\rangle\,|n_2\rangle\,|n_3\rangle \ldots |n_\ell\rangle \ldots \equiv |n_1, n_2, n_3, \ldots, n_\ell, \ldots\rangle,$$

where we have n_1 excitations in mode 1, n_2 excitations in mode 2, and so forth. Here we consider the individual modes to be independent of each other.

The physical meaning of the number states comes out most clearly when we calculate the expectation value of the total energy in the resonator in the state $|\Psi\rangle$.

We recall from Eq. (10.53) the expression

$$\hat{\mathcal{H}} \equiv \sum_\ell \hbar\Omega_\ell \hat{n}_\ell$$

for the Hamiltonian of the electromagnetic field in a resonator and neglect the zero-point energy $\hat{\mathcal{H}}_0$ to find

$$\langle\hat{\mathcal{H}}\rangle = \langle\{n_\ell\}|\hat{\mathcal{H}}|\{n_\ell\}\rangle = \sum_\ell \hbar\Omega_\ell \langle n_1, n_2, \ldots, n_\ell, \ldots|\hat{n}_\ell|n_1, n_2, \ldots, n_\ell, \ldots\rangle$$

where the number operator $\hat{n}_\ell$ acts only on the state of the ℓth mode, which in the present case is in the number state $|n_\ell\rangle$.

We use the eigenvalue equation Eq. (10.71) and arrive at

$$\langle\hat{\mathcal{H}}\rangle = \sum_\ell \hbar\Omega_\ell n_\ell \langle n_1, n_2, \ldots, n_\ell, \ldots|n_1, n_2, \ldots, n_\ell, \ldots\rangle.$$

Since the number states $|n_\ell\rangle$ are orthogonal and normalized, that is

$$\langle n_\ell|n_{\ell'}\rangle = \delta_{\ell,\ell'},$$

we find

$$\langle n_1, n_2, \ldots, n_\ell, \ldots|n_1, n_2, \ldots, n_\ell, \ldots\rangle = \langle n_1|n_1\rangle\langle n_2|n_2\rangle \ldots \langle n_\ell|n_\ell\rangle \ldots = 1.$$

Hence, the expectation value of the total energy reads

$$\langle\hat{\mathcal{H}}\rangle = \sum_\ell \hbar\Omega_\ell n_\ell.$$

The total energy of the electromagnetic field in the state $|\{n_\ell\}\rangle$ consisting of a direct product of photon number states is the sum of the energies $n_\ell \cdot \hbar\Omega_\ell$ corresponding to the energies in the individual modes. Since the ℓth mode has n_ℓ quanta in it, we arrive at the energy $n_\ell \cdot \hbar\Omega_\ell$.

10.6.3 Superposition and Entangled States

The state $|\psi_\ell\rangle$ of the ℓth mode does not have to be in a number state $|n_\ell\rangle$, but could be in a superposition

$$|\psi_\ell\rangle = \sum_{n_\ell} \psi_{n_\ell}|n_\ell\rangle$$

of number states with probability amplitudes ψ_{n_ℓ}. In this case the radiation field is a direct product

$$|\Psi\rangle \equiv |\psi_1\rangle|\psi_2\rangle|\psi_3\rangle \ldots |\psi_\ell\rangle \ldots$$

of superposition states $|\psi_\ell\rangle$.

Examples for such superposition states include coherent states, squeezed states and quadrature eigenstates as discussed for a mechanical oscillator in Chapter 4. We will discuss the field analogue of these states in more detail in the next chapter.

So far, we have only considered product states of the individual modes. However, these individual modes can be entangled, that is the states of the individual modes cannot be factorized into product states of these modes. An example for such an entangled multi-mode state is

$$|\Psi\rangle = \sum_{\ldots,n_i,n_j,n_k,\ldots} \Psi_{\ldots,n_i,n_j,n_k,\ldots} |\ldots,n_i,n_j,n_k\ldots\rangle$$

where the probability amplitudes $\Psi_{\ldots,n_i,n_j,n_k,\ldots}$ do not factor into products of functions of indices $\ldots,n_i,n_j,n_k,\ldots$ only.

The expression entangled states stems from Schrödinger's German expression "verschränkter Zustand". It is the true difference between classical wave physics and quantum mechanics. It is at the very heart of the Einstein-Podolsky-Rosen situation where Schrödinger coined this phrase. Most recently it has been used to achieve teleportation of quantum states, quantum communication and hopefully quantum computing. We will return to this topic of entangled states in Chapter 16.

Problems

10.1 Eigenvalues of the Laplacian

Show that the eigenvalues λ of the Laplacian defined by

$$\vec{\nabla}^2 u_j(\vec{r}) = \lambda u_j(\vec{r}) \tag{10.72}$$

are negative if the boundary conditions (10.24) and (10.25) hold.
Hint: Multiply Eq. (10.72) with u_j and sum over j. Then make use of Green's first identity.

10.2 Boundary Conditions on Electric and Magnetic Field

Use Maxwell's equations in integral form to show that the requirements, Eq. (10.25), on the magnetic field components are satisfied on the boundary provided the electric field constraints, Eq. (10.24), are fulfilled. This statement is valid for a resonator of arbitrary form.

10.3 Properties of Bernoulli Numbers

The Bernoulli numbers B_n are defined by the relation

$$\frac{z}{e^z - 1} = \sum_{n=0}^{\infty} \frac{B_n}{n!} z^n.$$

Prove that

$$\sum_{\nu=0}^{n-1} \binom{n}{\nu} B_\nu = 0 \qquad \text{for } n \geq 2 \tag{10.73}$$

and calculate B_n for $n = 0, 1, \ldots, 4$.
Hint: Multiply Eq. (10.73) with $e^z - 1$ and write the right-hand side as a Taylor series in z.

10.4 Field Quantization with Running Waves

(a) Show that in an appropriate gauge the electromagnetic field in the absence of charges and currents can *solely* be described by the vector potential $\vec{A}$. What are the equations of motion for $\vec{A}$?

(b) Show that in a box with edges L_1, L_2, L_3 the vector potential can be written in the form

$$\vec{A}(\vec{x},t) = \sum_{\vec{k}} \sum_{\sigma=1,2} \sqrt{\frac{\hbar}{2\omega_{\vec{k}}\varepsilon_0 V}}\, \vec{e}_{\vec{k}\sigma}\left[a_{\vec{k}\sigma}(t)e^{i\vec{k}\vec{x}} + a^*_{\vec{k}\sigma}(t)e^{-i\vec{k}\vec{x}}\right]$$

if we assume periodic boundary conditions and Coulomb gauge. Here, $\vec{e}_{\vec{k}\sigma}$, $\sigma = 1,2$ are the polarization vectors. What are the conditions for $\vec{k}$ and $\vec{e}_{\vec{k}\sigma}$? What is the time dependence of $a_{\vec{k}\sigma}(t)$ and $a^*_{\vec{k}\sigma}(t)$?
Remark: Here the prefactor, which initially is completely arbitrary, was chosen in such a way that after quantizing the field $a^*_{\vec{k}\sigma}$ and $a_{\vec{k}\sigma}$ become the usual creation and annihilation operators.

(c) Show that with a suitable choice of the polarization vectors $\vec{e}_{\vec{k}\sigma}$ the total energy of the free electromagnetic field can be written as

$$H = \frac{1}{2}\sum_{\vec{k},\sigma} \hbar\omega_{\vec{k}}\left[a_{\vec{k},\sigma}(t)a^*_{\vec{k},\sigma}(t) + a^*_{\vec{k},\sigma}(t)a_{\vec{k},\sigma}(t)\right].$$

Hint: Although the calculation is done with classical quantities it is advantageous to take care of the order of $a_{\vec{k},\sigma}(t)$ and $a^*_{\vec{k},\sigma}(t)$ since they will become operators when the field is quantized.

(d) Show that the total momentum $\vec{G}(t)$ of the free electromagnetic field is given by

$$\vec{G}(t) = \frac{1}{2}\sum_{\vec{k}}\sum_{\sigma=1,2} \hbar\vec{k}\left[a^*_{\vec{k}\sigma}(t)a_{\vec{k}\sigma}(t) + a_{\vec{k}\sigma}(t)a^*_{\vec{k}\sigma}(t)\right].$$

(e) The form of the above Hamiltonian and the time dependence of the quantities $a_{\vec{k}\sigma}(t)$ and $a^*_{\vec{k}\sigma}(t)$ suggest to consider each mode of the electromagnetic field as a harmonic oscillator. When we quantize the field in the Heisenberg picture $a^*_{\vec{k}\sigma}(t)$ and $a_{\vec{k}\sigma}(t)$ become the time dependent creation and annihilation operators $\hat{a}^\dagger_{\vec{k}\sigma}(t)$ and $\hat{a}_{\vec{k}\sigma}(t)$ of the respective mode and H becomes the Hamilton operator. One assumes the modes to be independent and postulates for $\hat{a}^\dagger_{\vec{k}\sigma}(t)$ and $\hat{a}_{\vec{k}\sigma}(t)$ the commutator relations

$$\left[\hat{a}_{\vec{k}\sigma}(t),\hat{a}_{\vec{k}'\sigma'}(t)\right] = \left[\hat{a}^\dagger_{\vec{k}\sigma}(t),\hat{a}^\dagger_{\vec{k}'\sigma'}(t)\right] = 0, \quad \left[\hat{a}_{\vec{k}\sigma}(t),\hat{a}^\dagger_{\vec{k}'\sigma'}(t)\right] = \delta_{\vec{k}\vec{k}'}\delta_{\sigma\sigma'}.$$

Show that then the Heisenberg equation of motion

$$\frac{d\hat{a}_{\vec{k}\sigma}}{dt} = \frac{i}{\hbar}\left[\hat{H},\hat{a}_{\vec{k}\sigma}\right] \qquad \frac{d\hat{a}^\dagger_{\vec{k}\sigma}}{dt} = \frac{i}{\hbar}\left[\hat{H},\hat{a}^\dagger_{\vec{k}\sigma}\right]$$

are identical to the classical equations of motion for $a^*_{\vec{k}\sigma}$ and $a_{\vec{k}\sigma}$.

(f) Express the operators $\hat{\vec{A}}(\vec{x},t)$, $\hat{\vec{E}}(\vec{x},t)$, and $\hat{\vec{B}}(\vec{x},t)$ for the vector potential and the fields in term of the creation and annihilation operators.

(g) What are the corresponding operators in the Schrödinger picture?

10.5 Field Commutation Relations

In a box with edges L_1, L_2, L_3 the operator for the vector potential can be written in the form

$$\hat{\vec{A}}(\vec{x}) = \sum_{\vec{k}} \sum_{\sigma=1,2} \sqrt{\frac{\hbar}{2\omega_{\vec{k}}\varepsilon_0 V}} \, \vec{e}_{\vec{k}\sigma} \left[\hat{a}_{\vec{k}\sigma} e^{i\vec{k}\vec{x}} + \hat{a}^\dagger_{\vec{k}\sigma} e^{-i\vec{k}\vec{x}} \right]$$

if we assume periodic boundary conditions and Coulomb gauge. (Notation as in the preceding problem.)

(a) What are the operators for $\hat{\vec{D}}(\vec{x})$ and $\hat{\vec{B}}(\vec{x})$?

(b) Show that the vectors $\vec{e}_{\vec{k}\sigma}$ and $\vec{k}$ with the cartesian components $\left(\vec{e}_{\vec{k}\sigma}\right)_i$ fulfill the completeness relation

$$\sum_{\sigma=1,2} \left(\vec{e}_{\vec{k}\sigma}\right)_i \left(\vec{e}_{\vec{k}\sigma}\right)_j + \frac{k_i k_j}{|\vec{k}|^2} = \delta_{ij}.$$

(c) Show that the operators $\hat{\vec{A}}(\vec{x})$, $\hat{\vec{D}}(\vec{x})$, and $\hat{\vec{B}}(\vec{x})$ obey the following commutation relations:

$$\left[\hat{A}_i(\vec{x}), \hat{A}_j(\vec{x}')\right] = 0$$
$$\left[\hat{D}_i(\vec{x}), \hat{D}_j(\vec{x}')\right] = 0$$
$$\left[\hat{B}_i(\vec{x}), \hat{B}_j(\vec{x}')\right] = 0$$
$$\left[\hat{A}_i(\vec{x}), \hat{D}_j(\vec{x}')\right] = -i\hbar \delta^T_{ij}(\vec{x}-\vec{x}')$$
$$\left[\hat{A}_i(\vec{x}), \hat{B}_j(\vec{x}')\right] = 0$$
$$\left[\hat{D}_i(\vec{x}), \hat{B}_j(\vec{x}')\right] = -i\hbar \, \epsilon_{ijk} \frac{\partial}{\partial x_k} \delta(\vec{x}-\vec{x}').$$

Here $\delta^T_{ij}(\vec{x}-\vec{x}')$ is the *transverse* delta function which can be defined by

$$\delta^T_{ij}(\vec{x}-\vec{x}') = \frac{1}{V} \sum_{\vec{k}} \left(\delta_{ij} - \frac{k_i k_j}{|\vec{k}|^2}\right) e^{i\vec{k}(\vec{x}-\vec{x}')}.$$

10.6 Properties of the Transverse δ-Function

The transverse delta function is defined by

$$\delta^T_{ij}(\vec{x}) = \frac{1}{V} \sum_{\vec{k}} \left(\delta_{ij} - \frac{k_i k_j}{|\vec{k}|^2}\right) e^{i\vec{k}\vec{x}}.$$

Show:

(a) $\delta_{ij}^T(\vec{x}) = \delta_{ji}^T(\vec{x})$, $\delta_{ij}^T(-\vec{x}) = \delta_{ij}^T(\vec{x})$

(b) $\sum_j \frac{\partial}{\partial x_j}\delta_{ij}^T(\vec{x}) = \sum_i \frac{\partial}{\partial x_i}\delta_{ij}^T(\vec{x}) = 0$

(c) $\delta_{ij}^T(\vec{x}) = \delta_{ij}\delta(\vec{x}) + \frac{1}{4\pi}\frac{\partial^2}{\partial x_i \partial x_j}\frac{1}{|\vec{x}|}$ for the transition to the continuum.

References

Quantization of the Field

The first quantum theory of radiation was developed by
M. Born and **P. Jordan**, *Zur Quantenmechanik*, Z. Physik **34**, 858–889 (1925)
and
M. Born, **W. Heisenberg** and **P. Jordan**, *Zur Quantenmechanik II*, Z. Physik **35**, 557–615 (1925)

The goals of these papers were to take over the classical formula for the emission of radiation of a dipole into the quantum theory and to evaluate the energy fluctuations in a field of black-body radiation, respectively.

The field of quantum electrodynamics started with
P.A.M. Dirac, *The Quantum Theory of the Emission and Absorption of Radiation*, Proc. Roy. Soc. A **114**, 243–265 (1927)

This paper is reprinted in
J. Schwinger, *Selected Papers on Quantum Electrodynamics*, Dover Publications, New York, 1958

For a particular nice treatment of quantum theory of radiation emphasizing the role of the mode function we refer to the classic paper
E. Fermi, *Quantum theory of radiation*, Rev. Mod. Phys. **4**, 87–132 (1932).

The material for this article came from a 1930 Theoretical Physics Summer School at the university of Michigan at Ann Arbor. In this paper Fermi used the quantum theory of radiation to treat Lippmann fringes. Moreover he showed that the radiation emitted by one atom and absorbed by another traveled with the speed of light. This problem has attracted recently again a lot of interest, because strictly speaking this model predicts instantaneous propagation as shown by
G.C. Hegerfeldt, *Causality problems for Fermi's two-atom system*, Phys. Rev. Lett. **72**, 596–599 (1994)

For the quantization of longitudinal fields and the method of Gupta-Bleuler see
J.M. Jauch and **F. Rohrlich**, *The Theory of Photons and Electrons*, Addison-Wesley, New York, 1955
C. Itzykson and **J.B. Zuber**, *Quantum Field Theory*, McGraw-Hill, New York, 1966

For a detailed discussion of the commutation relations of the field operators and the transverse delta function see for example
W.H. Louisell, *Quantum Statistical Properties of Radiation*, Wiley, New York, 1973

W. Vogel and D.-G. Welsch, *Lectures on Quantum Optics*, Akademie Verlag, Berlin, 1994

C. Cohen-Tannoudji, J. Dupont-Roc and G. Grynberg, *Photons and Atoms – Introduction to Quantum Electrodynamics*, Wiley, New York, 1989

L. Mandel and E. Wolf, *Optical Coherence and Quantum Optics*, Cambridge University Press, Cambride, 1995

Resonator Modes

For the discussion of the modes of various resonator configurations see for example
A.E. Siegman, *An Introduction to Lasers and Masers*, McGraw-Hill, New York, 1971

A.E. Siegman, *Lasers*, University Science Books, Mill Valley, 1986

P.W. Milonni and J.H. Eberly, *Lasers*, Wiley, New York, 1988

B.E.A. Saleh and M.C. Teich, *Fundamentals of Photonics*, Wiley, New York, 1991

Limits to the Measurement of the Electromagnetic Field

The question of the measurability of field operators was addressed in
N. Bohr and L. Rosenfeld, *Zur Frage der Meßbarkeit der elektromagnetischen Feldgrößen*, Kgl. Danske Videnskab. Selskab., Mat.-fys. Medd. **12**, no. 8, (1933)

N. Bohr and L. Rosenfeld, *Field and Charge Measurement in Quantum Electrodynamics*, Phys. Rev. **78**, 794–798 (1950)

These papers have been reprinted in
J.A. Wheeler and W.H. Zurek, *Quantum Theory and Measurement*, Princeton University Press, Princeton, 1983

Casimir Effect

For the original paper on the Casimir effect see
H.B.G. Casimir, *On the attraction between two perfectly conducting plates*, Proc. Kon. Ned. Akad. **51**, 793–795 (1948)

H.B.G. Casimir, *Introductory remarks on quantum electrodynamics*, Physica **XIX**, 846–849 (1953)

See also the treatment in
E.A. Power, *Introductory Quantum Electrodynamics*, Longmans, Green and Co., London, 1964

P.W. Milonni, *The Quantum Vacuum: An Introduction to Quantum Electrodynamics*, Academic, New York, 1994

A. Sarlemijn and M.J. Sparnaay, *Physics in the Making: Essays in Honor of H.B.G. Casimir on the Occasion of his 80th Birthday*, North Holland, Amsterdam, 1989

For the treatment of the Casimir effect using the Zeta function see
E. Elizalde, *Ten Physical Applications of Spectral Zeta Functions*, Springer, Berlin, 1995

The Casimir force was investigated for various arrangements of conducting plates, for example wedges and spheres. For a summary see
T.H. Boyer, *Quantum Electromagnetic Zero-Point Energy of a Conducting Spherical Shell and the Casimir Model for a Charged Particle*, Phys. Rev. **174**, 1764–1776 (1968)
E. Elizalde and **A. Romeo**, *Essentials of the Casimir effect and its computation*, Am. J. Phys. **59**, 711–719 (1991)
G. Barton, *New Aspects of the Casimir Effect: Fluctuations and Radiative Reaction*, in: *Cavity Quantum Electrodynamics*, edited by P.R. Berman, Adv. At. Mol. Opt. Phys. **28** Supplement 2, Academic Press, Boston, 1994
V.M. Mostepanenko and **N.N. Trunov**, *The Casimir Effect and Its Applications*, Oxford University Press, Oxford, 1997

There is an interesting "classical" Casimir effect. Two ships that lie parallel to each other in a harbor attract each other as discussed in
S.L. Boersma, *A maritime analogy of the Casimir effect*, Am. J. Phys. **64**, 539–541 (1996)

The first experiment on the Casimir effect was performed in 1958 in the Phillips Labs. See for example
M.J. Sparnaay, *Measurements of attractive forces between flat plates*, Physica **24**, 751–764 (1958)

For a most recent measurement of the Casimir effect see
S. Lamoreaux, *Demonstration of the Casimir Force in the 0.6 to 6 µm Range*, Phys. Rev. Lett. **78**, 5–8 (1997)

For the attraction of an atom by two parallel plates see
C.I. Sukenik, M.G. Boshier, D. Cho, V. Sandoghdar and **E.A. Hinds**, *Measurement of the Casimir-Polder Force*, Phys. Rev. Lett. **70**, 560–563 (1993)

Photon Wave Function and Localization of a Photon

For the question of the wave function of the photon we refer to the introduction of
M.O. Scully and **M.S. Zubairy**, *Quantum Optics*, Cambridge University Press, Cambridge, 1997

A nice review and novel approach can be found in
I. Bialynicki-Birula, *Photon Wave Function*, Progress in Optics, Volume XXXVI, North Holland, 1996, p. 245–294

For a detailed criticism of the concept of the photon and a nice historical overview see
W.E. Lamb Jr., *Anti-photon*, Appl. Phys. B **60**, 77–84 (1995)

The question of the position operator of a photon has been discussed by
T.D. Newton and **E.P. Wigner**, *Localized States for Elementary Systems*, Rev. Mod. Phys. **21**, 400–406 (1949)

The question of the localization of the photon in space was discussed in
E.R. Pike and **S. Sarkar**, *Spatial Dependence of Weakly Localized Single-Photon Wave Packets*, Phys. Rev. A **35**, 926–928 (1987)

For a more detailed discussion see

E.R. Pike and **S. Sarkar**, *The Quantum Theory of Radiation*, Oxford Science Publications, Oxford, 1995

The idea to use the correlation functions of the electric field operators as wave functions of the photon see

M.O. Scully and **K. Drühl**, *Quantum eraser: A proposed photon correlation experiment concerning observation and "delayed choice" in quantum mechanics*, Phys. Rev. A **25**, 2208–2213 (1982)

M.O. Scully, *Photon-Photon Correlations from Single Atoms*, in: Advances in Quantum Phenomena, edited by E.G. Beltrametti and J.-M. Lévy-Leblond, Plenum Press, New York, 1995

U.W. Rathe and **M.O. Scully**, *Theoretical Basis for a New Subnatural Spectroscopy Via Correlation Interferometry*, Lett. Math. Phys. **34**, 297–307 (1995)

Similar ideas have been pursued in the concept of the bi-photon to express entanglement between different modes by

D.N. Klyshko, *Photons and Nonlinear Optics*, Gordon & Breach, New York, 1988

M.H. Rubin, D.N. Klyshko, Y.H. Shih and **A.V. Sergienko**, *Theory of Two-photon Entanglement in Type-II Optical Parametric Down-conversion*, Phys. Rev. A **50**, 5122–5133 (1994)

A.V. Sergienko, Y.H. Shih and **M.H. Rubin**, *Experimental Evaluation of a Two-photon Wave Packet in Type-II Parametric Downconversion*, J. Opt. Soc. Am. **B12**, 859–862 (1995)

M.H. Rubin, *Transverse Correlations in optical spontaneous parametric down-conversion*, Phys. Rev. A **54**, 5349–5360 (1996)

T.E. Keller and **M.H. Rubin**, *Theory of two-photon entanglement for spontaneous parametric down-conversion driven by a narrow pump pulse*, Phys. Rev. A **56**, 1534–1541 (1997)

For another approach towards the wave function see

J.E. Sipe, *Photon Wave Functions*, Phys. Rev. A **52**, 1875–1883 (1995)

11 Field States

The main subject of this book is the interaction of matter with quantized light. The enormous progress of experimental quantum optics allows us to focus on situations where a single atom interacts with a single mode of the radiation field. In Chapter 14 we discuss a simple model for such an interaction. We therefore concentrate now on a single mode and suppress in the remainder of this chapter the mode index ℓ. The electric field operator then reads

$$\hat{\vec{E}}(\vec{r},t) = -\mathcal{E}_0\,\vec{u}(\vec{r})\,\hat{\pi}(t) = \mathcal{E}_0\,\vec{u}(\vec{r})\,\frac{i}{\sqrt{2}}\left(\hat{a}(t) - \hat{a}^\dagger(t)\right). \tag{11.1}$$

In this formulation the electric field is expressed in units of the vacuum electric field $\mathcal{E}_0$. Its spatial dependence enters via the spatial function $\vec{u}(\vec{r})$ which is solely determined by the geometry of the resonator and the boundary conditions, Eqs. (10.24)–(10.25). The operator nature of the electric field manifests itself in the creation and annihilation operators.

In the course of these notes we have already met many different quantum states of a mechanical harmonic oscillator and have discussed their properties. In particular, the states $|n\rangle$ of well-defined energy have turned out to be extremely useful. Moreover, we have also studied coherent and squeezed states. These states reappear in the context of the radiation field. In the present chapter we therefore return to a discussion of their properties. Here, we focus especially on the coherent states since they allow us to construct the formalism of generalized phase space distribution functions. Moreover, we discuss the non-classical features of Schrödinger cat states arising from the superposition principle of quantum mechanics.

11.1 Properties of the Quantized Electric Field

In the present section we show that the expectation value of the electric field operator in a photon number state vanishes. In contrast the intensity is determined by the number of photons. We then turn to eigenstates of the electric field operator consisting of an appropriate linear combination of creation and annihilation operators. Here the expectation value of the electric field is non-vanishing.

11.1.1 Photon Number States

We start our discussion with the analogue of the energy eigenstates, the so-called photon number states. In particular, we calculate the expectation value of the electric field operator $\hat{\vec{E}}(\vec{r}, t)$ of a single mode in a number state $|n\rangle$.

Average Electric Field

The average field in a number state reads

$$\langle \hat{\vec{E}} \rangle_{|n\rangle} \equiv \langle n| \hat{\vec{E}} |n\rangle = \frac{i}{\sqrt{2}} \mathcal{E}_0 \vec{u}(\vec{r}) \langle n| \hat{a} - \hat{a}^\dagger |n\rangle. \tag{11.2}$$

When we make use of the properties

$$\hat{a}|n\rangle = \sqrt{n}\,|n-1\rangle \quad \text{and} \quad \hat{a}^\dagger|n\rangle = \sqrt{n+1}\,|n+1\rangle$$

of the annihilation and creation operator this expectation value takes the form

$$\langle \hat{\vec{E}} \rangle_{|n\rangle} = \frac{i}{\sqrt{2}} \mathcal{E}_0 \vec{u}(\vec{r}) \left[\sqrt{n}\, \langle n|n-1\rangle - \sqrt{n+1}\, \langle n|n+1\rangle \right]. \tag{11.3}$$

Since the number states are mutually orthogonal we arrive at

$$\langle n| \hat{\vec{E}} |n\rangle = 0,$$

that is the average electric field in a number state vanishes.

Origin of the Vacuum Electric Field

We note, however, that the intensity which is proportional to the square of the electric field, does not vanish in a number state. Indeed, we find

$$\langle \hat{\vec{E}}^2 \rangle_{|n\rangle} \equiv \langle n| \hat{\vec{E}}^2 |n\rangle = -\frac{1}{2} \mathcal{E}_0^2 \vec{u}^2(\vec{r}) \langle n| \hat{a}^2 + \hat{a}^{\dagger 2} - \hat{a}^\dagger \hat{a} - \hat{a} \hat{a}^\dagger |n\rangle.$$

We recall that the operator $\hat{a}^2$ annihilates two quanta creating a number state $|n-2\rangle$ which is orthogonal on $|n\rangle$. Analogously, the operator $\hat{a}^{\dagger 2}$ creates a number state $|n+2\rangle$ which is again orthogonal to $|n\rangle$.

Hence, when we use the commutation relation $[\hat{a}, \hat{a}^\dagger] = 1$ the above equation reduces to

$$\langle n| \hat{\vec{E}}^2 |n\rangle = \frac{1}{2} \mathcal{E}_0^2 \vec{u}^2(\vec{r}) \langle n| 2\hat{n} + 1 |n\rangle = \mathcal{E}_0^2 \vec{u}^2(\vec{r}) \left(n + \frac{1}{2} \right). \tag{11.4}$$

The square of the electric field increases linearly with the number of quanta in the number state.

Since for $n = 0$, that is for the vacuum state $|0\rangle$, we find

$$\langle 0| \hat{\vec{E}}^2 |0\rangle = \frac{1}{2} \mathcal{E}_0^2 \vec{u}^2(\vec{r})$$

11.1 Properties of the Quantized Electric Field

the fluctuations $\langle \hat{\vec{E}}^2 \rangle_{|0\rangle}$ of the electric field, in the vacuum state are apart from the geometrical factor $\vec{u}^2(\vec{r})$, given by $\mathcal{E}_0^2$. This explains why we sometimes refer to $\mathcal{E}_0$ as the vacuum electric field.

We emphasize, however, that the average electric field in the vacuum state vanishes. We can interpret $\mathcal{E}_0$ only as the effective vacuum electric field

$$E_{\text{effec.}}/|\vec{u}(\vec{r})| \equiv \left[\langle 0|\hat{\vec{E}}^2|0\rangle / \vec{u}^2(\vec{r}) \right]^{1/2} = \frac{\mathcal{E}_0}{\sqrt{2}}.$$

This discussion brings to light that the number states $|n\rangle$ have very intuitive properties for the square of the electric field, that is for the intensity. However, the properties of the electric field operator in a number state are counterintuitive as exemplified by the vanishing of the expectation value. This feature just reflects the fact that the number states are eigenstates of the number operator $\hat{n} \equiv \hat{a}^\dagger \hat{a}$ which consists of the *product* of a creation and annihilation operator. In contrast, the electric field operator consists of a *difference* of creation and annihilation operators.

11.1.2 Electromagnetic Field Eigenstates

We now construct eigenstates of the electric or magnetic field operator. An electromagnetic field in such a state has therefore a well-defined amplitude. Since the electric field operator consists of a combination of the annihilation and creation operators, the states are analogous to the quadrature eigenstates discussed in the context of a mechanical oscillator in Sec. 4.4.

Definition of Radiation Field Operator

The electric field operator of a single mode reads

$$\hat{\vec{E}}(\vec{r},t) = \mathcal{E}_0\, \vec{u}(\vec{r})\, \frac{i}{\sqrt{2}} \left(\hat{a}(0)e^{-i\theta} - \hat{a}^\dagger(0)e^{i\theta} \right) \tag{11.5}$$

where $\theta \equiv \Omega t$. Apart from the mode function and the vacuum electric field amplitude it is therefore a linear combination of the annihilation and creation operators with phase factors. This feature suggests to introduce the radiation field operator

$$\hat{\mathcal{E}}_\theta \equiv \frac{1}{\sqrt{2}} \left(\hat{a} e^{-i\theta} + \hat{a}^\dagger e^{i\theta} \right) = \cos\theta\, \hat{x} + \sin\theta\, \hat{\pi}. \tag{11.6}$$

In the last step we have used the definitions Eqs. (10.64) and (10.67) of the dimensionless position and momentum operators $\hat{x}$ and $\hat{\pi}$.

Obviously, $\hat{\mathcal{E}}_\theta$ is a linear combination of $\hat{x}$ and $\hat{\pi}$. In particular, for $\theta = 0$ the operator $\hat{\mathcal{E}}_\theta$ is identical to the dimensionless position operator and hence, it corresponds to the magnetic field operator. For $\theta = \pi/2$ the operator reduces to the dimensionless momentum operator corresponding to the electric field operator.

Photon Number Expansion

The eigenstates $|\mathcal{E}_\theta\rangle$ of the rotated electric field operator satisfy the eigenvalue equation

$$\hat{\mathcal{E}}_\theta |\mathcal{E}_\theta\rangle = \mathcal{E}_\theta |\mathcal{E}_\theta\rangle. \tag{11.7}$$

We now derive an explicit representation of these states by expanding the radiation field eigenstate into photon number eigenstates, that is

$$|\mathcal{E}_\theta\rangle = \sum_{n=0}^{\infty} \langle n|\mathcal{E}_\theta\rangle |n\rangle.$$

In order to find the expansion coefficients $\langle n|\mathcal{E}_\theta\rangle$ we first consider the case $\theta = 0$. Here the operator $\hat{\mathcal{E}}_\theta$ reduces to the position operator of an oscillator. Hence, in complete analogy to the position representation $u_n(x) \equiv \langle x|n\rangle$ of an energy eigenstate, Eqs. (2.17), the expansion coefficients read

$$\langle n|\mathcal{E}_{\theta=0}\rangle \equiv \langle n|x\rangle = \pi^{-1/4} \frac{1}{\sqrt{2^n n!}} H_n(\mathcal{E}_{\theta=0}) \exp\left(-\frac{1}{2}\mathcal{E}_{\theta=0}^2\right).$$

Here H_n denotes the nth Hermite polynomial.

From the definition Eq. (11.6) we recognize that the states $|\mathcal{E}_\theta\rangle$ follow from the states $|\mathcal{E}_{\theta=0}\rangle$ by a phase space rotation of an angle θ. We are therefore led to the ansatz

$$|\mathcal{E}_\theta\rangle = \sum_{n=0}^{\infty} \langle n|x = \mathcal{E}_\theta\rangle e^{in\theta} |n\rangle,$$

that is

$$|\mathcal{E}_\theta\rangle = \pi^{-1/4} \exp\left(-\frac{1}{2}\mathcal{E}_\theta^2\right) \sum_{n=0}^{\infty} \frac{1}{\sqrt{2^n n!}} H_n(\mathcal{E}_\theta) e^{in\theta} |n\rangle \tag{11.8}$$

for the photon number representation of the field eigenstates $|\mathcal{E}_\theta\rangle$.

Indeed we can verify that the so-defined states satisfy the eigenvalue equation Eqs. (11.7) with (11.6). With the help of the well known relation

$$\left(\hat{a}e^{-i\theta} + \hat{a}^\dagger e^{i\theta}\right)|n\rangle = \sqrt{n}\, e^{-i\theta}|n-1\rangle + \sqrt{n+1}\, e^{i\theta}|n+1\rangle$$

the eigenvalue equation Eq. (11.7) takes the form

$$\hat{\mathcal{E}}_\theta |\mathcal{E}_\theta\rangle = \pi^{-1/4} \exp\left(-\frac{1}{2}\mathcal{E}_\theta^2\right)$$
$$\times \sum_{n=0}^{\infty} \frac{1}{\sqrt{2^n n!}} H_n(\mathcal{E}_\theta) e^{in\theta} \frac{1}{\sqrt{2}} \left[\sqrt{n}\, e^{-i\theta}|n-1\rangle + \sqrt{n+1}\, e^{i\theta}|n+1\rangle\right].$$

When we introduce the new summation indices $m \equiv n-1$ and $m \equiv n+1$ we arrive at

$$\hat{\mathcal{E}}_\theta |\mathcal{E}_\theta\rangle = \pi^{-1/4} \exp\left(-\frac{1}{2}\mathcal{E}_\theta^2\right)$$
$$\times \sum_{m=0}^{\infty} \frac{e^{im\theta}}{\sqrt{2^m m!}} |m\rangle \left[\frac{1}{2} H_{m+1}(\mathcal{E}_\theta) + m H_{m-1}(\mathcal{E}_\theta)\right].$$

We recall the recurrence relation

$$H_{m+1}(z) = 2zH_m(z) - 2mH_{m-1}(z)$$

for the Hermite polynomials and find

$$\hat{\mathcal{E}}_\theta \ket{\mathcal{E}_\theta} = \mathcal{E}_\theta \ket{\mathcal{E}_\theta}.$$

Hence the states $\ket{\mathcal{E}_\theta}$ defined by Eq. (11.8) satisfy the eigenvalue equation of the radiation field operator. An electric field in such a state has therefore the well-defined amplitude $\mathcal{E}_\theta$.

For more properties of the radiation field eigenstates we refer to the analogous discussion of quadrature eigenstates of the mechanical oscillator in Sec. 4.4.

11.2 Coherent States Revisited

An electromagnetic field in an eigenstate $\ket{\mathcal{E}_\theta}$ enjoys a well-defined electromagnetic field amplitude $\mathcal{E}_\theta$. The conjugate variable $\mathcal{E}_{\theta+\pi/2}$ is therefore completely uncertain. This fact stands out most clearly for the phase angle $\theta = 0$, where $\hat{\mathcal{E}}_{\theta=0} = \hat{x}$ and $\hat{\mathcal{E}}_{\theta=\pi/2} = \hat{\pi}$ and the uncertainty principle for these operators enforces the relation

$$\Delta \hat{x}\, \Delta \hat{\pi} \geq 1/2$$

for the fluctuations Δx and $\Delta \pi$. Since the fields $\hat{\vec{E}}$ and $\hat{\vec{H}}$ are proportional to the operators $\hat{\pi}$ and $\hat{x}$ and are hence conjugate variables, we have a uncertainty relation

$$\Delta \hat{E}\, \Delta \hat{H} \geq \mathcal{I}$$

for the fields. Here $\mathcal{I}$ is a constant determined by the definitions of the electric and magnetic field operators and the commutation relations of the operators $\hat{\pi}$ and $\hat{x}$.

Consequently, an electromagnetic field in an radiation field eigenstate does not behave in a way predicted by classical electrodynamics.

We are now interested in quantum states which, within the uncertainty relation, would allow for such a simultaneously well-defined electric and magnetic field. Such states are the coherent states. They have been discussed for the first time by E. Schrödinger as special solutions of the time dependent Schrödinger equation of a mechanical harmonic oscillator. Later R. Glauber investigated these states in more detail. They have become the foundations of Quantum Optics. We now briefly review some of their properties.

11.2.1 Eigenvalue Equation

Following R. Glauber we define coherent states $\ket{\alpha}$ as eigenstates of the annihilation operator $\hat{a}$, that is

$$\hat{a}\ket{\alpha} = \alpha\ket{\alpha}. \tag{11.9}$$

Since, $\hat{a}$ is not hermitian the eigenvalues α are not necessarily real. In general they are complex. Moreover, there are no boundary conditions which enforce the spectrum of $\hat{a}$ to be discrete. Hence, α can take on any complex value.

Expansion in Number States

We express the coherent state

$$|\alpha\rangle = \sum_{n=0}^{\infty} w_n |n\rangle, \qquad (11.10)$$

in terms of the photon number states $|n\rangle$. Here w_n denotes the probability amplitude to find n photons in the coherent state $|\alpha\rangle$. We recall that the term *n photons* stands for n *excitations* of that mode.

When we substitute the ansatz Eq. (11.10) into Eq. (11.9) and use

$$\hat{a}|n\rangle = \sqrt{n}\,|n-1\rangle$$

we arrive at

$$\hat{a}|\alpha\rangle = \sum_{n=0}^{\infty} w_n \sqrt{n}\,|n-1\rangle = \sum_{n=0}^{\infty} \alpha w_n |n\rangle.$$

We multiply from the left with $\langle m|$ and find with the help of the orthonormality property of the number states the two-term recurrence relation

$$w_{m+1}\sqrt{m+1} = \alpha w_m,$$

that is

$$w_{m+1} = \frac{\alpha}{\sqrt{m+1}}\, w_m.$$

When we repeatedly apply this formula we arrive at

$$w_m = \frac{\alpha}{\sqrt{m}}\, w_{m-1} = \frac{\alpha}{\sqrt{m}}\,\frac{\alpha}{\sqrt{m-1}}\cdots\frac{\alpha}{1}\, w_0,$$

that is

$$w_m = \frac{\alpha^m}{\sqrt{m!}}\, w_0.$$

Normalization and Special Class of Coherent States

We determine the coefficient w_0 from the normalization condition

$$1 = \langle \alpha | \alpha \rangle = \sum_{n=0}^{\infty} |w_n|^2$$

and find

$$1 = \sum_{n=0}^{\infty} \frac{|\alpha|^{2n}}{n!} |w_0|^2 = e^{|\alpha|^2} |w_0|^2$$

or

$$|w_0| = e^{-\frac{1}{2}|\alpha|^2}.$$

This still leaves the phase φ of

$$w_0 = |w_0| e^{i\varphi(\alpha)}$$

undetermined. Indeed, φ could also depend in a complicated way on the parameter α. However, in the standard definition of coherent states φ is chosen to be zero. Hence, with this choice of the phase a coherent state $|\alpha\rangle$ of complex-valued parameter α reads

$$|\alpha\rangle = e^{-\frac{1}{2}|\alpha|^2} \sum_{n=0}^{\infty} \frac{\alpha^n}{\sqrt{n!}} |n\rangle. \tag{11.11}$$

In the next section we show that this definition is consistent with the alternative definition of a coherent state as a displaced vacuum.

11.2.2 Coherent State as a Displaced Vacuum

In Sec. 4.2 we have introduced the coherent state as a displaced ground state. We now want to formulate this concept in an exact mathematical way. For this purpose we define the coherent state by the relation

$$|\alpha\rangle \equiv \hat{D}(\alpha)|0\rangle \equiv e^{\alpha \hat{a}^\dagger - \alpha^* \hat{a}}|0\rangle, \tag{11.12}$$

where $\hat{D}(\alpha)$ is the displacement operator.

We first convince ourselves that this definition of a coherent state is equivalent to the one based on the eigenvalue equation Eq. (11.9) of $\hat{a}$. To this end we use the Baker-Hausdorff formula

$$e^{\hat{A}+\hat{B}} = e^{\hat{A}} e^{\hat{B}} e^{-[\hat{A},\hat{B}]/2}$$

valid for two operators $\hat{A}$ and $\hat{B}$ which satisfy the commutations relations

$$[\hat{A},[\hat{A},\hat{B}]] = [\hat{B},[\hat{A},\hat{B}]] = 0.$$

We note that the annihilation and creation operators $\hat{a}$ and $\hat{a}^\dagger$ satisfy this condition since $[\hat{a},\hat{a}^\dagger] = 1$.

When we make use of the Baker-Hausdorff theorem we find from the definition of the coherent state Eq. (11.12) the formula

$$|\alpha\rangle = e^{\alpha \hat{a}^\dagger} e^{-\alpha^* \hat{a}} e^{-\frac{1}{2}[\alpha \hat{a}^\dagger, -\alpha^* \hat{a}]}|0\rangle,$$

that is

$$|\alpha\rangle = e^{-\frac{1}{2}|\alpha|^2} e^{\alpha \hat{a}^\dagger} e^{-\alpha^* \hat{a}}|0\rangle.$$

We then expand the exponential of $\alpha^* \hat{a}$ in powers of $\hat{a}$. The relation $\hat{a}|0\rangle = 0$ implies

$$e^{-\alpha^* \hat{a}}|0\rangle = (1 - \alpha^* \hat{a} + ...)|0\rangle = |0\rangle$$

and therefore

$$|\alpha\rangle = e^{-\frac{1}{2}|\alpha|^2} e^{\alpha \hat{a}^\dagger}|0\rangle. \tag{11.13}$$

When we now expand the operator $e^{\alpha \hat{a}^\dagger}$ we arrive at

$$|\alpha\rangle = e^{-\frac{1}{2}|\alpha|^2} \sum_{n=0}^{\infty} \frac{\alpha^n}{n!} \hat{a}^{\dagger n}|0\rangle = e^{-\frac{1}{2}|\alpha|^2} \sum_{n=0}^{\infty} \frac{\alpha^n}{n!} \sqrt{n!}|n\rangle,$$

or
$$|\alpha\rangle = e^{-\frac{1}{2}|\alpha|^2} \sum_{n=0}^{\infty} \frac{\alpha^n}{\sqrt{n!}} |n\rangle$$

in agreement with the representation Eq. (11.11) of the coherent state in terms of photon number states.

We conclude by emphasizing that the two definitions are only identical because we have chosen the phase of w_0 to be zero.

11.2.3 Photon Statistics of a Coherent State

In Sec. 4.2.2 we have discussed the Poissonian energy distribution of a coherent state of a mechanical oscillator. In particular, we have identified its origin by evaluating the overlap between the appropriate wave functions in position space. In the present section we focus on a single mode of the radiation field in a coherent state. In this case the photon statistics

$$W_n = |w_n|^2 = |\langle n|\alpha\rangle|^2 = \frac{|\alpha|^{2n}}{n!} e^{-|\alpha|^2},$$

is the probability $W_n \equiv |w_n|^2$ to find n photons in a coherent state.

Moments of Photon Distribution

We now calculate the average number

$$\langle \hat{n} \rangle \equiv \langle \alpha | \hat{n} | \alpha \rangle = \sum_{n=0}^{\infty} \langle \alpha | \hat{n} | n \rangle \langle n | \alpha \rangle = \sum_{n=0}^{\infty} n |\langle \alpha | n \rangle|^2 = \sum_{n=0}^{\infty} n W_n$$

of photons in a coherent state $|\alpha\rangle$. When we introduce the expression

$$\langle \hat{n} \rangle = \sum_{n=0}^{\infty} n \frac{|\alpha|^{2n}}{n!} e^{-|\alpha|^2} = |\alpha|^2 \sum_{n=1}^{\infty} \frac{|\alpha|^{2(n-1)}}{(n-1)!} e^{-|\alpha|^2}$$

the new summation index $\bar{n} \equiv n - 1$ and perform the summation we arrive at

$$\langle n \rangle = |\alpha|^2.$$

Hence, the parameter $|\alpha|^2$ is the average number of photons in this state.

In order to find the width of the Poisson distribution defined by the variance

$$\sigma^2 \equiv \langle \hat{n}^2 \rangle - \langle \hat{n} \rangle^2$$

we now calculate the second moment

$$\langle \hat{n}^2 \rangle \equiv \langle \alpha | \hat{n}^2 | \alpha \rangle = \sum_{n=0}^{\infty} \langle \alpha | \hat{n}^2 | n \rangle \langle n | \alpha \rangle = \sum_{n=0}^{\infty} n^2 W_n = \sum_{n=1}^{\infty} \frac{n-1+1}{n-1} \frac{|\alpha|^{2n}}{(n-2)!} e^{-|\alpha|^2},$$

that is,

$$\langle \hat{n}^2 \rangle = |\alpha|^4 \sum_{n=2}^{\infty} \frac{|\alpha|^{2(n-2)}}{(n-2)!} e^{-|\alpha|^2} + |\alpha|^2 \sum_{n=1}^{\infty} \frac{|\alpha|^{2(n-1)}}{(n-1)!} e^{-|\alpha|^2}.$$

11.2 Coherent States Revisited

When we introduce the summation indices $\bar{\bar{n}} \equiv n - 2$ and $\bar{n} \equiv n - 1$, respectively, and perform the summations we find

$$\langle \hat{n}^2 \rangle = |\alpha|^4 + |\alpha|^2. \tag{11.14}$$

Therefore, the variance of the Poissonian photon distribution of a coherent state reads

$$\sigma = \sqrt{|\alpha|^2} = \sqrt{\langle \hat{n} \rangle}.$$

For increasing average photon number the maximum located at $\langle n \rangle = |\alpha|^2$ and the width $\sigma = |\alpha|$ of the Poisson distribution increase. However, their ratio

$$\left(\frac{\text{width of Poisson distribution}}{\text{mean of Poisson distribution}}\right) \equiv \frac{\sigma}{\langle n \rangle} = \frac{1}{\sqrt{\langle \hat{n} \rangle}} \xrightarrow{\sqrt{\langle \hat{n} \rangle} \to \infty} 0$$

goes to zero with the square root of the average number of photons.

Variance from Operator Algebra

We conclude this section by discussing another possibility for calculating the average number $\langle \hat{n} \rangle$ of photons and the variance. Here we use the eigenvalue equation Eq. (11.9) of the coherent state rather than the summation over the photon statistics. This brings out most clearly the way quantum mechanics enters into the second moment.

When we make use of the eigenvalue equation Eq. (11.9) of the coherent states we find for the average number of photons

$$\langle \hat{n} \rangle = \langle \alpha | \hat{a}^\dagger \hat{a} | \alpha \rangle = \alpha^* \alpha \langle \alpha | \alpha \rangle = |\alpha|^2$$

where we have also used the formulas $\hat{a} | \alpha \rangle = \alpha | \alpha \rangle$ and $\langle \alpha | \hat{a}^\dagger = \alpha^* \langle \alpha |$.

The second moment then reads

$$\langle \hat{n}^2 \rangle = \langle \alpha | \hat{a}^\dagger \hat{a} \hat{a}^\dagger \hat{a} | \alpha \rangle = \langle \alpha | \hat{a}^\dagger (\hat{a}^\dagger \hat{a} + [\hat{a}, \hat{a}^\dagger]) \hat{a} | \alpha \rangle = \langle \alpha | \hat{a}^\dagger (\hat{a}^\dagger \hat{a} + 1) \hat{a} | \alpha \rangle,$$

or

$$\langle \hat{n}^2 \rangle = \langle \alpha | \hat{a}^{\dagger\, 2} \hat{a}^2 + \hat{a}^\dagger \hat{a} | \alpha \rangle = |\alpha|^4 + |\alpha|^2$$

in agreement with Eq. (11.14).

Hence, the nonzero variance $\sigma^2 \equiv \langle \hat{n}^2 \rangle - \langle \hat{n} \rangle^2 = |\alpha|^2$ of the photon distribution arises through the application of the commutation relation in the evaluation of the second moment of the photon number operator.

11.2.4 Electric Field Distribution of a Coherent State

We now calculate the probability distribution

$$W_{|\beta\rangle}(\mathcal{E}_\theta) \equiv |\langle \mathcal{E}_\theta | \beta \rangle|^2$$

to find the radiation field eigenstate $|\mathcal{E}_\theta\rangle$ in the coherent state $|\beta\rangle$. Hence we are left with the problem of calculating the scalar product between the radiation field state $|\mathcal{E}_\theta\rangle$ and the coherent state $|\beta\rangle$.

Scalar Product

For this purpose we use the number state representation

$$|\mathcal{E}_\theta\rangle = \pi^{-1/4} \exp\left(-\frac{1}{2}\mathcal{E}_\theta^2\right) \sum_{n=0}^{\infty} \frac{1}{\sqrt{2^n n!}} H_n(\mathcal{E}_\theta) e^{in\theta} |n\rangle \qquad (11.15)$$

of the radiation field eigenstate $|\mathcal{E}_\theta\rangle$, Eq. (11.8) and find

$$\langle \mathcal{E}_\theta | \beta \rangle = \pi^{-1/4} \exp\left(-\frac{1}{2}\mathcal{E}_\theta^2\right) \sum_{n=0}^{\infty} \frac{1}{\sqrt{2^n n!}} H_n(\mathcal{E}_\theta) e^{-in\theta} \langle n | \beta \rangle,$$

which with the help of the photon number probability amplitudes

$$\langle n | \beta \rangle = \frac{\beta^n}{\sqrt{n!}} \exp\left(-\frac{1}{2}|\beta|^2\right)$$

of the coherent state takes the form

$$\langle \mathcal{E}_\theta | \beta \rangle = \pi^{-1/4} \exp\left(-\frac{1}{2}\mathcal{E}_\theta^2\right) \sum_{n=0}^{\infty} \frac{\left(\beta e^{-i\theta}/\sqrt{2}\right)^n}{n!} H_n(\mathcal{E}_\theta) e^{-\frac{1}{2}|\beta|^2}.$$

The generating function

$$\sum_{n=0}^{\infty} \frac{z^n}{n!} H_n(\xi) = \exp(-z^2 + 2\xi z)$$

of the Hermite polynomials finally yields the expression

$$\langle \mathcal{E}_\theta | \beta \rangle = \pi^{-1/4} \exp\left[-\frac{1}{2}\mathcal{E}_\theta^2 - \frac{1}{2}\left(\beta e^{-i\theta}\right)^2 + \sqrt{2}\mathcal{E}_\theta \beta e^{-i\theta} - \frac{1}{2}|\beta|^2\right] \qquad (11.16)$$

for the scalar product of the electric field eigenstate and a coherent state.

Nonrotated Eigenstates

Of special importance are the probability amplitudes $\langle \mathcal{E}_{\theta=0} | \beta \rangle$. For the phase angle $\theta = 0$ the rotated electric field operator reduces to the position operator $\hat{x}$ and the expression Eq. (11.16) for the probability amplitude takes the form

$$\langle \mathcal{E}_{\theta=0} | \beta \rangle = \pi^{-1/4} \exp\left[-\frac{1}{2}\mathcal{E}_\theta^2 - \frac{1}{2}\beta^2 + \sqrt{2}\mathcal{E}_\theta \beta - \frac{1}{2}|\beta|^2\right],$$

or

$$\langle \mathcal{E}_{\theta=0} | \beta \rangle \equiv \langle x | \beta \rangle = \pi^{-1/4} \exp\left[\frac{1}{2}\left(\beta^2 - |\beta|^2\right)\right] \cdot \exp\left[-\frac{1}{2}\left(\mathcal{E}_{\theta=0} - \sqrt{2}\beta\right)^2\right]. \qquad (11.17)$$

Thus, the corresponding distribution is a Gaussian centered at $\mathcal{E}_{\theta=0} = \sqrt{2}\beta$. Only for real values of β the prefactor of the Gaussian cancels. In particular, when we consider in Sec. 11.3 the superposition of two different coherent states these phase factors become important.

Probability Distribution

We conclude this section by presenting the expression for the probability to find the electric field eigenstate in a coherent state. For this purpose we evaluate the absolute value squared of the above scalar product which reads

$$|\langle \mathcal{E}_\theta | \beta \rangle|^2 = \pi^{-1/2} \exp\left\{-\left[\mathcal{E}_\theta^2 + \frac{1}{2}\left(\beta e^{-i\theta}\right)^2 + \frac{1}{2}\left(\beta^* e^{i\theta}\right)^2 - \sqrt{2}\mathcal{E}_\theta\left(\beta e^{-i\theta} + \beta^* e^{i\theta}\right) + |\beta|^2\right]\right\}.$$

When we combine the terms in the exponent we find

$$W_{|\beta\rangle}(\mathcal{E}_\theta) = |\langle \mathcal{E}_\theta | \beta \rangle|^2 = \pi^{-1/2} \exp\left\{-\left[\mathcal{E}_\theta - \frac{1}{\sqrt{2}}\left(\beta e^{-i\theta} + \beta^* e^{i\theta}\right)\right]^2\right\}.$$

Hence a coherent state enjoys a Gaussian electric field distribution.

11.2.5 Over–completeness of Coherent States

Coherent states form an over-complete set, that is they are complete but not orthogonal. This leads to some interesting expansion properties which we now consider in more detail.

Completeness of Coherent States

We start our discussion by first proving that the coherent states are complete. For this purpose we study the operator

$$\hat{I} \equiv \frac{1}{\pi} \int d^2\alpha |\alpha\rangle\langle\alpha|.$$

We recall that the parameter $\alpha = |\alpha|e^{i\theta}$ describing the coherent state is a complex number which has a real and an imaginary part. The differential $d^2\alpha$ thus reads

$$d^2\alpha = d(\operatorname{Re}\alpha)\, d(\operatorname{Im}\alpha) = d\alpha_r\, d\alpha_i = |\alpha|\, d|\alpha|\, d\theta.$$

When we substitute the number state representation, Eq. (11.11), of the coherent state into the operator $\hat{I}$ we find

$$\hat{I} = \sum_{m,n=0}^{\infty} \frac{1}{\sqrt{m!n!}} \int_0^\infty d|\alpha|\, |\alpha|^{m+n+1} e^{-|\alpha|^2} \frac{1}{\pi} \int_{-\pi}^{\pi} d\theta\, e^{-i(m-n)\theta} |n\rangle\langle m|.$$

With the help of the relation

$$\frac{1}{2\pi} \int_{-\pi}^{\pi} d\theta\, e^{im\theta} = \delta_{m,0} \qquad (11.18)$$

we can perform the integration over θ and after introducing the integration variable $\xi \equiv |\alpha|^2$ the operator $\hat{I}$ takes the form

$$\hat{I} = \sum_{n=0}^{\infty} \frac{1}{n!} \int_0^{\infty} d|\alpha|^2 \, |\alpha|^{2n} e^{-|\alpha|^2} |n\rangle\langle n| = \frac{1}{n!} \int_0^{\infty} d\xi \, \xi^n e^{-\xi} |n\rangle\langle n|.$$

We recall the integral formula

$$\int_0^{\infty} d\xi \, \xi^n e^{-\xi} = \Gamma(n+1) = n!$$

for the Gamma function Γ and arrive at the completeness relation for Fock states

$$\hat{I} = \sum_{n=0}^{\infty} |n\rangle\langle n| = \mathbb{1},$$

where $\mathbb{1}$ denotes the identity operator.

Hence we have found a representation of the identity operator in terms of coherent states, namely

$$\mathbb{1} = \frac{1}{\pi} \int d^2\alpha \, |\alpha\rangle\langle\alpha|. \tag{11.19}$$

Nonorthogonality

In the preceding section we have derived the completeness relation for the coherent states. We now show that they are not orthogonal on each other.

To analyze this feature in more detail we consider two coherent states

$$|\alpha\rangle = e^{-\frac{1}{2}|\alpha|^2} \sum_{m=0}^{\infty} \frac{\alpha^m}{\sqrt{m!}} |m\rangle$$

and

$$|\beta\rangle = e^{-\frac{1}{2}|\beta|^2} \sum_{n=0}^{\infty} \frac{\beta^n}{\sqrt{n!}} |n\rangle.$$

When we make use of the orthogonality of the number states we find

$$\langle\alpha|\beta\rangle = e^{-\frac{1}{2}(|\alpha|^2+|\beta|^2)} \sum_{n=0}^{\infty} \frac{(\alpha^*\beta)^n}{n!},$$

or

$$\langle\alpha|\beta\rangle = e^{-\frac{1}{2}(|\alpha|^2+|\beta|^2)+\alpha^*\beta}. \tag{11.20}$$

We therefore arrive at

$$|\langle\alpha|\beta\rangle|^2 = e^{-|\alpha-\beta|^2}, \tag{11.21}$$

where we have used

$$-|\alpha|^2 - |\beta|^2 + \alpha^*\beta + \alpha\beta^* = -|\alpha-\beta|^2.$$

Hence the overlap between two coherent states is given by a Gaussian. The larger the separation the smaller the overlap.

11.2.6 Expansion into Coherent States

A rather remarkable consequence of this nonorthogonality of the coherent states is the possibility to expand a coherent state $|\alpha\rangle$ into a set of coherent states. Moreover, we can expand photon number states into coherent states, either in the complete complex plane or along a circle. This is the topic of the present section.

Coherent State Expanded into Coherent States

In order to find the corresponding expansion we multiply a coherent state by the completeness relation Eq. (11.19) and find

$$|\alpha\rangle = \mathbb{1}\,|\alpha\rangle = \frac{1}{\pi} \int d^2\beta\, |\beta\rangle\langle\beta|\alpha\rangle,$$

which when we apply Eq. (11.20) reads

$$|\alpha\rangle = \frac{1}{\pi} \int d^2\beta\, e^{-\frac{1}{2}(|\alpha|^2+|\beta|^2)+\alpha\beta^*}\,|\beta\rangle.$$

According to this formula we represent the coherent state $|\alpha\rangle$ by a continuous superposition of coherent states $|\beta\rangle$ over the whole space spanned by β_r and β_i. Each coherent state $|\beta\rangle$ brings in a weight factor

$$e^{-\frac{1}{2}(|\alpha|^2+|\beta|^2)+\alpha\beta^*} = e^{-\frac{1}{2}|\alpha-\beta|^2}\, e^{\frac{1}{2}(\alpha\beta^*-\alpha^*\beta)} \equiv e^{-\frac{1}{2}|\alpha-\beta|^2}\, e^{i\phi(\beta;\alpha)}$$

where $\phi(\beta;\alpha) \equiv \operatorname{Im}(\alpha\beta^*)$. Hence, the weight factor consists of a Gaussian which measures the separation of the coherent state $|\beta\rangle$ from the state to be expanded and the phase ϕ is the triangular area defined by the origin of complex space and the two complex numbers α and β.

Fock State into Plane of Coherent States

Due to the completeness of the coherent states it is obvious that a number state can be expanded in coherent states. Indeed, we find by inserting the completeness relation Eq. (11.19) the expression

$$|n\rangle = \frac{1}{\pi}\int d^2\beta\,|\beta\rangle\langle\beta|n\rangle = \frac{1}{\pi}\int d^2\beta\,\frac{e^{-\frac{1}{2}|\beta|^2}}{\sqrt{n!}}\,(\beta^*)^n|\beta\rangle,$$

which with $\beta \equiv |\beta|e^{i\theta}$ takes the form

$$|n\rangle = 2\int_0^\infty d|\beta|\,|\beta|^{n+1}\frac{e^{-\frac{1}{2}|\beta|^2}}{\sqrt{n!}}\frac{1}{2\pi}\int_{-\pi}^{\pi} d\theta\, e^{-in\theta}||\beta|e^{i\theta}\rangle.$$

This is a rather complicated representation of the number state as a superposition of coherent states over the whole complex space. Here we have parameterized complex space by circles of continuous radius. We emphasize, that the coherent states along a circle of fixed radius have a fixed phase relation $e^{in\theta}$ which depends on the number state $|n\rangle$. In addition to this superposition along a circle we have a superposition of many of these circles.

Fock State into Circle of Coherent States

However, due to the over-completeness of the coherent states the integration over the radius $|\beta|$ of the circles is not really necessary. We can derive the representation

$$|n\rangle = \mathcal{N}_n(|\beta|) \frac{1}{2\pi} \int_{-\pi}^{\pi} d\theta \, e^{-in\theta} \left||\beta|e^{i\theta}\right\rangle$$

of the number state which only involves coherent states along a single circle of arbitrary radius. We find the normalization constant $\mathcal{N}_n(|\beta|)$ by expanding the coherent state

$$\left||\beta|e^{i\theta}\right\rangle = e^{-\frac{1}{2}|\beta|^2} \sum_{m=0}^{\infty} \frac{|\beta|^m}{\sqrt{m!}} e^{im\theta} |m\rangle.$$

into number states and substituting this expression into the representation of $|n\rangle$ which yields

$$|n\rangle = \mathcal{N}_n e^{-\frac{1}{2}|\beta|^2} \sum_{m=0}^{\infty} \frac{|\beta|^m}{\sqrt{m!}} \frac{1}{2\pi} \int_{-\pi}^{\pi} d\theta \, e^{i(m-n)\theta} |m\rangle = \mathcal{N}_n e^{-\frac{1}{2}|\beta|^2} \frac{|\beta|^n}{\sqrt{n!}} |n\rangle,$$

that is

$$\frac{1}{\mathcal{N}_n} = e^{-\frac{1}{2}|\beta|^2} \frac{|\beta|^n}{\sqrt{n!}}.$$

Here we have used the integral formula Eq. (11.18).

We therefore arrive at the representation

$$|n\rangle = \frac{1}{2\pi} \int_{-\pi}^{\pi} d\theta \left[\frac{|\beta|^{2n} e^{-|\beta|^2}}{n!}\right]^{-1/2} e^{-in\theta} \left||\beta|e^{i\theta}\right\rangle$$

of the nth number state as a superposition of coherent states along a circle.

The radius of this circle is still arbitrary. This is counterintuitive when we recall from Sec. 8.2 the most elementary representation of a number state as a Planck-Bohr-Sommerfeld band in phase space. This band is centered around a radius $\sqrt{2(n+1/2)}$ in phase space.

We note, that a similar emphasis on a particular phase space radius β takes place in the above representation in the semiclassical limit, that is for $n \gg 1$. Indeed, in this case we recall from the asymptotic expansion of the Poisson distribution performed in Sec. 4.2 that the maximum of the expression in square brackets occurs at $|\beta| = \sqrt{n}$. The difference of a factor of $\sqrt{2}$ reflects the fact that in the present analysis we deal with α_r-α_i phase space rather than with x-p phase space.

The fact that the radius $|\beta|$ is still arbitrary although the asymptotic expansion of the Poissonian suggests a particular value reflects the over-completeness of the coherent states.

11.2.7 Electric Field Expectation Values

In Sec. 11.1.1 we have shown that the expectation value of the electric field operator in a number state vanishes. We now perform the analogous calculation for a coherent state.

From the electric field operator

$$\hat{\vec{E}}(\vec{r}, t) = \mathcal{E}_0 \, \vec{u}(\vec{r}) \, \frac{i}{\sqrt{2}} \left[\hat{a}(t) - \hat{a}^\dagger(t) \right]$$

we find with the help of the eigenvalue equation $\hat{a}|\alpha\rangle = \alpha|\alpha\rangle$ of the coherent state

$$\langle \hat{\vec{E}}(\vec{r}, t) \rangle = \mathcal{E}_0 \, \vec{u}(\vec{r}) \, \frac{i}{\sqrt{2}} \left[\alpha(t) - \alpha^*(t) \right] = -\sqrt{2} \, \mathcal{E}_0 \, \vec{u}(\vec{r}) \, \operatorname{Im} \alpha(t),$$

or

$$\langle \hat{\vec{E}}(\vec{r}, t) \rangle = -\sqrt{2} \, \mathcal{E}_0 \, \vec{u}(\vec{r}) \, |\alpha| \sin \theta(t). \tag{11.22}$$

This is the expression for a classical electric field with amplitude $\sqrt{2}\mathcal{E}_0 |\alpha| \vec{u}(\vec{r})$ and phase θ. In contrast to the expectation value in a number state this one is not vanishing.

We now consider the second moment

$$\langle \hat{\vec{E}}^2 \rangle = -\frac{1}{2} \mathcal{E}_0^2 \, \vec{u}^2(\vec{r}) \, \langle \alpha | \hat{a}^2 + \hat{a}^{\dagger 2} - 2\hat{a}^\dagger \hat{a} - 1 | \alpha \rangle$$

$$= -\frac{1}{2} \mathcal{E}_0^2 \, \vec{u}^2(\vec{r}) \left(\alpha^2 + \alpha^{*2} - 2\alpha^* \alpha - 1 \right)$$

of the electric field operator in a coherent state, that is

$$\langle \hat{\vec{E}}^2 \rangle = -\frac{1}{2} \mathcal{E}_0^2 \, \vec{u}^2(\vec{r}) \left[(\alpha - \alpha^*)^2 - 1 \right]. \tag{11.23}$$

From this equation and the expression Eq. (11.22) for $\langle \vec{E} \rangle$ we find for the variance

$$\sigma_E^2 \equiv \langle \hat{\vec{E}}^2 \rangle - \langle \hat{\vec{E}} \rangle^2$$

of the electric field

$$\sigma_E^2 = \frac{1}{2} \mathcal{E}_0^2 \, \vec{u}^2(\vec{r}).$$

Hence the fluctuations of the electric field in a coherent state are independent of the complex amplitude α. We emphasize that this result is quite different from the corresponding one for the number state: According to Eq. (11.4) the fluctuations are proportional to the number of quanta n contained in the state. However, for the case of the vacuum, that is for $n = 0$, the fluctuations in the vacuum and in the coherent state are identical: A coherent state only contains the vacuum fluctuations. Therefore, coherent states are closest to classical physics.

11.3 Schrödinger Cat State

The superposition principle lies at the very heart of quantum mechanics. As P.A.M. Dirac formulated it in the first chapter of his classical treatise "... any two or more states may be superposed to give a new state." Nowhere clearer comes the power of this principle to light than in the superposition of two coherent states. Indeed a single coherent state is a pseudo-classical state, that is a state closest to classical physics. However, a quantum mechanical superposition of two coherent states can already display highly non-classical properties as we show in the present section.

11.3.1 The Original Cat Paradox

In 1935 A. Einstein, B. Podolsky and N. Rosen published their seminal paper criticizing quantum mechanics by asking the question: Is quantum mechanics complete? This motivated E. Schrödinger to summarize the formalism of quantum mechanics in his, by now classic, paper published in the *Naturwissenschaften*. In this article he put forward the notation of entanglement of two quantum systems and described a Gedankenexperiment with a cat.

A box, which is opaque to the outside observer, contains a cat and a vial of poisonous gas. A mechanism can release a hammer which destroys the glass with the gas. This kills the cat. The mechanism for the release of the hammer is triggered by a radioactive decay and is hence determined by chance. After the experiment is set up and many half-times of radioactive decay have passed by, there exist two possibilities: When the hammer has not fallen down and the glass is still intact, the cat is still alive. This is summarized by a two system state vector $|a, i\rangle$ where a denotes the *alive* state of the cat and i the *intact* glass. In the other alternative the hammer has broken the glass and the cat is dead. This state is described by the vector $|d, b\rangle$ where d represents the *dead* cat and b the *broken* glass.

Since cat and glass are in the opaque box the outside observer is not included in this two-system state vector. Both alternatives exist simultaneously. Hence, the state vector Ψ of the two component system, cat and glass, is the quantum mechanical superposition of both state vectors. Their relative amplitudes are determined by the radioactive process. In the most elementary way we can write a superposition with equal weights, that is

$$|\Psi\rangle = \frac{1}{\sqrt{2}} \left(|a, i\rangle + |d, b\rangle \right).$$

Loosely speaking, the cat is in a superposition of two macroscopically distinguishable states, namely *dead* and *alive*. We can represent this state of the cat alone by

$$|\psi_c\rangle = \frac{1}{\sqrt{2}} \left(|a\rangle + |d\rangle \right).$$

Here, the paradox is not that the cat can be dead or alive. However, the paradox consists of the fact that such a macroscopic object as a cat is simultaneously dead *and* alive reflecting the interference between the two states $|a\rangle$ and $|d\rangle$.

However, the state $|\Psi\rangle$ given by Schrödinger is not this superposition state $|\psi_c\rangle$. The state $|\Psi\rangle$ is more complicated. It also involves the state of the glass. It is not a one-particle state, but a two-particle state. Moreover, the alive cat is correlated with the intact glass and the dead cat with the broken glass. This state carries the name *entangled* state.

11.3.2 Definition of the Field Cat State

In the context of the Jaynes-Cummings-Paul model describing the interaction of a two-level atom with a quantized light field we return to the entanglement of two quantum systems, namely atomic states and field states.

However, in the present context we confine ourselves to a single mode and define a Schrödinger cat state as the superposition of two macroscopically distinguishable states. We choose two coherent states that have two utterly different amplitudes or phases. When additionally they have large amplitudes, this quantum state involves two classically distinguishable states, such as dead or alive. Moreover, we emphasize that S. Haroche and his group have realized experimentally the states discussed theoretically in this section. We briefly outline his preparation scheme in Sec. 16.3.1, but focus now on the properties of these states.

We define the Schrödinger cat state

$$|\psi\rangle \equiv \mathcal{N}\frac{1}{\sqrt{2}}\left(|\alpha e^{i\varphi}\rangle + |\alpha e^{-i\varphi}\rangle\right) \tag{11.24}$$

as the quantum-mechanical superposition of two coherent states, $|\alpha e^{\pm i\varphi}\rangle$, of real, positive displacement $\sqrt{2}\,\alpha$ in dimensionless x-p oscillator phase space and real phase, $\pm\varphi$.

Since two coherent states are not orthogonal, the normalization constant $\mathcal{N}$ in Eq. (11.24) is not unity. In order to calculate it we recall from Eq. (11.20) the scalar product

$$\langle\beta|\gamma\rangle = \exp[-\tfrac{1}{2}(|\beta|^2 + |\gamma|^2) + \beta^*\gamma]$$

between two coherent states $|\beta\rangle$ and $|\gamma\rangle$.

With this expression we find for the normalization constant

$$\mathcal{N}^2(\varphi) = \frac{1}{1 + \cos[\alpha^2 \sin(2\varphi)]\exp(-p_\varphi^2)},$$

where

$$|p_\varphi| \equiv |\pm\sqrt{2}\,\alpha\sin\varphi)|.$$

In the definition of the state $|\psi\rangle$, Eq. (11.24), we have separated the normalization factor $1/\sqrt{2}$ which arises naturally in the superposition of two *orthogonal* states from the factor $\mathcal{N}$ measuring the *nonorthogonality* of the two interfering coherent states.

11.3.3 Wigner Phase Space Representation

In a most elementary way we can represent the state $|\psi\rangle$ by two circles in phase space located at the points with polar coordinates $(\sqrt{2}\,\alpha, \pm\varphi)$ as shown in Fig. 11.1. A more sophisticated treatment which brings out most clearly the consequences of the interference of the two states relies on the Wigner function

$$W_{|\psi\rangle}(x,p) \equiv \frac{1}{2\pi}\int_{-\infty}^{\infty}dy\, e^{-ipy}\,\psi^*(x-\tfrac{1}{2}y)\,\psi(x+\tfrac{1}{2}y) \tag{11.25}$$

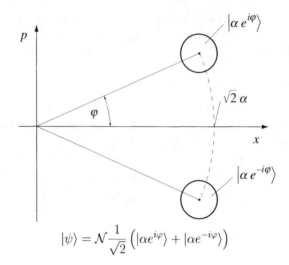

$$|\psi\rangle = \mathcal{N}\frac{1}{\sqrt{2}}\left(|\alpha e^{i\varphi}\rangle + |\alpha e^{-i\varphi}\rangle\right)$$

Fig. 11.1: In its most elementary version the quantum mechanical superposition of two coherent states of mean quantum number $\langle n \rangle = \alpha^2$ and phase difference φ can be visualized by two circles of radius unity displaced by an amount $\sqrt{2}\,\alpha$ from the origin and having the angle φ between them.

discussed in Chapter 3. Here $\psi(x) \equiv \langle x | \psi \rangle$ is the x representation of the state $|\psi\rangle$.

Since according to Eq. (11.17) the x-representation of a coherent state $|\beta\rangle$ takes the form

$$\langle x | \beta \rangle = \pi^{-1/4} \exp[\tfrac{1}{2}(\beta^2 - |\beta|^2)] \exp[-\tfrac{1}{2}(x - \sqrt{2}\,\beta)^2]$$

we obtain for the position representation of the state $|\psi\rangle$, Eq. (11.24), the formula

$$\psi(x) = \pi^{-1/4} \mathcal{N} \tfrac{1}{\sqrt{2}} \exp[-\alpha^2 \sin^2 \varphi] \exp[\tfrac{i}{2}\alpha^2 \sin(2\varphi)]$$
$$\times \left\{ \exp\left[-\tfrac{1}{2}\left(x - \sqrt{2}\,\alpha e^{i\varphi}\right)^2\right] + \exp[-i\alpha^2 \sin(2\varphi)] \exp\left[-\tfrac{1}{2}\left(x - \sqrt{2}\,\alpha e^{-i\varphi}\right)^2\right] \right\}.$$

When we substitute this expression into the definition, Eq. (11.25), of the Wigner function and perform the integration, we arrive after minor algebra at

$$W_{|\psi\rangle} = \frac{1}{2}\mathcal{N}^2 \left(W_{|\alpha \exp(i\varphi)\rangle} + W_{|\alpha \exp(-i\varphi)\rangle} + W_{\text{int}} \right). \tag{11.26a}$$

Here

$$W_{|\alpha \exp(\pm i\varphi)\rangle} \equiv \frac{1}{\pi} \exp\left[-\left(x - \sqrt{2}\,\alpha \cos\varphi\right)^2 - \left(p \mp \sqrt{2}\,\alpha \sin\varphi\right)^2\right] \tag{11.26b}$$

denotes the Wigner function of a single coherent state of displacement α and phase $\pm\varphi/2$. The interference term

$$W_{\text{int}}(x,p) \equiv \frac{2}{\pi} \cos\left[2\sqrt{2}\,\alpha \sin\varphi\,\left(x - \tfrac{1}{2}\sqrt{2}\,\alpha \cos\varphi\right)\right]$$
$$\times \exp\left[-\left(x - \sqrt{2}\,\alpha \cos\varphi\right)^2 - p^2\right] \tag{11.26c}$$

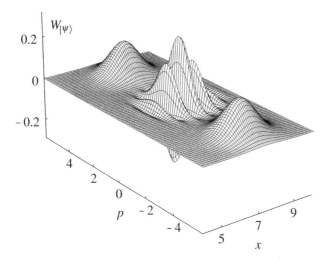

Fig. 11.2: A more sophisticated representation of the quantum mechanical superposition of two coherent states relies on the Wigner function. The latter does not only consist of two Gaussian bells located in x-p oscillator phase space at $x = \sqrt{2}\,\alpha \cos\varphi$ and $p = \pm\sqrt{2}\,\alpha \sin\varphi$ corresponding to two individual coherent states $|\alpha e^{i\varphi}\rangle$ and $|\alpha e^{-i\varphi}\rangle$ but involves an interference term located on the x axis. This contribution originates from the quantum mechanical superposition of the two coherent states and the bilinearity of the Wigner distribution Eq. (11.25) in the wave function. This interference bell can be narrower in the x-direction than the individual coherent state Gaussian bells giving rise to squeezing in the x-variable. It can even take on negative values and create in this way an oscillatory photon statistics. Here we have chosen $\alpha = 6$ and $\varphi = \pi/6$.

arises as a consequence of the bilinearity of the Wigner distribution, Eq. (11.24), in the wave function. Therefore, $W_{|\psi\rangle}$ is not the sum of the two Wigner functions, $W_{|\alpha\exp(\pm i\varphi)\rangle}$, Eq. (11.26b), of the two individual coherent states but involves the Gaussian bell W_{int}, Eq. (11.26c), located at the positive axis at $x = \sqrt{2}\,\alpha \cos\varphi$ and modulated by the oscillatory function of phase

$$2\phi_x = 2\sqrt{2}\,\alpha \sin\varphi\,(x - \tfrac{1}{2}\sqrt{2}\,\alpha \cos\varphi).$$

We note from Eqs. (11.26) that the local widths of the three peaks in the variable p are identical and equal to unity. Moreover, they are independent of φ. The same holds true for the width in the x-direction of $W_{|\alpha\exp(+i\varphi)\rangle}$.

In contrast the width of the interference term W_{int}, Eq. (11.26c) in x, strongly depends on φ. This is a consequence of the $\cos(2\phi_x)$ modulation. For appropriate values of φ, it gets narrower than the coherent state Gaussian bell, Eq. (11.26b), that is, it causes a squeezing of the Gaussian. In a quite different φ region, however, domains in phase space exist in which W_{int} takes on negative as well as positive values – ditches in phase space, as illustrated in Fig. 11.2. These Wigner wave crests and troughs are the origin of these non-classical features of the superposition state $|\psi\rangle$.

11.3.4 Photon Statistics

We now turn to the discussion of the probability W_m to find m quanta in this superposition state in its dependence on the phase difference 2φ between the two states. In particular, we show that there are various domains, in which the photon distribution is narrower than a Poissonian. In this case we call the statistics sub-Poissonian statistics. When the distribution is broader than a Poissonian, we call it super-Poissonian. We show that for the present example of the superposition of two coherent states with identical amplitudes but different phases there exist also domains of phase angles where the photon statistics is super-Poissonian, but displays oscillatory behavior. This is a consequence of interference in phase space discussed in Chapter 7 and is thus analogous to the oscillatory photon statistics of highly squeezed state.

Complete Description

The probability W_m to find m quanta in $|\psi\rangle$ is given by

$$W_m[|\psi\rangle] \equiv |\langle m|\psi\rangle|^2 = \mathcal{N}^2 \frac{1}{2} \left|\langle m|\alpha e^{i\varphi}\rangle + \langle m|\alpha e^{-i\varphi}\rangle\right|^2. \tag{11.27}$$

With the help of the explicit expression Eq. (4.13)

$$\langle m|\beta\rangle = \frac{\beta^m}{\sqrt{m!}} \exp(-\tfrac{1}{2}|\beta|^2), \tag{11.28}$$

for the photon number probability amplitude of a coherent state we find, that the photon statistics takes the form

$$W_m[|\psi\rangle] = |\mathcal{A}_m^{1/2} \exp(i\phi_m) + \mathcal{A}_m^{1/2} \exp(-i\phi_m)|^2, \tag{11.29}$$

or

$$W_m[|\psi\rangle] = 4\mathcal{A}_m \cos^2 \phi_m. \tag{11.30a}$$

In order to bring out the similarities and differences with the analogous expression Eq. (4.44) for the photon statistics of a highly squeezed state, we have introduced the quantity

$$\mathcal{A}_m \equiv \frac{1}{2} \mathcal{N}^2 \frac{\alpha^{2m}}{m!} e^{-\alpha^2} \tag{11.30b}$$

which up to a normalization factor $\mathcal{N}^2/2$ is the Poissonian photon statistics of a single coherent state. The phase

$$\phi_m \equiv m\varphi, \tag{11.30c}$$

is the area of the wedge-shaped phase space area enclosed by the inner circular edge of the mth photon number state and the two lines connecting the centers of the two coherent states with the origin of phase space.

According to Eqs. (11.27) and (11.28) the superposition of the two coherent states – two contributors of interfering probability amplitudes $\langle m|\alpha e^{\pm i\varphi}\rangle$ – creates the

11.3 Schrödinger Cat State

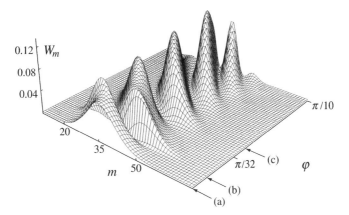

Fig. 11.3: The probability W_m for finding m quanta in the quantum mechanical superposition state $|\psi\rangle$ of Eq. (11.24) in its dependence on the relative phase φ. For increasing φ the Poissonian distribution for the case $\varphi = 0$ (a) narrows while its maximum moves towards smaller m values (b). This curved wave front suddenly breaks off to start a new front and yields a distribution broader than a Poissonian with more than one maximum (c). Here we have chosen $\alpha = 6$.

interference term $\cos^2 \phi_m$ which modulates the familiar Poissonian statistics of a single coherent state.

In Fig. 11.3 we analyze the consequences of this contribution in more detail by depicting the probability W_m of Eqs. (11.30) as a function of quantum number m and the phase φ. All curves here are plotted for definiteness for the same value $\alpha = 6$ of the displacement parameter.

For a vanishing phase angle, that is $\varphi = 0$, the two coherent states are on top of each other, that is, $|\psi\rangle$ is a single coherent state and W_m is a Poisson distribution (Fig. 11.3(a)). When we increase φ the probability distribution narrows having a slightly shifted, higher maximum as shown in Fig. 11.3(b). This narrowing effect stands out most clearly when we compare and contrast the initial Poisson distribution of $\varphi = 0$ (solid line in Fig. 11.4(a) and dashed lines in Fig. 11.4(b–d), to the distribution W_m for the specific φ value indicated in the lower part of Fig. 11.4 by (b).

Normalized Variance

Mathematically, we describe the narrowing effect by the normalized variance

$$\sigma^2 \equiv \frac{\langle m^2 \rangle}{\langle m \rangle} - \langle m \rangle, \tag{11.31}$$

where we have introduced the moments

$$\langle m \rangle \equiv \sum_{m=0}^{\infty} m W_m \quad \text{and} \quad \langle m^2 \rangle \equiv \sum_{m=0}^{\infty} m^2 W_m$$

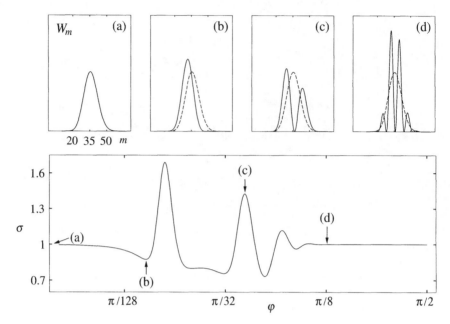

Fig. 11.4: The probability distribution W_m for the quantum mechanical superposition of two coherent states, Eq. (11.24), is (a) Poissonian, (b) sub-Poissonian, (c) super-Poissonian, or (d) oscillatory depending on the relative phase φ between the two coherent states. The Poissonian distribution for $\varphi = 0$ is plotted for comparison by a dashed line. The various domains of the phase angle in which the photon statistics shows these different behaviors come out most clearly in the normalized variance σ, Eq. (11.31) shown in the bottom. To emphasize the oscillations in σ, we have chosen a logarithmic scale for φ. Here we have chosen $\alpha = 6$.

of the distribution W_m, Eq. (11.30). Values $\sigma < 1$ define sub-Poisson statistics, whereas $\sigma > 1$ indicate super-Poisson statistics. For $\sigma = 1$ the variance is identical to that of a Poissonian distribution.

For the evaluation of these moments we refer to Problem 11.8 and present here only the results. The lower part of Fig. 11.4 depicts the so-calculated variance σ as a function of the phase φ for fixed displacement $\alpha = 6$. For $\varphi = 0$, that is for a coherent state with Poisson statistics, we find from Eq. (11.31) $\sigma(\varphi = 0) = 1$. A phase φ such that $2(\alpha\varphi)^2 \ll 1$ leads to

$$\sigma^2 \cong 1 - 2(\alpha\varphi)^2.$$

Consequently, there exists a range of φ values, shown in the lower part of Fig. 11.4, in which the distribution W_m of $|\psi\rangle$ exhibits a substantial amount of sub-Poissonian statistics. This is a remarkable result when we recall that the transition from the Poisson distribution of a coherent state to the sub-Poissonian was induced solely by superposing two coherent states. This example illustrates in a striking way the power of the superposition principle.

When we increase φ further, the first "wave front" of Fig. 11.3 bends to the left and abruptly a second wave train breaks off, giving rise to the two peaks in

W_m. As a result, the probability distribution is broader than a Poisson distribution, that is super-Poissonian with $\sigma > 1$ as shown in the lower part of Fig. 11.4. For even larger φ, this second wave front again gains height, indicating the recurrence of the narrowing of W_m to a sub-Poissonian distribution. However, this narrowing again gets abruptly interrupted by the sudden breakoff of the third wave front, at the phase value (c) of Fig. 11.3, again leading to super-Poissonian statistics. Hence the oscillations in σ only appear for φ values when the two coherent states are not distinguishable yet, that is,

$$|p_\varphi| = \sqrt{2}\,\alpha \sin \varphi < 1.$$

For larger φ values (in this specific example, $\varphi \geq \pi/8$), the wave fronts align themselves more and more parallel to the φ axis. This results in multipeaked probability distributions W_m, as shown in Fig. 11.4(d), that is, in an oscillatory distribution with a Poissonian envelope. This example demonstrates that $\sigma \cong 1$ does not automatically imply Poisson statistics since the second moment of W_m is not sensitive to the oscillations in W_m.

Interference in Phase Space

For $\varphi \geq \pi/4$ the two coherent states states are distinguishable and the oscillations in W_m merely reflect this fact. These oscillations are analogous to the ones in the quantum statistics of a highly squeezed state. This analogy stands out most clearly when we compare Eqs. (11.30) to (4.43): The two expressions look very similar. They only differ in their definitions of the amplitude $\mathcal{A}_m$ and the phase ϕ_m.

This suggests that there exists a common physical principle. Indeed, the expression, Eq. (11.30), for the photon statistics of the Schrödinger cat state has the same simple interpretation in the language of interference in phase space as the oscillatory photon statistics of a highly squeezed state. We find the probability W_m from the interfering areas of overlap between the mth band of the mth number state and the two Gaussian bells representing the two coherent states. The overlap of the band with a single Gaussian bell gives the Gaussian limit of the Poissonian, that is the Gaussian limit of $\mathcal{A}$, as shown in Sec. 8.3. Since there are two states, there are two overlaps which interfere and the phase difference ϕ_m is determined by the central lines of the two states of interest.

We conclude our discussion of the Schrödinger cat state by stating that the oscillations in σ are very reminiscent of those in the corresponding quantity in the photon statistics of the one-atom maser shown in Fig. 18.4.

Problems

11.1 Creation Operator Acting on a Coherent State

A coherent state is an eigenstate of the annihilation operator. We might wonder what is the result if we apply the creation operator $\hat{a}^\dagger$ on a coherent state. The resulting equation will be very important in the context of the Fokker-Planck equation and the damping of radiation.

Show that
$$\hat{a}^\dagger |\alpha\rangle = \left(\frac{\partial}{\partial \alpha} + \frac{1}{2}\alpha^*\right)|\alpha\rangle.$$

Hint: Use the relation Eq. (11.13) to find
$$\hat{a}^\dagger |\alpha\rangle = \hat{a}^\dagger e^{-\frac{1}{2}\alpha\alpha^*} e^{\alpha \hat{a}^\dagger}|0\rangle$$
and interpret α and α^* as independent variables.

11.2 Eigenvalue Problem for the Creation Operator

Solve the eigenvalue problem
$$\hat{a}^\dagger \varphi(\xi) = \alpha \varphi(\xi)$$
and show that for no complex value of α there exits a normalizable eigenfunction.

11.3 Forced Harmonic Oscillator

The Hamiltonian of the forced harmonic oscillator reads
$$H = \frac{p^2}{2M} + \frac{M\Omega^2}{2}x^2 - K(t)x.$$

(a) Show that this Hamiltonian yields the classical equations of motion of a forced harmonic oscillator. Solve these equations for the initial condition $x(0) \equiv x_0$ and $p(0) \equiv p_0$.

(b) Show that the Hamiltonian of this problem can be cast in the form
$$\hat{H} = \hbar\Omega\left(\hat{a}^\dagger \hat{a} + \frac{1}{2}\right) - \hbar f(t)(\hat{a}^\dagger + \hat{a}),$$
where $\hat{a}^\dagger$ and $\hat{a}$ are the creation and annihilation operators of the harmonic oscillator. Derive the relation between the function $f(t)$ and the force $K(t)$.

(c) Solve the Schrödinger equation
$$i\hbar\frac{d}{dt}|\psi\rangle = \hat{H}|\psi\rangle$$
with the ansatz
$$|\psi(t)\rangle = e^{i\varphi(t)}\exp\left[\alpha(t)\hat{a}^\dagger - \alpha^*(t)\hat{a}\right]|0\rangle,$$
where $\varphi(t)$ and $\alpha(t)$ are time dependent quantities and $|0\rangle$ is the ground state of the harmonic oscillator. How do we have to choose $\varphi(0)$ and $\alpha(0)$ in order for the ansatz to yield the time dependent solution with $|\alpha_0\rangle$ as initial condition?

(d) Show that only for real φ the state $|\psi\rangle$ is normalized. Does the differential equation for φ allow this?

Hint: P. Carruthers and M.M. Nieto (1965).

11.4 Properties of Field States

Consider a single-mode electromagnetic field. Calculate the expectation values of the operators $\hat{a}_1 \equiv (\hat{a} + \hat{a}^\dagger)/2$ and $\hat{n} \equiv \hat{a}^\dagger \hat{a}$ as well as the variances

(a) for the state $\sum_m c_m |m\rangle$ with $\sum_m |c_m|^2 = 1$,

(b) for the Fock state $|m\rangle$,

(c) for the state $\frac{1}{\sqrt{2}}(|m\rangle + e^{i\varphi}|m+1\rangle)$,

(d) for a coherent state.

11.5 Bogoliubov Transformation and Squeezed States

Consider the operators

$$\hat{A} \equiv \cosh\varepsilon\, \hat{a} + \sinh\varepsilon\, \hat{a}^\dagger$$
$$\hat{A}^\dagger \equiv \cosh\varepsilon\, \hat{a}^\dagger + \sinh\varepsilon\, \hat{a},$$

where ε is real.

(a) Show

$$\left[\hat{A}, \hat{A}^\dagger\right] = 1.$$

(b) Express $\hat{a}$ and $\hat{a}^\dagger$ in terms of $\hat{A}$ and $\hat{A}^\dagger$.

(c) Determine the eigenstates of $\hat{A}$

$$\hat{A}|\epsilon, \gamma\rangle = \gamma|\epsilon, \gamma\rangle$$

Hint: The calculation is most easily performed in the position representation where in scaled variables we have

$$\hat{a} = \frac{1}{\sqrt{2}}\left(x + \frac{d}{dx}\right) \qquad \hat{a}^\dagger = \frac{1}{\sqrt{2}}\left(x - \frac{d}{dx}\right).$$

(d) Calculate for a squeezed state the expectation values of the operators $\hat{a}_1 \equiv (\hat{a}+\hat{a}^\dagger)/2$ and $\hat{a}_2 \equiv (\hat{a}-\hat{a}^\dagger)/(2i)$, as well as their variances. Compare the squeezed state with a coherent state.

(e) Show the completeness relation

$$\frac{1}{\pi}\int |\gamma, \epsilon\rangle\langle\gamma, \epsilon|\, d^2\gamma = 1.$$

11.6 Interfering Coherent States Build a Number State

Show that a Fock state $|n\rangle$ can be expressed as a superposition of coherent states $|\alpha\rangle$ via

$$|n\rangle = \frac{\sqrt{n!}}{2\pi i} \oint_C \frac{e^{|\alpha|^2/2}}{\alpha^{n+1}} |\alpha\rangle \, d\alpha,$$

where C is an arbitrary integration path which circumvents the origin.

11.7 Completeness of Coherent States

Given a sequence of coherent states $|\alpha_k\rangle$ ($k = 0, 1, 2, \ldots, \infty$) with at least one accumulation point. We want to show that the states $|\alpha_k\rangle$ are complete. In order to prove this theorem we assume that the states $|\alpha_k\rangle$ are *not* complete and construct a contradiction.

(a) Show that there exists a normalizable state $|\psi\rangle \neq 0$ with

$$\langle \psi | \alpha_k \rangle = 0, \, k = 0, 1, 2, \ldots$$

if the states $|\alpha_k\rangle$ are not complete.

(b) Show that the function

$$f(\alpha) = e^{|\alpha|^2/2} \langle \psi | \alpha \rangle$$

is an entire function with $f(\alpha_k) = 0$.

(c) Show that from this it follows that $f(\alpha) \equiv 0$.

(d) Show that because of

$$\frac{1}{\pi} \int |\alpha\rangle\langle\alpha| \, d^2\alpha = 1$$

we also have

$$\frac{1}{\pi} \int e^{-|\alpha|^2} |f(\alpha)|^2 \, d^2\alpha = \langle \psi | \psi \rangle \neq 0.$$

11.8 Properties of an Amplitude-Cat State

We consider the Schrödinger cat state

$$|\psi\rangle \equiv \mathcal{N} \frac{1}{\sqrt{2}} \left(|\alpha e^{i\varphi}\rangle + |\alpha e^{-i\varphi}\rangle \right)$$

discussed in Sec. 11.3.

(a) Derive the Wigner function, Eqs. (11.26), of the Schrödinger cat.
Hint: To perform the integration use the formula

$$\pi^{-1/2} \int_{-\infty}^{\infty} dx \, x^n \, e^{-(x-z)^2} = \begin{cases} 1 & \text{for } n = 0 \\ z & \text{for } n = 1 \\ z^2 + \frac{1}{2} & \text{for } n = 2. \end{cases}$$

(b) Calculate the probability distributions for the quadrature variables x and p by integrating the Wigner function over the conjugate variables. Determine the parameter regimes for which squeezing of the quadrature variables exists.

(c) Derive the first two moments

$$\langle m \rangle = \mathcal{N}^2 \alpha^2 \left\{ 1 + \cos\left[\alpha^2 \sin(2\varphi) + 2\varphi\right] \exp\left(-2\alpha^2 \sin^2 \varphi\right) \right\}$$

and

$$\langle m^2 \rangle = \mathcal{N}^2 \alpha^2 \left\{ 1 + \cos\left(\alpha^2 \sin(2\varphi) + 2\varphi\right) \exp\left(-2\alpha^2 \sin^2 \varphi\right) \right. \\ \left. + \alpha^2 \left[1 + \cos\left(\alpha^2 \sin(2\varphi) + 4\varphi\right) \exp\left(-2\alpha^2 \sin^2 \varphi\right)\right] \right\}$$

of the photon distribution Eq. (11.30) corresponding to the Schrödinger cat state.

(d) Use these expressions to show that for $2(\alpha\varphi)^2 \ll 1$ the normalized variance σ^2, Eq. (11.31) reduces to

$$\sigma^2 \cong 1 - 2(\alpha\varphi)^2.$$

Hence, for appropriately small angles φ the state shows sub-Poissonian statistics and is therefore amplitude squeezed. This has led to the name *amplitude-cat*.

11.9 Phase Distribution of a Phase-Cat

The superposition state

$$|\psi\rangle \equiv \frac{\mathcal{N}}{\sqrt{2}} \left(|\alpha - \delta\alpha\rangle + |\alpha + \delta\alpha\rangle\right)$$

consisting of two coherent states of the same phase but different amplitudes $\alpha \pm \delta\alpha$ exhibits also interesting properties. In particular, depending on the separation $2\delta\alpha$ of the two coherent states the phase distribution can be narrower than that of a single coherent state. This has led to the name *phase-cat*. Calculate the London phase distribution discussed in Sec. 8.5 for this state.

Hint: See Schaufler et al., (1994)

References

Electromagnetic Field Eigenstates

Electromagnetic field eigenstates are discussed in
M. Schubert and **W. Vogel**, *Field Fluctuations of Two-Photon Coherent States*, Phys. Lett. **68A**, 321–322 (1978)
B. Yurke, D.F. Walls and **W.P. Schleich**, *Quantum Superpositions Generated by Quantum Nondemolition Measurements*, Phys. Rev. A **42**, 1703–1711 (1988)

Coherent States

For the original paper on a non-dispersive wave packet in a harmonic oscillator see
E. Schrödinger, *Der stetige Übergang von der Makro- zur Mikromechanik*, Naturwissenschaften **14**, 664–666 (1926),

E.H. Kennard, *Zur Quantenmechanik einfacher Bewegungstypen*, Z. Physik **44**, 326–352 (1927)

For a nice historical account of coherent states and non-dispersive wave packets see
M.M. Nieto, *The Discovery of Squeezed States – in 1927*, in: *Fifth International Conference on Squeezed States and Uncertainty Relations*, edited by D. Han, J. Janszky, Y.S. Kim and V.I. Man'ko, NASA Center for AeroSpace Information, Hanover, MD, 1997

The formalism of coherent states as the foundation of quantum optics was set up in
R.J. Glauber, *The Quantum Theory of Optical Coherence*, Phys. Rev. **130**, 2529–2539 (1963)

R.J. Glauber, *Coherent and Incoherent States of the Radiation Field*, Phys. Rev. **131**, 2766–2788 (1963)

R.J. Glauber, in: Quantum Optics and Electronics, Les Houches, edited by C. DeWitt, A. Blandin and C. Cohen-Tannoudji, Gordon and Breach, New York, 1965, pp. 331–381

R.J. Glauber, *Quantum Optics*, Proceedings of the 10th Session of Scottish Universities Summer School in Physics 1969, Academic Press, London, 1970

More on coherent states can also be found in
J.R. Klauder, *The Action Option and a Feynman Quantization of Spinor Fields in Terms of Ordinary c-numbers*, Ann. Phys. **11**, 123–168 (1960)

J.R. Klauder and **E.C.G. Sudarshan**, *Fundamentals of Quantum Optics*, Bendjamin, New York, 1968

P. Carruthers and **M.M. Nieto**, *Coherent States and the Forced Quantum Oscillator*, Am. J. Phys. **33**, 537–544 (1965)

A. Perelomov, *Generalized Coherent States and Their Applications*, Springer, Heidelberg, 1986

Schrödinger Cats

The original paper describing the cat paradox appeared in a series of papers by
E. Schrödinger, *Die gegenwärtige Situation der Quantenmechanik*, Naturwissenschaften **23**, 807–812, 823–828, 844–849 (1935)

The english translation by J.D. Trimmer
E. Schrödinger, *The Present Situation in Quantum Mechanics*, Proc. Am. Philos. Soc. **124**, 323–338 (1980)

is reprinted in
J.A. Wheeler and **W.H. Zurek**, *Quantum Theory and Measurement*, Princeton University Press, Princeton, 1983

The properties of a superposition of two coherent states of identical amplitudes but different phases are summarized in
W.P. Schleich, F. Le Kien and **M. Pernigo**, *Non-Classical States from Two Pseudo-Classical States*, Phys. Rev. A **44**, 2172–2187 (1991)

For the superposition of two coherent states of identical phases but different amplitudes see
S. Schaufler, M. Freyberger and **W.P. Schleich**, *The birth of a phase-cat*, J. Mod. Opt. **41**, 1765-1779 (1994)

Schrödinger cat states of field or mechanical oscillators have been realized experimentally in high-Q cavity, ion or Rydberg atom experiments, see
M. Brune, E. Hagley, J. Dreyer, X. Maitre, A. Maali, C. Wunderlich, J.M. Raimond and **S. Haroche**, *Observing the Progressive Decoherence of the "Meter" in a Quantum Measurement*, Phys. Rev. Lett. **77**, 4887–4891 (1996)
C. Monroe, D.M. Meekhof, B.E. King and **D.J. Wineland**, *A Schrödinger Cat Superposition State of an Atom*, Science **272**, 1131–1135 (1996)
M.W. Noel and **C.R. Stroud Jr.**, *Young's double-slit interferometry within an atom*, Phys. Rev. Lett. **75**, 1252-1255 (1995)

12 Phase Space Functions

In the last chapter we have discussed various examples of states of the electromagnetic field. Many of their properties stand out, when we visualize these states in phase space. The question, however, is in which phase space.

In classical mechanics we can associate the dynamics of a system with the dynamics of a distribution evolving in a phase space. For a system with a single degree of freedom two conjugate variables, such as position and momentum in the case of a mechanical oscillator, or electric and magnetic field in the case of a field oscillator span this phase space.

In quantum mechanics, however, Heisenberg's uncertainty principle prevents the notion of a system being characterized by a point in phase space. Only domains of minimum area $2\pi\hbar$ in phase space are allowed. It is therefore surprising that nevertheless a phase space approach towards quantum mechanics exists. We have already discussed extensively the Wigner function approach. In the present chapter we show that coherent states allow us to define other phase space distributions in quantum mechanics.

12.1 There is more than Wigner Phase Space

On first sight it is surprising to realize that there are many quantum phase space distribution functions, especially when we realize that there is an infinite amount of such functions. We wonder what their use is? In the earlier chapters we have mainly used the Wigner function to illustrate the properties of a given quantum state. In the present section we briefly discuss the use of generalized phase space distributions to calculate quantum mechanical averages and make the connection to the Wigner function.

12.1.1 Who Needs Phase Space Functions?

The uncertainty principle always looms in the back. All phase space distributions have some features quite different from a classical phase space distribution. For example, the Wigner function can assume negative values. In contrast, the Q-function

$$Q(\alpha) \equiv \frac{1}{\pi} \langle \alpha | \hat{\rho} | \alpha \rangle,$$

introduced in this chapter as the expectation value of the density operator in a coherent state, is always positive. However, it does not allow to calculate the marginals, that is, the probability distributions of one of the conjugate variables by integrating the phase space distribution over the other variable. Nevertheless it is very useful, because it allows us to calculate other quantum mechanical expectation values in a very convenient way, as we will show.

Another phase space distribution based on coherent states is the P-distribution. R.J. Glauber and E.C.G. Sudarshan have introduced this function independently from each other at the same time. We have therefore decided to call this phase space function Glauber-Sudarshan P-distribution where we subscribe to alphabetical order. The P-distribution can become highly singular, since it can involve derivatives of delta functions. Due to these alien features of quantum phase space distributions we refer to them as *pseudo*-probability distributions.

Moreover, each of these distributions refers to a specific choice of operator ordering. Indeed, the Q-, Wigner- or P-distributions are associated with antinormal, symmetric or normal ordering, respectively: We can evaluate expectation values of quantum mechanical operators as in statistical mechanics using a distribution function provided we have first ordered the operators appropriately.

To understand this we recall that a classical distribution function $P^{(cl)}(x,p)$ allows us to calculate averages of a function $O(x,p)$ of two conjugate variables by averaging them with the help of this distribution, that is

$$\langle O(x,p) \rangle = \int dx \int dp \, O(x,p) \, P^{(cl)}(x,p). \tag{12.1}$$

In quantum mechanics we now calculate the expectation value $\langle \hat{O}(\hat{x},\hat{p}) \rangle$ of the operator $\hat{O}$ using a quantum phase space distribution in a way analogous to the classical one. However, since the two operators $\hat{x}$ and $\hat{p}$ do not commute, it is not clear how we have to translate the operator $\hat{O}$ into the c-number representation appearing in the above integral. Obviously, we have to invoke some operator ordering. For every specific ordering there exists a given phase space distribution function as to obtain always the correct quantum mechanical expectation value. This statement is the major topic of the present chapter.

12.1.2 Another Description of Phase Space

So far we have based our discussion of the phase space of a mechanical oscillator of mass M and frequency Ω on the position x and the momentum p. However, in the context of the electromagnetic field oscillator, and especially in view of the coherent states, it is now advantageous to use the phase space variables $\alpha_r \equiv \operatorname{Re}\alpha$ and $\alpha_i \equiv \operatorname{Im}\alpha$, where α is the complex eigenvalue of the annihilation operator $\hat{a}$. Indeed, the Q- and the P-distribution are formulated in terms of α_r and α_i instead of x and p.

We can immediately make the transition between the two sets of variables when we recall the standard definition

$$x \equiv \sqrt{\frac{\hbar}{2M\Omega}}(\alpha + \alpha^*) = \frac{1}{\kappa}\sqrt{2}\,\alpha_r \tag{12.2a}$$

and
$$p \equiv \frac{1}{i}\sqrt{\frac{M\hbar\Omega}{2}}(\alpha - \alpha^*) = \hbar\kappa\sqrt{2}\,\alpha_i \qquad (12.2b)$$

of x and p in terms of complex-valued amplitudes α_r and α_i. Here we have used $\kappa \equiv \sqrt{M\Omega/\hbar}$. We have already met a special case of Eq. (12.2) in Eq. (10.47) where the mass of the field oscillator is set to unity.

Moreover, the phase space volume $dx \cdot dp$ is related to the corresponding one $d\alpha_r \cdot d\alpha_i$ via
$$dx \cdot dp = 2\hbar\, d\alpha_r \cdot d\alpha_i. \qquad (12.3)$$

These relations allow us a straightforward comparison of the various phase space distributions to the Wigner function.

For example, in Chapter 4 we have introduced the Wigner function
$$W(x,p) = \frac{1}{\pi\hbar}\exp\left[-\left(\kappa x - \sqrt{2}\,\alpha_{0r}\right)^2 - \left(\frac{p}{\hbar\kappa} - \sqrt{2}\,\alpha_{0i}\right)^2\right].$$

of a coherent state of displacement $\alpha_0 \equiv \alpha_{0r} + i\alpha_{0i}$. When the state is squeezed we have found the Wigner function
$$W(x,p) = \frac{1}{\pi\hbar}\exp\left[-s\left(\kappa x - \sqrt{2}\,\alpha_{0r}\right)^2 - \frac{1}{s}\left(\frac{p}{\hbar\kappa} - \sqrt{2}\,\alpha_{0i}\right)^2\right].$$

where $s > 0$ describes the squeezing parameter. For $s = 1$ we obtain the coherent state.

When we now make the transition to α-phase space these expressions take the forms
$$W(\alpha_r, \alpha_i) = \frac{2}{\pi}\exp\left[-2(\alpha_r - \alpha_{0r})^2 - 2(\alpha_i - \alpha_{0i})^2\right] \qquad (12.4a)$$
and
$$W(\alpha_r, \alpha_i) = \frac{2}{\pi}\exp\left[-2s(\alpha_r - \alpha_{0r})^2 - \frac{2}{s}(\alpha_i - \alpha_{0i})^2\right]. \qquad (12.4b)$$

Here we have used the transformations, Eqs. (12.2) and (12.3), of the variables and of the phase space volume, respectively.

Hence, in these dimensionless variables the widths $\Delta\alpha_r$ and $\Delta\alpha_i$ of the squeezed state defined by the contour line of exponential decay of the Wigner function read
$$\Delta\alpha_r^{(W)} \equiv \sqrt{\frac{1}{2s}} \quad \text{and} \quad \Delta\alpha_i^{(W)} \equiv \sqrt{\frac{s}{2}}.$$

The two uncertainties are up to a factor 2 the inverse of each other. In the next section we show that this is not the case when we represent the squeezed state using the Q-function.

12.2 The Husimi-Kano Q-Function

In Chapter 3 we have introduced the Wigner function as the visualization of a quantum state. We have also shown, that this distribution lives in phase space spanned by the phase space variables position and momentum. In the case of the electromagnetic field these variables are electric field and magnetic field.

However, the Wigner function is not the only phase space distribution. There exists an infinite family of distribution functions. In the present section we introduce the so-called Q-function, which has the nice property that it is positive everywhere in phase space. We first define the Q-function and illustrate it for various examples.

12.2.1 Definition of Q-Function

We define the Q-function of a pure quantum state $|\psi\rangle$ by

$$Q(\alpha_r, \alpha_i) \equiv \frac{1}{\pi} |\langle \alpha | \psi \rangle|^2. \tag{12.5}$$

We recall that two real numbers α_r and α_i (or $|\alpha|$, θ) describe the coherent state $|\alpha\rangle$. Hence the quantity Q depends on these two variables, that is α_r and α_i span the phase space of the Q-function.

The alternative form

$$Q(\alpha_r, \alpha_i) = \frac{1}{\pi} \langle \alpha | \psi \rangle \langle \psi | \alpha \rangle$$

of the above definition together with the representation

$$\hat{\rho} \equiv |\psi\rangle\langle\psi| \tag{12.6}$$

of a pure state $|\psi\rangle$ as a density operator leads immediately to the generalization

$$Q(\alpha_r, \alpha_i) \equiv \frac{1}{\pi} \langle \alpha | \hat{\rho} | \alpha \rangle \tag{12.7}$$

of the Q-function (12.5) to a mixed state described by a density operator $\hat{\rho}$.

Hence the Q-function is the expectation value of the *density operator* in a coherent state.

12.2.2 Q-Functions of Specific Quantum States

In the present section we calculate and discuss the Q-functions of specific quantum states. Moreover, we compare them to the corresponding Wigner functions.

Coherent State

Our first example is a coherent state $|\alpha_0\rangle$. When we use the nonorthogonality relation, Eq. (11.21), of two coherent states $|\alpha\rangle$ and $|\alpha_0\rangle$ we immediately find

$$Q(\alpha_r, \alpha_i) = \frac{1}{\pi} |\langle \alpha | \alpha_0 \rangle|^2 = \frac{1}{\pi} \exp\left[-|\alpha - \alpha_0|^2\right].$$

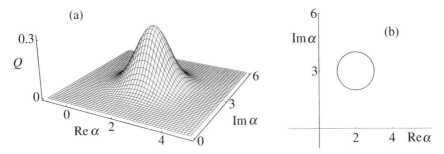

Fig. 12.1: (a) The Q-function of a coherent state $|\alpha_0\rangle$ is a Gaussian bell located at $\alpha_r = \alpha_{0r}$ and $\alpha_i = \alpha_{0i}$. (b) Contour line of the Q-function corresponding to the $1/e$ decay. Here we have chosen $\alpha_0 = 2 + 3i$.

The relation

$$|\alpha - \alpha_0|^2 = |(\alpha_r - \alpha_{0r}) + i(\alpha_i - \alpha_{0i})|^2 = (\alpha_r - \alpha_{0r})^2 + (\alpha_i - \alpha_{0i})^2$$

brings this expression into the form

$$Q(\alpha_r, \alpha_i) = \frac{1}{\pi} \exp\left[-(\alpha_r - \alpha_{0r})^2 - (\alpha_i - \alpha_{0i})^2\right]. \tag{12.8}$$

Thus the Q-function of a coherent state $|\alpha_0\rangle$ is a Gaussian bell located at $\alpha_r = \alpha_{0r}$ and $\alpha_i = \alpha_{0i}$ as shown in Fig. 12.1. Moreover, the Gaussian bell is symmetric: The contour line where the Gaussian has decayed to $1/e$ is a circle of radius one. Sometimes we simplify this representation by just displaying the circular contour line of exponential decay of the Q-function.

We note that the Q-function of the coherent state is always Gaussian independent of the parameters α_{0r} or α_{0i}. Moreover, the radius of the circular contour line is also independent of them. This reflects the fact that the fluctuations of the electric field operator in a coherent state are independent of the displacement α_0: The coherent state is a displaced ground state of a harmonic oscillator. Hence its fluctuations are only determined by the properties of the harmonic oscillator and not by the parameters of the displacement.

We conclude this subsection by noting that this distribution is broader than the corresponding Wigner function, Eq. (12.4).

Number State

We now concentrate on the Q-function of a number state $|n\rangle$. When we substitute the state $|n\rangle$ into the definition, Eq. (12.5), of the Q-function and recall that the photon distribution of a coherent state is a Poisson distribution, we arrive at

$$Q(\alpha_r, \alpha_i) = \frac{1}{\pi} |\langle \alpha | n \rangle|^2 = \frac{1}{\pi} \frac{|\alpha|^{2n}}{n!} e^{-|\alpha|^2}. \tag{12.9}$$

Note that in the discussion of the photon distribution $W_n(|\alpha\rangle)$ we have considered W_n as a function of the quantum number n for a fixed coherent state $|\alpha\rangle$. However in

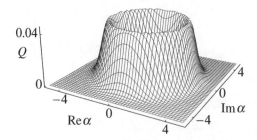

Fig. 12.2: The Q-function of a number state $|n\rangle$ is radially symmetric and does not display a preferred phase. Here we have chosen $n = 8$.

the present context, we keep the quantum number n fixed and consider the Poisson distribution in its dependence on α_r and α_i.

We first note that the Q-function of a number state only involves the absolute value $|\alpha| \equiv \sqrt{\alpha_r^2 + \alpha_i^2}$ of the phase space variable and not its phase θ. Therefore, the Q-function of a number state is radially symmetric as shown in Fig. 12.2.

Vacuum State

A particularly interesting case is the vacuum, which is the number state with $n = 0$. According to Eq. (12.9) the Q-function then reads

$$Q(\alpha_r, \alpha_i) = \frac{1}{\pi} \exp\left[-\alpha_r^2 - \alpha_i^2\right], \tag{12.10}$$

which is a Gaussian centered at the origin of phase space.

When we recall that the Q-function of a coherent state $|\alpha_0\rangle$ reads

$$Q(\alpha_r, \alpha_i) = \frac{1}{\pi} \exp\left[-(\alpha_r - \alpha_{0r})^2 - (\alpha_i - \alpha_{0i})^2\right]$$

the abstract definition

$$|\alpha_0\rangle \equiv \hat{D}(\alpha_0)|0\rangle = e^{\alpha_0 \hat{a}^\dagger - \alpha_0^* \hat{a}}|0\rangle$$

of a coherent state as the result of the action of the displacement operator $\hat{D}(\alpha_0)$ on the vacuum assumes an extremely intuitive picture: It is a displacement of the Gaussian bell, located at the origin of the (α_r, α_i) phase space, to a position $(\alpha_{0r}, \alpha_{0i})$.

Squeezed State

Another important state of the radiation field is the squeezed state, which we have discussed extensively in the context of the mechanical oscillator. We now briefly discuss its Q-function

$$Q(\alpha) = \frac{1}{\pi}|\langle\alpha|\psi_{sq}\rangle|^2 = \frac{2}{\pi}\frac{\sqrt{s}}{s+1}\exp\left[-\frac{2s}{s+1}(\alpha_r - \gamma_r)^2 - \frac{2}{s+1}(\alpha_i - \gamma_i)^2\right] \tag{12.11}$$

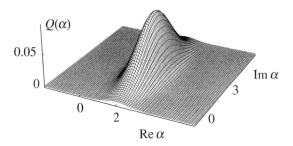

Fig. 12.3: In contrast to the example of a coherent state the Q-function of a squeezed state $|\psi_{sq}\rangle$ is asymmetric in phase space. It is squeezed in one direction. Here we have chosen $\gamma = 2 + 3i$ and $s = 10$.

calculated in Problem 12.1. Hence the Q-function is a Gaussian which has different widths in α_r- and α_i-direction as shown in Fig. 12.3.

We emphasize, however, that these widths

$$\Delta\alpha_r^{(Q)} = \sqrt{\frac{s+1}{2s}} \quad \text{and} \quad \Delta\alpha_i^{(Q)} = \sqrt{\frac{s+1}{2}}$$

are not the inverse of each other as it is the case for the widths $\Delta\alpha_r^{(W)}$ and $\Delta\alpha_i^{(W)}$ of the corresponding Wigner function, Eq. (12.5). This comes out most clearly for the case of a highly squeezed state, that is $s \ll 1$. Here, we find the widths

$$\Delta\alpha_r^{(Q)} \cong \sqrt{\frac{1}{2}} \quad \text{and} \quad \Delta\alpha_i^{(Q)} \cong \sqrt{\frac{s}{2}}$$

of the Q-function in contrast to the widths

$$\Delta\alpha_r^{(W)} \cong \sqrt{\frac{1}{2s}} \cong 0 \quad \text{and} \quad \Delta\alpha_i^{(W)} \cong \sqrt{\frac{s}{2}}.$$

Hence, in the limit of strong squeezing the width $\Delta\alpha_i^{(Q)}$ of the Q-function approaches that of the Wigner function. However, its width $\Delta\alpha_r^{(Q)}$ approaches a constant whereas the corresponding property of the Wigner function goes to zero. In other words, the Wigner function of a squeezed state will become infinitely long and infinitely thin while its area

$$\Delta\alpha_r^{(W)} \cdot \Delta\alpha_i^{(W)} = \frac{1}{2}$$

remains constant.

In contrast, the Q-function becomes infinitely long but always keeps a finite width. This manifests itself in the fact that the area in phase space

$$\Delta\alpha_r^{(Q)} \cdot \Delta\alpha_i^{(Q)} = \frac{1}{2}\frac{s+1}{\sqrt{s}}$$

goes to infinity as s increases. We can understand this feature when we recall in the context of Problem 12.5 that there is an additional averaging involved, to be precise, the Q-function is a Wigner function averaged over a Gaussian.

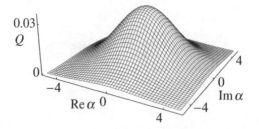

Fig. 12.4: The Q-function of a thermal light field is a Gaussian bell located at the origin of phase space. It is radially symmetric and does not display a preferred direction in phase space. For a non-vanishing temperature it is broader than the vacuum state. The average number $\langle \hat{n} \rangle$ of photons determines the width. Here we have chosen $\langle \hat{n} \rangle = 8$.

Thermal State

So far, we have discussed Q-functions of pure states. We now turn to a mixed state of a single mode. In Sec. 2.3.4 we have introduced a thermal field defined by the density operator

$$\hat{\rho} \equiv \frac{1}{\langle \hat{n} \rangle + 1} \sum_{n=0}^{\infty} \left(\frac{\langle \hat{n} \rangle}{\langle \hat{n} \rangle + 1} \right)^n |n\rangle\langle n|. \tag{12.12}$$

Here $\langle \hat{n} \rangle$ is the average number of photons in the field.

In order to calculate its Q-function, we now substitute this density operator, Eq. (12.12), into the definition of the Q-function, Eq. (12.7), and find with the help of the photon number probability amplitude

$$\langle n|\alpha \rangle = \frac{\alpha^n}{\sqrt{n!}} e^{-|\alpha|^2/2}$$

of a coherent state the expression

$$Q(\alpha_r, \alpha_i) = \frac{1}{\pi[\langle \hat{n} \rangle + 1]} \sum_{n=0}^{\infty} \left(\frac{\langle \hat{n} \rangle}{\langle \hat{n} \rangle + 1} \right)^n \frac{|\alpha|^{2n}}{n!} e^{-|\alpha|^2}.$$

When we recall the expansion of the exponential function, we can perform the summation and arrive at

$$Q(\alpha_r, \alpha_i) = \frac{1}{\pi[\langle \hat{n} \rangle + 1]} \exp\left[\left(\frac{\langle \hat{n} \rangle}{\langle \hat{n} \rangle + 1} - 1 \right) |\alpha|^2 \right],$$

which after minor algebra yields

$$Q(\alpha_r, \alpha_i) = \frac{1}{\pi[\langle \hat{n} \rangle + 1]} \exp\left(-\frac{|\alpha|^2}{\langle \hat{n} \rangle + 1} \right). \tag{12.13}$$

Hence the Q-function of a thermal state is a Gaussian bell centered at the origin of the α_r-α_i phase space as shown in Fig. 12.4. Since only the absolute value $|\alpha|$ of the phase space variable enters, the Gaussian is radially symmetric.

12.2 The Husimi-Kano Q-Function

The width of the Gaussian is determined by the average number of photons $\langle \hat{n} \rangle$. Note that for the vacuum, that is for $\langle \hat{n} \rangle = 0$, we find

$$Q(\alpha_r, \alpha_i) = \frac{1}{\pi} e^{-|\alpha|^2}$$

in agreement with the expressions Eqs. (12.8) and (12.10) for the Q-functions of the coherent state $|\alpha_0\rangle = |0\rangle$ or of the photon number state $|n=0\rangle$, respectively. For a non-vanishing photon number $\langle \hat{n} \rangle$ the Gaussian is broader than the vacuum state.

Thermal Phase State

The radial symmetry of the Q-function of a number state clearly shows that the number state does not have a preferred direction in phase space. It is a Poissonian doughnut shown in Fig. 12.2.

In contrast, the coherent state $||\alpha|e^{i\theta}\rangle$ is a Gaussian distribution located at the amplitude $|\alpha|$ and phase θ. Therefore, it displays a preferred direction. This is surprising, when we recall that a coherent state is a superposition of number states: Each number state has no preferred phase but the superposition of number states leads to a symmetry breaking in phase space, that is to a rather well-defined phase.

Thus, we expect that the Q-function of the thermal phase state discussed in Sec. 2.3.4 and defined by the operator

$$\hat{\rho} \equiv |\varphi_0\rangle\langle\varphi_0| \equiv \frac{1}{\langle\hat{n}\rangle + 1} \sum_{m,n=0}^{\infty} \left(\frac{\langle\hat{n}\rangle}{\langle\hat{n}\rangle + 1}\right)^{(n+m)/2} |n\rangle\langle m|$$

also displays a preferred phase direction. To investigate this in more detail we now calculate the Q-function

$$Q(\alpha_r, \alpha_i) = \frac{1}{\pi[\langle\hat{n}\rangle + 1]} \left| \sum_{n=0}^{\infty} \left(\frac{\langle\hat{n}\rangle}{\langle\hat{n}\rangle + 1}\right)^{n/2} \frac{\alpha^n e^{-|\alpha|^2/2}}{\sqrt{n!}} \right|^2 \quad (12.14)$$

of a thermal phase state.

Here we note a typical manifestation of quantum interference: We sum probability amplitudes rather than probabilities to obtain the total probability. Unfortunately this interference of probability amplitudes also brings in a mathematical complication. The photon probability amplitude of the coherent state contains the term $(n!)^{-1/2}$, which makes it difficult to perform the sum analytically. We therefore have to resort to a numerical evaluation of this sum.

In Fig. 12.5 we show the so-calculated Q-function. It is a wedge-shaped distribution aligned along the real axis of phase space. This confirms the idea of interference of number states giving rise to a preferred direction in phase space.

In order to get a feeling for the behavior of this Q-function, we calculate this sum in Problem 12.2 in an approximate but analytical way. We find

$$Q(\alpha_r, \alpha_i) \cong \sqrt{\frac{8}{\pi} \frac{|\alpha|}{\langle\hat{n}\rangle}} \exp\left[-\frac{|\alpha|^2}{\langle\hat{n}\rangle} - 2|\alpha|^2\theta^2\right]. \quad (12.15)$$

Hence, in contrast to the Q-function of a thermal state this expression not only involves the absolute value square $|\alpha|^2$ of the phase space variable, but also the angle θ.

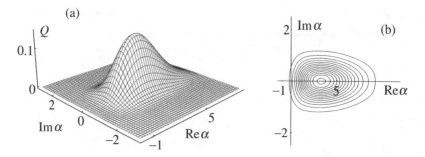

Fig. 12.5: The Q-function of a thermal phase state $|\varphi_0\rangle$ depicted in (a) by a 3D-plot and in (b) by contour lines clearly shows a preferred direction in phase space. The mean number of photons in this example is $\langle \hat{n} \rangle = 20$ and the preferred phase angle is zero.

12.3 Averages Using Phase Space Functions

In the preceding section we have discussed the Q-function of various quantum states. These distributions are another excellent visualization of quantum states. Moreover, in comparison to the Wigner functions, the Q-functions have the advantage that they are always positive. Hence we are lead to the question: Why not use the Q-function rather than the Wigner function?

We recall, that the Wigner function emphasizes the interference nature and hence the Wigner function is useful when we want to study interference phenomena. However, one question remains: Apart from a visualization of a quantum state, what is the use of a phase space distribution? In the present section we show that we can use the Q-function to evaluate expectation values of anti-normally ordered products of annihilation and creation operators.

12.3.1 Heuristic Argument

We first calculate the expectation values of simple combinations of creation and annihilation operators and then generalize the treatment in the next section. Throughout this section we confine ourselves to the treatment of a single mode.

We consider the dimensionless electric field operator

$$\hat{\mathcal{E}}(\hat{a}, \hat{a}^\dagger) \equiv \hat{\pi} = \frac{1}{\sqrt{2}i} (\hat{a} - \hat{a}^\dagger). \qquad (12.16)$$

We note from the representation Eq. (10.68) of the electric field operator $\hat{\vec{E}}$, that apart from the vacuum electric field $\mathcal{E}_0$ and the mode function u_ℓ this operator contains all the operator nature of $\hat{\vec{E}}$.

The classical quantity corresponding to $\hat{\mathcal{E}}$ reads

$$\mathcal{E}(\alpha, \alpha^*) = \frac{1}{\sqrt{2}i} (\alpha - \alpha^*) = \sqrt{2}\alpha_i. \qquad (12.17)$$

12.3 Averages Using Phase Space Functions

Here we have replaced the annihilation and creation operators $\hat{a}$ and $\hat{a}^\dagger$ by the complex c-numbers α and α^*, respectively. We sometimes refer to this procedure as the classical limit.

In Sec. 11.2.7 we have calculated the first two moments of the electric field operator $\hat{\vec{E}}$ in a coherent state $|\alpha_0\rangle$. With the definition Eq. (12.16) of the dimensionless electric field operator the results Eqs. (11.22) and (11.23) yield

$$\langle \hat{\mathcal{E}}(\hat{a}, \hat{a}^\dagger) \rangle = \sqrt{2}\alpha_{0i} \tag{12.18a}$$

and

$$\langle \hat{\mathcal{E}}^2(\hat{a}, \hat{a}^\dagger) \rangle = 2\alpha_{0i}^2 + \frac{1}{2}. \tag{12.18b}$$

In this derivation we have made use of the eigenvalue equation of the annihilation operator.

We now calculate expectation values of the electric field operator $\hat{\mathcal{E}}$ in a coherent state assuming that we can use the Q-function as a classical phase space distribution function. Hence, the Q-function serves as the weight function to integrate the classical representation $\mathcal{E}(\alpha, \alpha^*)$ of $\hat{\mathcal{E}}(\hat{a}, \hat{a}^\dagger)$ over the variables α and α^*.

First Moment of Electric Field

In this scheme we find the quantum mechanical expectation value $\langle \hat{\mathcal{E}}(\hat{a}, \hat{a}^\dagger) \rangle$ by evaluating the integral

$$\langle \hat{\mathcal{E}}(\hat{a}, \hat{a}^\dagger) \rangle \equiv \int d^2\alpha \, \mathcal{E}(\alpha, \alpha^*) \, Q(\alpha, \alpha^*) = \int_{-\infty}^{\infty} d\alpha_r \int_{-\infty}^{\infty} d\alpha_i \, \mathcal{E}(\alpha_r, \alpha_i) \, Q(\alpha_r, \alpha_i), \tag{12.19}$$

which with the help of the expressions, Eqs. (12.8) and (12.17), for the Q-function of the coherent state and the classical electric field, respectively, reduces to

$$\langle \hat{\mathcal{E}}(\hat{a}, \hat{a}^\dagger) \rangle = \sqrt{2} \, \frac{1}{\sqrt{\pi}} \int_{-\infty}^{\infty} d\alpha_r \, e^{-(\alpha_r - \alpha_{0r})^2} \, \frac{1}{\sqrt{\pi}} \int_{-\infty}^{\infty} d\alpha_i \, \alpha_i \, e^{-(\alpha_i - \alpha_{0i})^2}.$$

With the help of the integral relations

$$\frac{1}{\sqrt{\pi}} \int_{-\infty}^{\infty} d\zeta \, e^{-\zeta^2} = 1 \quad \text{and} \quad \frac{1}{\sqrt{\pi}} \int_{-\infty}^{\infty} d\zeta \, \zeta \, e^{-(\zeta - \zeta_0)^2} = \zeta_0$$

we therefore find indeed

$$\langle \hat{\mathcal{E}}(\hat{a}, \hat{a}^\dagger) \rangle = \sqrt{2}\,\alpha_{0i},$$

that is the correct quantum result, Eq. (12.18a).

Second Moment

So far the creation and annihilation operators have entered in a linear way. We now perform a similar calculation for the second moment

$$\hat{\mathcal{E}}^2(\hat{a}, \hat{a}^\dagger) = -\frac{1}{2}\left(\hat{a}^2 + \hat{a}^{\dagger\,2} - \hat{a}\hat{a}^\dagger - \hat{a}^\dagger\hat{a}\right)$$

of $\hat{\mathcal{E}}$.

The first question we have to address is the appropriate form of the corresponding classical quantity. We therefore have to discuss the question of operator ordering. We note that if we now take the classical limit by replacing the operators $\hat{a}$ and $\hat{a}^\dagger$ by the complex c-numbers α and α^*, we find

$$\mathcal{E}^2(\alpha, \alpha^*) = -\frac{1}{2}\left(\alpha^2 + \alpha^{*2} - \alpha\alpha^* - \alpha^*\alpha\right) = -\frac{1}{2}\left(\alpha^2 + \alpha^{*2} - 2\alpha^*\alpha\right).$$

However, we get a completely different result when we use the commutation relation $[\hat{a}, \hat{a}^\dagger] = 1$ before we perform the replacement. In this case we start either from

$$\mathcal{N}\left[\hat{\mathcal{E}}^2(\hat{a}, \hat{a}^\dagger)\right] = -\frac{1}{2}\left(\hat{a}^2 + \hat{a}^{\dagger 2} - 2\hat{a}^\dagger\hat{a} - 1\right)$$

or from

$$\mathcal{A}\left[\hat{\mathcal{E}}^2(\hat{a}, \hat{a}^\dagger)\right] = -\frac{1}{2}\left(\hat{a}^2 + \hat{a}^{\dagger 2} - 2\hat{a}\hat{a}^\dagger + 1\right).$$

Here, the symbol $\mathcal{N}$ denotes that products formed out of annihilation and creation operators are ordered in a way such that creation operators are always to the left of the annihilation operators. We call such expressions *normally ordered*. In contrast in $\mathcal{A}$ creation operators are always to the right of the annihilation operators. We refer to this type of ordering as *antinormally ordered*.

When we now replace $\hat{a}$ and $\hat{a}^\dagger$ by α and α^*, respectively, we find that the two classical expressions

$$\mathcal{E}^{2\,(n)}(\alpha, \alpha^*) = -\frac{1}{2}\left(\alpha^2 + \alpha^{*2} - 2\alpha^*\alpha - 1\right) = -\frac{1}{2}(\alpha - \alpha^*)^2 + \frac{1}{2} = 2\alpha_i^2 + \frac{1}{2}$$

and

$$\mathcal{E}^{2\,(a)}(\alpha, \alpha^*) = -\frac{1}{2}\left(\alpha^2 + \alpha^{*2} - 2\alpha^*\alpha + 1\right) = -\frac{1}{2}(\alpha - \alpha^*)^2 - \frac{1}{2} = 2\alpha_i^2 - \frac{1}{2},$$

corresponding to the normally ordered or antinormally ordered operators are indeed different.

Motivated by the success for the first moment and in order to decide whether the normal or antinormal ordering procedure provides the correct quantum mechanical result, we now calculate the integrals

$$I\binom{n}{a} \equiv \int d^2\alpha\, \mathcal{E}^{2\binom{n}{a}}(\alpha, \alpha^*)\, Q(\alpha, \alpha^*)$$

$$= \int_{-\infty}^{\infty} d\alpha_r \int_{-\infty}^{\infty} d\alpha_i \left(2\alpha_i^2 \pm \frac{1}{2}\right)\frac{1}{\pi}e^{-(\alpha_r - \alpha_{0r})^2}e^{-(\alpha_i - \alpha_{0i})^2}$$

corresponding to the classical phase space representations of the expectation value $\langle\hat{\mathcal{E}}^2(\hat{a}, \hat{a}^\dagger)\rangle$. When we perform the integrations we find

$$I\binom{n}{a} = \frac{2}{\sqrt{\pi}}\int_{-\infty}^{\infty}d\bar\zeta\,(\bar\zeta + \alpha_{0i})^2 e^{-\bar\zeta^2} \pm \frac{1}{2}$$

which with the help of the relations

$$\frac{1}{\sqrt{\pi}} \int_{-\infty}^{\infty} d\zeta \, e^{-\zeta^2} = 1 \quad \text{and} \quad \frac{1}{\sqrt{\pi}} \int_{-\infty}^{\infty} d\zeta \, \zeta^2 e^{-\zeta^2} = \frac{1}{2}$$

yields the explicit formulas

$$I\binom{n}{a} = 1 + 2\alpha_{0i}^2 \pm \frac{1}{2} = 2\alpha_{0i}^2 + \frac{3/2}{1/2}.$$

When we compare these expressions to the correct quantum mechanical result, Eq. (12.18b),

$$\langle \hat{\mathcal{E}}^2(\hat{a}, \hat{a}^\dagger) \rangle = 2\alpha_{0i}^2 + 1/2,$$

we note that only the antinormally ordered expression agrees. We are therefore led to the conclusion that the Q-function allows us to calculate antinormally ordered products of annihilation and creation operators.

This suggests the following procedure when we have to calculate the expectation value of the operator $\hat{O}(\hat{a}, \hat{a}^\dagger)$ consisting of products of annihilation and creation operators which are not in a particular order yet. We first antinormally order this operator $\hat{O}(\hat{a}, \hat{a}^\dagger)$ using the commutator relation $[\hat{a}, \hat{a}^\dagger] = 1$, that is we calculate $\mathcal{A}[\hat{O}(\hat{a}, \hat{a}^\dagger)]$. We then find the classical expression $O^{(a)}(\alpha, \alpha^*)$ of this antinormally ordered product by replacing $\hat{a}$ and $\hat{a}^\dagger$ by α and α^*. We perform the average by averaging the classical quantity $O^{(a)}$ using the Q-function, that is

$$\langle \hat{O}(\hat{a}, \hat{a}^\dagger) \rangle = \int d^2\alpha \, O^{(a)}(\alpha, \alpha^*) \, Q(\alpha, \alpha^*).$$

In the next section we prove this relation in a rigorous way.

12.3.2 Rigorous Treatment

In the preceding section we have shown, that we can evaluate the first and the second moment of the electric field operator in a coherent state using the Q-function. In the present section we want to generalize this to an operator $\hat{O}(\hat{a}, \hat{a}^\dagger)$ which contains an arbitrary combination of annihilation and creation operators.

The general problem then is to find a convenient way to evaluate the expectation value

$$\langle \hat{O} \rangle = \text{Tr}(\hat{O} \hat{\rho}) \tag{12.20}$$

of the operator $\hat{O}$ in the quantum state described by the density operator $\hat{\rho}$. There are many ways to do this and in the present section we highlight only two.

The example of the second moment of the electric field operator has shown that by making use of the commutation relations we can represent the same operator in many formally different versions, which however are equivalent to one another. Hence we can represent the operators $\hat{\rho}$ and $\hat{O}$ both in normal, or both in antinormal ordering. We could also have a mixed representation in which $\hat{\rho}$ is in normal order whereas $\hat{O}$ is in antinormal order, or vice versa. All these forms are equivalent. However, in the evaluation some are more convenient and, in particular, provide a link to classical phase space integration. They allow us to obtain the expectation value as a classical integration.

Expectation Values of Antinormally Ordered Operators

In this section we evaluate the expectation value using the normally ordered density operator $\hat{\rho}^{(n)}$ and the antinormally ordered operator $\hat{O}^{(a)}$. Moreover, we show that in this case the formula

$$\langle \hat{O} \rangle = \mathrm{Tr}(\hat{O}^{(a)} \hat{\rho}^{(n)}) \tag{12.21}$$

enjoys a classical phase space interpretation with the Q-function as the distribution function.

We first antinormally order the operator $\hat{O}$ using the commutation relations $[\hat{a}, \hat{a}^\dagger] = 1$ and find the antinormally ordered operator

$$\hat{O}^{(a)} \equiv \mathcal{A}\left[\hat{O}(\hat{a}, \hat{a}^\dagger)\right] \equiv \sum_{i,j} O^{(a)}_{ij} \hat{a}^i \, \hat{a}^{\dagger j} \tag{12.22}$$

with the c-number coefficients $O^{(a)}_{ij}$.

We now find a representation of the density operator $\hat{\rho}$ in terms of the annihilation and creation operators that leads directly to the Q-function. This representation is normally ordered, as we show now.

For this purpose we recall that the Q-function defined by

$$Q(\alpha_r, \alpha_i) \equiv \frac{1}{\pi} \langle \alpha | \hat{\rho} | \alpha \rangle \equiv Q(\alpha, \alpha^*) \tag{12.23}$$

depends on the real and imaginary part of α, or on α and α^*. We can therefore expand the function

$$Q(\alpha, \alpha^*) \equiv \frac{1}{\pi} \sum_{k,l} \rho^{(n)}_{kl} \alpha^{*k} \alpha^l \tag{12.24}$$

in a power series of α and α^*. The coefficients of this expansion are then the c-numbers $\rho^{(n)}_{kl}$.

Indeed, when we substitute the operator representation

$$\hat{\rho} = \sum_{k,l} \rho^{(n)}_{kl} \hat{a}^{\dagger k} \hat{a}^l \tag{12.25}$$

of the density operator into the definition, Eq. (12.23), of the Q-function and use the eigenvalue equation of the annihilation operator, we arrive at the above relation.

Now we are in a position to calculate the expectation value $\langle \hat{O} \rangle$ of the operator $\hat{O}$. When we insert the density operator, Eq. (12.25), in its normally ordered form, and the operator $\hat{O}$ in its antinormally ordered form, Eq. (12.22), we find

$$\langle \hat{O} \rangle = \mathrm{Tr}(\hat{O}^{(a)} \hat{\rho}^{(n)}) = \sum_{i,j,k,l} O^{(a)}_{ij} \rho^{(n)}_{kl} \mathrm{Tr}(\hat{a}^i \, \hat{a}^{\dagger j} \, \hat{a}^{\dagger k} \, \hat{a}^l).$$

With the help of the property

$$\mathrm{Tr}(\hat{A}\hat{B}) = \mathrm{Tr}(\hat{B}\hat{A})$$

of the trace operation valid for two operators $\hat{A}$ and $\hat{B}$ we can combine the individual powers of the annihilation and creation operators to yield

$$\langle \hat{O} \rangle = \sum_{i,j,k,l} O^{(a)}_{ij} \rho^{(n)}_{kl} \, \text{Tr}(\hat{a}^{\dagger\, j+k} \hat{a}^{i+l}).$$

We perform the trace using coherent states and arrive at

$$\langle \hat{O} \rangle = \sum_{i,j,k,l} O^{(a)}_{ij} \rho^{(n)}_{kl} \frac{1}{\pi} \int d^2\alpha \langle \alpha | \hat{a}^{\dagger\, j+k} \hat{a}^{i+l} | \alpha \rangle.$$

With help of the eigenvalue equation for coherent states we find

$$\langle \hat{O} \rangle = \sum_{i,j,k,l} O^{(a)}_{ij} \rho^{(n)}_{kl} \frac{1}{\pi} \int d^2\alpha \, \alpha^{*\, j+k} \alpha^{i+l},$$

or

$$\langle \hat{O} \rangle = \int d^2\alpha \left(\sum_{i,j} O^{(a)}_{ij} \alpha^i \alpha^{*j} \right) \left(\frac{1}{\pi} \sum_{k,l} \rho^{(n)}_{kl} \alpha^{*k} \alpha^l \right).$$

When we now introduce the c-number representation

$$O^{(a)}(\alpha) = \sum_{i,j} O^{(a)}_{ij} \alpha^i \alpha^{*j}$$

of the antinormally ordered operator $\hat{O}$ and the expansion Eq. (12.24) this expression takes the form

$$\langle \hat{O} \rangle = \int d^2\alpha \, O^{(a)}(\alpha) \, Q(\alpha).$$

Hence we obtain the expectation value of an operator $\hat{O}$ consisting of an arbitrary combination of creation and annihilation operators by integrating the classical representation of the antinormally ordered operator using the Q-function.

Expectation Values of Normally Ordered Operators

In the present section we evaluate the expectation value

$$\langle \hat{O} \rangle = \text{Tr}(\hat{O}^{(n)} \hat{\rho}^{(a)}) \tag{12.26}$$

of the normally ordered form $\hat{O}^{(n)}$ of the operator $\hat{O}$ using the antinormally ordered density operator $\hat{\rho}^{(a)}$. We show that in this case we also have a phase space interpretation using the so-called Glauber-Sudarshan P-function.

We express the operator $\hat{O}$ in normally ordered products of powers of annihilation and creation operators, that is

$$\hat{O}^{(n)} = \sum_{i,j} O^{(n)}_{ij} \hat{a}^{\dagger\, i} \hat{a}^j \tag{12.27}$$

with the c-number expansion coefficients $O^{(n)}_{ij}$.

In contrast, we express the density operator $\hat{\rho}$ in antinormally ordered products of powers of the annihilation and creation operators, that is

$$\hat{\rho}^{(a)} = \sum_{k,l} \rho_{kl}^{(a)} \hat{a}^k \hat{a}^{\dagger l}. \tag{12.28}$$

Here the coefficients $\rho_{kl}^{(a)}$ are c-numbers.

We are now able to calculate averages of $\hat{O}$ using these expressions and start from

$$\langle \hat{O} \rangle = \text{Tr}(\hat{O}^{(n)} \hat{\rho}^{(a)}) = \sum_{i,j,k,l} O_{ij}^{(n)} \rho_{kl}^{(a)} \text{Tr}(\hat{a}^{\dagger i} \hat{a}^j \hat{a}^k \hat{a}^{\dagger l})$$

where we have used Eqs. (12.27) and (12.28).

When we combine the annihilation and creation operators due to the trace operation we find

$$\langle \hat{O} \rangle = \sum_{i,j,k,l} O_{ij}^{(n)} \rho_{kl}^{(a)} \text{Tr}(\hat{a}^{\dagger i+l} \hat{a}^{j+k}).$$

We perform the trace using coherent states

$$\langle \hat{O} \rangle = \sum_{i,j,k,l} O_{ij}^{(n)} \rho_{kl}^{(a)} \frac{1}{\pi} \int d^2\alpha \langle \alpha | \hat{a}^{\dagger i+l} \hat{a}^{j+k} | \alpha \rangle$$

and arrive with the help of the eigenvalue equation for coherent states at

$$\langle \hat{O} \rangle = \sum_{i,j,k,l} O_{ij}^{(n)} \rho_{kl}^{(a)} \frac{1}{\pi} \int d^2\alpha \, \alpha^{*i+l} \alpha^{j+k}$$

or

$$\langle \hat{O} \rangle = \int d^2\alpha \left(\sum_{i,j} O_{ij}^{(n)} \alpha^{*i} \alpha^j \right) \left(\frac{1}{\pi} \sum_{k,l} \rho_{kl}^{(a)} \alpha^k \alpha^{*l} \right).$$

We notice that the term in the first parenthesis is the c-number representation

$$O^{(n)}(\alpha) = \sum_{i,j} O_{ij}^{(n)} \alpha^{*i} \alpha^j$$

of the normally ordered operator $\hat{O}^{(n)}$. This motivates us to introduce the new distribution function

$$P(\alpha) \equiv \frac{1}{\pi} \sum_{k,l} \rho_{kl}^{(a)} \alpha^k \alpha^{*l}.$$

We emphasize that this distribution is not the Q-function since the expansion coefficients $\rho_{kl}^{(a)}$ resulting from the antinormal expansion of the density operator are in general different from the expansion coefficients $\rho_{kl}^{(n)}$ of the normally ordered expansion. In the next section we will discuss this distribution in more detail.

Hence we find

$$\langle \hat{O} \rangle = \int d^2\alpha \, O^{(n)}(\alpha) P(\alpha)$$

that is, we can perform the average of the quantum mechanical operator $\hat{O}$ by first normally ordering it, replacing the operators by c-numbers and calculating a phase space integral using the P-distribution.

12.4 The Glauber-Sudarshan P-Distribution

In the preceding section we have introduced the P-function

$$P(\alpha) \equiv \frac{1}{\pi} \sum_{i,j} \rho_{ij}^{(a)} \alpha^i \alpha^{*j} \tag{12.29}$$

in order to evaluate the expectation value of a quantum mechanical operator by a phase space integration. Here $\rho_{ij}^{(a)}$ are the expansion coefficients of the antinormally density operator. We now show that the so-defined P-function allows us to represent the density operator in a diagonal form of coherent states. Moreover, we discuss its relation to the Q-function. We conclude this section by giving various examples.

12.4.1 Definition of P-Distribution

So far we have represented the density operator in terms of photon number states $|m\rangle$. We have found this representation by multiplying the completeness relation

$$\mathbb{1} = \sum_{m=0}^{\infty} |m\rangle\langle m| \tag{12.30}$$

of the photon number states onto the density operator from the left and from the right. From

$$\hat{\rho} = \mathbb{1} \cdot \hat{\rho} \cdot \mathbb{1} \tag{12.31}$$

we thus find

$$\hat{\rho} = \sum_{m,n=0}^{\infty} |m\rangle\langle m|\hat{\rho}|n\rangle\langle n| \equiv \sum_{m,n=0}^{\infty} \rho_{mn}|m\rangle\langle n|.$$

Likewise we can use the completeness relation

$$\mathbb{1} = \frac{1}{\pi} \int d^2\alpha |\alpha\rangle\langle\alpha|$$

for coherent states to find the representation

$$\hat{\rho} = \frac{1}{\pi^2} \int d^2\alpha \int d^2\beta \langle\alpha|\hat{\rho}|\beta\rangle |\alpha\rangle\langle\beta| \tag{12.32}$$

of the density operator.

Note that in complete analogy to the photon number representation this coherent state representation in general involves off-diagonal elements $\langle\alpha|\hat{\rho}|\beta\rangle$. Hence it contains projectors $|\alpha\rangle\langle\beta|$ in which α could be different from β. Moreover, this representation also involves two phase space integrations.

However, we can also express the density operator $\hat{\rho}$ in the diagonal representation

$$\hat{\rho} = \int d^2\alpha \, P(\alpha) |\alpha\rangle\langle\alpha| \tag{12.33}$$

of coherent states. Note that this is completely different from the coherent state representation in Eq. (12.32). First of all, it involves each coherent state $|\alpha\rangle$ only in the combination $|\alpha\rangle\langle\alpha|$, that is in a diagonal form. Moreover, it needs only one phase space integration. R. Glauber and E.C.G. Sudarshan have introduced this function independently from each other.

On first sight, this diagonal representation is quite surprising. However, we can prove the above relation by substituting the expansion Eq. (12.29) in terms of powers of α and α^* into the right hand side of Eq. (12.33). When we recall the eigenvalue equation of the annihilation operator we find

$$\int d^2\alpha\, P(\alpha)|\alpha\rangle\langle\alpha| = \sum_{k,l} \rho_{kl}^{(a)} \frac{1}{\pi} \int d^2\alpha \left(\hat{a}^k |\alpha\rangle\right) \left(\langle\alpha| \hat{a}^{\dagger l}\right)$$

which with the completeness relation of the coherent states immediately reduces to the antinormally ordered representation, Eq. (12.28), of the density operator.

We emphasize that the so-called Glauber-Sudarshan P-distribution is not a true probability distribution. This fact stands out most clearly in the example of a coherent state $|\alpha_0\rangle$. Since here the density operator reads

$$\hat{\rho} = |\alpha_0\rangle\langle\alpha_0|,$$

the P-distribution in the representation

$$\hat{\rho} = \int d^2\alpha\, P(\alpha)|\alpha\rangle\langle\alpha|$$

has to be a Dirac delta function, that is

$$P(\alpha) = \delta(\alpha - \alpha_0).$$

Hence the P-function has to be understood in a sense of a generalized distribution function. In the next section we show that indeed the P-function can become more singular than a delta function. For the mth number state it is the mth derivative of a delta function. For a squeezed state it even involves infinitely high derivatives.

12.4.2 Connection between Q- and P-Function

The P-distribution as well as the Q-function depends on the real and imaginary part of the amplitude α of a coherent state $|\alpha\rangle$. However, they have quite a different shape: For the example of a coherent state we have already seen, that the P-distribution is a delta function whereas the Q-function is a Gaussian bell of non-vanishing width. We now establish a relation between the two phase space distributions of an arbitrary density operator.

For this purpose we take the expectation value of the equation

$$\hat{\rho} = \int d^2\alpha\, P(\alpha)|\alpha\rangle\langle\alpha| \tag{12.34}$$

defining the P-distribution, and find

$$Q(\alpha) = \frac{1}{\pi}\langle\alpha|\hat{\rho}|\alpha\rangle = \frac{1}{\pi}\int d^2\beta\, P(\beta)\langle\alpha|\beta\rangle\langle\beta|\alpha\rangle$$

or
$$Q(\alpha) = \frac{1}{\pi}\int d^2\beta\, P(\beta)|\langle\alpha|\beta\rangle|^2.$$

When we recall the expression
$$|\langle\alpha|\beta\rangle|^2 = e^{-|\alpha-\beta|^2}$$

for the scalar product between two coherent states, we find
$$Q(\alpha) = \frac{1}{\pi}\int d^2\beta\, P(\beta) e^{-|\alpha-\beta|^2}. \tag{12.35}$$

Hence the Q-function is the P-distribution integrated over phase space weighted with a Gaussian. The latter is the Q-function of a coherent state $|\beta\rangle$. This relation suggests the following interpretation: The Q-function of a quantum state emerges, when we read out the P-distribution using a coherent state.

Note that Eq. (12.35) is consistent with the averaging procedure discussed in Sec. 12.3. Indeed, we recall that the Q-function is the expectation value of the density operator in a coherent state $|\alpha\rangle$. We can evaluate this expectation value by first antinormally ordering the density operator and then performing the classical limit that is replacing the operators by c-numbers. This yields the P-function, Eq. (12.29). We then perform a phase space integration with the Q-function Q_α of the quantum state of interest, that is of a coherent state $|\alpha\rangle$. We therefore arrive at

$$Q(\alpha) = \frac{1}{\pi}\int d^2\beta\, P(\beta) Q_\alpha(\beta).$$

Equation (12.35) also shows that the Q-function of a quantum state is always broader than the corresponding P-distribution, since it is averaged over a Gaussian distribution. The two distributions serve to evaluate expectation values of quantum mechanical operators. However, the ordering of the operators is different, which reflects itself in the distributions being different. Therefore, they also live in different phase spaces.

12.4.3 P-Function from Q-Function

The Q-function is the expectation value of the density operator in coherent states. Hence, the Q-function is defined explicitly. In contrast the P-function is defined implicitly. It is the diagonal representation of the density operator in coherent states. In the preceding section we have derived the relation

$$Q(\alpha) = \frac{1}{\pi}\int d^2\beta\, P(\beta) e^{-|\alpha-\beta|^2} \tag{12.36}$$

which allows us to find the Q-function, provided we already know the P-function. But how to invert this equation? This is the question we address in the present section.

We can establish a simple connection between the Fourier transforms of the two distribution functions. To bring this out most clearly we now define the Fourier transform

$$\tilde{P}(\xi) \equiv \tilde{P}(\xi_r, \xi_i) \equiv \int d\alpha_r \int d\alpha_i\, P(\alpha_r, \alpha_i)\, e^{-i(\xi_r \alpha_r + \xi_i \alpha_i)} \tag{12.37}$$

of the P-function and similarly

$$\tilde{Q}(\xi) \equiv \tilde{Q}(\xi_r, \xi_i) \equiv \int d\alpha_r \int d\alpha_i Q(\alpha_r, \alpha_i) \, e^{-i(\xi_r \alpha_r + \xi_i \alpha_i)} \tag{12.38}$$

denotes the Fourier transform of the Q-function.

We substitute the expression, Eq. (12.36), for the Q-function in terms of the P-function into the Fourier transform of the Q-function, Eq. (12.38), and find

$$\tilde{Q}(\xi_r, \xi_i) = \frac{1}{\pi} \int d\alpha_r \int d\alpha_i \int d\beta_r \int d\beta_i \, P(\beta_r, \beta_i) \, e^{-(\alpha_r - \beta_r)^2 - (\alpha_i - \beta_i)^2} e^{-i(\xi_r \alpha_r + \xi_i \alpha_i)}$$

or after interchanging the order of integrations

$$\tilde{Q}(\xi_r, \xi_i) = \frac{1}{\pi} \int d\beta_r \int d\beta_i P(\beta_r, \beta_i) \left(\int d\alpha_r \, e^{-(\alpha_r - \beta_r)^2 - i\xi_r \alpha_r} \right) \left(\int d\alpha_i \, e^{-(\alpha_i - \beta_i)^2 - i\xi_i \alpha_i} \right).$$

We evaluate the Gaussian integrals in braces by introducing the integration variable $\alpha'_i \equiv \alpha_i - \beta_i$ and by using the formula

$$\int_{-\infty}^{\infty} dx \, e^{-ax^2 + bx} = \sqrt{\frac{\pi}{a}} \exp\left(\frac{b^2}{4a}\right). \tag{12.39}$$

Hence, we obtain the connection formula

$$\tilde{Q}(\xi_r, \xi_i) = e^{-(\xi_r^2 + \xi_i^2)/4} \int d\beta_r \int d\beta_i P(\beta_r, \beta_i) \, e^{-i(\xi_r \beta_r + \xi_i \beta_i)} = e^{-(\xi_r^2 + \xi_i^2)/4} \tilde{P}(\xi_r, \xi_i), \tag{12.40}$$

or

$$\tilde{P}(\xi_r, \xi_i) = e^{(\xi_r^2 + \xi_i^2)/4} \, \tilde{Q}(\xi_r, \xi_i) \tag{12.41}$$

between the Fourier transforms $\tilde{Q}$ and $\tilde{P}$ of the Q- and the P-function.

We now substitute this formula into the inverse transformation

$$P(\alpha_r, \alpha_i) = \frac{1}{(2\pi)^2} \int d\xi_r \int d\xi_i \, \tilde{P}(\xi_r, \xi_i) \, e^{i(\xi_r \alpha_r + \xi_i \alpha_i)} \tag{12.42}$$

for the P-distribution and arrive at

$$P(\alpha_r, \alpha_i) = \frac{1}{(2\pi)^2} \int d\xi_r \int d\xi_i \, e^{(\xi_r^2 + \xi_i^2)/4} \, \tilde{Q}(\xi_r, \xi_i) e^{i(\xi_r \alpha_r + \xi_i \alpha_i)}. \tag{12.43}$$

This equation allows us to find the P-distribution of an arbitrary quantum state from its Q-function. For this purpose we first have to calculate the Fourier transform of the Q-function and then evaluate the above integral. In Appendix J we pursue this approach to derive the P-functions of various quantum states discussed in the next section.

Equation (12.43) also shows that there can be cases, where the P-function does not exist as an analytic function, but only in the sense of a distribution. Indeed, in this integral the two-dimensional anti-Gaussian $e^{(\xi_r^2 + \xi_i^2)/4}$ multiplies the Fourier transform of Q. Since this function has a positive sign in the exponent, the integral can only converge provided $\tilde{Q}(\xi_r, \xi_i)$ decays faster than the anti-Gaussian grows.

12.4 The Glauber-Sudarshan P-Distribution

However, the integral always exists as a distribution, as was shown by E.C.G. Sudarshan.

In contrast, the Q-function always exists. The normalization property

$$\int Q(\alpha)\, d^2\alpha = \frac{1}{\pi} \int \langle \alpha | \hat{\rho} | \alpha \rangle\, d^2\alpha = \operatorname{Tr} \hat{\rho} = 1$$

of the Q-function guarantees that it vanishes for $\xi_r, \xi_i \to \infty$.

Moreover, Eq. (12.43) shines more light on the fact that the Q-function is *always* broader than the P-function. Equation (12.40) shows that due to the Gaussian with the negative exponent the Fourier transform of the Q-function is always narrower than the Fourier transform of the P-function. Therefore, the inverse Fourier transform of $\tilde{Q}$, the Q-function itself, is always broader than the P-function.

12.4.4 Examples of P-Distributions

In the preceding section we have given a general prescription how to obtain the P-function of a quantum state from its Q-function. In the present section we discuss the properties of the P-function for various quantum states. In particular, we show that states with non-classical features such as photon number states or squeezed states have P-functions, that exist only in the sense of a distribution. They are only properly defined when they appear under an integral.

Thermal State

We start our discussion with the example of a thermal state. According to Appendix J its P-function reads

$$P(\alpha_r, \alpha_i) = \frac{1}{\pi \langle \hat{n} \rangle} e^{-|\alpha|^2 / \langle \hat{n} \rangle}.$$

When we compare this expression to the corresponding Q-function

$$Q(\alpha) = \frac{1}{\pi [\langle \hat{n} \rangle + 1]} \exp\left(-\frac{|\alpha|^2}{\langle \hat{n} \rangle + 1} \right), \tag{12.44}$$

we recognize that both functions are Gaussians. However, they have very different widths: The width of the P-function defined by the position where the Gaussian has decayed to its eth part is $\langle \hat{n} \rangle$ whereas the width of the Q-function is $\langle \hat{n} \rangle + 1$. Hence it is larger by one additional quantum. This makes our former statement that the Q-function is always broader than the corresponding P-function more precise.

In the limit $\langle \hat{n} \rangle \gg 1$, that is when this single quantum does not matter anymore, the difference ceases to exist. We can understand this phenomenon in physical terms, when we recall from Sec. 2.3.4 that the average number of photons is related to the temperature T of the thermal field and the Boltzmann constant k_B by

$$\langle \hat{n} \rangle = \left[\exp\left(\frac{\hbar \Omega}{k_B T} \right) - 1 \right]^{-1}. \tag{12.45}$$

Hence, when we assume that the thermal energy $k_B T$ is much larger than the quantized energy level separation $\hbar\Omega$ we find

$$\langle \hat{n} \rangle \cong \frac{k_B T}{\hbar\Omega}.$$

Thus the limit $\langle \hat{n} \rangle \gg 1$ corresponds to the large temperature limit, where the discreteness of photon numbers and the difference between n and $n+1$ is irrelevant.

The difference between the Q- and the P-functions stands out most clearly, when we take the limit $\langle \hat{n} \rangle \to 0$ of the two distribution functions of a thermal state. We then find for the Q-function a Gaussian

$$Q(\alpha_r, \alpha_i) = \frac{1}{\pi} e^{-|\alpha|^2}, \qquad (12.46)$$

whereas the P-function is a delta function, that is

$$P(\alpha_r, \alpha_i) = \delta(\alpha_r)\,\delta(\alpha_i) = \delta(\alpha). \qquad (12.47)$$

This is not surprising when we recall from Eq. (12.45) that the limit $\langle \hat{n} \rangle \to 0$ implies that the temperature approaches zero. In this limit, the thermal state becomes the vacuum, which is also a coherent state.

Number State

The P-function of a thermal state is a well-behaved Gaussian. Only in the limit of the vacuum state does it turn into a Dirac delta function. We now show that a photon number state is more singular than a vacuum state. Its P-function brings in higher derivatives.

In Appendix J we find that the P-function

$$P(\alpha) = L_n(-\tfrac{1}{4}\Delta)\,\delta(\alpha)$$

of the nth number state is characterized by an operator $L_n(\Delta)$ acting on a delta function. The operator is in terms of the two-dimensional Laplacian

$$\Delta \equiv \frac{\partial^2}{\partial \alpha_r^2} + \frac{\partial^2}{\partial \alpha_i^2}. \qquad (12.48)$$

Due to the appearance of the nth Laguerre polynomial L_n powers of the Laplacian up to the nth power emerge and act on on the delta function. Hence, the nth photon number state involves up to the $2n$th derivative of a delta function.

When we recall the explicit expressions $L_0(x) = 1$ and $L_1(x) = 1 - x$ for the two lowest Laguerre polynomials, we find

$$P_{|0\rangle}(\alpha) = \delta(\alpha)$$

for the vacuum state and

$$P_{|1\rangle}(\alpha) = (1 + \tfrac{1}{4}\Delta)\delta(\alpha)$$

for the number state of one photon.

Squeezed State

The nth number state enjoys a P-function that involves the $2n$th derivative of a delta function. A squeezed state is at least in one direction narrower than a coherent state. Hence its P-function must be more complicated than the delta function.

In Appendix J we find indeed that the P-function of the squeezed state reads

$$P(\alpha_r, \alpha_i) = \exp\left[\frac{1-s}{8s}\frac{\partial^2}{\partial \alpha_r^2} - \frac{1-s}{8}\frac{\partial^2}{\partial \alpha_i^2}\right]\delta(\alpha - \gamma).$$

Since it brings in an infinite amount of derivatives of delta functions, it is more singular than the P-function of a number state. Nevertheless, this expression is useful whenever it appears under an integral as it is usually the case for distribution functions.

Problems

12.1 Q-Function of a Squeezed State

Calculate the Q-function, Eq. (12.11), of a squeezed state.

Hint: Start from the Q-function amplitude

$$\langle \alpha | \psi_{sq} \rangle = \int_{-\infty}^{\infty} dx\, \langle \alpha | x \rangle \langle x | \psi_{sq} \rangle,$$

and use the position representations

$$\langle \alpha | x \rangle = \left(\frac{1}{\pi}\right)^{1/4} \exp\left[-\tfrac{1}{2}(x - \sqrt{2}\,\alpha_r)^2 - \sqrt{2}\,i\alpha_i x + i\alpha_r \alpha_i\right],$$

and

$$\langle x | \psi_{sq} \rangle = \left(\frac{s}{\pi}\right)^{1/4} \exp\left[-\frac{s}{2}(x - \sqrt{2}\,\gamma_r)^2 + \sqrt{2}\,i\gamma_i x - i\gamma_r \gamma_i\right].$$

of a coherent state $|\alpha\rangle$ and a squeezed state $|\psi_{sq}\rangle$, respectively. Here α_r and α_i are the real and imaginary parts of α, and $|\psi_{sq}\rangle$ is characterized by the squeezing parameter s and displacement $\gamma = \gamma_r + i\gamma_i$.

12.2 Q-Function of Thermal Phase State

Derive the analytical expression Eq. (12.15) for the Q-Function of the thermal phase state.

Hint: Use the Gaussian limit of the Poissonian distribution and replace summation by integration. See Däubler et al. (1993).

12.3 Normal Ordering

Given the operators $\hat{a}$ and $\hat{a}^\dagger$ with $[\hat{a}, \hat{a}^\dagger] = 1$. In the following $|\alpha\rangle$ is a coherent state, $|n\rangle$ and $|m\rangle$ are Fock states, and $\hat{F}$ is an arbitrary operator. Show:

(a) An operator $\hat{F}$ is uniquely defined by the function
$$f(\alpha^*, \alpha) = \langle \alpha | \hat{F} | \alpha \rangle.$$

(b) If $f(\alpha^*, \alpha)$ can be expanded in a power series, i. e.
$$f(\alpha^*, \alpha) = \sum_{n,m} f_{nm} (\alpha^*)^n \alpha^m,$$
then the operator $\hat{F}$ can be represented as
$$\hat{F} = \sum_{n,m} f_{nm} (\hat{a}^\dagger)^n \hat{a}^m.$$

(c) The operator $|0\rangle\langle 0|$ can be represented as
$$|0\rangle\langle 0| = \sum_{\ell=0}^{\infty} \frac{(-1)^\ell}{\ell!} (\hat{a}^\dagger)^\ell \hat{a}^\ell.$$

(d) The operator $e^{\lambda \hat{a}^\dagger \hat{a}}$ (λ complex) can be represented as
$$e^{\lambda \hat{a}^\dagger \hat{a}} = \sum_{\ell=0}^{\infty} \frac{(e^\lambda - 1)^\ell}{\ell!} (\hat{a}^\dagger)^\ell \hat{a}^\ell.$$

(e) The operator $|n\rangle\langle m|$ can be represented as
$$|n\rangle\langle m| = \sum_{\ell=0}^{\infty} \frac{(-1)^\ell}{\ell!} \frac{(\hat{a}^\dagger)^{\ell+n} \hat{a}^{\ell+m}}{\sqrt{n!m!}}.$$

(f) The operator $|\alpha\rangle\langle\alpha|$ can be represented as
$$|\alpha\rangle\langle\alpha| = e^{-\alpha^*\alpha} \sum_{\ell=0}^{\infty} \frac{(-1)^\ell}{\ell!} e^{\alpha \hat{a}^\dagger} (\hat{a}^\dagger)^\ell \hat{a}^\ell e^{\alpha^*\hat{a}}.$$

12.4 Antinormal Ordering

Throughout this problem the operators $\hat{a}$ and $\hat{a}^\dagger$ satisfy the commutation relation $[\hat{a}, \hat{a}^\dagger] = 1$ and $|\alpha\rangle$ denotes a coherent state.

(a) For a function $g(\alpha)$ which is analytic in α we have
$$\int g(\alpha) \exp\left(-z\alpha^*\alpha + \alpha^*\xi\right) \frac{d^2\alpha}{\pi} = \frac{g(\xi/z)}{z}.$$

(b) The trace of an operator $\hat{F}$ is
$$\text{Tr}\,\hat{F} = \int \langle \alpha | \hat{F} | \alpha \rangle \frac{d^2\alpha}{\pi}.$$

(c) For an arbitrary operator $\hat{F}$ the following identity holds:

$$\hat{F} = \int \mathrm{Tr}\left(\hat{F} e^{\xi \hat{a}^\dagger - \xi^* \hat{a}}\right) e^{-\xi \hat{a}^\dagger + \xi^* \hat{a}} \frac{d^2\xi}{\pi}.$$

(d) Define a function $f(\alpha^*, \alpha)$ via

$$\hat{F} = \int f(\alpha^*, \alpha)|\alpha\rangle\langle\alpha| \frac{d^2\alpha}{\pi}.$$

If $f(\alpha^*, \alpha)$ can be expanded in a power series

$$f(\alpha^*, \alpha) = \sum_{n,m} f_{nm} (\alpha^*)^n \alpha^m,$$

then

$$\hat{F} = \sum_{n,m} f_{nm} \hat{a}^m (\hat{a}^\dagger)^n.$$

(e) Determine with the help of (c) a function $f(\alpha^*, \alpha)$ corresponding to the operator $\hat{F}$ such that

$$\hat{F} = \int f(\alpha^*, \alpha)|\alpha\rangle\langle\alpha| \frac{d^2\alpha}{\pi}.$$

What can be said about the existence of the function $f(\alpha^*, \alpha)$?

(f) Determine a function $f(\alpha^*, \alpha)$ corresponding to the operator $e^{\lambda \hat{a}^\dagger \hat{a}}$ (λ complex) such that

$$e^{\lambda \hat{a}^\dagger \hat{a}} = \int f(\alpha^*, \alpha)|\alpha\rangle\langle\alpha| \frac{d^2\alpha}{\pi}.$$

(g) Show:

$$e^{\lambda \hat{a}^\dagger \hat{a}} = e^{-\lambda} \sum_{\ell=0}^{\infty} \frac{(1 - e^{-\lambda})^\ell}{\ell!} \hat{a}^\ell (\hat{a}^\dagger)^\ell.$$

(h) What can be said about the convergence of this expansion?

12.5 s-Parameterized Phase Space Distributions

We define the s-parameterized quasiprobability distributions via

$$W(\alpha, s) = \int \mathrm{Tr}\{\hat{\rho} e^{\xi \hat{a}^\dagger - \xi^* \hat{a} + s|\xi|^2/2}\} e^{-\xi \alpha^* + \xi^* \alpha} \frac{d^2\xi}{\pi^2}.$$

Show the following properties:

(a) $W(\alpha, s)$ is normalized:

$$\int W(\alpha, s) d^2\alpha = 1.$$

(b) $s = -1$ yields the Q-function:
$$W(\alpha, s = -1) = \frac{1}{\pi} \langle \alpha | \hat{\rho} | \alpha \rangle.$$

(c) $s = 1$ yields the P-function:
$$\int W(\alpha, s = 1) |\alpha\rangle\langle\alpha| \, d^2\alpha = \hat{\rho}.$$

(d) $s = 0$ yields the Wigner function in suitable variables:
$$W\left(\alpha = \sqrt{\frac{M\omega}{2\hbar}} x + \frac{i}{\sqrt{2\hbar M\omega}} p, s = 0\right) = \frac{1}{\pi} \int \langle x + \frac{y}{2} | \hat{\rho} | x - \frac{y}{2} \rangle e^{-ipy/\hbar} \, dy.$$

Hint: Express $\hat{a}$ and $\hat{a}^\dagger$ in terms of position and momentum operators and perform the trace using the eigenstates of the position operator.

(e) Derive the diffusion type of equation
$$\frac{\partial W}{\partial s} = -\frac{1}{8}\left(\frac{\partial^2}{\partial \alpha_r^2} + \frac{\partial^2}{\partial \alpha_i^2}\right) W.$$

(f) Calculate $W(\alpha, s)$ for a coherent state $|\alpha_0\rangle$.

(g) Evaluate $W(\alpha, s)$ for a Fock state $|n\rangle$:
  i. Derive the relation
  $$\text{Tr}\{|n\rangle\langle n| e^{\xi \hat{a}^\dagger - \xi^* \hat{a} + s|\xi|^2/2}\} = e^{(s-1)|\xi|^2/2} L_n(|\xi|^2),$$
  where
  $$L_n(x) = \sum_{\nu=0}^{n} \binom{n}{\nu} \frac{(-x)^\nu}{\nu!}$$
  are the Laguerre polynomials.
  Hint: Perform the trace using Fock states.

  ii. Calculate $W(\alpha, s)$.
  Result:
  $$W(\alpha, s) = \frac{2}{\pi} \frac{1}{1-s} \left(\frac{s+1}{s-1}\right)^n L_n\left(\frac{4\alpha\alpha^*}{1-s^2}\right) \exp\left(-\frac{2\alpha\alpha^*}{1-s}\right).$$

  Hint: First show that the Laguerre polynomials satisfy the following relation:
  $$\sum_{n=0}^{\infty} L_n(x) z^n = \frac{\exp\left(-\frac{zx}{1-z}\right)}{1-z}.$$
  With this formula the problem reduces to the integration of exponentials.

(h) Show for the Fock state that the limiting case $s \to -1$ yields the Q-function.

Hint: The formulas of Problems 12.4(a),(c) may be helpful.

12.6 Different Representation of the Wigner Function

Show that the Wigner function can be represented as two displacements and a parity operation. Indeed, prove the relation:

$$W(\alpha) = \frac{2}{\pi} Tr\left[\hat{\rho}\hat{D}^{\dagger}(-\alpha)\hat{P}\hat{D}(-\alpha)\right]$$

where

$$\hat{P} = \int_{-\infty}^{\infty} dx\, |-x\rangle\langle x|$$

is the parity operator and

$$\alpha = \frac{x + ip}{\sqrt{2}}.$$

References

The Q-function was first introduced by
K. Husimi, *Some Formal Properties of the Density Matrix*, Proc. Phys. Math. Soc. Jpn. **22**, 264–314 (1940)

It was later rediscovered by
Y. Kano, *A new phase-space distribution function in the statistical theory of the electromagnetic field*, J. Math. Phys. **6**, 1913–1915 (1965)

The P-distribution was introduced independently by
R.J. Glauber, *Photon correlations*, Phys. Rev. Lett. **10**, 84–86 (1963)

and by
E.C.G. Sudarshan, *Equivalence of semiclassical and quantum mechanical descriptions of statistical light beams*, Phys. Rev. Lett. **10**, 277–279 (1963)

The relation between the Husimi-Kano Q-function and the Glauber-Sudarshan P-distribution was discussed by
C.L. Metha and **E.C.G. Sudarshan**, *Relation between quantum and semiclassical description of optical coherence*, Phys. Rev. **138**, B274–B280 (1965)

The s-parameterized phase space distribution was introduced by
K.E. Cahill and **R.J. Glauber**, *Density Operators and Quasiprobability Distributions*, Phys. Rev. A **177**, 1882–1902 (1969)

For the most general formulation of phase space calculus we refer to
G.S. Agarwal and **E. Wolf**, *Calculus for Functions of Noncommuting Operators and General Phase-Space Methods in Quantum Mechanics. I. Mapping Theorems and Ordering of Functions of Noncommuting Operators*, Phys. Rev. D **2**, 2161–2186 (1970)

G.S. Agarwal and **E. Wolf**, *Calculus for Functions of Noncommuting Operators and General Phase-Space Methods in Quantum Mechanics. II. Quantum Mechanics in Phase Space*, Phys. Rev. D **2**, 2187–2205 (1970)

G.S. Agarwal and **E. Wolf**, *Calculus for Functions of Noncommuting Operators and General Phase-Space Methods in Quantum Mechanics. III. A Generalized Wick Theorem and Multitime Mapping*, Phys. Rev. D **2**, 2206–2225 (1970)

G.S. **Agarwal** and E. **Wolf**, *Quantum Dynamics in Phase Space*, Phys. Rev. Lett. **21**, 180–183 (1968)

For a discussion of thermal phase states see

B. **Däubler, Ch. Miller, H. Risken** and L. **Schoendorff**, *Quantum States with Minimum Phase Uncertainty for the Süssmann Measure*, Physica Scripta, **T48**, 119–123 (1993)

13 Optical Interferometry

How can we measure the internal structure of a quantum state, that is how can we measure phase space distributions? In the present chapter we introduce and analyze two approaches which achieve this goal: The method of quantum state tomography uses a single beam splitter to mix a field mode with a local oscillator and cuts the Wigner function into many slices. From the distributions of these slices for various local oscillator phases we can reconstruct the Wigner function with the help of the Radon transformation discussed in Sec. 4.5.1. Hence, this method measures directly the cut distributions and evaluates mathematically the Wigner function.

In contrast the eight-port homodyne detector uses four beam splitters and a phase shifter to make two simultaneous homodyne measurements on the light field. The count statistics in this case provides directly the scaled Q-function without any further mathematical operations. However, the simultaneous measurement of two conjugate variables has only been made possible by allowing additional noise entering through the open input port at the entrance beam splitter.

Both techniques heavily rely on one or many beam splitters. In order to gain insight into the question why the homodyne or the eight-port homodyne detector allow us to reconstruct phase space distributions we have to discuss the action of a beam splitter on the quantum states of the radiation field. In particular, we have to understand how quantum states transform at a beam splitter, that is how the states of the incident fields transform into the states of the outgoing fields.

For this purpose we analyze in Sec. 13.1 the action of a beam splitter. The behavior of the field states is especially simple when we make use of the Glauber-Sudarshan distribution discussed in the preceding chapter. We then in Sec. 13.2 turn to the analysis of the homodyne detector. Here we derive the photon statistics at the two exit ports and show that in the limit of a strong local oscillator field we can measure the electric field distribution. We devote Sec. 13.3 to the discussion of the eight-port homodyne detector and demonstrate that the measured photon count statistics is a scaled Q-function. In Sec. 13.4 we connect this approach with the old question of phase operators. Indeed, this multiport interferometer suggests to introduce measured phase operators which have strong ties to the Einstein-Podolsky-Rosen situation.

13.1 Beam Splitter

Chapter 10 briefly outlines how the quantum theory of radiation approaches a given optical setup using the example of a box-shaped resonator. We start from Maxwell's equations, separate the electromagnetic field expressed in Coulomb gauge by the vector potential into a time dependent part, defined by an oscillator equation, and a space dependent part governed by the Helmholtz equation. The boundary conditions provided by the resonator together with the Helmholtz equation fix the spatial part of the electromagnetic field. They define the modes. Quantum physics enters through the time dependent part as harmonic oscillator excitations of these modes.

When we want to understand a beam splitter and its quantum features in the framework of the quantum theory of radiation we have to follow the above prescription: First we have to find the modes and then quantize them as outlined in the preceding chapters. But what are in this example the boundary conditions determining the modes?

The boundary conditions arise from the beam splitter itself. We therefore need a model for it. The most elementary model is a dielectric medium localized in space. For the sake of simplicity we assume that it is a thin plate dividing the space of interest. Before we discuss quantized light fields and light impinging on and emerging from a beam splitter our first duty is to find the modes for this problem. For this purpose we have to solve the Helmholtz equation in the presence of the dividing dielectric medium subjected to the appropriate boundary conditions. In the space outside of the medium, that is, in free space the solutions of the Helmholtz equations are just plane waves $\exp(\pm i \vec{k} \cdot \vec{r})$. The form of the solutions inside the medium depend on the properties of the dielectric. The boundary conditions provide matching conditions between the solutions outside and inside of the beam splitter.

In this way the beam splitter couples the solutions on one side to the solutions on the other side. In the most elementary case of a linear dielectric medium we find a linear coupling of all modes. We can therefore express the amplitude $a_{\ell'}$ of the ℓ'th mode on one side of the beam splitter by the amplitudes a_ℓ of the ℓth mode on the other side. The matrix $T_{\ell'\ell}$ connects the two amplitudes according to

$$a_{\ell'} = \sum_l T_{\ell'\ell} a_\ell.$$

The matrix elements $T_{\ell'\ell}$ are determined by the dielectric medium and the boundary conditions. Moreover, conservation of energy puts some restrictions on this matrix.

We make the transition to a quantized field by quantizing the amplitude a_ℓ and replacing it by the operator $\hat{a}_\ell$. This leads to a transformation between the mode operators on the two sides. Hence we can describe the action of a beam splitter by a transformation of the operators. Likewise we can also describe it by a transformation of the states.

13.1.1 Classical Treatment

In the most elementary discussion we confine ourselves to two input modes and couple them to two output modes on the other side of the beam splitter as shown in

13.1 Beam Splitter   351

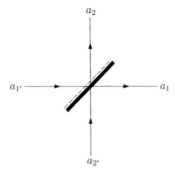

Fig. 13.1: Schematic sketch of a beam splitter. A dielectric medium transforms the amplitudes $a_{1'}$ and $a_{2'}$ of the ingoing modes 1' and 2' into the amplitudes a_1 and a_2 of the outgoing modes 1 and 2. The situation is not quite symmetric since energy conservation requires a phase shift of π in the reflection of one of the two modes. We express this asymmetry by the dashed line on one side of the mirror.

Fig. 13.1. We first derive the 2×2 transformation matrix from general considerations and then show how the operators and the quantum states transform. We want to express the amplitudes of the two input modes in terms of the amplitudes of the output modes. We denote the two modes on one side by 1' and 2' and abbreviate their amplitudes by $a_{1'}$ and $a_{2'}$. This notation is rather suggestive since in the later treatment these classical amplitudes become the annihilation operators of the modes 1' and 2'. We call the two modes on the other side of the beam splitter 1 and 2 and the corresponding amplitudes are a_1 and a_2.

We can now express the amplitudes a_1 and a_2 by the amplitudes $a_{1'}$ and $a_{2'}$ on the other side. When we recall that two paths contribute to the individual amplitudes a_1 and a_2 namely a path where the light is transmitted and reflected we find the transformation

$$\begin{aligned} a_1 &= t_1 a_{1'} + r_2 a_{2'} \\ a_2 &= t_2 a_{2'} + r_1 a_{1'} \end{aligned} \quad \text{or} \quad \begin{pmatrix} a_1 \\ a_2 \end{pmatrix} = \begin{pmatrix} t_1 & r_2 \\ r_1 & t_2 \end{pmatrix} \begin{pmatrix} a_{1'} \\ a_{2'} \end{pmatrix}. \quad (13.1)$$

Here the quantities t_i and r_i denote the transmission and the reflection coefficients of the respective beam. The specific details of the beam splitter, that is of the dielectric medium determine these numbers.

However, for a lossless beam splitter conservation of energy imposes the condition

$$|a_1|^2 + |a_2|^2 = |a_{1'}|^2 + |a_{2'}|^2$$

on the absolute value squares of these amplitudes, that is on the intensities in these modes. This equation holds true for arbitrary input amplitudes $a_{1'}$ and $a_{2'}$ and puts a constraint on the reflection and transmission coefficients. We find this constraint by substituting the amplitudes a_1 and a_2 from Eqs. (13.1) into this condition and obtain

$$\begin{aligned} |a_1|^2 + |a_2|^2 &= (t_1 a_{1'} + r_2 a_{2'})(t_1^* a_{1'}^* + r_2^* a_{2'}^*) + (t_2 a_{2'} + r_1 a_{1'})(t_2^* a_{2'}^* + r_1^* a_{1'}^*) \\ &= \left[|t_1|^2 + |r_1|^2\right] |a_{1'}|^2 + \left[|t_2|^2 + |r_2|^2\right] |a_{2'}|^2 \\ &\quad + (t_1 r_2^* + t_2^* r_1) a_{1'} a_{2'}^* + (t_1^* r_2 + t_2 r_1^*) a_{2'} a_{1'}^*. \end{aligned}$$

Since this relation must hold true for arbitrary amplitudes $a_{1'}$ and $a_{2'}$ we arrive at the relations

$$|t_1|^2 + |r_1|^2 = |t_2|^2 + |r_2|^2 = 1 \tag{13.2a}$$

$$t_1 r_2^* + t_2^* r_1 = t_1^* r_2 + t_2 r_1^* = 0. \tag{13.2b}$$

The first equation states that there is no absorption in the beam splitter. The second equation relates the different modes to each other and it brings in a minus sign. This comes out more clearly in the case of a completely symmetric beam splitter.

13.1.2 Symmetric Beam Splitter

In the following sections we will use for the sake of simplicity only 50 : 50-beam splitters. Here, one half of the *energy* is transmitted and the other half is reflected. In this case, the reflection and transmission coefficients, Eq. (13.2a), take the form

$$|t_1|^2 = |r_1|^2 = |t_2|^2 = |r_2|^2 = \frac{1}{2}. \tag{13.3}$$

The condition Eq. (13.2b) connects the phases of the reflection and transmission coefficients. These phases depend on the details of the beam splitter. Indeed, one or both sides of a beam splitter can be partial permeable mirrors, or the whole beam splitter can consist of several layers with different index of refraction.

Independent of these experimental details the condition Eq. (13.2b) puts a restriction on the relative phase. We can for example choose a completely symmetric situation

$$t_1 = t_2 = \frac{1}{\sqrt{2}}$$

for the transmission of the modes. However, Eq. (13.2b) dictates that the reflection is asymmetric, that is

$$r_1 = \frac{1}{\sqrt{2}} \qquad r_2 = e^{i\pi} \frac{1}{\sqrt{2}} = -\frac{1}{\sqrt{2}}.$$

In this case we deal with the situation depicted in Fig. 13.2. The beam which is reflected at the dark side of the beam splitter experiences a phase shift of π whereas the beam in the mode $1'$ is reflected from the side indicated by the dashed line without a phase shift.

Hence, we arrive at the amplitude relations

$$a_1 = \frac{1}{\sqrt{2}}(a_{1'} - a_{2'})$$

and

$$a_2 = \frac{1}{\sqrt{2}}(a_{1'} + a_{2'}),$$

or the transformation

$$\begin{pmatrix} a_1 \\ a_2 \end{pmatrix} = T \begin{pmatrix} a_{1'} \\ a_{2'} \end{pmatrix} = \frac{1}{\sqrt{2}} \begin{pmatrix} 1 & -1 \\ 1 & 1 \end{pmatrix} \begin{pmatrix} a_{1'} \\ a_{2'} \end{pmatrix}.$$

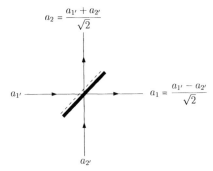

Fig. 13.2: The symmetric beam splitter creates out of two classical input fields their sum and differences. Moreover, the amplitudes get reduced by a factor of $\sqrt{2}$. The difference of the two amplitudes arises due to the reflection from the thicker medium and is a consequence of conservation of energy.

These equations predict the classical intensities

$$I_1 \equiv |a_1|^2 = \frac{1}{2}|a_{1'} - a_{2'}|^2$$

and

$$I_2 \equiv |a_2|^2 = \frac{1}{2}|a_{1'} + a_{2'}|^2$$

at the two output ports of the beam splitter.

13.1.3 Transition to Quantum Mechanics

We now make the transition to quantum mechanics by replacing the mode amplitudes $(a_{1'}, a_{2'})$ and (a_1, a_2) by operators $(\hat{a}_{1'}, \hat{a}_{2'})$ and $(\hat{a}_1, \hat{a}_2)$. The beam splitter transformation of the operators is then

$$\hat{a}_1 = \frac{1}{\sqrt{2}}(\hat{a}_{1'} - \hat{a}_{2'}) \tag{13.4a}$$

and

$$\hat{a}_2 = \frac{1}{\sqrt{2}}(\hat{a}_{1'} + \hat{a}_{2'}). \tag{13.4b}$$

Hence, the beam splitter creates linear combinations of the mode operators: The beam splitter transformation is linear.

13.1.4 Transformation of Quantum States

The transformation of the operators is only one possible point of view. Another but equivalent approach rests on quantum states. However, the situation in the state language is more complicated.

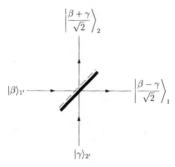

Fig. 13.3: Coherent states transform like classical amplitudes at a beam splitter. Hence, the input state $|\Psi_{\text{in}}\rangle \equiv |\beta\rangle_{1'}|\gamma\rangle_{2'}$ yields state $|\Psi_{\text{out}}\rangle \equiv |(\beta-\gamma)/\sqrt{2}\rangle_1|(\beta+\gamma)/\sqrt{2}\rangle_2$ of the two output ports.

Since, we have two input modes we have to specify the complete quantum state $|\Psi_{\text{in}}\rangle$ of the two modes. In the most elementary case the two quantum states $|\phi\rangle_{1'}$ and $|\psi\rangle_{2'}$ are independent from each other, that is they are a product state

$$|\Psi_{\text{in}}\rangle = |\phi\rangle_{1'}|\psi\rangle_{2'}.$$

However, the input states can be entangled with each other. In this case we cannot denote them by a product state. Moreover, states in the input mode could be mixed states and have therefore to be denoted by a density operator giving rise to a two-mode density operator $\hat{\rho}_{\text{in}}$. Hence, the most general input state is $\hat{\rho}_{\text{in}}$.

What is the quantum state of the two output modes? Since the beam splitter combines the two input modes into linear combinations we can expect the output modes to be entangled. In general, we cannot factorize the quantum states of the output into two product states. Obviously we can denote the quantum state by a two-mode density operator $\hat{\rho}_{\text{out}}$. But what is the connection between $\hat{\rho}_{\text{in}}$ and $\hat{\rho}_{\text{out}}$?

Transformation of Coherent States

In order to answer this question we first concentrate on the beam splitter transformation of two independent coherent states $|\beta\rangle_{1'}$ and $|\gamma\rangle_{2'}$ in the two modes $1'$ and $2'$, respectively. Since coherent states are eigenstates of the annihilation operators they transform like amplitudes. Therefore, we find that the input state

$$|\Psi_{\text{in}}\rangle \equiv |\beta\rangle_{1'}|\gamma\rangle_{2'} \qquad (13.5\text{a})$$

transforms into the output state

$$|\Psi_{\text{out}}\rangle \equiv \left|\frac{\beta-\gamma}{\sqrt{2}}\right\rangle_1 \left|\frac{\beta+\gamma}{\sqrt{2}}\right\rangle_2. \qquad (13.5\text{b})$$

The transformation of coherent states at a beam splitter is depicted in Fig. 13.3.

Transformation of an Arbitrary Density Matrix

So far we have only considered coherent states in the two input modes. We now generalize this treatment to the case when the radiation fields in the modes 1' and 2' are described by a density operator $\hat{\rho}_{\text{in}}$.

Since we know that coherent states remain coherent states under a beam splitter transformation it is advantageous to expand the density operator

$$\hat{\rho}_{\text{in}} = \int d^2\beta \int d^2\gamma \, P_{1',2'}(\beta,\gamma) \, |\beta\rangle_{1'}\langle\beta| \otimes |\gamma\rangle_{2'}\langle\gamma| \qquad (13.6)$$

into coherent states. Here $P_{1',2'}(\beta,\gamma)$ is the Glauber-Sudarshan distribution of the fields in the two modes 1' and 2'. Moreover, to simplify the notation we have included the mode index between the ket- and the bra-vector forming the density operator only once.

Now we are able to apply the beam splitter transformation

$$|\beta\rangle_{1'}|\gamma\rangle_{2'} \to \left|\frac{\beta-\gamma}{\sqrt{2}}\right\rangle_1 \left|\frac{\beta+\gamma}{\sqrt{2}}\right\rangle_2 \qquad (13.7)$$

to the two coherent states. Hence, the density matrix $\hat{\rho}_{\text{out}}$ of the two output ports 1 and 2 reads

$$\hat{\rho}_{\text{out}} = \int d^2\beta \int d^2\gamma \, P_{1',2'}(\beta,\gamma) \left|\frac{\beta-\gamma}{\sqrt{2}}\right\rangle_1\!\!\left\langle\frac{\beta-\gamma}{\sqrt{2}}\right| \otimes \left|\frac{\beta+\gamma}{\sqrt{2}}\right\rangle_2\!\!\left\langle\frac{\beta+\gamma}{\sqrt{2}}\right|. \qquad (13.8)$$

Special Examples of Input Fields

When both input modes are in coherent states $|\beta_0\rangle$ and $|\alpha\rangle$ that is, for

$$P_{1',2'}(\beta,\gamma) = \delta(\beta-\beta_0)\delta(\gamma-\alpha)$$

we can perform the integrations and find that the density operator

$$\hat{\rho}_{\text{out}} = \left|\frac{\beta_0-\alpha}{\sqrt{2}}\right\rangle_1\!\!\left\langle\frac{\beta_0-\alpha}{\sqrt{2}}\right| \otimes \left|\frac{\beta_0+\alpha}{\sqrt{2}}\right\rangle_2\!\!\left\langle\frac{\beta_0+\alpha}{\sqrt{2}}\right|$$

consists of the two independent coherent states $|(\beta_0-\alpha)/\sqrt{2}\rangle_1$ and $|(\beta_0+\alpha)/\sqrt{2}\rangle_2$ in the two modes 1 and 2.

Note, however, that when $P_{1',2'}(\beta,\gamma)$ is not a delta function in the two integration variables, the two modes are not independent anymore but are correlated with each other via the integration over the complex variables β and γ, that is via the integration over phase space. The two modes are entangled.

We conclude this section by presenting the result when one of the two input modes, say the mode 2', is in a coherent state $|\alpha\rangle$. This example is central to the analysis of the homodyne detector discussed in the next section. In this case we have

$$P_{1',2'}(\beta,\gamma) = P(\beta)\delta(\gamma-\alpha) \qquad (13.9)$$

where $P(\beta)$ denotes the Glauber-Sudarshan distribution for the input mode 1'.

The δ-function in the phase space variable γ allows us to perform one of the integrations and we arrive at the density operator

$$\hat{\rho}_{\text{out}} = \int d^2\beta \, P(\beta) \left|\frac{\beta-\alpha}{\sqrt{2}}\right\rangle_1 \left\langle\frac{\beta-\alpha}{\sqrt{2}}\right| \otimes \left|\frac{\beta+\alpha}{\sqrt{2}}\right\rangle_2 \left\langle\frac{\beta+\alpha}{\sqrt{2}}\right| \qquad (13.10)$$

of the radiation field in the two modes 1 and 2.

13.1.5 Count Statistics at the Exit Ports

In the preceding section we have derived the density operator for the radiation field in the two exit modes of a 50 : 50 beam splitter. We now calculate the probability

$$W(n_1, n_2) \equiv \langle n_1| \langle n_2| \hat{\rho}_{\text{out}} |n_1\rangle |n_2\rangle \qquad (13.11)$$

of finding n_1 and n_2 excitations of the modes 1 and 2, respectively. We recall that the states $|n_1\rangle$ and $|n_2\rangle$ are the photon number states of the mode 1 and 2. Hence, we can associate this distribution with the photon statistics in the two modes. Note, that $W(n_1, n_2)$ is a joint probability, that is, the probability to obtain n_1 excitations in the first mode and n_2 in the second.

When we use the expression Eq. (13.8) for the density operator $\hat{\rho}_{\text{out}}$ of the output modes we arrive at

$$W(n_1, n_2) = \int d^2\beta \int d^2\gamma P_{1',2'}(\beta,\gamma) K_{n_1,n_2}(\gamma,\beta).$$

The kernel

$$K_{n_1,n_2}(\gamma,\beta) \equiv \left|\left\langle n_1 \left| \frac{\beta-\gamma}{\sqrt{2}} \right\rangle_1 \right|^2 \left|\left\langle n_2 \left| \frac{\beta+\gamma}{\sqrt{2}} \right\rangle_2 \right|^2$$

is the probability of counting n_1 photons in mode 1 and n_2 photons in mode 2 when the modes 1' and 2' are in the coherent states $|\beta\rangle$ and $|\gamma\rangle$.

We recall the Poisson photon statistics

$$|\langle n|\alpha\rangle|^2 = \frac{|\alpha|^{2n}}{n!} e^{-|\alpha|^2} \qquad (13.12)$$

of a coherent state $|\alpha\rangle$ and find the explicit expression

$$K_{n_1,n_2}(\gamma,\beta) = \left|\frac{\beta-\gamma}{\sqrt{2}}\right|^{2n_1} \left|\frac{\beta+\gamma}{\sqrt{2}}\right|^{2n_2} \frac{1}{n_1! n_2!} \exp\left[-\frac{1}{2}(|\beta-\gamma|^2 + |\beta+\gamma|^2)\right] \qquad (13.13)$$

for the kernel.

Hence, the photon count statistics at the exit ports of a beam splitter is given as the phase space integration of the Glauber-Sudarshan distribution $P_{1',2'}$ of the fields in the modes 1' and 2' with respect to the kernel $K_{n_1,n_2}(\gamma,\beta)$.

We conclude this section by briefly discussing the case when one of the two modes, say the mode 2', is in a coherent state $|\alpha\rangle$ and the field in the mode 1' is described by a Glauber-Sudarshan P-distribution $P(\beta)$. Since in this case we have according to Eq. (13.9)

$$P_{1',2'}(\beta,\gamma) = P(\beta)\delta(\gamma-\alpha)$$

the count statistics reads

$$W(n_1, n_2) = \int d^2\beta P(\beta) K_{n_1,n_2}(\alpha, \beta), \tag{13.14}$$

where the kernel K_{n_1,n_2} is given by Eq. (13.13).

The same result emerges when we substitute the density operator $\hat{\rho}_{\text{out}}$ Eq. (13.10) into the expression Eq. (13.11) for the count statistics.

13.2 Homodyne Detector

In the preceding section we have derived expressions for the count statistics, that is for the probability of counting n_1 photons in the mode 1 and n_2 in the mode 2. A special type of measurement is to subtract the number of counts in the two modes and to make a histogram of the difference counts. This type of detection is called homodyne detection. We emphasize that here the two modes have identical frequencies. When they have different frequencies this detection scheme is called heterodyne detection.

13.2.1 Classical Considerations

In Sec. 13.1.2 we have provided a classical description of a symmetric beam splitter. Here, we have found that the classical intensities I_1 and I_2 in the two exit modes of the beam splitter follow from the electric fields $a_{1'} \equiv \mathcal{E}$ and $a_{2'} \equiv \alpha e^{i\theta}$ in the two input modes $1'$ and $2'$ via the relations

$$I_1 = \frac{1}{2}|a_{1'} - a_{2'}|^2 = \frac{1}{2}\left[|\mathcal{E}|^2 + \alpha^2 - \alpha\left(\mathcal{E}e^{-i\theta} + \mathcal{E}^*e^{i\theta}\right)\right]$$

and

$$I_2 = \frac{1}{2}|a_{1'} + a_{2'}|^2 = \frac{1}{2}\left[|\mathcal{E}|^2 + \alpha^2 + \alpha\left(\mathcal{E}e^{-i\theta} + \mathcal{E}^*e^{i\theta}\right)\right].$$

When we now subtract the two intensities on the two detectors the difference intensity

$$I_{21} \equiv I_2 - I_1 = \alpha\left(\mathcal{E}e^{-i\theta} + \mathcal{E}^*e^{i\theta}\right)$$

contains the phase θ of the electric field in mode $2'$. Moreover, the electric field $\mathcal{E}$ in mode $1'$ enters in a linear combination of $\mathcal{E}e^{-i\theta}$ and $\mathcal{E}^*e^{i\theta}$. This is very reminiscent of the quadrature operator

$$\hat{\mathcal{E}}_\theta \equiv \frac{1}{\sqrt{2}}\left(\hat{a}e^{-i\theta} + \hat{a}^\dagger e^{i\theta}\right) \tag{13.15}$$

of the electric field defined in Eq. (11.6). Indeed, when we replace in our classical treatment the amplitude $\mathcal{E}$ by the annihilation operator $\hat{a}$ and the complex conjugate $\mathcal{E}^*$ by the creation operator $\hat{a}^\dagger$ we arrive at the quadrature operator.

Hence the intensity difference I_{21} measures the quadrature operator of the electric field. This in turn allows us to reconstruct the Wigner function as discussed in Chapter 4.

13.2.2 Quantum Treatment

The intensity difference provides information about the electric field distribution. This is the message of the above classical considerations. They rely on the replacement of the classical fields by operators. Note however, that we have only replaced the field in one mode by operators but have kept the field in the other mode classical.

In order to replace this heuristic argument by a rigorous derivation we now consider the case when the radiation field in the mode $1'$ is described by a density matrix $\hat{\rho}$, and the field in the entrance mode $2'$ is a coherent state $|\alpha\rangle$. Moreover, we use quantum states rather than field operators to derive an analytical result for the statistics of photon count differences.

Outline of the Measurement Strategy

We first briefly summarize how to translate the intensity difference measurement into quantum language. Measuring the intensities I_1 and I_2 at the detectors D_1 and D_2 translates into measuring the number of photons n_1 and n_2. To subtract the two intensities I_1 and I_2 translates into subtracting the two photon numbers which yields the photon number difference $n_{21} \equiv n_1 - n_2$. Hence the intensity difference distribution corresponds to the distribution function $W(n_{21})$ for the photon count differences. Motivated by the classical argument we expect that in the limit of a coherent state of large amplitude $|\alpha| \gg 1$, the distribution $W(n_{21})$ is the distribution corresponding to the operator

$$\hat{\mathcal{E}}_\theta \equiv \frac{1}{\sqrt{2}} \left(\hat{a} e^{-i\theta} + \hat{a}^\dagger e^{i\theta} \right). \tag{13.16}$$

Therefore, in the experiment we count the number of photons in each detector and subtract them. We re-prepare the field states $\hat{\rho}$ and $|\alpha\rangle$, repeat the experiment and find again the photon number difference. We make a histogram of these counts. This histogram is then the distribution $W(n_{21})$. Such experiments have indeed been performed by the groups of M. Raymer in Eugene (USA) and J. Mlynek in Konstanz (Germany) and allow to measure the field state. For typical measured field distributions we refer to Fig. 4.11 in Chapter 4.

Derivation of Photon Count Statistics

We are interested in the distribution $W(n_{21})$ for the photon count differences $n_{21} \equiv n_2 - n_1$. When we do not store simultaneously the sum $\tilde{n}_{21} \equiv n_1 + n_2$ of the photon counts we have to take the trace of the distribution $W(n_1, n_2)$ over $\tilde{n}_{21}$. Hence we find the distribution $W(n_{21})$ of the photon count differences by expressing the photon number n_1 by the difference n_{21} and n_2 and sum over n_2, that is

$$W(n_{21}) \equiv \sum_{n_2=0}^{\infty} W(n_1 = n_2 - n_{21}; n_2) = \sum_{n_1=0}^{\infty} W(n_1; n_2 = n_{21} + n_1).$$

In the last step of this equation we have expressed n_2 by n_{21} and n_1 and summed over n_1.

When we substitute the explicit expression

$$W(n_1, n_2) = \int d^2\beta P(\beta) K_{n_1,n_2}(\alpha, \beta)$$

for the count statistics, Eq. (13.14), into this expression we arrive at

$$W(n_{21}) = \int d^2\beta\, P(\beta)\, K_{n_{21}}(\alpha, \beta) \tag{13.17}$$

where the kernel

$$K_{n_{21}}(\alpha, \beta) \equiv \sum_{n_1=0}^{\infty} K_{n_1, n_{21}+n_1}(\alpha, \beta)$$

$$= \sum_{n_1=0}^{\infty} \left|\left\langle n_1 \left| \frac{\beta-\alpha}{\sqrt{2}} \right\rangle_1\right|^2 \left|\left\langle n_{21}+n_1 \left| \frac{\beta+\alpha}{\sqrt{2}} \right\rangle_2\right|^2 \tag{13.18}$$

is the probability of counting the photon number difference n_{21} at the exit modes of a symmetric beam splitter when the fields in the two input modes are in coherent states $|\alpha\rangle$ and $|\beta\rangle$.

In Appendix K we evaluate this kernel in an analytic way and find the explicit expression

$$K_{n_{21}}(\alpha, \beta) = \left|\frac{\beta+\alpha}{\beta-\alpha}\right|^{n_{21}} I_{|n_{21}|}(|\beta^2 - \alpha^2|)\, e^{-|\alpha|^2 - |\beta|^2} \tag{13.19}$$

with the modified Bessel function I_ν.

Hence the statistics of the photon count differences of a homodyne detector is the Glauber-Sudarshan P-distribution of the input state integrated over β-phase space using the kernel $K_{n_{21}}(\alpha, \beta)$. Here, we have assumed that the other input field is in a coherent state $|\alpha\rangle$ of arbitrary amplitude.

Strong Local Oscillator Limit

In the limit of a coherent state $|\alpha\rangle = \left||\alpha|e^{i\theta}\right\rangle$ of large amplitude, that is for $|\alpha| \gg 1$, the expressions, Eqs. (13.17) and (13.19) for the count statistics simplify considerably. In particular, we now show that in this limit the photon count statistics reduces to the electric field distribution $W_{\text{el}}(\mathcal{E}_\theta)$ scaled in the amplitude of the local oscillator. The phase θ of the field distribution is the phase of the coherent state.

For this purpose we first establish the connection between the probability distribution

$$W_{\hat{\rho}}(\mathcal{E}_\theta) \equiv \langle \mathcal{E}_\theta | \hat{\rho} | \mathcal{E}_\theta \rangle$$

to find the quadrature state $|\mathcal{E}_\theta\rangle$ in the field state defined by the density operator $\hat{\rho}$ and the photon count statistics.

When we substitute the Glauber-Sudarshan representation

$$\hat{\rho} = \int d^2\beta\, P(\beta)\, |\beta\rangle\langle\beta|,$$

of the field in the input mode $1'$ into this equation we arrive at

$$W_{\hat{\rho}}(\mathcal{E}_\theta) = \int d^2\beta\, P(\beta) W_{|\beta\rangle}(\mathcal{E}_\theta). \tag{13.20}$$

Therefore, when we want to relate the photon difference count statistics $W(n_{21})$, Eq. (13.17), to the electric field distribution $W_{\hat{\rho}}(\mathcal{E}_\theta)$, Eq. (13.20), we have to relate the kernel $K_{n_{21}}$ given by Eq. (13.19) to the electric field distribution

$$W_{|\beta\rangle}(\mathcal{E}_\theta) \equiv |\langle \mathcal{E}_\theta | \beta \rangle|^2$$

of the coherent state. In Sec. 11.2.4 we have calculated this distribution and have found the Gaussian

$$W_{|\beta\rangle}(\mathcal{E}_\theta) \equiv |\langle \mathcal{E}_\theta | \beta \rangle|^2 = \pi^{-1/2} \exp\left\{-\left[\mathcal{E}_\theta - \frac{1}{\sqrt{2}}\left(\beta e^{-i\theta} + \beta^* e^{i\theta}\right)\right]^2\right\}$$

centered at $\beta e^{-i\theta} + \beta^* e^{i\theta}$.

When we compare this Gaussian field distribution $W_{|\beta\rangle}(\mathcal{E}_\theta)$ with the exact kernel $K_{n_{21}}$ containing modified Bessel functions the two expressions are obviously different. However, they become identical in the limit of a strong local oscillator, that is for a coherent state of large amplitude.

In order to prove this identity we are therefore left with the problem of calculating the appropriate asymptotic expansion of the kernel $K_{n_{21}}$. Here, we consider the limit when $n_{21}/|\alpha|$ remains finite as $|\alpha| \to \infty$. For this case we in Appendix K derive the asymptotic expression

$$K_{n_{21}}(|\alpha| \to \infty, \beta) \cong \frac{1}{\sqrt{2\pi|\alpha|^2}} \exp\left\{-\frac{1}{2}\left[\frac{n_{21}}{|\alpha|} - \left(\beta e^{-i\theta} + \beta^* e^{i\theta}\right)\right]^2\right\}, \quad (13.21)$$

which provides the connection

$$K_{n_{21}}(|\alpha| \to \infty, \beta) \cong \frac{1}{\sqrt{2}|\alpha|} \left|\left\langle \mathcal{E}_\theta = \frac{n_{21}}{\sqrt{2}|\alpha|} \Big| \beta \right\rangle\right|^2.$$

Hence, in the strong local oscillator limit the probability distribution $W(n_{21})$ for the photon count differences n_{21} is given by

$$W(n_{21}) \cong \frac{1}{\sqrt{2}|\alpha|} W_{\hat{\rho}}\left(\mathcal{E}_\theta = \frac{n_{21}}{\sqrt{2}|\alpha|}\right), \quad (13.22)$$

which is the rescaled electric field distribution of the incident quantum state.

Equation (13.22) is quite an interesting result from the point of view of a quantum measurement: Indeed, so far we have not discussed how to measure the electric field distribution of light in a single mode. The above relation shows that counting the excitations of the exit modes of a beam splitter and making a histogram of the count differences constitutes such a measurement. However, we emphasize that this strategy only works when we combine the field to be measured with a classical field, that is, a coherent state with a large amplitude.

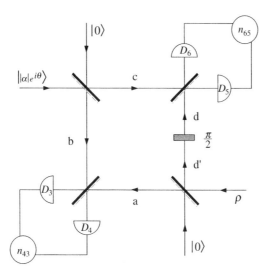

Fig. 13.4: An eight-port interferometer is built out of two simple homodyne schemes given by the detector pairs D_4/D_3 and D_6/D_5. They measure the photon count differences $n_{43} \equiv n_4 - n_3$ and $n_{65} \equiv n_6 - n_5$. Two additional beam splitters couple the resulting four input ports and one arm contains a $\pi/2$-phase shifter.

13.3 Eight-Port Interferometer

A single beam splitter transforms two input modes into two output modes. However, we can construct a more complicated setup by combining two homodyne setups together with a $\pi/2$-phase shifter as shown in Fig. 13.4. Note that in this case we have a device that transforms four input modes into four output modes. Since in this case there exist eight ports such an interferometer is called *eight-port homodyne detector* or *eight-port interferometer*.

13.3.1 Quantum State of the Output Modes

We now consider this setup in more detail and focus on the case when two of the four input modes are in a coherent state and a density matrix $\hat{\rho}$ as shown in Fig. 13.4. The two other input ports are in the vacuum states. In this case the density matrix of the input modes reads

$$\hat{\rho}_{\text{in}} = \int d^2\beta \, P(\beta) \, |\beta\rangle \langle\beta| \otimes |0\rangle \langle 0| \otimes |\alpha\rangle \langle\alpha| \otimes |0\rangle \langle 0|$$

where we have again represented the density matrix $\hat{\rho}$ of the input field by a Glauber-Sudarshan distribution $P(\beta)$.

We denote the four output modes by $3, 4, 5$ and 6 and find the density matrix $\hat{\rho}_{\text{out}}$ of these modes by transforming the individual coherent states according to the ordinary beam splitter transformation. For this purpose we now follow the actions of the individual beam splitters.

We first consider a coherent state $|\beta\rangle$ and the vacuum $|0\rangle$ entering the beam splitter at the lower right corner. According to Eq. (13.7) the beam splitter transforms the initial two mode states

$$|0\rangle \otimes |\beta\rangle$$

into the states

$$\left|-\frac{\beta}{\sqrt{2}}\right\rangle_{d'} \otimes \left|\frac{\beta}{\sqrt{2}}\right\rangle_{a}.$$

Here we have inserted a minus sign due to the reflection at the beam splitter whereas there is no such sign change for the transmitted beam. Moreover, we have assumed a 50:50 beam splitter which divides the intensity into half, that is the amplitude of the coherent state which is proportional to the electric field gets divided by a factor of $\sqrt{2}$.

Likewise the beam splitter at the upper left corner creates from

$$|\alpha\rangle \otimes |0\rangle$$

the states

$$\left|\frac{\alpha}{\sqrt{2}}\right\rangle_{b} \otimes \left|\frac{\alpha}{\sqrt{2}}\right\rangle_{c}.$$

Note that due to the orientation of the beam splitter – in contrast to the situation at the lower right beam splitter – there is no sign change in the coherent state since there is no reflection at the medium of larger index of refraction.

We now concentrate on the beam splitter at the lower left corner, which combines the two modes a and b and brings them together on the detectors D_3 and D_4. This beam splitter transforms the states

$$\left|\frac{\alpha}{\sqrt{2}}\right\rangle_{b} \otimes \left|\frac{\beta}{\sqrt{2}}\right\rangle_{a}$$

into

$$\left|\frac{1}{2}(\beta - \alpha)\right\rangle_{3} \otimes \left|\frac{1}{2}(\beta + \alpha)\right\rangle_{4}.$$

Here we have again included a minus sign in the mode 3 due to the reflection.

We now turn to the upper right beam splitter. Note that due to the $\pi/2$-phase shifter the coherent state

$$\left|-\frac{\beta}{\sqrt{2}}\right\rangle_{d'}$$

created by the beam splitter at the lower right corner at the entrance of the upper right beam splitter now reads

$$\left|-e^{i\pi/2}\frac{\beta}{\sqrt{2}}\right\rangle_{d} = \left|-i\frac{\beta}{\sqrt{2}}\right\rangle_{d}.$$

Hence we find from the states

$$\left|\frac{\alpha}{\sqrt{2}}\right\rangle_{c} \otimes \left|-i\frac{\beta}{\sqrt{2}}\right\rangle_{d}$$

in the modes c and d the states

$$\left|-\frac{1}{2}(i\beta - \alpha)\right\rangle_5 \otimes \left|-\frac{1}{2}(i\beta + \alpha)\right\rangle_6$$

in the modes 5 and 6 impinging on the photon detectors D_5 and D_6.

In summary we have found that the states

$$|0\rangle \otimes |\beta\rangle \otimes |\alpha\rangle \otimes |0\rangle$$

in the four input modes transform into the states

$$\left|\frac{1}{2}(\beta - \alpha)\right\rangle_3 \otimes \left|\frac{1}{2}(\beta + \alpha)\right\rangle_4 \otimes \left|-\frac{1}{2}(i\beta - \alpha)\right\rangle_5 \otimes \left|-\frac{1}{2}(i\beta + \alpha)\right\rangle_6$$

in the four output modes.

Since we have decomposed the field density operator $\hat{\rho}$ into coherent states $|\beta\rangle$ we can now evaluate the density operator $\hat{\rho}_{\text{out}}$ of the four output modes 3, 4, 5 and 6. Indeed we find

$$\hat{\rho}_{\text{out}} = \int d^2\beta\, P(\beta) \left|\frac{1}{2}(\beta - \alpha)\right\rangle_3 \left\langle\frac{1}{2}(\beta - \alpha)\right| \otimes \left|\frac{1}{2}(\beta + \alpha)\right\rangle_4 \left\langle\frac{1}{2}(\beta + \alpha)\right|$$
$$\otimes \left|-\frac{1}{2}(i\beta - \alpha)\right\rangle_5 \left\langle-\frac{1}{2}(i\beta - \alpha)\right| \otimes \left|-\frac{1}{2}(i\beta + \alpha)\right\rangle_6 \left\langle-\frac{1}{2}(i\beta + \alpha)\right|.$$

This density operator describes the quantum state of the fields in the four output modes of the eight-port interferometer. Due to the integration over the phase space variable β these four modes are entangled, since β appears in all modes. Only in the case of the Glauber-Sudarshan distribution P of the input field mode being a delta function, that is the field mode being in a coherent state, do we find no coupling between the modes, that is no entanglement.

13.3.2 Photon Count Statistics

We now can calculate the photon counts n_j at the four detectors D_j for $j = 3, 4, 5, 6$ using the corresponding photon number eigenstates $|n_j\rangle$. The probability $W(n_3, n_4, n_5, n_6)$ of finding n_3, n_4, n_5 and n_6 photons at the Detectors D_3, D_4, D_5 and D_6 then reads

$$W(n_3, n_4, n_5, n_6) = \langle n_3, n_4, n_5, n_6 | \hat{\rho}_{\text{out}} | n_3, n_4, n_5, n_6 \rangle$$
$$= \int d^2\beta\, P(\beta) K_{n_3,n_4}(\alpha, \beta) K_{n_5,n_6}(\alpha, \beta) \quad (13.23)$$

where the kernels

$$K_{n_3,n_4}(\alpha, \beta) \equiv \left|\left\langle n_3 \left|\frac{1}{2}(\beta - \alpha)\right\rangle\right|^2 \cdot \left|\left\langle n_4 \left|\frac{1}{2}(\beta + \alpha)\right\rangle\right|^2\right.\right.$$

and

$$K_{n_5,n_6}(\alpha, \beta) \equiv \left|\left\langle n_5 \left|-\frac{1}{2}(i\beta - \alpha)\right\rangle\right|^2 \cdot \left|\left\langle n_6 \left|-\frac{1}{2}(i\beta + \alpha)\right\rangle\right|^2\right.\right.$$

denote the count statistics in the output ports for a coherent state input port.

When we compare this expression with the corresponding one of a single beam splitter, Eq. (13.14), we note that due to the additional two exit ports we now have two kernels. They are again given by the photon statistics of coherent states. The amplitude of these coherent states are again linear combinations of coherent state amplitudes. The main differences compared to the case of a single beam splitter are the occurrence of a factor of 2 rather than $\sqrt{2}$ and a factor i due to the phase shift. The factor of 2 results from the fact that each state has to pass two beam splitters before it falls onto a detector.

Double Homodyne Statistics

When we now perform only a measurement of the photon number differences $n_{43} \equiv n_4 - n_3$ and $n_{65} \equiv n_6 - n_5$ we have to trace the expression Eq. (13.23) for the count statistics of the eight port interferometer over the sums $n_4 + n_3$ and $n_6 + n_5$ of the photon numbers for fixed differences n_{43} and n_{65}. Hence the probability $W(n_{43}, n_{65})$ of finding the photon differences n_{43} and n_{65} reads

$$W(n_{43}, n_{65}) = \int d^2\beta \, P(\beta) \, K_{n_{43}}(\alpha, \beta) \, K_{n_{65}}(\alpha, \beta), \tag{13.24}$$

where we have defined the kernels

$$K_{n_{43}}(\alpha, \beta) = \sum_{n_3=0}^{\infty} \left|\left\langle n_3 \left| \frac{1}{2}(\beta - \alpha) \right.\right\rangle\right|^2 \cdot \left|\left\langle n_{43} + n_3 \left| \frac{1}{2}(\beta + \alpha) \right.\right\rangle\right|^2$$

and

$$K_{n_{65}}(\alpha, \beta) = \sum_{n_5=0}^{\infty} \left|\left\langle n_5 \left| -\frac{1}{2}(i\beta - \alpha) \right.\right\rangle\right|^2 \cdot \left|\left\langle n_{65} + n_5 \left| -\frac{1}{2}(i\beta + \alpha) \right.\right\rangle\right|^2.$$

In the context of the homodyne detector we have already calculated sums of this type. Indeed, when we compare the above expressions for the kernels $K_{n_{43}}$ and $K_{n_{65}}$ with the expression Eqs. (13.18) for the kernel $K_{n_{21}}$ of the single beam splitter we can read off the result of the summation by using the formula Eq. (13.19) for the kernel $K_{n_{21}}$. We note, that the the substitutions $\alpha \to \alpha/\sqrt{2}$ and $\beta \to \beta/\sqrt{2}$ relate the kernel $K_{n_{21}}$ to $K_{n_{43}}$. Hence, we arrive with the help of Eq. (13.19) at

$$K_{n_{43}}(\alpha, \beta) = \left|\frac{\beta + \alpha}{\beta - \alpha}\right|^{n_{43}} I_{|n_{43}|}\left(\tfrac{1}{2}|\beta^2 - \alpha^2|\right) \exp\left[-\frac{1}{2}\left(|\beta|^2 + |\alpha|^2\right)\right].$$

Likewise, we find the substitutions $\alpha \to -\alpha/\sqrt{2}$ and $\beta \to -i\beta/\sqrt{2}$ for the kernel $K_{n_{65}}$ which results in

$$K_{n_{65}}(\alpha, \beta) = \left|\frac{i\beta + \alpha}{-i\beta + \alpha}\right|^{n_{65}} I_{|n_{65}|}\left(\tfrac{1}{2}|\beta^2 + \alpha^2|\right) \cdot \exp\left[-\frac{1}{2}\left(|\beta|^2 + |\alpha|^2\right)\right].$$

Here, I_ν denotes the modified Bessel function.

These two expressions together with Eq. (13.24) provide the exact expression for the photon count differences n_{43} and n_{65} at an eight-port homodyne detector when two of the four input modes are occupied by a coherent state and an arbitrary field, and the other two input modes contain the vacuum.

These exact expressions are rather complicated and resemble the result of a single homodyne detector. There we have found a rather simple expression when we have considered the limit of a coherent field of large amplitude, that is when we consider the case of a strong local oscillator. This serves as a motivation to now consider the limit of a strong local oscillator at the eight port homodyne detector.

Strong Local Oscillator Limit

We can find the asymptotic limit of $K_{n_{43}}$ by substituting $\alpha \to \alpha/\sqrt{2}$ and $\beta \to \beta/\sqrt{2}$ into the asymptotic expression Eq. (13.21) and arrive at

$$K_{n_{43}}(|\alpha| \to \infty, \beta) \cong \frac{1}{\sqrt{\pi |\alpha|^2}} \exp\left\{-\left[\frac{n_{43}}{|\alpha|} - \frac{1}{2}\left(\beta e^{-i\theta} + \beta^* e^{i\theta}\right)\right]^2\right\}. \qquad (13.25)$$

Here we have introduced $\alpha \equiv |\alpha| e^{i\theta}$.

Hence the detector pair (D_3, D_4) measures via the scaled photon count difference $n_{43}/|\alpha|$ the quadrature variable

$$x_\theta \equiv \frac{1}{2}\left(\beta e^{-i\theta} + \beta^* e^{i\theta}\right). \qquad (13.26)$$

When we compare this expression to the corresponding one Eq. (13.21) for a single beam splitter we find that the width and the center of the Gaussian have changed. The center is now at half the position of the single beam splitter case. Moreover, the width is now half of the width of the corresponding beam splitter expression.

In a similar way we obtain the asymptotic limit of $K_{n_{65}}$ by substituting $\alpha \to -\alpha/\sqrt{2}$ and $\beta \to -i\beta/\sqrt{2}$. This results in

$$K_{n_{65}}(|\alpha| \to \infty, \beta) \cong \frac{1}{\sqrt{\pi |\alpha|^2}} \exp\left\{-\left[\frac{n_{65}}{|\alpha|} - \frac{1}{2i}\left(\beta e^{-i\theta} - \beta^* e^{i\theta}\right)\right]^2\right\}. \qquad (13.27)$$

Hence, the detector pair (D_5, D_6) measures via the scaled photon count difference $n_{65}/|\alpha|$ the out-of-phase quadrature variable

$$p_\theta \equiv \frac{1}{2i}\left(\beta e^{-i\theta} - \beta^* e^{i\theta}\right). \qquad (13.28)$$

13.3.3 Simultaneous Measurement and EPR

The analysis of the preceding section suggests that the eight-port interferometer makes a simultaneous measurement of the variables x_θ and p_θ. However, the question arises to which operators does this measurement correspond. We now show that the underlying operators are two-mode rather than one-mode operators.

Naive Approach

In the most elementary approach we recall that β stems from the field operator $\hat{b}$ of the input mode. This suggests that the experiment makes a simultaneous measurement of the in-phase operators

$$\hat{x}_\theta \equiv \frac{1}{2}\left(\hat{b}e^{-i\theta} + \hat{b}^\dagger e^{i\theta}\right)$$

and

$$\hat{p}_\theta \equiv \frac{1}{2i}\left(\hat{b}e^{-i\theta} - \hat{b}^\dagger e^{i\theta}\right).$$

We note however, that these operators do not commute

$$[\hat{x}_\theta, \hat{p}_\theta] = \frac{i}{2}$$

and their measurement is forbidden by the postulates of quantum mechanics.

Two-Mode Operators

In the strong local oscillator limit the four input modes reduce to two inputs. Indeed, the two input modes at the beam splitter in the upper left corner of the interferometer shown in Fig. 13.4 are a coherent state of large amplitude and the vacuum. However, in this case the vacuum mode does not have to be treated quantum mechanically, since the fields of the two output ports of this beam splitter are dominated by the strong classical field of the coherent state. Therefore, in this case the four-mode input of the interferometer reduces to a two-mode input. These are the two modes at the beam splitter in the lower right corner. The annihilation and creation operators $\hat{b}$ and $\hat{b}^\dagger$ characterize the field described by the density operator $\hat{\rho}$ and $\hat{b}_0$ and $\hat{b}_0^\dagger$ the vacuum.

In Problem 13.3 we use the beam splitter transformation Eqs. 13.4 for the field operators to show that in the strong local oscillator limit the two detector pairs (D_4,D_3) and (D_6,D_5) of the eight-port interferometer measure the two-mode operators

$$\hat{\tilde{n}}_{43}(\theta) \equiv \frac{1}{2}\left(\hat{b}^\dagger + \hat{b}_0^\dagger\right)e^{i\theta} + \frac{1}{2}\left(\hat{b} + \hat{b}_0\right)e^{-i\theta} \tag{13.29a}$$

and

$$\hat{\tilde{n}}_{65}(\theta) \equiv \frac{i}{2}\left(\hat{b}^\dagger - \hat{b}_0^\dagger\right)e^{i\theta} - \frac{i}{2}\left(\hat{b} - \hat{b}_0\right)e^{-i\theta}. \tag{13.29b}$$

Here θ is the reference phase of the local oscillator.

For the special choice of $\theta = 0$ these operators reduce to the operator pair

$$\hat{\tilde{n}}_{43}(\theta = 0) = \frac{1}{\sqrt{2}}\left(\hat{x} + \hat{x}_0\right) \equiv \frac{1}{\sqrt{2}}\hat{X}$$

and

$$\hat{\tilde{n}}_{65}(\theta = 0) = \frac{1}{\sqrt{2}}\left(\hat{p} - \hat{p}_0\right) \equiv \frac{1}{\sqrt{2}}\hat{P},$$

where we have introduced the quadrature operators $(\hat{x}, \hat{p})$ and $(\hat{x}_0, \hat{p}_0)$ using the definition $\hat{b} \equiv 1/\sqrt{2}(\hat{x} + i\hat{p})$.

Hence, in the strong local oscillator limit and for the particular reference phase $\theta = 0$ the operators $\hat{\tilde{n}}_{43}$ and $\hat{\tilde{n}}_{65}$ corresponding to the difference in the photon numbers at the four detectors of the eight-port interferometer are the two-mode operators $\hat{X}$ and $\hat{P}$.

Hence, the main difference to the naive approach lies in the fact that we need two modes to describe a simultaneous measurement. Since the operators $\hat{x}$ and $\hat{p}$ of the incident mode do not commute they cannot be measured simultaneously with arbitrary accuracy. However, by coupling this mode to a second mode, namely the vacuum mode described $\hat{x}_0$ and $\hat{p}_0$ the operators

$$\hat{X} \equiv \hat{x} + \hat{x}_0 \quad \text{and} \quad \hat{P} \equiv \hat{p} - \hat{p}_0$$

do commute with each other, that is

$$\left[\hat{X}, \hat{P}\right] = [\hat{x}, \hat{p}] - [\hat{x}_0, \hat{p}_0] = 0.$$

This simultaneous measurement of $\hat{X}$ and $\hat{P}$ with arbitrary accuracy is therefore allowed by quantum mechanics. We emphasize that it is the special arrangement of the eight-port interferometer which enforces the special construction of the two-mode operators $\hat{X}$ and $\hat{P}$.

We recall that the operator pair $(\hat{X}, \hat{P})$ is the basis of the work of A. Einstein, B. Podolsky and N. Rosen on the completeness of quantum mechanics. Its importance was emphasized by N. Bohr in his reply to the EPR situation. This thought experiment considers the measurement of the position of the center-of-mass and the relative momentum of two correlated massive particles. Our operators $\hat{X}$ and $\hat{P}$ correspond to these variables. Each mode represents one particle.

To bring out the analogy most clearly we recall that two commuting operators have common eigenstates. Hence, the operators $\hat{X}$ and $\hat{P}$ have common eigenstates $|X, P\rangle$. In Problem 13.3 we derive the quadrature representation

$$|X, P\rangle = \frac{1}{\sqrt{2\pi}} \int_{-\infty}^{\infty} dz \int_{-\infty}^{\infty} dy \, \exp[-iPy] \, \delta(z + y - X) \, |y\rangle_1 |z\rangle_0$$

of these eigenstates. Here, $|y\rangle_1$ and $|z\rangle_0$ denote the quadrature eigenstates of the two input modes.

This representation shows the entanglement of the two modes through the δ-function and the well defined momentum due to the Fourier transform.

13.3.4 Q-Function Measurement

Due to the special arrangement of beam splitters and phase shifters the eight-port interferometer measures through its count statistics something close to two conjugate, noncommuting variables. However, we have to pay a price for this simultaneous measurement: When we mix the input mode with the vacuum mode, we bring in additional noise. Consequently we measure the Q-function rather than the Wigner function of the input field as we show now.

For this purpose we now substitute the asymptotic expression for the kernels $K_{n_{43}}$ and $K_{n_{65}}$, Eqs. (13.25) and (13.27) into the formula Eq. (13.24) for the photon count statistics. We therefore arrive at

$$W(n_{43}, n_{65}) = \frac{1}{\pi|\alpha|^2} \int d^2\beta P(\beta) \exp\left[-\left(\frac{n_{43}}{|\alpha|} - x_\theta\right)^2 - \left(\frac{n_{65}}{|\alpha|} + p_\theta\right)^2\right].$$

From the definition of the quadrature variables x_θ and p_θ we find the relation

$$\left(\frac{n_{43}}{|\alpha|} - x_\theta\right)^2 + \left(\frac{n_{65}}{|\alpha|} + p_\theta\right)^2$$

$$= \left(\frac{n_{43}}{|\alpha|}\right)^2 + \left(\frac{n_{65}}{|\alpha|}\right)^2 - 2\left(\frac{n_{43}}{|\alpha|}x_\theta - \frac{n_{65}}{|\alpha|}p_\theta\right) + x_\theta^2 + p_\theta^2$$

$$= \left(\frac{n_{43}}{|\alpha|}\right)^2 + \left(\frac{n_{65}}{|\alpha|}\right)^2 - \left(\frac{n_{43}}{|\alpha|} + i\frac{n_{65}}{|\alpha|}\right)\beta e^{-i\theta} - \left(\frac{n_{43}}{|\alpha|} - i\frac{n_{65}}{|\alpha|}\right)\beta^* e^{i\theta} + |\beta|^2$$

$$= \left|\left(\frac{n_{43}}{|\alpha|} - i\frac{n_{65}}{|\alpha|}\right)e^{i\theta} - \beta\right|^2.$$

Hence, in the strong local oscillator limit the count statistics of the photon number difference reads

$$W(n_{43}, n_{65}) = \frac{1}{\pi|\alpha|^2} \int d^2\beta\, P(\beta) \exp\left[-|z - \beta|^2\right], \tag{13.30}$$

where

$$z \equiv \left(\frac{n_{43}}{|\alpha|} - i\frac{n_{65}}{|\alpha|}\right) e^{i\theta}$$

is a complex number build out of the normalized count rates $n_{43}/|\alpha|$ and $n_{65}/|\alpha|$ and the phase θ of the local oscillator.

We gain deeper insight into the expression (13.30) for $W(n_{43}, n_{65})$ when we recall that the scalar product between the coherent state $|z\rangle$ associated with the number z reads according to Eq. (11.21)

$$|\langle z|\beta\rangle|^2 = \exp\left[-|z - \beta|^2\right],$$

which yields

$$W(n_{43}, n_{65}) = \frac{1}{\pi|\alpha|^2} \int d^2\beta\, P(\beta)\, |\langle z|\beta\rangle|^2.$$

We now express this formula in terms of the Q-function

$$Q(z) = \frac{1}{\pi} \langle z|\hat{\rho}|z\rangle$$

of the density operator

$$\hat{\rho} = \int d^2\beta\, P(\beta)\, |\beta\rangle\langle\beta|$$

13.3 Eight-Port Interferometer 369

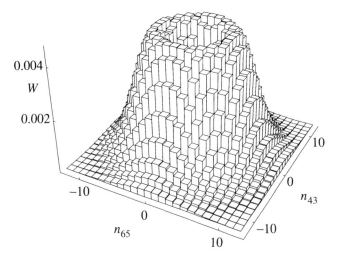

Fig. 13.5: Photon count probability $W(n_{43}, n_{65})$ of an eight-port interferometer in the strong local oscillator limit. Here, the incident field is in a one-photon Fock state.

of the incident field. Hence, after substituting $\hat{\rho}$ into the definition of Q we find

$$Q(z) = \frac{1}{\pi} \int d^2\beta \, P(\beta) \, |\langle z|\beta\rangle|^2$$

and thus the formula

$$W(n_{43}, n_{65}) = \frac{1}{|\alpha|^2} Q\left[z = \left(\frac{n_{43}}{|\alpha|} - i\frac{n_{65}}{|\alpha|}\right) e^{i\theta}\right]. \tag{13.31}$$

The statistics of the photon count differences is a scaled read out of the Q-function of the incident light field provided the coherent state incident on the other beam splitter has a large amplitude, that is the other light field is a strong local oscillator.

When we choose $\theta = 0$ for the local oscillator phase we find

$$W(n_{43}, n_{65}) = \frac{1}{|\alpha|^2} Q\left(z = \frac{n_{43}}{|\alpha|} - i\frac{n_{65}}{|\alpha|}\right).$$

Hence the scaled photon counts $n_{43}/|\alpha|$ and $n_{65}/|\alpha|$ represent the two quadrature components, that is the two phase space variables, as already suggested by the asymptotic limit of the kernels $K_{n_{43}}$ and $K_{n_{65}}$.

We conclude this section by briefly discussing how to measure the count probability $W(n_{43}, n_{65})$. We prepare the local oscillator state $|\alpha\rangle = ||\alpha|\rangle$ and the field state $\hat{\rho}$ to be measured. We count n_3, n_4, n_5 and n_6 photons at the four detectors and take the differences $n_{43} \equiv n_4 - n_3$ and $n_{65} \equiv n_6 - n_5$. We store these two numbers and repeat the experiment, that is we again prepare the field states, count the photon numbers n_j and find the differences. In this way we build a histogram for the differences n_{43} and n_{65} by plotting them in a diagram whose axis are $n_{43}/|\alpha|$ and $n_{65}/|\alpha|$ as shown in Fig. 13.5 for the case of a Fock state of one photon. This is indeed the scaled Q-function of the one-photon Fock state.

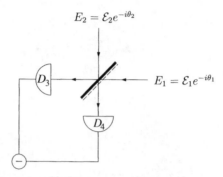

Fig. 13.6: Two classical electric fields enter a beam splitter and we measure the difference current at the two detectors D_3 and D_4.

13.4 Measured Phase Operators

The question of the operator corresponding to the phase of a harmonic oscillator is an old one and we have briefly discussed it in Sec. 8.5. It is interesting to note that the group of L. Mandel has used an eight-port interferometer to define new operators $\hat{C}_M$ and $\hat{S}_M$ for the cosine and the sine of the phase difference between the local oscillator field and a quantum field. Moreover, these operators are not just theoretical constructs but can be measured. We now briefly summarize this approach and in particular make a connection to a phase space interpretation using the Q-function. For this purpose we first present the classical analysis of an eight-port interferometer and then show how this analysis suggests new phase operators.

13.4.1 Measurement of Classical Trigonometry

We now briefly repeat a classical discussion of the intensity difference measurement at a beam splitter similar to Sec. 13.2.1. For this purpose we consider again two classical electric fields $E_j = \mathcal{E}_j e^{-i\theta_j}$ incident on a beam splitter as shown in Fig. 13.6. We find the intensities

$$I_3 = |E_1 - E_2|^2 = \mathcal{E}_1^2 + \mathcal{E}_2^2 - 2\mathcal{E}_1\mathcal{E}_2 \cos(\theta_2 - \theta_1)$$

and

$$I_4 = |E_1 + E_2|^2 = \mathcal{E}_1^2 + \mathcal{E}_2^2 + 2\mathcal{E}_1\mathcal{E}_2 \cos(\theta_2 - \theta_1)$$

at the two detectors and the difference current $I_{43} \equiv I_4 - I_3$ reads

$$I_{43} = 4\mathcal{E}_1\mathcal{E}_2 \cos(\theta_2 - \theta_1).$$

Hence we can express the cosine of the phase difference by the current difference I_{43} and the electric field amplitudes $\mathcal{E}_1$ and $\mathcal{E}_2$ via

$$\cos(\theta_2 - \theta_1) \equiv \frac{I_{43}}{4\mathcal{E}_1\mathcal{E}_2} = \frac{I_4 - I_3}{4\mathcal{E}_1\mathcal{E}_2}. \tag{13.32}$$

We emphasize that we *cannot* find $\cos(\theta_2 - \theta_1)$ by only measuring the difference I_{43}. We also need the product $\mathcal{E}_1\mathcal{E}_2$ of the electric field amplitudes.

13.4 Measured Phase Operators

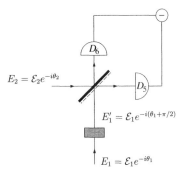

Fig. 13.7: Two classical electric fields enter a beam splitter where beam 1 passes a $\lambda/4$ wave plate which results in a phase shift of $\pi/2$. A measurement of the difference current at the two detectors D_5 and D_6 is performed.

Analogously we can measure the sine of the phase difference between the two classical waves by shifting the phase of one field by $\pi/2$ using a $\lambda/4$ wave plate before detection as indicated by Fig. 13.7. We find for the two intensities I_5 and I_6 at the detectors D_5 and D_6

$$I_5 = |E_1' - E_2|^2 = \mathcal{E}_1^2 + \mathcal{E}_2^2 - 2\mathcal{E}_1\mathcal{E}_2 \sin(\theta_2 - \theta_1)$$

and

$$I_6 = |E_1' + E_2|^2 = \mathcal{E}_1^2 + \mathcal{E}_2^2 + 2\mathcal{E}_1\mathcal{E}_2 \sin(\theta_2 - \theta_1),$$

that is for the difference current

$$I_{65} \equiv I_6 - I_5 = 4\mathcal{E}_1\mathcal{E}_2 \sin(\theta_2 - \theta_1).$$

Hence we can express the sine function via

$$\sin(\theta_2 - \theta_1) \equiv \frac{I_{65}}{4\mathcal{E}_1\mathcal{E}_2} = \frac{I_6 - I_5}{4\mathcal{E}_1\mathcal{E}_2}. \tag{13.33}$$

When we now combine both homodyne setups to an eight-port homodyne detector we can simultaneously measure sine and cosine. Using the trigonometric relation

$$1 = \cos^2(\theta_2 - \theta_1) + \sin^2(\theta_2 - \theta_1) = \frac{I_{43}^2 + I_{65}^2}{16\mathcal{E}_1^2\mathcal{E}_2^2}$$

and Eqs. (13.32) and (13.33) we can find the product $4\mathcal{E}_1\mathcal{E}_2$ of the electric field amplitudes $\mathcal{E}_1$ and $\mathcal{E}_2$ in terms of the current differences via

$$4\mathcal{E}_1\mathcal{E}_2 = \sqrt{I_{43}^2 + I_{65}^2} = \sqrt{(I_4 - I_3)^2 + (I_6 - I_5)^2}.$$

Now we are in a position to express sine and cosine in terms of the currents I_j only. Indeed we find

$$\cos(\theta_2 - \theta_1) = \frac{I_4 - I_3}{\sqrt{(I_4 - I_3)^2 + (I_6 - I_5)^2}}$$

and
$$\sin(\theta_2 - \theta_1) = \frac{I_6 - I_5}{\sqrt{(I_4 - I_3)^2 + (I_6 - I_5)^2}}.$$

Hence, we can find the cosine and the sine of the phase difference by measuring the intensities I_3, I_4, I_5 and I_6 at the four exit ports of the eight-port interferometer and substituting the measured data into the above expression.

13.4.2 Measurement of Quantum Trigonometry

We achieve the transition from classical to quantum fields by replacing the classical currents I_j by the photon number operators $\hat{n}_j$. This guiding principle leads us to the measured phase operators

$$\widehat{\cos}_M(\theta_2 - \theta_1) \equiv \frac{\hat{n}_4 - \hat{n}_3}{\sqrt{(\hat{n}_4 - \hat{n}_3)^2 + (\hat{n}_6 - \hat{n}_5)^2}} \tag{13.34a}$$

and

$$\widehat{\sin}_M(\theta_2 - \theta_1) \equiv \frac{\hat{n}_6 - \hat{n}_5}{\sqrt{(\hat{n}_4 - \hat{n}_3)^2 + (\hat{n}_6 - \hat{n}_5)^2}}. \tag{13.34b}$$

In Sec. 13.3.2 we have derived the joint count probability $W(n_{43}, n_{65})$ for the photon number differences $n_{43} = n_4 - n_3$ and $n_{65} = n_6 - n_5$ at the four detectors of the eight-port homodyne detector. Moreover, we can measure this distribution experimentally. We can therefore measure and analytically evaluate the expectation value

$$\langle f(\hat{C}_M, \hat{S}_M) \rangle \equiv \mathcal{N} \sum_{n_{43}, n_{65} \neq 0} f\left(\frac{n_{43}}{\sqrt{n_{43}^2 + n_{65}^2}}, \frac{n_{65}}{\sqrt{n_{43}^2 + n_{65}^2}}\right) W(n_{43}, n_{65}) \tag{13.35}$$

of any function f of the operators $\hat{C}_M$ and $\hat{S}_M$. In order to avoid the problems associated with the definitions of Eq. (13.35) for $n_{43} = n_{65} = 0$ the Mandel group suggests to disregard such measurements and renormalize the joint count probability via

$$\mathcal{N} \equiv \frac{1}{1 - W(0,0)}.$$

How is this approach related to phase space? In Sec. 13.3.4 we have shown that in the strong local oscillator limit the photon count statistics $W(n_{43}, n_{65})$ is a scaled Q-function of the input field state. Moreover, the measured cosine and sine operators satisfy the standard trigonometric relations. This suggests to define a phase distribution by integrating the Q-function of the quantum state over the radius. We now show that this is indeed the underlying phase distribution of these operationally defined phase operators.

In Sec. 13.3.4 we have shown that in the strong local oscillator limit the photon count differences loose their discreteness. This is due to the fact that they are scaled with the amplitude $|\alpha|$ of the local oscillator. We can therefore introduce the continuous variables $\xi \equiv n_{43}/|\alpha|$ and $\pi \equiv n_{65}/|\alpha|$ and replace in Eq. (13.35) summation by integration, which yields

$$\langle f(\hat{C}_M, \hat{S}_M) \rangle \cong \int_{-\infty}^{\infty} d\xi \int_{-\infty}^{\infty} d\pi \, f\left(\frac{\xi}{\sqrt{\xi^2 + \pi^2}}, \frac{\pi}{\sqrt{\xi^2 + \pi^2}}\right) |\alpha|^2 W(|\alpha|\xi, |\alpha|\pi). \quad (13.36)$$

Here we have used $\mathcal{N} \cong 1$ since in the strong local oscillator limit

$$W(0,0) \cong \frac{1}{\pi |\alpha|^2} \langle 0|\hat{\rho}|0\rangle \ll 1.$$

When we introduce the polar coordinates r and ϕ via the relations

$$\xi \equiv r\cos(\theta - \phi) \quad \text{and} \quad \pi \equiv r\sin(\theta - \phi)$$

the integrals in expression Eq. (13.36) for the expectation value take the form

$$\langle f(\hat{C}_M, \hat{S}_M) \rangle \cong \int_{-\pi}^{\pi} d\phi \, f[\cos(\theta - \phi), \sin(\theta - \phi)]$$

$$\times \int_{0}^{\infty} dr \, r |\alpha|^2 W[|\alpha| r \cos(\theta - \phi), |\alpha| r \sin(\theta - \phi)].$$

This separation into phase integration and radial integration suggests to introduce the phase distribution

$$W^Q(\theta - \phi) \equiv \int_{0}^{\infty} dr \, r |\alpha|^2 W[|\alpha| r \cos(\theta - \phi), |\alpha| r \sin(\theta - \phi)]$$

and the expectation value is obtained by integrating the c-number representation of f with respect to the phase distribution $W^Q(\phi)$, that is

$$\langle f(\hat{C}_M, \hat{S}_M) \rangle = \int_{-\pi}^{\pi} d\phi \, f[\cos(\theta - \phi), \sin(\theta - \phi)] W^Q(\theta - \phi).$$

We find the so-defined phase distribution by expressing the photon count statistics $W(n_{43}, n_{65})$ in polar coordinates and integrated over the radius. With the explicit formula Eq. (13.31)

$$W(n_{43}, n_{65}) = \frac{1}{|\alpha|^2} Q\left[z = \left(\frac{n_{43}}{|\alpha|} - i\frac{n_{65}}{|\alpha|}\right)e^{i\theta}\right] = \frac{1}{|\alpha|^2} Q\left[z = re^{i(\phi-\theta)} e^{i\theta}\right]$$

for the photon count distribution in the strong local oscillator limit this phase distribution reduces to

$$W^Q(\phi) = \int_{0}^{\infty} dr \, r Q(z = re^{i\phi}) = \frac{1}{\pi} \int_{0}^{\infty} dr \, r \left\langle re^{i\phi} \middle| \hat{\rho} \middle| re^{i\phi} \right\rangle.$$

Hence we have found an experimentally defined phase distribution which is connected to phase space. In the strong local oscillator limit the phase distribution $W^Q(\phi)$ corresponding to the measured phase operators follow from the Q-function. Indeed, when we express the Q-function in polar coordinates and integrate over the radius we find this phase distribution. With the help of this distribution we can calculate the expectation value of any function of operators by integrating them with respect to $W^Q(\phi)$.

13.4.3 Two-Mode Phase Operators

In Sec. 13.3.3 we have shown that in the limit of a strong local oscillator the eight-port interferometer measures the two-mode operators $\hat{X}$ and $\hat{P}$. Moreover, they commute. It is therefore possible to define trigonometric operators

$$\hat{\tilde{C}}_M \equiv \frac{\hat{X}}{\sqrt{\hat{X}^2 + \hat{P}^2}}$$

and

$$\hat{\tilde{S}}_M \equiv \frac{\hat{P}}{\sqrt{\hat{X}^2 + \hat{P}^2}}.$$

For the one-mode problem this definition is highly problematic since in this case $\hat{x}$ and $\hat{p}$ do not commute and an operator ordering problem arises. However, since the two-mode operators commute the definition is unique.

With the simultaneous eigenstates $|X, P\rangle$ of the operators $\hat{X}$ and $\hat{P}$ satisfying the eigenvalue equations

$$\hat{X}|X, P\rangle = X|X, P\rangle \quad \text{and} \quad \hat{P}|X, P\rangle = P|X, P\rangle$$

we find the spectral representations

$$\hat{\tilde{C}}_M(\theta = 0) \int_{-\infty}^{\infty} dX \int_{-\infty}^{\infty} dP \, \frac{X}{\sqrt{X^2 + P^2}} |X, P\rangle \langle X, P|$$

and

$$\hat{\tilde{S}}_M(\theta = 0) \int_{-\infty}^{\infty} dX \int_{-\infty}^{\infty} dP \, \frac{P}{\sqrt{X^2 + P^2}} |X, P\rangle \langle X, P|.$$

They allow us to calculate expectation values of $\hat{\tilde{C}}_M$ and $\hat{\tilde{S}}_M$.

Problems

13.1 Beam Splitter Transformation

Show that the beam splitter transformation Eqs. (13.4) of the field operators is a unitary transformation. Use this result to derive the transformation Eqs. (13.5) of the coherent states.

13.2 Beam Splitter and Number States

In Sec. 13.1.4 we have shown how coherent states transform at a 50/50 beam splitter. The transformation of arbitrary states can be reduced to the transformation of coherent states because they are complete. Alternatively one could try to explain the beam splitter in the particle picture, that is, with the help of the photon concept. In this picture we would describe a beam splitter as follows: A photon is transmitted with the probability 1/2 and it is reflected with the probability 1/2. Both models for the beam splitter will be investigated in the following examples.

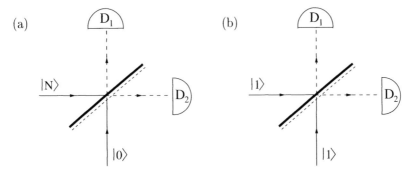

Fig. 13.8: Beam splitter an number states. (a) The vacuum state and a N-photon Fock state are incident on a beam splitter. (b) Two one-photon Fock states are incident on a beam splitter.

(a) At one input port of a beam splitter we inject a vacuum state, whereas at the other a Fock state with N photons, as shown in Fig. 13.8(a).

   i. How are the states $|0\rangle$ and $|N\rangle$ transformed at the beam splitter?
   ii. What is the probability to find n_1 photons at the detector D_1 and n_2 photons at the detector D_2?
   iii. Which result does one obtain with the particle picture?

(b) At both input ports of the beam splitter we inject Fock states $|1\rangle$, as shown in Fig. 13.8(b).

   i. How do the states transform at the beam splitter?
   ii. What is the probability to find n_1 photons at the detector D_1 and n_2 photons at the detector D_2?
   iii. Which result does one obtain with the particle picture?

(c) Explain what is common and what is different in the two pictures.

13.3 Two-Mode Operators in Eight-Port Interferometer

In Sec. 13.3 we have derived the photon count statistics of an eight-port interferometer using the transformation of field states at a beam splitter. Analogously we can derive this result using the transformation Eqs. (13.4) of the field operators at a beam splitter.

(a) Show that in the strong local oscillator limit the two detector pairs (D_4, D_3) and (D_6, D_5) of the eight-port interferometer measure the count rate

$$W(n_{43}, n_{65}) = \frac{1}{|\alpha|^2 (2\pi)^2} \int d\nu' \int d\nu''$$
$$\times \left\langle \Psi \left| \exp\left[i\left(\hat{n}_{43} - \frac{n_{43}}{|\alpha|}\right)\nu'\right] \exp\left[i\left(\hat{n}_{65} + \frac{n_{65}}{|\alpha|}\right)\nu''\right] \right| \Psi \right\rangle.$$

Here, we have introduced the two-mode operators

$$\hat{n}_{43}(\theta) \equiv \frac{1}{2}\left(\hat{a}^\dagger + \hat{a}_0^\dagger\right) e^{i\theta} + \frac{1}{2}\left(\hat{a} + \hat{a}_0\right) e^{-i\theta}$$

and

$$\hat{n}_{65}(\theta) \equiv \frac{i}{2}\left(\hat{a}^\dagger - \hat{a}_0^\dagger\right) e^{i\theta} - \frac{i}{2}\left(\hat{a} - \hat{a}_0\right) e^{-i\theta}$$

and $|\Psi\rangle$ denotes the two-mode input state of the field.

(b) Derive the quadrature representation

$$|X,P\rangle = \frac{1}{\sqrt{2\pi}} \int_{-\infty}^{\infty} dz \int_{-\infty}^{\infty} dy \, \exp\left[-iPy\right] \delta(z+y-X) |y\rangle_1 |z\rangle_0$$

of the two-mode state $|X,P\rangle$.

(c) Prove the relation

$$\left\langle \Psi \left| f\left[\hat{\tilde{C}}_M, \hat{\tilde{S}}_M\right] \right| \Psi \right\rangle = \int_{-\infty}^{\infty} dX \int_{-\infty}^{\infty} dP \, f\left[\frac{X}{\sqrt{X^2+P^2}}, \frac{P}{\sqrt{X^2+P^2}}\right] \mathcal{W}(X,P)$$

for the expectation value of a function f consisting of two-mode phase operators $\hat{\tilde{C}}_M$ and $\hat{\tilde{S}}_M$. Here we can interpret the probability density

$$\mathcal{W}(X,P) \equiv |\langle X,P|\Psi\rangle|^2$$

as a generalized, positive distribution in the phase space spanned by the simultaneously measurable (X,P) eigenvalue pair.

(d) Show how the distribution $\mathcal{W}$ reduces to the Q-function when only the vacuum state and an arbitrary state enter the interferometer at the lower right beam splitter.

Hint: See M. Freyberger *et al.* (1995).

13.4 Eight-Port Interferometer and Single Photons

In Sec. 13.3 we have investigated the eight-port interferometer and have derived an expression for the difference count rate $W(n_{43}, n_{65})$. We have also shown that in the limiting case of a strong local oscillator ($|\alpha| \gg 1$) the Q-function of the density operator $\hat{\rho}$ can be measured. In this problem we want to investigate the case $\hat{\rho} = |1\rangle\langle 1|$.

(a) Calculate $W(n_{43}, n_{65})$ for finite α and $\hat{\rho} = |1\rangle\langle 1|$, that is for a single-photon state.

Hint: We have already shown in Sec. J.2 that the P-function of a Fock state is given by

$$P(\alpha) = L_n\left(-\frac{1}{4}\left(\frac{\partial^2}{\partial \alpha_r^2} + \frac{\partial^2}{\partial \alpha_i^2}\right)\right) \delta(\alpha)$$

First show that this is equivalent to

$$P(\alpha) = L_n\left(-\frac{\partial^2}{\partial\alpha\partial\alpha^*}\right)\delta(\alpha).$$

(b) What is the result in the limiting case $|\alpha| \to 0$?

(c) Show that the limiting case $|\alpha| \gg 1$ allows to measure the Q-function.

13.5 Asymptotic of Bessel Functions

Show with the help of the saddle point method the following asymptotic expansions:

(a) The Stirling formula reads

$$n! = \int_0^\infty t^n e^{-t}\, dt \cong \sqrt{2\pi n}\, n^n\, e^{-n} \qquad \text{for } n \gg 1.$$

(b) The modified Bessel functions

$$I_n(x) = \frac{1}{\pi}\int_0^\pi e^{x\cos\theta} \cos(n\theta)\, d\theta$$

can be represented in the form

$$I_n(x) = \sum_{\nu=0}^\infty \frac{1}{(n+\nu)!\,\nu!}\left(\frac{x}{2}\right)^{2\nu+n}.$$

Hint: Expand the exponential in a power series and integrate each term.

(c) Similarly the modified Bessel function behaves as

$$I_n(x) \cong \frac{e^x}{\sqrt{2\pi x}} \qquad \text{for } x \gg n^2.$$

Hint: Split the integral in a part $0\ldots\pi/2$ and a second part $\pi/2\ldots\pi$ and show that the second part does not contribute in the limit $x \to \infty$.

(d)

$$I_n(x) \cong \frac{e^x}{\sqrt{2\pi x}} \exp\left(-\frac{n^2}{2x}\right) \qquad \text{for } x \to \infty,\ n \to \infty,\ \text{we keep } n/\sqrt{x}\ \text{fixed}.$$

Hint: The substitution $z = e^{i\theta}$ is useful.

References

Beam Splitter

For the theory of the quantum mechanical beam splitter and its different approaches we refer to the early work of

G. Richter, W. Brunner and **H. Paul**, *Elektrische Feldstärke und Interferenz von Laserstrahlen*, Ann. Physik (Leipzig) **14**, 239–261 (1964)

W. Brunner, H. Paul and **G. Richter**, *Absorption und Streuung quantenmechanisch kohärenter Strahlen II*, Ann. Physik (Leipzig) **15**, 17–29 (1965)

A summary of this early work can be found in

H. Paul, *Ein Beitrag zur Quantentheorie der optischen Kohärenz*, Fortschr. d. Physik **14**, 141–204 (1966)

A. Zeilinger, *General properties of lossless beam splitters in interferometry*, Am. J. Phys. **49**, 882–883 (1981)

S. Prasad, M.O. Scully and **W. Martienssen**, *A Quantum Description of the Beam Splitter*, Opt. Commun. **62**, 139–145 (1987)

K. Mølmer, *A particle description of the quantized field – the beam splitter*, J. Phys. B **21**, L573–L578 (1988)

U. Leonhardt, *Quantum Measurements of Light*, Cambridge University Press, 1997

Homodyne Detection

For a discussion of homodyne detection see

H.P. Yuen and **V.W.S. Chan**, *Noise in homodyne and heterodyne detection*, Opt. Lett. **8**, 177–179 (1983)

J.H. Shapiro and **S.S. Wagner**, *Phase and Amplitude Uncertainties in Heterodyne Detection*, IEEE J. Quantum Electron. QE **20**, 803–813 (1984)

An analysis of the count statistics in a homodyne experiment is presented in

S.L. Braunstein and **C.M. Caves**, *Phase and Statistics of Generalized Squeezed States*, Phys. Rev. A **42**, 4115–4119 (1990)

S.L. Braunstein, *Homodyne Statistics*, Phys. Rev. A **42**, 474–481 (1990)

Eight-Port Homodyne Detector

The eight-port homodyne detector has been used to measure Q-functions of coherent states and thermal states by

N.G. Walker and **J.E. Carroll**, *Simultaneous Phase and Amplitude Measurements on Optical Signals Using a Multiport Junction*, Electron. Lett. **20**, 981–983 (1984)

N.G. Walker and **J.E. Carroll**, *Multiport-Homodyne Detection near the Quantum Noise Limit*, Opt. Quantum Electron. **18**, 355–363 (1986)

For the use of the eight-port homodyne detector to analyze phase distributions we refer to

J.W. Noh, A. Fougeres and **L. Mandel**, *Measurement of the Quantum Phase by Photon Counting*, Phys. Rev. Lett. **67**, 1426–1429 (1991)

J.W. Noh, A. Fougères and **L. Mandel**, *Operational approach to phase operators based on classical optics*, Physica Scripta **T48**, 29–34 (1993)

J.R. Torgerson and **L. Mandel**, *Measuring the Quantum Phase of the E.M. Field by Interference as $\langle n \rangle \to 0$*, Physica Scripta **T76**, 110–114 (1998)

For an analysis of this experiment see
M. Freyberger and **W.P. Schleich**, *Photon counting, quantum phase, and phase-space distributions*, Phys. Rev. A **47**, R30-R33 (1993)
M. Freyberger, K. Vogel and **W.P. Schleich**, *From photon counts to quantum phase*, Phys. Lett. A **176**, 41–46 (1993)
B.-G. Englert and **K. Wódkiewicz**, *Intrinsic and operational observables in quantum mechanics*, Phys. Rev. A **51**, R2661–R2664 (1995)
B.-G. Englert, K. Wódkiewicz and **P. Riegler**, *Intrinsic phase operator of the Noh-Fougères-Mandel experiments*, Phys. Rev. A **52**, 1704–1711 (1995)

Eight-Port Homodyne Detector and EPR

For the original paper posing the EPR paradox and Bohr's reply see
A. Einstein, B. Podolsky and **N. Rosen**, *Can a Quantum-Mechanical Description of Physical Reality be Considered Complete?*, Phys. Rev. **47**, 777–780 (1935)
N. Bohr, *Can a Quantum-Mechanical Description of Physical Reality be Considered Complete?*, Phys. Rev. **48**, 696–702 (1935)

For a detailed discussion of the eight-port homodyne detector and the EPR-variables
M. Freyberger, M. Heni and **W.P. Schleich**, *Two-mode quantum phase*, Quantum. Semiclass. Opt. **7**, 187–203 (1995)
M.T. Fontenelle, M. Freyberger, M. Heni, W.P. Schleich and **M.S. Zubairy**, *Quantum phase, Photon Counting and EPR-Variables*, in: *The Dilemma of Einstein, Podolsky and Rosen - 60 Years later*, edited by A. Mann and M. Revzen, Institute of Physics Publishing, Bristol, 1996

The eight-port homodyne detector is a special case of a system where one measures simultaneously two conjugate variables by adding extra noise. For a general treatment see
E. Arthurs and **J.L. Kelly Jr**, *On the simultaneous measurement of a pair of conjugate observables*, Bell Syst. Tech. J. **44**, 725–729 (1965)
S. Stenholm, *Simultaneous measurement of conjugate variables*, Ann. Phys. (NY), **218**, 233–254 (1992)

Multiports

The concept of multiports, that is a network of beam splitters and phase shifters which creates n input ports and n output ports has been investigated theoretically by
N.G. Walker, *Quantum theory of multiport optical homodyning*, J. Mod. Opt. **34**, 15–60 (1987)
P. Törmä, S. Stenholm and **I. Jex**, *Hamiltonian theory of symmetric optical network transforms*, Phys. Rev. A **52**, 4853–4860 (1995)
P. Törmä and **I. Jex**, *Properties of Ising-type linear networks*, J. Opt. B **1**, 8–13 (1999)

K. Mattle, M. Michler, H. Weinfurter, A. Zeilinger and M. Zukowski, *Nonclassical statistics at multiport beam splitters*, Appl. Phys. B **60**, S111–S117 (1995)

It has been shown that any unitary matrix can be represented by a multiport assembled only from 4-ports

M. Reck, A. Zeilinger, H.J. Bernstein and P. Bertani, *Experimental Realization of Any Discrete Unitary Operator*, Phys. Rev. Lett. **73**, 58–61 (1994)

14 Atom-Field Interaction

So far we have only considered properties of quantized light fields in the absence of interaction with matter. We now concentrate on quantum effects that arise due to the interaction of the atoms with quantized radiation. In order to do this we first have to find the appropriate coupling between light and matter. Here we are guided by the gauge principle of H. Weyl. In 1918 he introduced the principle of gauge invariance to create a unified theory of gravitation and electricity. This principle was rather speculative as H. Weyl wrote later himself. However, in 1928, he recognized that gauge invariance leads to a coupling of electricity and matter. Gauge invariance is central to modern gauge theories such as quantum-electrodynamics or quantum-chromodynamics and creates the coupling between these fields.

In Sec. 14.1 we briefly outline the two most popular methods of coupling an atom to an electromagnetic wave and turn in Sec. 14.2 to a brief review of the principle of gauge invariance. In particular, we show how the Hamiltonian of minimal coupling arises. We then use this result to consider the Hamiltonian of a hydrogen atom interacting with an electromagnetic field. In the dipole approximation discussed in Sec. 14.3 this model leads to a rather compact interaction Hamiltonian. It couples the internal degrees of freedom of the atom to its center-of-mass motion and to the electromagnetic field.

Apart from the minimal coupling there exists another coupling scheme. It is based on the fact that the atom is a dipole which interacts with the electric field. As shown in Sec. 14.4 this approach creates an interaction Hamiltonian which contains the position of the electron and the electric field. We devote Sec. 14.5 to the discussion of the transition to quantum mechanics. Gauge transformations allow us to show in Secs. 14.6 and 14.7 and Appendix L that under appropriate conditions both coupling schemes are equivalent.

In Sec. 14.8 we reduce the problem of the atom-field interaction to the most elementary situation of a two-level atom interacting with a single mode of the radiation field. This model has been introduced by E.T. Jaynes together with F.W. Cummings, and independently by H. Paul in 1963. For a long time it has only been the favorite toy of theorists but due to the modern developments of quantum optics, in particular high-Q cavities for microwaves and optical radiation, it has been realized experimentally. Due to its simplicity, power of prediction and experimental implementation it has become the drosophila of quantum optics.

14.1 How to Construct the Interaction?

Two possibilities offer themselves:

The first approach relies on the minimal coupling between a charged particle of mass m_e and charge e at the position $\vec{r}_e$ and an electromagnetic field represented in Coulomb gauge by its vector potential $\vec{A}(\vec{r}_e, t)$. We replace the momentum $\vec{p}_e$ in the kinetic energy $\vec{p}_e^{\,2}/(2m_e)$ by the momentum $\vec{p}_e - e\vec{A}(\vec{r}_e, t)$. The square of this difference leads to the cross term

$$H_{cp} \equiv -\frac{e}{m_e}\vec{A}(\vec{r}_e, t) \cdot \vec{p}_e$$

which couples the momentum of the particle to the field.

We now consider this coupling for the most elementary atom consisting of a single electron and a single proton. Moreover, we recall the dipole approximation: The vector potential of an optical field does not change appreciably over the size of the atom. We then arrive at a coupling

$$H_{\vec{A}\cdot\vec{p}} \equiv -\frac{e}{\mu}\vec{A}(\vec{R}, t) \cdot \vec{p}$$

between the momentum $\vec{p}$ of the relative motion of the electron with reduced mass μ and the vector potential at the position $\vec{R}$ of the center-of-mass.

In the second approach we consider the atom as a dipole with a dipole moment $\vec{\wp} \equiv e\vec{r}$. Here $\vec{r}$ denotes the position of the electron relative to the proton. We recall again the dipole approximation: The electric field $\vec{E}$ of an optical electromagnetic field does not change considerably over the size of the atom. Hence the dipole has a potential energy

$$H_{\vec{r}\cdot\vec{E}} \equiv -e\vec{r} \cdot \vec{E}(\vec{R}, t) \tag{14.1}$$

in the electric field $\vec{E}(\vec{R}, t)$ evaluated at the position $\vec{R}$ of the center-of-mass.

In the present chapter we derive both interactions starting from the Hamiltonian of an electron and a proton in an electromagnetic field and using the dipole approximation. In the absence of the center-of-mass motion both Hamiltonians are equivalent, that is, there exists a gauge transformation which connects the two corresponding wave functions. However, this transformation is not straightforward in the presence of center-of-mass motion. In this case we have to go beyond the simple dipole approximation as discussed in great detail in Appendix L. As a consequence additional terms such as the Röntgen Hamiltonian arise.

14.2 Vector Potential-Momentum Coupling

In this section we first briefly review the essential ingredients of the minimal coupling scheme of a single particle and then turn to the case of an atom in an electromagnetic field. Throughout this section we focus on the complete Hamiltonian of the atom including the center-of-mass motion. The latter will be of great importance for the discussion of atom optics in quantized light fields.

14.2.1 Gauge Principle Determines Minimal Coupling

The standard approach of coupling a single, charged particle of mass m to an external field described by a vector potential $\vec{A} = \vec{A}(\vec{r},t)$ and a scalar potential $\Phi(\vec{r},t)$ relies on the minimal coupling scheme. Here we replace the canonical momentum $\vec{p}$ by the kinetic momentum $\vec{p} - e\vec{A}$. Moreover, we add the potential energy $V(\vec{r}) \equiv e\Phi(\vec{r},t)$. But what is the deeper reason for this procedure?

H. Weyl in 1928 provided the answer. He connected the gauge invariance of electrodynamics with quantum mechanics and showed that this results in the minimal coupling. Modern quantum field theories, in particular gauge theories such as quantum electrodynamics and quantum chromodynamics, rest on the same principle. Motivated by this importance, we now briefly review the principle of gauge invariance. Throughout this section we consider the electromagnetic field to be classical.

Global Transformations

In quantum mechanics we never directly observe the wave function of a quantum system. We rather observe probabilities for the outcome of a specific measurement. These probabilities follow from the absolute value square of the wave function in the appropriate representation. Therefore, the wave function is only determined up to an overall phase. We now start from the wave function $\psi(\vec{r},t)$ which satisfies the Schrödinger equation

$$i\hbar \frac{\partial \psi(\vec{r},t)}{\partial t} = \hat{H}\psi(\vec{r},t) \equiv \left[\frac{1}{2m}\left(\frac{\hbar}{i}\vec{\nabla}\right)^2 + U(\vec{r}) \right]\psi(\vec{r},t) \tag{14.2}$$

where $U(\vec{r})$ is an external potential.

The transformed wave function

$$\tilde{\psi}(\vec{r},t) \equiv e^{i\alpha}\psi(\vec{r},t) \tag{14.3}$$

satisfies the same Schrödinger equation

$$i\hbar \frac{\partial \tilde{\psi}(\vec{r},t)}{\partial t} = \left[\frac{1}{2m}\left(\frac{\hbar}{i}\vec{\nabla}\right)^2 + U(\vec{r}) \right]\tilde{\psi}(\vec{r},t),$$

provided the phase α is independent of position and time.

In the transformation Eq. (14.3) the change of the wave function ψ is independent of position and time. All points in space-time are treated equally. Therefore, this transformation is called a *global transformation*.

Local Transformations

We now consider a situation in which the phase does depend on position and time, and thus analyze the transformed wave function

$$\tilde{\psi}(\vec{r},t) \equiv \exp\left[\frac{i}{\hbar}e\Lambda(\vec{r},t)\right]\psi(\vec{r},t). \tag{14.4}$$

Here for convenience we have introduced Planck's constant and the electric charge e.

Since in the transformation Eq. (14.4) the phase depends on the point in space-time the transformation is called *local transformation*.

Due to this position and time dependence of Λ the wave function $\tilde{\psi}$ does not satisfy the Schrödinger equation, but the more complicated equation

$$i\hbar \frac{\partial \tilde{\psi}(\vec{r},t)}{\partial t} = \frac{1}{2m}\left[\frac{\hbar}{i}\vec{\nabla} - e\vec{\nabla}\Lambda(\vec{r},t)\right]^2 \tilde{\psi}(\vec{r},t) + \left[U(\vec{r}) - e\frac{\partial \Lambda(\vec{r},t)}{\partial t}\right]\tilde{\psi}(\vec{r},t). \quad (14.5)$$

We can convince ourselves that this equation is correct by substituting the wave function Eq. (14.4)

$$\psi(\vec{r},t) = \exp\left[-\frac{i}{\hbar}e\Lambda(\vec{r},t)\right]\tilde{\psi}(\vec{r},t)$$

into Eq. (14.2) and making use of the relations

$$i\hbar\frac{\partial \psi}{\partial t} = i\hbar\frac{\partial}{\partial t}\left[\exp\left(-\frac{i}{\hbar}e\Lambda\right)\tilde{\psi}\right] = \exp\left(-\frac{i}{\hbar}e\Lambda\right)\left[e\frac{\partial \Lambda}{\partial t}\tilde{\psi} + i\hbar\frac{\partial \tilde{\psi}}{\partial t}\right]$$

and

$$\frac{\hbar}{i}\vec{\nabla}\psi = \frac{\hbar}{i}\vec{\nabla}\left[\exp\left(-\frac{i}{\hbar}e\Lambda\right)\tilde{\psi}\right] = \exp\left(-\frac{i}{\hbar}e\Lambda\right)\left(\frac{\hbar}{i}\vec{\nabla} - e\vec{\nabla}\Lambda\right)\tilde{\psi}$$

which implies

$$\left(\frac{\hbar}{i}\vec{\nabla}\right)^2\left[\exp\left(-\frac{i}{\hbar}e\Lambda\right)\tilde{\psi}\right] = \frac{\hbar}{i}\vec{\nabla}\left[\exp\left(-\frac{i}{\hbar}e\Lambda\right)\left(\frac{\hbar}{i}\vec{\nabla} - e\vec{\nabla}\Lambda\right)\tilde{\psi}\right]$$
$$= \exp\left(-\frac{i}{\hbar}e\Lambda\right)\left(\frac{\hbar}{i}\vec{\nabla} - e\vec{\nabla}\Lambda\right)^2\tilde{\psi}.$$

Thus we arrive indeed at the Schrödinger equation Eq. (14.5) for $\tilde{\psi}$.

Gauge Invariance of Electrodynamics

Despite the additional terms in Eq. (14.5) there is still a similarity to the Schrödinger equation: The momentum has been replaced by the momentum minus the gradient of the phase, and the potential energy has been modified by the time derivative of the phase. This is reminiscent of the gauge freedom of electrodynamics discussed in Sec. 10.1.2.

We recall, that we can solve the homogeneous Maxwell's equations by introducing the vector potential $\vec{A}$ and the scalar potential Φ. Since the magnetic flux density follows from the curl of the vector potential $\vec{A}$, that is

$$\vec{B} \equiv \vec{\nabla} \times \vec{A},$$

the vector potential is only determined up to a gradient of a scalar field. Indeed, the transformed vector potential

$$\vec{A}' \equiv \vec{A} + \vec{\nabla}\Lambda \quad (14.6)$$

creates the same magnetic flux.

14.2 Vector Potential-Momentum Coupling

Similarly, since the electric field $\vec{E}$ follows from

$$\vec{E} \equiv -\vec{\nabla}\Phi - \frac{\partial \vec{A}}{\partial t}$$

we can include the time derivative $\dot{\Lambda}$ in the scalar potential Φ. The so-transformed scalar potential

$$\Phi' \equiv \Phi - \frac{\partial \Lambda}{\partial t} \tag{14.7}$$

creates the same electric field

$$\vec{E}' \equiv -\vec{\nabla}\Phi' - \frac{\partial \vec{A}'}{\partial t} = \vec{E}.$$

We emphasize however, that we are restricted in our choice of the gauge potential Λ. Indeed the Coulomb gauge condition $\vec{\nabla} \cdot \vec{A}$ puts the constraint $\Delta\Lambda = 0$ on Λ.

Interaction – A Consequence of the Invariance Postulate

We now postulate that the Schrödinger equation should stay invariant under local transformations such as Eq. (14.4), that is the wave equation for the transformed wave function $\tilde{\psi}$ should take the same form as the Schrödinger equation for the original wave function ψ. We therefore have to start from a Schrödinger equation which already contains electromagnetic potentials. They allow us to absorb the extra terms $\vec{\nabla}\Lambda$ and $\dot{\Lambda}$ which we obtain when we make the local transformation. We start from the Schrödinger equation

$$i\hbar \frac{\partial \psi(\vec{r},t)}{\partial t} = \frac{1}{2m}\left[\left(\frac{\hbar}{i}\vec{\nabla}\right) - e\vec{A}(\vec{r},t)\right]^2 \psi(\vec{r},t) + [U(\vec{r}) + e\Phi(\vec{r},t)]\,\psi(\vec{r},t).$$

When we now perform the transformation

$$\tilde{\psi}(\vec{r},t) \equiv \exp\left[\frac{i}{\hbar}e\Lambda(\vec{r},t)\right]\psi(\vec{r},t)$$

the new wave function $\tilde{\psi}$ satisfies the Schrödinger equation

$$i\hbar \frac{\partial \tilde{\psi}(\vec{r},t)}{\partial t} = \frac{1}{2m}\left[\left(\frac{\hbar}{i}\vec{\nabla}\right) - e\vec{A}'(\vec{r},t)\right]^2 \tilde{\psi}(\vec{r},t) + [U(\vec{r}) + e\Phi'(\vec{r},t)]\,\tilde{\psi}(\vec{r},t),$$

which is formally identical to the first one.

Here we have used the gauge freedom of electrodynamics to absorb the additional terms in the modified potentials, Eqs. (14.6) and (14.7). Due to this connection to gauge transformations, the transformations Eqs. (14.3) and (14.4) are called *global* or *local gauge transformations*, respectively.

Hamiltonian of a Charged Particle in an Electromagnetic Field

We are now in a position to write the Hamiltonian of a charged particle without internal degrees, that is without spin in an electromagnetic field. We perform minimal coupling and arrive at the Hamiltonian

$$H \equiv \frac{1}{2m}\left(\vec{p} - e\vec{A}\right)^2 + e\Phi + \mathcal{H}$$

where we have also included the Hamiltonian

$$\mathcal{H} \equiv \int d^3r \left[\frac{1}{2}\varepsilon_0 \vec{E}^2(\vec{r},t) + \frac{1}{2}\mu_0 \vec{H}^2(\vec{r},t)\right] \tag{14.8}$$

of the free field, Eq. (10.39).

When we evaluate the square in the kinetic energy we find the two cross-terms $\vec{A}\cdot\vec{p}$ and $\vec{p}\cdot\vec{A}$. Since in the transition to quantum mechanics we have to replace $\vec{p}$ by the operator $\hat{\vec{p}} \equiv \frac{\hbar}{i}\vec{\nabla}$ the order of the operators is important.

We recall the relation

$$\hat{\vec{p}}\cdot\left[\vec{A}(\vec{r})\psi(\vec{r})\right] = \frac{\hbar}{i}\vec{\nabla}\cdot\left[\vec{A}(\vec{r})\psi(\vec{r})\right] = \frac{\hbar}{i}\left[\psi(\vec{r})\vec{\nabla}\cdot\vec{A} + \vec{A}\cdot\vec{\nabla}\psi(\vec{r})\right]$$

valid for any arbitrary position dependent function $\psi(\vec{r})$. Since in Coulomb gauge we have $\vec{\nabla}\cdot\vec{A} = 0$ we find

$$\hat{\vec{p}}\cdot\vec{A} = \vec{A}\cdot\hat{\vec{p}}.$$

Hence the quantum mechanical Hamiltonian

$$\hat{H} = \frac{\hat{\vec{p}}^2}{2m} - \frac{e}{m}\vec{A}(\hat{\vec{r}},t)\cdot\hat{\vec{p}} + \frac{e^2}{2m}\vec{A}^2(\hat{\vec{r}},t) + e\Phi(\hat{\vec{r}},t) + \mathcal{H} \tag{14.9}$$

is, indeed, identical to the classical one.

14.2.2 Interaction of an Atom with a Field

We now consider the situation of an atom in a field. For simplicity we take a hydrogen atom with a proton of mass m_p at position $\vec{r}_p$ and an electron of mass m_e at $\vec{r}_e$ as shown in Fig. 14.1. Throughout this section we suppress the Hamiltonian $\mathcal{H}$ of the free field.

With the Coulomb interaction

$$V(\vec{r}) \equiv -\frac{1}{4\pi\varepsilon_0}\frac{e^2}{|\vec{r}|}$$

between electron and proton the total Hamiltonian of the hydrogen atom in the electromagnetic field reads

$$H = H_e + H_p - V(\vec{r}_e - \vec{r}_p). \tag{14.10}$$

Here we have followed Eq. (14.9) and have defined the Hamiltonians

$$H_e \equiv \frac{\hat{\vec{p}}_e^2}{2m_e} - \frac{e}{m_e}\vec{A}(\vec{r}_e,t)\cdot\vec{p}_e + \frac{e^2}{2m_e}\vec{A}^2(\vec{r}_e,t) \tag{14.11}$$

14.2 Vector Potential-Momentum Coupling

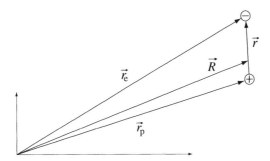

Fig. 14.1: The simplest model of a hydrogen atom consists of a proton at position $\vec{r}_p$ and an electron at $\vec{r}_e$. Due to the non-vanishing mass ratio between electron and proton the center-of-mass $\vec{R} \equiv (m_p + m_e)^{-1}(m_e\vec{r}_e + m_p\vec{r}_p)$ is slightly shifted from the proton to the electron.

and

$$H_p \equiv \frac{\vec{p}_p^{\,2}}{2m_p} + \frac{e}{m_p}\vec{A}(\vec{r}_p,t) \cdot \vec{p}_p + \frac{e^2}{2m_p}\vec{A}^{\,2}(\vec{r}_p,t) \tag{14.12}$$

of the electron and the proton in the external field.

In order to bring out most clearly the influence of the dipole approximation discussed in the next section we have included the position dependence of the vector potential. Moreover, for the sake of simplicity we have suppressed the operator hats on the positions and momenta of the electron and the proton, as well as on the Hamiltonians.

Center-of-Mass and Relative Coordinates

We now introduce the center-of-mass

$$\vec{R} \equiv \frac{m_e\vec{r}_e + m_p\vec{r}_p}{m_p + m_e} = \frac{m_e}{M}\vec{r}_e + \frac{m_p}{M}\vec{r}_p$$

of the two-particle system of total mass

$$M \equiv m_e + m_p = m_p\left(1 + \frac{m_e}{m_p}\right) \tag{14.13}$$

and the relative coordinate

$$\vec{r} \equiv \vec{r}_e - \vec{r}_p$$

between electron and proton.

We can then express the position

$$\vec{r}_e = \vec{R} + \frac{m_p}{M}\vec{r} \cong \left(\vec{R} - \frac{m_e}{m_p}\vec{r}\right) + \vec{r} \tag{14.14}$$

of the electron and the position

$$\vec{r}_p = \vec{R} - \frac{m_e}{M}\vec{r} \cong \vec{R} - \frac{m_e}{m_p}\vec{r} \tag{14.15}$$

of the proton in terms of the center-of-mass $\vec{R}$ and relative coordinate $\vec{r}$. Note, that in the last step of the two equations we have made use of Eq. (14.13) for the total mass of the system.

We emphasize that due to the non-vanishing mass ratio $m_e/m_p \cong 1/1836 \ll 1$ the position of the proton is not identical with the center-of-mass. In particular, the latter is slightly shifted towards the electron as indicated in Fig. 14.1.

We introduce the momentum

$$\vec{P} \equiv \vec{p}_e + \vec{p}_p$$

associated with the motion of the total mass M, and the momentum

$$\vec{p} \equiv \frac{m_p}{M} \vec{p}_e - \frac{m_e}{M} \vec{p}_p \qquad (14.16)$$

associated with the relative motion of the reduced mass

$$\mu \equiv \frac{m_e m_p}{m_e + m_p} = m_e \left(1 + \frac{m_e}{m_p}\right)^{-1}. \qquad (14.17)$$

These definitions allow us to express the momentum

$$\vec{p}_e = \frac{m_e}{M} \vec{P} + \vec{p} \qquad (14.18)$$

of the electron and the momentum

$$\vec{p}_p = \frac{m_p}{M} \vec{P} - \vec{p} \qquad (14.19)$$

of the proton in terms of $\vec{P}$ and $\vec{p}$. When we use these relations to evaluate the kinetic energies of the electron and the proton the cross-terms cancels in the sum and we find

$$\frac{\vec{p}_e^{\,2}}{2m_e} + \frac{\vec{p}_p^{\,2}}{2m_p} = \frac{\vec{P}^{\,2}}{2M} + \frac{\vec{p}^{\,2}}{2\mu}, \qquad (14.20)$$

which is just the sum of the kinetic energies of the center-of-mass and the relative motion.

Hamiltonian of Atom in an Electromagnetic Field

We are now in a position to express the Hamiltonian of the hydrogen atom in the electromagnetic field, Eq. (14.10) in the center-of-mass and relative coordinates. When we make use of the relations Eqs. (14.19), (14.14), (14.15), (14.18) and (14.20) expressing the various quantities in the center-of-mass and relative coordinate variables we arrive at

$$\begin{aligned} H = & \frac{\vec{P}^{\,2}}{2M} + \frac{\vec{p}^{\,2}}{2\mu} - \frac{1}{4\pi\varepsilon_0} \frac{e^2}{|\vec{r}|} \\ & - \frac{e}{\mu} \left\{ \frac{\mu}{m_e} \vec{A}\left[\vec{R} + \frac{m_p}{M}\vec{r}, t\right] + \frac{\mu}{m_p} \vec{A}\left[\vec{R} - \frac{m_e}{M}\vec{r}, t\right] \right\} \vec{p} \\ & - \frac{e}{M} \left\{ \vec{A}\left[\vec{R} + \frac{m_p}{M}\vec{r}, t\right] - \vec{A}\left[\vec{R} - \frac{m_e}{M}\vec{r}, t\right] \right\} \vec{P} \\ & + \frac{e^2}{2m_e} \vec{A}^{\,2}\left[\vec{R} + \frac{m_p}{M}\vec{r}, t\right] + \frac{e^2}{2m_p} \vec{A}^{\,2}\left[\vec{R} - \frac{m_e}{M}\vec{r}, t\right]. \end{aligned} \qquad (14.21)$$

The Hamiltonian H contains the Hamiltonian

$$H_\mathrm{H} \equiv \frac{\vec{P}^2}{2M} + \frac{\vec{p}^{\,2}}{2\mu} - \frac{1}{4\pi\varepsilon_0}\frac{e^2}{|\vec{r}|} \tag{14.22}$$

of the hydrogen atom in the absence of an electromagnetic field. This Hamiltonian consists of the kinetic energies of the center-of-mass motion and the relative motion and the Coulomb interaction between electron and proton.

Moreover, the complete Hamiltonian Eq. (14.21) of the atom in the field includes a coupling of the atom to the vector potential. We note two types of couplings: The momentum $\vec{p}$ of the relative motion couples to the sum of the vector potentials evaluated at the positions of the electron and the proton. In contrast, the momentum $\vec{P}$ of the center-of-mass motion couples to the difference of the two vector potentials.

The last two terms containing the squares of the vector potential do not couple to any momenta but do contain center-of-mass motion as indicated by $\vec{R}$ as well as the internal degrees of freedom summarized by $\vec{r}$.

14.3 Dipole Approximation

The exact Hamiltonian Eq. (14.21) of a hydrogen atom in an electromagnetic field simplifies considerably when we apply the so-called dipole approximation. In this section we develop this approximation in great detail.

14.3.1 Expansion of Vector Potential

We recall that the position dependence of $\vec{A}$ in Eq. (14.21) is of the form $\vec{A}(\vec{R} + \delta\vec{r})$ where $\delta\vec{r}$ is either given by $\delta\vec{r} \equiv -m_e/M\vec{r}$ or $\delta\vec{r} \equiv m_p/M\vec{r}$. Hence when we expand the vector potential $\vec{A}(\vec{R} + \delta\vec{r})$ into a Taylor series around $\vec{R}$ we arrive at

$$\vec{A}(\vec{R} + \delta\vec{r}) \cong \vec{A}(\vec{R}) + (\delta\vec{r} \cdot \vec{\nabla}_{\vec{R}})\vec{A}(\vec{R}) + \ldots. \tag{14.23}$$

The vector potential changes on a characteristic length determined by the wave vector $\vec{k}$. When we recall that $m_e/m_p \ll 1$ and $\vec{r}$ is of the size of the atom we can estimate the linear correction term $(\delta\vec{r} \cdot \vec{\nabla}_{\vec{R}})\vec{A}(\vec{R})$ relative to $\vec{A}(\vec{R})$ and find

$$\frac{\left|(\delta\vec{r} \cdot \vec{\nabla}_{\vec{R}})\vec{A}(\vec{R})\right|}{\left|\vec{A}(\vec{R})\right|} \cong \delta\vec{r} \cdot \vec{k} \lesssim \vec{r} \cdot \vec{k} \cong 2\pi \frac{\text{size of the atom}}{\text{wavelength of light}} \ll 1. \tag{14.24}$$

Here we have used that the wave vector is inversely proportional to the wavelength of the light.

Hence the vector potential does not change considerably over the size of the atom: The electron and the proton both feel essentially the same vector potential, namely the one at the center-of-mass coordinate, that is

$$\vec{A}(\vec{r}_e) \cong \vec{A}(\vec{r}_p) \cong \vec{A}(\vec{R}). \tag{14.25}$$

This approximation carries the name dipole approximation, which will become clear once we have discussed the $\vec{E} \cdot \vec{r}$-formulation of the interaction.

14.3.2 $\vec{A} \cdot \vec{p}$-Interaction

The Taylor expansion Eq. (14.23) simplifies the Hamiltonian Eq. (14.21) of the hydrogen atom in an electromagnetic field considerably. In lowest order we replace the position arguments of the vector potential by the center-of-mass coordinate $\vec{R}$ and arrive at

$$H^{(0)} \equiv \frac{\vec{P}^2}{2M} + \frac{\vec{p}^2}{2\mu} - \frac{1}{4\pi\varepsilon_0}\frac{e^2}{|\vec{r}|} - \frac{e}{\mu}\vec{A}(\vec{R},t)\,\vec{p} + \frac{e^2}{2\mu}\vec{A}^2(\vec{R},t). \tag{14.26}$$

Hence in this formulation of the interaction the relative coordinate $\vec{r}$ does not couple to the field. Since $\vec{r}$ is associated with the dipole moment $\vec{\wp} \equiv e\vec{r}$ of the atom we are tempted to call this level of the approximation the zeroth dipole approximation. However, this notation is misleading. The problem stands out most clearly in the $\vec{r} \cdot \vec{E}$-formulation, Eq. (14.1), of the interaction. Indeed, the Hamiltonian Eq. (14.26) in zeroth approximation states that it is the momentum $\vec{p}$ of the relative motion which couples to the vector potential. In Sec. 14.6 we will show that in the absence of the center-of-mass motion the two interactions are equivalent. Hence the zeroth approximation of the Taylor series Eq. (14.23) which is independent of $\vec{r}$ does already bring in the dipole through the relative momentum $\vec{p}$.

14.3.3 Various Forms of the $\vec{A} \cdot \vec{p}$ Interaction

We gain a deeper understanding of the form of the Hamiltonian $H^{(0)}$ when we now rearrange its various terms.

Emphasizing the Role of the Electron

One such representation combines the relative momentum and the vector potential to the square and reads

$$H^{(0)} = H_{\text{kin}} + H_\mu = \frac{\vec{P}^2}{2M} + \frac{1}{2\mu}\left[\vec{p} - e\vec{A}(\vec{R},t)\right]^2 - \frac{1}{4\pi\varepsilon_0}\frac{e^2}{|\vec{r}|}. \tag{14.27}$$

Hence the Hamiltonian of the atom in the electromagnetic field consists of the sum of the kinetic energy

$$H_{\text{kin}} \equiv \frac{\vec{P}^2}{2M}$$

of the center-of-mass motion and the Hamiltonian

$$H_\mu = \frac{1}{2\mu}\left[\vec{p} - e\vec{A}(\vec{R},t)\right]^2 + V(\vec{r})$$

of a charged particle of reduced mass μ in a Coulomb potential and in an electromagnetic wave of vector potential $\vec{A}$. The latter is evaluated at the position $\vec{R}$ of the center-of-mass. This form of the Hamiltonian emphasizes the role of the electron in the atom.

Emphasizing the Role of the Atom

Another representation emphasizes the separation of center-of-mass motion and relative motion. The latter couples to the electromagnetic field through the vector potential at the position $\vec{R}$.

We start from the Hamiltonian

$$H_{\text{atom}} \equiv \frac{\vec{p}^2}{2\mu} - \frac{1}{4\pi\varepsilon_0}\frac{e^2}{|\vec{r}|} \tag{14.28}$$

of an infinitely heavy atom located at the position $\vec{R}$ in the electromagnetic wave. The atom interacts with the vector potential $\vec{A}$ via the interaction Hamiltonian

$$H_{\vec{A}\cdot\vec{p}} \equiv -\frac{e}{\mu}\vec{A}(\vec{R},t)\cdot\vec{p}.$$

We note that this so-called $\vec{A}\cdot\vec{p}$ interaction term couples the electromagnetic field via its vector potential to the momentum $\vec{p}$ of the electron. We hence arrive at the decomposition

$$H^{(0)} = H_{\text{cm}} + H_{\text{atom}} + H_{\vec{A}\cdot\vec{p}} \tag{14.29}$$

where we have introduced the Hamiltonian

$$H_{\text{cm}} \equiv \frac{\vec{P}^2}{2M} + \frac{e^2}{2\mu}\vec{A}^2(\vec{R},t) \tag{14.30}$$

which only involves center-of-mass coordinates. It hence describes the motion of the center-of-mass in the ponderomotive potential

$$U_{\text{pp}}(\vec{R}) \equiv \frac{e^2}{2\mu}\vec{A}^2(\vec{R},t)$$

which is the square of the vector potential. This potential plays an important role in plasma physics, atoms in strong fields and the free electron laser.

Summary

We conclude this section by noting that the Hamiltonian Eq. (14.26) follows directly from Eq. (14.10). Indeed, when we approximate in the Hamiltonians H_e and H_p of the electron and the proton, Eqs. (14.11) and (14.12), their positions $\vec{r}_e$ und $\vec{r}_p$ by $\vec{r}$ and $\vec{R}$ and use the dipole approximation Eq. (14.25) we arrive at

$$H_e \cong \frac{\vec{p}_e^2}{2m_e} - \frac{e}{m_e}\vec{A}(\vec{R},t)\cdot\vec{p}_e + \frac{e^2}{2m_e}\vec{A}^2(\vec{R},t) \tag{14.31}$$

and

$$H_p \cong \frac{\vec{p}_p^2}{2m_p} + \frac{e}{m_p}\vec{A}(\vec{R},t)\cdot\vec{p}_p + \frac{e^2}{2m_p}\vec{A}^2(\vec{R},t) \tag{14.32}$$

which yields

$$H = \frac{\vec{p}_e^2}{2m_e} + \frac{\vec{p}_p^2}{2m_p} - e\vec{A}(\vec{R},t)\cdot\left(\frac{1}{m_e}\vec{p}_e - \frac{1}{m_p}\vec{p}_p\right) + \frac{e^2}{2\mu}\vec{A}^2(\vec{R},t) - \frac{1}{4\pi\varepsilon_0}\frac{e^2}{|\vec{r}|}. \tag{14.33}$$

Together with Eqs. (14.16) and (14.20) we find Eq. (14.26).

14.3.4 Higher Order Corrections

The linear term of the Taylor expansion Eq. (14.23) brings in the relative coordinate $\vec{r}$. Since the zeroth order approximation contains already the dipole we call this level of approximation the higher order dipole approximation. Here the treatment is slightly more complicated. To focus on the essential ideas we summarize the first order correction in Appendix L and find

$$H^{(1)} = \frac{1}{2M}\left[\vec{P} - e\left(\vec{r}\cdot\vec{\nabla}_{\vec{R}}\right)\vec{A}\right]^2 + \frac{1}{2\mu}\left[\vec{p} - e\vec{A} - e\frac{\Delta m}{M}\left(\vec{r}\cdot\vec{\nabla}_{\vec{R}}\right)\vec{A}\right]^2 + V(\vec{r}) \quad (14.34)$$

where

$$\Delta m \equiv m_p - m_e$$

is the difference in masses between the proton and the electron.

There exists a coupling between the momentum $\vec{P}$ of the center-of-mass motion and the directional derivative $\left(\vec{r}\cdot\vec{\nabla}_{\vec{R}}\right)\vec{A}$ of the vector potential giving rise to the interaction Hamiltonian

$$H' \equiv -\frac{e}{M}\left(\vec{r}\cdot\vec{\nabla}_{\vec{R}}\right)\vec{A}\cdot\vec{P}.$$

This term brings in a quantity closely related to the magnetic field. Indeed, from the vector relation

$$\vec{r}\times\vec{B} = \vec{r}\times\vec{\nabla}_{\vec{R}}\times\vec{A} = \vec{\nabla}_{\vec{R}}\left(\vec{r}\cdot\vec{A}\right) - \left(\vec{r}\cdot\vec{\nabla}_{\vec{R}}\right)\vec{A}$$

we recognize that H' is closely related to the so-called Röntgen Hamiltonian

$$H_{RM} \equiv \frac{e}{M}\left(\vec{r}\times\vec{B}\right)\cdot\vec{P}$$

discussed in more detail in the next section.

We emphasize however, that H' is not identical to the Röntgen Hamiltonian since the term $\vec{\nabla}_{\vec{R}}\left(\vec{r}\cdot\vec{A}\right)$ is missing. In Appendix L we show that by an appropriate gauge transformation we can indeed obtain the missing term.

We conclude this discussion by noting that a similar Röntgen term arises for the relative motion. The gauge transformation derived in Appendix L brings the Hamiltonian

$$H'' \equiv -\frac{e}{\mu}\frac{\Delta m}{M}\left(\vec{r}\cdot\vec{\nabla}_{\vec{r}}\right)\vec{A}\cdot\vec{p}$$

arising from the cross term in the square in the kinetic energy of the relative motion into the gauge invariant form

$$H_{Re} \equiv \frac{e}{2\mu}\frac{\Delta m}{M}\left(\vec{r}\times\vec{B}\right)\cdot\vec{p}.$$

Here an additional factor 2 appears which is closely related to the Thomas factor.

The subscripts in the Röntgen Hamiltonians H_{RM} and H_{Re} will become clear when we discuss the interaction between an electric dipole and an electric field. Similarly, the name Röntgen Hamiltonian will become apparent.

14.4 Electric Field-Dipole Interaction

Apart from the description of the interaction between light and matter based on $\vec{A} \cdot \vec{p}$ there exists one which rests on the coupling of a dipole to an electric field. In the present section we focus on this approach.

14.4.1 Dipole Approximation

We recall that an atom consisting of an electron and proton separated by a vector $\vec{r}$ enjoys a dipole moment $\vec{\wp} \equiv e\vec{r}$. Again we note that the electric field does not change considerably over the size of the atom. Hence the electric field $\vec{E}(\vec{r}_e, t)$ at the position $\vec{r}_e$ of the electron is almost identical to the electric field $\vec{E}(\vec{r}_p, t)$ at the position $\vec{r}_p$ of the proton or to the one at the center-of-mass, that is

$$\vec{E}(\vec{r}_e, t) \cong \vec{E}(\vec{r}_p, t) \cong \vec{E}(\vec{R}, t).$$

The dipole in the electric field $\vec{E}$ experiences a potential energy

$$H_{\vec{r} \cdot \vec{E}} \equiv -\vec{\wp} \cdot \vec{E}(\vec{R}, t) = -e\vec{r} \cdot \vec{E}(\vec{R}, t). \tag{14.35}$$

This argument suggests that the total Hamiltonian of the atom in the electromagnetic field in dipole approximation reads

$$\widetilde{H}^{(0)} \equiv \frac{\vec{P}^2}{2M} + \frac{\vec{p}^2}{2\mu} + V(\vec{r}) - e\vec{r} \cdot \vec{E}(\vec{R}, t). \tag{14.36}$$

We emphasize that for many applications this Hamiltonian suffices.

14.4.2 Röntgen Hamiltonians and Others

However, this ad hoc procedure can lead to inconsistent results. These problems manifest themselves when we take into account the center-of-mass motion. In particular, it has been shown that the conventional dipole coupling Eq. (14.35) leads to an unphysical atomic velocity dependence in the angular distribution of photons spontaneously emitted by a moving atom.

The resolution of these inconsistency problem emerges from the correction terms $\left(\vec{r} \cdot \vec{\nabla}_{\vec{R}}\right) \vec{A}$ in the Hamiltonian $H^{(1)}$, Eq. (14.34), Indeed, in Appendix L we show that the gauge invariant Hamiltonian $\widetilde{H}^{(1)}$ derived from $H^{(1)}$ reads

$$\widetilde{H}^{(1)} \equiv \frac{1}{2M} \left[\vec{P} + e\left(\vec{r} \times \vec{B}\right)\right]^2 + \frac{1}{2\mu} \left[\vec{p} + \frac{e}{2}\frac{\Delta m}{M}\left(\vec{r} \times \vec{B}\right)\right]^2 + V(\vec{r})$$
$$- e\vec{r} \cdot \vec{E} - \frac{e}{2}\frac{\Delta m}{M}\left(\vec{r} \cdot \vec{\nabla}_{\vec{R}}\right)\left(\vec{r} \cdot \vec{E}\right). \tag{14.37}$$

In contrast to the expression Eq. (14.34) this formula now contains fields rather than the vector potential. Moreover, when we neglect the contributions containing the magnetic field and the gradient of the electric field the Hamiltonian reduces to $\widetilde{H}^{(0)}$, Eq. (14.36).

The physics of the individual contributions to $\widetilde{H}^{(1)}$ stands out most clearly when we calculate the squares in the kinetic energies which results in

$$\widetilde{H}^{(1)} = \widetilde{H}^{(0)} + H_{RM} + H_{Re} + H_{(\vec{r}\times\vec{B})^2} + H_{\vec{\nabla}(\vec{\wp}\cdot\vec{E})}.$$

Hence we have derived the Röntgen term

$$H_{RM} \equiv \frac{e}{M} \left(\vec{r} \times \vec{B}\right) \cdot \vec{P}.$$

We gain deeper insight into this term by recalling from Eq. (L.12) the relation

$$\vec{P} = M\dot{\vec{R}} - e\left(\vec{r} \times \vec{B}\right)$$

connecting the canonical momentum $\vec{P}$ and the kinetic momentum $M\dot{\vec{R}}$. The Röntgen Hamiltonian then takes the form

$$H_{RM} = e\left(\vec{r} \times \vec{B}\right) \cdot \dot{\vec{R}} - \frac{e^2}{M}\left(\vec{r} \times \vec{B}\right)^2 = -e\vec{r}\cdot\left(\dot{\vec{R}} \times \vec{B}\right) - \frac{e^2}{M}\left(\vec{r} \times \vec{B}\right)^2,$$

or

$$H_{RM} \equiv H_{\vec{r}\cdot\vec{E}_{cm}} - \frac{e^2}{M}\left(\vec{r} \times \vec{B}\right)^2$$

and contains two contributions. The first one results from the motion of the dipole $\vec{\wp} \equiv e\vec{r}$ with velocity $\dot{\vec{R}}$ through a magnetic field. In the moving frame the dipole sees an electric field $\vec{E}_{cm} \equiv \dot{\vec{R}} \times \vec{B}$. A moving observer does not only feel a magnetic field but also an electric field. This fact was already known to W. Röntgen in 1888. For this reason the contribution to the Hamiltonian carries his name. The second contribution involves the square of the magnetic field. A similar term arises from the squares of the momenta and will be discussed at the end of the section.

Similarly, we find a Röntgen term

$$H_{Re} \equiv \frac{1}{\mu}\frac{\Delta m}{2M} e \left(\vec{r} \times \vec{B}\right) \cdot \vec{p}$$

for the relative motion. When we recall the relation Eq. (L.13) for the canonical momentum

$$\vec{p} = \mu\dot{\vec{r}} - \frac{e}{2}\frac{\Delta m}{M}\left(\vec{r} \times \vec{B}\right)$$

of the relative motion we arrive at

$$H_{Re} = \frac{\Delta m}{2M} e \left(\vec{r} \times \vec{B}\right) \cdot \dot{\vec{r}} - \frac{1}{\mu}\left(\frac{e}{2}\frac{\Delta m}{M}\right)^2 \left(\vec{r} \times \vec{B}\right)^2$$

or

$$H_{Re} = -\vec{\wp}\cdot\vec{E}_e - \frac{1}{\mu}\left(\frac{e}{2}\frac{\Delta m}{M}\right)^2 \left(\vec{r} \times \vec{B}\right)^2.$$

Indeed, since the electron is moving around the nucleus, the dipole is moving with a velocity $\dot{\vec{r}}$. Consequently, the dipole $\vec{\wp} \equiv e\vec{r}$ sees a motion-induced electric field

$\vec{E}_e \equiv \frac{\Delta m}{2M}\left(\dot{\vec{r}} \times \vec{B}\right)$. However, in comparison to the analogous effect in the center-of-mass motion the field is now more complicated: It contains a factor $1/2$, which is the familiar Thomas factor and the mass ratio

$$\frac{\Delta m}{M} = \frac{m_p - m_e}{m_p + m_e} = \frac{1 - \frac{m_e}{m_p}}{1 + \frac{m_e}{m_p}} \cong 1 - 2\frac{m_e}{m_p},$$

which is close to unity.

More insight into the physics of the term H_{Re} springs from the form

$$H_{Re} = -\frac{e}{\mu}\frac{\Delta m}{2M} \cdot \vec{B} \cdot (\vec{r} \times \vec{p})$$

which brings out the coupling of the orbital angular momentum $\vec{r} \times \vec{p}$ to the magnetic field. The Hamiltonian H_{Re} is therefore closely related to the familiar paramagnetic term of an electron in a magnetic field.

We now turn to the term

$$H_{(\vec{r} \times \vec{B})^2} \equiv \left[\frac{1}{2M} + \frac{1}{2\mu}\left(\frac{\Delta m}{2M}\right)^2\right] e^2 \left(\vec{r} \times \vec{B}\right)^2 = \frac{e^2}{8\mu}\left(\vec{r} \times \vec{B}\right)^2.$$

In the last step we have combined the two mass terms resulting from the two squares using the definitions Eqs. (14.13) and (14.17) of the total and reduced mass, respectively. We recognize that in the contribution $H_{(\vec{r} \times \vec{B})^2}$ only the reduced mass μ appears.

Moreover, the magnetic field and the relative coordinate appear quadratically. Therefore, the term is associated with the so-called diamagnetic term in atomic physics.

When we recall the vector relation

$$\left(\vec{r} \times \vec{B}\right)^2 = \vec{r}^{\,2}\vec{B}^{\,2} - \left(\vec{r} \cdot \vec{B}\right)^2$$

the Hamiltonian $H_{(\vec{r} \times \vec{B})^2}$ takes the alternative form

$$H_{(\vec{r} \times \vec{B})^2} = \frac{1}{8\mu}\left[\vec{\wp}^{\,2} \cdot \vec{B}^2 - \left(\vec{\wp} \cdot \vec{B}\right)^2\right]$$

where we have expressed $e\vec{r}$ by the dipole moment $\vec{\wp}$.

The last term

$$H_{\vec{\nabla}(\vec{\wp} \cdot \vec{E})} \equiv -\frac{e}{2}\frac{\Delta m}{M}\left(\vec{r} \cdot \vec{\nabla}_{\!\vec{R}}\right)\vec{r} \cdot \vec{E}$$

is a correction to the familiar dipole interaction: It measures the inhomogenity of the field over the size of the atom.

14.5 Subsystems, Interaction and Entanglement

So far most of our consideration have been of classical nature. We now briefly address the quantum problem.

The complete quantum system consisting of atom plus field is described by a state vector $|\Phi\rangle$. Its change as a function of time follows from the Schrödinger equation

$$i\hbar \frac{d|\Phi\rangle}{dt} = \hat{H}|\Phi\rangle.$$

Here $|\Phi\rangle$ consists of the state $|\psi_{\text{cm}}\rangle$ of the center-of-mass motion, the internal state $|\psi_{\text{atom}}\rangle$ of the atom, and the state $|\psi_{\text{field}}\rangle$ of the radiation field. In the remainder of this book we use capital greek letters such as Φ and Ψ to denote the state vector of a combined system, whereas small greek letters such as ψ represent the state vectors of subsystems. In the following chapters we neglect for a moment the center-of-mass motion. In this case we denote the state vector of the combined system of field plus internal degrees of freedom by $|\Psi\rangle$.

The operators $\hat{\vec{R}}$ and $\hat{\vec{P}}$ corresponding to the position and the momentum of the center-of-mass motion act on the state $|\psi_{\text{cm}}\rangle$, the operators $\hat{\vec{r}}$ and $\hat{\vec{p}}$ corresponding to the position and momentum of the electron act on the internal state $|\psi_{\text{atom}}\rangle$, and the operators of the vector potential $\hat{\vec{A}}$, the electric field $\hat{\vec{E}}$ and the magnetic field $\hat{\vec{B}}$, given by Eqs. (10.61), (10.65) and (10.69) and consisting of linear combinations of creation and annihilation operators $\hat{a}_\ell$ and $\hat{a}_\ell^\dagger$ act on the field state $|\psi_{\text{field}}\rangle$. Moreover, all field operators also act on the center-of-mass-motion since they contain the mode functions $\vec{u}_\ell(\hat{\vec{R}})$ which involve the position operator $\hat{\vec{R}}$ of the center-of-mass-motion.

We note, that both interaction Hamiltonians $\hat{H}_{\vec{A}\cdot\vec{p}}$ and $\hat{H}_{\vec{r}\cdot\vec{E}}$ involve all three degrees of freedom. Therefore, even when the initial state of the system is a direct product of the three state vectors the interaction creates a highly entangled state – a state which cannot be factorized again in a product of states of the three subsystems. This entanglement is of great importance in the topics discussed in great detail in the context of the Jaynes-Cummings-Paul model, the concepts of state engineering and the field of atom optics in quantized light fields.

14.6 Equivalence of $\vec{A}\cdot\vec{p}$ and $\vec{r}\cdot\vec{E}$

Two ways of coupling an atom to radiation offer themselves: The first couples the *momentum* of the electron and the *vector potential* whereas the second uses the *position* of the electron and the *electric field*. On first sight it is not clear that both methods lead to identical results. This will be the topic of the present section.

In 1931 M. Göppert-Mayer published the results of her PhD thesis with M. Born in Göttingen. Her thesis dealt with two-photon transitions in atoms based on Dirac's quantum electrodynamics. In the beginning of her analysis she decides to work with the interaction Hamiltonian $H_{\vec{r}\cdot\vec{E}}$ rather than $H_{\vec{A}\cdot\vec{p}}$. For this purpose she adds a complete time derivative to the classical Lagrangian. This addition allows her to show that both interactions are equivalent.

We now briefly review this argument and then turn to the quantum case using a special example of the local gauge transformation discussed in Sec. 14.2.1. We then show that there is still a subtlety left in evaluating matrix elements. Throughout this section we neglect the center-of-mass motion and only deal with the relative motion of the electron.

14.6.1 Classical Transformation of Lagrangian

We start our analysis by considering the classical Hamiltonian

$$H^{(0)} = \frac{1}{2\mu}\left[\vec{p} - e\vec{A}(\vec{R}, t)\right]^2 + V(\vec{r}) \tag{14.38}$$

of an electron of reduced mass μ in the vector potential $\vec{A} = \vec{A}(\vec{R}, t)$ and the potential $V(\vec{r}) = -e^2/(4\pi\varepsilon_0|\vec{r}|)$.

In order to make the connection to the interaction Hamiltonian

$$\widetilde{H}^{(0)} = \frac{\vec{p}^{\,2}}{2\mu} + V(\vec{r}) - e\vec{r}\cdot\vec{E}(\vec{R}, t) \tag{14.39}$$

of a dipole $e\vec{r}$ in the electric field $\vec{E}$ we use the fact that two Lagrangians are equivalent when they only differ in a complete time derivative.

Equivalence of Lagrangians

We obtain the equations of motion from the variation of the action integral. Indeed, the space time trajectory $\vec{r} = \vec{r}(t)$ of a classical particle makes the variation

$$\delta I = \delta \left\{\int_{t_1}^{t_2} dt\, L(\vec{r}, \dot{\vec{r}}, t)\right\} = 0$$

of the action integral containing the Lagrangian an extremum. Here we do not include the end-points of the integral in the variation, that is $\delta\vec{r}(t_1) = \delta\vec{r}(t_2) \equiv 0$.

When we now consider a Lagrangian

$$\tilde{L}(\vec{r}, \dot{\vec{r}}, t) \equiv L(\vec{r}, \dot{\vec{r}}, t) + \frac{d}{dt}f(\vec{r}, t)$$

which only differs from L by a complete time derivative of a scalar function f we find for the action

$$I' \equiv \int_{t_1}^{t_2} dt\, L'(\vec{r}, \dot{\vec{r}}, t) = \int_{t_1}^{t_2} dt\, L(\vec{r}, \dot{\vec{r}}, t) + f(\vec{r}, t)\Big|_{t_1}^{t_2} = I + f(\vec{r}(t_2), t_2) - f(\vec{r}(t_1), t_1).$$

Since the additional terms due to f involve only the end-points the variation of these contributions vanishes. The two Lagrangians yield the same equation of motion and are thus equivalent.

Complete Time Derivative

We find the Lagrangian from the Hamiltonian by the Legendre transformation

$$L^{(0)} = \dot{\vec{r}}\cdot\vec{p} - H^{(0)}$$

where we have to express the momentum with the help of the Hamilton equation of motion

$$\dot{\vec{r}} = \frac{\partial H^{(0)}}{\partial \vec{p}} = \frac{1}{\mu}\left[\vec{p} - e\vec{A}(\vec{R},t)\right].$$

Therefore, the Lagrangian corresponding to the Hamiltonian Eq. (14.38) reads

$$L^{(0)} = \frac{\mu}{2}\dot{\vec{r}}^2 - V(\vec{r}) + e\dot{\vec{r}} \cdot \vec{A}(\vec{R},t).$$

Since we can subtract the complete time derivative $d/dt(\vec{r}\cdot\vec{A})$ we find the Lagrangian

$$\tilde{L}^{(0)} \equiv L^{(0)} - e\frac{d}{dt}(\vec{r}\cdot\vec{A}) = \frac{\mu}{2}\dot{\vec{r}}^2 - V(\vec{r}) - e\vec{r}\frac{d}{dt}\vec{A}(\vec{R},t).$$

In the present treatment we neglect the center-of-mass motion. Hence the position argument $\vec{R}$ of the vector potential is time independent. This fact allows us to replace the total time derivative by the partial time derivative, that is

$$\frac{d}{dt}\vec{A}(\vec{R},t) = \frac{\partial}{\partial t}\vec{A}(\vec{R},t) = -\vec{E}(\vec{R},t).$$

Here we have used that in Coulomb gauge the electric field follows from the vector potential according to

$$\vec{E} = -\frac{\partial \vec{A}}{\partial t}.$$

We therefore arrive at the Lagrangian

$$\tilde{L}^{(0)} = \frac{\mu}{2}\dot{\vec{r}}^2 - V(\vec{r}) + e\vec{r}\cdot\vec{E}(\vec{R},t).$$

which corresponds to the Hamiltonian Eq. (14.39).

Influence of Center-of-Mass Motion

We emphasize, that a crucial ingredient of this derivation is the dipole approximation, that is the fact that the vector potential does not depend on the position of the electron. Moreover, it is important that the position argument $\vec{R}$ is not time dependent. When we include the center-of-mass motion, that is when we allow $\vec{R}$ to depend on time, we obtain an additional term since

$$\frac{d}{dt}\vec{A}(\vec{R},t) = \left(\frac{d\vec{R}(t)}{dt}\cdot\vec{\nabla}_{\vec{R}}\right)\vec{A}(\vec{R},t) + \frac{\partial}{\partial t}\vec{A}(\vec{R},t).$$

Hence, in the presence of center-of-mass motion the equivalence of the Hamiltonians does not hold anymore. This is the deeper reason for the additional terms in the Hamiltonian Eq. (14.37), such as the Röntgen Hamiltonians and others.

14.6.2 Quantum Mechanical Treatment

We now turn to the quantum mechanical treatment of the equivalence of the two Hamiltonians $H^{(0)}$ and $\widetilde{H}^{(0)}$, Eqs. (14.38) and (14.39). We again neglect the center-of-mass motion and consider only the wave function $\psi(\vec{r}, t)$ of the electron. In particular, we focus on the Schrödinger equation

$$i\hbar \frac{\partial \widetilde{\psi}(\vec{r}, t)}{\partial t} = \frac{1}{2\mu} \left[\hat{\vec{p}} - e\vec{A}(\vec{R}, t) \right]^2 \widetilde{\psi}(\vec{r}, t) + V(\vec{r})\widetilde{\psi}(\vec{r}, t).$$

In quantum mechanics the two Hamiltonians are connected via the gauge transformation

$$\widetilde{\psi}(\vec{r}, t) \equiv \exp\left[i\frac{e}{\hbar} \vec{r} \cdot \vec{A}(\vec{R}, t) \right] \psi(\vec{r}, t). \tag{14.40}$$

We note that this gauge transformation is a special case of the one discussed in Sec. 14.2.1, that is here we have

$$\Lambda(\vec{r}, t) \equiv \vec{r} \cdot \vec{A}(\vec{R}, t). \tag{14.41}$$

From Eqs. (14.2) and (14.5) we immediately find the Schrödinger equation for the wave function $\psi(\vec{r}, t)$ which contains the $\vec{r} \cdot \vec{E}$ interaction. Indeed, we arrive at

$$i\hbar \frac{\partial \psi(\vec{r}, t)}{\partial t} = \left[\frac{\hat{p}^2}{2\mu} + V(\vec{r}) + e\vec{r} \cdot \frac{\partial}{\partial t} \vec{A}(\vec{R}, t) \right] \psi(\vec{r}, t).$$

We recall that in Coulomb gauge $\vec{E} = -\frac{\partial \vec{A}}{\partial t}$ which yields

$$i\hbar \frac{\partial \psi(\vec{r}, t)}{\partial t} = \left[\frac{\hat{p}^2}{2\mu} + V(\vec{r}) \right] \psi(\vec{r}, t) - e\vec{r} \cdot \vec{E}(\vec{R}, t) \psi(\vec{r}, t).$$

When the wave function $\widetilde{\psi}(\vec{r}, t)$ satisfies the Schrödinger equation in the minimal coupling, that is with the $\vec{A} \cdot \vec{p}$-coupling, then the wave function $\psi(\vec{r}, t)$ satisfies the Schrödinger equation with the $\vec{r} \cdot \vec{E}$-interaction term. Again we emphasize that this equivalence of the two interactions is based on the gauge transformation Eq. (14.40) in which the relative position $\vec{r}$ of the electron does not enter in the vector potential and the center-of-mass $\vec{R}$ is not time dependent.

14.6.3 Matrix elements of $\vec{A} \cdot \vec{p}$ and $\vec{r} \cdot \vec{E}$

It is crucial that we are consistent in using the wave function ψ or $\widetilde{\psi}$ when we evaluate matrix elements of the two interaction Hamiltonians. In particular, the matrix elements

$$\langle j | \hat{H}_{\vec{A} \cdot \vec{p}} | j' \rangle = -e\vec{A}(\vec{R}, t) \frac{1}{\mu} \langle j | \hat{\vec{p}} | j' \rangle$$

and

$$\langle j | \hat{H}_{\vec{r} \cdot \vec{E}} | j' \rangle = -e\vec{E}(\vec{R}, t) \langle j | \hat{\vec{r}} | j' \rangle$$

of the operators
$$\hat{H}_{\vec{A}\cdot\vec{p}} \equiv -\frac{e}{\mu}\vec{A}(\vec{R},t)\cdot\hat{\vec{p}}$$
and
$$\hat{H}_{\vec{r}\cdot\vec{E}} \equiv -e\hat{\vec{r}}\cdot\vec{E}(\vec{R},t)$$

are different. We emphasize that here we evaluate both matrix elements using the same atomic energy eigenstates $|j\rangle$ and $|j'\rangle$ defined by

$$\hat{H}_{\text{atom}}|j\rangle \equiv \left[\frac{\hat{\vec{p}}^2}{2\mu}+V(\hat{\vec{r}})\right]|j\rangle = E_j|j\rangle. \tag{14.42}$$

We therefore do not transform the states using a gauge transformation.

Since $\vec{E}$ and $\vec{A}$ only depend on the position $\vec{R}$ of the nucleus it is enough to analyze the matrix elements $\langle j|\hat{\vec{r}}|j'\rangle$ and $\langle j|\hat{\vec{p}}|j'\rangle$. For this purpose we recall the Heisenberg equation of motion

$$\frac{\hat{\vec{p}}}{\mu} = \frac{d\hat{\vec{r}}}{dt} = \frac{i}{\hbar}\left[\hat{H}_{\text{atom}},\hat{\vec{r}}\right] = \frac{i}{\hbar}\left(\hat{H}_{\text{atom}}\hat{\vec{r}} - \hat{\vec{r}}\hat{H}_{\text{atom}}\right)$$

which yields the matrix elements

$$\frac{1}{\mu}\langle j|\hat{\vec{p}}|j'\rangle = \frac{i}{\hbar}\langle j|\hat{H}_{\text{atom}}\hat{\vec{r}} - \hat{\vec{r}}\hat{H}_{\text{atom}}|j'\rangle.$$

When we use the definition Eq. (14.42) of the energy eigenstates $|j\rangle$ and $|j'\rangle$ we find

$$\frac{1}{\mu}\langle j|\hat{\vec{p}}|j'\rangle = \frac{i}{\hbar}(E_j - E_{j'})\langle j|\hat{\vec{r}}|j'\rangle = i\omega\langle j|\hat{\vec{r}}|j'\rangle$$

where $\omega \equiv (E_j - E_{j'})/\hbar$. Hence we arrive at the relation

$$\langle j|\hat{H}_{\vec{A}\cdot\vec{p}}|j'\rangle = -ei\omega\vec{A}(\vec{R},t)\langle j|\hat{\vec{r}}|j'\rangle.$$

We note however, that this matrix element is not identical to $\langle j|\hat{H}_{\vec{r}\cdot\vec{E}}|j'\rangle$ corresponding to the dipole-electric field interaction.

14.7 Equivalence of Hamiltonians $H^{(1)}$ and $\widetilde{H}^{(1)}$

In the preceding sections and in particular, in Appendix L we have derived the Hamiltonian $\widetilde{H}^{(1)}$, Eq. (14.37) of an atom moving in an electromagnetic field characterized by $\vec{E}$ and $\vec{B}$. In this derivation we have started from the minimal coupling Hamiltonian

$$H = \frac{1}{2m_e}\left[\vec{p}_e - e\vec{A}(\vec{r}_e)\right]^2 + \frac{1}{2m_p}\left[\vec{p}_p + e\vec{A}(\vec{r}_p)\right]^2 + V(\vec{r}) \tag{14.43}$$

14.7 Equivalence of Hamiltonians $H^{(1)}$ and $\tilde{H}^{(1)}$

of an electron and a proton in Coulomb gauge. We have then introduced center-of-mass and relative coordinates and have expanded the vector potential into a Taylor series around the center-of-mass coordinate. A gauge transformation

$$\Phi(\vec{r}, \vec{R}, t) \equiv \exp\left[\frac{i}{\hbar} e \Lambda(\vec{r}, \vec{R}, t)\right] \tilde{\Phi}(\vec{r}, \vec{R}, t)$$

with the potential

$$\Lambda(\vec{r}, \vec{R}, t) \equiv \vec{r} \cdot \vec{A}(\vec{R}, t) + \frac{\Delta m}{2M} \left(\vec{r} \cdot \vec{\nabla}_{\vec{R}}\right) \left[\vec{r} \cdot \vec{A}(\vec{R}, t)\right] \quad (14.44)$$

transforms the approximate Hamiltonian $H^{(1)}$, Eq. (14.34) into the gauge invariant Hamiltonian $\tilde{H}^{(1)}$, Eq. (14.37).

One might wonder if it is possible to perform a gauge transformation directly on the exact Hamiltonian Eq. (14.43) and then make the dipole approximation. Indeed, the gauge transformation

$$\Phi(\vec{r}_e, \vec{r}_p, t) \equiv \exp\left[\frac{i}{\hbar} e \Lambda(\vec{r}_e, \vec{r}_p, t)\right] \tilde{\Phi}(\vec{r}_e, \vec{r}_p, t) \quad (14.45)$$

with the Dirac-Heisenberg line gauge potential

$$\Lambda(\vec{r}_e, \vec{r}_p, t) \equiv \int_0^1 d\lambda\, (\vec{r}_e - \vec{r}_p) \cdot \vec{A}[\lambda \vec{r}_e + (1-\lambda)\vec{r}_p, t] \quad (14.46)$$

achieves this goal. In Problem 2 we derive the Schrödinger equation for $\tilde{\Phi}$ assuming that Φ satisfies the Schrödinger equation with the Hamiltonian Eq. (14.43). We then make the dipole approximation and arrive at the Hamiltonian $\tilde{H}^{(1)}$, Eq. (14.37).

The Dirac-Heisenberg gauge potential Λ is quite a remarkable object. It is a generalization of the expression $\vec{r} \cdot \vec{A}$, Eq. (14.41) for a single particle, to two particles. Such a generalization is not quite straightforward since the vector potential $\vec{A}$ only depends on one space variable. The line integral avoids this problem. Indeed, for $\lambda = 0$ the argument

$$\vec{r}' \equiv \lambda \vec{r}_e + (1-\lambda)\vec{r}_p = \vec{r}_p + \lambda(\vec{r}_e - \vec{r}_p) = \vec{r}_p + \lambda \vec{r}$$

of the vector potential reduces to the position $\vec{r}_p$ of the proton, whereas for $\lambda = 1$ we recover the position $\vec{r}_e$ of the electron. Hence we go continuously from the position of one of the particles to the other.

Moreover, the fact that the line integral, Eq. (14.46) is a generalization of $\vec{r} \cdot \vec{A}'$ stands out most clearly in the representation

$$\Lambda(\vec{r}_e, \vec{r}_p, t) = \int_{\vec{r}_p}^{\vec{r}_e} d\vec{r}' \cdot \vec{A}(\vec{r}', t).$$

Due to the relation $\vec{\nabla} \times \vec{A} = \vec{B} \neq 0$ this line integral depends on the path we choose. In agreement with the definition Eq. (14.46) we choose the line connecting the two charges.

We conclude this section by connecting the Dirac-Heisenberg line gauge factor, Eq. (14.46) with the gauge potential Λ, Eq. (14.44), used in Appendix L to show quantum mechanically the equivalence of the Hamiltonians $H^{(1)}$ and $\widetilde{H}^{(1)}$, Eqs. (14.34) and (14.37) in the presence of the center-of-mass motion.

For this purpose we make use of Eqs. (14.14) and (14.15) to express the positions $\vec{r}_e$ and $\vec{r}_p$ of the electron and proton in the definition Eq. (14.46) of Λ by the center-of-mass and relative coordinates $\vec{R}$ and $\vec{r}$, respectively which yields

$$\Lambda = \vec{r} \cdot \int_0^1 d\lambda\, \vec{A}\left[\vec{R} - \left(\frac{m_e}{M} - \lambda\right)\vec{r}, t\right] \cong \vec{r} \cdot \int_0^1 d\lambda\left[\vec{A}(\vec{R}, t) - \left(\frac{m_e}{M} - \lambda\right)(\vec{r} \cdot \vec{\nabla}_{\vec{R}}) \cdot \vec{A}(\vec{R}, t)\right]$$

In the last step we have expanded the vector potential into a Taylor series following Eq. (14.23).

When we perform the integral over λ we arrive at

$$\Lambda \cong \vec{r} \cdot \vec{A} - \left(\frac{m_e}{M} - \frac{1}{2}\right)(\vec{r} \cdot \vec{\nabla}_{\vec{R}}) \cdot \vec{A}.$$

After combining the mass terms we find indeed the potential Λ given by Eq. (14.44).

14.8 Simple Model for Atom-Field Interaction

In Sec. 14.3 we have derived the Hamiltonian for a hydrogen atom in a quantized electromagnetic field including the center-of-mass motion. This Hamiltonian contains many atomic levels and all modes of the radiation field. In the present section we simplify this Hamiltonian considerably by allowing only two atomic levels to be in resonance with the field. Moreover, we consider only a single mode of the cavity field. We still allow the center-of-mass motion of the atom.

This elementary model summarized in Fig. 14.2 still contains enough physics to describe most phenomena in cavity quantum electrodynamics (CQED) and atom optics. It is considered the drosophila of quantum optics. When we neglect the center-of-mass motion, that is consider only the interaction between the quantized cavity field and the two levels we call this model the Jaynes-Cummings-Paul model.

14.8.1 Derivation of the Hamiltonian

In this section we reduce the dipole-field interaction Hamiltonian

$$H_{\vec{r}\cdot\vec{E}} = -\vec{\wp} \cdot \vec{E}(\vec{R}, t) = -e\vec{r} \cdot \vec{E}(\vec{R}, t)$$

to the case of a two-level atom interacting with a single mode of the radiation field. We start by expressing the internal degrees of freedom of the atom by Pauli spin matrices.

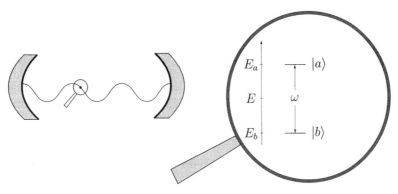

Fig. 14.2: Simple model for atom-field interaction including center-of-mass motion. A single mode in a perfect resonator characterized by a mode function $\vec{u}(\vec{r})$ displayed here in its simplest form as a sine-function interacts with an atom of total mass M located at a position $\vec{R}$. The atom performs a center-of-mass motion with a kinetic energy $\vec{P}^2/2M$. We only consider two internal degrees of freedom of the atom, that is, an excited state $|a\rangle$ and ground state $|b\rangle$ with energies $E_a \equiv \hbar\omega_a$ and $E_b \equiv \hbar\omega_b$, respectively. The transition frequency is $\omega \equiv \omega_a - \omega_b$. A further simplification arises when we consider the atom to be fixed at a given position, that is we neglect the center-of-mass motion. In this case the model referred to in the literature as the Jaynes-Cummings-Paul model describes the interaction between the two internal levels and a single mode of the quantized cavity field.

Internal States

It is clear that there are no two-level atoms in nature. However, by optical pumping we can create a situation in which effectively only two energy levels of an atom are involved. We denote the upper level of energy $E_a \equiv \hbar\omega_a$ by $|a\rangle$ and the lower level of energy $E_b \equiv \hbar\omega_b$ by $|b\rangle$. These states are eigenstates of the Hamiltonian H_{atom} of the atom.

Since we can represent any atomic operator $\hat{O}$ in terms of these states via

$$\hat{O} = \mathbf{1} \cdot \hat{O} \cdot \mathbf{1} = \sum_{j,j'=a,b} |j\rangle \langle j| \hat{O} |j'\rangle \langle j'| \qquad (14.47)$$

we find for the Hamiltonian

$$\hat{H}_{\text{atom}} = E_a |a\rangle \langle a| + E_b |b\rangle \langle b|.$$

With the vector representation

$$|a\rangle \equiv \begin{pmatrix} 1 \\ 0 \end{pmatrix}, \qquad \langle a| \equiv (1,0)$$

and

$$|b\rangle \equiv \begin{pmatrix} 0 \\ 1 \end{pmatrix}, \qquad \langle b| \equiv (0,1)$$

the Hamiltonian of the atom takes the matrix representation

$$\hat{H}_{\text{atom}} \equiv \begin{pmatrix} E_a & 0 \\ 0 & E_b \end{pmatrix} = \begin{pmatrix} \frac{1}{2}(E_a+E_b) + \frac{1}{2}(E_a-E_b) & 0 \\ 0 & \frac{1}{2}(E_a+E_b) - \frac{1}{2}(E_a-E_b) \end{pmatrix},$$

or
$$\hat{H}_{\text{atom}} = E\mathbb{1} + \frac{1}{2}\hbar\omega \begin{pmatrix} 1 & 0 \\ 0 & -1 \end{pmatrix}$$

where $E \equiv \frac{1}{2}(E_a + E_b)$ and $\omega \equiv \omega_a - \omega_b$ denotes the transition frequency.

Since the constant energy in the Hamiltonian does not contribute we suppress it from now on and consider
$$\hat{H}_{\text{atom}} \equiv \frac{1}{2}\hbar\omega\hat{\sigma}_z$$
where we have introduced the Pauli spin matrix
$$\hat{\sigma}_z \equiv \begin{pmatrix} 1 & 0 \\ 0 & -1 \end{pmatrix}.$$

Obviously, the Hamiltonian of the atom in energy representation is diagonal.

Dipole Operator

We now use the completeness relation Eq. (14.47) to express the position operator $\hat{\vec{r}}$ in energy eigenstates. Since energy eigenstates of wave function $\psi_j(\vec{r})$ have a well-defined parity, diagonal elements vanish, that is,
$$\langle j|\hat{\vec{r}}|j\rangle = \int d^3r \, |\psi_j(\vec{r})|^2 \vec{r} = 0.$$

Indeed, since $|\psi_j|^2$ is a symmetric function and $\vec{r}$ is an anti–symmetric function, the integrand is anti–symmetric.

The off-diagonal elements read
$$e\langle a|\hat{\vec{r}}|b\rangle = e\int d^3r \, \psi_a^*(\vec{r})\vec{r}\psi_b(\vec{r}) \equiv \vec{\wp}$$
and
$$e\langle b|\hat{\vec{r}}|a\rangle = e\int d^3r \, \psi_b^*(\vec{r})\vec{r}\psi_a(\vec{r}) \equiv \vec{\wp}^*$$
and hence the dipole operator $e\hat{\vec{r}}$ takes the form
$$e\hat{\vec{r}} = \vec{\wp}\,|a\rangle\langle b| + \vec{\wp}^*\,|b\rangle\langle a|.$$

We note that this operator describes transitions from the ground state $|b\rangle$ to the excited state $|a\rangle$ and vice versa. In order to bring this feature out most clearly we apply the operator $e\hat{\vec{r}}$ to the state $|b\rangle$ and find
$$e\hat{\vec{r}}|b\rangle = \vec{\wp}\,|a\rangle\langle b|b\rangle + \vec{\wp}^*\,|b\rangle\langle a|b\rangle = \vec{\wp}\,|a\rangle,$$
and likewise
$$e\hat{\vec{r}}|a\rangle = \vec{\wp}\,|a\rangle\langle b|a\rangle + \vec{\wp}^*\,|b\rangle\langle a|a\rangle = \vec{\wp}^*\,|b\rangle.$$

Hence the operator
$$\hat{\sigma}^{\dagger} \equiv |a\rangle\langle b| = \begin{pmatrix} 0 & 1 \\ 0 & 0 \end{pmatrix} \tag{14.48}$$

creates an atom in the excited state and acts as a creation operator for atoms in the excited state $|a\rangle$.

In contrast the operator

$$\hat{\sigma} \equiv |b\rangle\langle a| = \begin{pmatrix} 0 & 0 \\ 1 & 0 \end{pmatrix} \qquad (14.49)$$

destroys an atom in the excited state and is therefore an annihilation operator for excited atoms.

In the language of the Pauli spin matrices $\hat{\sigma}^\dagger$ and $\hat{\sigma}$, Eqs. (14.48) and (14.49), the dipole operator reads

$$\hat{e\vec{r}} = \vec{\wp}\,\hat{\sigma}^\dagger + \vec{\wp}^*\,\hat{\sigma}. \qquad (14.50)$$

Field States

We now turn to the single mode of the radiation field in the cavity. From Eq. (10.65) we recall

$$\hat{\vec{E}}(\vec{R}, t) = \mathcal{E}_0\,\vec{u}(\vec{R})\,i\,(\hat{a} - \hat{a}^\dagger), \qquad (14.51)$$

where in contrast to the original definition Eq. (10.66) of the vacuum electric field we now include the square root of two, that is

$$\mathcal{E}_0 \equiv \sqrt{\frac{\hbar\Omega}{2\epsilon_0 V}}.$$

Moreover, $\vec{u}(\vec{R})$ is the mode function of the resonator. Since the atom is at the position $\vec{R}$ we have to evaluate the mode function at this position.

According to Eq. (10.52) the Hamiltonian of a single mode of the radiation field of frequency Ω reads

$$\hat{H}_{\text{field}} = \hbar\Omega\,\hat{a}^\dagger\hat{a}$$

where we have neglected the zero–point energy.

Interaction Hamiltonian

We now turn to the interaction Hamiltonian

$$\hat{H}_{\vec{r}\cdot\vec{E}} = -\hat{e\vec{r}}\cdot\hat{\vec{E}}(\vec{R}, t).$$

When we combine the two expressions Eqs. (14.50) and (14.51) for the dipole operator $\hat{e\vec{r}}$ and the electric field operator $\hat{\vec{E}}(\vec{R}, t)$ we arrive at

$$\hat{H}_{\vec{r}\cdot\vec{E}} = -\mathcal{E}_0\,i\left[\vec{\wp}\cdot\vec{u}(\vec{R})\,\hat{\sigma}^\dagger + \vec{\wp}^*\cdot\vec{u}(\vec{R})\,\hat{\sigma}\right](\hat{a} - \hat{a}^\dagger).$$

The scalar product $\vec{\wp}\cdot\vec{u}$ between the dipole moment $\vec{\wp}$ and the mode function $\vec{u}$ yields a complex number

$$\vec{\wp}\cdot\vec{u} = |\vec{\wp}\cdot\vec{u}|\,e^{i\varphi}$$

with phase φ. Hence the interaction Hamiltonian reduces to

$$\hat{H}_{\vec{r}\cdot\vec{E}} = \hbar g(\hat{\vec{R}})(-i)\left(\hat{\sigma}^\dagger e^{i\varphi} + \hat{\sigma} e^{-i\varphi}\right)\left(\hat{a} - \hat{a}^\dagger\right),$$

where we have introduced the position dependent vacuum Rabi frequency

$$g(\vec{R}) \equiv \frac{|\vec{\wp}\cdot\vec{u}(\vec{R})|}{\hbar}\mathcal{E}_0. \tag{14.52}$$

The physics of this quantity and its name will become clear when discuss the dynamics of the Jaynes-Cummings-Paul model in Chapter 16.

When we choose the phase $\varphi = \pi/2$ we find

$$\hat{H}_{\vec{r}\cdot\vec{E}} = \hbar g(\hat{\vec{R}})\left(\hat{\sigma}^\dagger - \hat{\sigma}\right)\left(\hat{a} - \hat{a}^\dagger\right).$$

Total Hamiltonian

We combine all results and arrive at the total Hamiltonian of the atom-field system

$$\hat{H} = \frac{\hat{\vec{P}}^2}{2M} + \hbar\Omega\,\hat{a}^\dagger\hat{a} + \frac{1}{2}\hbar\omega\hat{\sigma}_z + \hbar g(\hat{\vec{R}})\left(\hat{\sigma}^\dagger - \hat{\sigma}\right)\left(\hat{a} - \hat{a}^\dagger\right).$$

This Hamiltonian consists of the kinetic energy operator $\hat{\vec{P}}^2/2M$ of the center-of-mass motion of the atom, the Hamiltonian $\hbar\hat{a}^\dagger\hat{a}$ of the free field and the Hamiltonian $\frac{1}{2}\hbar\omega\hat{\sigma}_z$ corresponding to the internal states. The interaction between these degrees of freedom results from the position dependent coupling operator $g(\hat{\vec{R}})$ which brings in the Pauli spin matrices $\hat{\sigma}$ and $\hat{\sigma}^\dagger$ and the field operator $\hat{a} - \hat{a}^\dagger$. The Pauli spin matrices are off-diagonal in the energy representation.

14.8.2 Rotating-Wave Approximation

We now continue to simplify the interaction Hamiltonian

$$\hat{H}_{\vec{r}\cdot\vec{E}} = \hbar g\left(\hat{\sigma}\hat{a}^\dagger + \hat{\sigma}^\dagger\hat{a} - \hat{\sigma}\hat{a} - \hat{\sigma}^\dagger\hat{a}^\dagger\right).$$

Two arguments offer themselves: The first relies on the fact that the terms $\hat{\sigma}\hat{a}$ and $\hat{\sigma}^\dagger\hat{a}^\dagger$ violate energy conservation. The second is a more mathematical argument and is based on the method of averaging: We transform into the interaction picture defined by the internal states and the light field and neglect the rapidly oscillating terms. This has led to the name of *rotating-wave approximation*.

Violation of Energy Conservation

The terms $\hat{\sigma}\hat{a}$ and $\hat{\sigma}^\dagger\hat{a}^\dagger$ violate energy conservation. We recognize this fact when we note that for example the term $\hat{\sigma}\hat{a}$ destroys an excited atom while at the same time destroys a field excitation. Similarly the term $\hat{\sigma}^\dagger\hat{a}^\dagger$ creates a field excitation while also exciting an atom from the ground state to the excited state.

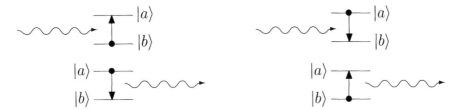

Fig. 14.3: Rotating-wave approximation. Excitation of an atom by absorption of a photon (top left) and de-excitation by emission of a photon (bottom left). While these processes are resonant the ones shown on the right hand side of the figure are nonresonant. They do not conserve energy: An atom is de-excited and simultaneously absorbs a photon (top right) and an atom is excited while it emits a photon (bottom right).

In contrast the terms $\hat{\sigma}\hat{a}^\dagger$ and $\hat{\sigma}^\dagger\hat{a}$ either destroy an excited atom while creating a field excitation or create an excited atom by annihilating a field excitation as shown in Fig. 14.3.

Hence, when we neglect the terms $\hat{\sigma}\hat{a}$ and $\hat{\sigma}^\dagger\hat{a}^\dagger$ we arrive at the final expression

$$\hat{H}_{\vec{r}\cdot\vec{E}} \cong \hat{H}_{\text{int}} = \hbar g \left(\hat{\sigma}\hat{a}^\dagger + \hat{\sigma}^\dagger\hat{a} \right) \tag{14.53}$$

for our interaction Hamiltonian.

The approximation to neglect the terms $\hat{\sigma}\hat{a}$ and $\hat{\sigma}^\dagger\hat{a}^\dagger$ is referred to as *rotating-wave approximation*. The name only becomes clear in the more mathematical treatment of this problem.

Method of Averaging

To bring out most clearly the underlying mathematics of the rotating wave approximation, which so far has only be motivated heuristically, we transform the Hamiltonian

$$\hat{H} = \frac{\hat{P}^2}{2M} + \hat{H}_0 + \hat{H}_{\vec{r}\cdot\vec{E}} = \frac{\hat{P}^2}{2M} + \hat{H}_0 + \hbar g(\vec{R})(\hat{\sigma}^\dagger - \hat{\sigma})(\hat{a} - \hat{a}^\dagger)$$

of the complete system into the interaction picture of the internal states and the light field. Here

$$\hat{H}_0 \equiv \hbar\Omega\,\hat{a}^\dagger\hat{a} + \frac{1}{2}\hbar\omega\,\hat{\sigma}_z$$

denotes the Hamiltonian of the free atom and the free field.

For this purpose we define the state $\left|\Phi^{(1)}(t)\right\rangle$ by the ansatz

$$|\Phi(t)\rangle \equiv \exp\left[-\frac{i}{\hbar}\hat{H}_0 t\right] \left|\Phi^{(1)}(t)\right\rangle. \tag{14.54}$$

Note that here we have not included the free center-of-mass motion of the atom in the interaction picture, that is we have not transformed the kinetic energy. Since the Rabi frequency g depends on the position $\vec{R}$ of center-of-mass such a transformation would have led to a complicated Rabi frequency.

When we substitute the ansatz Eq. (14.54) for the state $|\Phi^{(I)}(t)\rangle$ in the interaction picture into the Schrödinger equation

$$i\hbar \frac{d|\Phi(t)\rangle}{dt} = \left(\frac{\hat{P}^2}{2M} + \hat{H}_0 + \hat{H}_{\vec{r}\cdot\vec{E}} \right) |\Phi(t)\rangle$$

we arrive at

$$i\hbar \frac{d|\Phi^{(I)}(t)\rangle}{dt} = \left[\frac{\hat{P}^2}{2M} + \hat{H}_{\vec{r}\cdot\vec{E}}^{(I)} \right] |\Phi^{(I)}(t)\rangle$$

where

$$\hat{H}_{\vec{r}\cdot\vec{E}}^{(I)} \equiv \exp\left[\frac{i}{\hbar}\hat{H}_0 t\right] \hat{H}_{\vec{r}\cdot\vec{E}} \exp\left[-\frac{i}{\hbar}\hat{H}_0 t\right] \tag{14.55}$$

is the interaction Hamiltonian in the interaction picture.

We now find an explicit expression for the interaction Hamiltonian $\hat{H}_{\vec{r}\cdot\vec{E}}^{(I)}$. For this purpose we substitute the Hamiltonian $\hat{H}_0$ into Eq. (14.55) and note that the atomic and the field operators commute. We arrive at

$$\hat{H}_{\vec{r}\cdot\vec{E}}^{(I)} = \hbar g \exp\left(\frac{i}{2}\omega t\, \hat{\sigma}_z\right) \left(\hat{\sigma}^\dagger - \hat{\sigma}\right) \exp\left(-\frac{i}{2}\omega t\, \hat{\sigma}_z\right) e^{i\hat{a}^\dagger \hat{a}\Omega t} \left(\hat{a} - \hat{a}^\dagger\right) e^{-i\hat{a}^\dagger \hat{a}\Omega t}.$$

We start evaluating this expression by considering first the atomic part: Since the Pauli matrix $\hat{\sigma}_z$ is diagonal we find

$$\exp\left(\frac{i}{2}\omega t\, \hat{\sigma}_z\right) = \begin{pmatrix} e^{i\omega t/2} & 0 \\ 0 & e^{-i\omega t/2} \end{pmatrix}$$

which yields

$$\exp\left(\frac{i}{2}\omega t\hat{\sigma}_z\right) \hat{\sigma} \exp\left(-\frac{i}{2}\omega t\hat{\sigma}_z\right) = \begin{pmatrix} e^{i\omega t/2} & 0 \\ 0 & e^{-i\omega t/2} \end{pmatrix} \begin{pmatrix} 0 & 0 \\ 1 & 0 \end{pmatrix} \begin{pmatrix} e^{-i\omega t/2} & 0 \\ 0 & e^{i\omega t/2} \end{pmatrix}$$

$$= \begin{pmatrix} 0 & 0 \\ e^{-i\omega t} & 0 \end{pmatrix} = \hat{\sigma}\, e^{-i\omega t}. \tag{14.56}$$

Here we have used the matrix representation Eq. (14.49) of $\hat{\sigma}$. Consequently the atomic part reads

$$\exp\left(\frac{1}{2}i\omega t\hat{\sigma}_z\right) \left(\hat{\sigma}^\dagger - \hat{\sigma}\right) \exp\left(-\frac{1}{2}i\omega t\hat{\sigma}_z\right) = \hat{\sigma}^\dagger e^{i\omega t} - \hat{\sigma}\, e^{-i\omega t}.$$

We now turn to the evaluation of the field operators. For this purpose we recall from Sec. 10.5.1 the time evolution

$$\hat{a}(t) = e^{i\hat{a}^\dagger \hat{a}\Omega t}\, \hat{a}(0)\, e^{-i\hat{a}^\dagger \hat{a}\Omega t} = \hat{a}(0) e^{-i\Omega t}$$

in the Heisenberg picture which yields

$$e^{i\hat{a}^\dagger \hat{a}\Omega t} \left(\hat{a} - \hat{a}^\dagger\right) e^{-i\hat{a}^\dagger \hat{a}\Omega t} = \hat{a}\, e^{-i\Omega t} - \hat{a}^\dagger e^{i\Omega t}.$$

Hence, the Hamiltonian $\hat{H}^{(I)}_{\vec{r}\cdot\vec{E}}$ in the interaction picture reads

$$\hat{H}^{(I)}_{\vec{r}\cdot\vec{E}} = \hbar g \left(\hat{\sigma}^\dagger e^{i\omega t} - \hat{\sigma} e^{-i\omega t} \right) \left(\hat{a} e^{-i\Omega t} - \hat{a}^\dagger e^{i\Omega t} \right)$$

or

$$\hat{H}^{(I)}_{\vec{r}\cdot\vec{E}} = \hbar g \left(\hat{\sigma}\hat{a}^\dagger e^{i(\Omega-\omega)t} + \hat{\sigma}^\dagger \hat{a} e^{-i(\Omega-\omega)t} - \hat{\sigma}\hat{a} e^{-i(\Omega+\omega)t} - \hat{\sigma}^\dagger \hat{a}^\dagger e^{i(\Omega+\omega)t} \right).$$

We note that the terms $\hat{\sigma}\hat{a}$ and $\hat{\sigma}^\dagger \hat{a}^\dagger$ which do not conserve the energy are multiplied by oscillatory terms which involve the sum of the frequencies of the cavity and the atomic transition.

In contrast the contributions $\hat{\sigma}\hat{a}^\dagger$ and $\hat{\sigma}^\dagger \hat{a}$ involve terms that only contain the frequency difference, that is the detuning $\Delta \equiv \Omega - \omega$. Whereas the terms $\hat{\sigma}\hat{a}$ and $\hat{\sigma}^\dagger \hat{a}^\dagger$ oscillate roughly with twice the optical frequency, the terms $\hat{\sigma}\hat{a}^\dagger$ and $\hat{\sigma}^\dagger \hat{a}$ are slowly varying.

Since the Schrödinger equation is a differential equation of first order in time we have to integrate in time. This time integration brings the frequency sum and difference into the denominator. Hence the dominant contribution must come from the slowly varying part. We can therefore approximate the interaction Hamiltonian in the interaction picture by

$$\hat{H}^{(I)}_{\vec{r}\cdot\vec{E}} \cong \hat{H}_{\text{int}} \equiv \hbar g \left(\hat{\sigma}\hat{a}^\dagger e^{i\Delta t} + \hat{\sigma}^\dagger \hat{a} e^{-i\Delta t} \right). \tag{14.57}$$

This method can be made rigorous by the method of averaging. In particular, we can calculate higher order contributions.

This Hamiltonian corresponds in the original Schrödinger picture to the interaction Hamiltonian

$$\hat{H}_{\vec{r}\cdot\vec{E}} \equiv \hbar g \left(\hat{\sigma}\hat{a}^\dagger + \hat{\sigma}^\dagger \hat{a} \right).$$

Summary

In conclusion we have shown that in the rotating wave approximation our simple model for the interaction of a two-level atom with a single mode of the radiation field is summarized by the Hamiltonian

$$\hat{H} = \frac{\hat{P}^2}{2M} + \hbar\Omega\, \hat{a}^\dagger \hat{a} + \frac{1}{2}\hbar\omega\hat{\sigma}_z + \hbar g(\hat{\vec{R}}) \left(\hat{\sigma}\hat{a}^\dagger + \hat{\sigma}^\dagger \hat{a} \right).$$

In the next two chapters we investigate this model in more detail. In particular, we first consider the atoms to be stationary, that is we neglect the center-of-mass motion. In the last chapters we then focus on the influence of the kinetic energy operator.

Problems

14.1 Construction of an Interaction Hamiltonian

We consider an atom with a singly charged nucleus at position $\vec{r}_p$ and a single electron at position $\vec{r}_e$ in a static electric field described by the potential $\phi(\vec{x})$.

(a) What is the Hamiltonian for this system?

(b) Express the Hamiltonian in terms of the center-of-mass and relative coordinates and expand the potential up to second order in the relative coordinates.

(c) Now consider the motion of an electron in the potential of a proton sitting at the position $\vec{R}$ and an additional external field described by the potential $\phi(\vec{x})$. What is the Hamiltonian in this case?

(d) Compare the results.

(e) What changes in (b) if the nucleus consists of not only one proton but Z protons?

(f) Consider an ion having a nucleus with Z protons and a single electron. Show that for the center-of-mass of the ion there doesn't exist a stable equilibrium if in the surrounding of the ion there are no further charges or currents.

14.2 Dipole Approximation for Two Particles

Use the gauge transformation Eq. (14.45) with the Dirac-Heisenberg line gauge potential Eq. (14.46) to derive from the Hamiltonian Eq. (14.43) of two oppositely charged particles in an electromagnetic field in Coulomb gauge the Hamiltonian Eq. (14.37). It is convenient to first perform the gauge transformation and then to introduce center-of-mass and relative coordinates. Conclude by making use of the dipole approximation. Is it in this approach necessary to retain the first derivative of the vector potential in the Taylor expansion Eq. (14.23)?

References

Gauge Invariance as an Agent for Coupling to Fields

For the original proposal by H. Weyl to use the principle of gauge invariance to create a unified theory of gravitation and electricity see

H. Weyl, *Raum, Zeit, Materie*, Springer, Berlin, 1918

For the use of gauge invariance to create a coupling of electricity and matter see

H. Weyl, *Gruppentheorie und Quantenmechanik*, Hirzel, Leipzig, 1928

For a pedagogical introduction to gauge theories see

F. Hayot, *Introduction to gauge theories*, in: *New trends in atomic physics*, Les Houches 1982, ed. by G. Grynberg and R. Stora, North-Holland, Amsterdam, 1984

$\vec{A}\cdot\vec{p}$ versus $\vec{r}\cdot\vec{E}$

The equivalence of the two interaction Hamiltonians $\vec{A}\cdot\vec{p}$ and the $\vec{r}\cdot\vec{E}$ was discussed for the first time in

M. Göppert-Mayer, *Über Elementarakte mit zwei Quantensprüngen*, Ann. Phys. (Leipzig) **9**, 273–294 (1931)

For a particular clear treatment of this problem see

E.A. Power and **S. Zienau**, *Coulomb Gauge in Non-Relativistic Quantum Electrodynamics and the Shape of Spectral Lines*, Philos. Tran. Roy. Soc. London A **251**, 427–454 (1959)

C. Cohen-Tannoudji, J. Dupont-Roc and **G. Grynberg**, *Photons and Atoms, Introduction to Quantum-Electrodynamics*, Wiley, New York, 1989

In the context of the Lamb shift it was pointed out, that the calculations based on $\vec{p} \cdot \vec{A}$ are not in agreement with the experiments. However, the $\vec{r} \cdot \vec{E}$ Hamiltonian is in agreement as discussed in

W.E. Lamb Jr., *Fine Structure of the Hydrogen Atom*, Phys. Rev. **85**, 259–276 (1952)

For a resolution of this paradox see for example

W.E. Lamb Jr., R.R. Schlicher and **M.O. Scully**, *Matter-field interaction in atomic physics and quantum optics*, Phys. Rev. A **36**, 2763–2772 (1987)

M.O. Scully and **M.S. Zubairy**, *Quantum Optics*, Cambridge University Press, 1997

For pedagogical treatment of the gauge problem see

D.H. Kobe, *Gauge transformations and the electric dipole approximation*, Am. J. Phys. **50**, 128–133 (1982)

D.H. Kobe and **R.D. Gray**, *Operator Gauge Transformations in Nonrelativistic Quantum Electrodynamics: Application to the Multipolar Hamiltonian*, Il Nuovo Cimento **86 B**, 155–169 (1985)

D.H. Kobe, in: *Ode to a Quantum Physicist*, ed. by W.P. Schleich, H. Walther and W.E. Lamb, Elsevier, Amsterdam, 2000

Atom in Field

A treatment similar to the one of Sec. 14.2.2 was suggested by

J.H. Eberly, *Note on the Electric Dipole Interaction*, Internal Memorandum, M. L. No. 808, W. W. Hansen Laboratories of Physics, Stanford, 1961

and in the context of laser cooling by

S. Stenholm, *The semiclassical theory of laser cooling*, Rev. Mod. Phys. **58**, 699–739 (1986)

For a discussion of the appropriate interaction Hamiltonian including the center-of-mass motion using the $\vec{E} \cdot \vec{r}$ interaction see

W.C. Röntgen, *Über die durch Bewegung eines im homogenen electrischen Felde befindlichen Dielectricums hervorgerufene electrodynamische Kraft*, Ann. Phys. Chem. **35**, 264–270 (1888)

C. Baxter, M. Babiker and **R. Loudon**, *Canonical approach to photon pressure*, Phys. Rev. A **47**, 1278–1287 (1993)

M. Wilkens, *Significance of Röntgen current in quantum optics: Spontaneous emission of moving atoms*, Phys. Rev. A **48**, 570–573 (1994)

Jaynes-Cummings-Paul Model

The original papers on the Jaynes-Cummings-Paul model are

E.T. Jaynes and **F.W. Cummings**, *Comparison of Quantum and Semiclassical Radiation Theories with Applications to the Beam Maser*, Proc. IEEE **51**, 89–109 (1963)

H. Paul, *Induzierte Emission bei starker Einstrahlung*, Ann. Phys. (Leipzig) **11**, 411–412 (1963)

Similar ideas have also been put forward by

C.R. Willis, *A Model of Interacting Radiation and Matter*, J. Math. Phys. **5**, 1241–1252 (1964)

R.H. Picard and **C.R. Willis**, *Coherence in a Model of Interacting Radiation and Matter*, Phys. Rev. **139**, A10–A15 (1965)

For an exposition of the method of averaging see for example

N.N. Bogoljubow and **J.A. Mitropolski**, *Asymptotic Methods in Non Linear Oscillations*, Hindustan Publ., Delhi, 1961

15 Jaynes-Cummings-Paul Model: Dynamics

The Jaynes-Cummings-Paul model introduced in the preceding section is the most elementary model for the interaction of an atom with a quantized field: it considers only a single two-level atom and a single mode of the radiation field. Moreover, one can also include the center-of-mass motion. In the present chapter, however, we neglect for the time being the motion of the atom: This is the original version of the Jaynes-Cummings-Paul model. We return to the problem of the quantum treatment of the center-of-mass motion in Chapters 19 and 20.

In the present chapter we solve the Schrödinger equation

$$i\hbar \frac{d\left|\Psi(t)\right\rangle}{dt} = \hat{H}_{\text{int}} \left|\Psi(t)\right\rangle \tag{15.1}$$

for the state vector $\left|\Psi\right\rangle$ where the Hamiltonian in the interaction picture reads

$$\hat{H}_{\text{int}} \equiv \hbar g \left(\hat{\sigma} \hat{a}^\dagger e^{i\Delta t} + \hat{\sigma}^\dagger \hat{a} e^{-i\Delta t}\right). \tag{15.2}$$

We consider the dynamics in two limits: *(i)* In the resonant case the Hamiltonian is time independent and we derive in Sec. 15.1 an explicit expression for the time evolution operator using operator algebra. *(ii)* In the off-resonant case discussed in Sec. 15.2 we make an ansatz for the state vector and then solve in Sec. 15.3 the resulting equations using a Laplace transformation. The strongly off-resonant case is of particular importance since it conserves the number of photons and the populations in the atomic levels. Nevertheless, it creates phase shifts which depend on the number of photons and the atomic state. In this way we can create Schrödinger cat states of the field as discussed in Sec. 15.4.

15.1 Resonant Jaynes-Cummings-Paul Model

In order to gain some insight into the dynamics of the Jaynes-Cummings-Paul model we first consider the resonant case $\Delta = 0$. In this limit the interaction Hamiltonian

$$\hat{H}_{\text{int}} = \hbar g(\hat{\sigma} \hat{a}^\dagger + \hat{\sigma}^\dagger \hat{a})$$

is not explicitly time dependent. We can therefore use the formal solution

$$|\Psi(t)\rangle = \exp\left(-\frac{i}{\hbar}\hat{H}_{\text{int}}t\right)|\Psi(0)\rangle = \exp\left[-igt(\hat{\sigma}\hat{a}^\dagger + \hat{\sigma}^\dagger \hat{a})\right]|\Psi(0)\rangle \equiv \hat{\mathcal{U}}(t, t_0 = 0)|\Psi(0)\rangle$$

of the Schrödinger equation discussed in Sec. 2.4.2.

In the present section we discuss the dynamics by evaluating the time evolution operator

$$\hat{\mathcal{U}}(t, t_0 = 0) \equiv \exp\left[-igt\left(\hat{\sigma}\hat{a}^\dagger + \hat{\sigma}^\dagger \hat{a}\right)\right]$$

of the resonant Jaynes-Cummings-Paul model.

The initial state vector

$$|\Psi(0)\rangle \equiv |\psi_{\text{field}}\rangle \otimes |\psi_{\text{atom}}\rangle \equiv \sum_{n=0}^{\infty} w_n[c_a |a\rangle + c_b |b\rangle]|n\rangle = \sum_{n=0}^{\infty} w_n \begin{pmatrix} c_a \\ c_b \end{pmatrix} |n\rangle, \quad (15.3)$$

is the direct product of the field state

$$|\psi_{\text{field}}\rangle \equiv \sum_{n=0}^{\infty} w_n |n\rangle$$

and the atomic state

$$|\psi_{\text{atom}}\rangle \equiv c_a |a\rangle + c_b |b\rangle.$$

In the last step of Eq. (15.3) we have represented the atomic states by the vectors

$$|a\rangle \equiv \begin{pmatrix} 1 \\ 0 \end{pmatrix} \quad \text{and} \quad |b\rangle \equiv \begin{pmatrix} 0 \\ 1 \end{pmatrix},$$

which is convenient when we consider the action of the atom-field Hamiltonian on the initial state.

15.1.1 Time Evolution Operator Using Operator Algebra

We first cast the operator $\hat{\mathcal{U}}$ into a different form and start from the definition

$$e^{-\lambda \hat{A}} \equiv \sum_{n=0}^{\infty} \frac{(-\lambda)^n}{n!} \hat{A}^n$$

of the exponential function as a power series for a scalar λ and an operator $\hat{A}$. We decompose the sum into even and odd terms, that is,

$$\hat{\mathcal{U}}(t, t_0 = 0) = \exp[-igt(\hat{\sigma}\hat{a}^\dagger + \hat{\sigma}^\dagger \hat{a})] = \hat{c} - i\hat{s},$$

where

$$\hat{c} \equiv \sum_{n=0}^{\infty} \frac{(-1)^n (gt)^{2n}}{(2n)!} (\hat{\sigma}\hat{a}^\dagger + \hat{\sigma}^\dagger \hat{a})^{2n} \quad (15.4)$$

and

$$\hat{s} \equiv \sum_{n=0}^{\infty} \frac{(-1)^n (gt)^{2n+1}}{(2n+1)!} (\hat{\sigma}\hat{a}^\dagger + \hat{\sigma}^\dagger \hat{a})^{2n} (\hat{\sigma}\hat{a}^\dagger + \hat{\sigma}^\dagger \hat{a}). \quad (15.5)$$

15.1 Resonant Jaynes-Cummings-Paul Model

Even Powers

We first consider the operator

$$(\hat{\sigma}\hat{a}^\dagger + \hat{\sigma}^\dagger \hat{a})^2 = \hat{\sigma}^2 \hat{a}^{\dagger 2} + \hat{\sigma}^{\dagger 2} \hat{a}^2 + \hat{\sigma}\hat{\sigma}^\dagger \hat{a}^\dagger \hat{a} + \hat{\sigma}^\dagger \hat{\sigma} \hat{a} \hat{a}^\dagger$$

and recall the relations

$$\hat{\sigma}^{\dagger 2} = (|a\rangle \langle b|)^2 = |a\rangle \langle b| |a\rangle \langle b| = |a\rangle \langle b| \cdot \langle b| a\rangle = \hat{0}$$

and

$$\hat{\sigma}^2 = (|b\rangle \langle a|)^2 = |b\rangle \langle a| |b\rangle \langle a| = \hat{0}$$

reflecting the fact that in a two-level atom we can apply the raising and lowering operators for the atomic levels only once.

Moreover, we have the operator

$$\hat{\sigma}\hat{\sigma}^\dagger = (|b\rangle \langle a|)(|a\rangle \langle b|) = |b\rangle \langle b| = \begin{pmatrix} 0 & 0 \\ 0 & 1 \end{pmatrix} \tag{15.6}$$

corresponding to transitions from the ground state $|b\rangle$ to the excited state $|a\rangle$ and back to $|b\rangle$.

Analogously, the operator

$$\hat{\sigma}^\dagger \hat{\sigma} = (|a\rangle \langle b|)(|b\rangle \langle a|) = |a\rangle \langle a| = \begin{pmatrix} 1 & 0 \\ 0 & 0 \end{pmatrix} \tag{15.7}$$

describes transitions from $|a\rangle$ to $|b\rangle$ and back to $|a\rangle$.

The atomic operator $\hat{\sigma}\hat{\sigma}^\dagger$ is accompanied by the field operator $\hat{a}^\dagger \hat{a}$, that is, the atomic path $|b\rangle \to |a\rangle \to |b\rangle$ is accompanied by the annihilation of a photon and its re-creation. Likewise, the operator $\hat{\sigma}^\dagger \hat{\sigma}$ is connected with $\hat{a}\hat{a}^\dagger$, that is the path $|a\rangle \to |b\rangle \to |a\rangle$ goes together with the creation of a photon and its annihilation. Thus the operator

$$(\hat{\sigma}\hat{a}^\dagger + \hat{\sigma}^\dagger \hat{a})^2 = \hat{\sigma}\hat{\sigma}^\dagger \hat{a}^\dagger \hat{a} + \hat{\sigma}^\dagger \hat{\sigma} \hat{a} \hat{a}^\dagger = \begin{pmatrix} \hat{a}\hat{a}^\dagger & 0 \\ 0 & \hat{a}^\dagger \hat{a} \end{pmatrix}$$

following from Eqs. (15.6) and (15.7) corresponds to a two-photon process giving rise to the representation

$$(\hat{\sigma}\hat{a}^\dagger + \hat{\sigma}^\dagger \hat{a})^{2n} = \begin{pmatrix} (\hat{a}\hat{a}^\dagger)^n & 0 \\ 0 & (\hat{a}^\dagger \hat{a})^n \end{pmatrix} = \begin{pmatrix} \sqrt{\hat{a}\hat{a}^\dagger}^{2n} & 0 \\ 0 & \sqrt{\hat{a}^\dagger \hat{a}}^{2n} \end{pmatrix}. \tag{15.8}$$

Hence we arrive for the operator $\hat{c}$, defined in Eq. (15.4), at the expression

$$\hat{c} = \begin{pmatrix} \sum_{n=0}^{\infty} \frac{(-1)^n (gt)^{2n}}{(2n)!} \sqrt{\hat{a}\hat{a}^\dagger}^{2n} & 0 \\ 0 & \sum_{n=0}^{\infty} \frac{(-1)^n (gt)^{2n}}{(2n)!} \sqrt{\hat{a}^\dagger \hat{a}}^{2n} \end{pmatrix},$$

or

$$\hat{c} = \begin{pmatrix} \cos(gt\sqrt{\hat{a}\hat{a}^\dagger}) & 0 \\ 0 & \cos(gt\sqrt{\hat{a}^\dagger \hat{a}}) \end{pmatrix}. \tag{15.9}$$

Odd Powers

From the definition Eq. (15.5) of the operator $\hat{s}$ we find

$$\hat{s} = \begin{pmatrix} \sum_{n=0}^{\infty} \frac{(-1)^n (gt)^{2n+1}}{(2n+1)!} \frac{\sqrt{\hat{a}\hat{a}^\dagger}^{2n+1}}{\sqrt{\hat{a}\hat{a}^\dagger}} & 0 \\ 0 & \sum_{n=0}^{\infty} \frac{(-1)^n (gt)^{2n+1}}{(2n+1)!} \frac{\sqrt{\hat{a}^\dagger\hat{a}}^{2n+1}}{\sqrt{\hat{a}^\dagger\hat{a}}} \end{pmatrix} (\hat{\sigma}\hat{a}^\dagger + \hat{\sigma}^\dagger \hat{a}),$$

where we have made use of the representation Eq. (15.8) of the square of the operator

$$\hat{\sigma}\hat{a}^\dagger + \hat{\sigma}^\dagger \hat{a} = \begin{pmatrix} 0 & \hat{a} \\ \hat{a}^\dagger & 0 \end{pmatrix}.$$

When we use the above relation we arrive at the formula

$$\hat{s} = \begin{pmatrix} \frac{\sin(gt\sqrt{\hat{a}\hat{a}^\dagger})}{\sqrt{\hat{a}\hat{a}^\dagger}} & 0 \\ 0 & \frac{\sin(gt\sqrt{\hat{a}^\dagger\hat{a}})}{\sqrt{\hat{a}^\dagger\hat{a}}} \end{pmatrix} \begin{pmatrix} 0 & \hat{a} \\ \hat{a}^\dagger & 0 \end{pmatrix} = \begin{pmatrix} 0 & \frac{\sin(gt\sqrt{\hat{a}\hat{a}^\dagger})}{\sqrt{\hat{a}\hat{a}^\dagger}} \hat{a} \\ \frac{\sin(gt\sqrt{\hat{a}^\dagger\hat{a}})}{\sqrt{\hat{a}^\dagger\hat{a}}} \hat{a}^\dagger & 0 \end{pmatrix}. \tag{15.10}$$

We emphasize that in contrast to $\hat{c}$ the operator $\hat{s}$ is off-diagonal in the atomic state basis.

Time Evolution Operator

We can now combine the formulas Eqs. (15.9) and (15.10) to find an explicit expression for the time evolution operator. Before we do this we express the operator $\hat{a}\hat{a}^\dagger$ in terms of the photon number operator $\hat{n}$ using the relation $\hat{a}\hat{a}^\dagger = \hat{a}^\dagger\hat{a} + 1 = \hat{n} + 1$. Hence the time evolution operator of the Jaynes-Cummings-Paul model takes the form

$$\hat{\mathcal{U}}(t, t_0 = 0) = \exp[-igt(\hat{\sigma}\hat{a}^\dagger + \hat{\sigma}^\dagger \hat{a})]$$

$$= \begin{pmatrix} \cos(gt\sqrt{\hat{n}+1}) & 0 \\ 0 & \cos(gt\sqrt{\hat{n}}) \end{pmatrix} - i \begin{pmatrix} 0 & \frac{\sin(gt\sqrt{\hat{n}+1})}{\sqrt{\hat{n}+1}} \hat{a} \\ \frac{\sin(gt\sqrt{\hat{n}})}{\sqrt{\hat{n}}} \hat{a}^\dagger & 0 \end{pmatrix}. \tag{15.11}$$

This operator relation provides the dynamics of this model.

15.1.2 Interpretation of Time Evolution Operator

The structure of the expression Eq. (15.11) for the time evolution operator is rather transparent: The atom-field dynamics consists of an even or odd number of exchanges of quanta of excitation between the field and the atom corresponding to the two contributions in this equation.

Even Powers

Since an annihilation of a field quantum is associated with the creation of an atomic excitation and a creation of a field quantum with the annihilation of an atomic excitation, an even number of exchanges of quanta leaves the atomic states invariant. Hence the operator corresponding to the exchange of even number of photons is diagonal.

Moreover, due to the unitary time evolution creating the exponential we find that the time dependence is governed by cosine and sine functions. Since the diagonal terms do not describe transitions and at $t = 0$ the initial state must emerge, this diagonal term can only contain cosine functions.

The emergence of the operators $\sqrt{\hat{n}}$ and $\sqrt{\hat{n}+1}$ becomes clear when we recall that the element in the upper left corner reflects $2n$-photon transitions starting from the excited state and ending again there. Hence we have to first create a photon and then annihilate it. This process takes place n times, that is

$$\underbrace{(\hat{a}\hat{a}^\dagger)(\hat{a}\hat{a}^\dagger)\cdots(\hat{a}\hat{a}^\dagger)}_{n \text{ times}} = (\hat{a}\hat{a}^\dagger)^n = (\hat{a}^\dagger\hat{a}+1)^n = \sqrt{\hat{n}+1}^{2n}.$$

Likewise, the element in the lower right corner corresponds to a $2n$-photon process starting from the ground state and ending there. Hence we first annihilate a photon and then create it again and this process takes place n times, that is

$$\underbrace{(\hat{a}^\dagger\hat{a})(\hat{a}^\dagger\hat{a})\cdots(\hat{a}^\dagger\hat{a})}_{n \text{ times}} = (\hat{a}^\dagger\hat{a})^n = \sqrt{\hat{n}}^{2n}.$$

Odd Powers

The second term contains the exchange of an odd number of photons. Hence the operator cannot be diagonal – it has to be off-diagonal. When we start from the ground state and end up finally in the excited state we must first annihilate a photon and then go through a $2n$-photon process which creates a photon and annihilates it n times, that is

$$\underbrace{(\hat{a}\hat{a}^\dagger)(\hat{a}\hat{a}^\dagger)\cdots(\hat{a}\hat{a}^\dagger)}_{n \text{ times}} \hat{a} = (\hat{a}\hat{a}^\dagger)^n \hat{a} = \sqrt{\hat{n}+1}^{2n} \hat{a}.$$

In order to create a sine function from the odd terms in the expansion of the exponential we have to multiply and divide by $\sqrt{\hat{n}+1}$. This explains the second column of the second term in Eq. (15.11).

Analogously, we can explain the first column as corresponding to the $2n + 1$ photon transition from the excited state ending up in the ground state. Here we first create a photon and then go through a $2n$-photon process of annihilation and creation and so on, that is

$$\underbrace{(\hat{a}^\dagger\hat{a})(\hat{a}^\dagger\hat{a})\cdots(\hat{a}^\dagger\hat{a})}_{n \text{ times}} \hat{a}^\dagger = (\hat{a}^\dagger\hat{a})^n \hat{a}^\dagger = \sqrt{\hat{n}}^{2n} \hat{a}^\dagger.$$

Since there are no transitions for $t = 0$, the time dependence is determined by a sine-function.

15.1.3 State Vector of Combined System

We are now in a position to apply the time evolution operator to our initial state vector

$$|\Psi(t=0)\rangle = \sum_{n=0}^{\infty} w_n \begin{pmatrix} c_a \\ c_b \end{pmatrix} |n\rangle.$$

When we note the relations

$$\frac{\sin(gt\sqrt{\hat{n}+1})}{\sqrt{\hat{n}+1}} \hat{a} |n\rangle = \frac{\sin(gt\sqrt{\hat{n}+1})}{\sqrt{\hat{n}+1}} \sqrt{n} |n-1\rangle = \sin(gt\sqrt{n}) |n-1\rangle$$

and

$$\frac{\sin(gt\sqrt{\hat{n}})}{\sqrt{\hat{n}}} \hat{a}^\dagger |n\rangle = \frac{\sin(gt\sqrt{\hat{n}})}{\sqrt{\hat{n}}} \sqrt{n+1} |n+1\rangle = \sin(gt\sqrt{n+1}) |n+1\rangle$$

we find

$$|\Psi(t)\rangle = \sum_{n=0}^{\infty} w_n \left[\begin{pmatrix} \cos(gt\sqrt{n+1}) & 0 \\ 0 & \cos(gt\sqrt{n}) \end{pmatrix} \begin{pmatrix} c_a \\ c_b \end{pmatrix} |n\rangle \right.$$
$$\left. - i \begin{pmatrix} 0 & \sin(gt\sqrt{n}) |n-1\rangle \\ \sin(gt\sqrt{n+1}) |n+1\rangle & 0 \end{pmatrix} \begin{pmatrix} c_a \\ c_b \end{pmatrix} \right].$$

Hence we arrive at

$$|\Psi(t)\rangle = \sum_{n=0}^{\infty} w_n \left[\begin{pmatrix} c_a \cos(gt\sqrt{n+1}) |n\rangle \\ c_b \cos(gt\sqrt{n}) |n\rangle \end{pmatrix} - i \begin{pmatrix} c_b \sin(gt\sqrt{n}) |n-1\rangle \\ c_a \sin(gt\sqrt{n+1}) |n+1\rangle \end{pmatrix} \right],$$

or

$$|\Psi(t)\rangle = \sum_{n=0}^{\infty} w_n \{ c_a [\cos(gt\sqrt{n+1}) |a,n\rangle - i \sin(gt\sqrt{n+1}) |b,n+1\rangle]$$
$$+ c_b [\cos(gt\sqrt{n}) |b,n\rangle - i \sin(gt\sqrt{n}) |a,n-1\rangle] \}. \quad (15.12)$$

Here we have introduced the notation $|a\rangle |n\rangle \equiv |a,n\rangle$ and $|b\rangle |n+1\rangle \equiv |b,n+1\rangle$.

15.1.4 Dynamics Represented in State Space

It is instructive to represent the dynamics of the Jaynes-Cummings-Paul model as expressed by Eq. (15.12) for the most elementary case of an atom in the superposition state $|\psi_{\text{atom}}\rangle \equiv c_a |a\rangle + c_b |b\rangle$ interacting with a Fock state $|n\rangle$. In this case the time evolution transforms the initial state

$$|\Psi(t=0)\rangle \equiv |\psi_{\text{atom}}\rangle \otimes |\psi_{\text{field}}\rangle \equiv [c_a |a\rangle + c_b |b\rangle] \otimes |n\rangle,$$

that is,

$$|\Psi(t=0)\rangle = c_a |a,n\rangle + c_b |b,n\rangle$$

into the state

$$|\Psi(t)\rangle = c_a [C_n |a,n\rangle - i S_n |b,n+1\rangle] + c_b [C_{n-1} |b,n\rangle - i S_{n-1} |a,n-1\rangle] \quad (15.13)$$

15.1 Resonant Jaynes-Cummings-Paul Model

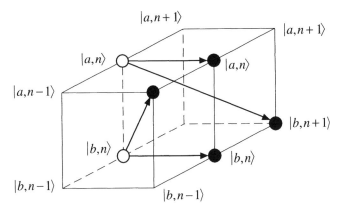

Fig. 15.1: Dynamics of the Jaynes-Cummings-Paul model represented in an atom-field space. Each corner of the cubes depicted either by an empty or filled circle denotes an atom-field state $|j, m\rangle$ where $|j\rangle = |a\rangle$ or $|b\rangle$ and $|m\rangle$ is the mth Fock state. The upper plane represents states in which the atom is in the excited state and m photons are present whereas the lower plane contains the states in which the atom is in the ground state. The axes into the board denote the number of photons. Since the Jaynes-Cummings-Paul model allows only for an exchange of a single quantum of excitation between atom and field the state $|a, n\rangle$ either stays unaltered represented by the arrow in the upper plane or it transforms into $|b, n+1\rangle$ depicted by the arrow pointing to the right corner in the lower plane in the background. The probability amplitudes associated with these processes are $C_n \equiv \cos(\sqrt{n+1}\,gt)$ and $-iS_n \equiv -i\sin(\sqrt{n+1}\,gt)$, respectively. Likewise, the state $|b, n\rangle$ can stay unchanged indicated by the arrow in the lower plane or it transforms into the state $|a, n-1\rangle$ shown by the arrow pointing to the right corner in the upper plane in the foreground. The corresponding probability amplitudes are C_{n-1} and $-iS_{n-1}$, respectively. The complete state of atom plus field is the interference between the four states $|a, n\rangle$, $|a, n-1\rangle$, $|b, n\rangle$ and $|b, n+1\rangle$, that is the four black corners. We emphasize however that this picture does not include or represent correctly the interference between these states.

at time t.

Here we have introduced the abbreviations

$$C_n \equiv \cos\left(\sqrt{n+1}\,gt\right) \qquad (15.14\text{a})$$

and

$$S_n \equiv \sin\left(\sqrt{n+1}\,gt\right). \qquad (15.14\text{b})$$

In Fig. 15.1 we represent this dynamics in a three-dimensional lattice. Each corner of the cubes represents a state of the combined atom-field system. On the upper floor we have the states $|a, m\rangle$ that is the Fock states in the presence of an atom in the excited state. In the lower floor we depict the states $|b, m\rangle$, that is, the photon states with atoms in the ground state.

The corners on the left side mark the states *before* the interaction, whereas the corners on the right side represent the states *after* the interaction. We depict the dynamics of the model by arrows connecting the corners.

Since at most one excitation can be exchanged between the atom and the field the state $|a, n\rangle$ can either stay, shown by the horizontal arrow, or transform to $|b, n+1\rangle$, as depicted by the arrow pointing towards the right corner. The transition probability amplitude corresponding to these transitions are C_n and $-iS_n$, respectively.

Likewise the state $|b, n\rangle$ can either remain unchanged or go to $|a, n-1\rangle$ as indicated by the horizontal and the diagonal arrow pointing up to the right corner in the foreground. The transition probability amplitudes for this evolution are C_{n-1} and $-iS_{n-1}$.

When the atom is initially in a coherent superposition of excited and ground state the above mentioned transition probability amplitudes must be multiplied by the probability amplitude c_a to be in the state $|a\rangle$ and the probability amplitude c_b to be in the state $|b\rangle$. Thererfore, the total probability amplitudes for a transition from the upper to the lower atomic level read $c_a C_n$ and $-ic_a S_n$. Similarly, the total probability amplitudes for a transition from the lower to the upper level are given by $c_b C_{n-1}$ and $-ic_b S_{n-1}$. In this case the state vector $|\Psi\rangle$ of the combined atom-field system is the interference of all four state vectors $|a, n\rangle$, $|a, n-1\rangle$, $|b, n\rangle$ and $|b, n-1\rangle$, that is, the interference of all black corners.

So far we have only considered the evolution of the state vector of a system consisting of a two-level atom interacting with a single Fock state. In the case of an arbitrary initial field state

$$|\psi_{\text{field}}\rangle = \sum_{n=0}^{\infty} w_n |n\rangle$$

every corner on the left side serves as a starting point of the arrows indicating transitions. Now we have to attach to each corner a probability amplitude w_n as shown in Fig. 15.2.

15.2 Role of Detuning

We now turn to the more complicated case of the Jaynes-Cummings-Paul model with a non-vanishing detuning. Here the interaction Hamiltonian Eq. (15.2) is explicitly time dependent and we cannot use the formal solution presented in the preceding section. However, we could, in principle, use the time ordered product discussed in Sec. 2.4. Fortunately, this is not necessary.

We recall that the Hamiltonian $\hat{H}_{\text{int}}$ couples the subsystems of the atomic and the field states. We therefore make an ansatz for the state vector containing appropriate combinations of field and atomic states. From the Schrödinger equation we derive the equations of motion for the expansion coefficients.

15.2.1 Atomic and Field States

The state vector $|\Psi\rangle$ consists of field and atomic states. However, due to the particular atom-field coupling $\hat{\sigma}\hat{a}^\dagger$ and $\hat{\sigma}^\dagger\hat{a}$ the atomic and field states only enter in a specific combination.

Indeed, atoms in the excited state $|a\rangle$ get annihilated and create a field excitation. Therefore, the Hamiltonian $\hat{H}_{\text{int}}$ transforms the state $|a\rangle |n\rangle \equiv |a, n\rangle$ where $|n\rangle$

15.2 Role of Detuning

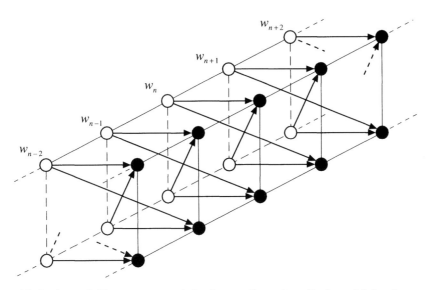

Fig. 15.2: Atom-field state space of the Jaynes-Cummings-Paul model for the case of a superposition of Fock states with probability amplitudes w_n. Every empty circle at the corners on the left side represents one component of the initial state weighted with the probability amplitude w_n. The arrows indicate for each component of the initial state the different possibilities of evolution which contribute to the components of the final state represented by the filled circles.

denotes the nth Fock state of the field into

$$\hat{H}_{\text{int}} |a, n\rangle = \hbar g \left(\hat{\sigma} \hat{a}^\dagger e^{i\Delta t} + \hat{\sigma}^\dagger \hat{a} e^{-i\Delta t} \right) |a, n\rangle = \hbar g \sqrt{n+1}\, e^{i\Delta t} |b, n+1\rangle. \quad (15.15)$$

Here we have made use of the property

$$\hat{a}^\dagger |n\rangle = \sqrt{n+1}\, |n+1\rangle$$

of the creation operator and the relations

$$\hat{\sigma} |a\rangle = |b\rangle$$

and

$$\hat{\sigma}^\dagger |a\rangle = |a\rangle \langle b|a\rangle = |a\rangle \langle b|a\rangle = 0$$

of the atomic operator. The last formula reflects the fact that the excited state cannot be excited further.

Similarly, atoms in the ground state $|b\rangle$ get excited at the expense of a field excitation. Hence, the Hamiltonian $\hat{H}_{\text{int}}$ transforms the state $|b\rangle |n+1\rangle \equiv |b, n+1\rangle$ into

$$\hat{H}_{\text{int}} |b, n+1\rangle = \hbar g \left(\hat{\sigma} \hat{a}^\dagger e^{i\Delta t} + \hat{\sigma}^\dagger \hat{a} e^{-i\Delta t} \right) |b, n+1\rangle = \hbar g \sqrt{n+1}\, e^{-i\Delta t} |a, n\rangle. \quad (15.16)$$

Here we have used the relation

$$\hat{a} |n\rangle = \sqrt{n}\, |n-1\rangle$$

and the properties
$$\hat{\sigma}^\dagger |b\rangle = |a\rangle$$
and
$$\hat{\sigma} |b\rangle = |b\rangle \langle a| |b\rangle = |b\rangle \langle a|b\rangle = 0.$$

The last formula reflects the fact that we cannot go below the ground state.

The state $|b, 0\rangle$ which corresponds to $n = -1$ in Eq. (15.16) plays a special role: According to Eq. (15.16) we find

$$\hat{H}_{\text{int}} |b, 0\rangle = 0. \tag{15.17}$$

The vacuum cannot excite an atom initially in the ground state. The state $|b, 0\rangle$ decouples from the rest of the states.

15.2.2 Rabi Equations

We are now in a position to discuss the dynamics of the nonresonant Jaynes-Cummings-Paul model. For this purpose we make an ansatz for the state vector.

Ansatz for State Vector

Since the Hamiltonian $\hat{H}_{\text{int}}$ couples only the states $|b, n+1\rangle$ and $|a, n\rangle$ we try the ansatz

$$|\Psi(t)\rangle = \sum_{n=0}^{\infty} \left[\Psi_{a,n}(t) |a, n\rangle + \Psi_{b,n+1}(t) |b, n+1\rangle \right] + \Psi_{b,0}(t) |b, 0\rangle \tag{15.18}$$

for the state vector.

We find the equations of motion for the time dependent probability amplitudes $\Psi_{a,n}$ and $\Psi_{b,n+1}$ by substituting this ansatz into the Schrödinger equation Eq. (15.1) which yields

$$i\hbar \sum_{n=0}^{\infty} \left[\dot{\Psi}_{a,n}(t) |a, n\rangle + \dot{\Psi}_{b,n+1}(t) |b, n+1\rangle \right] + i\hbar \, \dot{\Psi}_{b,0} |b, 0\rangle$$
$$= \sum_{n=0}^{\infty} \left[\Psi_{a,n}(t) \hat{H}_{\text{int}} |a, n\rangle + \Psi_{b,n+1}(t) \hat{H}_{\text{int}} |b, n+1\rangle \right] + \Psi_{b,0} \hat{H}_{\text{int}} |b, 0\rangle,$$

or

$$i\hbar \sum_{n=0}^{\infty} \left[\dot{\Psi}_{a,n}(t) |a, n\rangle + \dot{\Psi}_{b,n+1}(t) |b, n+1\rangle \right] + i\hbar \, \dot{\Psi}_{b,0} |b, 0\rangle$$
$$= \sum_{n=0}^{\infty} \left[\Psi_{a,n}(t) \hbar g \sqrt{n+1} \, e^{i\Delta t} |b, n+1\rangle + \Psi_{b,n+1}(t) \hbar g \sqrt{n+1} \, e^{-i\Delta t} |a, n\rangle \right]$$

where we have used the transformations Eqs. (15.15)–(15.17) of the states $|a, n\rangle$, $|b, n+1\rangle$ and $|b, 0\rangle$.

When we project on the states $|a, n\rangle$, $|b, n+1\rangle$ and $|b, 0\rangle$, and note that they are mutually orthogonal we arrive at the so-called Rabi equations

$$\dot{\Psi}_{a,n}(t) = -ig\sqrt{n+1}\, e^{-i\Delta t}\Psi_{b,n+1}(t)$$
$$\dot{\Psi}_{b,n+1}(t) = -ig\sqrt{n+1}\, e^{i\Delta t}\Psi_{a,n}(t). \qquad (15.19)$$

We recognize that indeed the state $|b, 0\rangle$ decouples from the dynamics since

$$\dot{\Psi}_{b,0}(t) = 0,$$

in complete accordance with the second of the two Rabi equations for $n = -1$.

Initial Conditions

The Rabi equations Eq. (15.19) are two coupled ordinary differential equations of first order and thus require two initial conditions: These are the probability amplitudes $\Psi_{a,n}(t=0)$ and $\Psi_{b,n}(t=0)$ at time $t = 0$. They follow from the assumption that at time $t = 0$ the atom is uncorrelated with the field and the state vector $|\Psi(t=0)\rangle$ is a direct product

$$|\Psi(t=0)\rangle \equiv |\psi_{\text{atom}}\rangle \otimes |\psi_{\text{field}}\rangle$$
$$\equiv [c_a |a\rangle + c_b |b\rangle] \otimes \sum_{n=0}^{\infty} w_n |n\rangle \equiv \sum_{n=0}^{\infty}[w_n c_a |a, n\rangle + w_n c_b |b, n\rangle],$$

which in the notation of the ansatz Eq. (15.18) reads

$$|\Psi(t=0)\rangle = \sum_{n=0}^{\infty}[w_n c_a |a, n\rangle + w_{n+1} c_b |b, n+1\rangle] + w_0 c_b |b, 0\rangle. \qquad (15.20)$$

Here c_a, c_b and w_n are the probability amplitudes to be at time $t = 0$ in the atomic state $|a\rangle$ or $|b\rangle$, or to find the nth Fock state in the radiation field, respectively.

When we compare the expression Eq. (15.20) for $|\Psi(t=0)\rangle$ with the ansatz Eq. (15.18) for the state vector $|\Psi\rangle$ we find the initial conditions

$$\Psi_{a,n}(t=0) = w_n c_a \qquad (15.21a)$$
$$\Psi_{b,n+1}(t=0) = w_{n+1} c_b \qquad (15.21b)$$

and

$$\Psi_{b,0}(t=0) = \Psi_{b,0}(t) = w_0 c_b. \qquad (15.21c)$$

Hence, we have to solve the Rabi equations Eq. (15.19) subjected to these initial conditions.

15.3 Solution of Rabi Equations

The Rabi equations Eq. (15.19) are two coupled first order differential equations. There exist many techniques to decouple such equations. One possibility consists of differentiating both equations with respect to time and substituting one into the other. However, in the present section we find it more convenient to use a Laplace transformation. In this way we can automatically include the initial conditions.

15.3.1 Laplace Transformation

The Laplace transformation allows us to cast the Rabi equations into a set of algebraic equations which we can solve in a straightforward way. We then perform the inverse Laplace transformation to find the probability amplitudes $\Psi_{a,n}(t)$ and $\Psi_{b,n+1}(t)$.

Derivation of Algebraic Equations

We start by introducing the Laplace transformed probability amplitudes

$$\overline{\Psi}_{a,n}(s) \equiv \int_0^\infty dt\, e^{-st} e^{i\Delta t/2} \Psi_{a,n}(t)$$

and

$$\overline{\Psi}_{b,n+1}(s) \equiv \int_0^\infty dt\, e^{-st} e^{-i\Delta t/2} \Psi_{b,n+1}(t).$$

Here we have transformed the probability amplitudes $e^{i\Delta t/2}\Psi_{a,n}$ and $e^{-i\Delta t/2}\Psi_{b,n+1}$, that is, we have included part of the time dependence of the Hamiltonian due to the detuning. This approach allows us to obtain simple equations for $\overline{\Psi}_{a,n}(s)$ and $\overline{\Psi}_{b,n+1}(s)$.

Indeed, when we multiply the two Rabi equations Eq. (15.19) by $e^{-st}e^{i\Delta t/2}$ and $e^{-st}e^{-i\Delta t/2}$ and integrate over time we arrive at

$$\int_0^\infty dt\, e^{-st} e^{i\Delta t/2} \dot{\Psi}_{a,n}(t) = -ig\sqrt{n+1}\,\overline{\Psi}_{b,n+1}(s)$$

and

$$\int_0^\infty dt\, e^{-st} e^{-i\Delta t/2} \dot{\Psi}_{b,n+1}(t) = -ig\sqrt{n+1}\,\overline{\Psi}_{a,n}(s).$$

After integrating by parts we find the inhomogeneous system

$$\mathbf{U}\begin{pmatrix} \overline{\Psi}_{a,n}(s) \\ \overline{\Psi}_{b,n+1}(s) \end{pmatrix} \equiv \begin{pmatrix} s - i\frac{\Delta}{2} & ig\sqrt{n+1} \\ ig\sqrt{n+1} & s + i\frac{\Delta}{2} \end{pmatrix} \begin{pmatrix} \overline{\Psi}_{a,n}(s) \\ \overline{\Psi}_{b,n+1}(s) \end{pmatrix} = \begin{pmatrix} \Psi_{a,n}(t=0) \\ \Psi_{b,n+1}(t=0) \end{pmatrix}$$

of two coupled algebraic equations.

Solution of Algebraic Equations

The solution of this system reads

$$\begin{pmatrix} \overline{\Psi}_{a,n}(s) \\ \overline{\Psi}_{b,n+1}(s) \end{pmatrix} = \mathbf{U}^{-1} \begin{pmatrix} \Psi_{a,n}(0) \\ \Psi_{b,n+1}(0) \end{pmatrix} \tag{15.22}$$

where the inverse matrix of $\mathbf{U}$ is given by

$$\mathbf{U}^{-1} = \frac{1}{\det \mathbf{U}} \begin{pmatrix} s + i\frac{\Delta}{2} & -ig\sqrt{n+1} \\ -ig\sqrt{n+1} & s - i\frac{\Delta}{2} \end{pmatrix} = \frac{1}{\det \mathbf{U}} \mathbf{U}^\dagger. \tag{15.23}$$

The determinant

$$\det \mathbf{U} = s^2 + \left(\frac{\Delta}{2}\right)^2 + g^2(n+1) = (s + i\lambda_n)(s - i\lambda_n) \tag{15.24}$$

of the matrix $\mathbf{U}$ enjoys the zeros $s = \pm i\lambda_n$ with

$$\lambda_n \equiv \sqrt{\left(\frac{\Delta}{2}\right)^2 + g^2(n+1)},$$

which allow us to find the probability amplitudes $\Psi_{a,n}(t)$ and $\Psi_{b,n+1}(t)$.

15.3.2 Inverse Laplace Transformation

We now invert the Laplace transform via the relation

$$\begin{pmatrix} e^{i\Delta t/2}\Psi_{a,n}(t) \\ e^{-i\Delta t/2}\Psi_{b,n+1}(t) \end{pmatrix} = \frac{1}{2\pi i} \int_C ds\, e^{st} \begin{pmatrix} \overline{\Psi}_{a,n}(s) \\ \overline{\Psi}_{b,n+1}(s) \end{pmatrix}$$

where the path of integration C has to be chosen in such a way as to include all poles. When we substitute the solution Eq. (15.22) together with the expression Eq. (15.23) for $\mathbf{U}^{-1}$ into this inversion formula we recognize from the resulting expression

$$\begin{pmatrix} e^{i\Delta t/2}\Psi_{a,n}(t) \\ e^{-i\Delta t/2}\Psi_{b,n+1}(t) \end{pmatrix} = \frac{1}{2\pi i} \int_C ds\, \frac{e^{st}}{\det \mathbf{U}} \mathbf{U}^\dagger \begin{pmatrix} \Psi_{a,n}(t=0) \\ \Psi_{b,n+1}(t=0) \end{pmatrix}$$

that the poles are the zeros of the determinant Eq. (15.24) of the matrix $\mathbf{U}$. Indeed, when we note the relation

$$\frac{1}{\det \mathbf{U}} = \frac{1}{(s+i\lambda_n)(s-i\lambda_n)} = \frac{1}{2i\lambda_n}\left(\frac{1}{s-i\lambda_n} - \frac{1}{s+i\lambda_n}\right)$$

we arrive at

$$\begin{pmatrix} e^{i\Delta t/2}\Psi_{a,n}(t) \\ e^{-i\Delta t/2}\Psi_{b,n+1}(t) \end{pmatrix} = \frac{1}{2\pi i} \oint ds \left(\frac{1}{s-i\lambda_n} - \frac{1}{s+i\lambda_n}\right) \frac{e^{st}}{2i\lambda_n} \mathbf{U}^\dagger(s) \begin{pmatrix} \Psi_{a,n}(0) \\ \Psi_{b,n+1}(0) \end{pmatrix}.$$

We now recall from complex analysis the *Cauchy integral formula*

$$\frac{1}{2\pi i} \oint dz\, \frac{f(z)}{z-a} = f(a)$$

valid for an analytic function f. Here the closed path of integration circumvents the simple pole at $z = a$ in the counterclockwise direction.

This formula reduces the above expression for the probability amplitudes to

$$\begin{pmatrix} e^{i\Delta t/2}\Psi_{a,n}(t) \\ e^{-i\Delta t/2}\Psi_{b,n+1}(t) \end{pmatrix} = \frac{1}{2i\lambda_n}\left[e^{i\lambda_n t}\mathbf{U}^\dagger(s=i\lambda_n) - e^{-i\lambda_n t}\mathbf{U}^\dagger(s=-i\lambda_n)\right] \begin{pmatrix} \Psi_{a,n}(0) \\ \Psi_{b,n+1}(0) \end{pmatrix}.$$

When we use the definition of $\mathbf{U}^\dagger$, Eq. (15.23), we find

$$\frac{1}{2i\lambda_n}\left[e^{i\lambda_n t}\mathbf{U}^\dagger(s=i\lambda_n) - e^{-i\lambda_n t}\mathbf{U}^\dagger(s=-i\lambda_n)\right]$$

$$= \frac{1}{2\lambda_n}\left[e^{i\lambda_n t}\begin{pmatrix}\lambda_n+\frac{\Delta}{2} & -g\sqrt{n+1} \\ -g\sqrt{n+1} & \lambda_n-\frac{\Delta}{2}\end{pmatrix} - e^{-i\lambda_n t}\begin{pmatrix}-\lambda_n+\frac{\Delta}{2} & -g\sqrt{n+1} \\ -g\sqrt{n+1} & -\lambda_n-\frac{\Delta}{2}\end{pmatrix}\right]$$

which after combining the appropriate terms reads

$$\begin{pmatrix}e^{i\Delta t/2}\Psi_{a,n}(t) \\ e^{-i\Delta t/2}\Psi_{b,n+1}(t)\end{pmatrix} = $$

$$\begin{pmatrix}\cos(\lambda_n t)+i\frac{\Delta}{2\lambda_n}\sin(\lambda_n t) & -i\frac{g\sqrt{n+1}}{\lambda_n}\sin(\lambda_n t) \\ -i\frac{g\sqrt{n+1}}{\lambda_n}\sin(\lambda_n t) & \cos(\lambda_n t)-i\frac{\Delta}{2\lambda_n}\sin(\lambda_n t)\end{pmatrix}\begin{pmatrix}\Psi_{a,n}(0) \\ \Psi_{b,n+1}(0)\end{pmatrix}.$$

Result

We therefore arrive at the explicit expressions

$$\Psi_{a,n}(t) = e^{-i\Delta t/2}\left\{\left[\cos(\lambda_n t)+i\frac{\Delta}{2\lambda_n}\sin(\lambda_n t)\right]\Psi_{a,n}(0)\right.$$

$$\left. - i\frac{g\sqrt{n+1}}{\lambda_n}\sin(\lambda_n t)\Psi_{b,n+1}(0)\right\} \qquad (15.25a)$$

and

$$\Psi_{b,n+1}(t) = e^{i\Delta t/2}\left\{-i\frac{g\sqrt{n+1}}{\lambda_n}\sin(\lambda_n t)\Psi_{a,n}(0)\right.$$

$$\left. + \left[\cos(\lambda_n t)-i\frac{\Delta}{2\lambda_n}\sin(\lambda_n t)\right]\Psi_{b,n+1}(0)\right\} \qquad (15.25b)$$

for the probability amplitudes $\Psi_{a,n}(t)$ or $\Psi_{b,n+1}(t)$ to find the atom in the excited state $|a\rangle$ and n photons, or the atom in the ground state $|b\rangle$ and $n+1$ photons.

We emphasize that these expressions are in the interaction picture defined by the internal states and the field states.

15.4 Discussion of Solution

We now discuss the physical content of the solutions of the Jaynes-Cummings-Paul model. In particular, we analyze them in two limiting cases: *(i)* When the light field is resonant with the atomic transition and *(ii)* when it is far detuned. In the first case we make the connection with the results of the first section derived using operator algebra. In the latter case the time evolution of the state vector $|\Psi\rangle$ of the complete system of atom plus field follows from an effective Hamiltonian which conserves the atomic populations and the photon statistics. This Hamiltonian is of great importance in the field of cavity quantum electrodynamics and atom optics.

15.4.1 General Considerations

In the preceding section we have realized that the Jaynes-Cummings-Paul interaction Hamiltonian couples only the two probability amplitudes $\Psi_{a,n}$ and $\Psi_{b,n+1}$. Since we restrict ourselves to unitary time evolution, that is, we consider the closed system of atom and field and since quantum mechanics is linear the time dependent probability amplitudes are linear combinations of the probability amplitudes at time $t = 0$.

Moreover, since the interaction Hamiltonian creates an atomic excitation while annihilating a photon and vice versa the time evolution of this quantum system can only be a periodic exchange of field and atomic excitation. Hence, the probability amplitudes $\Psi_{a,n}$ and $\Psi_{b,n+1}$ have to be periodic functions. The most elementary periodic functions are sine and cosine and indeed the expressions Eqs. (15.25) involve linear combinations of the initial probability amplitudes with periodic coefficients. The period of this exchange between atom and field is given by the generalized Rabi frequency

$$\lambda_n \equiv \sqrt{\left(\frac{\Delta}{2}\right)^2 + g^2(n+1)} \tag{15.26}$$

which, as we have seen in the preceding section, is a zero of the determinant of **U**.

We now discuss how the trigonometric functions sine and cosine enter into the state vector. In order to make a transition from the state $|b, n+1\rangle$ to $|a, n\rangle$ or vice versa we need a non-vanishing field. Hence, the probability amplitude to make this transition has to be at least proportional to the electric field, that is to the Rabi frequency λ_n. Consequently, the contribution to $\Psi_{a,n}$ resulting from the transition from $\Psi_{b,n+1}$ has to be proportional to a sine rather than a cosine function since in lowest order the latter is independent of the Rabi frequency. Likewise, the contribution to $\Psi_{b,n+1}$ resulting from the transition from $\Psi_{a,n}$ has to contain a sine function.

Moreover, the probability to remain in the same state has to be proportional to a cosine function in order to guarantee that in the short time limit or in the weak field limit there are no transitions.

The prefactors of the trigonometric functions only become transparent when we consider a resonant or a far off-resonant situation.

15.4.2 Resonant Case

We first consider the resonant case, that is, $\Delta \equiv \Omega - \omega = 0$. In this limit the generalized Rabi frequency Eq. (15.26) reduces to

$$\lambda_n = g\sqrt{n+1}.$$

The time dependent probability amplitudes then read

$$\Psi_{a,n}(t) = \cos(\sqrt{n+1}\, gt)\, \Psi_{a,n}(0) - i\sin(\sqrt{n+1}\, gt)\, \Psi_{b,n+1}(0) \tag{15.27a}$$

and

$$\Psi_{b,n+1}(t) = -i\sin(\sqrt{n+1}\, gt)\, \Psi_{a,n}(0) + \cos(\sqrt{n+1}\, gt)\, \Psi_{b,n+1}(0). \tag{15.27b}$$

Again we note the periodic exchange of excitation between atom and field. Moreover, we recognize that the sine-functions always contain the prefactor $-i$ which we can trace back to the prefactor $-i$ in the Schrödinger equation, that is to the factor $-i$ in the Rabi equations, Eq. (15.19).

Vacuum Rabi Oscillations

A special case emerges when the electromagnetic field is initially in the ground state, that is in the vacuum state. In this case the above equations reduce to

$$\Psi_{a,0}(t) = \cos(gt)\,\Psi_{a,0}(0) - i\sin(gt)\,\Psi_{b,1}(0)$$

and

$$\Psi_{b,1}(t) = -i\sin(gt)\,\Psi_{a,0}(0) + \cos(gt)\,\Psi_{b,1}(0).$$

They indicate a periodic exchange between the atomic excitation and the field excitation even in the case of the vacuum. The frequency of this exchange is the vacuum Rabi frequency g, Eq. (14.52).

When the atom is initially in the ground state we have the initial conditions $\Psi_{b,0}(0) \equiv 1$ and $\Psi_{b,1}(0) \equiv 0$ and $\Psi_{a,0}(0) \equiv 0$. The above equations predict that the atom stays in the ground state for all times, that is $\Psi_{b,0}(t) = 1$ and $\Psi_{b,1}(t) = 0$ and $\Psi_{a,0}(t) = 0$. Hence, the atom is not able to go into the excited state if the field is in the vacuum state.

However, when the atom is initially in the excited state then there is a periodic exchange between field and atom. Indeed, in this case we have the initial conditions $\Psi_{b,1}(0) \equiv 0$ and $\Psi_{a,0}(0) \equiv 1$ which selects the time dependent solutions

$$\Psi_{a,0}(t) = \cos(gt) \quad (15.28a)$$

and

$$\Psi_{b,1}(t) = -i\sin(gt). \quad (15.28b)$$

This periodic exchange between the vacuum state of the cavity and the atom is apparent in Fig. 16.8.

Connection with Operator Algebra Approach

We conclude by presenting the state vector of the resonant Jaynes-Cummings-Paul model for arbitrary initial conditions in the field and in the atom. For this purpose we substitute the initial conditions Eqs. (15.21) into the probability amplitudes, Eqs. (15.27), and find

$$\Psi_{a,n}(t) = C_n w_n c_a - i S_n w_{n+1} c_b \quad (15.29a)$$

and

$$\Psi_{b,n+1}(t) = -i S_n w_n c_a + C_n w_{n+1} c_b \quad (15.29b)$$

where we have recalled the abbreviations Eqs. (15.14) for C_n and S_n.

With the help of these expressions for the probability amplitudes $\Psi_{a,n}$ and $\Psi_{b,n+1}$ the state vector

$$|\Psi(t)\rangle = \sum_{n=0}^{\infty} \left(\Psi_{a,n}\,|a,n\rangle + \Psi_{b,n+1}\,|b,n+1\rangle \right) + \Psi_{b,0}\,|b,0\rangle$$

takes the form

$$|\Psi(t)\rangle = \sum_{n=0}^{\infty} w_n c_a[C_n |a,n\rangle - iS_n |b,n+1\rangle]$$

$$+ \sum_{n=0}^{\infty} w_{n+1} c_b[C_n |b,n+1\rangle - iS_n |a,n\rangle] + w_0 c_b |b,0\rangle.$$

When we change the summation index in the second sum via $n' = n+1$ and combine it with the last term we arrive at

$$|\Psi(t)\rangle = \sum_{n=0}^{\infty} w_n \{ c_a[C_n |a,n\rangle - iS_n |b,n+1\rangle]$$
$$+ c_b[C_{n-1} |b,n\rangle - iS_{n-1} |a,n-1\rangle]\}, \qquad (15.30)$$

in complete accordance with Eq. (15.12).

15.4.3 Far Off-Resonant Case

More insight into the time dependence of the probability amplitudes $\Psi_{a,n}$ and $\Psi_{b,n+1}$ springs from the limit of large detuning, that is, for

$$\frac{2g\sqrt{n+1}}{|\Delta|} \ll 1. \qquad (15.31)$$

We now analyze the solutions Eqs. (15.25) of the Jaynes-Cummings-Paul model in this case.

Probability Amplitudes

For this purpose we first note that in the case of strong detuning defined by Eq. (15.31) the generalized Rabi frequency Eq. (15.26) reduces to

$$\lambda_n = \frac{\Delta}{2}\left(1 + \frac{4g^2(n+1)}{\Delta^2}\right)^{1/2} \cong \frac{\Delta}{2}\left(1 + \frac{2g^2(n+1)}{\Delta^2}\right) = \frac{\Delta}{2} + \frac{g^2(n+1)}{\Delta}$$

and we find from Eq. (15.25a)

$$\Psi_{a,n}(t) \cong e^{-i\Delta t/2}[\cos(\lambda_n t) + i\sin(\lambda_n t)]\Psi_{a,n}(0) = \exp\left[-i\left(\frac{\Delta}{2} - \lambda_n\right)t\right]\Psi_{a,n}(0),$$

or

$$\Psi_{a,n}(t) \cong \exp\left(i\frac{g^2(n+1)}{\Delta}t\right)\Psi_{a,n}(0). \qquad (15.32)$$

Here we have neglected transitions from the ground state since

$$\left|\frac{g\sqrt{n+1}}{\lambda_n}\right| = \frac{2g\sqrt{n+1}}{|\Delta|[1 + 4g^2(n+1)/\Delta^2]^{1/2}} < \frac{2g\sqrt{n+1}}{|\Delta|} \ll 1.$$

Indeed, the strong detuning prevents the atom from making a transition: It has to stay in its initial atomic state.

Similarly, we find

$$\Psi_{b,n+1}(t) \cong \exp\left[i\left(\frac{\Delta}{2} - \lambda_n\right)t\right] \Psi_{b,n+1}(0) = \exp\left(-i\frac{g^2(n+1)}{\Delta}t\right) \Psi_{b,n+1}(0). \quad (15.33)$$

We note that the two probability amplitudes $\Psi_{a,n}$ and $\Psi_{b,n+1}$ rotate differently: Whereas $\Psi_{b,n+1}$ rotates clockwise, $\Psi_{a,n}$ rotates counterclockwise.

Effective Hamiltonian

We are now in a position to simplify the formulas Eqs. (15.18) and (15.25) for the state vector of the combined atom-field system. When we substitute the approximate expressions Eqs. (15.32) and (15.33) for the probability amplitudes $\Psi_{a,n}$ and $\Psi_{b,n+1}$ into the ansatz Eq. (15.18) for the state vector we find

$$|\Psi(t)\rangle = \sum_{n=0}^{\infty}\left[\exp\left(i\frac{g^2(n+1)}{\Delta}t\right) \Psi_{a,n}(0)|a,n\rangle \right.$$
$$\left. + \exp\left(-i\frac{g^2(n+1)}{\Delta}t\right) \Psi_{b,n+1}(0)|b,n+1\rangle\right] + \Psi_{b,0}(0)|b,0\rangle.$$

We shift the summation over the Fock states in the contribution of the ground state atoms and arrive at

$$|\Psi(t)\rangle = \sum_{n=0}^{\infty}\left[\exp\left(i\frac{g^2(n+1)}{\Delta}t\right) \Psi_{a,n}(0)|a,n\rangle + \exp\left(-i\frac{g^2 n}{\Delta}t\right) \Psi_{b,n}(0)|b,n\rangle\right]. \quad (15.34)$$

We recall the relations

$$e^{i\varphi\hat{\sigma}_z \hat{a}^\dagger \hat{a}}|a,n\rangle = e^{i\varphi n}|a,n\rangle \quad \text{and} \quad e^{i\varphi\hat{\sigma}_z \hat{a}^\dagger \hat{a}}|b,n\rangle = e^{-i\varphi n}|b,n\rangle$$

for an arbitrary real-valued phase φ and

$$\frac{1}{2}(\hat{\sigma}_z + \mathbb{1})|a\rangle = |a\rangle \quad \text{and} \quad \frac{1}{2}(\hat{\sigma}_z + \mathbb{1})|b\rangle = 0,$$

which allow us to cast the above expression for the state vector into the form

$$|\Psi(t)\rangle \cong \exp\left\{\frac{i}{\hbar}\frac{\hbar g^2}{\Delta}\left[\hat{\sigma}_z \hat{a}^\dagger \hat{a} + \frac{1}{2}(\hat{\sigma}_z + \mathbb{1})\right]t\right\} \sum_{n=0}^{\infty}\left[\Psi_{a,n}(0)|a,n\rangle + \Psi_{b,n}(0)|b,n\rangle\right],$$

or

$$|\Psi(t)\rangle = \exp\left(-\frac{i}{\hbar}\hat{H}_{\text{eff}} t\right)|\Psi(0)\rangle.$$

Here we have introduced the effective Hamiltonian

$$\hat{H}_{\text{eff}} \equiv -\frac{\hbar g^2}{\Delta}\left[\hat{\sigma}_z \hat{a}^\dagger \hat{a} + \frac{1}{2}(\hat{\sigma}_z + \mathbb{1})\right]. \quad (15.35)$$

In Appendix M we use a second order perturbation expansion of the time evolution operator $\hat{\mathcal{U}}$ to rederive this result.

State Vector

In the limit of strong detuning the dynamics of the Jaynes-Cummings-Paul model follows from the effective Hamiltonian Eq. (15.35). The resulting dynamics conserves the photon statistics and the atomic populations. It only introduces phase shifts

$$\varphi_n^{(a)}(t) \equiv \frac{g^2(n+1)}{\Delta} t$$

and

$$\varphi_n^{(b)}(t) \equiv \frac{g^2 n}{\Delta} t$$

into the state vectors $|a, n\rangle$ and $|b, n\rangle$.

We recognize these features when we substitute the initial conditions, Eq. (15.21), $\Psi_{a,n}(0) = w_n c_a$ and $\Psi_{b,n}(0) = w_n c_b$ into Eq. (15.34) and find

$$|\Psi(t)\rangle = \sum_{n=0}^{\infty} w_n \left(e^{i\varphi_n^{(a)}(t)} c_a |a\rangle + e^{-i\varphi_n^{(b)}(t)} c_b |b\rangle \right) |n\rangle . \qquad (15.36)$$

Since the phase shifts $\varphi_n^{(a)}$ and $\varphi_n^{(b)}$ depend on the photon number and are different for the ground and the excited state the atomic states do not factorize from the field state: The effective Hamiltonian which does not cause any transitions in the system creates an entanglement between the atomic and the field states. We will study this entanglement in more detail in the next section and conclude by emphasizing that the effective Hamiltonian, Eq. (15.35), has been used extensively in the context of the generation of Schrödinger cats and atom optics in quantized light fields.

Problems

15.1 Semiclassical Jaynes-Cummings-Paul Model

Consider an atom (nucleus at $\vec{x} = 0$) interacting with a classical electromagnetic field $\vec{E}(\vec{x}, t) = E_0 \vec{e}_z \cos\left(\vec{k}\vec{x} - \nu t\right)$.

(a) Show that under certain approximations (dipole approximation, rotating wave approximation, no permanent dipole moment, only transitions are important which correspond approximately to the frequency of the electromagnetic field) the atom can be described by the state

$$|\psi\rangle = c_a e^{-iE_a t/\hbar}|a\rangle + c_b e^{-iE_b t/\hbar}|b\rangle.$$

Here $|a\rangle$ and $|b\rangle$ are the states whose transition frequency is approximately the same as the frequency of the electromagnetic field, and c_a and c_b obey the equations of motion

$$\dot{c}_a = ig e^{i\Delta t} c_b$$
$$\dot{c}_b = ig e^{-i\Delta t} c_a ,$$

where
$$g \equiv \frac{E_0 e}{2\hbar} \langle a|z|b \rangle$$
and
$$\Delta \equiv \omega_{ab} - \nu \equiv \frac{E_a - E_b}{\hbar} - \nu.$$

In the framework of semiclassical quantum mechanics, the spontaneous decay from $|a\rangle$ or $|b\rangle$ into the ground state (or other states of the atom) has to be included in the form of phenomenological damping terms:

$$\dot{c}_a = -\frac{\gamma_a}{2} c_a + ige^{i\Delta t} c_b$$
$$\dot{c}_b = -\frac{\gamma_b}{2} c_b + ige^{-i\Delta t} c_a. \quad (15.37)$$

Quantum mechanically the form of Eq. (15.37) can be justified in the Weisskopf-Wigner theory of spontaneous emission discussed in Sec. 18.5.6.

(b) What is the probability to find the atom at time t in the state $|b\rangle$ if at time $t = 0$ it was in the state $|a\rangle$? Calculate the result exactly as well as in first order time dependent perturbation theory. Take into account the damping terms in Eq. (15.37).
Hint: First perform the transformation
$$c_a = e^{-\gamma_a t/2} \tilde{c}_a$$
$$c_b = e^{-\gamma_b t/2} \tilde{c}_b.$$

Result: The exact result is
$$W_b(t) = g^2 \left| \frac{\sin(\mu t/2)}{\mu/2} \right|^2 e^{-(\gamma_a + \gamma_b)t/2},$$
where μ is given by
$$\mu^2 \equiv \left[\Delta - i\frac{\gamma_a - \gamma_b}{2} \right]^2 + 4g^2.$$

First order time dependent perturbation theory yields
$$W_b^{(1)}(t) = \frac{g^2}{\Delta^2 + \left(\frac{\gamma_a - \gamma_b}{2}\right)^2} \left[e^{-\gamma_a t} + e^{-\gamma_b t} - 2 e^{-(\gamma_a + \gamma_b)t/2} \cos(\Delta t) \right].$$

(c) Determine from $W_b(t)$ the probability that between $t = 0$ and $t = \infty$ a photon is spontaneously emitted from state $|b\rangle$, and discuss your result.

(d) Determine from $W_b^{(1)}(t)$ the probability that between $t = \tau$ and $t = \infty$ a photon is spontaneously emitted from state $|b\rangle$, and discuss your result. Of particular interest are large values of τ.

15.2 Raman-Coupled Jaynes-Cummings-Paul Model

In the Raman-coupled Jaynes-Cummings-Paul model the interaction between a two-level atom and a quantized mode of the electromagnetic field is described by the Hamiltonian

$$\hat{H} \equiv \hbar\omega\,\hat{a}^\dagger\hat{a} + \hbar\lambda\,\hat{a}^\dagger\hat{a}\,\hat{\sigma}_x,$$

where

$$\hat{\sigma}_x \equiv \begin{pmatrix} 0 & 1 \\ 1 & 0 \end{pmatrix}.$$

(a) Show that the time evolution of an arbitrary state $|\Psi_0\rangle$ is given by

$$|\Psi(t)\rangle = \exp(-i\omega t\,\hat{a}^\dagger\hat{a})\left[\cos(\lambda t\,\hat{a}^\dagger\hat{a}) - i\hat{\sigma}_x \sin(\lambda t\,\hat{a}^\dagger\hat{a})\right]|\Psi_0\rangle.$$

(b) Evaluate the state $|\Psi(t)\rangle$ provided at time $t = 0$ the field is in the Fock state $|n\rangle$ and the atom in the state $|a\rangle$, or the field is in the coherent state $|\alpha\rangle$ and the atom in the state $|a\rangle$.

(c) What is the probability to find the atom at time $t > 0$ in the state $|a\rangle$ if at time $t = 0$ the field was in the coherent state $|\alpha\rangle$ and the atom in the state $|a\rangle$?

15.3 Jaynes-Cummings-Paul Model: Time Dependent Coupling Strength

The interaction of a two-level system with a resonant mode of the electromagnetic field is described in the rotating wave approximation by the Hamiltonian

$$\hat{H}_{\text{int}} \equiv \hbar g\left(\hat{\sigma}\hat{a}^\dagger + \hat{\sigma}^\dagger\hat{a}\right).$$

If the atom moves in the resonator g becomes time dependent (why?).

(a) Solve for a time dependent coupling strength $g = g(t)$ the corresponding Schrödinger equation for the two initial conditions $|\Psi(0)\rangle \equiv |n, a\rangle$, or $|\Psi(0)\rangle \equiv |n, b\rangle$.

(b) Why is it impossible to find an (analytical) solution in the case of a non-vanishing detuning?

15.4 Time Evolution of Mixed States

A two-level atom interacts with an electromagnetic field in a resonator according to the Jaynes-Cummings-Paul model. At time $t = 0$ the field is in a thermal state

$$\hat{\rho}_F(0) \equiv \frac{1}{\bar{n}+1}\sum_{n=0}^{\infty}\left(\frac{\bar{n}}{\bar{n}+1}\right)^n |n\rangle\langle n|$$

and the atom is in an incoherent superposition

$$\hat{\rho}_A(0) \equiv \cos^2\theta\,|a\rangle\langle a| + \sin^2\theta\,|b\rangle\langle b|.$$

of excited state $|a\rangle$ and ground state $|b\rangle$.

(a) Calculate the density operator of the total system after the time t for the case that atom and field are resonant.

(b) Find the density operator for the atom and for the field after the time t.

(c) Calculate the probability to find the atom at time t in the excited state.

(d) What is the density operator for the field after the atom has been found in the excited state?

(e) What is the probability to find n photons in the field after the time t?

(f) What is the density operator for the atom after the detection of n photons in the field?

(g) What happens for $\cot^2\theta = \bar{n}/(\bar{n}+1)$? How can this result be interpreted?

References

Jaynes-Cummings-Paul Model

The original papers on the Jaynes-Cummings-Paul model are
E.T. Jaynes and **F.W. Cummings**, *Comparison of Quantum and Semiclassical Radiation Theories with Applications to the Beam Maser*, Proc. IEEE **51**, 89–109 (1963)
H. Paul, *Induzierte Emission bei starker Einstrahlung*, Ann. Phys. (Leipzig) **11**, 411–412 (1963)

For a topical review of the Jaynes-Cummings-Paul model and its many applications in quantum optics see
B.W. Shore, *The Theory of Coherent Atomic Excitations*, Wiley, New York, 1990
B.W. Shore and **P.L. Knight**, *Topical Review: The Jaynes-Cummings model*, J. Mod. Opt. **40**, 1195–1238 (1993)

16 State Preparation and Entanglement

In the preceeding chapter we have solved the Schrödinger equation for the state vector of the Jaynes-Cummings-Paul model describing the internal states of a single atom interacting with a single mode of the cavity field. In the present chapter we focus on the feature of entanglement between these two degrees of freedom.

The word entanglement expresses the fact that due to their interaction the two quantum systems – atom and field – cannot be separated anymore. Even after the interaction has taken place the complete system can only be described by a single state vector and not by a product state. This on first sight seemingly innocent statement is the central lesson of quantum mechanics.

The importance of entanglement was recognized by Erwin Schrödinger in 1936 who called it *Verschränktheit zweier Quantensysteme*. We now show that the phenomenon of entanglement allows us to obtain information about the radiation field by making measurements on the internal degrees of freedom. Moreover, we can use entanglement to create arbitrary field states in a resonator. We call this scheme *Quantum State Engineering*. Entanglement and quantum state engineering are thus the main topics of the present chapter.

16.1 Measurements on Entangled Systems

The Jaynes-Cummings-Paul model is the most elementary model which demonstrates the entanglement between the field and the atomic degrees of freedom. We can observe this entanglement by performing various measurements, either on the atom or the field, or a joint measurement on both quantum systems. Such joint measurements have been at the center of interest of quantum optics over the last years. In the present section we first discuss the mathematical formalism to extract from the state vector the relevant probabilities and then briefly describe the experimental procedures necessary to make the appropriate measurements on the atom, or the field, or on both.

16.1.1 How to Get Probabilities

The state vector contains the complete information about the quantum system. In particular, it allows us to obtain the probabilities for the outcome of various measurements. We now discuss how to derive these probabilities from the state vector.

Joint Atom-Field Measurement

Our quantum system consists of an atom and a single mode of the radiation field. We can therefore perform joint measurements on the atom and the field. The experiment proceeds as follows: We prepare the initial field state $|\psi_{\text{field}}\rangle$, and the atomic state $|\psi_{\text{atom}}\rangle$, let the two systems interact for a given time and then test if the internal degrees of freedom are in the atomic reference state $|\tilde{\psi}_{\text{atom}}\rangle$ and the field state is in the reference state $|\tilde{\psi}_{\text{field}}\rangle$. If the two subsystems are in their respective reference states we record this event; when they are not we discard this event. We then repeat the procedure many times.

We recall that quantum mechanics cannot predict the outcome of each individual measurement act, but it can give the probability for this event. Indeed, the probability W to measure the atom in the state $|\tilde{\psi}_{\text{atom}}\rangle$ and the field in the state $|\tilde{\psi}_{\text{field}}\rangle$ given the complete system is in $|\Psi(t)\rangle$ reads

$$W(t;|\tilde{\psi}_{\text{atom}}\rangle,|\tilde{\psi}_{\text{field}}\rangle) \equiv \begin{pmatrix} \text{Probability to find at the} \\ \text{time } t \text{ the atom in the state} \\ |\tilde{\psi}_{\text{atom}}\rangle \text{ and the field in } |\tilde{\psi}_{\text{field}}\rangle \end{pmatrix} = |\langle\tilde{\psi}_{\text{field}}|\langle\tilde{\psi}_{\text{atom}}|\Psi(t)\rangle|^2. \tag{16.1}$$

Measurements of this type, that is joint measurements of quantum systems bring out the entanglement of the subsystems. To demonstrate this we now use the representation

$$|\Psi(t)\rangle = \sum_{n=0}^{\infty} \Psi_{a,n}(t)\,|a,n\rangle + \Psi_{b,n}(t)\,|b,n\rangle = \sum_{n=0}^{\infty}\sum_{j=a,b} \Psi_{j,n}(t)\,|j,n\rangle \tag{16.2}$$

of the state vector, Eq. (15.18). Here we have included the term $\Psi_{b,0}(t)\,|b,0\rangle$ in the summation.

When we substitute this expression into the definition Eq. (16.1) of the joint probability we find

$$W(t;|\tilde{\psi}_{\text{atom}}\rangle,|\tilde{\psi}_{\text{field}}\rangle) = \left|\sum_{n=0}^{\infty}\sum_{j=a,b} \Psi_{j,n}(t)\,\langle\tilde{\psi}_{\text{atom}}|j\rangle\,\langle\tilde{\psi}_{\text{field}}|n\rangle\right|^2. \tag{16.3}$$

We note, that in this joint measurement we obtain a maximum of quantum interference. Indeed, we first sum over all Fock states and over all atomic states before we calculate the absolute value square. We sum the product of the probability amplitudes $\Psi_{j,n}$ and the amplitudes of the reference states of atom and field over all photon states and internal states. This double sum creates a lot of interference terms.

Measurement of the Internal Atomic State

The most elementary measurement on a quantum system consisting of an atom and a single mode field is the measurement of the internal state of the atom independent of the field state. In this case the probability W to find the atomic state $|\tilde{\psi}_{\text{atom}}\rangle$ independent of the field state follows from the state vector $|\Psi(t)\rangle$ as

$$W(t;|\tilde{\psi}_{\text{atom}}\rangle) \equiv \begin{pmatrix} \text{Probability to find at time } t \\ \text{the atom in the state } |\tilde{\psi}_{\text{atom}}\rangle \\ \text{independent of field state} \end{pmatrix} = \sum_{n=0}^{\infty} |\langle n|\langle \tilde{\psi}_{\text{atom}}|\Psi(t)\rangle|^2. \quad (16.4)$$

Since we have not measured the field state after the interaction we have to trace over the field. For this trace we have used Fock states.

With the help of the ansatz Eq. (16.2) for the state vector $|\Psi(t)\rangle$ of the system we find

$$W(t;|\tilde{\psi}_{\text{atom}}\rangle) = \sum_{n=0}^{\infty} \left| \sum_{j=a,b} \Psi_{j,n}(t) \langle \tilde{\psi}_{\text{atom}}|j\rangle \right|^2. \quad (16.5)$$

When we compare this expression to the corresponding one for the joint measurement Eq. (16.3) we note that we have lost interference. Whereas in the joint measurement we sum the probability amplitudes of both subsystems before we take the square, we now, in the single system measurement, only sum over one subsystem and then take the square. For a reference state in a coherent superposition of the two internal states the probability $W(t;|\tilde{\psi}_{\text{atom}}\rangle)$ involves the sum of probability amplitudes $\Psi_{a,n}(t)$ and $\Psi_{b,n}(t)$. The weight factor of each of these contributions is the amplitude of finding the atomic states $|a\rangle$ or $|b\rangle$ in the reference state $|\tilde{\psi}_{\text{atom}}\rangle$.

We emphasize that the probability $W(t;|\tilde{\psi}_{\text{atom}}\rangle)$, Eq. (16.5), still contains a sum over the field states. However, this sum now only involves probabilities. The concept of the joint measurement allows us to interpret these probabilities. Indeed, the probability to find the atom in the reference state $|\tilde{\psi}_{\text{atom}}\rangle$ independent of the field state is the sum

$$W(t;|\tilde{\psi}_{\text{atom}}\rangle) = \sum_{n=0}^{\infty} W(t;\left|\tilde{\psi}_{\text{atom}}\right\rangle, |n\rangle)$$

of joint probabilities

$$W(t;\left|\tilde{\psi}_{\text{atom}}\right\rangle, |n\rangle) \equiv \left| \sum_{j=a,b} \Psi_{j,n}(t) \langle \tilde{\psi}_{\text{atom}}|j\rangle \right|^2$$

to find the atom in the reference state $\left|\tilde{\psi}_{\text{atom}}\right\rangle$ and the field in the nth Fock state.

We conclude this discussion by presenting the expressions for the case when the reference state is one of the two internal states. In this case there is no interference left and the probability to find the atom in the excited or the ground state reads

$$W(t;|a\rangle) = \sum_{n=0}^{\infty} |\Psi_{a,n}(t)|^2,$$

or

$$W(t;|b\rangle) = \sum_{n=0}^{\infty} |\Psi_{b,n}(t)|^2.$$

Indeed, in order to arrive at the probabilities $W(t; |a\rangle)$ and $W(t; |b\rangle)$ we have to first evaluate the absolute square of the individual probability amplitudes $\Psi_{a,n}$ and $\Psi_{b,n}$ to be in the atomic state $|a\rangle$ or $|b\rangle$ and in an individual Fock state. Then we have to sum these probabilities over all Fock states.

It is interesting to note that these expressions have been measured experimentally for the case of the Jaynes-Cummings-Paul model and the one-atom maser. We discuss these experiments in more detail in Sec. 16.2.

Measurement of the Field State

We now turn to the discussion of a measurement of the field alone, that is, we do not measure the atomic state. We therefore ask for the probability $W(t; |\tilde{\psi}_{\text{field}}\rangle)$ to find the field state $|\tilde{\psi}_{\text{field}}\rangle$ independent of the atomic state. This probability reads

$$W(t; |\tilde{\psi}_{\text{field}}\rangle) \equiv \begin{pmatrix} \text{Probability to find at} \\ \text{time } t \text{ the field in the} \\ \text{state } |\tilde{\psi}_{\text{field}}\rangle \text{ independent} \\ \text{of the atomic state} \end{pmatrix} = \sum_{j=a,b} |\langle j|\langle \tilde{\psi}_{\text{field}}|\Psi(t)\rangle|^2.$$

Here we have performed the trace over the two unobserved internal states.

The ansatz Eq. (16.2) for the state vector $|\Psi(t)\rangle$ yields

$$W(t; |\tilde{\psi}_{\text{field}}\rangle) = \sum_{j=a,b} \left| \sum_{n=0}^{\infty} \Psi_{j,n}(t) \langle \tilde{\psi}_{\text{field}}|n\rangle \right|^2. \tag{16.6}$$

We emphasize that this expression is completely analogous to the one for the measurement of the atomic states. The role of the summation indices j and n are interchanged. We first sum the product of the probability amplitude $\Psi_{j,n}$ and the probability amplitude $\langle n|\tilde{\psi}_{\text{field}}\rangle$ of finding n photons in the reference state over all Fock states. Since we do not measure the internal state we have to take the trace over the atomic state, that is, sum over the probabilities in the two states.

With the help of the concept of joint measurements we can interpret the probability $W(t; |\tilde{\psi}_{\text{field}}\rangle)$ to find the field state $|\tilde{\psi}_{\text{field}}\rangle$ independent of the atomic state as the sum of two joint probabilities. Indeed, it is the sum

$$W(t; |\tilde{\psi}_{\text{field}}\rangle) = W(t; |a\rangle, |\tilde{\psi}_{\text{field}}\rangle) + W(t; |b\rangle, |\tilde{\psi}_{\text{field}}\rangle)$$

of the joint probability

$$W(t; |a\rangle, |\tilde{\psi}_{\text{field}}\rangle) \equiv |\langle \tilde{\psi}_{\text{field}}|\langle a|\Psi(t)\rangle|^2 = \left| \sum_{n=0}^{\infty} \Psi_{a,n}(t) \langle \tilde{\psi}_{\text{field}}|n\rangle \right|^2$$

to be in the atomic state $|a\rangle$ and find the field state $|\tilde{\psi}_{\text{field}}\rangle$, and the joint probability

$$W(t; |b\rangle, |\tilde{\psi}_{\text{field}}\rangle) \equiv |\langle \tilde{\psi}_{\text{field}}|\langle b|\Psi(t)\rangle|^2 = \left| \sum_{n=0}^{\infty} \Psi_{b,n}(t) \langle \tilde{\psi}_{\text{field}}|n\rangle \right|^2$$

to be in the atomic state $|b\rangle$ and find the field state $|\tilde{\psi}_{\text{field}}\rangle$.

16.1 Measurements on Entangled Systems

Summary

We conclude this discussion of measurements on entangled quantum systems by summarizing our main results:

In the case of an atomic measurement of the state $|a\rangle$ *together* with a field measurement $|\tilde{\psi}_{\text{field}}\rangle$ we add the probability amplitudes $\Psi_{a,n}$ of being in the state $|a\rangle$ and n photons in the resonator multiplied by the probability amplitude $\langle n|\tilde{\psi}_{\text{field}}\rangle^*$ of finding n photons in the reference state. In contrast, in the case of a measurement on $|a\rangle$ and no field measurement we have to add the probabilities $|\Psi_{a,n}|^2$ of finding the atomic state $|a\rangle$ and n photons. Therefore, in the discussion of measurement using a reference state we have quantum mechanical interference of complex probability amplitudes, whereas in the second case no such interference terms exist.

16.1.2 State of the Subsystem after a Measurement

In the preceding sections we have derived expressions for the probabilities to find a specific atomic state and/or a specific field state. In the present section we address a completely different question: What is the quantum state of one subsystem after we have made a measurement on the other subsystem? What is, for example, the field state after we have made a measurement on the atom? What is the state of the atom after we have made a measurement on the field?

We start the discussion with the case of the field state prepared by finding after a measurement the atomic state in the reference state $|\tilde{\psi}_{\text{atom}}\rangle$. The resulting field state is then

$$|\psi_{\text{field}}\rangle \equiv \mathcal{N} \langle \tilde{\psi}_{\text{atom}}|\Psi\rangle.$$

We emphasize that this object is indeed a field state: The quantum system described by the state vector $|\Psi\rangle$ consists of the two subsystems atomic levels and field states. We project onto the atomic subsystem and the resulting state is therefore a field state.

However, due to this projection the new field state is not normalized. To remove this problem we have introduced the normalization constant $\mathcal{N}$ which follows from the condition

$$1 = \langle \psi_{\text{field}}|\psi_{\text{field}}\rangle = |\mathcal{N}|^2 \sum_{n=0}^{\infty} \left|\langle n|\langle \tilde{\psi}_{\text{atom}}|\Psi\rangle\right|^2.$$

From Eq. (16.4) we recall that the sum is the probability $W(|\tilde{\psi}_{\text{atom}}\rangle)$ to find the atomic reference state $|\tilde{\psi}_{\text{atom}}\rangle$ in the state $|\Psi\rangle$. Hence the normalized field state reads

$$|\psi_{\text{field}}\rangle = \frac{1}{\sqrt{W(|\tilde{\psi}_{\text{atom}}\rangle)}} \langle \tilde{\psi}_{\text{atom}}|\Psi\rangle. \qquad (16.7)$$

This expression has a very simple physical meaning: The quantum state $|\psi_{\text{field}}\rangle$ only emerges when we find the atom in the reference state $|\tilde{\psi}_{\text{atom}}\rangle$. This, however, only happens with the probability $W(|\tilde{\psi}_{\text{atom}}\rangle)$. Therefore, we have to normalize the field state with respect to this probability.

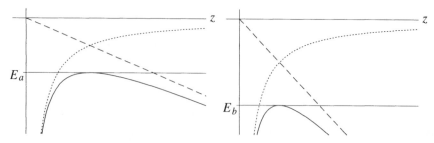

Fig. 16.1: Stark effect and ionization. The Stark potential V_S associated with a constant electric field (dashed curve) deforms the Coulomb potential (dotted curve) into a potential with a maximum (solid line). An electron of energy E_a, originally bound in the Coulomb potential, is now free (left). For a larger field strength the Stark potential is steeper and a lower lying state of energy E_b can be ionized (right).

We now briefly discuss the case of the atomic state given we have found the field in a reference state $|\widetilde{\psi}_{\text{field}}\rangle$. Following the above arguments we arrive at

$$|\psi_{\text{atom}}\rangle = \frac{1}{\sqrt{W(|\widetilde{\psi}_{\text{field}}\rangle)}} \langle \widetilde{\psi}_{\text{field}} | \Psi \rangle.$$

The normalization is again determined by the probability $W(|\widetilde{\psi}_{\text{field}}\rangle)$ to find the reference state $|\widetilde{\psi}_{\text{field}}\rangle$ in the state $|\Psi\rangle$ of the complete system.

In Sec. 16.3 we use the concept of joint measurements to prepare various states of the radiation field such as Schrödinger cat states and phase states.

16.1.3 Experimental Setup

The entanglement between atomic and field variables manifests itself in the probabilities to measure these quantities. We now describe methods that in principle, and in practice, allow us to perform such a measurement on the combined system of atom and field.

Measurement of Internal States

We start with the simplest case, namely the measurement of an atomic state $|a\rangle$ or $|b\rangle$. A popular technique is the ionization method. Here we apply an electric field $\vec{E} = E_0 \vec{e}_z$ in the z-direction to the atom. The resulting potential energy

$$V_S \equiv -e\vec{E} \cdot \vec{r} = -eE_0 z$$

shown in Fig. 16.1 by the dashed curve deforms the Coulomb potential

$$V_C \equiv -\frac{e^2}{4\pi\epsilon_0} \frac{1}{\sqrt{x^2 + y^2 + z^2}}$$

displayed by a dotted curve into the effective atomic potential shown by the solid curve. Consequently, an energy eigenstate of the Coulomb problem of energy E_a

16.1 Measurements on Entangled Systems 441

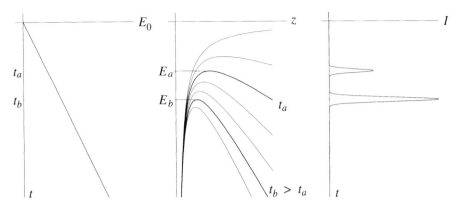

Fig. 16.2: Measurement of internal states using the method of field ionization. A linear electric field ramp ionizes different atomic levels at different times. An atom in this ramp feels a linearly growing electric field strength $E_0(t)$ (left). This field creates a potential that is linear in the position variable z. Its steepness increases linearly with time (left). The electron in the atom feels the combination of the Coulomb potential and the linear potential from the electric field (middle). At time t_a the field strength is such that the maximum of the combined Coulomb-Stark potential is equal to the energy E_a of the excited state $|a\rangle$. At this moment the excited state ionizes and gives rise to a peak (right) in the ionization current I. At a later time t_b the Coulomb-Stark potential is even steeper which leads to the ionization of the state $|b\rangle$ and a second peak in the ionization current (right). By comparing the areas underneath of the two peaks we find the populations of the individual levels.

which lies just barely above the maximum of the combined Coulomb-Stark potential is not bound anymore and ionizes.

We note, that a larger electric field strength creates a steeper Stark potential. Thus we can reach lower lying energy levels of the Coulomb throat. In Fig. 16.1 we have chosen a field strength such that we have ionized the energy level E_b. Hence ionization field strength translates itself into energy of atomic level.

Therefore, the most effective way of performing such a measurement is to increase the electric field strength as a function of time and observe the ionization current $I(t)$. Maxima in this current, their height and their location relative to the field strength provide information about the atomic levels, their population and their eigenenergies, as indicated in Fig. 16.2.

This technique has been extensively used in the context of Rydberg atoms since these highly excited atoms of principal numbers $n \cong 60$ are easy to ionize and are extremely sensitive to electric fields. In particular, in the context of the one-atom maser discussed in Sec. 18.4 the ionization method is used to measure the population of different atomic energy levels.

Measurement of Atomic Dipole

This version of the field ionization technique only allows us to measure the populations in the individual energy levels. For the case of a two-level atom this measurement translates into projections using the atomic states $|a\rangle$ or $|b\rangle$. However, it does not constitute a reference superposition state $|\tilde{\psi}_{\text{atom}}\rangle \equiv \tilde{c}_a |a\rangle + \tilde{c}_b |b\rangle$ necessary to make a measurement on the atomic state $|\psi_{\text{atom}}\rangle \equiv c_a |a\rangle + c_b |b\rangle$ corresponding to the projection

$$\langle \tilde{\psi}_{\text{atom}} | \psi_{\text{atom}} \rangle = \tilde{c}_a^* c_a + \tilde{c}_b^* c_b.$$

How can we achieve such a measurement?

To answer this question we recall that a classical, resonant electromagnetic field of given amplitude $\mathcal{E}_0$, and interaction time t transforms an atomic state

$$|\psi_{\text{atom}}\rangle \equiv c_a |a\rangle + c_b |b\rangle$$

into

$$|\psi'_{\text{atom}}\rangle = c'_a |a\rangle + c'_b |b\rangle$$

where the new probability amplitudes

$$c'_a = c'_a(c_a, c_b) = \cos(gt)\, c_a - i \sin(gt)\, c_b$$

and

$$c'_b = c'_b(c_a, c_b) = \cos(gt)\, c_b - i \sin(gt)\, c_a$$

are functions of the initial probability amplitudes c_a and c_b. Here $g \equiv \wp \mathcal{E}_0 / \hbar$ denotes the classical Rabi frequency.

Throughout this book we have focused on the case of the interaction of an atom with a quantized rather than a classical light field. We note however, that the latter is equivalent to the interaction with a quantized light field, summarized by the Rabi equations Eqs. (15.19) when the field is in a photon number state. The classical electric field strength $\mathcal{E}_0$ is then determined by the vacuum electric field strength times the square root of the number of photons.

When we now perform on the rotated field state $|\psi'_{\text{atom}}\rangle$ shown in Fig. 16.3 a measurement of the atomic state $|a\rangle$ we find

$$\langle a | \psi'_{\text{atom}} \rangle = \cos(gt)\, c_a - i \sin(gt)\, c_b$$

which we have to compare with the envisioned projection

$$\langle \tilde{\psi}_{\text{atom}} | \psi_{\text{atom}} \rangle = \tilde{c}_a^* c_a + \tilde{c}_b^* c_b$$

corresponding to a measurement using the reference state $|\tilde{\psi}_{\text{atom}}\rangle$.

Hence by appropriately choosing the parameters of the classical electromagnetic field we can achieve a situation in which $\tilde{c}_a^* = \cos(gt)$ and $\tilde{c}_b^* = -i \sin(gt)$. We emphasize that in order to satisfy this relation we, in general, have to allow for a nonresonant classical light wave.

In summary, we can realize a measurement of the atomic states using a reference state $|\tilde{\psi}_{\text{atom}}\rangle$ by sending the atom through a classical field followed by an atomic measurement using the ionization method.

16.1 Measurements on Entangled Systems

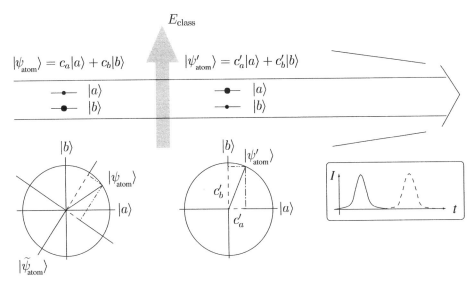

Fig. 16.3: Measurement of the atomic reference state $|\widetilde{\psi}_{\text{atom}}\rangle$ using a classical field and a measurement of the population in one atomic level. The upper part of the figure shows in a schematic way the arrangement of the classical field and the detector for the atom. The lower part of the figure depicts the state of the atom as a vector in the two-dimensional Hilbert space spanned by the basis vectors $|a\rangle$ and $|b\rangle$. (For the purpose of this figure we neglect the fact that the probability amplitudes are in general complex.) We want to project the initial state vector $|\psi_{\text{atom}}\rangle$ onto the vector $|\widetilde{\psi}_{\text{atom}}\rangle$. In the experiment we can only measure whether the atom is in $|a\rangle$ or in $|b\rangle$ which corresponds to a projection onto the axes of the vector space. The interaction with a classical field provides a rotation in the vector space in such a way that the state $|\widetilde{\psi}_{\text{atom}}\rangle$ is transformed into $|b\rangle$. A measurement of the atom in state $|b\rangle$ is now equivalent with an effective projection onto the state $|\widetilde{\psi}_{\text{atom}}\rangle$.

Field Measurements

So far we have considered measurements of the atomic variables. We conclude this section by briefly discussing the measurement of the field in a reference state $|\widetilde{\psi}_{\text{field}}\rangle$.

The most prominent reference state is a photon number state $|n\rangle$. Here we measure the number of photons in the cavity in a specific experiment. To be precise, we really do not count photons but photo electrons in a detector. At this point we do not want to go into the details of the theory of photo-detection. For this topic we refer to the Les Houches lectures by R. Glauber or the textbook by L. Mandel and E. Wolf. We only state that counting photons is equivalent to projecting the reference state $|\widetilde{\psi}_{\text{field}}\rangle \equiv |n\rangle$ on the state vector.

We now turn to the experimental realization of homodyne states as reference states of the field. In the discussion of Sec. 13.2 we have found that a homodyne detector shown in Fig. 16.4 measures the electromagnetic field eigenstate $|\mathcal{E}_\theta\rangle$. This procedure corresponds to a projection using the state $|\mathcal{E}_\theta\rangle$. In this case our reference state is $|\widetilde{\psi}_{\text{field}}\rangle \equiv |\mathcal{E}_\theta\rangle$.

We conclude by emphasizing again that the probabilities $W(t; |\widetilde{\psi}_{\text{atom}}\rangle, |\widetilde{\psi}_{\text{field}}\rangle)$, $W(t; |\widetilde{\psi}_{\text{field}}\rangle)$ and $W(t; |\widetilde{\psi}_{\text{atom}}\rangle)$ are probabilities for the outcomes of measurements

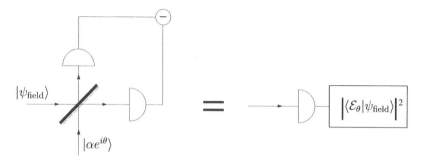

Fig. 16.4: A balanced homodyne detector with a coherent state $|\alpha\rangle \equiv ||\alpha|e^{i\theta}\rangle$ of large amplitude $|\alpha| \gg 1$ is equivalent to a measurement of the electromagnetic field distribution $W(\mathcal{E}_\theta) = |\langle \mathcal{E}_\theta | \psi_{\text{field}} \rangle|^2$. Hence it serves as an example of an electromagnetic field eigenstate $|\mathcal{E}_\theta\rangle$ as a reference state for the measurement of the electric field state $|\psi_{\text{field}}\rangle$.

on identically prepared systems. In particular, we first prepare the initial atomic and field states, let the atom interact with the field for a given time t, and then make the measurement of either atom and field, or field, or atom alone. We record the result, repeat the experiment many times and build up a histogram.

16.2 Collapse, Revivals and Fractional Revivals

In the preceding section we have shown how to obtain probabilities for measuring a specific atomic state and/or a specific field state in the Jaynes-Cummings-Paul model. In the present section we concentrate on the time evolution of atomic variables and briefly discuss the experiments in this field.

16.2.1 Inversion as Tool for Measuring Internal Dynamics

The atomic inversion $\mathcal{I}$ is a quantity of central interest since it is easily accessible in experiments. We first derive an explicit expression for the inversion in the Jaynes-Cummings-Paul model and then discuss its time evolution.

Inversion

The inversion is defined as

$$\mathcal{I}(t) \equiv W(t; |a\rangle) - W(t; |b\rangle),$$

and therefore involves the probability

$$W(t; |a\rangle) \equiv \sum_{n=0}^{\infty} |\Psi_{a,n}(t)|^2$$

to be after a measurement at time t in the excited state $|a\rangle$, and the probability

$$W(t; |b\rangle) \equiv \sum_{n=0}^{\infty} |\Psi_{b,n}(t)|^2$$

16.2 Collapse, Revivals and Fractional Revivals

to be in the ground state $|b\rangle$.

The atomic inversion measures the difference in the populations of the two levels. It therefore plays an important role in laser theory. When $\mathcal{I}$ is positive the probability to find an atom in the excited state is higher than in the ground state. For an ensemble of atoms this fact implies that there are more atoms in the excited than in the ground state. This is usually a condition for the onset of laser oscillations. However, recently also lasers without inversion have been suggested. We will return to the concept of inversion in Chapter 18.

From the solution Eq. (15.29) of the resonant Jaynes-Cummings-Paul model we find the explicit expressions

$$\Psi_{a,n}(t) = -iS_n w_{n+1} = -i\sin(\sqrt{n+1}gt)w_{n+1}$$

and

$$\Psi_{b,n+1}(t) = C_n w_{n+1} = \cos(\sqrt{n+1}gt)w_{n+1}.$$

Here we have assumed that initially the atom was in its ground state.

We can shift the summation in the probability

$$W(t;|a\rangle) = \sum_{n=0}^{\infty} \sin^2\left(\sqrt{n+1}gt\right)|w_{n+1}|^2 = \sum_{n=0}^{\infty} \sin^2\left(\sqrt{n}gt\right)W_n,$$

and similarly in

$$W(t;|b\rangle) = \sum_{n=0}^{\infty} \cos^2\left(\sqrt{n}gt\right)W_n,$$

where $W_n \equiv |w_n|^2$ denotes the photon statistics of the cavity field before the interaction with the atom.

With the help of the trigonometric relation

$$\cos^2\alpha - \sin^2\alpha = \cos(2\alpha)$$

the inversion reads

$$\mathcal{I}(t) = -\sum_{n=0}^{\infty} W_n \cos(\sqrt{n}2gt). \tag{16.8}$$

Time Evolution of Inversion

The summation over n is not trivial since it involves $\sqrt{n}$. This is reminiscent of our general problem of wave packet dynamics discussed in Chapter 9. We can therefore immediately apply all the results derived in that chapter and understand the behavior of the sum.

In Fig. 16.5 we display the time evolution of the inversion for the case of a cavity field whose photon statistics W_n is centered around a rather large average photon number $\bar{n} \gg 1$. This figure has been obtained by numerically evaluating the sum Eq. (16.8) defining the inversion in the Jaynes-Cummings-Paul model. We note that indeed the inversion displays the same behavior as the wave packets discussed in Chapter 9: After an oscillatory regime with decaying amplitude the inversion vanishes over an extended period but revives periodically. This periodic revival of

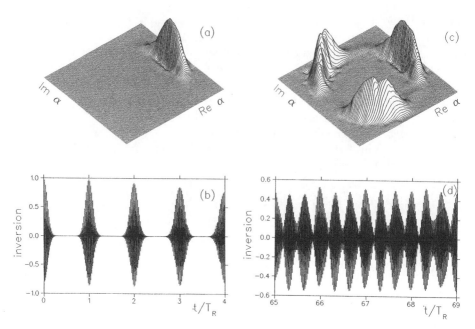

Fig. 16.5: Dynamics of the Jaynes-Cummings-Paul model represented by the Q-function of the field (top) and the atomic population inversion (bottom) for two time regions: At the initial stage (left column), the Q-function of the field rotates in phase space giving rise to a periodic inversion. This phenomenon corresponds to the classical, periodic motion of a wave packet of a mechanical oscillator. In the language of the Jaynes-Cummings-Paul model this periodic behavior is called revival. We note that in the domain of fractional revivals (right column) near $t \cong (1/3)T_2/2$ the Q-function of the field displays more peaks and the periodicity of the inversion has changed. Taken from I. Sh. Averbukh, Phys. Rev. A **46**, R2205 (1992).

the inversion is referred to in the literature by the name of *Jaynes-Cummings revival*. These revivals get broader and their amplitudes decrease. Eventually the revivals overlap and new structures develop. These are the well-known fractional revivals.

We emphasize, however, that the Jaynes-Cummings revivals are analogous to the classical oscillation of an atomic or molecular wave packet between the turning points as discussed in Chapter 9. They occur in the early stage of the evolution at integer multiples of a time scale $T_1 \equiv \bar{n}^{1/2} 2\pi/g$ where $\bar{n}$ is the average number of photons in the field state.

In order to understand this scenario we make use of the fact that the photon statistics is localized around $\bar{n} \gg 1$. This feature allows us to expand

$$n^{1/2} = \bar{n}^{1/2} \left[1 + \frac{n - \bar{n}}{\bar{n}} \right]^{1/2} \cong \bar{n}^{1/2} \left[1 + \frac{1}{2} \frac{n - \bar{n}}{\bar{n}} - \frac{1}{8} \left(\frac{n - \bar{n}}{\bar{n}} \right)^2 \right]$$

into a Taylor series about $\bar{n}$ and the inversion reads

$$\mathcal{I}(t) = -\frac{1}{2} \mathcal{S}(t) - c.c. \quad .$$

Here we have introduced the time dependent signal

$$\mathcal{S}(t) \equiv \exp(i\sqrt{\bar{n}}2gt) \sum_{m=-\bar{n}}^{\infty} W_{m+\bar{n}} \exp\left\{2\pi i \left[\frac{m}{T_1}t - \frac{m^2}{T_2}t + \cdots\right]\right\}$$

and the time scales

$$T_1 \equiv \bar{n}^{1/2} \frac{2\pi}{g}$$

and

$$T_2 \equiv 4\bar{n}^{3/2} \frac{2\pi}{g} = 4\bar{n}T_1.$$

Since $\bar{n} \gg 1$ the time scales T_1 and T_2 satisfy the hierarchy $T_1 \ll T_2$, that is they separate. Moreover, this approximate expression for the inversion brings out most clearly that the inversion oscillates with the effective Rabi frequency $2g\sqrt{\bar{n}}$.

Following the recipe given in Chapter 9 we can neglect the quadratic and higher contributions in the short time limit. When we then replace the summation over m by integration we find the collapse of the inversion, that is the decaying envelope of the inversion around the origin of time. Moreover, with the help of the Poisson summation formula we can find the revivals of the inversion at integer multiples of T_1. Due to the quadratic contribution in $\mathcal{S}(t)$ the revivals spread and neighboring peaks at T_1 start to overlap. This creates the fractional revivals.

16.2.2 Experiments on Collapse and Revivals

We conclude this section by emphasizing that quantities such as the inversion have been measured in experiments. In particular, the collapse and the first revival have been observed in the context of the one-atom maser, high-Q microwave cavity experiments and ion traps. Unfortunately, fractional revivals have not been detected in these systems yet.

The revival experiments provide clear indication that the radiation field is of granular structure, that is, that the photon number n is discrete. Indeed, the revivals, that is the periodic recurrence of the inversion at integer multiples of T_1 would not occur if n would not be discrete. We recall from the above discussion and from the analysis of the wave packet dynamics of Chapter 9 that the collapse arises when we replace the summation over n by integration. The revivals only emerge when we save the discreteness by using the Poisson summation formula.

To bring this out most clearly we now briefly discuss the experiments related to the Jaynes-Cummings revivals. They measure the probability

$$W(t; |b\rangle) = \sum_{n=0}^{\infty} |\Psi_{b,n}(t)|^2$$

to find the atom in the ground state when before the interaction it was in the excited state. In this case we find from Eq. (15.29) the probability amplitude

$$\Psi_{b,n}(t) = -iS_{n-1}w_{n-1},$$

and hence

$$W(t; |b\rangle) = \sum_{n=0}^{\infty} W_n \sin^2(\sqrt{n+1}gt). \quad (16.9)$$

448   16 State Preparation and Entanglement

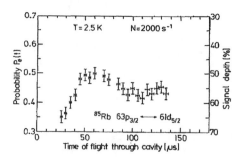

Fig. 16.6: Collapse of the occupation probability in the one-atom maser. The probability for an atom to remain in the excited state (63p$_{3/2}$ of ^{85}Rb) as a function of the interaction time, expressed here as time of flight through the cavity approaches a steady state. We also recognize a last Rabi oscillation at the end of the collapse leading to the steady state. The flux N of the atoms is $N = 2000 s^{-1}$ and the temperature of the black body radiation is $T = 2.5K$. Taken from G. Rempe et al., Phys. Rev. Lett. **58**, 353 (1987).

One-Atom Maser

The first experiment to test this relation and, in particular, to observe the collapse and the revivals was performed using the one-atom maser. We discuss this amazing maser in Sec. 18.4 in more detail. Here we only focus on the feature of the collapse and revivals shown in Fig. 16.5.

In the one-atom maser many atoms pass through a high-quality resonator and build up a stationary single mode maser field. The photon statistics $W_n(t)$ of this field depends on the interaction time t of each atom. We can use one of the atoms driving the maser field as a test atom to probe the field. For this purpose we ask for the probability of finding the atom in the ground state after it has interacted with the maser field. The interaction time for this atom is also t. Therefore, this probability reads

$$W(t; |b\rangle) = \sum_{n=0}^{\infty} W_n(t) \sin^2(\sqrt{n+1}gt).$$

Indeed, the probability involves the sum of Rabi oscillations. It therefore displays the phenomena of collapse and revivals shown in Figs. 16.6 and 16.7. As outlined above the existence of the revival is a clear indication of the granular structure of the radiation.

Cavity Experiments with Injected Microwaves

However, the maser situation is more sophisticated than the dynamics of the Jaynes-Cummings-Paul model. The latter describes the interaction of an atom with an identically prepared field. In particular, when we change the interaction time the atom still interacts with the same initial field. Moreover, the photon statistics of the field before the interaction does not depend on the interaction time. In contrast, the atoms of the one-atom maser are used to prepare the field and to read out the dynamics. A change of the interaction time therefore causes a change of the stationary field.

16.2 Collapse, Revivals and Fractional Revivals

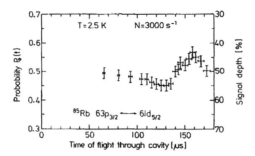

Fig. 16.7: Revival of the occupation probability in the one-atom maser. After the collapse and a quiet period the probability for an atom to be in the excited state revives and oscillates again. In contrast to Fig. 16.6 the flux of atoms is increased to $N = 3000s^{-1}$. Taken from G. Rempe et al., Phys. Rev. Lett. **58**, 353 (1987).

We can circumvent this problem by injecting a field from outside into the resonator. The results of this approach are shown in Fig. 16.8. Here coherent states of four different amplitudes have been injected into the cavity and the resulting atomic dynamics have been observed. The left column shows the probability to make a transition to the ground state as a function of interaction time. In (A) there is no injected field, but the atom undergoes Rabi oscillations as predicted by Eq. (15.28). In (C) and (D) the injected coherent state is such that a clear collapse and the onset of a revival become apparent.

In the expression Eq. (16.9) for the probability to be in the ground state the photon statistics and time enter. The above mentioned experiments observe the time dependence of this probability. Thus we can use a Fourier transform to express the photon statistics in terms of an integral over the measured probabilities. This allows us to find the photon statistics in the field.

The middle and the right column of Fig. 16.8 display the so-calculated Fourier transforms and the photon statistics, respectively. We note that indeed for the top case the photon statistics is centered at zero photons corresponding to the vacuum state whereas in the bottom case it has a maximum which is shifted away from the vacuum. The Fourier transform displays maxima at $\sqrt{n}$ which correspond to the Rabi frequencies. These maxima are rather localized around the frequencies $\sqrt{n}$. This is an indication that the electromagnetic field is quantized since only discrete frequencies enter. Therefore, this experiment is another clear indication that the electromagnetic field involves discrete excitations.

Motion of Ion in Paul Trap

We conclude this section by mentioning that the Jaynes-Cummings-Paul dynamics has also been observed in the context of a single ion moving in a Paul trap and interacting with a classical light field. We discuss this situation in more detail in the next chapter. Here it suffices to say that the quantized energy levels of the center-of-mass motion of the ion in the trap play the role of the photon number states.

450   16 State Preparation and Entanglement

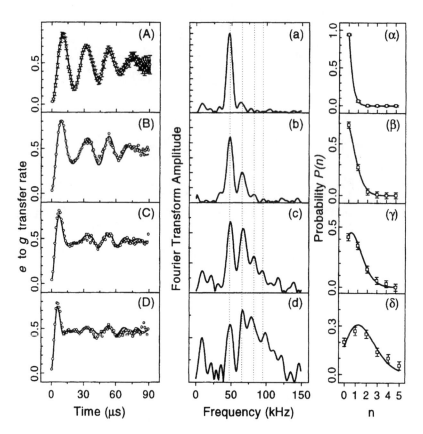

Fig. 16.8: Measured Rabi nutation signal (left column), inferred Fourier transform amplitude (middle column) and photon statistics (right column). A coherent state is injected into an initially empty high-Q microwave cavity. An atom in the excited state passes through the cavity and interacts with the field for a time t. In the left column we show the probability for finding a ground state atom for fields with increasing amplitudes. (A) No injected field and 0.06($\pm$0.01) thermal photons on average; (B), (C), and (D) correspond to coherent fields with 0.4($\pm$0.02), 0.85($\pm$0.04), and 1.77($\pm$0.15) photons on average. The points are experimental [error bars in (A) only for clarity]; the solid lines correspond to theoretical fits. The Fourier transform (middle column) of this probability provides the Rabi frequencies $\nu = 47$ kHz, $\nu\sqrt{2}$, $\nu\sqrt{3}$, and 2ν indicated by vertical dotted lines. Vertical scales are proportional to 4, 3, 1.5, and 1 from (a) to (d). The right column depicts the corresponding photon number distributions inferred from the experimental signal (points). Solid lines show the theoretical thermal (α) or coherent [(β), (γ), (δ)] distributions which provide the optimal fit of the data. Taken from M. Brune *et al.*, Phys. Rev. Lett. **76**, 1800 (1996).

In Fig. 16.9 we show the dynamics of the internal levels of a two-level ion moving in a Paul trap and interacting with a classical light wave. Initially the motional state of the ion is described by a coherent state and the corresponding dynamics of the internal states shows a collapse and a clear revival.

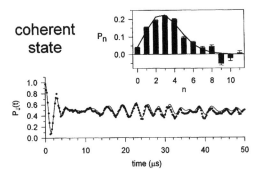

Fig. 16.9: Collapse and revival of the internal dynamics of an ion stored in a Paul trap. A single ion moving according to the laws of quantum mechanics in a harmonic oscillator potential and interacting with a standing classical light wave can be described by the Jaynes-Cummings-Paul model as discussed in Sec. 17.4. In this case the energy eigenstates of the oscillator potential play the role of the photon states. The initial state of center-of-mass motion is a coherent state of average number of phonons $\bar{n} = 3.1 \pm 0.1$. The internal dynamics displays a clear collapse with Rabi oscillations and a revival. The insert shows the decomposition of the data onto the expected phonon number states fitted by a Poissonian distribution of average number $\bar{n} = 2.9 \pm 0.1$. Taken from D.M. Meekhof et al., Phys. Rev. Lett. **76**, 1796 (1996).

16.3 Quantum State Preparation

So far we have always assumed that the radiation field is in a specific state $|\psi_{\text{field}}\rangle$. However, we have not given any prescription for the preparation of such a state. In the present section we first briefly review a method to obtain a superposition of two coherent states of different phases – a quantum state commonly referred to as Schrödinger cat state. Such a state has been recently produced experimentally using this prescription.

We then suggest in Sec. 16.4 a general method to produce an arbitrary but finite superposition of Fock states. In both cases we make use of the Jaynes-Cummings-Paul dynamics of the two-level atom interacting with a single quantized mode.

There exist many more methods to prepare states and the special issue of *Journal of Modern Optics* devoted to quantum state preparation and measurement provides an overview over this area of quantum optics.

16.3.1 State Preparation with a Dispersive Interaction

In Sec. 16.1.2 we have shown that the measurement of an atomic variable in an entangled atom-field system prepares a quantum state of the field. In the present section we illustrate this preparation scheme using the example of a strongly detuned atom interacting according to the Jaynes-Cummings-Paul model. We show that in this way we can create a coherent superposition of two coherent states of different phase. This scheme has been realized experimentally as shown in Figs. 16.10 and 16.11.

Fig. 16.10: Generation of a Schrödinger cat state in a resonator field. A classical light field creates out of a two-level atom initially in its ground state an atom in a coherent superposition $|\psi_{\text{atom}}\rangle \equiv c_a |a\rangle + c_b |b\rangle$ of ground and excited state. The so-prepared atom traverses a single mode cavity which is strongly detuned with respect to the atomic transition. This dispersive interaction described by the effective Hamiltonian Eq. (15.35) produces the entangled state Eq. (16.11) between the atom and the field. A measurement of the atomic dipole $|\widetilde{\psi}_{\text{atom}}\rangle \equiv \tilde{c}_a |a\rangle + \tilde{c}_b |b\rangle$ using another classical field and an ionization detector generates the field state $|\psi_{\text{field}}\rangle$, Eq. (16.12). This two-field configuration is reminiscent of the Ramsey method in high-resolution radio-frequency spectroscopy. Since not every atom leaving the cavity is in the reference state $|\widetilde{\psi}_{\text{atom}}\rangle$ the state $|\psi_{\text{field}}\rangle$ is not created in every run. In the case of a coherent state $|\alpha\rangle$ as the initial field state the final field state is a superposition $|\psi\rangle = \frac{\mathcal{N}}{\sqrt{2}} [|\alpha e^{i\varphi}\rangle + |\alpha e^{-i\varphi}\rangle]$ of two coherent states of identical amplitude α but different phases $+\varphi$ and $-\varphi$ – an amplitude cat.

We first consider the interaction of an atom in the atomic state

$$|\psi_{\text{atom}}\rangle \equiv c_a |a\rangle + c_b |b\rangle \tag{16.10}$$

with a strongly detuned quantized light field in the state

$$|\psi_{\text{field}}\rangle = \sum_{n=0}^{\infty} w_n |n\rangle.$$

An appropriate classical field can prepare such a coherent atomic superposition Eq. (16.10) as indicated in Fig. 16.10.

According to Eq. (15.36) the quantum state of the atom-field system after an interaction time t reads

$$|\Psi(t)\rangle = \sum_{n=0}^{\infty} w_n \left(e^{i\varphi_1(t)} e^{i\varphi_n(t)} c_a |a\rangle + e^{-i\varphi_n(t)} c_b |b\rangle \right) |n\rangle \tag{16.11}$$

where we have chosen the decomposition

$$\varphi_n(t) \equiv \varphi(t)\, n$$

of the phase with $\varphi(t) \equiv g^2 t/\Delta$.

We now perform a measurement on the atom and, in particular, ask whether the atom is in the atomic reference state

$$|\widetilde{\psi}_{\text{atom}}\rangle \equiv \tilde{c}_a |a\rangle + \tilde{c}_b |b\rangle.$$

We can realize such a measurement using an appropriate classical field and an ionization detector as discussed in Sec. 16.1.

16.3 Quantum State Preparation   453

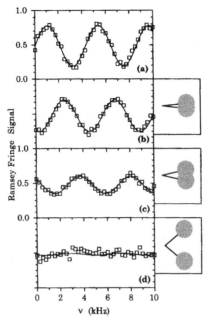

Fig. 16.11: Experimental generation of a Schrödinger cat state based on a joint measurement. An atom passing through the Ramsey setup of Fig. 16.10 exhibits fringes in the transition probability. Whereas in (a) the cavity field is in the vacuum, the cases (b)–(d) correspond to a coherent field with $|\alpha| = 3.1$. The detuning between the transition and the field frequency is different in (b)–(d). This gives rise to three different angles φ between the two coherent states forming the cat as shown by the phase space representation of the field components displayed in the right column. Points are experimental values and curves are sinusoidal fits. Taken from M. Brune et al., Phys. Rev. Lett. **77**, 4887 (1996).

When the atom is indeed in this atomic state $|\tilde{\psi}_{\text{atom}}\rangle$, then according to Eq. (16.7) the state of the field reads

$$|\psi_{\text{field}}(t)\rangle = \frac{1}{\sqrt{W(t;|\tilde{\psi}_{\text{atom}}\rangle)}}\left[c_a\tilde{c}_a^* e^{i\varphi_1(t)}\sum_{n=0}^{\infty} w_n e^{i\varphi_n(t)}|n\rangle + c_b\tilde{c}_b^* \sum_{n=0}^{\infty} w_n e^{-i\varphi_n(t)}|n\rangle\right]$$
(16.12)

where $W(t;|\tilde{\psi}_{\text{atom}}\rangle)$ is the probability to find after the interaction time t the atomic reference state $|\tilde{\psi}_{\text{atom}}\rangle$.

We can simplify this expression by choosing the atomic probability amplitudes $\tilde{c}_a$, $\tilde{c}_b$ and c_a, c_b of the reference and initial state such that

$$c_a\tilde{c}_a^* e^{i\varphi_1(t)} = c_b\tilde{c}_b^* = \frac{1}{\sqrt{2}}.$$

In this case the so-prepared state $|\psi\rangle$ of the field reads

$$|\psi\rangle = \frac{\mathcal{N}}{\sqrt{2}}\left[\sum_{n=0}^{\infty} w_n e^{i\varphi_n(t)}|n\rangle + \sum_{n=0}^{\infty} w_n e^{-i\varphi_n(t)}|n\rangle\right],$$
(16.13)

where $\mathcal{N}$ is the normalization constant.

We note that the initial photon number amplitudes w_n have been modified by the phase factors $\exp[\pm i\,\varphi_n(t)]$. Therefore, the field state has changed considerably due to the interaction with and the measurement of the atom.

16.3.2 Generation of Schrödinger Cats

An extremely important application of this result arises when we consider a coherent state

$$|\alpha\rangle \equiv \sum_{n=0}^{\infty} w_n \,|n\rangle \equiv \sum_{n=0}^{\infty} \frac{\alpha^n}{\sqrt{n!}} e^{-\alpha^2/2}\,|n\rangle \tag{16.14}$$

as an initial field state.

We recall that the phases $\varphi_n(t) \equiv \varphi(t)\,n$ are linear in n. Consequently, we can combine the phases φ_n with the probability amplitudes w_n and perform the summation in Eq. (16.13), that is

$$|\psi\rangle = \frac{\mathcal{N}}{\sqrt{2}} \left[\sum_{n=0}^{\infty} \frac{(\alpha e^{i\varphi})^n}{\sqrt{n!}} \exp(-|\alpha e^{i\varphi}|^2/2)\,|n\rangle + \sum_{n=0}^{\infty} \frac{(\alpha e^{-i\varphi})^n}{\sqrt{n!}} \exp(-|\alpha e^{-i\varphi}|^2/2)\,|n\rangle\right],$$

and arrive at the superposition state

$$|\psi\rangle = \frac{\mathcal{N}}{\sqrt{2}} \left[\left|\alpha e^{i\varphi}\right\rangle + \left|\alpha e^{-i\varphi}\right\rangle\right]. \tag{16.15}$$

This state is the Schrödinger amplitude cat discussed in Sec. 11.3.

In summary, we have used the dispersive interaction of the strongly detuned Jaynes-Cummings-Paul model to create from an atomic superposition and a single coherent state a superposition of two coherent states of identical amplitudes but different phases. In this process we have transfered the coherence from the atoms to the field. This method allows us not only to prepare states of the type Eq. (16.15) but can be generalized to prepare any finite superposition of Fock states as will be discussed in the next section.

16.4 Quantum State Engineering

We now discuss a method to produce an arbitrary but finite superposition of Fock states. We make use of the Jaynes-Cummings-Paul dynamics of the two-level atom interacting resonantly with a single mode quantized field and rely on joint measurements.

16.4.1 Outline of the Method

Our ultimate desire is to engineer a quantum state

$$|\psi_d\rangle \equiv \sum_{n=0}^{N} d_n\,|n\rangle = d_0\,|0\rangle + d_1\,|1\rangle + \ldots \tag{16.16}$$

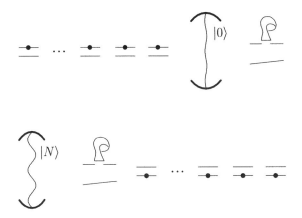

Fig. 16.12: Generation of a Fock state $|N\rangle$ from the vacuum $|0\rangle$. We inject one by one N excited two-level atoms into a resonator with a single mode initially in the vacuum. Provided all atoms are in the ground state after their interaction with the resonant field, we have created in the cavity the Fock state $|N\rangle$. This condition requires that we make a measurement of the internal levels using an atomic state sensitive detector.

of a single mode of a resonator. This desired field state is a superposition of the first $N+1$ Fock states with given complex-valued coefficients d_n. How to obtain such a superposition in the resonator starting from the vacuum?

Fock State Generation

Since the desired field state contains the Fock state $|N\rangle$ of N photons and we begin from the vacuum it is clear that we have to transfer excitation into the cavity. One obvious method relies on N excited two-level atoms. Following this first naive idea we inject N excited atoms one by one into the cavity. Here and in the remainder of this section we assume for the sake of simplicity that there is only one single atom in the cavity at a time. Provided all N atoms transfer their excitation to the field we have, indeed, engineered the Fock state $|N\rangle$ as indicated in Fig. 16.12.

We emphasize that such an experiment has just been performed using the high-Q microwave resonator of the one-atom maser in Garching. Starting from the vacuum state the experimentalists have created successively a one-photon and a two-photon state. They have probed this state with an additional atom and have measured the Rabi oscillations of this atom due to the so-prepared field as shown in Fig. 16.13.

Superposition State

However, this state is not the desired field state $|\psi_d\rangle$, that is, it is not the coherent superposition Eq. (16.16) of the first $N+1$ Fock states. In order to arrive at a coherent superposition of field states it is necessary to bring into the cavity not only excitation but also coherence, that is, quantum mechanical interference.

The most elementary interference is the superposition of the atomic states in a single atom. Hence, in order to transfer excitation as well as coherence onto the field we inject N resonant two-level atoms into the cavity, where now each atom is

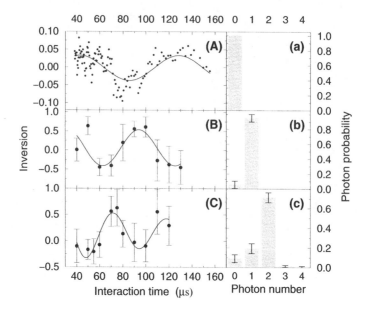

Fig. 16.13: Fock state preparation and measurement in the one-atom maser. A sequence of two-level atoms passing a high-Q cavity can prepare and probe a one-photon or a two-photon Fock state by giving the atomic excitation into the field and by measuring the resulting Rabi oscillations. When we start from a cavity field in the vacuum state (a), a single atom making a transition from its excited to the ground state prepares a one-photon Fock state (b). Two such atoms depositing their internal excitation create a two-photon Fock state (c). An additional atom probes the number of photons by recording the corresponding Rabi oscillations shown in the left column. Taken from B.T.H. Varcoe et al., Nature **403**, 743 (2000).

in an appropriate superposition between the ground and the excited state as shown in Fig. 16.14.

Let us now analyze the buildup of the cavity field by the appropriately prepared atoms. Initially the cavity is in the vacuum mode. We now send the first atom in an appropriate superposition

$$|\varphi_{\text{atom}}^{(1)}\rangle = |a\rangle + i\epsilon_1 |b\rangle$$

into the resonator. Note that for the time being we use states that are not normalized and we perform the normalization at the end.

After the interaction of the atom with the vacuum field we make a measurement on the atomic state. In case the atom is in the excited state we have to stop this procedure and abort this run. Why? The desired field state $|\psi_d\rangle$, Eq. (16.16), contains the Fock state $|N\rangle$ of N photons. Since we have only N atoms, that is, N excitations all atoms have to exit in the ground state.

Hence, we now consider the situation when the atom does indeed exit in the ground state. In this case there exist two possibilities for the field state in the cavity. Since the atom is initially in a coherent superposition of ground and excited state

16.4 Quantum State Engineering   457

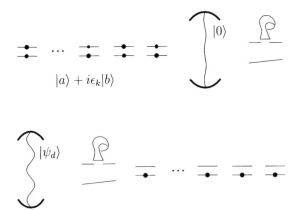

Fig. 16.14: Quantum state engineering: The setup. We engineer an arbitrary desired field state $|\psi_d\rangle$ of the $N+1$ first Fock states by injecting N appropriately prepared resonant two-level atoms into the resonator. The field is initially in the vacuum. The kth atom is in a superposition $|a\rangle + i\epsilon_k |b\rangle$ of its atomic states where ϵ_k is a complex parameter which depends on and is determined by the desired field state $|\psi_d\rangle$. Provided all atoms exit the cavity in the ground state the cavity mode is in the desired field state $|\psi_d\rangle$.

and afterwards in the ground state the field state is a superposition

$$|\varphi^{(1)}\rangle = \varphi_0^{(1)} |0\rangle + \varphi_1^{(1)} |1\rangle$$

of the vacuum state $|0\rangle$ and the one photon Fock state $|1\rangle$ as indicated in Fig. 16.15. The complex-valued probability amplitudes $\varphi_0^{(1)}$ and $\varphi_1^{(1)}$ are determined by the dynamics of the atom-field interaction and will be calculated later in this section.

We emphasize that due to the reduction of the complete system of atom plus field to the field system only, the field state $|\varphi^{(1)}\rangle$ is not normalized. We therefore denote unnormalized field states by $|\varphi^{(k)}\rangle$. Moreover, the notation used throughout this section describes the number of the atom as a superscript and the number of the Fock state by a subscript.

The second atom continues this process and transforms the vacuum contribution of $|\varphi^{(1)}\rangle$ into a superposition of the vacuum $|0\rangle$ and the one photon state $|1\rangle$. Likewise, it transforms the state $|1\rangle$ of $|\varphi^{(1)}\rangle$ into a superposition of $|1\rangle$ and the next higher Fock state $|2\rangle$. Hence the field state after the interaction of the second atom and its detection in the ground state is the superposition

$$|\varphi^{(2)}\rangle = \varphi_0^{(2)} |0\rangle + \varphi_1^{(2)} |1\rangle + \varphi_2^{(2)} |2\rangle$$

of the Fock states $|0\rangle$, $|1\rangle$ and $|2\rangle$ with probability amplitudes $\varphi_0^{(2)}$, $\varphi_1^{(2)}$ and $\varphi_2^{(2)}$.

Figure 16.15 provides more information about the character of these probability amplitudes. We note that only one path leads from the vacuum $|0\rangle$ to the Fock state $|2\rangle$. Likewise only one path leaves the vacuum state unchanged. In contrast there exist *two* paths connecting $|0\rangle$ with $|1\rangle$: It was either the first or the second atom which brought in this excitation. The two paths interfere and hence the probability amplitude $\varphi_1^{(2)}$ is the sum of the probability amplitudes corresponding to the two paths.

458   16 State Preparation and Entanglement

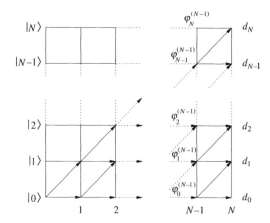

Fig. 16.15: Quantum state engineering: Evolution of field probability amplitudes due to the interaction of successive atoms with the cavity field. On the vertical axis we show the development of the photon number states of the field and on the horizontal axis we depict time governed by the number of atoms. We assume that after N two-level atoms have interacted resonantly with a single mode initially in the vacuum they are all in the ground state. Before each atom enters the cavity it is in an appropriately chosen superposition of ground and excited state. The atom can therefore either increase the number of photons by one or leave it unchanged as indicated by diagonal or horizontal arrows. Each elementary cell of this lattice can be thought of as the lower plane of Fig. 15.1 since we only consider atoms leaving the resonator in the ground state.

The next atom continues the excitation process, and after N atoms have passed the cavity, we find, indeed, the superposition

$$|\varphi^{(N)}\rangle = \sum_{n=0}^{N} \varphi_n^{(N)} |n\rangle = \varphi_0^{(N)} |0\rangle + \varphi_1^{(N)} |1\rangle + \ldots + \varphi_{N-1}^{(N)} |N-1\rangle + \varphi_N^{(N)} |N\rangle$$

of the first $N+1$ Fock states with the probability amplitudes $\varphi_n^{(N)}$.

We note, however, that in general these probability amplitudes are not identical to the expansion coefficients d_n of the desired field state. This stands out most clearly when we recognize that many different paths contribute to a single Fock state. Figure 16.15 shows that there are two exceptions to this rule: There exists only a single path which leads from the vacuum to the Fock state $|N\rangle$. Likewise there is only one path which leaves the vacuum unchanged. However, for all other Fock states there exist many paths which lead from $|0\rangle$ to $|n\rangle$. All the paths connecting $|0\rangle$ with $|n\rangle$ interfere and the corresponding probability amplitudes $\varphi_n^{(N)}$ are the sum of the probability amplitudes corresponding to these interfering paths.

16.4.2 Inverse Problem

Due to the multitude of interfering paths the inverse problem namely to determine the appropriate atomic superposition parameters ϵ_k leading to the desired field amplitudes, that is, $\varphi_n^{(N)} = d_n$ seems to be hopeless. However, the triangular structure

16.4 Quantum State Engineering

of the field-atom diagram, Fig. 16.15, allows to solve the inverse problem step by step.

Recurrence Relation for Amplitudes

For this purpose we consider the effect of the kth atom. We describe the interaction by the resonant Jaynes-Cummings-Paul model discussed in detail in Sec. 15.1.

Before the injection of the kth atom in an atomic superposition $|a\rangle + i\epsilon_k |b\rangle$ the cavity field is in a state

$$|\varphi^{(k-1)}\rangle = \sum_{n=0}^{k-1} \varphi_n^{(k-1)} |n\rangle.$$

After the interaction, when the atom has left the cavity, the state $|\Phi^{(k)}\rangle$ of the combined atom-field system reads

$$|\Phi^{(k)}\rangle = \sum_{n=0}^{k-1} \varphi_n^{(k-1)} \left[C_n^{(k)} |a,n\rangle - i S_n^{(k)} |b,n+1\rangle + i\epsilon_k C_{n-1}^{(k)} |b,n\rangle + \epsilon_k S_{n-1}^{(k)} |a,n-1\rangle \right].$$

Here we have recalled the time evolution Eq. (15.13) of the Jaynes-Cummings-Paul model and have defined $C_n^{(k)} \equiv \cos\left(\sqrt{n+1}\, g\tau_k\right)$ and $S_n^{(k)} \equiv \sin\left(\sqrt{n+1}\, g\tau_k\right)$. Moreover, we have allowed the individual atoms to have different interaction times τ_k.

When we now make a measurement on the atomic state and find the atom in the ground state $|b\rangle$ the state $|\varphi^{(k)}\rangle$ of the field reads

$$|\varphi^{(k)}\rangle \equiv \langle b|\Phi^{(k)}\rangle = -i \sum_{n=0}^{k-1} \varphi_n^{(k-1)} \left[S_n^{(k)} |n+1\rangle - \epsilon_k C_{n-1}^{(k)} |n\rangle \right]$$

$$= -i \sum_{n=0}^{k} \left[S_{n-1}^{(k)} \varphi_{n-1}^{(k-1)} - \epsilon_k C_{n-1}^{(k)} \varphi_n^{(k-1)} \right] |n\rangle,$$

or

$$|\varphi^{(k)}\rangle = -i \sum_n \varphi_n^{(k)} |n\rangle.$$

Here the new probability amplitudes $\varphi_n^{(k)}$ follow from the recurrence relation

$$\varphi_n^{(k)} = S_{n-1}^{(k)} \varphi_{n-1}^{(k-1)} - \epsilon_k C_{n-1}^{(k)} \varphi_n^{(k-1)}. \tag{16.17}$$

They are defined in terms of the field probability amplitudes $\varphi_n^{(k-1)}$, before the kth atom has entered the cavity. We emphasize that we have factored out the common factor $-i$.

Formulation of the Inverse Problem

We are now in a position to address the inverse problem, that is, to find the atomic superposition $|a\rangle + i\epsilon_N |b\rangle$ and the field distribution $\varphi_n^{(N-1)}$ such that after the passage of the Nth atom the field is in a desired field state

$$|\varphi^{(N)}\rangle = |\psi_d\rangle = \sum_{n=0}^{N} d_n |n\rangle.$$

For this purpose we cast the recurrence relation

$$\varphi_n^{(N)} = S_{n-1}^{(N)} \varphi_{n-1}^{(N-1)} - \epsilon_N C_{n-1}^{(N)} \varphi_n^{(N-1)}$$

following from Eq. (16.17) for the $k = N$th atom into a system of equations for the individual Fock states. We note that $\varphi_N^{(N-1)} = 0$ since only the Nth atom brings in the Nth Fock state. Hence together with $C_{-1}^{(k)} = 1$ and $S_{-1}^{(k)} = 0$ we find

$$S_{N-1}^{(N)} \varphi_{N-1}^{(N-1)} = d_N$$
$$\vdots$$
$$S_{n-1}^{(N)} \varphi_{n-1}^{(N-1)} - \epsilon_N C_{n-1}^{(N)} \varphi_n^{(N-1)} = d_n \qquad (16.18)$$
$$\vdots$$
$$- \epsilon_N \varphi_0^{(N-1)} = d_0.$$

Note that the right hand side of this system of $N+1$ equations, namely the coefficients d_n of the desired field state are given. The unknown quantities are the N probability amplitudes $\varphi_n^{(N-1)}$ of the field state before the passage of the Nth atom and the superposition parameter ϵ_N. Hence we have, indeed, $N+1$ equations for $N+1$ variables. This is the deeper reason why we can solve the inverse problem.

Solution of the Inverse Problem

We start by solving the first equation, that is

$$\varphi_{N-1}^{(N-1)} = d_N / S_{N-1}^{(N)}$$

and substitute this result into the next equation determining $\varphi_{N-2}^{(N-1)}$ and so on. We obtain the solution

$$\varphi_n^{(N-1)} = \sum_{\nu=1}^{N-n} \left[\prod_{\mu=n}^{n+\nu-2} \frac{C_\mu^{(N)}}{S_\mu^{(N)}} \right] \frac{d_{n+\nu}}{S_{n+\nu-1}^{(N)}} \epsilon_N^{\nu-1}. \qquad (16.19)$$

Substitution of the so-obtained coefficient $\varphi_0^{(N-1)}$ into the last equation of the set Eq. (16.18) yields

$$d_0 + \sum_{\nu=1}^{N} \left[\prod_{\mu=0}^{\nu-2} \frac{C_\mu^{(N)}}{S_\mu^{(N)}} \right] \frac{d_\nu}{S_{\nu-1}^{(N)}} \epsilon_N^\nu = 0 \qquad (16.20)$$

as the characteristic equation for ϵ_N.

We note that Eq. (16.20) is a polynomial of degree N. It therefore has N solutions. We solve the characteristic equation numerically and choose one value ϵ_N from the N roots of Eq. (16.20). Equation (16.19) immediately gives us the corresponding coefficients $\varphi_n^{(N-1)}$ of the state $|\varphi^{(N-1)}\rangle$. We have thus determined the field state and the atomic superposition necessary to achieve the desired field state after the Nth atom has passed the cavity.

We now take this field state $|\varphi^{(N-1)}\rangle$ as a new desired state which we have to obtain by sending $N-1$ atoms through the cavity. For the state $|\varphi^{(N-1)}\rangle$ we perform

k	$\lvert\epsilon_k\rvert$	β_k/π	$P_b^{(k)}$
1	0.5412	−0.5075	0.4938
2	0.5730	0.5102	0.3616
3	0.6951	−0.7585	0.6477
4	0.8283	−0.9977	0.8106
5	1.0562	0.7783	0.7368
6	1.3334	0.5141	0.4918
7	1.5002	−0.5389	0.4086

Tab. 16.1: Internal state $|a\rangle + i|\epsilon_k| e^{i\beta_k}|b\rangle$ of the kth atom needed to obtain the truncated phase state, Eq. (16.21), for a fixed interaction parameter $g\tau = \pi/5$. The right column gives the probability $P_b^{(k)}$, Eq. (16.25), to find the kth atom in state $|b\rangle$ after its interaction with the cavity field provided all earlier atoms have been detected in the state $|b\rangle$. The probability $\mathcal{P}_7$, Eq. (16.27), to find all atoms in the ground state is $\mathcal{P}_7 = P_b^{(1)} \cdot P_b^{(2)} \cdots P_b^{(7)} = 0.01388$.

the same calculations as for the state $|\psi_d\rangle$ and obtain the parameter ϵ_{N-1} and state $|\varphi^{(N-2)}\rangle$ with $N-1$ coefficients $\varphi_n^{(N-2)}$.

We repeat this procedure until we end up with the vacuum state. A string of complex numbers ϵ_1, ϵ_2, ..., ϵ_N defines the internal states of a sequence of N atoms we have to inject into the cavity in order to obtain the desired state $|\psi_d\rangle$ from the vacuum state.

16.4.3 Example: Preparation of a Phase State

We illustrate this method by creating the truncated phase state

$$|\psi_d\rangle \equiv \frac{1}{\sqrt{8}} \sum_{n=0}^{7} |n\rangle. \qquad (16.21)$$

In this case the probability amplitudes of the desired field state are all equal and real. We substitute them into the expressions Eq. (16.19) for the field amplitudes before the Nth atom enters the cavity and calculate the superposition parameter ϵ_N from Eq. (16.20). We then repeat the procedure for the next $N-2$ field states until we arrive at the vacuum state. In Table 16.1 we give the so-calculated values ϵ_1, ϵ_2, ..., ϵ_l for identical interaction parameters $g\tau_k = \pi/5$.

In order to give an impression about the individual steps of the evolution of the field state from the vacuum state to the truncated phase state, Eq. (16.21), we show in Fig. 16.16 the Q-function for the field state $|\varphi^{(k)}\rangle$ after the kth atom has passed through the cavity and has been detected in the ground state.

Success Probability

But what is the probability to create the state, that is, what is the probability $\mathcal{P}_N$ to find all atoms in the ground state after they have passed through the cavity? So

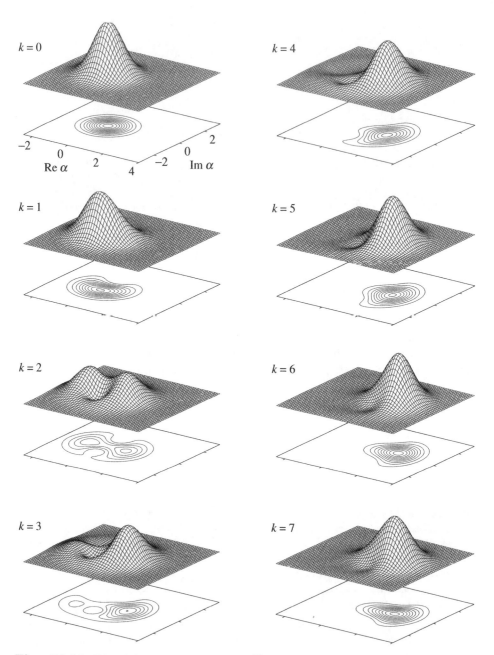

Fig. 16.16: The Q-function $Q(\alpha) \equiv |\langle\alpha|\varphi^{(k)}\rangle|^2/\pi$ for the field state $|\varphi^{(k)}\rangle$ after the kth atom has interacted with the field and has been detected in the ground state. The parameters for the internal states of the incoming atoms are given in Tab. 16.1. The contour lines are at $Q = 0.025, 0.050, 0.075, \ldots$.

16.4 Quantum State Engineering

far we have used unnormalized states for the atoms and the field because it was convenient for calculating the superposition parameters ϵ_k and the amplitudes $\varphi_n^{(k)}$. However, when we need probabilities we have to use normalized field states

$$|\psi^{(k)}\rangle \equiv \sum_{n=0}^{k} \psi_n^{(k)} |n\rangle \qquad (16.22)$$

and normalized atomic field states $(|a\rangle + i\epsilon_k |b\rangle)/\sqrt{1 + |\epsilon_k|^2}$.

For the coefficients $\psi_n^{(k)}$ we obtain equations similar to Eqs. (16.18) which read

$$\psi_k^{(k)} = \mathcal{N}_k S_{k-1}^{(k)} \psi_{k-1}^{(k-1)}$$
$$\vdots \qquad \vdots$$
$$\psi_n^{(k)} = \mathcal{N}_k \left[S_{n-1}^{(k)} \psi_{n-1}^{(k-1)} - \epsilon_k C_{n-1}^{(k)} \psi_n^{(k-1)} \right] \qquad (16.23)$$
$$\vdots \qquad \vdots$$
$$\psi_0^{(k)} = -\mathcal{N}_k \epsilon_k \psi_0^{(k-1)}.$$

Here the normalization constant

$$\mathcal{N}_k \equiv \frac{1}{\sqrt{P_b^{(k)} (1 + |\epsilon_k|^2)}} \qquad (16.24)$$

consists of two parts: The factor $1/\sqrt{1 + |\epsilon_k|^2}$ which takes into account the normalization of the internal state of the kth atom, and the factor $1/\sqrt{P_b^{(k)}}$ which is due to the normalization of the field state after the state reduction.

We can calculate the probability to find the kth atom in the ground state from Eq. (16.24) and arrive at

$$P_b^{(k)} = \sum_{n=0}^{k} \left| S_{n-1}^{(k)} \psi_{n-1}^{(k-1)} - \epsilon_k C_{n-1}^{(k)} \psi_n^{(k-1)} \right|^2 (1 + |\epsilon_k|^2). \qquad (16.25)$$

We are now in a position to calculate the success probability, that is the probability

$$\mathcal{P}_N = \prod_{k=1}^{N} P_b^{(k)}$$

to find all N atoms leaving the cavity in the ground state. For this purpose we recognize that from the first equation of the set Eq. (16.24) follows

$$\psi_N^{(N)} = \prod_{k=1}^{N} \left(\mathcal{N}_k S_{k-1}^{(k)} \right) \psi_0^{(0)}. \qquad (16.26)$$

Since we start from the vacuum state we have $\psi_0^{(0)} = 1$. Moreover, we note that for a normalized desired state we have $\psi_n^{(N)} = d_n$ and Eq. (16.26) reduces to

$$|d_N|^2 = \prod_{k=1}^{N} \left(\mathcal{N}_k S_{k-1}^{(k)} \right)^2.$$

k	$\|\epsilon_k\|$	β_k/π	$g\tau_k/\pi$	$P_b^{(k)}$
1	0.7462	−0.6016	0.5000	1.0000
2	0.8513	0.5569	0.3370	0.5655
3	0.8543	0.7427	0.2780	0.7435
4	0.9972	−0.6821	0.2477	0.6265
5	1.2000	−0.5256	0.2363	0.5196
6	1.3198	0.5097	0.1937	0.4366
7	1.1003	1.0000	0.1524	0.8690

Tab. 16.2: Internal state $|a\rangle + i|\epsilon_k| e^{i\beta_k}|b\rangle$ of the kth atom needed to obtain the truncated phase state, Eq. (16.21). Here we have optimized the interaction parameters $g\tau_k$ as to maximize the probability $\mathcal{P}_7$, Eq. (16.27), to find all atoms in the ground state. The right column gives the probability $P_b^{(k)}$, Eq. (16.25), to find the kth atom in state $|b\rangle$ after its interaction with the cavity field provided all earlier atoms have been detected in the state $|b\rangle$. In this case we have $\mathcal{P}_7 = 0.05193$.

We substitute the normalization constant $\mathcal{N}_k$ from Eq. (16.24) into this equation and obtain the probability $\mathcal{P}_N$ to find all N atoms in the ground state by solving the equation

$$|d_N|^2 = \prod_{k=1}^{N} \frac{1}{P_b^{(k)}} \frac{(S_{k-1}^{(k)})^2}{1+|\epsilon_k|^2} = \left(\prod_{k=1}^{N} \frac{1}{P_b^{(k)}}\right)\left(\prod_{k=1}^{N} \frac{(S_{k-1}^{(k)})^2}{1+|\epsilon_k|^2}\right) = \frac{1}{\mathcal{P}_N}\left(\prod_{k=1}^{N} \frac{(S_{k-1}^{(k)})^2}{1+|\epsilon_k|^2}\right)$$

for $\mathcal{P}_N$ which yields

$$\mathcal{P}_N = \frac{1}{|d_N|^2} \prod_{k=1}^{N} \left[\frac{(S_{k-1}^{(k)})^2}{1+|\epsilon_k|^2}\right]. \qquad (16.27)$$

The probability $\mathcal{P}_N$ depends on the choice of roots of the characteristic equation, Eq. (16.20), and the interaction times τ_k. Can we use these "degrees of freedom" to optimize the probability $\mathcal{P}_N$?

To get an idea about the possibilities of this optimization let us consider the simplest case of identical interaction times $\tau_k = \tau$ for the example of the truncated phase state, Eq. (16.21). The dependence of the probability $\mathcal{P}_7$ on the interaction parameter $g\tau$ is shown in Fig. 16.17. For this curve we have chosen for each atom the parameter ϵ_k with the smallest absolute value. We note that $\mathcal{P}_7$ increases for increasing interaction parameter $g\tau$ and reaches its maximum $\mathcal{P}_7 \cong 0.02067$ at $g\tau \cong 0.2445\pi$ and then decreases. Moreover, trapping states, that is, interaction parameters $g\tau$ with $\sin(g\tau\sqrt{n}) = 0$ $(n = 1, 2, \ldots, 7)$, manifest themselves in vanishing probabilities $\mathcal{P}_7$ as apparent from Eq. (16.27). As a general rule the maximum value for the probability occurs for interaction parameters smaller than those corresponding to trapping states.

In the next step of the optimization we allow each atom to have its individual interaction time τ_k with the cavity field. In Table 16.2 we have chosen τ_k such that the probability $\mathcal{P}_7$ to find all seven atoms in the ground state has a maximum. Using this strategy we increase $\mathcal{P}_7$ up to the value $\mathcal{P}_7 \cong 0.05193$.

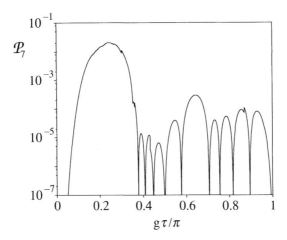

Fig. 16.17: Probability $\mathcal{P}_7$ to find all seven atoms in the ground state as a function of the interaction parameter $g\tau$ for the truncated phase state, Eq. (16.21). Here we have chosen ϵ_k with the smallest absolute value. Note the occurrence of trapping states where, according to Eq. (16.27), the probability $\mathcal{P}_7$ vanishes.

Conclusion

In conclusion we emphasize that we can construct *any* superposition of the first $N+1$ number states from the vacuum state by injecting N appropriately prepared atoms into a cavity and detecting all of them in the ground state. Furthermore, we note that the Jaynes-Cummings-Paul Hamiltonian is not crucial for this method. Similar interactions between field and atom can also be used provided that they allow for energy exchange between field and atoms.

Problems

16.1 Collapse and Revivals in the Jaynes-Cummings-Paul Model

For a resonant atom initially in its ground state the inversion $\mathcal{I}(t)$ reads

$$\mathcal{I}(t) = -\sum_{n=0}^{\infty} W_n \cos(\sqrt{n}2gt),$$

where W_n is the photon statistics of the field state at time $t = 0$.

(a) Show with the help of the Poisson summation formula

$$\sum_{n=0}^{\infty} f_n = \sum_{\nu=-\infty}^{+\infty} \int_0^{\infty} dn\, f(n)\, e^{2\pi i \nu n} + \frac{1}{2} f_0,$$

that $\mathcal{I}(t)$ can be written in the form

$$\mathcal{I}(t) = \sum_{\nu=-\infty}^{+\infty} \mathcal{I}_\nu(t) - \frac{1}{2} W_0,$$

where $\mathcal{I}_\nu(t)$ is given by

$$\mathcal{I}_\nu(t) \equiv -\mathrm{Re}\left[\int_0^\infty dn\, W(n) e^{2\pi i \nu n - 2igt\sqrt{n}}\, dn\right].$$

(b) When the photon statistics W_n is slowly varying compared to the phase factor the integral can be evaluated approximately with the method of stationary phase. Show that for $\nu \geq 1$ the quantity $\mathcal{I}_\nu$ takes the form

$$\mathcal{I}_\nu(t) \cong -W\left(n = \frac{g^2 t^2}{4\pi^2 \nu^2}\right) \frac{gt}{\pi\sqrt{2\nu^3}} \cos\left(\frac{g^2 t^2}{2\pi\nu} - \frac{\pi}{4}\right),$$

whereas for $\nu \leq -1$ we find

$$\mathcal{I}_\nu(t) \cong 0.$$

Hint: In the case $\nu = 0$ the method of stationary phase cannot be applied. Here we simply have

$$\mathcal{I}_0(t) = -\int_0^\infty dn\, W(n)\cos(2gt\sqrt{n}).$$

(c) Discuss the form of the inversion for the case that the field is at time $t = 0$ in a coherent state with large amplitude. What is the condition to have nice revivals? Are there revivals when the initial field state is a thermal state?

Hint: For a coherent state $|\alpha\rangle$ with $\bar{n} = |\alpha|^2 \gg 1$ one can use the asymptotic expansion Eq. (4.15).

(d) In the case of an oscillatory photon statistics such as arising for a highly squeezed state the revivals have oscillations as discussed by Satyanarayana et al. (1989). Use the appropriate asymptotics of the photon statistics and the method of stationary phase to bring out the ringing revivals. Is there an interpretation in terms of interference in phase space?

Hint: See Fleischhauer and Schleich (1993).

16.2 State Engineering

The interaction of a two-level atom with a classical field can be described by

$$\left.\begin{array}{l}|a\rangle \to \cos\theta\, |a\rangle + \sin\theta\, e^{i\varphi}\, |b\rangle \\ |b\rangle \to \cos\theta\, |b\rangle - \sin\theta\, e^{-i\varphi}\, |a\rangle.\end{array}\right\} \qquad (16.28)$$

The interaction of the two-level atom with the quantized mode of the electromagnetic field is described in the interaction picture by the Hamiltonian

$$\hat{H} = -\frac{\hbar g^2}{\Delta}\left[\hat{\sigma}_z \hat{a}^\dagger \hat{a} + \frac{1}{2}(\hat{\sigma}_z + \mathbb{1})\right].$$

At time $t = 0$ the atom is in the state $|a\rangle$ and the field is in a coherent state $|\alpha_0\rangle$. The interaction with the fields consists in the three steps:

1. Interaction with a classical field according to Eq. (16.28)
2. interaction with the quantized field (interaction time τ)
3. Interaction with a classical field according to Eq. (16.28) but with φ replaced by $-\varphi$.

After this interaction a measurement on the atom is performed. We assume that the atom is always found in the excited state.

(a) What is the state of the field after the measurement?
(b) Show that with an appropriate choice of the parameters θ und φ any superposition of the states $|\alpha_0 e^{-ig^2\tau/\Delta}\rangle$ and $|\alpha_0 e^{ig^2\tau/\Delta}\rangle$ can be generated.
(c) Show that with the help of N atoms with appropriately chosen parameters θ_k, φ_k ($k = 1, \ldots, N$) and $\tau < \pi\Delta/(g^2 N)$ a given field state

$$|\psi\rangle = \sum_{n=0}^{N} d_n |\alpha_0 e^{-ig^2\tau/\Delta\,(2n-N)}\rangle \qquad (16.29)$$

can be generated if *all* atoms are found in the excited state.

Hint: Pursue a similar approach as in Sec. 16.4: Derive a recurrence relation for the field coefficients $c_n^{(k-1)}$ before the interaction with the kth atom and the coefficients $c_n^{(k)}$ after the interaction with the kth atom. Show that for any state Eq. (16.29) at least one solution exists. Do no worry too much about normalization.

References

Theory of Photon Counting

For expositions of the theory of photon counting see the classic Les Houches lectures by
R.J. Glauber, in: *Quantum Optics and Electronics*, Les Houches, edited by C. DeWitt, A. Blandin and C. Cohen-Tannoudji, Gordon and Breach, New York, 1965, pp. 331–381
L. Mandel and **E. Wolf**, *Optical Coherence and Quantum Optics*, Cambridge U.P., New York, 1995

Cavity Quantum Electrodynamics

For a review of the field of cavity quantum electrodynamics see
S. Haroche and **D. Kleppner**, *Cavity Quantum Electrodynamics*, Phys. Today **42**, 24–30 (1989)
E.A. Hinds, *Perturbative Cavity Quantum Electrodynamics*, in: *Cavity Quantum Electrodynamics*, edited by P. Berman, Adv. At. Mol. Opt. Phys., Supplement 2, Academic Press, Boston, 1994
H. Yokoyama and **K. Ujihara**, *Spontaneous Emission and Laser Oscillation in Microcavities*, CRC, Boca Raton, 1995

The periodic exchange of excitation between an atom and a cavity field has been observed in the microwave domain by

G. Rempe, H. Walther and **N. Klein**, *Observation of Quantum Collapse and Revival in a One-Atom Maser*, Phys. Rev. Lett. **58**, 353–356 (1987)

M. Brune, F. Schmidt-Kaler, A. Maali, J. Dreyer, E. Hagley, J.M. Raimond and **S. Haroche**, *Quantum Rabi Oscillations: A Direct Test of Field Quantization in a Cavity*, Phys. Rev. Lett. **76**, 1800–1803 (1996)

B.T.H. Varcoe, S. Brattke, M. Weidinger and **H. Walther**, *Preparing pure photon number states of the radiation field*, Nature **403**, 743–746 (2000)

The first results in the optical domain were reported as a conference abstract by

R.J. Brecha, L.A. Orozco, M.G. Raizen, M. Xiao and **H.J. Kimble**, *Mode Splitting for Two-Level Atoms Inside an Optical Resonator*, J. Opt. Soc. Am. B, 238 (1986)

and published much later by

M.G. Raizen, R.J. Thompson, R.J. Brecha, H.J. Kimble and **H.J. Carmichael**, *Normal-Mode Splitting and Linewidth Averaging for Two-State Atoms in an Optical Cavity*, Phys. Rev. Lett. **63**, 240–243 (1989)

Ramsey Method

For the original work on the method of separated oscillatory fields in spectroscopy see

N.F. Ramsey, *Molecular Beams*, Oxford University Press, New York, 1956

For the application of this method to the one-atom maser we refer to

J. Krause, M.O. Scully and **H. Walter**, *Quantum theory of the micromaser: Symmetry breaking via off-diagonal atomic injection*, Phys. Rev. A **34**, 2032–2037 (1986)

M. Brune, S. Haroche, V. Lefevre, J.M. Raimond and **N. Zagury**, *Quantum Nondemolition Measurement of Small Photon Numbers by Rydberg-Atom Phase-Sensitive Detection*, Phys. Rev. Lett. **65**, 976–979 (1990)

Revivals in the Jaynes-Cummings-Paul Model

The time evolution of the mean number of photons in the Jaynes-Cummings-Paul model has been calculated numerically by

J. Frahm, *Quantenmechanische Streuung der elektrischen Feldstärke einer gestörten Resonator-Eigenschwingung*, Ann. Phys. (Leipzig) **18**, 205–208 (1966). The author found a decay and various recurrences of the Rabi oscillations. However, the paper does not discuss the recurrence phenomenon in more detail.

For a detailed theoretical analysis of the revivals in the Jaynes-Cummings-Paul model see

J.H. Eberly, N.B. Narozhny and **J.J. Sanchez-Mondragon**, *Periodic Spontaneous Collapse and Revival in a Simple Quantum Model*, Phys. Rev. Lett. **44**, 1323–1326 (1980)

M. Venkata Satyanarayana, P. Rice, R. Vyas and **H.J. Carmichael**, *Ringing revivals in the interaction of a two-level atom with squeezed light*, J. Opt. Soc. Am. B **6**, 228–237 (1989)

M. Fleischhauer and **W.P. Schleich**, *Revivals made simple: Poisson summation formula as a key to the revivals in the Jaynes-Cummings model*, Phys. Rev. A **47**, 4258–4269 (1993)

For the experimental observation of the granular nature of the radiation field as manifested in the Rabi oscillations and the Jaynes-Cummings revivals see

G. Rempe, H. Walther and **N. Klein**, *Observation of Quantum Collapse and Revival in a One-Atom Maser*, Phys. Rev. Lett. **58**, 353–356 (1987)

M. Brune, F. Schmidt-Kaler, A. Maali, J. Dreyer, E. Hagley, J.M. Raimond and **S. Haroche**, *Quantum Rabi Oscillations: A Direct Test of Field Quantization in a Cavity*, Phys. Rev. Lett. **76**, 1800–1803 (1996)

For the similarities and differences between the collapse and revivals in the Jaynes-Cummings-Paul model and the one-atom maser see

E.M. Wright and **P. Meystre**, *Collapse and revival in the micromaser*, Opt. Lett. **14**, 177–179 (1989)

There exists a close analogy between cavity QED and the quantized motion of an ion in a Paul trap as discussed in more detail in Chapter 17. The revivals predicted by the Jaynes-Cummings-Paul model have been observed experimentally in the Paul trap by

D.M. Meekhof, C. Monroe, B.E. King, W.M. Itano and **D.J. Wineland**, *Generation of Nonclassical Motional States of a Trapped Atom*, Phys. Rev. Lett. **76**, 1796–1799 (1996)

For a theoretical discussion of fractional revivals and full revivals in the Jaynes-Cummings-Paul model analogous to the ones in mechanical oscillators such as atoms and molecules, that is, at fractions of T_2, see, for example,

I.Sh. Averbukh, *Fractional revivals in the Jaynes-Cummings model*, Phys. Rev. A **46**, R2205-R2208 (1992)

State Preparation

For an early approach towards quantum state preparation see

W.E. Lamb, *An operational interpretation of non-relativistic quantum mechanics*, Phys. Today **22**(4), 23–28 (1969)

For the problem of state preparation in the one-atom maser and, in particular, the generation of Fock states see

J. Krause, M.O. Scully and **H. Walther**, *State Reduction and $|n\rangle$ State Preparation in a High-Q Micromaser*, Phys. Rev. A **36**, 4647–4550 (1987)

P. Meystre, *Repeated Quantum Measurements on a Single Harmonic Oscillator*, Opt. Lett. **12**, 669–671 (1987)

P. Meystre, *Cavity Quantum Optics and the Quantum Measurement Process*, Prog. Opt. **30**, 261–355 (1992)

F. De Martini, G. Di Giuseppe and **M. Marrocco**, *Single-Mode Generation of Quantum Photon States by Excited Single Molecules in a Microcavity Trap*, Phys. Rev. Lett. **76**, 900–903 (1996)

For a recent experiment on the preparation of a two-photon Fock state using the one-atom maser see
B.T.H. Varcoe, S. Brattke, M. Weidinger and **H. Walther**, *Preparing pure photon number states of the radiation field*, Nature **403**, 743–746 (2000)

For theoretical proposals of the preparation of a Schrödinger cat of a cavity field see
S. Song, C.M. Caves and **B. Yurke**, *Generation of superpositions of classically distinguishable quantum states from optical back-action evasion*, Phys. Rev. A **41**, 5261–5264 (1990)

B. Yurke, W.P. Schleich and **D.F. Walls**, *Quantum superpositions generated by quantum nondemolition measurements*, Phys. Rev. A **42**, 1703–1711 (1990)

C.M. Savage, S.L. Braunstein and **D.F. Walls**, *Macroscopic quantum superpositions by means of single-atom dispersion*, Opt. Lett. **15**, 628–630 (1990)

The latter suggestion has been pursued in the experimental realization in
M. Brune, E. Hagley, J. Dreyer, X. Maitre, A. Maali, C. Wunderlich, J.M. Raimond and **S. Haroche**, *Observing the Progressive Decoherence of the "Meter" in a Quantum Measurement*, Phys. Rev. Lett. **77**, 4887–4891 (1996)

For the experimental realization of a Schrödinger cat in the center-of-mass motion of an ion see
C. Monroe, D.M. Meekhof, B.E. King and **D.J. Wineland**, *A Schrödinger Cat Superposition State of an Atom*, Science **272**, 1131–1135 (1996)

For the concept of quantum state engineering see
K. Vogel, V.M. Akulin and **W.P. Schleich**, *Quantum State Engineering of the Radiation Field*, Phys. Rev. Lett. **71**, 1816–1819 (1993)

see also the approach by
A.S. Parkins, P. Marte, P. Zoller and **H.J. Kimble**, *Synthesis of Arbitrary Quantum States via Adiabatic Transfer of Zeeman Coherence*, Phys. Rev. Lett. **71**, 3095–3098 (1993)

For an overview over various methods of state preparation see
W.P. Schleich and **M. Raymer**, Special Issue on *Quantum State Preparation and Measurement*, J. Mod. Opt. **44**, (11/12) (1997)

For a QND-measurement of a single photon in a cavity see
G. Nogues, A. Rauschenbeutel, S. Osnaghi, M. Brune, J. M. Raimond and **S. Haroche**, *Seeing a single photon without destroying it*, Nature **400**, 239–242 (1999)

Entanglement Between Atoms

Consecutive atoms passing through a cavity are entangled as discussed in
B.-G. Englert, M. Löffler, O. Benson, B. Varcoe, M. Weidinger and **H. Walther**, *Entangled Atoms in Micromaser Physics*, Fortschr. Phys. **46**, 897–926 (1998)

E. Hagley, X. Maitre, G. Nogues, C. Wunderlich, M. Brune, J.M. Raimond and **S. Haroche**, *Generation of Einstein-Podolsky-Rosen Pairs of Atoms*, Phys. Rev. Lett. **79**, 1–5 (1997)

A. Rauschenbeutel, G. Nogues, S. Osnaghi, P. Bertet, M. Brune, J.M. Raimond and **S. Haroche**, *Step-by-Step Engineered Multiparticle Entanglement*, Science **288**, 2024–2028 (2000)

Four ions stored in a trap were entangled by

C.A. Sackett, D. Kielpinski, B.E. King, C. Langer, V. Meyer, C. J. Myatt, M. Rowe, Q.A. Turchette, W.M. Itano, D.J. Wineland and **C. Monroe**, *Experimental entanglement of four particles*, Nature **404**, 256–259 (2000)

17 Paul Trap

Trapping of single ions over a long period of time offers numerous new possibilities in laser spectroscopy. Moreover, a single trapped ion represents a unique system for testing fundamental concepts of quantum mechanics. For example, the dynamics of an ion in a Paul trap has put stringent limits on nonlinear versions of quantum mechanics. Quantum jumps, which have been so central to the early discussions of quantum mechanics between Bohr and Schrödinger, have been observed directly and are nowadays used to monitor the internal dynamics of the ion. Recently, a single ion stored in a Paul trap has been used to implement a quantum gate and many ions in a linear trap arranged in a string seem to be a promising tool to realize a quantum computer. Moreover, the experimental generation of non-classical states of the motion of an ion in a Paul trap has propelled the field of quantum state preparation into a new era. Due to the importance of the Paul trap illustrated by these examples we devote this chapter to the discussion of the physics of this remarkable device.

The chapter is organized as follows: In Sec. 17.1 we briefly review the basics of trapping ions. In particular, we show that we cannot trap charged particles in three dimensions based on static electric fields only. We need time dependent electric fields. We then in Sec. 17.2 give a brief introduction into laser cooling. This field has expanded rapidly over the last years and there exists a huge literature on this topic. Space does not allow us to go deeper into this exciting branch of quantum optics and we therefore refer to the references at the end of this chapter.

In Sec. 17.3 we briefly discuss dynamical features of the motion of a single ion in the Paul trap. In particular, we show that the temporal evolution of the ion, which due to the explicit time dependence of the trapping potential is rather complicated, can be understood in terms of a sequence of a rotation, a squeezing and a second rotation operation in phase space. Moreover, we focus on the so-called Floquet solution of a harmonic oscillator with time dependent but periodic frequency. We show, that in the basis of the initial values of the Floquet states the temporal evolution of the system simplifies considerably.

In Sec. 17.4 we introduce a model describing the coupling of the center-of-mass motion of the ion to its internal levels by a classical electromagnetic field. This leads to a multi-phonon Jaynes-Cummings-Paul Hamiltonian.

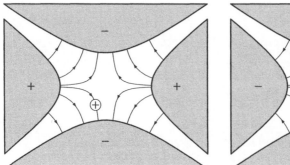
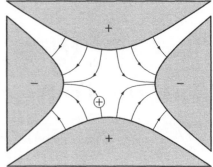

Fig. 17.1: Unstable electrostatic trap configurations consisting of a ring electrode and two end-caps. A positively charged particle is repelled by the ring electrode but attracted by the end-caps (left). A different polarity of the electrodes (right) does not change the situation: Now the end-caps provide binding but the particle is attracted by the ring electrodes.

17.1 Basics of Trapping Ions

In contrast to neutral atoms, ions can easily be influenced by electromagnetic fields because of their charge. We first show that the Laplace equation does not allow trapping in three dimensions based on static electric fields only. We then turn to the discussion of dynamical trapping due to time dependent electric fields.

17.1.1 No Static Trapping in Three Dimensions

In a first attempt towards trapping we consider static electric fields only and start from a ring electrode with two end-caps as shown in Fig. 17.1. We apply a constant voltage such that the ring electrode becomes positively charged and the end-caps negatively charged. However, this arrangement does not achieve three-dimensional confinement. To understand this, we follow a positively charged ion in its course along the electric field lines. The positive ring repels the positive ion and pushes it towards the center of the trap. Hence, there is binding in the radial direction. However, in the vertical (or orthogonal) direction the particle is attracted by the negatively charged end-caps and so the motion is unstable.

Such a motion results from a saddle-point potential

$$\Phi(\vec{r}) \equiv f_0 \cdot \left(x^2 + y^2 - 2z^2\right) \tag{17.1}$$

which causes stable motion in the radial direction and unstable motion in the orthogonal direction. Here f_0 is a constant.

It is tempting to think that another, more sophisticated arrangement of electrodes would achieve three-dimensional confinement but the laws of electrostatics, and more precisely the Laplace equation

$$\Delta \Phi = \left(\frac{\partial^2}{\partial x^2} + \frac{\partial^2}{\partial y^2} + \frac{\partial^2}{\partial z^2}\right) \Phi = 0,$$

prevent this.

Indeed, when we substitute the saddle-point potential Eq. (17.1) into the Laplace equation we find
$$\Delta \Phi = f_0 \cdot (2 + 2 - 2 \cdot 2) = 0.$$
Hence, the factor of 2 and the minus sign in the potential along the z-direction are dictated by the Laplace equation.

There exist at least two techniques which offer a way out of this saddle-point problem. The first one is based on static electric and magnetic fields and results in the Penning trap. The second one utilizes time dependent electric fields and results in the Paul trap discussed in the next section in more detail.

In the Penning trap we start from the configuration shown on the right hand side of Fig. 17.1 and superimpose on the electric field a constant homogeneous magnetic field aligned along the symmetry axis. This device achieves the goal of three-dimensional confinement – the vertical magnetic field forces the ion to perform a circular motion in the plane orthogonal to the magnetic field, which overrules the radial instability caused by the electrostatic field.

17.1.2 Dynamical Trapping

The Paul trap is solely based on the use of time varying electric fields. Its mode of operation is easily understood when we consider the above mentioned saddle potential. When the particle starts to roll down the unstable hillside, the sign of the voltage at the electrodes is reversed, so that the particle suddenly sees a rising rather than a decreasing potential. However, now the other coordinate has become unstable. Therefore, we have to reverse again the polarity on the electrodes. This application of an alternating voltage creates a dynamical binding.

Mathieu Equation

In this situation the potential reads
$$\Phi(\vec{r}, t) \equiv f(t) (x^2 + y^2 - 2z^2), \qquad (17.2)$$
where
$$f(t) \equiv \frac{U + V \cos(\omega_{\rm rf} t)}{r_0^2 + 2z_0^2}.$$
Here U and V are the amplitudes of the DC and AC voltages, r_0 is the radius of the ring electrode, $2z_0$ is the distance between the end-caps, and $\omega_{\rm rf} = 2\pi/T$ is the (radio) frequency of the AC voltage.

The classical motion of an ion in the potential Eq. (17.2) is governed by the equation of motion
$$M\ddot{\vec{r}} = -e\vec{\nabla}\Phi(\vec{r}, t),$$
where M and e are mass and charge of the ion. When we substitute the potential Eq. (17.2) into this equation of motion we find
$$M\ddot{\vec{r}} + 2\,e\,f(t)\,(\vec{r} - 3\vec{z}) = 0,$$

where $\vec{z} = (0, 0, z)^T$.

We transform this equation into the Mathieu equation

$$\frac{\partial^2 x}{\partial \tau^2} + [a + 2q\cos(2\tau)]x = 0, \tag{17.3}$$

where we have defined

$$\tau \equiv \tfrac{1}{2}\omega_{\text{rf}}t, \quad a \equiv \frac{8eU}{M\omega_{\text{rf}}^2 (r_0^2 + 2z_0^2)}, \quad \text{and} \quad q \equiv \frac{4eV}{M\omega_{\text{rf}}^2 (r_0^2 + 2z_0^2)}. \tag{17.4}$$

The equation for the y-coordinate is identical to Eq. (17.3) whereas for the z-coordinate the parameters a and q are multiplied by a factor of -2.

Domains of Stability

The Mathieu equation Eq. (17.3) is a linear differential equation with time-periodic coefficients. We can therefore make use of the Floquet theorem and the general solution of Eq. (17.3) reads

$$x(\tau) = A e^{+i\mu\tau} \phi(\tau) + B e^{-i\mu\tau} \phi(-\tau). \tag{17.5}$$

Here A and B are constants and

$$\phi(\tau) = \phi(\tau + \pi) = \sum_{n=-\infty}^{\infty} c_n e^{2in\tau} \tag{17.6}$$

is a periodic function.

When the characteristic exponent μ is purely real the variable $x(\tau)$ is bound and consequently the motion is stable. However, when μ has an imaginary part the function $x(\tau)$ contains an exponentially growing contribution. The motion is unstable. The parameters a and q, that is the voltages applied to the trap, determine if the motion is stable or unstable. When $\mu \equiv 0$ the solution $x(\tau)$ is strictly periodic.

In order to determine the stability regime of the Mathieu equation, that is the Paul trap, we substitute the Floquet ansatz

$$x^{(F)}(\tau) = e^{i\mu\tau} \phi(\tau) = \sum_n c_n e^{i(2n+\mu)\tau}$$

into the Mathieu equation Eq. (17.3) and arrive at

$$\sum_n \left\{ \left[-(2n+\mu)^2 + a \right] c_n e^{i(2n+\mu)\tau} + q c_n e^{i[2(n+1)+\mu]\tau} + q c_n e^{i[2(n-1)+\mu]\tau} \right\} = 0.$$

We shift the indices in the last two terms and find the three-term recurrence relation

$$\left[a - (2n+\mu)^2 \right] c_n + q (c_{n-1} + c_{n+1}) = 0, \tag{17.7}$$

which takes the form of a vector equation

$$M\vec{c} \equiv \begin{pmatrix} \ddots & \vdots & \vdots & \vdots & \vdots & \vdots \\ \cdots & q & a-(-2+\mu)^2 & q & 0 & 0 & \cdots \\ \cdots & 0 & q & a-\mu^2 & q & 0 & \cdots \\ \cdots & 0 & 0 & q & a-(2+\mu)^2 & q & \cdots \\ & \vdots & \vdots & \vdots & \vdots & \vdots & \ddots \end{pmatrix} \begin{pmatrix} \vdots \\ c_{-2} \\ c_{-1} \\ c_0 \\ c_1 \\ c_2 \\ \vdots \end{pmatrix} = 0$$

with a tri-diagonal matrix M.

In order for $\vec{c}$ to be a nontrivial vector the determinant of M, that is the Hill determinant, has to vanish

$$\det M = \det \begin{pmatrix} \ddots & \vdots & \vdots & \vdots & \vdots & \vdots \\ \cdots & q & a-(-2+\mu)^2 & q & 0 & 0 & \cdots \\ \cdots & 0 & q & a-\mu^2 & q & 0 & \cdots \\ \cdots & 0 & 0 & q & a-(2+\mu)^2 & q & \cdots \\ & \vdots & \vdots & \vdots & \vdots & \vdots & \ddots \end{pmatrix} = 0. \quad (17.8)$$

It is not trivial to evaluate for a given set of parameters a and q this determinant analytically since the matrix M has infinite dimensions. However, in various textbooks this determinant has been calculated analytically and the condition of a vanishing Hill determinant translates into a highly nonlinear equation.

In general due to this nonlinearity we have to find the characteristic exponent numerically as discussed in more detail in Sec. 17.3.4. In the stability diagram of Fig. 17.2 we indicate by light and dark grey areas the domains of the dimensionless DC and AC voltage amplitudes a and q where the r- and z-motion is stable. Due to the factor of 2 and the minus sign in the equation for the z-coordinate the domains of stability for the z-motion follow from the ones of the r-motion by a reduction and imaging at the q-axis.

In order to find trapping in three dimensions the r- as well as the z-motion have to be stable. Hence, we achieve dynamical, three-dimensional confinement only in the overlap of the light and the dark grey domains. Many such domains exist as indicated in Fig. 17.2 by the black areas. However, the one close to the origin is the one used most frequently in experiments.

Once we have determined the characteristic exponent for a set of parameters a and q we can find from the vector equation $M\vec{c} = 0$ the corresponding coefficients c_n. They play an important role in the quantum treatment of the Paul trap. Therefore, we devote Sec. 17.3.4 to a detailed discussion of the properties of c_n.

Approximate Stability Analysis

We gain some insight into the stability diagram when we consider the neighborhood of the origin. Indeed, in the limiting case $|a| \ll 1$ and $q < 1$ we can confine ourselves

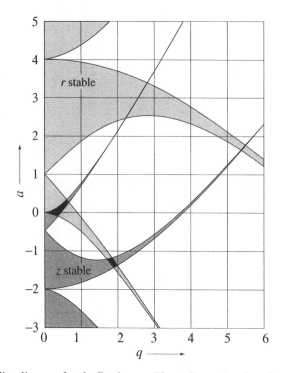

Fig. 17.2: Stability diagram for the Paul trap. Three-dimensional confinement is achieved for voltages a and q from the black domains where both the radial and the vertical motion are stable.

to a 3×3 matrix around $n = 0$. When we evaluate the corresponding determinant and make use of $|a| \ll 1$ and $q < 1$ we find for the characteristic exponent the formula

$$\mu^2 \cong \tfrac{1}{2} q^2 + a.$$

Only when μ is real the motion is stable, otherwise it is unstable. This implies the condition $\tfrac{1}{2} q^2 + a > 0$. The boundary of stability is determined by the requirement

$$\tfrac{1}{2} q^2 + a = 0$$

that is,

$$a = -\tfrac{1}{2} q^2.$$

In a diagram of stability spanned by a and q this curve is a parabola starting at the origin and bending downwards. Since the motion in the z-direction differs by a minus sign and a factor of 2 the corresponding curve for the z-motion bends upwards and is steeper as indicated by Fig. 17.2.

However, these curves do not define completely the area of stability. For this purpose we need the almost straight boundary emerging from $a = 1$ going downwards and the corresponding curve starting at $a = -1/2$ going upwards. An approximate analytical calculation similar to the one for $n = 0$ is possible and provides this curve.

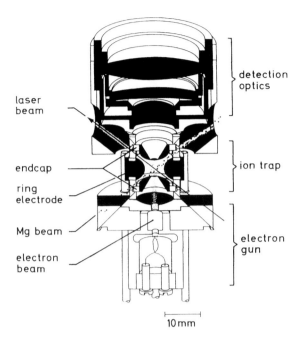

Fig. 17.3: Schematic illustration of a Paul trap with electron gun (bottom) and detection optics (top). A beam of neutral atoms, in this case Magnesium atoms, traverses the ion trap at the center of the picture through the openings between the ring and the end-cap electrodes. An electron beam propagation upwards through a hole in the lower end-cap transforms the atoms on their path through the trap into ions. They are now confined in three dimensions by the electric fields. A laser beam cools the motion of the ions and detects their position through resonance fluorescence. The emitted light is observed through a hole in the upper end-cap electrode by a microscope. Taken from F. Diedrich and H. Walther, Phys. Rev. Lett. **58**, 203 (1987).

17.2 Laser Cooling

So far we have learned how to keep an ion trapped, but how do we get it into the trap in the first place? The most obvious solution is to create it in there by, for example, ionizing a neutral atomic beam via electron collision at the trap center. Unfortunately, the resulting trapped ion has considerable kinetic energy rendering it useless for most applications, such as spectroscopy. Thus the next task is to extract kinetic energy from the ion. Radiation pressure exerted by a laser is the tool used to overcome this obstacle.

For this purpose we consider an atom or ion with momentum p traveling against an electromagnetic wave of frequency ω_L. We choose ω_L to be smaller than the atomic resonance transition frequency ω_a, so that the energy of the photon is not sufficient to excite the ion. Crudely speaking, the ion can extract the missing energy from its motion and thus reduce its kinetic energy. In other words, the atomic velocity Doppler-shifts the atom into resonance to bridge the detuning gap, $\Delta \equiv \omega_L - \omega_a$, and the atom absorbs the photon of momentum $\hbar k = \hbar \omega_L / c$. After the absorption process the momentum of the atom is reduced from p to $p - \hbar k$ as required by conservation of momentum.

The absorbed photon is reemitted spontaneously and its direction and hence the resulting recoil momentum are random. When we average over many scattering photons this recoil momentum vanishes, whereas the average momentum of the atom along the propagation axis of the electromagnetic wave gets reduced by the number N of scattering events, times the momentum of the photons. The number N is determined by the internal quantum physics of the atom and can be as large as 10^8 s^{-1}, allowing an ion to cool from room temperature to the milli-Kelvin region within seconds.

It is tempting to be side-tracked by the effects connected with laser cooling, to discuss how the temperatures of a single laser-cooled ion are being pushed towards ever lower temperatures or to mention the quite different game of trapping and cooling neutral atoms by radiation pressure to achieve an optical "molasses". Unfortunately we have to rush on and can only refer the reader to the ever growing literature on this subject. We should nevertheless mention that the ultimate temperature achievable by laser cooling is the micro-Kelvin region.

17.3 Motion of an Ion in a Paul Trap

In general the motion of an ion in a Paul trap contains two time scales: There is a slow motion – the so-called secular motion – governed by the frequency of the time averaged binding potential, and a fast motion – the micromotion – governed by the radio frequency of the AC voltage applied to the trap. To gain deeper insight into the quantum mechanical counterpart of this motion we first analyze the time evolution of the position and momentum operators of a harmonic oscillator with time dependent frequency. This oscillator serves as a model for the Paul trap. We show, that three real-valued parameters characterize the quantum motion and correspond to a rotation, squeezing, and rotation in phase space. Moreover, we show that by an appropriate choice of the coordinate system the time dependence of these parameters becomes simple and in some cases even periodic. This choice corresponds to a Floquet description.

The time evolution of the state vector $|\psi_{\text{cm}}(t)\rangle$ for the center-of-mass motion of a single ion in the Paul trap follows from the Schrödinger equation

$$i\hbar \frac{d}{dt}|\psi_{\text{cm}}(t)\rangle = \hat{H}_{\text{cm}}(t) |\psi_{\text{cm}}(t)\rangle .$$

Here the Hamiltonian

$$\hat{H}_{\text{cm}}(t) \equiv \frac{\hat{p}^2}{2M} + \frac{1}{2} M \omega^2(t) \hat{x}^2 \tag{17.9}$$

governs the one-dimensional motion of the ion with mass M, momentum p, and position x in a harmonic potential with time dependent steepness

$$\omega^2(t) = \tfrac{1}{4}\omega_{\text{rf}}^2 \left[a + 2q \cos(\omega_{\text{rf}} t)\right] . \tag{17.10}$$

The dimensionless parameters a and q are defined in Eq. (17.4).

17.3 Motion of an Ion in a Paul Trap

We can formally solve the time dependent Schrödinger equation by introducing the time evolution operator

$$\hat{U}_{\text{cm}}(t) \equiv \hat{\mathcal{T}} \exp\left[-\frac{i}{\hbar}\int_0^t dt'\, \hat{H}_{\text{cm}}(t')\right] \qquad (17.11)$$

where $\hat{\mathcal{T}}$ is the time ordering operator and thus the state vector reads

$$|\psi_{\text{cm}}(t)\rangle = \hat{U}_{\text{cm}}(t)\,|\psi_{\text{cm}}(0)\rangle$$

We now analyze the action of this time evolution operator in state space and Wigner phase space.

17.3.1 Reduction to Classical Problem

In order to gain some insight into the dynamics of the harmonic oscillator with time dependent frequency we first consider the time evolution of the operators in the Heisenberg picture. Throughout this chapter we use a reference oscillator with a time independent frequency ω_r. The initial operators, or in the Schrödinger picture the initial states, are defined with respect to this reference oscillator. The frequency ω_r is a real-valued parameter which is free at our disposal. Glauber has shown that the static harmonic oscillator with the frequency ω_r serves as a reference oscillator whose energy eigenstates $|n\rangle$ form a convenient complete basis set. Without any loss of generality this reference frequency may be chosen such that the states $|n\rangle$ become the initial values of the Floquet states, as we show in Sec. 17.3.4.

We now define the position and momentum operator in terms of the annihilation and creation operators $\hat{b}$ and $\hat{b}^\dagger$ of the static reference oscillator, that is

$$\hat{x} \equiv \sqrt{\frac{\hbar}{2M\omega_r}}\,(\hat{b} + \hat{b}^\dagger)$$

and

$$\hat{p} \equiv \sqrt{\frac{\hbar\omega_r M}{2}}\,\frac{1}{i}\,(\hat{b} - \hat{b}^\dagger).$$

We now consider the time dependence of these operators under the time evolution operator $\hat{U}_{\text{cm}}(t)$. Hence, we are interested in the operators

$$\tilde{\hat{x}}(t) \equiv \hat{U}_{\text{cm}}^\dagger(t)\,\hat{x}\,\hat{U}_{\text{cm}}(t)$$

and

$$\tilde{\hat{p}}(t) \equiv \hat{U}_{\text{cm}}^\dagger(t)\,\hat{p}\,\hat{U}_{\text{cm}}(t)$$

Here, the tilde denotes operators in the Heisenberg picture.

We now show that these Heisenberg operators can be expressed as time dependent linear combinations of the annihilation and creation operators $\hat{b}$ and $\hat{b}^\dagger$ of the static reference oscillator. Indeed, we can prove the relations

$$\hat{U}_{\text{cm}}^\dagger(t)\,\hat{x}\,\hat{U}_{\text{cm}}(t) = \sqrt{\frac{\hbar}{2M\omega_r}}\,\left[\epsilon^*(t)\,\hat{b} + \epsilon(t)\,\hat{b}^\dagger\right] \qquad (17.12a)$$

and

$$\hat{U}^\dagger_{cm}(t)\,\hat{p}\,\hat{U}_{cm}(t) = \sqrt{\frac{\hbar M}{2\omega_r}}\left[\dot{\epsilon}^*(t)\,\hat{b} + \dot{\epsilon}(t)\,\hat{b}^\dagger\right]. \tag{17.12b}$$

The complex-valued function $\epsilon(t)$ obeys the classical Mathieu differential equation

$$\ddot{\epsilon}(t) + \omega^2(t)\,\epsilon(t) = 0 \tag{17.13}$$

with the initial conditions

$$\epsilon(0) = 1 \quad \text{and} \quad \dot{\epsilon}(0) = i\omega_r. \tag{17.14}$$

We prove Eqs. (17.12) by showing that the operator $\hat{\tilde{x}}(t)$ as defined in terms of $\hat{b}$ and $\hat{b}^\dagger$ on the right hand side of Eqs. (17.12) and by the unitary transformation of the left hand side satisfies the same differential equation subjected to the same initial conditions. Indeed, when we differentiate the operator $\hat{\tilde{x}}(t)$ as defined by the unitary transformation we find using the relation

$$\frac{d\hat{U}_{cm}(t)}{dt} = -\frac{i}{\hbar}\,\hat{H}_{cm}(t)\,\hat{U}_{cm}(t)$$

following from Eq. (17.11) the equation of motion

$$\dot{\hat{\tilde{x}}}(t) = \frac{i}{\hbar}\,\hat{U}^\dagger_{cm}(t)\,[\hat{H}_{cm}(t),\hat{x}]\,\hat{U}_{cm}(t) = \frac{1}{M}\,\hat{U}^\dagger_{cm}(t)\,\hat{p}\,\hat{U}_{cm}(t) \equiv \frac{1}{M}\,\hat{\tilde{p}}(t).$$

In complete analogy to $\hat{\tilde{x}}(t)$ we find for the momentum operator

$$\dot{\hat{\tilde{p}}}(t) = -M\omega^2(t)\,\hat{\tilde{x}}(t).$$

Hence the operator $\hat{\tilde{x}}(t)$ satisfies the second order differential equation

$$\ddot{\hat{\tilde{x}}}(t) + \omega^2(t)\,\hat{\tilde{x}}(t) = 0.$$

We note that due to Eq. (17.13) the right hand side of Eqs. (17.12) also satisfies this equation. Moreover, due to Eq. (17.14), they agree in their initial conditions

$$\hat{\tilde{x}}(t=0) = \hat{x} = \sqrt{\frac{\hbar}{2M\omega_r}}\,(\hat{b}+\hat{b}^\dagger)$$

and

$$\dot{\hat{\tilde{x}}}(t=0) = \frac{1}{M}\,\hat{\tilde{p}}(t=0) = \frac{1}{M}\,\hat{p} = \sqrt{\frac{\hbar\omega_r}{2M}}\,i\,(\hat{b}^\dagger - \hat{b}).$$

Equations (17.12) are the key equations for the harmonic oscillator with time dependent frequency. They show that in this case time dependence appears only through the classical function $\epsilon(t)$. All the quantum mechanics is contained in the operators $\hat{b}$ and $\hat{b}^\dagger$ of the static reference oscillator.

We conclude this subsection by noting an important property of the function ϵ and its complex conjugate ϵ^*. They form the Wronskian

$$\mathcal{W}[\epsilon(t),\dot{\epsilon}(t)] \equiv \dot{\epsilon}(t)\epsilon^*(t) - \epsilon(t)\dot{\epsilon}^*(t),$$

which is constant. Indeed, when we differentiate $\mathcal{W}$ with respect to t and use the fact, that $\epsilon(t)$ and $\epsilon^*(t)$ satisfy the differential equation

$$\ddot{\epsilon}(t) + \omega^2(t)\,\epsilon(t) = 0\,,$$

we find $\dot{\mathcal{W}}(\epsilon, \dot{\epsilon}) \equiv 0$. With the initial conditions $\epsilon(0) = 1$ and $\dot{\epsilon}(0) = i\,\omega_r$, Eq. (17.14), we therefore arrive at

$$\mathcal{W}[\epsilon(t), \dot{\epsilon}(t)] \equiv \mathcal{W}[\epsilon(0), \dot{\epsilon}(0)] \equiv 2\,i\,\omega_r\,. \tag{17.15}$$

Moreover, provided that $\omega_r \neq 0$ the function $\epsilon(t)$ never vanishes, otherwise the Wronskian would be zero.

17.3.2 Motion as a Sequence of Squeezing and Rotations

In the Heisenberg picture the position and momentum operators $\hat{\tilde{x}}(t)$ and $\hat{\tilde{p}}(t)$ are time dependent linear combinations of the annihilation and creation operators $\hat{b}$ and $\hat{b}^\dagger$ of the static reference oscillator. This combination is very reminiscent of the squeezing transformation, introduced in Problem 11.5. Indeed, two-photon states are eigenstates of a linear combination of annihilation and creation operators. We now analyze the action of the unitary time evolution operator $\hat{U}_{\rm cm}$ on $\hat{b}$ from this perspective.

In particular, we show that $\hat{U}_{\rm cm}(t)$ is a sequence of a rotation operator

$$\hat{R}(\vartheta) \equiv \exp\left[-i\,\vartheta\left(\hat{b}^\dagger \hat{b} + \tfrac{1}{2}\right)\right] \tag{17.16}$$

followed by a squeezing operator

$$\hat{S}(r) \equiv \exp\left[\tfrac{r}{2}\left(\hat{b}^{\dagger 2} - \hat{b}^2\right)\right] \tag{17.17}$$

and a second rotation operator. The total time evolution then reads

$$\hat{U}_{\rm cm}(t) \equiv \hat{R}[\Theta(t)]\,\hat{S}[r(t)]\,\hat{R}[\vartheta(t)]\,.$$

We determine explicit expressions for the squeezing parameter r and the two rotation angles Θ and ϑ.

Generalized Squeezing Transformation

We first show that the time evolution operator $\hat{U}_{\rm cm}$ transforms the annihilation operator $\hat{b}$ of the reference oscillator into a linear combination of annihilation and creation operators.

When we recall the definition

$$\hat{b} = \sqrt{\frac{M\,\omega_r}{2\,\hbar}}\,\hat{x} + i\,\sqrt{\frac{1}{2\,\hbar\,M\,\omega_r}}\,\hat{p} \tag{17.18}$$

of the operator $\hat{b}$ in terms of the position and momentum operators and the transformation formulas

$$\hat{U}^\dagger_{\rm cm}(t)\,\hat{x}\,\hat{U}_{\rm cm}(t) = \sqrt{\frac{\hbar}{2\,M\,\omega_r}}\left[\epsilon^*(t)\,\hat{b} + \epsilon(t)\,\hat{b}^\dagger\right]$$

and
$$\hat{U}^\dagger_{cm}(t)\,\hat{p}\,\hat{U}_{cm}(t) = \sqrt{\frac{\hbar M}{2\omega_r}}\left[\dot{\epsilon}^*(t)\,\hat{b} + \dot{\epsilon}(t)\,\hat{b}^\dagger\right] \tag{17.19}$$

we find
$$\hat{U}^\dagger_{cm}(t)\,\hat{b}\,\hat{U}_{cm}(t) = c(t)\,\hat{b} - s(t)\,\hat{b}^\dagger. \tag{17.20}$$

Here the coefficients
$$c(t) \equiv \frac{1}{2}\left[\epsilon^*(t) + i\frac{1}{\omega_r}\dot{\epsilon}^*(t)\right] \tag{17.21}$$

and
$$s(t) \equiv -\frac{1}{2}\left[\epsilon(t) + i\frac{1}{\omega_r}\dot{\epsilon}(t)\right] \tag{17.22}$$

satisfy the equation
$$|c(t)|^2 - |s(t)|^2 = \frac{-i}{2\omega_r}\left[\dot{\epsilon}(t)\epsilon^*(t) - \epsilon(t)\dot{\epsilon}^*(t)\right]. \tag{17.23}$$

When we recall the relation Eq. (17.15)
$$\mathcal{W}[\epsilon(t),\dot{\epsilon}(t)] \equiv \dot{\epsilon}(t)\epsilon^*(t) - \epsilon(t)\dot{\epsilon}^*(t) = 2\,i\,\omega_r,$$

for the Wronskian we find
$$|c(t)|^2 - |s(t)|^2 \equiv 1. \tag{17.24}$$

This allows us to express the absolute values of c and s as $\cosh r$ and $\sinh r$, that is
$$c(t) = \cosh r(t)\,\exp[i\,\gamma(t)] \tag{17.25}$$

and
$$s(t) = \sinh r(t)\,\exp[i\,\beta(t)]. \tag{17.26}$$

For the parameter $r(t)$ we find with the help of Eqs. (17.21) and (17.22)
$$\exp[r(t)] = \cosh r(t) + \sinh r(t) = |c(t)| + |s(t)|$$
$$= \frac{1}{2}\left[\sqrt{|\epsilon(t)|^2 + |\dot{\epsilon}(t)|^2/\omega_r^2 + 2} + \sqrt{|\epsilon(t)|^2 + |\dot{\epsilon}(t)|^2/\omega_r^2 - 2}\right] \tag{17.27}$$

and similarly
$$\exp[-r(t)] = \frac{1}{2}\left[\sqrt{|\epsilon(t)|^2 + |\dot{\epsilon}(t)|^2/\omega_r^2 + 2} - \sqrt{|\epsilon(t)|^2 + |\dot{\epsilon}(t)|^2/\omega_r^2 - 2}\right]. \tag{17.28}$$

In the representations (17.25) and (17.26) for $c(t)$ and $s(t)$ Eq. (17.20) reads
$$\hat{U}^\dagger_{cm}(t)\,\hat{b}\,\hat{U}_{cm}(t) = \cosh r(t)e^{i\,\gamma(t)}\hat{b} - \sinh r(t)e^{i\,\beta(t)}\hat{b}^\dagger.$$

Identification of Rotations and Squeezing

We note, that the three parameters r, γ and β determine the time evolution of the annihilation operator. We now show that the time evolution can be represented by a sequence of a rotation, a squeezing and another rotation.

For this purpose we recall the properties

$$\hat{R}^\dagger(\varphi)\,\hat{b}\,\hat{R}(\varphi) \equiv \hat{b}\,\exp(-i\,\varphi)$$

of the rotation operator $\hat{R}$ and the action

$$\hat{S}^\dagger(r)\,\hat{b}\,\hat{S}(r) \equiv \cosh r\,\hat{b} - \sinh r\,\hat{b}^\dagger$$

of the squeezing operator $\hat{S}$. This allows us to establish the relation

$$\hat{R}^\dagger\left(\tfrac{\beta-\gamma}{2}\right)\hat{S}^\dagger(r)\left[\hat{R}^\dagger\left(-\tfrac{\gamma+\beta}{2}\right)\hat{b}\hat{R}\left(-\tfrac{\gamma+\beta}{2}\right)\right]\hat{S}(r)\hat{R}\left(\tfrac{\beta-\gamma}{2}\right)$$

$$= \hat{R}^\dagger\left(\tfrac{\beta-\gamma}{2}\right)\left[\hat{S}^\dagger(r)\,\hat{b}\,\exp\left(i\tfrac{\gamma+\beta}{2}\right)\hat{S}(r)\right]\hat{R}\left(\tfrac{\beta-\gamma}{2}\right)$$

$$= \hat{R}^\dagger\left(\tfrac{\beta-\gamma}{2}\right)\left[\cosh r\,e^{i\frac{\gamma+\beta}{2}}\,\hat{b} - \sinh r\,e^{i\frac{\gamma+\beta}{2}}\,\hat{b}^\dagger\right]\hat{R}\left(\tfrac{\beta-\gamma}{2}\right)$$

$$= \cosh r\,e^{i\gamma}\,\hat{b} - \sinh r\,e^{i\beta}\,\hat{b}^\dagger$$

$$= \hat{U}_{\text{cm}}^\dagger(t)\,\hat{b}\,\hat{U}_{\text{cm}}(t)\,.$$

So far we have shown that the operator $\hat{U}_{\text{cm}}(t)$ acting on $\hat{b}$ is identical to a sequence of rotation, squeezing and rotation operators. Obviously, the same holds for $\hat{b}^\dagger$ and hence for any operator. Therefore, we identify the action of the evolution operator $\hat{U}_{\text{cm}}(t)$ as a subsequent application of a rotation, a squeezing, and a second rotation,

$$\hat{U}_{\text{cm}}(t) \equiv \hat{R}[\Theta(t)]\,\hat{S}[r(t)]\,\hat{R}[\vartheta(t)]\,,$$

where the angles

$$\Theta(t) \equiv -\frac{\gamma(t) + \beta(t)}{2} \tag{17.29}$$

and

$$\vartheta(t) \equiv \frac{\beta(t) - \gamma(t)}{2} \tag{17.30}$$

are defined via

$$\cos[2\,\Theta(t)] = \cos\gamma(t)\cos\beta(t) - \sin\gamma(t)\sin\beta(t)$$
$$= \frac{\operatorname{Re} c(t)\operatorname{Re} s(t) - \operatorname{Im} c(t)\operatorname{Im} s(t)}{|c(t)|\,|s(t)|}$$
$$= \frac{|\dot{\epsilon}(t)|^2/\omega_r^2 - |\epsilon(t)|^2}{\sqrt{(|\epsilon(t)|^2 + |\dot{\epsilon}(t)|^2/\omega_r^2)^2 - 4}} \tag{17.31}$$

and

$$\sin[2\,\Theta(t)] = -\frac{\tfrac{2}{\omega_r}\operatorname{Re}[\epsilon(t)\,\dot{\epsilon}^*(t)]}{\sqrt{(|\epsilon(t)|^2 + |\dot{\epsilon}(t)|^2/\omega_r^2)^2 - 4}}\,. \tag{17.32}$$

Similarly, we find

$$\cos[2\vartheta(t)] = -\frac{\operatorname{Re}\left[\epsilon(t)^2 + \dot{\epsilon}(t)^2/\omega_r^2\right]}{\sqrt{(|\epsilon(t)|^2 + |\dot{\epsilon}(t)|^2/\omega_r^2)^2 - 4}} \tag{17.33}$$

and

$$\sin[2\vartheta(t)] = -\frac{\operatorname{Im}\left[\epsilon(t)^2 + \dot{\epsilon}(t)^2/\omega_r^2\right]}{\sqrt{(|\epsilon(t)|^2 + |\dot{\epsilon}(t)|^2/\omega_r^2)^2 - 4}}. \tag{17.34}$$

We note, that in the limit of $t \to 0$ the operator $\hat{U}_{\text{cm}}(t)$ approaches the unity operator. Indeed, from the initial conditions $\epsilon(0) = 1$ and $\dot{\epsilon}(0) = i\omega_r$, we find from Eqs. (17.21) and (17.22) the properties $c(0) \equiv 1$ and $s(0) \equiv 0$, which with the help of Eqs. (17.25) and (17.26) immediately yield $r(0) \equiv 0$ and $\gamma(0) \equiv 0$. Hence, the squeezing operator $\hat{S}(0)$ is identical to the unity operator and the two remaining rotations combine to a single rotation by the angle $\Theta(0) + \vartheta(0) = -\gamma(0) \equiv 0$, as can be seen from Eqs. (17.29) and (17.30).

The limit of the angles $\Theta(t)$ and $\vartheta(t)$ is more complicated, since their definitions Eqs. (17.29) and (17.30) both involve the angle $\beta(t)$, which due to $s(0) \equiv 0$ is not defined for $t = 0$. Hence, we have to consider higher derivatives of $s(t)$ at $t = 0$. A careful analysis yields $\Theta(0+) = \pi/4 = -\vartheta(0+)$ for $\omega_r > \omega(0)$, and $\Theta(0+) = -\pi/4 = -\vartheta(0+)$ for $\omega_r < \omega(0)$. For $\omega_r \equiv \omega(0)$ as we have chosen in Fig. 17.5, it is the first non-vanishing derivative of $\omega^2(t)$ at $t = 0$, which determines $\Theta(0+)$ and $\vartheta(0+)$. In this case of $\omega_r \equiv \omega(0)$ and for the specific $\omega^2(t)$ defined by Eq. (17.10), we find $\Theta(0+) = \pi/4 = -\vartheta(0+)$.

17.3.3 Dynamics in Wigner Phase Space

So far, we have analyzed the dynamics of a single ion in an explicitly time dependent Paul trap using the Heisenberg picture. We now return to the Schrödinger picture and consider the time dependence of the state given by

$$|\psi_{\text{cm}}(t)\rangle = \hat{U}_{\text{cm}}(t)|\psi_{\text{cm}}(0)\rangle \equiv \hat{\mathcal{T}}\exp\left[-\frac{i}{\hbar}\int_0^t dt'\,\hat{H}_{\text{cm}}(t')\right]|\psi_{\text{cm}}(0)\rangle. \tag{17.35}$$

We can visualize this dynamics of these quantum states with the help of the Wigner function

$$W(x,p;t) \equiv \frac{1}{2\pi\hbar}\int_{-\infty}^{\infty} d\xi\,\psi_{\text{cm}}^*(x-\tfrac{1}{2}\xi,t)\,\psi_{\text{cm}}(x+\tfrac{1}{2}\xi,t)\,e^{-ip\xi/\hbar}. \tag{17.36}$$

In Chapter 3 we have shown, that for all Hamiltonians, which are quadratic in the position and momentum operators, the Wigner function satisfies the classical Liouville equation

$$\left[\frac{\partial}{\partial t} + \frac{p}{M}\frac{\partial}{\partial x} - M\omega^2(t)\,x\frac{\partial}{\partial p}\right]W(x,p;t) = 0.$$

Hence we can express the Wigner function

$$W(x,p;t) = W[\bar{x}(x,p;t),\bar{p}(x,p;t);t=0] \tag{17.37}$$

17.3 Motion of an Ion in a Paul Trap   487

at an arbitrary time t in terms of the initial Wigner function $W(x, p; 0)$ at $t = 0$, when we replace the initial coordinate and the initial momentum by

$$\begin{pmatrix} \bar{x}(x, p; t) \\ \bar{p}(x, p; t) \end{pmatrix} = \begin{pmatrix} \frac{1}{\omega_r} \operatorname{Im} \dot{\epsilon}(t) & -\frac{1}{M\omega_r} \operatorname{Im} \epsilon(t) \\ -M \operatorname{Re} \dot{\epsilon}(t) & \operatorname{Re} \epsilon(t) \end{pmatrix} \begin{pmatrix} x \\ p \end{pmatrix}. \tag{17.38}$$

Dynamics from Mixing Phase Space Coordinates

We now study this dynamics in more detail. We note that the transformation Eq. (17.38) relates the coordinate x and momentum p of the Wigner function at time t to the phase space coordinates $\bar{x}$ and $\bar{p}$ of the Wigner function at time $t = 0$. In order to connect this phase space transformation to the propagator $\hat{U}_{\mathrm{cm}}(t)$ defined by Eq. (17.35), we use the inverse transformation

$$\begin{pmatrix} x(t) \\ p(t) \end{pmatrix} = \begin{pmatrix} \operatorname{Re} \epsilon(t) & \frac{1}{M\omega_r} \operatorname{Im} \epsilon(t) \\ M \operatorname{Re} \dot{\epsilon}(t) & \frac{1}{\omega_r} \operatorname{Im} \dot{\epsilon}(t) \end{pmatrix} \begin{pmatrix} \bar{x} \\ \bar{p} \end{pmatrix}, \tag{17.39}$$

that is we express the phase space coordinates (x, p) of the final distribution by the original ones. Here we have used the fact that according to Eqs. (17.23) and (17.24) the determinant of the phase space transformation Eq. (17.38) is unity.

We can therefore interpret the time evolution operator $\hat{U}_{\mathrm{cm}}$ acting in phase space as

$$\begin{pmatrix} x(t) \\ p(t) \end{pmatrix} \equiv U_{\mathrm{cm}}(t) \begin{pmatrix} \bar{x} \\ \bar{p} \end{pmatrix}.$$

We recall that $\hat{U}_{\mathrm{cm}}$ consists of a squeezing operation and two rotations. We now show that this can be read off directly from the matrix $\hat{U}_{\mathrm{cm}}$.

Indeed, in phase space we can write the rotation operation

$$R(\varphi) = \begin{pmatrix} \cos \varphi & \frac{1}{M\omega_r} \sin \varphi \\ -M \omega_r \sin \varphi & \cos \varphi \end{pmatrix}$$

and the squeezing operation

$$S(r) = \begin{pmatrix} \exp(-r) & 0 \\ 0 & \exp(r) \end{pmatrix}$$

in terms of 2×2 matrices. Note, that according to our definition Eq. (17.16) of the rotation operator $\hat{R}(\varphi)$ a positive angle φ yields a clockwise rotation. Hence, the unitary evolution operator $\hat{U}_{\mathrm{cm}}(t)$ acts in phase space as

$$U_{\mathrm{cm}}(t) = R[\Theta(t)] \, S[r(t)] \, R[\vartheta(t)],$$

which with the help of the explicit expressions for $r(t)$, $\Theta(t)$, and $\vartheta(t)$, Eqs. (17.27) and (17.28), Eqs. (17.31) and (17.32), and Eqs. (17.33) and (17.34) respectively, simplifies after minor algebra to

$$U_{\mathrm{cm}}(t) \equiv \begin{pmatrix} \operatorname{Re} \epsilon(t) & -\frac{1}{M\omega_r} \operatorname{Im} \epsilon(t) \\ M \operatorname{Re} \dot{\epsilon}(t) & \frac{1}{\omega_r} \operatorname{Im} \dot{\epsilon}(t) \end{pmatrix}.$$

Note, that this matrix is the phase space transformation Eq. (17.39) relating the initial Wigner function at $t = 0$ to the Wigner function at some later time $t > 0$.

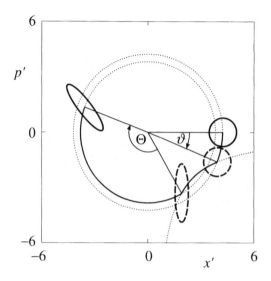

Fig. 17.4: Time evolution of a vibratory quantum state of an ion in the Paul trap as a sequence of rotation, squeezing and rotation. According to Eq.(17.37) an initial coherent state $|\psi_{\text{cm}}(0)\rangle = |\alpha = 3\rangle$ of the ion displayed here by a solid circular Wigner function contour line evolves into a squeezed state $|\psi_{\text{cm}}(\tau)\rangle$ shown by a solid elliptical contour line of its Wigner function. The first transformation rotates the initial Gaussian Wigner function centered at the scaled phase space coordinates $x' \equiv \sqrt{M\omega_r/\hbar}\, x = \sqrt{2}\,\alpha$ and $p' \equiv 1/\sqrt{\hbar M \omega_r}\, p = 0$ by the angle $\vartheta(\tau) = 0.387$ rad defined by Eqs. (17.33) and (17.34). The squeezing operation subsequently squeezes the resulting dashed circular contour line of the Wigner function along the two Cartesian axes of phase space. At the same time the center of the Gaussian moves on the hyperbola $x' \cdot p' = \text{const} \equiv \alpha^2 \sin[2\,\vartheta(\tau)]$. The final rotation rotates the ellipse of the squeezed Gaussian by the angle $\Theta(\tau) = 2.454$ rad defined in Eqs. (17.31) and (17.32), which results in the Wigner function of the final state $|\psi_{\text{cm}}(\tau)\rangle$. Note, that we have defined the rotation operator Eq. (17.16) such that positive rotation angles correspond to rotations in the clockwise direction. Here we have chosen the parameters $a = 0$, $q = 0.4$ and $\omega_r = \omega(0)$ and the time $\omega_{\text{rf}}\tau = 9/\sqrt{0.2}$.

Example of a Gaussian Wave Packet

The time dependence of ϑ, r, and Θ results from the dependence of these quantities on the complex-valued functions $\epsilon(t)$ and $\dot{\epsilon}(t)$. In Fig. 17.4 we illustrate the effect of all three transformations on the coherent state $|\psi_{\text{cm}}(0)\rangle = |\alpha = 3\rangle$, that is we start with the initial Gaussian Wigner function

$$W(x,p,0) = \frac{1}{\pi \hbar} \exp\left[-\frac{M\omega_r}{\hbar}(x - \sqrt{2}\,\alpha)^2 - \frac{1}{\hbar M \omega_r} p^2\right]$$

and construct the Wigner function $W(x, p, \tau)$ at the later time $\tau = 9/(\sqrt{0.2}\,\omega_{\text{rf}})$. The first transformation $R(\vartheta)$ only rotates the Wigner function by an angle $\vartheta(\tau) = 0.387$ rad around the origin of phase space. Hence the Wigner function is now centered at

$$x_\vartheta \sqrt{\frac{M\omega_r}{\hbar}} = \sqrt{2}\,\alpha \cos \vartheta(\tau)$$

17.3 Motion of an Ion in a Paul Trap   489

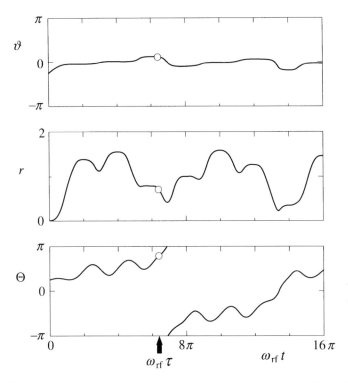

Fig. 17.5: Time dependence of the three transformation parameters $\vartheta(t)$, $r(t)$, and $\Theta(t)$ characterizing the temporal evolution of a single ion in the Paul trap. Time is scaled in terms of the frequency $\omega_{\rm rf}$ of the AC voltage. The arrow indicates the value of τ used in Fig. 17.4 and the circles are the corresponding values of ϑ, r, and Θ at τ. The remaining parameters are as in Fig. 17.4.

and
$$p_\vartheta \sqrt{\tfrac{1}{\hbar M \omega_r}} = \sqrt{2}\,\alpha \sin \vartheta(\tau).$$

The squeezing operation $S(r)$ then squeezes the resulting Wigner function with respect to the x-axes of the phase space, since the parameter $r(\tau) = 0.719$ is real. At the same time the center of the Gaussian moves on the hyperbola $x \cdot p = x_\vartheta \cdot p_\vartheta \equiv \alpha^2\,\hbar \sin[2\,\vartheta(\tau)]$ by an amount which depends on r. Finally the third transformation $R(\Theta)$ rotates the ellipse of the squeezed state by the angle $\Theta(\tau) = 2.454$ rad around the origin of phase space. Hence, the axes of the ellipse are rotated by $\Theta(\tau)$ with respect to the fixed coordinate system. In this figure we have taken the trap parameters $a = 0$, $q = 0.4$, and the reference frequency $\omega_r = \omega(0)$.

We emphasize that the three parameters $\vartheta(t)$, $r(t)$, and $\Theta(t)$ characterize the complete time evolution of any Wigner function. In Fig. 17.5 we display the time dependence of ϑ, r, and Θ using the same trap parameters as in Fig. 17.4. Here we have evaluated the functions $\epsilon(t)$ and $\dot\epsilon(t)$ by numerically integrating the Mathieu equation (17.13). We note the complicated time dependence of the transformation parameters $\vartheta(t)$, $r(t)$, and $\Theta(t)$ which results in a complicated temporal evolution of the system.

17.3.4 Floquet Solution

In the last section we have shown that the three parameters $\vartheta(t)$, $r(t)$, and $\Theta(t)$ describe the time evolution of any quantum state of motion in the Paul trap. We note that they depend on the choice of the reference frequency ω_r via the functions $\epsilon(t)$ and $\dot{\epsilon}(t)$. We now discuss a special choice of ω_r which simplifies the classical solution $\epsilon(t)$ and hence the time evolution of these parameters.

Secular Motion Determined by Characteristic Exponent

For this purpose we solve the classical differential equation (17.13) for $\epsilon(t)$ subjected to the initial condition (17.14). According to the Floquet theorem the general solution reads

$$\epsilon(t) = A\, e^{i\mu t}\, \phi(t) + B\, e^{-i\mu t}\, \phi(-t) \tag{17.40}$$

where A and B are constants and

$$\phi(t) = \phi(t+T) = \sum_{n=-\infty}^{\infty} c_n \exp\left(in\omega_{\mathrm{rf}} t\right) \tag{17.41}$$

is a periodic function.

We find the expansion coefficients c_n and the characteristic exponent μ by substituting the Floquet solution

$$\epsilon^{(F)}(t) = e^{i\mu t}\, \phi(t) \tag{17.42}$$

into Eq. (17.13). This yields for the case of the time dependent frequency Eq. (17.10) the three-term recurrence relation Eq. (17.7) which via the Hill determinant, Eq. (17.8), provides us with the characteristic exponent μ. The coefficients c_n follow from the resulting linear set of equations.

In Table 17.1 we present the coefficients c_n for $|n| \leq 3$ for three different pairs (a, q) of trap parameters from the first stable region of the Mathieu equation. We note that within this region the coefficients c_n are real. This is not by accident as we show now.

Due to the special symmetry $\omega^2(-t) \equiv \omega^2(t)$ of the time dependent frequency, Eq. (17.10), the functions $\epsilon^*(t)$ and $\epsilon(-t)$ satisfy the same differential equation with the same initial conditions. Hence, both functions are identical, that is

$$\epsilon^*(t) = \epsilon(-t) . \tag{17.43}$$

When we now substitute the ansatz Eq. (17.40) into Eq. (17.43) and use the fact that in the stable region of the Mathieu equation the secular frequency μ is real, we find the relation $\phi^*(t) = \phi(-t)$, which with the help of Eq. (17.41) yields

$$\sum_{n=-\infty}^{\infty} c_n^*\, e^{-in\omega_{\mathrm{rf}} t} = \sum_{n=-\infty}^{\infty} c_n\, e^{-in\omega_{\mathrm{rf}} t}. \tag{17.44}$$

Hence the coefficients c_n have to be real.

Tab. 17.1: The coefficients c_n for $|n| \leq 3$ for three different pairs (a, q) of trap parameters from the first stable regime of the Mathieu equation. Moreover we show the ratio $\Omega_n/\Omega_n^{(\mu)} \equiv c_0(\mu/\omega_r^{(F)})^{1/2}$ between the Rabi frequencies of the two Hamiltonians $\hat{\tilde{H}}_{\text{int}}^{(s)}$ and $\hat{\tilde{H}}_{\text{int}}^{(\mu)}$ including and discarding the influence of the micromotion together with the squeezing parameter $s \equiv \mu/\omega_r^{(F)}$, that is the ratio of the frequencies μ and $\omega_r^{(F)}$. In the last row we present the ratio ω_{rf}/μ between the trap frequency and the secular frequency.

	$a = 0, q = 0.1$	$a = 0, q = 0.4$	$a = 0.2, q = 0.6$
c_0	0.9519165	0.8198607	0.6540799
c_1	0.0222005	0.0624992	0.0551381
c_{-1}	0.0255825	0.1129407	0.2734725
c_2	0.0001340	0.0013571	0.0015066
c_{-2}	0.0001657	0.0032879	0.0154435
c_3	0.0000004	0.0000137	0.0000202
c_{-3}	0.0000005	0.0000404	0.0003334
$c_0 \left(\mu/\omega_r^{(F)}\right)^{1/2}$	1.001890	1.034380	1.190388
$s = \mu/\omega_r^{(F)}$	1.107758	1.591780	3.312210
ω_{rf}/μ	28.23	6.84	2.82

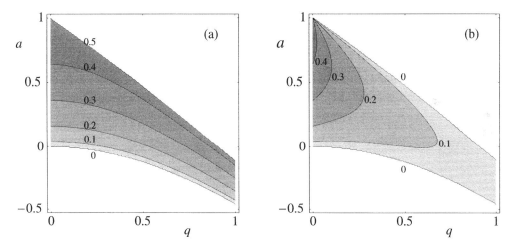

Fig. 17.6: Dependence of the characteristic exponent μ (left) and of the Floquet reference frequency $\omega_r^{(F)}$ (right) on the trap parameters a and q. Here we show the contour lines of $\mu = \mu(a, q)$ and $\omega_r^{(F)} = \omega_r^{(F)}(a, q)$ in the first stable region of the Mathieu equation. Both quantities are scaled in units of the radio frequency ω_{rf}.

In Fig. 17.6(a) we display the characteristic exponent μ in units of the frequency ω_{rf} in the first stable region of the Mathieu equation as a function of the trap parameters a and q. In this domain the characteristic exponent is purely real and gives the frequency of the secular motion of the ion.

Choice of Reference Oscillator

We now focus on the Floquet solution Eq. (17.42). We satisfy the condition $A = 1$ and $B = 0$ by appropriately choosing the reference frequency ω_r. To find this reference frequency $\omega_r^{(F)}$ we substitute the Floquet solution Eq. (17.42) into the initial conditions Eq. (17.14) and find

$$\epsilon^{(F)}(0) = 1 = \phi(0) = \sum_{n=-\infty}^{\infty} c_n \qquad (17.45)$$

and $\dot{\epsilon}^{(F)}(0) = i\omega_r^{(F)} = i\mu + \dot{\phi}(0)$ which together with Eq. (17.45) yields

$$\omega_r^{(F)} = \mu + \omega_{\text{rf}} \sum_{n=-\infty}^{\infty} n\, c_n \,. \qquad (17.46)$$

Hence, the characteristic exponent μ and the coefficients c_n determine via Eq. (17.46) the particular reference frequency $\omega_r^{(F)}$ such as to enforce the Floquet solution Eq. (17.42).

In Fig. 17.6(b) we show the dependence of $\omega_r^{(F)}$ on a and q again for the first stable region of the Mathieu equation. We note that for $q \equiv 0$, i.e. for a time-independent oscillator, the two frequencies μ and $\omega_r^{(F)}$ are identical.

Example of a Gaussian Wave Packet

In Fig. 17.7 we show the time dependence of ϑ, r, and Θ for the same trap parameters $a = 0$ and $q = 0.4$ as in Fig. 17.5. However, this time we have chosen the particular reference frequency $\omega_r^{(F)}$ enforcing a Floquet solution

$$\epsilon(t) \equiv \epsilon^{(F)}(t) = e^{i\mu t}\phi(t),$$

where $\phi(t+T) = \phi(t)$.

We note from this figure that the angle Θ and the squeezing parameter r are periodic with period T of the trap potential.

Indeed, this periodicity follows from our equations. Since the definitions of $r(t)$, Eqs. (17.27) and (17.28), and of $\Theta(t)$, Eqs. (17.31) and (17.32), only involve the absolute values of $\epsilon^{(F)}(t)$ and $\dot{\epsilon}^{(F)}(t)$ or the product of $\epsilon^{(F)}(t)$ and $[\dot{\epsilon}^{(F)}(t)]^*$, it is obvious that the parameters $r(t)$ and $\Theta(t)$ are periodic in t with period T, that is

$$r(t+T) = r(t)$$

and

$$\Theta(t+T) = \Theta(t) \,.$$

Moreover, since $\epsilon(t)$ and $\dot{\epsilon}(t)$ enter the definition of ϑ quadratically, we find

$$\vartheta(t+T) = \vartheta(t) + \mu T \,.$$

These periodicity properties imply for the transformation matrix

$$U_{\text{cm}}(t+T) = R[\Theta(t+T)]\, S[r(t+T)]\, R[\vartheta(t+T)]$$

17.3 Motion of an Ion in a Paul Trap

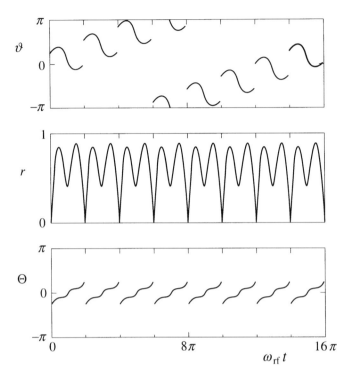

Fig. 17.7: Time dependence of the transformation parameters $\vartheta(t)$, $r(t)$ and $\Theta(t)$ for the reference frequency $\omega_r \equiv \omega_r^{(F)}$ enforcing the Floquet solution. Note, that in contrast to Fig. 17.5 the squeezing parameter r and the rotation angle Θ are periodic, that is $r(t+T) \equiv r(t)$ and $\Theta(t+T) \equiv \Theta(t)$, whereas for ϑ we find $\vartheta(t+T) \equiv \vartheta(t) + \mu T$. Here as in Fig. 17.5 the trap parameters are $a = 0$ and $q = 0.4$.

at time $t + T$ the expression

$$U_{\text{cm}}(t+T) = \left(R[\Theta(t)]\, S[r(t)]\, R[\vartheta(t)]\right) R^{-1}[\vartheta(t)]\, R[\vartheta(t+T)]$$
$$= U_{\text{cm}}(t)\, R^{-1}[\vartheta(t)]\, R[\vartheta(t+T)].$$

Here we have made use of the relation $R^{-1} R = 1$.

We combine the rotation by an angle $\vartheta(t+T)$ and the inverse rotation by the angle $\vartheta(t)$ to obtain a single rotation by the angle $\vartheta(t+T) - \vartheta(t) = \mu T$ and hence arrive at

$$U_{\text{cm}}(t+T) = U_{\text{cm}}(t)\, R(\mu T). \tag{17.47}$$

Therefore, in the basis of the Floquet oscillator the Wigner function of an arbitrary state at time $t + T$ is only rotated in phase space by the angle μT with respect to the Wigner function at time t.

We gain deeper insight into this special choice of the reference frequency when we consider the time evolution of the energy eigenstates $|n\rangle$ of the time independent harmonic oscillator of frequency $\omega_r^{(F)}$. From Eqs. (17.35) and (17.47) we find for the initial state $|\psi_{\text{cm}}(0)\rangle = |n\rangle$ after one period T the relation

$$|\psi_{\text{cm}}(T)\rangle = \hat{U}_{\text{cm}}(T)|n\rangle = \hat{R}(\mu T)|n\rangle = e^{-i(n+\frac{1}{2})\mu T}|n\rangle,$$

and hence the energy eigenstates $|n\rangle$ of the reference oscillator with frequency $\omega_r^{(F)}$ are initial values of the Floquet states at time t.

17.4 Model Hamiltonian

So far, we have shown that due to the explicit time dependence of the Paul trap the motion of a single ion in the Paul trap is rather complicated. We emphasize that this is not due to quantum mechanics but solely results from the time dependence of the trapping potential. Indeed, since we are dealing with a harmonic oscillator classical and quantum dynamics are identical as expressed by the Liouville equation for the Wigner function.

In the present section we couple this motion of the ion to its internal dynamics. We achieve this coupling via the interaction with a classical laser field. In this situation the state vector $|\Psi(t)\rangle$ of the complete system contains both the internal states and the center-of-mass motion and follows from the Schrödinger equation

$$i\hbar \frac{d}{dt}|\Psi(t)\rangle = \hat{H}(t)|\Psi(t)\rangle. \tag{17.48}$$

The full Hamiltonian

$$\hat{H}(t) \equiv \hat{H}_{\text{cm}}(t) + \hat{H}_a + \hat{H}_{\text{int}}(t) \tag{17.49}$$

is the sum of the Hamiltonians $\hat{H}_{\text{cm}}(t)$, $\hat{H}_a$, and $\hat{H}_{\text{int}}(t)$ describing the center-of-mass motion Eq. (17.9), the internal states and the interaction with the laser field, respectively.

For the sake of simplicity we consider only two internal states of the atom and write

$$\hat{H}_a \equiv \tfrac{1}{2}\hbar\omega_a\hat{\sigma}_z,$$

where ω_a denotes the transition frequency between the two levels and $\hat{\sigma}_z$ the Pauli matrix.

We can envision two types of laser fields interacting with the ion: (i) a running wave or (ii) a standing wave. In the case of a running wave the interaction Hamiltonian reads

$$\hat{H}_{\text{int}}^{(r)}(t) = \hbar g\left\{\hat{\sigma}^\dagger \exp[-i(\omega_L t - k\hat{x})] + \text{h.c.}\right\} \tag{17.50}$$

Here, g gives the interaction strength and ω_L and k are the laser frequency and wave number, respectively. The abbreviation h.c. denotes *hermitian conjugate*.

In the case of a standing laser field the interaction Hamiltonian takes the form

$$\hat{H}_{\text{int}}^{(s)}(t) = 2\hbar g \cos(\omega_L t)\left(\hat{\sigma} + \hat{\sigma}^\dagger\right)\sin(k\hat{x}). \tag{17.51}$$

In this section we concentrate on the running wave but also consider the Lamb-Dicke limit of Eq. (17.51).

17.4.1 Transformation to Interaction Picture

In order to gain insight into the dynamics of the combined system we work in the interaction picture. We therefore transform the interaction Hamiltonian Eq. (17.50) into the interaction picture via the transformation

$$\tilde{H}_{int}(t) = \hat{U}^\dagger_{cm}(t)\, \hat{U}^\dagger_a(t)\, \hat{H}_{int}(t)\, \hat{U}_a(t)\, \hat{U}_{cm}(t). \quad (17.52)$$

Here we have introduced the unitary transformation

$$\hat{U}_a(t) \equiv \exp\left(-\frac{i}{2} \omega_a \hat{\sigma}_z t\right)$$

of the internal levels and the time evolution operator $\hat{U}_{cm}(t)$ of the center-of-mass motion is defined by Eq. (17.35).

Lamb-Dicke Parameter

We recall the relation Eq. (14.56)

$$\hat{U}^\dagger_a\, \hat{\sigma}\, \hat{U}_a = \hat{\sigma}\, e^{-i\omega_a t}. \quad (17.53)$$

Moreover, from the transformation Eq. (17.12a), of the position operator $\hat{x}$ we find

$$\hat{U}^\dagger_{cm}(t)\, \exp(ik\hat{x})\, \hat{U}_{cm}(t) = \exp\left[ik\sqrt{\frac{\hbar}{2M\omega_r}}\, \left[\epsilon^*(t)\, \hat{b} + \epsilon(t)\, \hat{b}^\dagger\right]\right].$$

The definition Eq. (11.12) of the displacement operator

$$\hat{D}(\alpha) \equiv \exp\left(\alpha \hat{b}^\dagger - \alpha^* \hat{b}\right)$$

allows us to write this equation in the more compact form

$$\hat{U}^\dagger_{cm}(t)\, \exp(ik\hat{x})\, \hat{U}_{cm}(t) = \hat{D}[\alpha(t)]. \quad (17.54)$$

Here we have introduced the complex-valued time dependent displacement

$$\alpha(t) \equiv i\eta\epsilon(t)$$

and the Lamb-Dicke parameter

$$\eta \equiv k\sqrt{\frac{\hbar}{2M\omega_r}} \equiv \sqrt{2\pi}\, \frac{\Delta x}{\lambda}.$$

Indeed, the displacement α is proportional to the classical function $\epsilon(t)$ which follows from the Mathieu differential equation Eq. (17.13). As discussed in Sec. (17.3) the solutions of this differential equation displays a rather complicated time dependence.

Moreover, we recognize that the Lamb-Dicke parameter η measures the extension

$$\Delta x \equiv \sqrt{\frac{\hbar}{M\omega_r}}$$

of the ground state of the reference oscillator with respect to the wave length $\lambda \equiv 2\pi/k$ of the light field.

Hamiltonian in the Interaction Picture

When we substitute the relations Eqs. (17.53) and (17.54) for the unitary evolution of $\hat{\sigma}$ and the running wave into the expression for the interaction Hamiltonian $\hat{H}_{\text{int}}^{(r)}$ in the interaction picture we find

$$\hat{\tilde{H}}_{\text{int}}^{(r)}(t) = \hbar g \left[\hat{\sigma}^\dagger e^{-i\Delta t} \hat{D}(\alpha(t)) + \sigma e^{i\Delta t} \hat{D}(-\alpha(t)) \right]. \qquad (17.55)$$

Here we have introduced the detuning $\Delta \equiv \omega_L - \omega_a$ between the laser frequency and the atomic transition frequency. Moreover, we have used the fact that $\hat{D}^\dagger(\alpha) = \hat{D}(-\alpha)$.

The interaction Hamiltonian contains time explicitly. It enters through the detuning and the time dependent displacement $\alpha(t)$. The latter is quite remarkable since the Hamiltonian is the sum of two displacement operators. Hence, the action of the interaction Hamiltonian of a given state at a given instant of time is to displace the quantum state of a center-of-mass motion in complex space to a position $\alpha(t)$ *and* to $-\alpha(t)$. Moreover, these displacements are correlated and entangled with internal transitions. A displacement with $+\alpha$ is accompanied by a transition to the excited state and a displacement $-\alpha$ by a transition to the ground state. Consequently, this interaction Hamiltonian creates Schrödinger cats whose internal degrees are entangled with the center-of-mass motion.

We conclude this section by presenting the corresponding result for the interaction with a standing wave. Here we find

$$\hat{\tilde{H}}_{\text{int}}^{(s)}(t) = \hbar g \cos(\omega_L t) \left(\hat{\sigma} e^{-i\omega_a t} + \hat{\sigma}^\dagger e^{i\omega_a t} \right) \frac{1}{i} \left(\hat{D}(\alpha(t)) - \hat{D}(-\alpha(t)) \right). \qquad (17.56)$$

Again, the Hamiltonian consists of a combination of displacement operators with parameter $+\alpha$ and $-\alpha$. However, in contrast to the example of a running wave now the time dependence is much more complicated.

17.4.2 Lamb-Dicke Regime

We gain deeper insight into these interaction Hamiltonians when we confine our discussion to the Lamb-Dicke limit, that is when the characteristic length scale of the quantum mechanical state of the motion is small compared to the wave length $\lambda = 2\pi/k$ of the light field, that is when $\eta \ll 1$. Throughout this section we focus on the case of the standing wave.

Reduction of the Hamiltonian

Since for $\eta \ll 1$ we can expand the displacement operator as

$$\hat{D}(\alpha) = \hat{D}(i\eta\epsilon) = \exp[i\eta\epsilon \hat{b}^\dagger + i\eta\epsilon^* \hat{b}] \cong 1 + i\eta(\epsilon^* \hat{b} + \epsilon \hat{b}^\dagger).$$

the expression Eq. (17.56) for the Hamiltonian in the interaction picture reduces to

$$\hat{\tilde{H}}_{\text{int}}^{(s)}(t) \cong \eta \hbar g \left(1 + e^{-2i\omega_L t} \right) \hat{\sigma} \, e^{i\Delta t} \left[\epsilon^*(t) \hat{b} + \epsilon(t) \hat{b}^\dagger \right] + h.c. \qquad (17.57)$$

17.4 Model Hamiltonian

We can now simplify this interaction Hamiltonian considerably when we make use of the Floquet solution Eq. (17.42), that is when we choose the reference frequency $\omega_r \equiv \omega_r^{(F)}$ according to Eq. (17.46). In this case the explicit solution Eq. (17.42) reads

$$\epsilon^{(F)}(t) = e^{i\mu t}\phi(t) = e^{i\mu t}\sum_{n=-\infty}^{\infty} c_n \exp(in\omega_{\rm rf} t)$$

where μ is the secular frequency and c_n are the expansion coefficients.

We now substitute this expression into the interaction Hamiltonian (17.57) and obtain

$$\tilde{\hat{H}}_{\rm int}^{(s)}(t) = \eta \hbar g \left(1 + e^{-2i\omega_L t}\right)$$
$$\times \left\{ \sum_{n=-\infty}^{\infty} c_n \exp\left[i(\Delta - \mu - n\omega_{\rm rf})t\right] \hat{\sigma}\hat{b} \right.$$
$$\left. + \sum_{n=-\infty}^{\infty} c_n \exp\left[i(\Delta + \mu + n\omega_{\rm rf})t\right] \hat{\sigma}\hat{b}^\dagger \right\} + h.c. \quad (17.58)$$

where we have used that the coefficients c_n are real.

Hence, this Hamiltonian involves the annihilation operator of the internal states as well as the annihilation and creation operator of the vibratory motion. In particular, we note the appearance of the combinations of $\hat{\sigma}\hat{b}^\dagger$ and $\hat{\sigma}\hat{b}$. The first term corresponds to a internal transition from the excited state of the ion to its ground state while at the same time the vibratory motion gets excited by one quantum. This is in complete analogy with the ordinary Jaynes-Cummings-Paul model.

However, the second term involves two annihilation operators, namely for the vibrational and internal states of the ion. Hence, it shows that the ion can undergo a transition on the vibrational ladder while at the same time it goes from its excited state to the ground state. Such a combination we have neglected in the discussion of the ordinary Jaynes-Cummings-Paul model. It is sometimes referred to as the anti-Jaynes-Cummings-Paul model. We now show that by choosing appropriately the detuning Δ we can create either the Jaynes-Cummings-Paul model or its anti version. The latter is not possible in cavity electrodynamics.

Detuning Determines Type of Jaynes-Cummings-Paul Model

For this purpose we now consider Δ such that one of the terms in the sums in Eq. (17.58) is slowly varying whereas all the others are rapidly oscillating. This happens for example when $(-\Delta - \mu)/\omega_{\rm rf}$ is an integer n_0.

In order to achieve a large coupling to the field we choose Δ such that the coefficient c_{n_0} is the largest of all c_n. For the parameter pairs (a, q) displayed in Table 17.1 it is the coefficient c_0 which provides the largest coupling. Hence we choose $\Delta = -\mu$ provided $\Delta - \mu \equiv -2\mu$ is not a multiple integer of the trap frequency $\omega_{\rm rf}$. Note that Δ is of the order of the secular frequency μ and therefore much smaller than the laser frequency ω_L.

With the help of the rotating wave approximation we therefore arrive at

$$\tilde{\hat{H}}_{\rm int}^{(s)}(t) = \eta c_0 \hbar g \left(\hat{\sigma}\hat{b}^\dagger + \hat{\sigma}\hat{b}\right), \quad (17.59)$$

which is the ordinary Jaynes-Cummings-Paul Hamiltonian.

Note, that with the choice $\Delta = \mu$ we get the anti-Jaynes-Cummings Hamiltonian.

17.4.3 Multi-Phonon Jaynes-Cummings-Paul Model

So far, we have shown that in the Lamb-Dicke regime when the spatial extension of the center-of-mass wave function is small compared to the periodicity of the electromagnetic field, the interaction Hamiltonians reduces to the one-phonon Jaynes-Cummings-Paul model. We now drop this restriction and show that in this case we arrive at a multi-phonon Jaynes-Cummings-Paul model. To bring this out most clearly we concentrate on the case of the running wave.

Interaction Hamiltonian in Energy Representation

In Sec. 17.4.1 we have represented the interaction Hamiltonian $\hat{\tilde{H}}_{\text{int}}^{(r)}(t)$ in abstract operator language without resorting to a specific basis. However, in the present section we represent the displacement operators in the energy eigenstates $|n\rangle$ of the time dependent reference oscillator. This brings out most clearly that the atom undergoes transitions between the vibrational levels of the trap while it makes internal transitions.

For this purpose we introduce the matrix elements

$$\Omega^{(n,n+s)} \equiv g e^{-i\Delta t} \langle n| \hat{D}(\alpha(t)) |n+s\rangle$$

between the energy eigenstate $|n\rangle$ of the reference oscillator. Hence, the interaction Hamiltonian reads

$$\hat{\tilde{H}}_{\text{int}}^{(r)}(t) = \sum_{n=0}^{\infty} \sum_{s=-n}^{\infty} \hbar \Omega^{(n,n+s)}(t) \hat{\sigma}^\dagger |n\rangle \langle n+s| + h.c.$$

which demonstrates that the ion goes into its excited state at the expense of s vibrational quanta. The interaction strength with this s phonon transitions follows from the matrix element $\Omega^{(n,n+s)}$. We note, that this quantity is explicitly time dependent. This time dependence results from the detuning Δ and from the displacement $\alpha(t)$.

We can evaluate these matrix elements explicitly and find for $s \geq 0$

$$\Omega^{(n,n+s)}(t) = g \left[\frac{n!}{(n+s)!} \right]^{1/2} \exp(-i\Delta t)[i\eta\epsilon^*(t)]^s \exp\left[-\frac{\eta^2}{2}|\epsilon(t)|^2\right] L_n^s\left(\eta^2|\epsilon(t)|^2\right). \tag{17.60}$$

Here L_n^s denotes the generalized Laguerre polynomial. Similar relations hold for $s < 0$.

Choice of Reference Oscillator

We note, that the solution ϵ of the Mathieu differential equation enters the expression for $\Omega^{(n,n+s)}$, Eq. (17.60) in a rather complicated way. Indeed, we recall from Sec. 17.3.4 that the general solution of ϵ consists of the sum

$$\epsilon(t) = A\, e^{i\mu t}\, \phi(t) + B\, e^{-i\mu t}\, \phi(-t)$$

of two contributions where A and B are constants, and

$$\phi(t) \equiv \phi(t+T) \equiv \sum_{n=-\infty}^{\infty} c_n \exp\left(in\omega_{\mathrm{rf}}t\right)$$

is a periodic function.

However, the proper choice of the frequency of the reference oscillator simplifies the above solution. It selects from the two contributions only one. Indeed, as shown in Sec. 17.3.4 we then find

$$\epsilon^{(F)}(t) = e^{i\mu t}\,\phi(t).$$

We note from the expression Eq. (17.60) that $\epsilon^{(F)}$ enters mostly as $|\epsilon|$. Hence, at these places only the periodic function $\phi(t)$ enters. An exception is the term where the index s of the phonon transition appears as a power. Here, the characteristic exponent μ, that is the secular frequency appears. This creates the sth harmonic of the secular frequency and we arrive at

$$\Omega^{(n,n+s)}(t) = \sum_{l=-\infty}^{\infty} w_l^{(n,n+s)} \exp[i(l\omega_{rf} - s\mu - \Delta)t]$$

where the coefficients $w_l^{(n,n+s)}$ follow by expanding the part of $\Omega^{(n,n+s)}$ in Fourier series, which is periodic in $T = 2\pi/\omega_{rf}$.

When we now use this expression for $\Omega^{(n,n+s)}$ the interaction Hamiltonian in the interaction picture reads

$$\hat{H}_{\mathrm{int}}^{(\mathrm{r})}(t) = \sum_{n=0}^{\infty} \sum_{s=-n}^{\infty} \sum_{l=-\infty}^{\infty} \hbar w_l^{(n,n+s)} \exp[i(l\omega_{rf} - s\mu - \Delta)t]\, \hat{\sigma}^{\dagger}\, |n\rangle\langle n+s| + h.c. \quad (17.61)$$

We emphasize that this representation is exact. It shows that the time dependence of $\hat{H}_{\mathrm{int}}^{(\mathrm{r})}(t)$ is governed by the specific combination $l\omega_{rf} - s\mu - \Delta$ of all harmonics of the radio frequency ω_{rf} and of the secular frequency μ and the detuning Δ. This feature allows us to perform a time average of $\hat{H}_{\mathrm{int}}^{(\mathrm{r})}(t)$ in order to obtain a time independent s_0-phonon Hamiltonian.

Time Average of the Hamiltonian

So far, we have not specified the detuning Δ. We now choose it in such a way that one of the terms in the sum Eq. (17.61) is slowly varying whereas all the others are rapidly oscillating. This happens when

$$s_0\mu + \Delta = l_0\omega_{rf}.$$

Note that this condition leads to an interesting number theoretical problem.

In order to achieve a large coupling to the field we choose Δ such that only the term with the largest coefficient $w_{l_0}^{(n,n+s)}$ survives the time averaging. Hence, we choose

$$\Delta = l_0\omega_{rf} - s_0\mu$$

provided that $\Delta + s\mu$ is not a multiple integer of the frequency ω_{rf} for all $s \neq s_0$.

With the help of the rotating wave approximation we therefore arrive at the time averaged Hamiltonian

$$\hat{\tilde{H}}_{\text{int}}^{(r)} = \sum_{n=0}^{\infty} \hbar \omega_{l_0}^{(n,n+s_0)} \hat{\sigma}^{\dagger} |n\rangle \langle n + s_0| + h.c.$$

which is the s_0-phonon Jaynes-Cummings-Paul Hamiltonian.

17.5 Effective Potential Approximation

We emphasize that our treatment of the Paul trap as discussed in the preceding section takes into account the full time dependence Eq. (17.10) of the harmonic trap potential. In contrast, the effective potential approximation neglects the micromotion of the ion and describes the Paul trap by a time independent harmonic oscillator

$$\hat{H}_{\text{cm}}^{(\mu)} \equiv \frac{\hat{p}^2}{2M} + \frac{1}{2} M \mu^2 \hat{x}^2 \tag{17.62}$$

with the secular frequency μ.

The interaction Hamiltonian $\hat{\tilde{H}}_{\text{int}}^{(\mu)}$ in the rotating wave approximation takes on the form

$$\hat{\tilde{H}}_{\text{int}}^{(\mu)} \equiv \hbar \Omega \sqrt{\tfrac{\hbar}{2M\mu}} \left(\hat{\sigma} \, \hat{b}_{\mu}^{\dagger} + \hat{\sigma}^{\dagger} \, \hat{b}_{\mu} \right) . \tag{17.63}$$

Note, that the annihilation operator $\hat{b}_{\mu}$ and the creation operator $\hat{b}_{\mu}^{\dagger}$ belong to the oscillator with frequency μ.

Hence, the Rabi frequencies read

$$\Omega_n^{(\mu)} \equiv \Omega \sqrt{\tfrac{\hbar}{2M\mu}} \sqrt{n+1}$$

and we have to use the eigenstates $|n\rangle_{\mu}$ of the harmonic oscillator with frequency μ.

When we compare the effective potential model to the one which includes the influence of the micromotion, two differences stand out: (i) the two Rabi frequencies Ω_n and $\Omega_n^{(\mu)}$ differ by the factor $c_0 \, (\mu/\omega_r^{(F)})^{1/2}$; (ii) the operator $\hat{b}$ corresponds to a reference oscillator with the frequency $\omega_r^{(F)}$, whereas the operator $\hat{b}_{\mu}$ describes an oscillator with the secular frequency μ. Hence, the Fock states of the two oscillators are squeezed with respect to each other by an amount

$$s = \frac{\mu}{\omega_r^{(F)}},$$

where s is related to the squeezing parameter r in the definition of the squeezing operator Eq. (17.17) by the expression $s = \exp(2\,r)$.

In Table 17.1 we show the ratios $\Omega_n/\Omega_n^{(\mu)} = c_0 \, (\mu/\omega_r^{(F)})^{1/2}$ of the Rabi frequencies of the two Hamiltonians and the parameter $s = \mu/\omega_r^{(F)}$ for three different pairs (a, q) of trap parameters. We note that the larger the values of a and q the larger are these ratios and hence the differences in the two models. However for $a, |q| \ll 1$ they are almost identical.

Problems

17.1 Time Evolution of Wigner Function with Time Dependent Frequency

Consider a harmonic oscillator with time dependent frequency $\omega(t)$. The equation of motion for this oscillator reads

$$\ddot{x} + \omega^2(t)x = 0. \tag{17.64}$$

From Problem 3.1 we know that the time evolution of the Wigner function for such an oscillator is given by

$$W(x,p;t) = W_0\left(x_0(x,p;t), p_0(x,p;t)\right).$$

Here, $W_0(x,p)$ is the Wigner function at time $t=0$, and $(x_0(x,p;t), p_0(x,p;t))$ is the phase space point at which a *classical* particle would have to start at time $t=0$ in order to reach the point (x,p) at time t. Convince yourself that the time dependent frequency does not change this statement.

Show that $x_0(x,p;t)$ and $p_0(x,p;t)$ for a harmonic oscillator with time dependent frequency can be written in the form

$$\begin{pmatrix} x_0(x,p;t) \\ p_0(x,p;t) \end{pmatrix} = \begin{pmatrix} \frac{1}{\omega_r}\operatorname{Im}\dot{\epsilon}(t) & -\frac{1}{M\omega_r}\operatorname{Im}\epsilon(t) \\ -M\operatorname{Re}\dot{\epsilon}(t) & \operatorname{Re}\epsilon(t) \end{pmatrix} \begin{pmatrix} x \\ p \end{pmatrix}$$

where ω_r is an arbitrary but constant frequency and $\epsilon(t)$ satisfies Eq. (17.64) with the initial condition

$$\epsilon(0) = 1, \qquad \dot{\epsilon}(0) = i\omega_r.$$

17.2 Paul Trap Endoscopy

The goal is to reconstruct the initial vibrational state

$$|\psi_{\text{cm}}(0)\rangle \equiv \sum_{n=0}^{\infty} w_n |n\rangle$$

of the ion in the Paul trap from the measured time evolution of its internal states. Here we have expanded the state $|\psi_{\text{cm}}(0)\rangle$ into eigenstates $|n\rangle$ of the reference oscillator with frequency $\omega_r^{(F)}$.

Since we want to reconstruct the probability amplitudes w_n not only in their absolute values but also in their phases, we need a probe with a reference phase. We use a coherent superposition

$$|\psi_{\text{atom}}(0)\rangle \equiv \tfrac{1}{\sqrt{2}}\left(|b\rangle + e^{i\varphi}|a\rangle\right)$$

of the ground state $|b\rangle$ and the excited state $|a\rangle$ as the initial internal state $|\psi_{\text{atom}}(0)\rangle$.

(a) Propagate the total state vector

$$|\Psi(0)\rangle \equiv |\psi_{\text{atom}}(0)\rangle \otimes |\psi_{\text{cm}}(0)\rangle$$

in time using the interaction Hamiltonian Eq. (17.59) and show that the probability

$$W_a(t;\varphi) \equiv \sum_{n=0}^{\infty} \langle n| \langle a| e^{-i\hat{H}_{\text{int}}\,t/\hbar} |\Psi(0)\rangle\langle\Psi(0)| e^{i\hat{H}_{\text{int}}\,t/\hbar} |a\rangle |n\rangle$$

of finding the ion at time t in the excited state reads

$$W_a(t;\varphi) = \tfrac{1}{2} - \tfrac{1}{4}|w_0|^2 + \tfrac{1}{4}\sum_{n=0}^{\infty} \cos(2\Omega_n t)\left(|w_n|^2 - |w_{n+1}|^2\right)$$
$$- \tfrac{1}{2}\sum_{n=0}^{\infty} \sin(2\Omega_n t)\,\text{Im}\left(w_n w_{n+1}^* e^{-i\varphi}\right). \tag{17.65}$$

Here the Rabi frequencies Ω_n are given by

$$\Omega_n \equiv \Omega\, c_0 \sqrt{\frac{\hbar}{2M\omega_r^{(F)}}} \sqrt{n+1}.$$

(b) Use the time dependence of the probability $W_a(t;\varphi)$, Eq. (17.65), to infer the expansion coefficients w_n and hence the initial vibrational state $|\psi_{\text{cm}}(0)\rangle$. For this purpose we measure the probability $W_a(t;\varphi)$ for two different phases φ and N interaction times t.

Hint: See Schrade *et al.* (1995), (1997) and Bardroff *et al.* (1996).

References

Physics of Paul Trap

The Paul trap was originally proposed and experimentally demonstrated by
W. Paul, O. Osberghaus and **E. Fischer**, Forschungsberichte des Wirtschaftsministeriums Nordrhein-Westfalen Nr. 415 (1958)
E. Fischer, *Die dreidimensionale Stabilisierung von Ladungsträgern in einem Vierpolfeld*, Z. Physik **156**, 1–26 (1959)

A more recent review can be found in the Nobel prize lecture by
W. Paul, *Electromagnetic traps for charged and neutral particles*, Rev. Mod. Phys. **62**, 531–540 (1990)

The physics of ions in traps is reviewed in the special issues
R. Blatt, P. Gill and **R.C. Thompson**, *Current perspectives on the physics of trapped ions*, J. Mod. Opt. **39**, 193–220 (1992)
R. Blatt and **W. Neuhauser**, *High-Resolution Laser Spectroscopy*, Appl. Phys. B **59** (2 and 3) (1994)

Mathieu Equation

The Floquet theorem is explained in
E.T. Whittaker and **G.N. Watson**, *A Course of Modern Analysis*, Cambridge University Press, Cambridge, 1927

For a summary of the properties of Mathieu functions see
M. Abramowitz and **I.A. Stegun**, *Handbook of Mathematical Functions*, U.S. Government Printing Office, Washington, 1964

N.W. McLachlan, *Theory and application of Mathieu functions*, Clarendon Press, Oxford, 1947

J. Meixner and **F.W. Schäfke**, *Mathieusche Funktionen und Sphäroidfunktionen*, Springer, Heidelberg, 1954

Linear Paul Traps

Linear Paul traps have become very popular in the last years due to the potential use in a realization of a quantum computer. They originated from the mass filter invented by
W. Paul and **H. Steinwedel**, *Ein neues Massenspektrometer ohne Magnetfeld*, Z. Naturforsch. A**8**, 448–450 (1953)

Linear traps have been built in the form of a race track or in a linear shape, see for example
G. Birkl, **S. Kassner** and **H. Walther**, *Multiple-shell structures of laser-cooled $^{24}Mg^+$ ions in a quadrupole storage ring*, Nature **357**, 310–313 (1992)

M.G. Raizen, **J.M. Gilligan**, **J.C. Bergquist**, **W.M. Itano** and **D.J. Wineland**, *Linear trap for high-accuracy spectroscopy of stored ions*, J. Mod. Opt. **39**, 233–242 (1992)

Laser Cooling

For the original papers on laser cooling, see
T.W. Hänsch and **A.L. Schawlow**, *Cooling of Gases by Laser Radiation*, Opt. Commun. **13**, 68–69 (1975)

D.J. Wineland and **H. Dehmelt**, *Proposed $10^{14}\Delta\nu < \nu$ Laser Fluorescence Spectroscopy on Tl^+ Mono-Ion Oscillator III*, Bull. Am. Phys. Soc. **20**, 637 (1975)

D.J. Wineland and **W.M. Itano**, *Laser cooling of atoms*, Phys. Rev. A **20**, 1521–1540 (1979)

For a review of Doppler cooling with an exhaustive list of references we refer to the article by
S. Stenholm, *The semiclassical theory of laser cooling*, Rev. Mod. Phys. **58**, 699–739 (1986)

and the book
V.G. Minogin and **V.S. Lethokov**, *Laser Light Pressure on Atoms*, Gordon and Breach, New York, 1987

For an excellent introduction into the sophisticated techniques of Sisyphus and sub-recoil cooling we refer to
C. Cohen-Tannoudji, *Atomic Motion in Laser Light*, in: *Fundamental Systems in Quantum Optics*, edited by J. Dalibard, J.-M. Raimond and J. Zinn-Justin, North-Holland, Amsterdam, 1992

Quantum Treatment of Paul Trap

For the quantum motion of ions in traps see
M. Combescure, *A quantum particle in a quadrupole radio-frequency trap*, Ann. Inst. Henri Poincare **44**, 293–314 (1986)
L.S. Brown, *Quantum Motion in a Paul Trap*, Phys. Rev. Lett. **66**, 527–529 (1991)
G.S. Agarwal and S. Arun Kumar, *Exact Quantum-Statistical Dynamics of an Oscillator with Time-Dependent Frequency and Generation of Nonclassical States*, Phys. Rev. Lett. **67**, 3665–3668 (1991)
R.J. Glauber, *Laser Manipulation of Atoms and Ions*, Proc. Int. School of Physics "Enrico Fermi", Course CXVIII (1991), edited by E. Arimondo, W.D. Phillips and F. Strumia, North-Holland, Amsterdam, 1992

For a Wigner function treatment of this problem see
G. Schrade, V.I. Man'ko, W.P. Schleich and R.J. Glauber, *Wigner functions in the Paul trap*, Quantum Semiclass. Opt. **7**, 307–325 (1995)
G. Schrade, P.J. Bardroff, R.J. Glauber, C. Leichtle, V. Yakovlev and W.P. Schleich, *Endoscopy in the Paul trap: The influence of the micromotion*, Appl. Phys. B **64**, 181–191 (1997)

Analogy to Cavity QED

For the analogy between cavity QED and the quantized motion of an ion in a Paul trap see
C.A. Blockley, D.F. Walls and H. Risken, *Quantum collapses and revivals in a quantized trap*, Europhys. Lett. **17**, 509–514 (1992)
J.I. Cirac, R. Blatt, A.S. Parkins and P. Zoller, *Quantum collapse and revival in the motion of a single trapped ion*, Phys. Rev. A **49**, 1202–1207 (1994)
W. Vogel and R.L. de Matos Filho, *Nonlinear Jaynes-Cummings dynamics of a trapped ion*, Phys. Rev. A **52**, 4214–4217 (1995)

For a treatment of the nonlinear Jaynes-Cummings-Paul dynamics of a trapped ion in the presence of micro motion see
P.J. Bardroff, C. Leichtle, G. Schrade and W.P. Schleich, *Endoscopy in the Paul trap: measurement of the vibratory quantum state of a single ion*, Phys. Rev. Lett. **77**, 2198–2201 (1996)
P.J. Bardroff, C. Leichtle, G. Schrade and W.P. Schleich, *Paul Trap Multi-Quantum Interactions*, Acta Phys. Slovaca **46**, 231–240 (1996)

In these papers also a prescription for a measurement of the quantum state has been given.

The revivals in the Paul trap have been observed by
D.M. Meekhof, C. Monroe, B.E. King, W.M. Itano and **D.J. Wineland**, *Generation of Nonclassical Motional States of a Trapped Atom*, Phys. Rev. Lett. **76**, 1796–1799 (1996)

Applications of Paul Traps

Quantum jumps in atoms have been observed by
W. Nagourney, J. Sandberg and **H. Dehmelt**, *Shelved optical electron amplifier: observation of quantum jumps*, Phys. Rev. Lett. **56**, 2797–2799 (1986)
Th. Sauter, W. Neuhauser, R. Blatt and **P.E. Toschek**, *Observation of quantum jumps*, Phys. Rev. Lett. **57**, 1696–1698 (1986)

For a discussion of phase transitions between ordered structures and chaotic clouds of ions in a Paul trap see
F. Diedrich, E. Peik, J.M. Chen, W. Quint and **H. Walther**, *Observation of a phase transition of stored laser-cooled ions*, Phys. Rev. Lett. **59**, 2931–2934 (1987)
D.J. Wineland, J.C. Bergquist, W.M. Itano, J.J Bollinger and **C.H. Manney**, *Atomic-ion Coulomb clusters in an ion trap*, Phys. Rev. Lett. **59**, 2935–2938 (1987)
R. Blümel, J.M. Chen, E. Peik, W. Quint, W.P. Schleich, Y.R. Shen and **H. Walther**, *Phase Transitions of Stored Laser-Cooled Ions*, Nature **334**, 309–313 (1988)

Arrangements of ions in cylindrical shells in a Penning trap have been observed by
S.L. Gilbert, J.J. Bollinger and **D.J. Wineland**, *Shell-structure phase of magnetically confined strongly coupled plasmas*, Phys. Rev. Lett. **60**, 2022-2025 (1988)

For a nonlinear extension of quantum mechanics see
S. Weinberg, *Testing Quantum Mechanics*, Ann. Phys. **194**, 336–386 (1989)

For an experimental test of this extension see
J.J. Bollinger, D.J. Heinzen, W.M. Itano, S.L. Gilbert and **D.J. Wineland**, *Test of the Linearity of Quantum Mechanics by rf Spectroscopy of the $^9Be^+$ Ground State*, Phys. Rev. Lett. **63**, 1031–1034 (1989)

A quantum computer based on a linear trap was proposed by
J.I. Cirac and **P. Zoller**, *Quantum Computations with Cold Trapped Ions*, Phys. Rev. Lett. **74**, 4091–4094 (1995)

For the resonance fluorescence of a single ion stored in a Paul trap see
F. Diedrich and **H. Walther**, *Nonclassical Radiation of a Single Stored Ion*, Phys. Rev. Lett. **58**, 203–206 (1987)

Four ions stored in a trap were entangled by
C.A. Sackett, D. Kielpinski, B.E. King, C. Langer, V. Meyer, C.J. Myatt, M. Rowe, Q.A. Turchette, W.M. Itano, D.J. Wineland and **C. Monroe**, *Experimental entanglement of four particles*, Nature **404**, 256–259 (2000)

18 Damping and Amplification

The Jaynes-Cummings-Paul model describes the interaction of a two-level atom with a single quantized mode of the radiation field. The corresponding dynamics is governed by the Schrödinger equation for the state vector of the combined system consisting of atom plus field. Due to the entanglement of the two subsystems it is impossible to write an equation of motion for the state vector of a single subsystem.

When we are interested in the dynamics of the atom or the field itself we can still write equations of motion. However, they are not for the state vectors of the subsystems, but for the density operators of the field or the atom alone. The reduction of two interacting quantum systems to a subsystem inevitably leads to a description in terms of the density operator.

A quantum system damped by a reservoir also relies on the notion of the density operator. A reservoir consists of a large number of degrees of freedom, that is, many subsystems each of which interacts with the quantum system of interest. Since the number of subsystems is large we cannot make measurements on each individual one and can therefore not keep track of the entanglement between the quantum system and the reservoir systems. This fact forces us into the density operator formalism.

Nevertheless, it is possible to describe problems of this type using wave functions rather than the density operator. The so-called *Monte Carlo wave function technique* relies on stochastic wave functions and has been developed in the context of laser cooling to treat the mechanical action of light on atoms. It is an extremely useful technique to analyze damping of quantum systems.

In Sec. 18.1 we illustrate a model which allows us to describe quantum mechanically damping and amplification of a cavity field. We then in Sec. 18.2 present a general formulation of the dynamics of a small system coupled to a large reservoir. In Sec. 18.3 we put this formalism to work and derive a master equation for the model of a cavity field interacting with a stream of two-level atoms. Here we follow a perturbative and an exact approach. The exact one immediately yields the master equation of the one-atom maser – the topic of Sec. 18.4. In Sec. 18.5 we turn to the problem of a two-level atom coupled to a thermal reservoir. We find spontaneous emission and the Lamb shift.

Space does not allow us to cover other interesting approaches towards damping and amplification such as the Langevin method. For more details we refer to the literature at the end of this chapter.

18.1 Damping and Amplification of a Cavity Field

In the present chapter we study the time evolution of a single mode of the radiation field due to its interaction with a beam of two-level atoms as shown in Fig. 18.1. Here we do not make a measurement on the atoms after they have left the cavity. Since we are only interested in the dynamics of the field due to the atoms, that is, the dynamics of the subsystem, the state of the field mode can only be described by a density operator.

We can envision various scenarios depending on the initial state of the atoms. We first consider atoms which are not in a coherent superposition of their excited state $|a\rangle$ and their ground state $|b\rangle$, but in a classical statistical mixture of $|a\rangle$ and

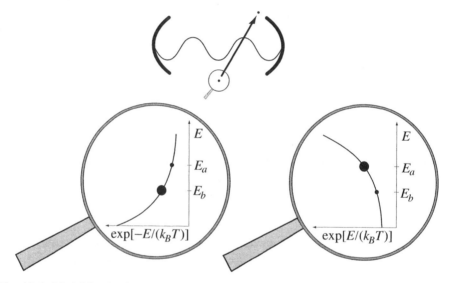

Fig. 18.1: Model for the damping or amplification of a single mode of a cavity. A stream of resonant two-level atoms passes through the cavity (top) at an anti-node of the field mode. The atoms are in a statistical mixture of excited and ground state, that is, the populations in the states $|a\rangle$ and $|b\rangle$ follow a Boltzmann distribution of temperature T. When there are more atoms in the ground state than in the excited state (left) then on average more ground state atoms get excited by the cavity field and thus take field excitation, that is, photons out of the resonator than excited atoms get de-excited and leave their excitation in the field. Hence, on average the cavity field looses photons, that is, the field is damped by the reservoir of two-level atoms of temperature T. Note that no measurement on the internal state of the atoms exiting the cavity is made. Only the population in the states governed by the temperature is known as appropriate for a reservoir. When there is an inversion in the atom, that is, there are more atoms in the excited state than in the ground state (right) then the atoms transfer excitation into the field, that is, they amplify the field. In this case the population of the levels is described by a Boltzmann distribution of "negative temperature".

$|b\rangle$, that is, the atoms are described by a density operator

$$\hat{\rho}_{\text{at}} \equiv \begin{pmatrix} \rho_{aa} & 0 \\ 0 & \rho_{bb} \end{pmatrix}.$$

Here ρ_{aa} and ρ_{bb} are the populations in the excited and the ground state, respectively.

The cavity field can excite the ground state atoms during their flight and they carry field excitation out of the resonator. Consequently, the field is damped by these atoms.

In contrast, atoms in the excited state can make a transition to the ground state transferring their excitation into the field. In this case the field is amplified.

When we have a beam of temperature T, that is, a beam in which there are more atoms in the ground state than in the excited state according to a Boltzmann distribution

$$\frac{\rho_{aa}}{\rho_{bb}} = \exp\left(-\frac{E_a - E_b}{k_B T}\right) = e^{-\hbar\omega/(k_B T)}, \tag{18.1}$$

the net effect of the reservoir of atoms is a damping of the cavity mode.

When on the other hand we have an inversion, that is, when there are more atoms in the excited state than in the ground state à la a Boltzmann distribution

$$\frac{\rho_{aa}}{\rho_{bb}} = e^{\hbar\omega/(k_B T)},$$

of "negative temperature", we find a net amplification of the cavity field. This is the principle of the laser and maser and will be discussed in more detail in the following sections.

The fact that we can describe the state of the field mode only by a density matrix rather than by a state vector originates from the lack of a measurement on the atom after the interaction, that is, lack of the measurement on the reservoir. Even if we injected all atoms in the excited (or in the ground state) we would still have to resort to a density operator formalism.

18.2 Density Operator of a Subsystem

The state of a quantum system such as a field mode interacting with a quantum reservoir consisting of two-level atoms is described by a density operator $\hat{\rho}_f$. What is the equation of motion for $\hat{\rho}_f$?

18.2.1 Coarse-Grained Equation of Motion

In the present section we answer this question by recalling the main results of Sec. 2.4.4. In order to keep our treatment rather general we consider the interaction of a quantum system denoted by subscript s with a reservoir denoted by r. The system could be, for example, a cavity field mode and the reservoir could be a beam of two-level atoms.

In Sec. 2.4.4 we have derived the formal expression

$$\hat{\rho}_{s+r}(t) = \hat{\rho}_{s+r}(t_0) + \sum_{n=1}^{\infty} \left(-\frac{i}{\hbar}\right)^n$$
$$\times \int_{t_0}^{t} dt_n \int_{t_0}^{t_n} dt_{n-1} \cdots \int_{t_0}^{t_2} dt_1 \, [\hat{H}(t_n), [\hat{H}(t_{n-1}), \cdots, [\hat{H}(t_1), \hat{\rho}_{s+r}(t_0)]\cdots]] \quad (18.2)$$

for the density operator $\hat{\rho}_{s+r}$ of the system plus reservoir. Here $\hat{H}$ denotes the interaction Hamiltonian in the interaction picture. In general $\hat{H}$ is time dependent.

We now derive the equation of motion for the system by tracing Eq. (18.2) over the reservoir variables. We define the density operator $\hat{\rho}_s$ of the system by taking the trace of $\hat{\rho}_{s+r}$ over the reservoir, that is

$$\hat{\rho}_s \equiv \text{Tr}_r(\hat{\rho}_{s+r}).$$

Moreover, we assume that before the interaction the system and the reservoir are uncoupled. Thus the density operator $\hat{\rho}_{s+r}$ of the combined system at time t_0 is a direct product

$$\hat{\rho}_{s+r}(t_0) = \hat{\rho}_s(t_0) \otimes \hat{\rho}_r(t_0)$$

of the density operator $\hat{\rho}_s(t_0)$ and $\hat{\rho}_r(t_0)$ of the system and the reservoir.

When we perform the trace over the reservoir and recall that $\text{Tr}_r \hat{\rho}_r(t_0) = 1$ we find

$$\hat{\rho}_s(t) = \hat{\rho}_s(t_0) + \sum_{n=1}^{\infty} \left(-\frac{i}{\hbar}\right)^n \int_{t_0}^{t} dt_n \int_{t_0}^{t_n} dt_{n-1} \cdots \int_{t_0}^{t_2} dt_1$$
$$\times \text{Tr}_r [\hat{H}(t_n), [\hat{H}(t_{n-1}), \cdots, [\hat{H}(t_1), \hat{\rho}_{s+r}(t_0)]\cdots]]. \quad (18.3)$$

The density operator of the system at time t is given by nested commutators of the interaction Hamiltonian and the initial density operator of the system plus reservoir traced over the reservoir. Therefore, even if we start from pure states of the system and the reservoir, due to the trace over the reservoir the state of the system is given by a density operator. This is due to the fact that we have chosen a setup in which we do not perform a measurement on the reservoir.

We now discuss the time evolution of the system due to the interaction with the reservoir for a time τ. Here the interaction of the system begins at t and ends at $t + \tau$. When we identify in Eq. (18.3) $t_0 \to t$ and $t \to t + \tau$ the change

$$\frac{1}{\tau}[\hat{\rho}_s(t+\tau) - \hat{\rho}_s(t)] \equiv \frac{d}{dt}\hat{\rho}_s(t) \quad (18.4)$$

of the density operator of the system due to this interaction is

$$\frac{d}{dt}\hat{\rho}_s(t) = \frac{1}{\tau}\sum_{n=1}^{\infty}\left(-\frac{i}{\hbar}\right)^n \int_{t}^{t+\tau} dt_n \int_{t}^{t_n} dt_{n-1} \cdots \int_{t}^{t_2} dt_1$$
$$\times \text{Tr}_r [\hat{H}(t_n), [\hat{H}(t_{n-1}), \cdots, [\hat{H}(t_1), \hat{\rho}_{s+r}(t)]\cdots]]. \quad (18.5)$$

The time derivative Eq. (18.4) is sometimes referred to as coarse-grained derivative since we consider the slow change of the system due its interaction with the reservoir.

18.2.2 Time Independent Hamiltonian

When the Hamiltonian is time independent we can immediately perform the integration over time using the relation

$$\int_{t_0}^{t} dt_n \int_{t_0}^{t_n} dt_{n-1} \cdots \int_{t_0}^{t_2} dt_1 = \frac{1}{n!}(t-t_0)^n. \tag{18.6}$$

We prove this formula by considering the step from n to $n+1$ via

$$\int_{t_0}^{t} dt_{n+1} \int_{t_0}^{t_{n+1}} dt_n \cdots \int_{t_0}^{t_2} dt_1 = \int_{t_0}^{t} dt_{n+1} \frac{1}{n!}(t_{n+1}-t_0)^n$$

$$= \frac{1}{(n+1)!}(t_{n+1}-t_0)^{n+1}\bigg|_{t_{n+1}=t_0}^{t} = \frac{1}{(n+1)!}(t-t_0)^{n+1},$$

in complete agreement with Eq. (18.6). In the first step of this equation we have assumed that Eq. (18.6) holds true for n.

When we note that the equation holds trivially for $n = 1$ we have proven the integral formula Eq. (18.6) by complete induction.

With the help of Eq. (18.6) we therefore arrive at the equation of motion

$$\frac{d}{dt}\hat{\rho}_s(t) = \frac{1}{\tau}\sum_{n=1}^{\infty}\frac{1}{n!}\left(-\frac{i\tau}{\hbar}\right)^n \text{Tr}_r [\hat{H},[\hat{H},\cdots,[\hat{H},\hat{\rho}_{s+r}(t)]\cdots]]_n. \tag{18.7}$$

for the density operator $\hat{\rho}_s(t)$ of the system. Here we have included a subscript n at the commutator as to remind ourselves that the nth order term contains n commutators and thus the interaction Hamiltonian appears n times.

18.3 Reservoir of Two-Level Atoms

We now illustrate the role of the nested commutators in Eq. (18.5) for the example shown in Fig. 18.1 of a cavity mode interacting with a beam of two-level atoms. Here we use the resonant interaction Hamiltonian

$$\hat{H} = \hbar g \left(\hat{\sigma}\hat{a}^\dagger + \hat{\sigma}^\dagger \hat{a}\right) = \hbar g \begin{pmatrix} 0 & \hat{a} \\ \hat{a}^\dagger & 0 \end{pmatrix} \equiv \hbar g \hat{A} \tag{18.8}$$

of the Jaynes-Cummings-Paul model. In the last step we have recalled that

$$\hat{\sigma} \equiv \begin{pmatrix} 0 & 0 \\ 1 & 0 \end{pmatrix} \quad \text{and} \quad \hat{\sigma}^\dagger \equiv \begin{pmatrix} 0 & 1 \\ 0 & 0 \end{pmatrix}$$

and have introduced the off-diagonal matrix $\hat{A}$ containing the field operators $\hat{a}$ and $\hat{a}^\dagger$.

The atom enters the cavity at time t and is therefore uncorrelated with the field. Consequently, the density operator $\hat{\rho}_{f+at}$ of field plus atom factorizes, that is,

$$\hat{\rho}_{f+at}(t) = \hat{\rho}_f(t) \otimes \hat{\rho}_{at}(t) = \hat{\rho}_f \otimes \begin{pmatrix} \rho_{aa} & \rho_{ab} \\ \rho_{ba} & \rho_{bb} \end{pmatrix} = \begin{pmatrix} \hat{\rho}_f \rho_{aa} & \hat{\rho}_f \rho_{ab} \\ \hat{\rho}_f \rho_{ba} & \hat{\rho}_f \rho_{bb} \end{pmatrix}.$$

18.3.1 Approximate Treatment

In the present discussion we confine ourselves to a case in which the initial atomic density operator $\hat{\rho}_{at}$ is diagonal, that is

$$\hat{\rho}_{at}(t) \equiv \begin{pmatrix} \rho_{aa} & 0 \\ 0 & \rho_{bb} \end{pmatrix}$$

and therefore

$$\hat{\rho}_{f+at}(t) = \begin{pmatrix} \hat{\alpha} & 0 \\ 0 & \hat{\beta} \end{pmatrix} = \begin{pmatrix} \hat{\rho}_f \rho_{aa} & 0 \\ 0 & \hat{\rho}_f \rho_{bb} \end{pmatrix} \equiv \hat{R}.$$

Since the Hamiltonian Eq. (18.8) of the resonant interaction is time independent we can immediately apply the equation of motion Eq. (18.7) and arrive at

$$\frac{d}{dt}\hat{\rho}_f(t) = \frac{1}{\tau}\sum_{n=1}^{\infty} \frac{(-ig\tau)^n}{n!}\, \text{Tr}_{at}\,[\hat{A},[\hat{A},\cdots,[\hat{A},\hat{R}]\cdots]]_n. \tag{18.9}$$

In order to find the explicit form of this density operator equation we have to calculate the commutators of 2×2 matrices which contain in the matrix elements the annihilation and creation operators of the field.

Single Commutator

To evaluate the commutators we first note that the product $\hat{A}\hat{R}$ of a diagonal matrix $\hat{R}$ and an off-diagonal matrix $\hat{A}$ leads to an off-diagonal matrix, that is,

$$\hat{A}\hat{R} = \begin{pmatrix} 0 & \hat{a} \\ \hat{a}^\dagger & 0 \end{pmatrix}\begin{pmatrix} \hat{\alpha} & 0 \\ 0 & \hat{\beta} \end{pmatrix} = \begin{pmatrix} 0 & \hat{a}\hat{\beta} \\ \hat{a}^\dagger\hat{\alpha} & 0 \end{pmatrix}.$$

Here we have been careful about the ordering of the operators $\hat{\alpha} \equiv \hat{\rho}_f \rho_{aa}$, $\hat{\beta} \equiv \hat{\rho}_f \rho_{bb}$, $\hat{a}$ and $\hat{a}^\dagger$ since the density operator $\hat{\rho}_f$ of the field contained in $\hat{\alpha}$ and $\hat{\beta}$ does not commute with the field operators $\hat{a}$ and $\hat{a}^\dagger$.

Likewise, we find

$$\hat{R}\hat{A} = \begin{pmatrix} \hat{\alpha} & 0 \\ 0 & \hat{\beta} \end{pmatrix}\begin{pmatrix} 0 & \hat{a} \\ \hat{a}^\dagger & 0 \end{pmatrix} = \begin{pmatrix} 0 & \hat{\alpha}\hat{a} \\ \hat{\beta}\hat{a}^\dagger & 0 \end{pmatrix}$$

which makes the commutator between $\hat{A}$ and $\hat{R}$ into the off-diagonal matrix

$$\hat{C} \equiv \hat{A}\hat{R} - \hat{R}\hat{A} = [\hat{A}, \hat{R}] = \begin{pmatrix} 0 & \hat{a}\hat{\beta} - \hat{\alpha}\hat{a} \\ \hat{a}^\dagger\hat{\alpha} - \hat{\beta}\hat{a}^\dagger & 0 \end{pmatrix} \equiv \begin{pmatrix} 0 & \hat{\gamma} \\ \hat{\delta} & 0 \end{pmatrix}. \tag{18.10}$$

Since the trace of this off-diagonal matrix vanishes, that is

$$\text{Tr}[\hat{A}, \hat{R}] = 0,$$

the first term in the perturbative expansion Eq. (18.9) of the field-density operator vanishes.

Double Commutator

The product $\hat{A}\hat{C}$ of two off-diagonal matrices such as $\hat{A}$ and $\hat{C}$ leads to a diagonal matrix. Indeed, since

$$\hat{A}\hat{C} = \begin{pmatrix} 0 & \hat{a} \\ \hat{a}^\dagger & 0 \end{pmatrix} \begin{pmatrix} 0 & \hat{\gamma} \\ \hat{\delta} & 0 \end{pmatrix} = \begin{pmatrix} \hat{a}\hat{\delta} & 0 \\ 0 & \hat{a}^\dagger\hat{\gamma} \end{pmatrix}$$

and

$$\hat{C}\hat{A} = \begin{pmatrix} 0 & \hat{\gamma} \\ \hat{\delta} & 0 \end{pmatrix} \begin{pmatrix} 0 & \hat{a} \\ \hat{a}^\dagger & 0 \end{pmatrix} = \begin{pmatrix} \hat{\gamma}\hat{a}^\dagger & 0 \\ 0 & \hat{\delta}\hat{a} \end{pmatrix}$$

we find

$$[\hat{A}, \hat{C}] = [\hat{A}, [\hat{A}, \hat{R}]] = \begin{pmatrix} \hat{a}\hat{\delta} - \hat{\gamma}\hat{a}^\dagger & 0 \\ 0 & \hat{a}^\dagger\hat{\gamma} - \hat{\delta}\hat{a} \end{pmatrix} \quad (18.11)$$

and the trace reads

$$\text{Tr}[\hat{A}, [\hat{A}, \hat{R}]] = \hat{a}\hat{\delta} - \hat{\gamma}\hat{a}^\dagger + \hat{a}^\dagger\hat{\gamma} - \hat{\delta}\hat{a}. \quad (18.12)$$

Hence, the first non-vanishing contribution to Eq. (18.9) is of second order, that is, proportional to $(g\tau)^2$.

We substitute the expressions for $\hat{\delta}$ and $\hat{\gamma}$ from Eq. (18.10) into Eq. (18.12) and arrive at

$$\text{Tr}[\hat{A}, [\hat{A}, \hat{R}]] = \hat{a}\hat{a}^\dagger\hat{\alpha} - \hat{a}\hat{\beta}\hat{a}^\dagger - \hat{a}\hat{\beta}\hat{a}^\dagger + \hat{\alpha}\hat{a}\hat{a}^\dagger + \hat{a}^\dagger\hat{a}\hat{\beta} - \hat{a}^\dagger\hat{\alpha}\hat{a} - \hat{a}^\dagger\hat{\alpha}\hat{a} + \hat{\beta}\hat{a}^\dagger\hat{a},$$

or

$$\text{Tr}[\hat{A}, [\hat{A}, \hat{R}]] = \rho_{aa} \left(\hat{a}\hat{a}^\dagger\hat{\rho}_f + \hat{\rho}_f\hat{a}\hat{a}^\dagger - 2\hat{a}^\dagger\hat{\rho}_f\hat{a} \right) + \rho_{bb} \left(\hat{a}^\dagger\hat{a}\hat{\rho}_f + \hat{\rho}_f\hat{a}^\dagger\hat{a} - 2\hat{a}\hat{\rho}_f\hat{a}^\dagger \right). \quad (18.13)$$

Here we have made use of the definitions of $\hat{\alpha}$ and $\hat{\beta}$.

Since the commutator $[\hat{A}, [\hat{A}, \hat{R}]]$ is a diagonal matrix the third order contribution to the perturbative expansion Eq. (18.9) of the field-density operator vanishes. Moreover, all odd orders vanish; only the even orders survive. In the follog discussion however, we confine ourselves to the terms up to $(g\tau)^2$.

Coarse Graining

We inject atoms in the excited or ground state with the rate r and let them interact with the field for the time τ. Since the Hamiltonian Eq. (18.8) describes only the interaction between a single atom with the field mode we must restrict ourselves to the case where the injection rate r is less or equal to $1/\tau$. Now suppose that at time t an atom enters the cavity and leaves it at time $t+\tau$. The next atom will enter not before $t+1/r$ which may be later. However, the density operator of the field does not change during the time interval with no atom in the cavity since we only consider the interaction part of the evolution. Therefore, we can define a coarse-grained derivative

$$\frac{d}{dt}\hat{\rho}_f(t) \equiv r\left[\hat{\rho}_f(t+1/r) - \hat{\rho}_f(t)\right] = r\left[\hat{\rho}_f(t+\tau) - \hat{\rho}_f(t)\right]$$

and with the definitions

$$\mathcal{R}_a \equiv r\rho_{aa}(g\tau)^2 \quad \text{and} \quad \mathcal{R}_b \equiv r\rho_{bb}(g\tau)^2 \qquad (18.14)$$

we find from Eq. (18.9) with the help of Eq. (18.13) the so-called master equation

$$\frac{d}{dt}\hat{\rho}_\mathrm{f} = -\frac{1}{2}\mathcal{R}_a\left(\hat{a}\hat{a}^\dagger\hat{\rho}_\mathrm{f} + \hat{\rho}_\mathrm{f}\hat{a}\hat{a}^\dagger - 2\hat{a}^\dagger\hat{\rho}_\mathrm{f}\hat{a}\right) - \frac{1}{2}\mathcal{R}_b\left(\hat{a}^\dagger\hat{a}\hat{\rho}_\mathrm{f} + \hat{\rho}_\mathrm{f}\hat{a}^\dagger\hat{a} - 2\hat{a}\hat{\rho}_\mathrm{f}\hat{a}^\dagger\right). \qquad (18.15)$$

We emphasize that this equation is rather complicated. It is an operator equation, that is, an equation for the density operator $\hat{\rho}_\mathrm{f}$. Moreover, it also contains the field operators $\hat{a}$ and $\hat{a}^\dagger$. Their order relative to $\hat{\rho}_\mathrm{f}$ is important, since $\hat{\rho}_\mathrm{f}$ contains field operators as well.

It is extremely difficult to solve the master equation directly. However, there exist many techniques for deriving a solution. The trick consists of converting the master equation Eq. (18.15) into a c-number equation by using a specific representation of the density operator such as phase space distribution functions or the photon number representation. This is the topic of the next section.

18.3.2 Density Operator in Number Representation

The equation of motion Eq. (18.15) for the density operator describes the damping or amplification of a cavity mode. Indeed, the first expression in brackets results from atoms in the excited state and should lead to amplification, whereas the second term originates from atoms in the ground state and should take excitations out of the cavity giving rise to damping.

Translation into Photon Number Space

To investigate damping and amplification in more detail we now consider the density operator of the field in number state representation

$$\hat{\rho}_\mathrm{f} \equiv \sum_{m,n=0}^{\infty} \langle m|\hat{\rho}_\mathrm{f}|n\rangle\, |m\rangle\langle n| \equiv \sum_{m,n=0}^{\infty} \rho_{m,n}\, |m\rangle\langle n|.$$

When we take matrix elements of the density operator equation Eq. (18.15) we arrive at

$$\frac{d}{dt}\rho_{m,n} = -\frac{1}{2}\mathcal{R}_a\left[\langle m|\hat{a}\hat{a}^\dagger\hat{\rho}_\mathrm{f}|n\rangle + \langle m|\hat{\rho}_\mathrm{f}\hat{a}\hat{a}^\dagger|n\rangle - 2\langle m|\hat{a}^\dagger\hat{\rho}_\mathrm{f}\hat{a}|n\rangle\right] \\ -\frac{1}{2}\mathcal{R}_b\left[\langle m|\hat{a}^\dagger\hat{a}\hat{\rho}_\mathrm{f}|n\rangle + \langle m|\hat{\rho}_\mathrm{f}\hat{a}^\dagger\hat{a}|n\rangle - 2\langle m|\hat{a}\hat{\rho}_\mathrm{f}\hat{a}^\dagger|n\rangle\right].$$

We recall that

$$\hat{a}|n\rangle = \sqrt{n}\,|n-1\rangle \quad \text{and} \quad \hat{a}^\dagger|n\rangle = \sqrt{n+1}\,|n+1\rangle$$

and

$$\langle m|\hat{a}^\dagger = \sqrt{m}\,\langle m-1| \quad \text{and} \quad \langle m|\hat{a} = \sqrt{m+1}\,\langle m+1|$$

18.3 Reservoir of Two-Level Atoms

and thus find

$$\frac{d}{dt}\rho_{m,n} = -\frac{1}{2}\mathcal{R}_a\left[(m+1+n+1)\rho_{m,n} - 2\sqrt{mn}\,\rho_{m-1,n-1}\right]$$
$$-\frac{1}{2}\mathcal{R}_b\left[(m+n)\rho_{m,n} - 2\sqrt{(m+1)(n+1)}\,\rho_{m+1,n+1}\right],$$

or

$$\frac{d}{dt}\rho_{m,n} = \mathcal{R}_a\sqrt{mn}\,\rho_{m-1,n-1}$$
$$-\frac{1}{2}\left[\mathcal{R}_a(m+1+n+1) + \mathcal{R}_b(m+n)\right]\rho_{m,n} + \mathcal{R}_b\sqrt{(m+1)(n+1)}\,\rho_{m+1,n+1}. \tag{18.16}$$

We therefore arrive at a first order differential equation for the matrix elements $\rho_{m,n}$. Moreover, they are coupled via a three-term recurrence relation to their nearest neighbors along the diagonals as indicated in the matrix

$$\begin{pmatrix} \ddots & & \ddots & & \ddots & & \\ & \rho_{m-1,m-2} & & \rho_{m-1,m-1} & & \rho_{m-1,m} & \\ & & \rho_{m,m-1} & & \rho_{m,m} & & \rho_{m,m+1} \\ & & & \rho_{m+1,m} & & \rho_{m+1,m+1} & & \rho_{m+1,m+2} \\ & & & & \ddots & & \ddots & & \ddots \end{pmatrix}$$

by the dots.

This coupling suggests to renumber the matrix elements in terms of the off-diagonals, that is, to introduce the quantity

$$\rho_{m,m+k} \equiv \langle m|\hat{\rho}_f|m+k\rangle \equiv \rho_m^{(k)}.$$

In this notation the equation of motion Eq. (18.15) for the density operator reads

$$\frac{d}{dt}\rho_m^{(k)} = \mathcal{R}_a\sqrt{m(m+k)}\,\rho_{m-1}^{(k)} - \left[\mathcal{R}_a\left(m+\frac{k}{2}+1\right) + \mathcal{R}_b\left(m+\frac{k}{2}\right)\right]\rho_m^{(k)}$$
$$+ \mathcal{R}_b\sqrt{(m+1)(m+k+1)}\,\rho_{m+1}^{(k)}. \tag{18.17}$$

Since the subscript k does not change in this equation it is not a matrix, but rather a scalar recurrence relation.

Time Evolution of Photon Statistics

We now consider the time evolution of the photon statistics. We therefore focus on the matrix elements along the main diagonal $m = n$, that is, $k = 0$. When we introduce the probability

$$W_m \equiv \langle m|\hat{\rho}_f|m\rangle = \rho_{m,m} = \rho_m^{(k=0)}$$

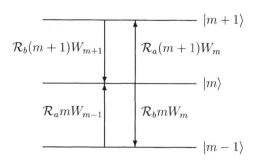

Fig. 18.2: Flow diagram for the probability W_m to find m photons in the cavity field interacting with a reservoir of two-level atoms. The level $|m\rangle$ is populated by the upper niveau $|m+1\rangle$ and the lower level $|m-1\rangle$. This corresponds to absorption of a photon from $m+1$ photons by a ground state atom and to stimulated emission of a photon from an excited atom and $m-1$ photons. Since it is absorption by ground state atoms from $m+1$ photons the transition rate is proportional to the photon number $m+1$, the rate $\mathcal{R}_b$ of ground state atoms injected into the cavity and the probability W_{m+1} to find $m+1$ photons. Emission of an excited atom consists of two parts: stimulated and spontaneous emission. Stimulated emission of an excited atom is proportional to the number of photons, whereas spontaneous emission contributes unity independent of the photon number in the field. Hence, emission of an atom in the presence of $m-1$ photons is proportional to $(m-1)+1 = m$, the rate $\mathcal{R}_a$ of excited atoms and the probability W_{m-1} to find the initial photon state $|m-1\rangle$. On the other hand, the level is depopulated by ground state atoms absorbing a photon and reducing the photon number from m to $m-1$. Moreover, atoms in the excited state can emit a photon and change the photon number from m to $m+1$. The corresponding rates associated with the absorption and the emission are $\mathcal{R}_b\, m\, W_m$ and $\mathcal{R}_a\,(m+1)\,W_m$.

to find m photons in the field, Eq. (18.17) reduces to

$$\frac{d}{dt} W_m = \mathcal{R}_a m\, W_{m-1} - [\mathcal{R}_a(m+1) + \mathcal{R}_b m]\, W_m + \mathcal{R}_b(m+1)\, W_{m+1}. \qquad (18.18)$$

We can represent this dynamics of the photon statistics using the flow diagram of Fig. 18.2. In order to gain some insight we study the time dependence of the mean photon number

$$\langle m \rangle \equiv \sum_{m=0}^{\infty} m W_m.$$

For this purpose we multiply the equation of motion Eq. (18.18) for the photon statistics with m and sum over all m which yields

$$\frac{d}{dt}\langle m \rangle = \mathcal{R}_a \sum_m m^2\, W_{m-1} - \sum_m [\mathcal{R}_a m(m+1) + \mathcal{R}_b m^2]\, W_m + \mathcal{R}_b \sum_m m(m+1)\, W_{m+1}.$$

In the first sum we introduce the new summation index $m' \equiv m - 1$ and in the last sum $m'' \equiv m + 1$ which yields

$$\frac{d}{dt}\langle m\rangle = \mathcal{R}_a \sum_{m'}(m'+1)^2 W_{m'} - \sum_m [\mathcal{R}_a m(m+1) + \mathcal{R}_b m^2] W_m + \mathcal{R}_b \sum_{m''}(m''-1)m'' W_{m''}.$$

Since the sums containing the square of the summation index m cancel we arrive at the differential equation

$$\frac{d}{dt}\langle m\rangle = -(\mathcal{R}_b - \mathcal{R}_a)\langle m\rangle + \mathcal{R}_a$$

for the mean number of photons.

The solution of this equation reads

$$\langle m(t)\rangle = e^{-(\mathcal{R}_b - \mathcal{R}_a)t}\langle m(0)\rangle + \frac{\mathcal{R}_a}{\mathcal{R}_b - \mathcal{R}_a}\left(1 - e^{-(\mathcal{R}_b - \mathcal{R}_a)t}\right). \tag{18.19}$$

Damping of the Field

When $\mathcal{R}_b > \mathcal{R}_a$ the mean number of photons $\langle m(t)\rangle$ decays from its original value $\langle m(0)\rangle$ to a stationary value. The decay rate

$$\gamma \equiv \mathcal{R}_b - \mathcal{R}_a = r(g\tau)^2(\rho_{bb} - \rho_{aa}). \tag{18.20}$$

is determined by the population difference and the interaction strength $(g\tau)^2$.

The steady state value reads

$$\langle m(t \to \infty)\rangle \equiv \frac{\mathcal{R}_a}{\mathcal{R}_b - \mathcal{R}_a} = \frac{\rho_{aa}}{\rho_{bb} - \rho_{aa}}.$$

The condition $\mathcal{R}_b > \mathcal{R}_a$ for a stationary state translates into $\rho_{bb} > \rho_{aa}$, that is, there must be more atoms in the ground state than in the excited state. On average the atoms are taking excitation out of the field until there is an equilibrium between the excitations in the atoms and the field: The field gets damped by this type of reservoir. Moreover, the field state eventually approaches a thermal state with a temperature determined by the population difference in the two levels. Indeed, when the populations in the two levels are distributed according to a Boltzmann distribution

$$\frac{\rho_{aa}}{\rho_{bb}} = \exp\left(-\frac{\hbar\omega}{k_B T}\right)$$

of temperature T we find

$$\langle m(t \to \infty)\rangle = \frac{1}{\frac{\rho_{bb}}{\rho_{aa}} - 1} = \frac{1}{\exp[\hbar\omega/(k_B T)] - 1} \equiv n_{\text{th}}. \tag{18.21}$$

Here we have introduced in the last step the average number n_{th} of photons in a thermal state of temperature T.

We can even show that in the limit of $t \to \infty$ all the off-diagonal elements $\rho_{m,n}$ of the density operator decay to zero and we obtain indeed, a thermal state.

Damping Equation in New Form

We can now use the definitions Eqs. (18.20) and (18.21) for the decay rate γ and the average number of thermal photons to express the coefficients $\mathcal{R}_a$ and $\mathcal{R}_b$ in Eq. (18.15). With the help of the formulas

$$\mathcal{R}_a = \gamma n_{\text{th}} \quad \text{and} \quad \mathcal{R}_b = \gamma(n_{\text{th}} + 1)$$

the equation of motion for the mean photon number takes the rather compact form

$$\frac{d}{dt}\langle m \rangle = -\gamma \langle m \rangle + \gamma n_{\text{th}}. \tag{18.22}$$

Moreover, the master equation Eq. (18.15) takes the standard form

$$\frac{d}{dt}\hat{\rho}_{\text{f}} = -\frac{\gamma}{2}(n_{\text{th}} + 1)\left(\hat{a}^\dagger \hat{a} \hat{\rho}_{\text{f}} + \hat{\rho}_{\text{f}} \hat{a}^\dagger \hat{a} - 2\hat{a}\hat{\rho}_{\text{f}}\hat{a}^\dagger\right) - \frac{\gamma}{2}n_{\text{th}}\left(\hat{a}\hat{a}^\dagger \hat{\rho}_{\text{f}} + \hat{\rho}_{\text{f}}\hat{a}\hat{a}^\dagger - 2\hat{a}^\dagger \hat{\rho}_{\text{f}}\hat{a}\right) \tag{18.23}$$

used to describe the damping of a field mode in a cavity of temperature T and decay constant γ.

This equation consists of two parts. The operators in the two brackets are multiplied by the decay constant $\gamma/2$ and the average number of thermal photons. However, the first term contains one additional quantum, that is, $n_{\text{th}} + 1$ rather than n_{th}. This difference is obviously important when we consider zero temperature, that is, when all atoms of the reservoir are in the ground state. In this case the average number of photons is zero. Hence, the master equation at zero temperature reduces to

$$\frac{d}{dt}\hat{\rho}_{\text{f}} = -\frac{\gamma}{2}\left(\hat{a}^\dagger \hat{a} \hat{\rho}_{\text{f}} + \hat{\rho}_{\text{f}}\hat{a}^\dagger \hat{a} - 2\hat{a}\hat{\rho}_{\text{f}}\hat{a}^\dagger\right).$$

A field coupled to a reservoir at zero temperature still experiences a damping.

Amplification of a Cavity Field

In the case $\mathcal{R}_b < \mathcal{R}_a$, the mean number of photons grows exponentially and we have amplification of the field. This condition for amplification translates into $\rho_{bb} < \rho_{aa}$, that is there are more atoms in the excited state than in the ground state, a situation referred to as inversion of the medium. In this case the atoms do not act as a reservoir which provides damping but on the contrary pump the cavity and deposit on the average more energy in the cavity than they take out.

Of course, in reality this amplification process cannot continue forever since there is always a counter-acting damping mechanism which eventually leads to a stationary mean photon number. We will discuss this problem in more detail in Sec. 18.4.

Phase Space Distributions

So far, we have only discussed the time evolution of one single parameter characterizing the quantum state, namely the average number of photons. However, a quantum state is determined either by a distribution function of continuous variables such as the Glauber-Sudarshan P-function, or the discrete photon number distribution. In the problems at the end of this chapter we discuss how the density operator equation Eq. (18.23) can be translated into a c-number representation and be solved using these tools. This approach allows us to discuss the influence of damping or amplification on the quantum state. In particular, we find that amplification always adds additional noise and the phase space distributions broaden.

18.3.3 Exact Master Equation

Before we discuss the properties of the master equation Eq. (18.23) in more detail we first sum the perturbative expansion in Eq. (18.9) and obtain an exact equation for the density operator of the field. Two approaches offer themselves: We can calculate all the commutators in Eq. (18.9), or use the time evolution operator Eq. (15.11) of the Jaynes-Cummings-Paul model. In the present section pursue the second approach.

Density Operator of Field from Time Evolution Operator

We start from the formal solution

$$\hat{\rho}_{f+at}(t+\tau) = \hat{\mathcal{U}}(t+\tau;t)\,\hat{\rho}_{f+at}(t)\,\hat{\mathcal{U}}^\dagger(t+\tau;t)$$

of the von Neumann equation given by Eq. (2.42) where

$$\hat{\mathcal{U}}(t+\tau;t) \equiv \exp\left[-\frac{i}{\hbar}\hat{H}\tau\right] \equiv \exp\left[-ig\tau\left(\hat{\sigma}\hat{a}^\dagger + \sigma^\dagger\hat{a}\right)\right] \qquad (18.24)$$

denotes the time evolution operator of the Jaynes-Cummings-Paul Hamiltonian.

Again we obtain the density operator of the field by tracing over the atoms, that is

$$\hat{\rho}_f(t+\tau) = \text{Tr}_{at}\left[\hat{\mathcal{U}}(t+\tau;t)\,\hat{\rho}_{f+at}(t)\,\hat{\mathcal{U}}^\dagger(t+\tau;t)\right].$$

From Eq. (15.11) we recall the expression

$$\hat{\mathcal{U}}(t+\tau;t) \equiv \begin{pmatrix} \hat{\mathcal{U}}_{aa} & \hat{\mathcal{U}}_{ab} \\ \hat{\mathcal{U}}_{ba} & \hat{\mathcal{U}}_{bb} \end{pmatrix} = \begin{pmatrix} \hat{C}_n & 0 \\ 0 & \hat{C}_{n-1} \end{pmatrix} - i\begin{pmatrix} 0 & \frac{\hat{S}_n}{\sqrt{\hat{n}+1}}\hat{a} \\ \frac{\hat{S}_{n-1}}{\sqrt{\hat{n}}}\hat{a}^\dagger & 0 \end{pmatrix}$$

for the time evolution operator with

$$\hat{C}_n(\tau) \equiv \cos(g\tau\sqrt{\hat{n}+1}) \quad \text{and} \quad \hat{S}_n(\tau) \equiv \sin(g\tau\sqrt{\hat{n}+1}).$$

We assume that at time t when the atom starts to interact with the field the density operator $\hat{\rho}_{f+at}$ factorizes into density operators $\hat{\rho}_f$ and $\hat{\rho}_{at}$ of the field and the atom, that is,

$$\hat{\rho}_{f+at}(t) \equiv \begin{pmatrix} \hat{\rho}_f\rho_{aa} & \hat{\rho}_f\rho_{ab} \\ \hat{\rho}_f\rho_{ba} & \hat{\rho}_f\rho_{bb} \end{pmatrix}.$$

In contrast to the preceding section we now also allow for off-diagonal elements ρ_{ab} in the atomic density operator.

Hence the density operator of the field reads

$$\hat{\rho}_f(t+\tau) = \mathrm{Tr}\left[\begin{pmatrix} \hat{U}_{aa} & \hat{U}_{ab} \\ \hat{U}_{ba} & \hat{U}_{bb} \end{pmatrix}\begin{pmatrix} \hat{\rho}_f\rho_{aa} & \hat{\rho}_f\rho_{ab} \\ \hat{\rho}_f\rho_{ba} & \hat{\rho}_f\rho_{bb} \end{pmatrix}\begin{pmatrix} \hat{U}_{aa}^\dagger & \hat{U}_{ba}^\dagger \\ \hat{U}_{ab}^\dagger & \hat{U}_{bb}^\dagger \end{pmatrix}\right], \quad (18.25)$$

and when we multiply the matrices and take the trace we arrive at

$$\hat{\rho}_f(t+\tau) =$$
$$\left[\hat{C}_n\,\hat{\rho}_f\,\hat{C}_n + \frac{\hat{S}_{n-1}}{\sqrt{\hat{n}}}\,\hat{a}^\dagger\,\hat{\rho}_f\,\hat{a}\,\frac{\hat{S}_{n-1}}{\sqrt{\hat{n}}}\right]\rho_{aa} + \left[\hat{C}_{n-1}\,\hat{\rho}_f\,\hat{C}_{n-1} + \frac{\hat{S}_n}{\sqrt{\hat{n}+1}}\,\hat{a}\,\hat{\rho}_f\,\hat{a}^\dagger\,\frac{\hat{S}_n}{\sqrt{\hat{n}+1}}\right]\rho_{bb}$$
$$-i\left[\left(\frac{\hat{S}_n}{\sqrt{\hat{n}+1}}\,\hat{a}\,\hat{\rho}_f\,\hat{C}_n - \hat{C}_{n-1}\hat{\rho}_f\hat{a}\,\frac{\hat{S}_{n-1}}{\sqrt{\hat{n}}}\right)\rho_{ba} - \left(\hat{C}_n\,\hat{\rho}_f\,\hat{a}^\dagger\,\frac{\hat{S}_n}{\sqrt{\hat{n}+1}} - \frac{\hat{S}_{n-1}}{\sqrt{\hat{n}}}\,\hat{a}^\dagger\,\hat{\rho}_f\,\hat{C}_{n-1}\right)\rho_{ab}\right].$$
$$(18.26)$$

This expression is exact and provides the density operator for the field at time $t+\tau$ given the field operator and the atomic density operator at time t.

Coarse Graining

We now turn to an atomic beam rather than a single atom and derive a coarse-grained equation of motion. For this purpose we subtract from Eq. (18.26) the density operator $\hat{\rho}_f(t)$ at time t and multiply the resulting equation by the rate r of atoms injected into the cavity. This procedure yields the master equation

$$\frac{d}{dt}\hat{\rho}_f(t) \equiv r[\hat{\rho}_f(t+\tau) - \hat{\rho}_f(t)]$$
$$= \left[\hat{C}_n(\tau)\,\hat{\rho}_f(t)\,\hat{C}_n(\tau) - \hat{\rho}_f(t) + \frac{\hat{S}_{n-1}(\tau)}{\sqrt{\hat{n}}}\,\hat{a}^\dagger\hat{\rho}_f(t)\,\hat{a}\,\frac{\hat{S}_{n-1}(\tau)}{\sqrt{\hat{n}}}\right]r\rho_{aa}$$
$$+ \left[\hat{C}_{n-1}(\tau)\,\hat{\rho}_f(t)\,\hat{C}_{n-1}(\tau) - \hat{\rho}_f(t) + \frac{\hat{S}_n(\tau)}{\sqrt{\hat{n}+1}}\,\hat{a}\,\hat{\rho}_f(t)\,\hat{a}^\dagger\,\frac{\hat{S}_n(\tau)}{\sqrt{\hat{n}+1}}\right]r\rho_{bb}$$
$$- i\left\{\left[\frac{\hat{S}_n(\tau)}{\sqrt{\hat{n}+1}}\,\hat{a}\,\hat{\rho}_f(t)\,\hat{C}_n(\tau) - \hat{C}_{n-1}(\tau)\,\hat{\rho}_f(t)\,\hat{a}\,\frac{\hat{S}_{n-1}(\tau)}{\sqrt{\hat{n}}}\right]r\rho_{ba}\right.$$
$$\left. - \left[\hat{C}_n(\tau)\,\hat{\rho}_f(t)\,\hat{a}^\dagger\,\frac{\hat{S}_n(\tau)}{\sqrt{\hat{n}+1}} - \frac{\hat{S}_{n-1}(\tau)}{\sqrt{\hat{n}}}\,\hat{a}^\dagger\,\hat{\rho}_f(t)\,\hat{C}_{n-1}(\tau)\right]r\rho_{ab}\right\}.$$
$$(18.27)$$

Here we have used the fact that $\rho_{aa} + \rho_{bb} = 1$.

Master Equation in Photon Number Representation

To gain more insight into Eq. (18.27) we now calculate the equation of motion for the matrix element

$$\rho_m^{(k)} \equiv \langle m|\,\hat{\rho}_f\,|m+k\rangle.$$

Here we number the diagonals by the superscript k which brings out most clearly the coupling scheme of the matrix elements.

From Eq. (18.27) we immediately find

$$\begin{aligned}\frac{d}{dt}\rho_m^{(k)} &= \left[(C_m C_{m+k} - 1)\rho_m^{(k)} + S_{m-1} S_{m+k-1} \rho_{m-1}^{(k)}\right] r\rho_{aa} \\ &+ \left[(C_{m-1} C_{m+k-1} - 1)\rho_m^{(k)} + S_m S_{m+k} \rho_{m+1}^{(k)}\right] r\rho_{bb} \\ &- i\left[\left(S_m C_{m+k}\rho_{m+1}^{(k-1)} - C_{m-1} S_{m+k-1}\rho_m^{(k-1)}\right) r\rho_{ba}\right. \\ &\left.- \left(C_m S_{m+k}\rho_m^{(k+1)} - S_{m-1} C_{m+k-1}\rho_{m-1}^{(k+1)}\right) r\rho_{ab}\right].\end{aligned}$$

The atomic populations ρ_{aa} and ρ_{bb} only couple matrix elements along a single diagonal, that is, they couple $\rho_m^{(k)}$ to its nearest neighbors $\rho_{m+1}^{(k)}$ and $\rho_{m-1}^{(k)}$. In contrast, the off-diagonal elements $\rho_{ab} = \rho_{ba}^*$ couple the diagonals with each other, that is,

$$\begin{array}{ccc} \rho_{m-1}^{(k)} & & \rho_{m-1}^{(k+1)} \\ \searrow & \updownarrow & \\ \rho_m^{(k-1)} & \leftrightarrow \rho_m^{(k)} \leftrightarrow & \rho_m^{(k+1)} \\ & \updownarrow & \searrow \\ \rho_{m+1}^{(k-1)} & & \rho_{m+1}^{(k)}. \end{array}$$

Reduction to only Populations

We now make the connection with the master equation Eq. (18.15) and consider the special case

$$\hat{\rho}_{\text{at}}(t) \equiv \begin{pmatrix} \rho_{aa} & 0 \\ 0 & \rho_{bb} \end{pmatrix},$$

of an atomic density operator without coherence, that is, we set $\rho_{ab} = \rho_{ba} = 0$. Hence the master equation Eq. (18.27) reduces to

$$\begin{aligned}\frac{d}{dt}\hat{\rho}_{\text{f}}(t) &= r\rho_{aa}\left[\hat{C}_n(\tau)\hat{\rho}_{\text{f}}(t)\hat{C}_n(\tau) - \hat{\rho}_{\text{f}}(t) + \frac{\hat{S}_{n-1}(\tau)}{\sqrt{\hat{n}}}\hat{a}^\dagger \hat{\rho}_{\text{f}}(t)\hat{a}\frac{\hat{S}_{n-1}(\tau)}{\sqrt{\hat{n}}}\right] \\ &+ r\rho_{bb}\left[\hat{C}_{n-1}(\tau)\hat{\rho}_{\text{f}}(t)\hat{C}_{n-1}(\tau) - \hat{\rho}_{\text{f}}(t) + \frac{\hat{S}_n(\tau)}{\sqrt{\hat{n}+1}}\hat{a}\hat{\rho}_{\text{f}}(t)\hat{a}^\dagger\frac{\hat{S}_n(\tau)}{\sqrt{\hat{n}+1}}\right].\end{aligned}$$

This equation is an important ingredient for the quantum theory of the one-atom maser discussed in the next section.

In the present section, however, we simplify this equation in the limit of small interaction times, that is, for $g\tau\sqrt{\langle m\rangle + 1} \ll 1$, where $\langle m\rangle$ is the average number of photons in the field state. Then we can expand the sine and cosine operators

$$\hat{C}_n(\tau) = \cos\left(g\tau\sqrt{\hat{a}^\dagger\hat{a}+1}\right) = \cos\left(g\tau\sqrt{\hat{a}\hat{a}^\dagger}\right) \cong 1 - \tfrac{1}{2}(g\tau)^2\hat{a}\hat{a}^\dagger$$

and
$$\hat{S}_n(\tau) = \sin\left(g\tau\sqrt{\hat{n}+1}\right) \cong g\tau\sqrt{\hat{n}+1}.$$

Here we have made use of the commutation relation $\hat{a}\hat{a}^\dagger = \hat{a}^\dagger\hat{a} + 1$.

When we retain terms quadratic in $g\tau$ we arrive at

$$\frac{d}{dt}\hat{\rho}_{\text{f}}(t) = -\frac{1}{2}\mathcal{R}_a\left[\hat{a}\,\hat{a}^\dagger\,\hat{\rho}_{\text{f}}(t) + \hat{\rho}_{\text{f}}(t)\,\hat{a}\,\hat{a}^\dagger - 2\hat{a}^\dagger\,\hat{\rho}_{\text{f}}(t)\,\hat{a}\right]$$
$$- \frac{1}{2}\mathcal{R}_b\left[\hat{a}^\dagger\,\hat{a}\,\hat{\rho}_{\text{f}}(t) + \hat{\rho}_{\text{f}}(t)\,\hat{a}^\dagger\,\hat{a} - 2\hat{a}\,\hat{\rho}_{\text{f}}(t)\,\hat{a}^\dagger\right],$$

where $\mathcal{R}_a \equiv r\,\rho_{aa}\,(g\tau)^2$ and $\mathcal{R}_b \equiv r\,\rho_{bb}\,(g\tau)^2$.

Indeed, this equation is identical to Eq. (18.15) derived in Sec. 18.3.1 by a perturbative technique.

18.3.4 Summary

In this section we have derived a master equation for the density operator $\hat{\rho}_{\text{f}}$ of the field in a cavity driven by a beam of resonant two-level atoms characterized by a density operator $\hat{\rho}_{\text{at}}$. We have used two methods:

(i) A formal solution of the von Neumann equation in terms of multiple commutators allowed us to perform the trace over the atomic reservoir and a coarse graining provided us with the equation of motion.

(ii) The exact solution of the Jaynes-Cummings-Paul model expressed in terms of the density operator $\hat{\rho}_{\text{f+at}}$ of the field plus atom enabled us to avoid the evaluation of the nested commutators. In this way we have found an exact expression for the density operator $\hat{\rho}_{\text{f}}$.

Whereas the first method is perturbative the second approach is exact. Why should we even bother with the perturbative treatment when we have an exact method? However, for the exact treatment it is essential that the interaction between the field and the atom is not explicitly time dependent. Since the field is resonant with the atoms this condition is satisfied. However, for a nonresonant interaction which involves time dependent phase factors due to the detuning this method can probably not be applied.

18.4 One-Atom Maser

In the last section we have derived the essential ingredients of the quantum theory of the one-atom maser or micromaser. In the present section we discuss this device in more detail and analyze the unique properties of its field.

The very heart of the one-atom maser is a high-Q cavity. A very weak beam of excited two-level atoms traverses the cavity and interacts resonantly with a single mode of the radiation field. The flux of the atomic beam is so small that only one atom at a time interacts with the field mode. The atom can deposit its excitation in the cavity and in this way amplify the field. The atoms also serve a different purpose:

They probe the field. A detector after the resonator measures the population of the internal levels. When we tune the cavity over the transition frequency the number of atoms in the excited state drops indicating an onset of maser action. In Fig. 1.14 we show the first experimentally measured maser resonance.

18.4.1 Density Operator Equation

When we describe this interaction by the resonant Jaynes-Cummings-Paul Hamiltonian the dynamics of the field follows from the equation of motion Eq. (18.27). We assume, that the time interval between two successive atoms is much larger than the interaction time of a single atom. In this time interval we take into account the damping of the field mode by an equation of the form Eq. (18.23).

We obtain an equation of motion for the density operator of the field which includes both the resonant interaction with the stream of two-level atoms and the damping of cavity field by adding to the exact master equation Eq. (18.27) the right-hand side of the damping equation Eq. (18.23).

In the standard configuration of the one-atom maser the atoms enter the cavity in the excited state. This implies $\rho_{aa} = 1$ and $\rho_{bb} = \rho_{ab} = \rho_{ba} = 0$.

When we add the so-simplified Eq. (18.27) to the damping equation Eq. (18.23) we find the master equation

$$\frac{d}{dt}\hat{\rho}_{\mathrm{f}}(t) = r[\hat{C}_n(\tau)\,\hat{\rho}_{\mathrm{f}}(t)\,\hat{C}_n(\tau) - \hat{\rho}_{\mathrm{f}}(t)] + r\frac{\hat{S}_{n-1}(\tau)}{\sqrt{\hat{n}}}\,\hat{a}^\dagger \hat{\rho}_{\mathrm{f}}(t)\,\hat{a}\,\frac{\hat{S}_{n-1}(\tau)}{\sqrt{\hat{n}}}$$
$$-\frac{\gamma}{2}(n_{\mathrm{th}}+1)\left(\hat{a}^\dagger \hat{a}\hat{\rho}_{\mathrm{f}} + \hat{\rho}_{\mathrm{f}}\hat{a}^\dagger \hat{a} - 2\hat{a}\hat{\rho}_{\mathrm{f}}\hat{a}^\dagger\right)$$
$$-\frac{\gamma}{2}n_{\mathrm{th}}\left(\hat{a}\hat{a}^\dagger \hat{\rho}_{\mathrm{f}} + \hat{\rho}_{\mathrm{f}}\hat{a}\hat{a}^\dagger - 2\hat{a}^\dagger \hat{\rho}_{\mathrm{f}}\hat{a}\right), \qquad (18.28)$$

of the one-atom maser.

We transform this operator equation into a c-number representation by expressing the density operator

$$\hat{\rho}_{\mathrm{f}} = \sum_{m,k} \rho_m^{(k)} |m\rangle\langle m+k|$$

in photon number states.

From the master equation Eq. (18.28) we find the differential-recurrence equation

$$\frac{d}{dt}\rho_m^{(k)} = r\left[(C_m C_{m+k} - 1)\rho_m^{(k)} + S_{m-1}S_{m+k-1}\rho_{m-1}^{(k)}\right]$$
$$+ \frac{\gamma}{2}(n_{\mathrm{th}}+1)\left[2\sqrt{(m+1)(m+k+1)}\,\rho_{m+1}^{(k)} - (2m+k)\rho_m^{(k)}\right]$$
$$+ \frac{\gamma}{2}n_{\mathrm{th}}\left[2\sqrt{m(m+k)}\,\rho_{m-1}^{(k)} - (2m+k+2)\rho_m^{(k)}\right]. \qquad (18.29)$$

which only couples nearest neighbors along the diagonal. There is no coupling in the superscript k. In the remainder of this section we discuss the properties of this rather complicated recurrence relation.

18.4.2 Equation of Motion for the Photon Statistics

We first analyze the properties of the photon statistics

$$W_m \equiv \langle m| \hat{\rho}_\text{f} |m\rangle = \rho_{m,m} = \rho_m^{(k=0)}.$$

For this purpose we consider the diagonal elements of the density operator of the one-atom maser, that is the case $k = 0$ of Eq. (18.29) and arrive at

$$\frac{d}{dt}W_m = -r\sin^2(g\tau\sqrt{m+1})\,W_m + r\sin^2(g\tau\sqrt{m})\,W_{m-1}$$
$$+ \gamma(n_\text{th}+1)[(m+1)W_{m+1} - mW_m] + \gamma n_\text{th}[mW_{m-1} - (m+1)W_m]. \quad (18.30)$$

Steady-State Solution

The steady-state photon distribution follows from Eq. (18.30) when we set

$$\frac{d}{dt}W_m = 0$$

which after rearranging the terms yields

$$0 = -r\sin^2(g\tau\sqrt{m+1})\,W_m + \gamma(n_\text{th}+1)(m+1)W_{m+1} - \gamma n_\text{th}(m+1)W_m$$
$$+ r\sin^2(g\tau\sqrt{m})\,W_{m-1} - \gamma(n_\text{th}+1)mW_m + \gamma n_\text{th}mW_{m-1}. \quad (18.31)$$

We certainly have found a solution of this equation if both lines in Eq. (18.31) vanish separately. This condition is called *detailed balance* and reads

$$\left(r\sin^2(g\tau\sqrt{m}) + \gamma n_\text{th}m\right)W_{m-1} = \gamma(n_\text{th}+1)mW_m,$$

or

$$W_m = \left[\frac{n_\text{th}}{n_\text{th}+1} + \frac{r\sin^2(g\tau\sqrt{m})}{\gamma(n_\text{th}+1)m}\right]W_{m-1}. \quad (18.32)$$

This two-term recurrence relation has the solution

$$W_m = W_0 \prod_{l=1}^{m}\left[\frac{n_\text{th}}{n_\text{th}+1} + \frac{r\sin^2(g\tau\sqrt{l})}{\gamma(n_\text{th}+1)l}\right], \quad (18.33)$$

where W_0 is a normalization constant determined from

$$\sum_{m=0}^{\infty} W_m = 1.$$

Mean Photon Number

The stationary photon statistics Eq. (18.33) allows us to calculate the mean photon number

$$\langle m \rangle \equiv \sum_{m=0}^{\infty} m\,W_m.$$

18.4 One-Atom Maser

in the steady state. However, we now pursue a different approach and use the equation of motion Eq. (18.30) for the photon statistics to derive an equation of motion for $\langle m \rangle$.

For this purpose we multiply Eq. (18.30) by m and sum over m which yields

$$\frac{d}{dt}\langle m \rangle = -r \sum_{m=0}^{\infty} m \sin^2\left(g\tau\sqrt{m+1}\right) W_m + r \sum_{m=0}^{\infty} m \sin^2\left(g\tau\sqrt{m}\right) W_{m-1} + \gamma(n_{\text{th}} - \langle m \rangle)$$

where we have used Eq. (18.22) for the contribution from the damping part.

When we introduce the new summation index $m' \equiv m - 1$ in the second sum, the terms linear in the summation index cancel and we obtain

$$\frac{d}{dt}\langle m \rangle = r \sum_{m=0}^{\infty} \sin^2(g\tau\sqrt{m+1}) W_m + \gamma(n_{\text{th}} - \langle m \rangle). \tag{18.34}$$

In order to find a closed equation for $\langle m \rangle$ we assume that the photon statistics is well-localized around its mean value $\langle m \rangle$. This assumption allows us to replace m under the sum by $\langle m \rangle$ and we obtain the nonlinear differential equation

$$\frac{d}{dt}\langle m \rangle \cong r \sin^2\left(g\tau\sqrt{\langle m \rangle + 1}\right) + \gamma(n_{\text{th}} - \langle m \rangle). \tag{18.35}$$

Since this equation is still quite complicated, we now restrict ourselves to values of the interaction strength $g\tau$ such that $g\tau\sqrt{\langle m \rangle + 1} \ll 1$. In this limit we can expand the sine in Eq. (18.34) and only keep the leading term which yields

$$\frac{d}{dt}\langle m \rangle \cong (r(g\tau)^2 - \gamma)\langle m \rangle + r(g\tau)^2 + \gamma n_{\text{th}}.$$

The stationary solution of this equation reads

$$\langle m \rangle \equiv \frac{\frac{r}{\gamma}(g\tau)^2 + n_{\text{th}}}{1 - \frac{r}{\gamma}(g\tau)^2} = \frac{\theta^2 + n_{\text{th}}}{1 - \theta^2}, \tag{18.36}$$

where we have introduced the dimensionless interaction time

$$\theta \equiv \sqrt{\frac{r}{\gamma}} g\tau.$$

When there is no interaction, that is $\theta = 0$ the average number of photons in the field is only determined by the number n_{th} of thermal photons. For increasing θ the number of photons in the cavity increases. We note that it increases for two reasons: (i) The θ^2-dependence in the numerator creates a growth in $\langle m \rangle$. (ii) The θ^2-dependence in the denominator adds to this growth. When θ approaches unity the mean photon number increases drastically since here the denominator tends towards zero and $\langle m \rangle$ becomes singular.

This singularity is a clear indication that the linear approximation is no longer valid. We have to go beyond this approximation. However, this cannot be done analytically. We therefore resort to a numerical solution.

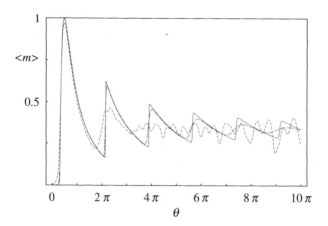

Fig. 18.3: Normalized mean photon number of the one-atom maser in steady state as a function of the dimensionless interaction time θ. We show three curves for different pump parameters: $r/\gamma = 20$ (short dashed line), $r/\gamma = 200$ (long dashed line), and $r/\gamma = 2000$ (solid line).

In Fig. 18.3 we show the mean photon number in steady state using the exact stationary photon statistics, Eq. (18.33). Indeed, we see a steep increase of $\langle m \rangle$ as θ approaches unity which is the maser threshold. A similar effect occurs in every laser.

However, due to the presence of the sine function in Eq. (18.34) more thresholds appear. However, they are not as pronounced as the first one. We also notice that when we increase the pump parameter the thresholds become sharper.

Variance

In the context of the Schrödinger cat state discussed in Sec. 11.3 we have introduced the normalized variance

$$\sigma^2 \equiv \frac{\langle m^2 \rangle - \langle m \rangle^2}{\langle m \rangle}$$

of the steady-state photon number distribution.

When $\sigma = 1$ we have Poisson statistics, whereas for $\sigma > 1$ or $\sigma < 1$ we have super-Poissonian or sub-Poissonian statistics, respectively. We now apply this measure of non-classicality of a radiation field to the one-atom maser field.

We start again from the equation of motion Eq. (18.30) for the photon statistics, multiply it by m^2 and sum over m, which results in

$$\frac{d}{dt}\langle m^2 \rangle = -r \sum_m \sin^2\left(g\tau\sqrt{m+1}\right) m^2 W_m + r \sum_m \sin^2\left(g\tau\sqrt{m}\right) m^2 W_{m-1}$$
$$+ \gamma(n_{\text{th}} + 1)\left[\sum_m m^2(m+1)W_{m+1} - \sum_m m^3 W_m\right]$$
$$+ \gamma n_{\text{th}}\left[\sum_m m^3 W_{m-1} - \sum_m m^2(m+1)W_m\right].$$

When we introduce the summation indices $m' \equiv m-1$ and $m'' \equiv m+1$ in order to bring the terms $W_{m\pm 1}$ to W_m, various terms cancel, that is,

$$\frac{d}{dt}\langle m^2 \rangle = r\sum_m \sin^2\left(g\tau\sqrt{m+1}\right)(2m+1)W_m$$
$$-\gamma(n_{\text{th}}+1)\sum_m (2m^2 - m)W_m + \gamma n_{\text{th}}\sum_m (2m^2 + 3m + 1)W_m$$

and we arrive at the exact equation

$$\frac{d}{dt}\langle m^2 \rangle = r\sum_{m=0}^{\infty} \sin^2(g\tau\sqrt{m+1})(2m+1)W_m - 2\gamma\langle m^2\rangle + \gamma(4n_{\text{th}}+1)\langle m\rangle + \gamma n_{\text{th}}.$$

We assume the photon distribution to be localized in the vicinity of $\langle m \rangle$ and restrict ourselves to short interaction times so that we can expand the sine function, that is

$$\frac{d}{dt}\langle m^2 \rangle = 2(r(g\tau)^2 - \gamma)\langle m^2 \rangle + \left[3r(g\tau)^2 + 4\gamma(4n_{\text{th}}+1)\right]\langle m \rangle + r(g\tau)^2 + \gamma n_{\text{th}}.$$

For the stationary solution we obtain

$$\langle m^2 \rangle = \frac{3\theta^2 + 4n_{\text{th}} + 1}{2(1-\theta^2)}\langle m \rangle + \frac{1}{2}\langle m \rangle = \frac{\theta^2 + 2n_{\text{th}} + 1}{1-\theta^2}\langle m \rangle.$$

The normalized variance

$$\sigma^2 = \frac{\theta^2 + 2n_{\text{th}} + 1}{1-\theta^2} - \langle m \rangle$$

reads after substituting Eq. (18.36)

$$\sigma^2 = \frac{n_{\text{th}} + 1}{1-\theta^2}. \tag{18.37}$$

For a coherent state with a Poissonian photon distribution the normalized variance σ is unity. Equation (18.37) predicts an increase of σ to values larger than unity when the interaction parameter θ comes close to unity. In this situation the one-atom maser is going through its threshold and hence the photon statistics is super-Poissonian. Indeed, the numerical result for σ, shown in Fig. 18.4, confirms this prediction for short interaction times.

However, for larger values of θ the photon statistics clearly shows sub-Poissonian characteristics, but also super-Poissonian peaks around values of θ, which are integer multiples of 2π.

Trapping States

The photon statistics becomes particularly interesting when there are no thermal photons present, that is for $n_{\text{th}} \equiv 0$ or $T \equiv 0$. In this case the stationary photon distribution given by Eq. (18.33) reduces to

$$W_m = W_0 \prod_{l=1}^{m} \frac{r\sin^2(g\tau\sqrt{l})}{\gamma l}.$$

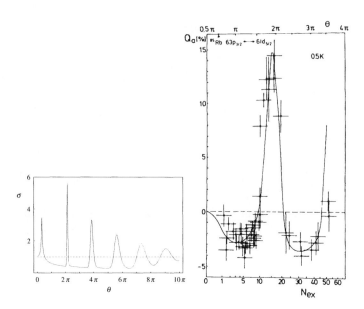

Fig. 18.4: Left: Normalized variance σ of the photon statistics of the one-atom maser in steady state. The pump parameter is $r/\gamma = 200$ and the dashed line indicates the value $\sigma = 1$ for a coherent state with Poissonian statistics. Right: Measured variance $Q_a \equiv \sigma^2 - 1$ of the atoms in the lower maser level as a function $N_{\text{ex}} \equiv r/\gamma$ above threshold. Taken from G. Rempe et al., Phys. Rev. Lett. **64**, 2783 (1990).

We now choose the interaction time θ such that

$$g\tau\sqrt{n_q + 1} = q\pi, \tag{18.38}$$

where q and n_q are positive integers. In this case we find from the recurrence relation Eq. (18.32) for the photon statistics

$$W_{n_q+1} = \frac{r \sin^2(g\tau\sqrt{n_q+1})}{\gamma(n_q+1)} W_{n_q} = 0,$$

and hence the probability for photon numbers higher then n_q vanishes.

Indeed, the steady state photon number distribution is of the form

$$W_m = \begin{cases} W_0 \prod_{l=1}^{m} \dfrac{r \sin^2(g\tau\sqrt{l})}{\gamma l} & \text{for } 0 \leq m \leq n_q \\ 0 & \text{for } m > n_q, \end{cases}.$$

Since the number states for $n > n_q$ are not populated the mean photon number is smaller than n_q and can be considerably smaller close to interaction times determined by the condition Eq. (18.38) – the mean photon number is trapped. We demonstrate this phenomenon in Fig. 18.5 where we compare the mean photon number as a function of the dimensionless interaction time θ for temperatures corresponding to $n_{\text{th}} = 0.1$ and $n_{\text{th}} = 10^{-7}$.

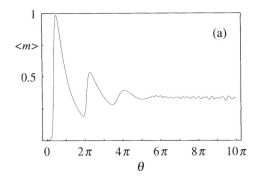 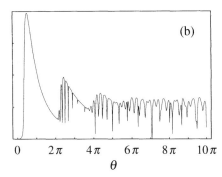

Fig. 18.5: Emergence of the trapping states in the steady state mean photon number for a pump parameter $r/\gamma = 50$ and mean number of thermal photons $n_{\text{th}} = 0.1$ (left) and $n_{\text{th}} = 10^{-7}$ (right). In case (b) we clearly observe the trapping states: For certain values of the interaction time θ the mean photon number is significantly reduced compared to the situation in (a).

An intuitive explanation of this phenomenon is based on the fact that an excited atom interacting with the state $|n_q\rangle$ cannot deposit its excitation in the cavity because for this particular interaction time the atom undergoes an integer number of Rabi cycles and therefore leaves the cavity in the excited state. Small initial populations in states with $m > n_q$ as well as nonzero temperature disturb the appearance of trapping states.

We conclude this section by noting that the trapping states have been observed experimentally as shown in Fig. 18.6.

18.4.3 Phase Diffusion

So far, we have only discussed properties of the one-atom maser related to the diagonal elements of the density operator. We now turn to the off-diagonal elements and study the time evolution of the first off-diagonal. These matrix elements determine the phase diffusion in the one-atom maser.

London Phase Operator

In Sec. 8.5 we have analyzed the problem of a hermitian phase operator and have introduced the London phase states. Our present approach is based on the corresponding phase operator

$$\widehat{e^{i\varphi}} \equiv \int_0^{2\pi} d\varphi\, e^{i\varphi}\, |\varphi\rangle\langle\varphi|, \qquad (18.39)$$

following from the phase states

$$|\varphi\rangle \equiv \frac{1}{\sqrt{2\pi}} \sum_{n=0}^{\infty} e^{i(n+1/2)\varphi} |n\rangle \qquad (18.40)$$

defined in Eq. (8.35).

530   18 Damping and Amplification

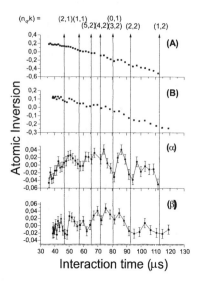

Fig. 18.6: Experimental observation of the trapping states and the one-atom maser. Atomic inversion as a function of interaction time for pump parameters $N_{\text{ex}} = 7$ and $N_{\text{ex}} = 10$, respectively. In plots (α) and (β) we represent the data of (A) and (B) after the linear trend was removed. The vertical lines indicate the theoretical positions of all low order trapping states over the range of interaction times of the plot. Dips in the inversion can be identified with the position of the trapping states. Taken from M. Weidinger *et al.*, Phys. Rev. Lett. **82**, 3795 (1999).

When we substitute the definition Eq. (18.40) of the phase state into the definition (18.39) of the phase operator and perform the integration we find

$$\widehat{e^{i\varphi}} = \sum_{n=0}^{\infty} |n\rangle\langle n+1|. \tag{18.41}$$

In the present section we calculate the expectation value

$$\langle \widehat{e^{i\varphi}} \rangle = \text{Tr}(\widehat{e^{i\varphi}} \, \hat{\rho}_{\text{f}}) = \sum_m \langle m|\widehat{e^{i\varphi}} \, \hat{\rho}_{\text{f}}|m\rangle$$

of the London phase operator in the maser field described by a density operator $\hat{\rho}_f$. For this purpose we substitute the expression Eq. (18.41) for the phase operator in number representation into this trace and arrive at

$$\langle \widehat{e^{i\varphi}} \rangle = \sum_n \langle n+1|\hat{\rho}_{\text{f}}|n\rangle = \sum_n \rho_{n+1,n} \equiv \sum_n (\rho_n^{(1)})^*.$$

Hence, the sum over the first off-diagonal of the density operator determines the expectation value of the phase.

Equation of Motion for Phase Operator

In order to derive the time evolution of this quantity we return to the equation of motion for the density operator, Eq. (18.29). We start from this equation for $k = 1$

18.4 One-Atom Maser

and sum over m which yields

$$\frac{d}{dt}\langle\widehat{e^{i\varphi}}\rangle = \sum_m \left(\frac{d}{dt}\rho_m^{(1)}\right)^* = r\left[\sum_m (C_m C_{m+1} - 1)\rho_m^{(1)} + \sum_m S_{m-1} S_m \rho_{m-1}^{(1)}\right]$$
$$+ \frac{\gamma}{2}(n_{\text{th}} + 1)\left[2\sum_m \sqrt{(m+1)(m+2)}\,\rho_{m+1}^{(1)} - \sum_m (2m+1)\rho_m^{(1)}\right]$$
$$+ \frac{\gamma}{2} n_{\text{th}}\left[2\sum_m \sqrt{m(m+1)}\,\rho_{m-1}^{(1)} - \sum_m (2m+3)\rho_m^{(1)}\right].$$

After appropriately shifting the summation indices we find

$$\frac{d}{dt}\langle\widehat{e^{i\varphi}}\rangle = \sum_n \left\{ r\left[\cos(g\tau\sqrt{n+2})\cos(g\tau\sqrt{n+1}) - 1 + \sin(g\tau\sqrt{n+2})\sin(g\tau\sqrt{n+1})\right] \right.$$
$$\left. - \gamma(n_{\text{th}}+1)\left[n + \tfrac{1}{2} - \sqrt{n(n+1)}\right] - \gamma n_{\text{th}}\left[n + \tfrac{3}{2} - \sqrt{(n+1)(n+2)}\right]\right\}(\rho_n^{(1)})^*.$$

The trigonometric relation

$$\cos\alpha\cos\beta - 1 + \sin\alpha\sin\beta = \cos(\alpha - \beta) - 1 = -2\sin^2\left[\tfrac{1}{2}(\alpha - \beta)\right]$$

simplifies the equation

$$\frac{d}{dt}\langle\widehat{e^{i\varphi}}\rangle = -\sum_n \mu_n (\rho_n^{(1)})^*, \tag{18.42}$$

where

$$\mu_n \equiv 2r\sin^2\left(g\tau\frac{\sqrt{n+2} - \sqrt{n+1}}{2}\right)$$
$$+ \gamma(n_{\text{th}}+1)\left[n + \tfrac{1}{2} - \sqrt{n(n+1)}\right] + \gamma n_{\text{th}}\left[n + \tfrac{3}{2} - \sqrt{(n+1)(n+2)}\right]. \tag{18.43}$$

We obtain a closed but approximate equation for $\langle\widehat{e^{i\varphi}}\rangle$ when we recognize that μ_n is a slowly varying function of n. In this case we can replace μ_n by its mean value

$$\langle\mu_n\rangle \equiv \sum_n \mu_n W_n \tag{18.44}$$

and obtain

$$\frac{d}{dt}\langle\widehat{e^{i\varphi}}\rangle \cong -\langle\mu_n\rangle \sum_n (\rho_n^{(1)})^* = -\langle\mu_n\rangle \langle\widehat{e^{i\varphi}}\rangle.$$

When we further assume that W_n is significantly different from zero only for $n \gg 1$ we may expand the square roots in the definition of μ_n, Eq. (18.43), for large n and obtain

$$D \equiv \langle\mu_n\rangle \cong \left\langle 4r\sin^2\left(\frac{g\tau}{4\sqrt{n}}\right)\right\rangle + \left\langle \frac{\gamma(2n_{\text{th}}+1)}{4n}\right\rangle \tag{18.45}$$

as an approximation for the phase diffusion constant D.

We can directly relate D to the average photon number when we assume that $\langle \mu_n \rangle \cong \mu_{n=\langle n \rangle}$. We then obtain from Eq. (18.45)

$$D \cong 4r \sin^2\left(\frac{g\tau}{4\langle n \rangle}\right) + \frac{\gamma(2n_{\text{th}}+1)}{4\langle n \rangle}.$$

The quantity D has the dimension of a frequency and can be interpreted as the linewidth of the one-atom maser: In the absence of diffusion the phase of the field would not change since we are working in the interaction picture. In the Schrödinger picture and with phase diffusion we have

$$\frac{d}{dt} e^{i\omega t + i\varphi(t)} = i(\omega + \dot{\varphi}) e^{i\omega t + i\varphi(t)} = i(\omega + \Delta\omega) e^{i\omega t + i\varphi(t)}.$$

This shows that a diffusion in phase leads to an uncertainty $\Delta\omega = \dot{\varphi}$ in frequency – nothing else but a linewidth.

18.5 Atom–Reservoir Interaction

In the present section we reverse the situation discussed in Sec. 18.3. Whereas there the field mode was the system and the reservoir consisted of two-level atoms, now our system is a single two-level atom and the reservoir consists of infinitely many field modes. This coupling of the atom to the field reservoir leads to spontaneous emission of the atoms, as well as, level shifts. In contrast to the examples discussed before we now do not want to derive an equation of motion for the density operator of the *field*, but for the *atom*.

18.5.1 Model and Equation of Motion

We consider the Hamiltonian

$$\hat{H} \equiv \hat{H}_{\text{at}} + \hat{H}_{\text{f}} + \hat{H}_{\text{int}} \tag{18.46}$$

of the complete system containing the Hamiltonian

$$\hat{H}_{\text{at}} \equiv \frac{1}{2}\hbar\omega\hat{\sigma}_z \tag{18.47}$$

of the free atom and

$$\hat{H}_{\text{f}} \equiv \sum_l \hbar\Omega_l \hat{a}_l^\dagger \hat{a}_l,$$

corresponding to the field modes of frequency Ω_l, and the interaction term

$$\hat{H}_{\text{int}} \equiv \sum_l \hbar g_l (\hat{\sigma}\hat{a}_l^\dagger + \hat{\sigma}^\dagger \hat{a}_l),$$

where

$$g_l(\vec{R}) \equiv \frac{\vec{\wp} \cdot \vec{u}_l(\vec{R})}{\hbar} \mathcal{E}_l \tag{18.48}$$

denotes the coupling strength of the atom at the position $\vec{R}$ to the lth mode.

In the context of atom optics – a topic discussed in chapter 19 – we consider the motion of the atom and hence $\vec{R}$ will turn into an operator. In this case the coupling strength $g_l(\vec{R})$ will also become an operator. Moreover, the density operator of the atom will consist not only of the atomic degrees of freedom but also involve the position of the atom. It is therefore important to keep track of the ordering of $\hat{g}_l(\vec{R})$ and the density operator of the atom. However, in the present section we neglect the motion of the atom and $\vec{R}$ is thus just a parameter.

Since our formal expression Eq. (18.3) for the density operator $\hat{\rho}_s$ of the system is in the interaction picture we transform the Hamiltonian $\hat{H}$, Eq. (18.46), into the interaction picture and find in complete analogy to Sec. 14.8 the expression

$$\hat{H}_{\text{int}}^{(I)} \equiv \hat{\mathcal{V}}(t) \equiv \hbar \sum_l g_l \left(\hat{\sigma} \hat{a}_l^\dagger e^{i\Delta_l t} + \hat{\sigma}^\dagger \hat{a}_l e^{-i\Delta_l t} \right) \equiv \hbar \bar{g} \left[\hat{\sigma} \hat{\mathcal{A}}^\dagger(t) + \hat{\sigma}^\dagger \hat{\mathcal{A}}(t) \right],$$

where $\Delta_l \equiv \Omega_l - \omega$ denotes the detuning of the lth field mode from the atom. Moreover, in order to simplify the notation we have introduced the multi-mode field operator

$$\hat{\mathcal{A}}(t) \equiv \frac{1}{\bar{g}} \sum_l g_l \hat{a}_l e^{-i\Delta_l t}$$

where $\bar{g}$ is a constant with the units of frequency. Its value is arbitrary since we have only introduced it as to cast the interaction Hamiltonian into the form of the Jaynes-Cummings-Paul Hamiltonian.

In contrast to the examples presented in Secs. 18.3 and 18.4 now the interaction Hamiltonian is explicitly time dependent and we have to resort to a perturbative approach. According to Eq. (18.5) the coarsed-grained equation of motion for the density operator $\hat{\rho}_{\text{at}}$ of the atom up to second order in the coupling strength reads

$$\frac{d}{dt} \hat{\rho}_{\text{at}} = -\frac{i}{\hbar} \frac{1}{\tau} \int_t^{t+\tau} dt_1 \, \text{Tr}_f \left\{ \left[\hat{\mathcal{V}}_1, \hat{\rho}(t) \right] \right\} - \frac{1}{\hbar^2} \frac{1}{\tau} \int_t^{t+\tau} dt_2 \int_t^{t+\tau} dt_1 \, \text{Tr}_f \left\{ \left[\hat{\mathcal{V}}_2, \left[\hat{\mathcal{V}}_1, \hat{\rho}(t) \right] \right] \right\}$$

$$\equiv \dot{\hat{\rho}}_{\text{at}}^{(1)} + \dot{\hat{\rho}}_{\text{at}}^{(2)} \tag{18.49}$$

where $\hat{\mathcal{V}}_n \equiv \hat{\mathcal{V}}(t_n)$ and $\hat{\rho}$ denotes the density operator for the atom and field.

We now evaluate the commutator, and the double commutator between $\hat{\mathcal{V}}_n$ and $\hat{\rho}$ and take the trace over the field. Since this is a rather lengthy and cumbersome procedure we use the first order term corresponding to the commutator to illustrate the method and then discuss the second order term corresponding to the double commutator in Appendix N.

18.5.2 First Order Contribution

In this section we evaluate the first order contribution to the density operator equation. For this purpose we find the commutator between $\mathcal{V}$ and $\hat{\rho}$ and perform the relevant time average.

Equation of Motion

We begin by calculating the contribution $\dot{\hat{\rho}}_{\text{at}}^{(1)}$ resulting from the commutator

$$\left[\hat{\mathcal{V}}_1, \hat{\rho}\right] = \hat{\mathcal{V}}_1\hat{\rho} - \hat{\rho}\hat{\mathcal{V}}_1 = \hbar\bar{g}\left[\hat{\sigma}\hat{\mathcal{A}}_1^\dagger\hat{\rho} + \hat{\sigma}^\dagger\hat{\mathcal{A}}_1\hat{\rho} - \hat{\rho}\hat{\sigma}\hat{\mathcal{A}}_1^\dagger - \hat{\rho}\hat{\sigma}^\dagger\hat{\mathcal{A}}_1\right]$$

between $\hat{\rho}$ and the multi-mode operators $\hat{\mathcal{A}}_1$ and $\hat{\mathcal{A}}_1^\dagger$ which yields

$$\text{Tr}_{\text{f}}\left\{\left[\hat{\mathcal{V}}_1, \hat{\rho}\right]\right\} = \hbar\bar{g}\left[\left(\hat{\sigma}\hat{\rho}_{\text{at}} - \hat{\rho}_{\text{at}}\hat{\sigma}\right)\text{Tr}\left(\hat{\mathcal{A}}_1^\dagger\hat{\rho}_{\text{f}}\right) + \left(\hat{\sigma}^\dagger\hat{\rho}_{\text{at}} - \hat{\rho}_{\text{at}}\hat{\sigma}^\dagger\right)\text{Tr}\left(\hat{\mathcal{A}}_1\hat{\rho}_{\text{f}}\right)\right].$$

Here we have decomposed the density operator $\hat{\rho}$ into the one of the atom (at) and the field (f), that is $\hat{\rho} = \hat{\rho}_{\text{at}} \otimes \hat{\rho}_{\text{f}}$. Moreover, we have made use of the fact that the field operators $\hat{\mathcal{A}}_1$ and $\hat{\mathcal{A}}_1^\dagger$ commute with the atomic operators $\hat{\rho}_{\text{at}}$, $\hat{\sigma}$ and $\hat{\sigma}^\dagger$ and have taken advantage of the relation

$$\text{Tr}\left(\hat{A} \cdot \hat{B}\right) = \text{Tr}\left(\hat{B} \cdot \hat{A}\right)$$

to interchange the order of $\hat{\mathcal{A}}_1$ or $\hat{\mathcal{A}}_1^\dagger$ with $\hat{\rho}_{\text{f}}$. This fact has allowed us to combine two terms of the commutator.

Hence in first order we arrive at the equation of motion

$$\dot{\hat{\rho}}_{\text{at}} = -ig^*(t)\left(\hat{\sigma}\hat{\rho}_{\text{at}} - \hat{\rho}_{\text{at}}\hat{\sigma}\right) - ig(t)\left(\hat{\sigma}^\dagger\hat{\rho}_{\text{at}} - \hat{\rho}_{\text{at}}\hat{\sigma}^\dagger\right) \tag{18.50}$$

for the density operator $\hat{\rho}_{\text{at}}$ of the atom and

$$g(t) \equiv \bar{g}\frac{1}{\tau}\int_t^{t+\tau} dt_1\, \text{Tr}\left\{\hat{\mathcal{A}}_1\hat{\rho}_{\text{f}}\right\} = \sum_l g_l\frac{1}{\tau}\int_t^{t+\tau} dt_1\, e^{-i\Delta_l t_1}\,\text{Tr}_{\text{f}}\left(\hat{a}_l\hat{\rho}_{\text{f}}\right) \tag{18.51}$$

denotes the Rabi frequency.

Effective Rabi Frequency

The quantity $g(t)$ contains the time averaged expectation value of the annihilation operator $\hat{\mathcal{A}}_1$ of the reservoir. This time average extends over the interaction time τ and contains a phase factor which involves the detuning Δ_l of the mode l of the reservoir from the atomic transition and the interaction time τ. We now show that despite this time average g still depends on time.

Moreover, the expectation value over the reservoir involves the density operator $\hat{\rho}_{\text{f}}$ of the multi-mode field. When the modes are not entangled with each other $\hat{\rho}_{\text{f}}$ is the direct product

$$\hat{\rho}_{\text{f}} \equiv \prod_m \hat{\rho}_m$$

of the density operators $\hat{\rho}_m$ of the individual modes. Hence we find for the trace

$$\text{Tr}\left(\hat{a}_l\hat{\rho}_{\text{f}}\right) = \text{Tr}_{\text{f}}\left(\hat{a}_l\prod_m \hat{\rho}_m\right) = \text{Tr}\left(\hat{a}_l\hat{\rho}_l\right)\prod_{m\neq l}\text{Tr}\left(\hat{\rho}_m\right),$$

that is

$$\text{Tr}\left(\hat{a}_l\hat{\rho}_{\text{f}}\right) = \text{Tr}\left(\hat{a}_l\hat{\rho}_l\right)$$

where in the last step we have used $\text{Tr}(\hat{\rho}_l) = 1$.

Hence the effective Rabi frequency reads

$$g(t) = \sum_l g_l e^{-i\Delta_l t} e^{-i\theta_l} \frac{\sin \theta_l}{\theta_l} \, \text{Tr}(\hat{a}_l \hat{\rho}_l) \tag{18.52}$$

where we have also performed the integration over the variable t_1 using the relation

$$\frac{1}{\tau} \int_t^{t+\tau} dt_1 \, e^{-i\Delta_l t_1} = e^{-i\Delta_l t} \frac{e^{-i\Delta_l \tau} - 1}{-i\Delta_l \tau} = e^{-i\Delta_l t} e^{-i\Delta_l \tau/2} \frac{e^{+i\Delta_l \tau/2} - e^{-i\Delta_l \tau/2}}{i\Delta_l \tau},$$

and have introduced the parameter $\theta_l \equiv \Delta_l \tau / 2$.

Resonance Approximation

The expression Eq. (18.52) for g clearly shows that only modes of the reservoir with a non-vanishing expectation value of the annihilation operator contribute to g. Moreover, the detuning Δ_l of this mode from the resonance frequency of the atom is of importance. Indeed, Eq. (18.48) involves the function

$$\mathcal{S}(\theta) \equiv \frac{\sin \theta}{\theta}$$

which has a dominant maximum at $\theta = 0$. In the summation in Eq. (18.52) over the modes the function $\mathcal{S}$ therefore selects only those modes which are resonant, that is $\Delta_l \tau / 2 \ll 1$. In this sense the function S acts as a delta-function.

Consequently, we find

$$g(t) \cong g_0 \, e^{-i\Delta_0 t} \, \text{Tr}(\hat{a}_0 \hat{\rho}_0)$$

where we have denoted the resonant mode by the index 0.

18.5.3 Bloch Equations

We now turn to the discussion of the dynamics resulting from the equation of motion Eq. (18.50). For this purpose we first translate the operator equation into a c-number equation by taking matrix elements and then connect the corresponding equations with the dynamics of a point on the sphere.

Equations for Matrix Elements

From Eq. (18.50) we find the equations

$$\dot{\rho}_{aa} \equiv \langle a| \dot{\hat{\rho}}_{\text{at}} |a\rangle = -ig^* \langle a| \hat{\sigma} \hat{\rho}_{\text{at}} - \hat{\rho}_{\text{at}} \hat{\sigma} |a\rangle - ig \langle a| \hat{\sigma}^\dagger \hat{\rho}_{\text{at}} - \hat{\rho}_{\text{at}} \hat{\sigma}^\dagger |a\rangle$$
$$\dot{\rho}_{bb} \equiv \langle b| \dot{\hat{\rho}}_{\text{at}} |b\rangle = -ig^* \langle b| \hat{\sigma} \hat{\rho}_{\text{at}} - \hat{\rho}_{\text{at}} \hat{\sigma} |b\rangle - ig \langle b| \hat{\sigma}^\dagger \hat{\rho}_{\text{at}} - \hat{\rho}_{\text{at}} \hat{\sigma}^\dagger |b\rangle$$
$$\dot{\rho}_{ab} \equiv \langle a| \dot{\hat{\rho}}_{\text{at}} |b\rangle = -ig^* \langle a| \hat{\sigma} \hat{\rho}_{\text{at}} - \hat{\rho}_{\text{at}} \hat{\sigma} |b\rangle - ig \langle a| \hat{\sigma}^\dagger \hat{\rho}_{\text{at}} - \hat{\rho}_{\text{at}} \hat{\sigma}^\dagger |b\rangle$$

and

$$\dot{\rho}_{ba} \equiv \langle b| \dot{\hat{\rho}}_{\text{at}} |a\rangle = -ig^* \langle b| \hat{\sigma} \hat{\rho}_{\text{at}} - \hat{\rho}_{\text{at}} \hat{\sigma} |a\rangle - ig \langle b| \hat{\sigma}^\dagger \hat{\rho}_{\text{at}} - \hat{\rho}_{\text{at}} \hat{\sigma}^\dagger |a\rangle.$$

18 Damping and Amplification

When we recall the relations

$$\hat{\sigma}\,|a\rangle = |b\rangle \quad \text{and} \quad \hat{\sigma}^\dagger\,|a\rangle = 0$$
$$\hat{\sigma}\,|b\rangle = 0 \quad \text{and} \quad \hat{\sigma}^\dagger\,|b\rangle = |a\rangle \tag{18.53}$$

the above set of equations takes the form

$$\dot{\rho}_{aa} = ig^*\rho_{ab} - ig\rho_{ba} \tag{18.54a}$$

$$\dot{\rho}_{bb} = -ig^*\rho_{ab} + ig\rho_{ba} \tag{18.54b}$$

$$\dot{\rho}_{ab} = -ig(\rho_{bb} - \rho_{aa}) \tag{18.54c}$$

and

$$\dot{\rho}_{ba} = -ig^*(\rho_{aa} - \rho_{bb}).$$

Since the density operator is hermitian the diagonal elements ρ_{aa} and ρ_{bb} must be real and $\rho_{ab} = \rho_{ba}^*$. Hence the forth equation is just the complex conjugate of the third equation.

Inversion and Polarization

Equations (18.54) describe the dynamics of a two-level atom due to a reservoir of modes. These so-called Bloch equations are the basis of semi-classical laser theory. It is therefore worthwhile to analyze them briefly.

When we add the first two equations we find

$$\dot{\rho}_{aa} + \dot{\rho}_{bb} = 0,$$

that is the probability $\rho_{aa} + \rho_{bb} = 1$ is conserved.

We now cast Eq. (18.54) in a slightly different form. For this purpose we introduce the inversion

$$w \equiv \rho_{aa} - \rho_{bb},$$

that is the difference between the populations in the excited and the ground level, and twice the real part

$$u \equiv \rho_{ab} + \rho_{ba} = \rho_{ab} + \rho_{ab}^* = 2\,\mathrm{Re}\,\rho_{ab}$$

and imaginary part

$$v \equiv i(\rho_{ab} - \rho_{ba}) = i(\rho_{ab} - \rho_{ab}^*) = -2\,\mathrm{Im}\,\rho_{ab}$$

of ρ_{ab}.

In this notation the Bloch equations read

$$\dot{u}(t) = -2g_i(t)\,w(t)$$
$$\dot{v}(t) = -2g_r(t)\,w(t)$$
$$\dot{w}(t) = 2g_i(t)\,u(t) + 2g_r(t)\,v(t)$$

where we have set $g(t) \equiv g_r(t) + i\,g_i(t)$.

Bloch Sphere

When we introduce the three-component column vectors $\vec{S} \equiv (u, v, w)$ and $\vec{B} \equiv (2g_r, -2g_i, 0)$ the Bloch equations take the form

$$\frac{d}{dt}\vec{S} = \vec{B} \times \vec{S},$$

familiar from the precession of an angular momentum vector $\vec{S}$ in a magnetic field.

Hence, the dynamics of a two-level atom in the presence of a field is identical to that of a spin in a magnetic field. Since this problem has been discussed extensively by D. Bloch, these equations carry his name.

There is a nice graphical representation of this dynamics as the motion of a point on a sphere. Indeed, we recognize that the length of the vector $\vec{S}$ is conserved, since

$$\frac{d}{dt}\vec{S}^2 = 2\vec{S}\cdot\dot{\vec{S}} = 2\left[u\,\dot{u} + v\,\dot{v} + w\,\dot{w}\right] = 2\left[-2\,g_i\,u\,w - 2\,g_r\,v\,w + 2\,g_i\,u\,w + 2\,g_r\,v\,w\right] = 0.$$

Hence the tip of the vector moves on the sphere.

18.5.4 Second Order Contribution

We now turn to the second order contribution $\dot{\rho}_{\text{at}}^{(2)}$, Eq. (18.49). We first present the resulting master equation and then discuss the physical significance of the individual parameters.

Master Equation

According to Appendix N we find

$$\dot{\rho}_{\text{at}} = -\frac{i}{\hbar}\left[\Delta\hat{H}, \hat{\rho}_{\text{at}}\right]$$
$$-(\Gamma_r + \tilde{G}_r)\left[\hat{\sigma}^\dagger\hat{\sigma}\hat{\rho}_{\text{at}} + \hat{\rho}_{\text{at}}\hat{\sigma}^\dagger\hat{\sigma} - 2\hat{\sigma}\hat{\rho}_{\text{at}}\hat{\sigma}^\dagger\right] - \Gamma_r\left[\hat{\sigma}\hat{\sigma}^\dagger\hat{\rho}_{\text{at}} + \hat{\rho}_{\text{at}}\hat{\sigma}\hat{\sigma}^\dagger - 2\hat{\sigma}^\dagger\hat{\rho}_{\text{at}}\hat{\sigma}\right]$$
$$+2\beta^*\hat{\sigma}\hat{\rho}_{\text{at}}\hat{\sigma} + 2\beta\hat{\sigma}^\dagger\hat{\rho}_{\text{at}}\hat{\sigma}^\dagger, \qquad (18.55)$$

with the abbreviations

$$\Gamma \equiv \Gamma_r + i\Gamma_i \equiv \sum_l |g_l|^2\,\bar{n}_l\,I(\Delta_l, \Delta_l) + \sum_{l\neq m} g_l^* g_m\,\langle\hat{a}_l^\dagger\rangle\langle\hat{a}_m\rangle\,I(\Delta_l, \Delta_m), \qquad (18.56)$$

and

$$\tilde{G} \equiv \tilde{G}_r + i\tilde{G}_i \equiv \sum_l |g_l|^2\,I(\Delta_l, \Delta_l) \qquad (18.57)$$

and

$$\beta \equiv \beta_r + i\beta_i \equiv \sum_l g_l^2\,\langle\hat{a}_l^2\rangle\,I(-\Delta_l, \Delta_l) + \sum_{l\neq m} g_l g_m\,\langle\hat{a}_l\rangle\langle\hat{a}_m\rangle\,I(-\Delta_l, \Delta_m).$$

Here the quantity I denotes the double time integral

$$I(\Delta_l, \Delta_m) \equiv \frac{1}{\tau}\int_t^{t+\tau} dt_2 \int_t^{t_2} dt_1\,e^{i(\Delta_l t_2 - \Delta_m t_1)}.$$

Moreover, we have introduced the Hamiltonian

$$\Delta \hat{H} \equiv -\hbar \left(\Gamma_i + \frac{1}{2} \tilde{G}_i \right) \hat{\sigma}_z, \qquad (18.58)$$

corresponding to a shift of the levels.

Discussion of Parameters

The state of the reservoir determines the value of Γ. Indeed, Γ depends on the average number $\bar{n}_l$ of photons in the individual reservoir modes, as well as the expectation value of the creation and of the annihilation operator.

For example, a reservoir in which each field mode is in a thermal state described by a diagonal density operator

$$\hat{\rho}_l \equiv \sum_n W_n^{(l)} |n\rangle \langle n|$$

of photon statistics $W_n^{(l)}$ yields immediately

$$\langle \hat{a}_l \rangle = \sum_n W_n^{(l)} \langle n| \hat{a}_l |n\rangle = \sum_n W_n^{(l)} \sqrt{n} \langle n| n-1 \rangle = 0,$$

and

$$\langle \hat{a}_l^\dagger \rangle = 0.$$

Thus for a thermal reservoir the expression for Γ reduces to

$$\Gamma_{\text{th}} \equiv \sum_l |g_l|^2 \, \bar{n}_l \, I(\Delta_l, \Delta_l). \qquad (18.59)$$

A particularly interesting reservoir is the vacuum, that is a thermal state at zero-temperature. Here all modes are in the ground state, that is the average number of quanta of excitation in each mode vanishes, that is $\bar{n}_l = 0$. Consequently, the parameter Γ vanishes, that is

$$\Gamma_{\text{vac}} = 0.$$

We now turn to the parameter $\tilde{G}$. From the definition Eq. (18.57) of $\tilde{G}$ we recognize that the value of

$$\tilde{G} = \sum_l |g_l|^2 I(\Delta_l, \Delta_l) \qquad (18.60)$$

is independent of the state of the reservoir, since it emerges from the commutator of the field operators giving rise to the c-number G, Eq. (N.13). Hence it is nonzero even for a reservoir in the vacuum state, that is

$$\tilde{G}_{\text{vac}} \neq 0.$$

Moreover, when we compare the expressions Eqs. (18.59) and (18.60) we note that $\tilde{G}$ is identical to Γ for the case of a thermal reservoir with $\bar{n}_l = 1$, that is for a thermal reservoir with one quantum of excitation in each mode.

As in the case of Γ the value of β depends crucially on the state of the reservoir: The necessary requirement for a non-vanishing value of β is that at least one of the reservoir modes has a non-vanishing expectation value of the annihilation operator, or of the square of the annihilation operator.

For a thermal reservoir we find

$$\langle \hat{a}_l^2 \rangle = \sum_n W_n^{(l)} \langle n| \hat{a}_l^2 |n\rangle = \sum_n W_n^{(l)} \sqrt{n(n-1)} \langle n| n-2 \rangle = 0,$$

and since $\langle \hat{a}_l \rangle = \langle \hat{a}_l^\dagger \rangle = 0$ we arrive at

$$\beta_{\text{th}} = 0,$$

that is the parameter β vanishes for a thermal reservoir.

However, there are cases when β is non-vanishing. For example a squeezed vacuum enjoys a non-vanishing expectation value $\langle \hat{a}_l^2 \rangle$ and thus leads to a non-vanishing value of β. This has important consequences for the decay of the atom as discussed in Sec. 18.5.6.

18.5.5 Lamb Shift

The expressions for the quantities $\Gamma, \tilde{G}$ and β are rather complicated and can be even infinite. Before we discuss these complications we first gain insight into the physical meaning of the individual terms of the master equation Eq. (18.55).

The Hamiltonian

$$\Delta \hat{H} = -\hbar \left(\Gamma_i + \tfrac{1}{2} \tilde{G}_i \right) \hat{\sigma}_z$$

in the master equation Eq. (18.55) is a correction to the Hamiltonian, Eq. (18.47),

$$H_{\text{at}} = \tfrac{1}{2} \hbar \omega \hat{\sigma}_z$$

of the free atom. Hence the interaction of the atom with a reservoir of field modes leads to a level shift of frequency

$$\Delta \omega \equiv -2 \left(\Gamma_i + \frac{1}{2} \tilde{G}_i \right). \tag{18.61}$$

This level shift carries the name Lamb shift. It has been discovered experimentally in 1947 by W. E. Lamb and his graduate student R. C. Retherford in hydrogen. The Dirac equation predicts that the $2^2S_{1/2}$ and $2^2P_{1/2}$ energy levels are degenerated. However, the experiment showed clearly that the two levels are shifted relative to each other. This shift is a manifestation of the quantization of the electromagnetic field. Indeed, we recognize from Eq. (18.61) that there are two shifts: The first arises from the imaginary part Γ_i of Γ. According to Eq. (18.56) the parameter Γ is essentially determined by the average number of photons in the reservoir modes. Hence this contribution to the level shift is analogous to the familiar second order Stark-effect of an atom in a static electric field.

In contrast, the contribution $\tilde{G}_i$ arises from the commutation relations of the field operators discussed in Appendix N, Eq. (N.12). It is therefore a pure quantum effect of the field and is nonzero even when all modes of the reservoir are in the ground state, that is for the vacuum.

18.5.6 Weisskopf-Wigner Decay

Apart from the shift of the levels the reservoir has another dramatic effect. It forces the atom to decay, that is the populations ρ_{aa} and ρ_{bb} in the two levels and the polarization ρ_{ab} decay. This phenomenon carries the name of Weisskopf-Wigner decay.

Equation of Motion for Matrix Elements

In order to bring this phenomenon out most clearly we now take matrix elements of the density operator equation Eq. (18.55). Since we focus on the decay aspect of the problem we neglect in the present discussion the Lamb shift.

When we recall the relations Eq. (18.53) for the action of the atomic operators on the internal states we arrive at the set of equations

$$\dot{\rho}_{aa} = -2\left(\Gamma_r + \tilde{G}_r\right)\rho_{aa} + 2\Gamma_r \rho_{bb}$$
$$\dot{\rho}_{bb} = +2\left(\Gamma_r + \tilde{G}_r\right)\rho_{aa} - 2\Gamma_r \rho_{bb}$$
$$\dot{\rho}_{ab} = -\left(2\Gamma_r + \tilde{G}_r\right)\rho_{ab} + 2\beta\,\rho_{ba}$$

and

$$\dot{\rho}_{ba} = -\left(2\Gamma_r + \tilde{G}_r\right)\rho_{ba} + 2\beta^*\,\rho_{ab}.$$

We take the sum of the first two equations and find

$$\dot{\rho}_{aa} + \dot{\rho}_{bb} = 0,$$

that is the normalization

$$\rho_{aa} + \rho_{bb} = 1,$$

is preserved.

We now use the normalization condition to express ρ_{bb} and ρ_{aa} in the two equations for ρ_{aa} and ρ_{bb} which yields

$$\dot{\rho}_{aa} = -2\left(2\Gamma_r + \tilde{G}_r\right)\rho_{aa} + 2\Gamma_r$$

and

$$\dot{\rho}_{bb} = -2\left(2\Gamma_r + \tilde{G}_r\right)\rho_{bb} + 2\left(\Gamma_r + \tilde{G}_r\right).$$

So far we have not yet derived explicit expressions for Γ_r nor for $\tilde{G}_r$. However, we already recognize that provided $\Gamma_r > 0$ and $\tilde{G}_r > 0$ the populations decay with rates $2(2\Gamma_r + \tilde{G}_r)$ due to the coupling of the atom to the reservoir. In steady state, that is for $\dot{\rho}_{aa} = \dot{\rho}_{bb} = 0$ we find

$$\rho_{aa} = \frac{\Gamma_r}{2\Gamma_r + \tilde{G}_r}$$

and

$$\rho_{bb} = \frac{\Gamma_r + \tilde{G}_r}{2\Gamma_r + \tilde{G}_r}.$$

18.5 Atom–Reservoir Interaction

We emphasize that again there are two contributions. One is due to Γ_r which contains the average number of photons. It is therefore induced decay, corresponding to induced emission.

The other contribution arises from $\widetilde{G}_r$ and thus from the quantum nature of field. It is even present if Γ_r vanishes. It corresponds to spontaneous emission.

Two Decay Constants For Polarization

We now turn to the two last equations, that is the two equations for the off-diagonal elements of the density operator.

When we recall the real and imaginary parts

$$u \equiv \frac{1}{2}\left(\rho_{ab} + \rho_{ab}^*\right)$$

and

$$v \equiv \frac{1}{2i}\left(\rho_{ab} - \rho_{ab}^*\right)$$

of ρ_{ab}, that is

$$\rho_{ab} = u + iv$$

and

$$\rho_{ab}^* = u - iv$$

we find

$$\dot{u} + i\dot{v} = -\Gamma_r(u + iv) + 2(\beta_r + i\beta_i)(u - iv)$$

that is

$$\dot{u} = -\left(\Gamma_r - 2\beta_r\right)u + 2\beta_i v$$

and

$$\dot{v} = 2\beta_i u - \left(\Gamma_r + 2\beta_r\right)v.$$

We gain insight into the dynamics of u and v from the eigenvalues λ following from

$$0 = \det\begin{pmatrix} \lambda - (\Gamma_r - 2\beta_r) & 2\beta_i \\ 2\beta_i & \lambda - (\Gamma_r - 2\beta_r) \end{pmatrix} = \lambda^2 - 2\lambda\Gamma_r + \Gamma_r^2 - 4(\beta_r^2 + \beta_i^2),$$

that is

$$[\lambda - (\Gamma_r + 2|\beta|)][\lambda - (\Gamma_r - 2|\beta|)] = 0.$$

Hence we find the two eigenvalues

$$\lambda_\pm \equiv \Gamma_r \pm 2|\beta|.$$

Consequently, the off-diagonal elements of the atomic density operator display two different decay constants provided $|\beta| \neq 0$: Indeed there is a decay constant $\Gamma_r - 2|\beta|$ which is smaller than $\Gamma_r + 2|\beta|$.

Problems

18.1 Density Operator of Field Plus Atom

Show that for the resonant Jaynes-Cummings-Paul model the density operator

$$\hat{\rho}_{\text{f+at}}(t) = \hat{\mathcal{U}}(t,t_0)\hat{\rho}_{\text{f+at}}(t_0)\hat{\mathcal{U}}^\dagger(t,t_0)$$

of the field plus atom at time t reads

$$\hat{\rho}(t) = \begin{pmatrix} \hat{C}_n \hat{\rho}_{\text{f}} \hat{C}_n \rho_{aa} & \hat{C}_n \hat{\rho}_{\text{f}} \hat{C}_{n-1} \rho_{ab} \\ \hat{C}_{n-1} \hat{\rho}_{\text{f}} \hat{C}_n \rho_{ba} & \hat{C}_{n-1} \hat{\rho}_{\text{f}} \hat{C}_{n-1} \rho_{bb} \end{pmatrix}$$

$$+ \begin{pmatrix} \frac{\hat{S}_n}{\sqrt{\hat{n}+1}} \hat{a} \hat{\rho}_{\text{f}} \hat{a}^\dagger \frac{\hat{S}_n}{\sqrt{\hat{n}+1}} \rho_{bb} & \frac{\hat{S}_n}{\sqrt{\hat{n}+1}} \hat{a} \hat{\rho}_{\text{f}} \hat{a} \frac{\hat{S}_{n-1}}{\sqrt{\hat{n}}} \rho_{ba} \\ \frac{\hat{S}_{n-1}}{\sqrt{\hat{n}}} \hat{a}^\dagger \hat{\rho}_{\text{f}} \hat{a}^\dagger \frac{\hat{S}_n}{\sqrt{\hat{n}+1}} \rho_{ab} & \frac{\hat{S}_{n-1}}{\sqrt{\hat{n}}} \hat{a}^\dagger \hat{\rho}_{\text{f}} \hat{a} \frac{\hat{S}_{n-1}}{\sqrt{\hat{n}}} \rho_{aa} \end{pmatrix}$$

$$-i \begin{pmatrix} \frac{\hat{S}_n}{\sqrt{\hat{n}+1}} \hat{a}\hat{\rho}_{\text{f}} \hat{C}_n \rho_{ba} - \hat{C}_n \hat{\rho}_{\text{f}} \hat{a}^\dagger \frac{\hat{S}_n}{\sqrt{\hat{n}+1}} \rho_{ab} & \frac{\hat{S}_n}{\sqrt{\hat{n}+1}} \hat{a}\hat{\rho}_{\text{f}} \hat{C}_{n-1} \rho_{bb} - \hat{C}_n \hat{\rho}_{\text{f}} \hat{a} \frac{\hat{S}_{n-1}}{\sqrt{\hat{n}}} \rho_{aa} \\ \frac{\hat{S}_{n-1}}{\sqrt{\hat{n}}} \hat{a}^\dagger \hat{\rho}_{\text{f}} \hat{C}_n \rho_{aa} - \hat{C}_{n-1} \hat{\rho}_{\text{f}} \hat{a}^\dagger \frac{\hat{S}_n}{\sqrt{\hat{n}+1}} \rho_{bb} & \frac{\hat{S}_{n-1}}{\sqrt{\hat{n}}} \hat{a}^\dagger \hat{\rho}_{\text{f}} \hat{C}_{n-1} \rho_{ab} - \hat{C}_{n-1} \hat{\rho}_{\text{f}} \hat{a} \frac{\hat{S}_{n-1}}{\sqrt{\hat{n}}} \rho_{ba} \end{pmatrix}.$$

Hint: Multiply the matrices $\hat{\mathcal{U}}$ and $\hat{\mathcal{U}}^\dagger$ given by Eq. (15.11) and the initial condition

$$\hat{\rho}_{\text{f+at}}(t_0) \equiv \begin{pmatrix} \hat{\rho}_{\text{f}} \rho_{aa} & \hat{\rho}_{\text{f}} \rho_{ab} \\ \hat{\rho}_{\text{f}} \rho_{ba} & \hat{\rho}_{\text{f}} \rho_{bb} \end{pmatrix}.$$

Here the operators $\hat{C}_n$ and $\hat{S}_n$ are evaluated at $t - t_0$.

18.2 Operator Algebra

Let $\hat{A}$ be an arbitrary, not necessarily Hermitian operator, $\hat{a}^\dagger$ and a are the creation and annihilation operator for the harmonic oscillator, and $|\alpha\rangle$ be a coherent state.

Show the following relations:

$$\langle\alpha|\hat{a}^\dagger \hat{A}|\alpha\rangle = \alpha^*\langle\alpha|\hat{A}|\alpha\rangle$$

$$\langle\alpha|\hat{a}\hat{A}|\alpha\rangle = \left(\alpha + \frac{\partial}{\partial\alpha^*}\right)\langle\alpha|\hat{A}|\alpha\rangle$$

$$\langle\alpha|\hat{A}\hat{a}|\alpha\rangle = \alpha\langle\alpha|A|\alpha\rangle$$

$$\langle\alpha|\hat{A}\hat{a}^\dagger|\alpha\rangle = \left(\alpha^* + \frac{\partial}{\partial\alpha}\right)\langle\alpha|\hat{A}|\alpha\rangle.$$

18.3 Q-Function of Damped Harmonic Oscillator

Consider a harmonic oscillator of frequency Ω coupled to a heat bath consisting of a continuum of harmonic oscillators. The equation of motion for the density operator reads

$$\dot{\rho} = -i\Omega[\hat{a}^\dagger \hat{a}, \rho] - \frac{\gamma}{2}(n_{\text{th}} + 1)\left(\hat{a}^\dagger \hat{a} \rho + \rho \hat{a}^\dagger \hat{a} - 2\hat{a}\rho \hat{a}^\dagger\right)$$
$$- \frac{\gamma}{2} n_{\text{th}} \left(\hat{a}\hat{a}^\dagger \rho + \rho \hat{a}\hat{a}^\dagger - 2\hat{a}^\dagger \rho \hat{a}\right), \quad (18.62)$$

where n_{th} is the defined in Eq. (18.21).

(a) Start from Eq. (18.62) to derive an equation of motion for the Q-function and make use of the relations derived in the preceding problem.
Result:

$$\frac{\partial Q}{\partial t} = \frac{\partial}{\partial \alpha}\left(\frac{\gamma}{2} + i\Omega\right)\alpha Q + \frac{\partial}{\partial \alpha^*}\left(\frac{\gamma}{2} - i\Omega\right)\alpha^* Q + \gamma(n_{\text{th}} + 1)\frac{\partial}{\partial \alpha^*}\frac{\partial}{\partial \alpha}Q.$$

(b) Transform the equation of motion for the Q-function into polar coordinates $\alpha \equiv re^{i\varphi}$ and $\alpha^* \equiv re^{-i\varphi}$.

(c) Show that the time evolution of the Q-function of a damped harmonic oscillator in the *interaction picture* reads

$$Q(\alpha^*, \alpha, t) = \frac{1}{\pi(n_{\text{th}} + 1)(1 - e^{-\gamma t})}$$
$$\times \int \exp\left[-\frac{|\alpha - e^{-\gamma t/2}\beta|^2}{(n_{\text{th}} + 1)(1 - e^{-\gamma t})}\right] Q(\beta^*, \beta, t = 0)\, d^2\beta. \quad (18.63)$$

Hint: Solve the equation of motion for the Q-function with the (unphysical) initial condition $Q(\alpha^*, \alpha, t = 0) = \delta(\alpha - \beta)$. This yields the Green's function for the time evolution of an arbitrary Q-function. In the coordinates α and α^* it is a Gaussian with time dependent mean value and variance.

(d) Calculate with the help of Eq. (18.63) the time evolution of the Q-function if at time $t = 0$ the oscillator is in a coherent state. Show that this solution approaches the Q-function of a thermal state, Eq. (12.13), for $t \to \infty$.

18.4 Damping of Schrödinger Cat

(a) Calculate $Q(\alpha^*, \alpha, t)$ for the case that at time $t = 0$ the oscillator is in the Schrödinger cat state

$$|\psi\rangle \equiv \mathcal{N}\frac{1}{\sqrt{2}}\left[|\alpha e^{i\varphi}\rangle + |\alpha e^{-i\varphi}\rangle\right]$$

introduced in Sec. 11.3. Discuss your result and focus on the decay of the interference term in the Q-function.

(b) Show by using the master equation Eq. (18.62) that for $n_{\text{th}} = 0$ the time evolution of the photon statistics is given by

$$\dot{W}_m = \gamma(m+1)W_{m+1} - \gamma m W_m.$$

(c) Show that the time dependent solution of this equation is given by

$$W_m(t) = e^{-m\gamma t} \sum_{\nu=0}^{\infty} \binom{m+\nu}{m} \left(1 - e^{-\gamma t}\right)^{\nu} W_{m+\nu}(t=0). \qquad (18.64)$$

(d) Calculate with the help of Eq. (18.64) the time dependent photon statistics of a damped Schrödinger cat state.

18.5 Time Evolution of P-Distribution

We describe a damped harmonic oscillator by the master equation Eq. (18.62). For $n_{th} = 0$, that is for $T = 0$, we now investigate this equation in more detail. Show:

(a) For a time dependent coherent state $|\alpha(t)\rangle$ we have

$$\frac{d}{dt}|\alpha(t)\rangle = \left(-\frac{1}{2}\dot\alpha\alpha^* - \frac{1}{2}\alpha\dot\alpha^* + \dot\alpha \hat a^\dagger\right)|\alpha\rangle.$$

(b) The density operator

$$|\alpha_0 e^{-(\gamma/2+i\Omega)t}\rangle\langle\alpha_0 e^{-(\gamma/2+i\Omega)t}|$$

obeys the equation of motion for the damped harmonic oscillator.

(c) If at time $t = 0$ the density operator is given by

$$\hat\rho(t=0) = \int d^2\alpha\, P(\alpha, \alpha^*, t=0)\, |\alpha\rangle\langle\alpha|,$$

where $P(\alpha, \alpha^*, t=0)$ is the P-function at time $t = 0$, then at times $t > 0$ the P-function is given by

$$P(\alpha, \alpha^*, t) = e^{\gamma t} P\left(\alpha e^{(\gamma/2+i\omega)t}, \alpha^* e^{(\gamma/2-i\omega)t}, t = 0\right).$$

18.6 Quantum Jump Method

At low enough temperatures a quantum mechanical system coupled to a reservoir can be approximated by the master equation

$$\dot{\hat\rho} = -\frac{i}{\hbar}[\hat H, \hat\rho] + \frac{\gamma}{2}(2\hat a\hat\rho\hat a^\dagger - \hat a^\dagger\hat a\hat\rho - \hat\rho\hat a^\dagger\hat a). \qquad (18.65)$$

One possibility for solving this equation is the method of quantum trajectories which we now want to investigate. We assume that at time t the system is in the state $|\psi(t)\rangle$. For the time evolution there are two possibilities:

(i) The time evolution is governed by the non-Hermitian Hamiltonian

$$\hat{\tilde H} \equiv \hat H - i\hbar\gamma/2\,\hat a^\dagger\hat a$$

which yields

$$|\psi(t+dt)\rangle = \frac{\left[1 - i\hat{\tilde H}dt/\hbar\right]|\psi(t)\rangle}{\sqrt{1 - \gamma\langle\psi(t)|\hat a^\dagger\hat a|\psi(t)\rangle\,dt}}.$$

(ii) The system makes a jump corresponding to the destruction of a photon leading to

$$|\psi(t+dt)\rangle = \frac{\hat{a}|\psi(t)\rangle}{\sqrt{\langle\psi(t)|\hat{a}^\dagger\hat{a}|\psi(t)\rangle}}.$$

For every time interval dt we choose by chance one of the two possibilities. The probability Pdt that between t and $t+dt$ a jump occurs is given by

$$Pdt = \gamma\langle\psi(t)|\hat{a}^\dagger\hat{a}|\psi(t)\rangle\, dt.$$

Since the time evolution is not unitary we have to renormalize the state after each step.

Show that the density operator

$$\hat{\rho}(t) = \overline{|\psi(t)\rangle\langle\psi(t)|}$$

which is averaged over all trajectories is a solution of the master equation Eq. (18.65).

Hint: See Dalibard *et al.* (1992), Mølmer *et al.* (1993) and Carmichael (1993).

References

Master Equations

M.O. Scully and **W.E. Lamb**, *Quantum Theory of an Optical Maser. I. General Theory*, Phys. Rev. **159**, 208–226 (1967)

M.O. Scully, *Varenna lecture*, in: *Proceedings of the international school of physics "Enrico Fermi", course XL2*, edited by R. Glauber, Academic Press, New York, 1969

For the model of a harmonic oscillator damped by a reservoir of two-level atoms see

W.E. Lamb, *Approach to Thermodynamic Equilibrium (and other Stationary States)*, in: *The Physicist's Conception of Nature*, edited by J. Mehra, D. Reidel, Dordrecht, Holland, 1973

This paper is also reprinted in

G.T. Moore and **M.O. Scully**, *Frontiers of Nonequilibrium Statistical Physics*, Plenum Press, New York, 1985

For various techniques to derive master equations see

F. Haake, *Statistical Treatment of Open Systems by Generalized Master Equations*, Springer Verlag, Heidelberg and Berlin, 1974

G.S. Agarwal, *Quantum Statistical Theories of Spontaneous Emission and their Relation to Other Approaches*, Springer Verlag, Heidelberg and Berlin, 1974

N.G. van Kampen, *Stochastic Processes in Physics and Chemistry*, North-Holland, Amsterdam, 1981

C.W. Gardiner, *Handbook of Stochastic Processes*, Springer Verlag, Heidelberg and Berlin, 1989

H. Risken, *The Fokker-Planck Equation*, Springer Verlag, Heidelberg and Berlin, 1989

Wave Function Approach Towards Dissipation

The method of the quantum trajectories was developed by

J. Dalibard, Y. Castin and **K. Mølmer**, *Wave-Functions approach to Dissipative Processes in Quantum Optics*, Phys. Rev. Lett. **68**, 580–583 (1992)

K. Mølmer, Y. Castin and **J. Dalibard**, *Monte Carlo Wave Function Method in Quantum Optics*, J. Opt. Soc. Am. B **10**, 524–538 (1993)

H. Carmichael, *An Open Systems Approach to Quantum Optics*, Springer Verlag, Heidelberg and Berlin, 1993

Quantum Langevin Equations

The field of quantum Langevin equations was pioneered by M. Lax and the school of H. Haken. For a summary see

M. Lax, *Quantum Noise. IV. Quantum Theory of Noise Sources*, Phys. Rev. **145**, 110–129 (1966)

H. Haken, *Laser Theory*, Springer Verlag, Heidelberg and Berlin, 1984.

Appeared originally in *Encyclopedia of Physics*, edited by S. Flügge, Springer, Heidelberg, 1970

W.H. Louisell, *Quantum Statistical Properties of Radiation*, Wiley, New York, 1973

For recent applications see Chapter 9 of

M.O. Scully and **M.S. Zubairy**, *Quantum Optics*, Cambridge U.P., New York, 1996

G.W. Ford, J.T. Lewis and **R.F. O'Connell**, *Quantum Langevin Equation*, Phys. Rev. A **37**, 4419–4428 (1988)

Decoherence

E. Joos and **H.D. Zeh**, *The Emergence of Classical Properties Through Interaction with the Environment*, Z. Phys. B **59**, 223–243 (1983)

W.H. Zurek, *Decoherence and the Transition from Quantum to Classical*, Physics Today **44**, 36–44 (1991)

D. Giulini, E. Joos, C. Kiefer, J. Kupsch, I.-O. Stamatescu and **H.D. Zeh**, *Decoherence and the Appearance of a Classical World in Quantum Theory*, Springer, Heidelberg, 1996

For the role of noise in amplification and damping see

R.J. Glauber, *Amplifiers, Attenuators, and Schrödinger's Cat*, in: *New Techniques and Ideas in Quantum Measurement Theory*, Vol. 480, ed. by D.M. Greenberger, New York Academy of Sciences, New York, 1986, p. 336–372

Decoherence of a Schrödinger cat state of a cavity field or the center-of-mass motion has been observed experimentally by

M. Brune, E. Hagley, J. Dreyer, X. Maitre, A. Maali, C. Wunderlich, J.M. Raimond and **S. Haroche**, *Observing the Progressive Decoherence of the Meter in a Quantum Measurement*, Phys. Rev. Lett. **77**, 4887–4891 (1996)

C.J. Myatt, B.E. King, Q.A. Turchette, C.A. Sackett, D. Kielpinski, W.M. Itano, C. Monroe and **D.J. Wineland**, *Decoherence of quantum superpositions through coupling to engineered reservoirs*, Nature **403**, 269–273 (2000)

One-Atom Maser

For the original experiment on the one-atom maser see
D. Meschede, H. Walther and **G. Müller**, *One-Atom Maser*, Phys. Rev. Lett. **54**, 551–554 (1985)
The experiment on a two-photon maser is described in
M. Brune, J.M. Raimond, P. Goy, L. Davidovich and **S. Haroche**, *Realization of a Two-Photon Maser Oscillator*, Phys. Rev. Lett. **59**, 1899–1902 (1987)
For the theory of the one-atom maser see
P. Filipowicz, J. Javanainen and **P. Meystre**, *Theory of a Microscopic Maser*, Phys. Rev. A **34**, 3077–3087 (1986)
L.A. Lugiato, M.O. Scully and **H. Walther**, *Connection between microscopic and macroscopic maser theory*, Phys. Rev. A **36**, 740–743 (1987)
P. Filipowicz, J. Javanainen and **P. Meystre**, *Quantum and Semiclassical States of a Kicked Cavity Mode*, J. Opt. Soc. Am. B **3**, 906–910 (1986)
For the measurement of the photon statistics in the one-atom maser and bistability in the photon statistics see
G. Rempe, F. Schmidt-Kaler and **H. Walther**, *Observation of sub-Poissonian photon statistics in a micromaser*, Phys. Rev. Lett. **64**, 2783–2786 (1990)
O. Benson, G. Raithel and **H. Walther**, *Quantum jumps of the micromaser field: dynamic behavior close to phase transition points*, Phys. Rev. Lett. **72**, 3506–3509 (1994)
The phase diffusion in a one-atom maser is discussed in
M.O. Scully, H. Walther, G.S. Agarwal, T. Quang and **W.P. Schleich**, *Micromaser spectrum*, Phys. Rev. A **44**, 5992–5996 (1991)
T. Quang, G.S. Agarwal, J. Bergou, M.O. Scully, H. Walther, K. Vogel and **W.P. Schleich**, *Calculation of the micromaser spectrum. I. Green's-function approach and approximate analytical techniques*, Phys. Rev. A **48**, 803–812 (1993)
K. Vogel, W.P. Schleich, M.O. Scully and **H. Walther**, *Calculation of the micromaser spectrum. II. Eigenvalue Approach*, Phys. Rev. A **48**, 813–817 (1993)
W.C. Schieve and **R.R. McGowan**, *Phase distribution and linewidth in the micromaser*, Phys. Rev. A **48**, 2315–2323 (1993)
G. Raithel, C. Wagner, H. Walther, L. Narducci and **M.O. Scully**, *The Micromaser: A Proving Ground for Quantum Physics*, in: *Cavity Quantum Electrodynamics*, edited by P.R. Berman, Adv. At. Mol. Opt. Phys., Supplement 2, Academic Press, Boston, 1994
N. Lu, *Natural linewidth of a micromaser*, Opt. Comm. **103**, 315–325 (1993)
N. Lu, *Micromaser spectrum: Trapped states*, Phys. Rev. A **47**, 1347–1357 (1993)
For the theoretical prediction and the observation of the trapping states see
P. Meystre, G. Rempe and **H. Walther**, *Very-low-temperature behavior of a micromaser*, Opt. Lett. **13**, 1078–1080 (1988)
M. Weidinger, B.T.H. Varcoe, R. Heerlein and **H. Walther**, *Trapping States in the Micromaser*, Phys. Rev. Lett. **82**, 3795–3798 (1999)
For a summary of the physics of entanglement in the one-atom maser see
B.-G. Englert, M. Löffler, O. Benson, B. Varcoe, M. Weidinger and **H. Walther**, *Entangled Atoms in Micromaser Physics*, Fortschr. Phys. **46**, 897–926 (1998)

Bloch Equations

The original paper on nuclear magnetic resonance appeared by
F. Bloch, *Nuclear Induction*, Phys. Rev. **70**, 460–474 (1946)

For a nice exposition of the Bloch equations see
L. Allen and **J.H. Eberly**, *Optical Resonance and Two-Level Atoms*, Wiley, New York, 1975, available as a reprinted version by Dover Publishing

For the concept of the Bloch sphere see
R.P. Feynman, F.L. Vernon and **R.W. Hellwarth**, *Geometrical Representation of the Schrödinger Equation for Solving Maser Problems*, J. Appl. Phys. **28**, 49–52 (1957)

Spontaneous Emission and Lamb Shift

V. Weisskopf and **E. Wigner**, *Calculation of the Natural Linewidth Based on Dirac's Theory of Light*, Z. Phys. **63**, 54–73 (1930)

W.E. Lamb and **R.C. Retherford**, *Fine Structure of the Hydrogen Atom by a Microwave Method*, Phys. Rev. **72**, 241–243 (1947)

For the first calculation of the Lamb shift, see
H.A. Bethe, *The Electromagnetic Shift of Energy Levels*, Phys. Rev. **72**, 339–341 (1947)

For the spontaneous emission into a squeezed vacuum see
C.W. Gardiner, *Inhibition of Atomic Phase Decays by Squeezed Light: A Direct Effect of Squeezing*, Phys. Rev. Lett. **56**, 1917–1920 (1986)

A similar effect exists in the Jaynes-Cummings-Paul model as pointed out by
Y. Ben-Aryeh, C.A. Miller, H. Risken and **W.P. Schleich**, *Inhibition of atomic dipole collapses by squeezed light: a Jaynes-Cummings model treatment*, Opt. Comm. **90**, 259–264 (1992)

19 Atom Optics in Quantized Light Fields

So far we have considered the interaction of an atom with a quantized light field with the atom at rest at a position $\vec{R}$, that is we have neglected the center-of-mass motion. In the present chapter we focus on the influence of quantized light on the motion of the atom. Hence we now treat not only the internal degrees of freedom and the light field quantum mechanically but also the center-of-mass motion. In this way we bring out the wave nature of matter, that is, the wave properties of atoms giving rise to the name atom optics. For the sake of simplicity we consider only a single mode of the radiation field.

19.1 Formulation of Problem

Figure 19.1 summarizes our setup. An atomic wave of a two-level atom propagates through a resonator and interacts resonantly with a single mode of the radiation field via the familiar Jaynes-Cummings Hamiltonian, Eq. (14.57). Hence, in the interaction picture we have the Hamiltonian

$$\hat{H} = \frac{\hat{\vec{p}}^2}{2M} + \vec{\wp} \cdot \vec{u}(\hat{\vec{r}}) \, \mathcal{E}_0 \left(\hat{\sigma} \hat{a}^\dagger + \hat{\sigma}^\dagger \hat{a} \right), \tag{19.1}$$

where $\hat{\vec{p}}^2/(2M)$ describes the kinetic energy of the center-of-mass motion and $\vec{u}(\hat{\vec{r}})$ is the mode function at the position $\vec{r}$ of the atom.

We emphasize that in contrast to Chapter 14 we now denote the center-of-mass variables $\vec{r}$ and $\vec{p}$ again by small rather than capital letters, since we express the dynamics of the relative coordinates in the dipole moment $\vec{\wp}$ and the flip operators $\hat{\sigma}$ and $\hat{\sigma}^\dagger$. Moreover, the small letter coordinates also agree with the notation of the general coordinate system shown in Fig. 19.1.

19.1.1 Dynamics

Since we are now treating the center-of-mass motion quantum mechanically the position $\vec{r}$ and the momentum $\vec{p}$ are conjugate operators and obey the commutator relation

$$[r_j, p_k] = i\hbar \delta_{jk}.$$

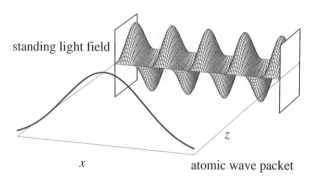

Fig. 19.1: Schematic illustration of the setup for atom optics. An atomic wave propagates through a resonator and interacts resonantly with a single mode of the standing light field. Here we show the case where the atomic wave packet covers many wavelengths of the light field and can therefore be considered a plane wave in z direction. We only show a cut through the probability distribution for one value of z.

State Vector of Complete System

Hence the dynamics of the state vector $|\Phi\rangle$ describing the combined system of center of mass motion, internal states of the atom and the states of the electromagnetic field is governed by the Schrödinger equation

$$i\hbar \frac{d|\Phi\rangle}{dt} = \hat{H} |\Phi\rangle. \tag{19.2}$$

We expand the state vector $|\Phi\rangle$ as

$$|\Phi(t)\rangle = \sum_{m=0}^{\infty} \int d^3r' \left[\Phi_{a,m-1}(\vec{r}',t) |a, m-1\rangle + \Phi_{b,m}(\vec{r}',t) |b, m\rangle \right] |\vec{r}'\rangle, \tag{19.3}$$

where

$$\Phi_{a,m-1}(\vec{r},t) \equiv \begin{pmatrix} \text{probability amplitude to find at time } t \text{ the} \\ \text{atom at position } \vec{r} \text{ and in the internal} \\ \text{state } |a\rangle \text{ with } m-1 \text{ photons in the field} \end{pmatrix}$$

and

$$\Phi_{b,m}(\vec{r},t) \equiv \begin{pmatrix} \text{probability amplitude to find at time } t \text{ the} \\ \text{atom at position } \vec{r} \text{ and in the internal} \\ \text{state } |b\rangle \text{ with } m \text{ photons in the field} \end{pmatrix}.$$

Generalized Rabi Equations

When we substitute this ansatz into the Schrödinger equation Eq. (19.2) and use the Hamiltonian Eq. (19.1) we find

$$\sum_{m=0}^{\infty} \int d^3r' \left[i\hbar \frac{\partial \Phi_{a,m-1}}{\partial t} |a, m-1\rangle + i\hbar \frac{\partial \Phi_{b,m}}{\partial t} |b, m\rangle \right] |\vec{r}'\rangle$$

$$= \sum_{m=0}^{\infty} \int d^3r' \left[\Phi_{a,m-1} |a, m-1\rangle + \Phi_{b,m} |b, m\rangle \right] \frac{\hat{\vec{p}}^2}{2M} |\vec{r}'\rangle$$

19.1 Formulation of Problem

$$+ \sum_{m=0}^{\infty} \int d^3r' \, \vec{\wp} \cdot \vec{u}(\vec{r}') \, \mathcal{E}_0 \sqrt{m} \left[\Phi_{a,m-1} \left| b, m \right\rangle + \Phi_{b,m} \left| a, m-1 \right\rangle \right] \left| \vec{r}' \right\rangle. \quad (19.4)$$

Here we have used the relations

$$\hat{\sigma} \left| b \right\rangle = \hat{\sigma}^{\dagger} \left| a \right\rangle = 0, \qquad \hat{a}^{\dagger} \left| m - 1 \right\rangle = \sqrt{m} \left| m \right\rangle \qquad \text{and} \qquad \hat{a} \left| m \right\rangle = \sqrt{m} \left| m - 1 \right\rangle.$$

We derive the equations of motion for the probability amplitudes $\Phi_{a,m-1}$ and $\Phi_{b,m}$ by multiplying Eq. (19.4) either by $\langle a, n-1, \vec{r} |$ or $\langle b, n, \vec{r} |$ and using the orthonormality relations $\langle a | b \rangle = 0$, $\langle a | a \rangle = \langle b | b \rangle = 1$ for the internal states and $\langle \vec{r} | \vec{r}' \rangle = \delta(\vec{r} - \vec{r}')$ for the center-of-mass motion.

We hence arrive at

$$i\hbar \frac{\partial \Phi_{a,n-1}(\vec{r}, t)}{\partial t} = \int d^3r' \, \Phi_{a,n-1}(\vec{r}', t) \left\langle \vec{r} \left| \frac{\hat{\vec{p}}^2}{2M} \right| \vec{r}' \right\rangle + \vec{\wp} \cdot \vec{u}(\vec{r}) \, \mathcal{E}_0 \sqrt{n} \, \Phi_{b,n}(\vec{r}, t) \quad (19.5a)$$

and

$$i\hbar \frac{\partial \Phi_{b,n}(\vec{r}, t)}{\partial t} = \int d^3r' \, \Phi_{b,n}(\vec{r}', t) \left\langle \vec{r} \left| \frac{\hat{\vec{p}}^2}{2M} \right| \vec{r}' \right\rangle + \vec{\wp} \cdot \vec{u}(\vec{r}) \, \mathcal{E}_0 \sqrt{n} \, \Phi_{a,n-1}(\vec{r}, t). \quad (19.5b)$$

When we recall from Problem 2.3 the expression

$$\langle \vec{r} | \hat{\vec{p}}^2 | \vec{r}' \rangle = \left(\frac{\hbar}{i} \vec{\nabla} \right)^2 \delta(\vec{r} - \vec{r}')$$

for the matrix element of the square of the momentum operator in position representation we can rewrite

$$\int d^3r' \, \Phi_{a,n-1}(\vec{r}', t) \left\langle \vec{r} \left| \frac{\hat{\vec{p}}^2}{2M} \right| \vec{r}' \right\rangle = \frac{1}{2M} \left(\frac{\hbar}{i} \vec{\nabla} \right)^2 \int d^3r' \, \Phi_{a,n-1}(\vec{r}', t) \, \delta(\vec{r} - \vec{r}')$$

$$= \frac{1}{2M} \left(\frac{\hbar}{i} \vec{\nabla} \right)^2 \Phi_{a,n-1}(\vec{r}, t) = \frac{\hat{\vec{p}}^2}{2M} \Phi_{a,n-1}(\vec{r}, t).$$

Similarly we derive the formula

$$\int d^3r' \, \Phi_{b,n}(\vec{r}', t) \left\langle \vec{r} \left| \frac{\hat{\vec{p}}^2}{2M} \right| \vec{r}' \right\rangle = \frac{\hat{\vec{p}}^2}{2M} \Phi_{b,n}(\vec{r}, t).$$

Consequently, the equations of motion (19.5) for the probability amplitudes $\Phi_{a,n-1}$ and $\Phi_{b,n}$ take the form

$$i\hbar \frac{\partial \Phi_{a,n-1}(\vec{r}, t)}{\partial t} = \frac{\hat{\vec{p}}^2}{2M} \Phi_{a,n-1}(\vec{r}, t) + \vec{\wp} \cdot \vec{u}(\vec{r}) \, \mathcal{E}_0 \sqrt{n} \, \Phi_{b,n}(\vec{r}, t) \quad (19.6a)$$

and

$$i\hbar \frac{\partial \Phi_{b,n}(\vec{r}, t)}{\partial t} = \frac{\hat{\vec{p}}^2}{2M} \Phi_{b,n}(\vec{r}, t) + \vec{\wp} \cdot \vec{u}(\vec{r}) \, \mathcal{E}_0 \sqrt{n} \, \Phi_{a,n-1}(\vec{r}, t). \quad (19.6b)$$

These equations are the Rabi equations Eq. (15.19) of the familiar Jaynes-Cummings-Paul model with the kinetic energy operator

$$\frac{\hat{\vec{p}}^2}{2M} \equiv -\frac{\hbar^2}{2M} \Delta.$$

Moreover, the mode function now depends on the position $\vec{r}$ of the atom, that is on the variable conjugate to momentum.

19.1.2 Time Evolution of Probability Amplitudes

In the preceding section we have derived two coupled equations for the probability amplitudes $\Phi_{a,n-1}$ and $\Phi_{b,n}$. In the present section we introduce linear combinations of them which correspond to dressed states and decouple in this way the two equations. The probability amplitudes then evolve like a particle in a given potential provided by the electromagnetic field. We derive these potentials and the corresponding initial conditions.

Effective Potentials from Dressed States

We decouple the generalized Rabi equations by considering the linear combinations

$$\Phi_n^{(\pm)} \equiv \Phi_{b,n} \pm \Phi_{a,n-1} \tag{19.7}$$

which results in

$$i\hbar \frac{\partial \Phi_n^{(\pm)}(\vec{r},t)}{\partial t} = \left[\frac{\hat{\vec{p}}^2}{2M} + U_n^{(\pm)}(\vec{r}) \right] \Phi_n^{(\pm)}(\vec{r},t). \tag{19.8}$$

Here we have defined the potentials

$$U_n^{(\pm)}(\vec{r}) \equiv \pm \vec{\wp} \cdot \vec{u}(\vec{r}) \, \mathcal{E}_0 \sqrt{n}. \tag{19.9}$$

Hence the probability amplitudes $\Phi_n^{(\pm)}$ satisfy a Schrödinger equation corresponding to a particle of mass M moving in a potential $U_n^{(\pm)}$. This potential is formed by the scalar product of the dipole moment $\vec{\wp}$ and the mode function $\vec{u}(\vec{r})$. Moreover, it scales with the vacuum electric field strength $\mathcal{E}_0$ and the square root of the photon number.

Special Example for Mode Functions

We illustrate these potentials $U_n^{(\pm)}$ using the modes of a box-shaped resonator discussed in Chapter 10. The mode

$$\vec{u}(\vec{r}) = \vec{e}_y \sin(k_x x) \sin(\pi z / L_z) \tag{19.10}$$

following from Eq. (10.35) for $l_z = 1$ gives rise to the potentials

$$U_n^{(\pm)} = \pm \wp \mathcal{E}_0 \sqrt{n} \sin(k_x x) \sin(\pi z / L_z) \tag{19.11}$$

where $\wp \equiv \vec{\wp} \cdot \vec{e}_y$.

In Fig. 19.2 we show the potential $U_n^{(+)}$ for $n = 1$. To bring out the influence of the photon number via the square root of n we show in the same figure a cut of this potential along x at $z = L_z/2$ for various values of n. Since the nodes of the potential are independent of the photon number and the amplitude of the modulation in space is proportional to $\sqrt{n}$, the potentials $U_n^{(\pm)}$ get steeper as n increases. The potential $U_n^{(-)}$ is just the negative of $U_n^{(+)}$.

Hence the probability amplitudes $\Phi_n^{(\pm)}$ evolve in the potentials $U_n^{(\pm)}$ according to the Schrödinger equation. But what is their initial condition?

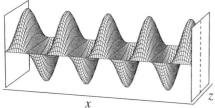 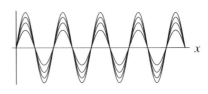

Fig. 19.2: Mode function in the cavity (left) and dressed state potential (right). The mode function $\vec{u}(\vec{r})$, Eq. (19.10), displays many oscillations along the x-axis but has only a single maximum along the z-axis. The dressed state potential, Eq. (19.11), reflects the mode function with an amplitude determined by the square root of the photon number n. We show a cut of this potential along x at $z = L_z/2$, that is at the center of the cavity, for different values of n.

Initial Conditions

We can answer this question by considering the initial condition for $|\Phi\rangle$. The state $|\Phi\rangle$ at time $t = 0$, that is, before the interaction, is a direct product of the atomic state

$$|\psi_{\text{atom}}\rangle \equiv |b\rangle$$

which we take to be the ground state, the field state

$$|\psi_{\text{field}}\rangle \equiv \sum_{n=0}^{\infty} w_n |n\rangle$$

as a superposition of photon number states with probability amplitudes w_n and the center-of-mass state

$$|\psi_{\text{cm}}\rangle \equiv \int d^3r' \, \mathcal{F}(\vec{r}') \, |\vec{r}'\rangle$$

with the normalization

$$\langle \psi_{\text{cm}} | \psi_{\text{cm}} \rangle = \int d^3r \, |\mathcal{F}(\vec{r})|^2 = 1.$$

Hence $|\Phi(t = 0)\rangle$ reads

$$|\Phi(t = 0)\rangle \equiv |\psi_{\text{atom}}\rangle \otimes |\psi_{\text{field}}\rangle \otimes |\psi_{\text{cm}}\rangle =$$
$$= |b\rangle \otimes \sum_{n=0}^{\infty} w_n |n\rangle \otimes \int d^3r' \, \mathcal{F}(\vec{r}') \, |\vec{r}'\rangle,$$

$$|\Phi(t = 0)\rangle = \sum_{n=0}^{\infty} \int d^3r' \mathcal{F}(\vec{r}') w_n \, |b, n\rangle \, |\vec{r}'\rangle.$$

When we compare this expression for the initial state with the ansatz Eq. (19.3) we find

$$\Phi_{b,n}(\vec{r}', 0) = \mathcal{F}(\vec{r}') w_n$$

and

$$\Phi_{a,n-1}(\vec{r}', 0) = 0$$

which translates via Eq. (19.7) into the initial conditions

$$\Phi_n^{(\pm)}(\vec{r}, 0) = \mathcal{F}(\vec{r}) \, w_n. \tag{19.12}$$

Hence the dynamics of the state vector $|\Phi\rangle$ of the combined system consisting of the three degrees of freedom, internal state of the atom, state of the field mode and state of the center-of-mass motion follows from the dynamics of the wave functions $\Phi_n^{(\pm)}$ moving in the potentials $U_n^{(\pm)}$ subject to the initial condition Eq. (19.12). Since in general the potential is three-dimensional and can be highly complicated it is a nontrivial task to solve the Schrödinger equation Eq. (19.8) for $\Phi_n^{(\pm)}$.

19.2 Reduction to One-Dimensional Scattering

We now consider a situation in which the atomic beam propagates initially orthogonal to the wave vector of the field in the cavity. In the remainder of the section we call this the x-direction. Since it is orthogonal to the atomic beam we sometimes refer to it as the transverse direction.

The motion along the z-axis, that is the longitudinal motion, we treat classically since we assume that the initial kinetic energy $p_z^2/(2M)$ in z-direction is much larger than the change in the longitudinal momentum due to the interaction with the light field.

19.2.1 Slowly Varying Approximation

These ideas motivate the ansatz

$$\Phi_n^{(\pm)}(\vec{r}, t) = \exp[-i(Et/\hbar - k_z z)] \, \tilde{\Phi}_n^{(\pm)}(x, y, z) \tag{19.13}$$

consisting of a plane wave with a slowly varying amplitude $\tilde{\Phi}_n^{(\pm)}(x, y, z)$. The wave propagates along the z-axis with the momentum

$$p_z \equiv Mv_z \equiv \hbar k_z$$

and the kinetic energy

$$E \equiv (\hbar k_z)^2/2M. \tag{19.14}$$

When we substitute this ansatz into the equation of motion Eq. (19.8) for the dressed-state probability amplitudes we find

$$E\Phi_n^{(\pm)} = \frac{1}{2M} \left[(\hbar k_z)^2 \tilde{\Phi}_n^{(\pm)} - \hbar^2 2ik_z \frac{\partial \tilde{\Phi}_n^{(\pm)}}{\partial z} - \hbar^2 \frac{\partial^2 \tilde{\Phi}_n^{(\pm)}}{\partial z^2} \right] \exp[-i(Et/\hbar - k_z z)]$$

$$+ \left[\frac{\hat{p}_x^2 + \hat{p}_y^2}{2M} + U_n^{(\pm)} \right] \Phi_n^{(\pm)}.$$

We recall that $\tilde{\Phi}_n^{(\pm)}$ is slowly varying as a function of z, that is,

$$\left| \frac{\partial^2 \tilde{\Phi}_n^{(\pm)}}{\partial z^2} \right| \ll k_z \left| \frac{\partial \tilde{\Phi}_n^{(\pm)}}{\partial z} \right|$$

19.2 Reduction to One-Dimensional Scattering

which allows us to neglect the third term in the bracket compared to the second term. With the help of the relation Eq. (19.14) for the energy E we arrive at

$$i\hbar \frac{\partial \tilde{\Phi}_n^{(\pm)}(x,y,z)}{\partial \left(z\frac{M}{\hbar k_z}\right)} = \left[\frac{\hat{p}_x^2 + \hat{p}_y^2}{2M} + U_n^{(\pm)}(x,y,z)\right] \tilde{\Phi}_n^{(\pm)}(x,y,z),$$

or

$$i\hbar \frac{\partial \tilde{\Phi}_n^{(\pm)}(x,y,v_z t)}{\partial t} = \left[\frac{\hat{p}_x^2 + \hat{p}_y^2}{2M} + U_n^{(\pm)}(x,y,v_z t)\right] \tilde{\Phi}_n^{(\pm)}(x,y,v_z t).$$

Here we have introduced the time

$$t \equiv zM/(\hbar k_z) = z/v_z$$

as "measured" by the position z reached by the wave of velocity v_z.

Hence by treating the motion in z-direction classically we have reduced the problem from a three-dimensional problem to a two-dimensional one.

19.2.2 From Two Dimensions to One

To simplify this scattering problem even further we now consider the case of the mode function Eq. (10.35) creating the potentials

$$U_n^{(\pm)}(x,y,z) = \pm \wp \mathcal{E}_0 \sqrt{n} \sin(k_x x) \sin(\pi z/L_z).$$

Since in this case the quantities $U_n^{(\pm)}$ do not depend on the y-coordinate the propagation in the y-direction can be separated via the ansatz

$$\tilde{\Phi}_n^{(\pm)}(x,y,v_z t) \equiv \Theta(y,t) \Psi_n^{(\pm)}(x,t), \qquad (19.15)$$

where

$$i\hbar \frac{\partial \Theta(y,t)}{\partial t} = \frac{\hat{p}_y^2}{2M} \Theta(y,t) \qquad (19.16)$$

and

$$i\hbar \frac{\partial \Psi_n^{(\pm)}(x,t)}{\partial t} = \left[\frac{\hat{p}_x^2}{2M} \pm \wp \mathcal{E}_0 \sqrt{n} \sin(\pi v_z t/L_z) \sin(k_x x)\right] \Psi_n^{(\pm)}(x,t). \qquad (19.17)$$

We have therefore reduced the three-dimensional scattering situation to the problem of solving a one-dimensional time dependent Schrödinger equation. We emphasize, however, that this problem is nontrivial, since due to the motion of the atom through the resonator – the interaction switches on and switches off via the mode function $\sin(\pi v_z t/L_z)$ – the potential is explicitly time dependent.

Moreover, the potential in x-direction is periodic and can allow for rather complicated solutions. Even in the absence of the time dependence the Schrödinger equation

$$i\hbar \frac{\partial \psi(x,t)}{\partial t} = \left[\frac{\hat{p}^2}{2M} + b \sin(kx)\right] \psi(x,t)$$

for the wave function ψ of a particle of mass M moving in a periodic potential cannot be solved in an analytic way. This equation appears in many physical phenomena in solid state physics and gives rise to Bloch waves and band gaps. Similar effects also make their appearance in the context of atom optics.

19.2.3 State Vector

We conclude this section by combining all results to rewrite the state vector $|\Phi(t)\rangle$, Eq. (19.3), of the combined system. For this purpose we express $\Phi_{a,m-1}$ and $\Phi_{b,m}$ in terms of $\Phi_n^{(\pm)}$ via Eq. (19.7) and use Eqs. (19.13) and (19.15) to cast the result

$$\Phi_{a,m-1}(\vec{r},t) = \frac{1}{2}\left(\Phi_m^{(+)} - \Phi_m^{(-)}\right)$$
$$= \exp[-i(Et/\hbar - k_z z)]\,\Theta(y,t)\,\frac{1}{2}\left[\Psi_m^{(+)}(x,t) - \Psi_m^{(-)}(x,t)\right]$$

and

$$\Phi_{b,m}(\vec{r},t) = \frac{1}{2}\left(\Phi_m^{(+)} + \Phi_m^{(-)}\right)$$
$$= \exp[-i(Et/\hbar - k_z z)]\,\Theta(y,t)\,\frac{1}{2}\left[\Psi_m^{(+)}(x,t) + \Psi_m^{(-)}(x,t)\right]$$

in terms of the plane wave in z-direction and the wave functions $\Theta(y,t)$ and $\Psi_n^{(\pm)}(x,t)$.

When we substitute these formulas into the expression Eq. (19.3) for the state vector $|\Phi\rangle$ we arrive at

$$|\Phi(t)\rangle = \int dz'\, e^{-i(Et/\hbar - k_z z')}\,|z'\rangle \otimes \int dy'\,\Theta(y',t)\,|y'\rangle \otimes |\Psi(t)\rangle,$$

where

$$|\Psi(t)\rangle \equiv \sum_{n=0}^{\infty}\int dx'\,\left\{\frac{1}{2}\left[\Psi_n^{(+)}(x',t) - \Psi_n^{(-)}(x',t)\right]|a,n-1\rangle\right.$$
$$\left. + \frac{1}{2}\left[\Psi_n^{(+)}(x',t) + \Psi_n^{(-)}(x',t)\right]|b,n\rangle\right\}|x'\rangle \quad (19.18)$$

denotes the state vector of the combined system in x-direction.

Now we are left with the problem of solving the Schrödinger equations Eqs. (19.16) and (19.17) subject to the initial condition Eq. (19.12) which according to Eqs. (19.13) and (19.15) read

$$\Phi_n^{(\pm)}(\vec{r},0) = e^{ik_z z}\,\tilde{\Phi}_n^{(\pm)}(\vec{r},0) = e^{ik_z z}\,\Theta(y,0)\,\Psi_n^{(\pm)}(x,0) = \mathcal{F}(\vec{r})\,w_n.$$

When we factorize the initial probability amplitude

$$\mathcal{F}(\vec{r}) = f(x)\,g(y)\,h(z)$$

of the center-of-mass motion we can read off the initial conditions

$$h(z) = e^{ik_z z}$$

$$g(y) = \Theta(y, t=0)$$

and

$$f(x)\,w_n = \Psi_n^{(\pm)}(x, t=0). \quad (19.19)$$

With the help of the propagator of a free particle derived in the Problem 2.4 we find the expression

$$\Theta(y,t) = \sqrt{\frac{M}{2\pi i\hbar t}} \int dy'\, \Theta(y', t=0) \exp\left[\frac{iM}{2\hbar}\frac{(y-y')^2}{t}\right]$$

for the wave function.

We still have to solve Eq. (19.17) for $\Psi_n^{(\pm)}$ subject to the initial condition Eq. (19.19). This is the topic of the next sections.

19.3 Raman-Nath Approximation

In the present discussion we confine ourselves to an approximate but analytical analysis of Eq. (19.17). Here we consider the limit when the atom does not move considerably during its passage through the standing wave. Nevertheless, there is a momentum transfer. This approximation is called the Raman-Nath approximation.

19.3.1 Heuristic Arguments

We gain some insight into this approximation by recalling that the atomic beam enters the resonator orthogonal to the wave vector of the electromagnetic field. For this situation we now present a classical estimate of the kinetic energy gained by the atom due to the interaction with the light field.

Estimate of Transfered Momentum

We estimate the classical force

$$\text{force} \equiv \left(\frac{\text{momentum change}}{\text{interaction time}}\right) \equiv \frac{\Delta p}{\tau}$$

on the particle by

$$\text{force} \equiv |\vec{\nabla} \text{potential}| \sim \left(\frac{\text{change of potential energy}}{\text{reduced wavelength } \lambda}\right) \sim \wp\mathcal{E}_0\sqrt{n}\, k_x.$$

Hence the classical momentum Δp transferred by the light to the atom reads

$$\Delta p \sim \wp\mathcal{E}_0\sqrt{n}\, k_x \tau.$$

Small Displacements

We can neglect the kinetic energy term $p_x^2/(2M)$ in Eq. (19.17) against the potential energy which we estimate by $\wp\mathcal{E}_0\sqrt{n}$, that is,

$$\frac{p_x^2}{2M} \sim \frac{(\Delta p)^2}{2M} \sim \frac{(\wp\mathcal{E}_0\sqrt{n}\, k_x\tau)^2}{2M} \ll \wp\mathcal{E}_0\sqrt{n}$$

provided
$$\frac{\wp \mathcal{E}_0 \sqrt{n}\, \tau^2 k_x^2}{2M} \ll 1. \qquad (19.20)$$

When we compare the distance Δx traveled by the atom to the wave length of the light, that is

$$\left(\frac{\text{distance traveled by atom due to momentum change } \Delta p}{\text{reduced wavelength } \lambdabar}\right) = \frac{\Delta x}{\lambdabar} \sim \frac{\tau \Delta p}{M} k_x = \frac{\wp \mathcal{E}_0 \sqrt{n}\, \tau^2 k_x^2}{M}$$

the condition (19.20) of neglecting the transverse kinetic energy translates into

$$\Delta x \ll \lambdabar.$$

In the Raman-Nath approximation the atom is allowed only to move over a distance which is much less than a wavelength of the light.

Small Recoil Action

We can also cast the condition Eq. (19.20) in a slightly different form when we recall that $\wp \mathcal{E}_0 \tau \sqrt{n}/\hbar$ measures the number of Rabi cycles the atom has undergone in a field of n photons during the interaction time τ.

We then find

$$\left(\frac{\wp \mathcal{E}_0 \tau}{\hbar} \sqrt{n}\right) \times \left(\frac{(\hbar k_x)^2}{2M} \tau \frac{1}{\hbar}\right) = \left(\begin{array}{c}\text{number of}\\ \text{Rabi cycles}\end{array}\right) \times \left(\frac{\text{action of recoil energy}}{\hbar}\right) \ll 1.$$

Due to the appearance of the mass of the atom and the interaction time the condition Eq. (19.20) can be always satisfied by going to heavier atoms or shorter interaction times.

19.3.2 Probability Amplitudes

We have shown that when the displacement caused by the electromagnetic field is smaller than its wavelength we can neglect the transverse kinetic energy term. In this approximation the Schrödinger equation (19.17) reduces to

$$i\hbar \frac{\partial \Psi_n^{(\pm)}(x,t)}{\partial t} = \pm \wp \mathcal{E}_0 \sqrt{n}\, \sin(\pi v_z t/L)\, \sin(k_x x)\, \Psi_n^{(\pm)}(x,t)$$

which, when integrated, yields

$$\Psi_n^{(\pm)}(x,t) = \Psi_n^{(\pm)}(x,0) \exp\left[\mp i \frac{\wp \mathcal{E}_0}{\hbar} \int_0^t dt'\, \sin(\pi v_z t'/L) \sqrt{n}\, \sin(k_x x)\right].$$

Since we are interested in the wave function at the exit of the cavity we can set $v_z \tau = L$. With the help of the relation

$$\int_0^{L/v_z} dt\, \sin(\pi v_z t/L) = -\frac{L}{\pi v_z} \cos(\pi v_z t/L)\bigg|_0^{L/v_z} = \frac{2L}{\pi v_z} = \frac{2}{\pi}\tau$$

and with the initial condition $\Psi_n^{(\pm)}(x,0) = f(x)w_n$, Eq. (19.19), we arrive at

$$\Psi_n^{(\pm)}(x,t) = w_n f(x) \exp\left[\mp i\kappa\sqrt{n}\sin(k_x x)\right].$$

Here we have introduced the dimensionless interaction parameter

$$\kappa \equiv \frac{2}{\pi}\frac{\wp\mathcal{E}_0}{\hbar}\tau.$$

We are now in a position to write the state vector $|\Psi\rangle$, Eq. (19.18), explicitly and find

$$|\Psi(t)\rangle = \sum_{n=0}^{\infty} w_n \int dx'\, f(x') \left\{\cos[\kappa\sqrt{n}\,\sin(k_x x')]\,|b,n\rangle\right.$$
$$\left. - i\sin[\kappa\sqrt{n}\,\sin(k_x x')]\,|a,n-1\rangle\right\}|x'\rangle. \qquad (19.21)$$

We note that, indeed, the position dependent interaction of the atom with the quantized light field has created a strong entanglement between the transverse motion, the field and the energy levels of the atom.

19.4 Deflection of Atoms

In the present section we consider the deflection of the atom, that is, the momentum transfer from the field to the atom. In particular, we study the scattering in the Raman-Nath approximation. In this regime the field does not displace the atom significantly but still changes the momentum.

19.4.1 Measurement Schemes and Scattering Conditions

In order to calculate the momentum transfer we express the state vector $|\Psi\rangle$ calculated in the preceding section in the momentum representation. For this purpose we insert a complete set of momentum states $|p'\rangle$ and find

$$|\Psi(t)\rangle = \sum_{n=0}^{\infty} w_n \int dp'\, [c_n(p')\,|b,n\rangle - is_n(p')\,|a,n-1\rangle]\,|p'\rangle, \qquad (19.22)$$

where

$$c_n(p) \equiv \frac{1}{\sqrt{2\pi\hbar}} \int dx\, f(x) \cos\left[\kappa\sqrt{n}\sin(k_x x)\right] e^{-ipx/\hbar} \qquad (19.23a)$$

and

$$s_n(p) \equiv \frac{1}{\sqrt{2\pi\hbar}} \int dx\, f(x) \sin\left[\kappa\sqrt{n}\sin(k_x x)\right] e^{-ipx/\hbar}. \qquad (19.23b)$$

This state vector $|\Psi\rangle$ of the combined system allows us to answer questions concerning the momentum distribution of the scattered atoms, especially when we consider joint measurements between the transverse motion and the quantum field in the cavity.

Joint Measurements

In this situation the atom traverses the cavity prepared in a given field state $|\psi_{\text{field}}\rangle$, interacts with it and as a consequence gets deflected. After the atom has left the cavity we observe the field and measure the momentum. We reprepare the complete atom-field system and repeat the experiment. Quantum mechanics can predict the resulting joint probability distribution of momentum and field.

In Sec. 16.1 we have discussed joint measurements on entangled systems. In particular, we have focused on measurements of internal states and the state of the cavity field. In the context of atom optics we also have to include the center-of-mass motion. It is straightforward to generalize the concepts developed in Sec. 16.1 to include this degree of freedom. We can therefore calculate the probability

$$W(p,|\tilde{\psi}_{\text{field}}\rangle) \equiv \sum_{j=a,b} \left| \langle j | \langle p | \langle \tilde{\psi}_{\text{field}} | \Psi \rangle \right|^2$$

to find the momentum p given that the field is in the reference state

$$|\tilde{\psi}_{\text{field}}\rangle \equiv \sum_{n=0}^{\infty} \tilde{\psi}_n |n\rangle.$$

Here we deal with a case where we have not made a measurement of the internal states $|j\rangle = |a\rangle, |b\rangle$ and therefore take the trace over them.

When we substitute the photon number representation of the reference state into the above expression for W we find

$$W(p,|\tilde{\psi}_{\text{field}}\rangle) = \sum_{j=a,b} \left| \sum_{n=0}^{\infty} \tilde{\psi}_n^* \langle j | \langle p | \langle n | \Psi \rangle \right|^2. \tag{19.24}$$

In complete analogy to the joint atom-field measurement we in the case of a joint motion-field measurement first sum over all probability amplitudes and then take the square of the resulting expression. Hence, this probability distribution originates from the coherent sum, that is the interference of many probability amplitudes.

When we now make use of the explicit expression Eq. (19.22) the probability distribution reads

$$W(p,|\tilde{\psi}_{\text{field}}\rangle) = \left| \sum_{n=0}^{\infty} \tilde{\psi}_n^* w_n c_n(p) \right|^2 + \left| \sum_{n=0}^{\infty} \tilde{\psi}_{n+1}^* w_{n+1} s_{n+1}(p) \right|^2,$$

or

$$W(p,|\tilde{\psi}_{\text{field}}\rangle) = \left| \sum_{n=0}^{\infty} \tilde{\psi}_n^* w_n c_n(p) \right|^2 + \left| \sum_{n=0}^{\infty} \tilde{\psi}_n^* w_n s_n(p) \right|^2. \tag{19.25}$$

Here we have changed the summation index and have used the fact that $s_0(p) \equiv 0$ as indicated by Eq. (19.23b).

Indeed, the two internal levels contribute in an incoherent way. In contrast, the field states represented by the probability amplitudes $\tilde{\psi}_n^*$ and w_n enter in a coherent way.

Averaged Measurements

We now consider a completely different experiment. The atom traverses the cavity and we only measure the momentum of the atom. We therefore ignore the change of the field due to the atom. In this case we take the trace over the cavity state. When we use photon number states to perform this trace, the resulting probability reads

$$W(p) = \sum_{j=a,b} \sum_{n=0}^{\infty} |\langle j | \langle p | \langle n | \Psi \rangle|^2.$$

In contrast to Eq. (19.24) here we first square and then take the sum. The resulting probability distribution therefore originates from an incoherent sum, that is a sum of probabilities.

When we substitute the explicit representation Eq. (19.22) of the state vector into the above expression for the averaged momentum distribution we arrive at

$$W(p) = \sum_{n=0}^{\infty} |w_n|^2 |c_n(p)|^2 + \sum_{n=0}^{\infty} |w_{n+1}|^2 |s_{n+1}(p)|^2.$$

Again we can shift the summation index in the second sum making use of $s_0(p) = 0$ and can combine the two terms to yield

$$W(p) = \sum_{n=0}^{\infty} |w_n|^2 \left[|c_n(p)|^2 + |s_n(p)|^2 \right]. \tag{19.26}$$

Here we only sum probabilities.

Scattering Situations

One of the initial conditions in the scattering process is the transverse position amplitude $f(x)$ of the atoms. According to Eqs. (19.23) the probability amplitudes c_n and s_n for finding the momentum p are Fourier transforms of the product of the initial position amplitude $f(x)$ and trigonometric functions of the mode function $\sin(kx)$ of the electromagnetic field. We can therefore distinguish two characteristic cases for these Fourier integrals: *(i)* The initial position distribution $|f(x)|^2$ of the atoms is broad compared to the period of the standing wave or *(ii)* the distribution is narrow.

In the first case, which is usually referred to as the Kapitza-Dirac regime, we can assume the distribution of atoms, and hence f, to be essentially constant. The resulting integrals for c_n and s_n are then Bessel functions as shown in Appendix O. Moreover, the periodicity of the standing wave leads to a discreteness of the transferred momentum.

The opposite case when the initial position distribution is narrow compared to the period of the electromagnetic wave is referred to as the Stern-Gerlach regime. Due to the narrowness we can linearize the field mode around the maximum of $f(x)$. Therefore, the atom feels only the gradient of the mode function. Again discrete momenta, that is discrete deflection angles occur. However, they now result from the discreteness of the electromagnetic field.

Whereas in the present chapter we focus on the Kapitza-Dirac regime the next chapter analyses the Stern-Gerlach regime.

19.4.2 Kapitza-Dirac Regime

We now consider the case where the initial atomic position distribution of width L reaches over N periods λ of the standing wave. For the sake of simplicity we assume it is constant, that is,

$$|f(x)|^2 = \frac{1}{L} = \frac{1}{N\lambda}.$$

In this case we can evaluate the functions c_n and s_n explicitly. Moreover, we can calculate the resulting momentum distributions.

Momentum Probability Amplitudes

We start our analysis with c_n which now reads

$$c_n(p) = \frac{1}{\sqrt{2\pi\hbar L}} \int_0^L dx \, \cos\left[\kappa\sqrt{n}\sin(kx)\right] \exp(-ipx/\hbar).$$

When we introduce the new integration variable $\theta = kx$ we find

$$c_n(p) = \frac{1}{\sqrt{\hbar k}} \frac{1}{\sqrt{N}} \frac{1}{2\pi} \int_0^{2\pi N} d\theta \, \cos\left[\kappa\sqrt{n}\sin\theta\right] \exp\left[-i\frac{p}{\hbar k}\theta\right]$$

where we have used the relation $kL = 2\pi N$. We now decompose the integration into N periods, that is

$$c_n(p) = \frac{1}{\sqrt{\hbar k}} \frac{1}{\sqrt{N}} \sum_{\nu=0}^{N-1} \frac{1}{2\pi} \int_{2\pi\nu}^{2\pi(\nu+1)} d\theta \, \cos\left[\kappa\sqrt{n}\sin\theta\right] \exp\left[-i\frac{p}{\hbar k}\theta\right]$$

and shift the integrations with the help of the new variable $\bar{\theta} \equiv \theta - 2\pi\nu$. This yields

$$c_n(p) = \frac{1}{\sqrt{\hbar k}} \frac{1}{\sqrt{N}} \sum_{\nu=0}^{N-1} \frac{1}{2\pi} \int_0^{2\pi} d\bar{\theta} \, \cos\left[\kappa\sqrt{n}\sin\bar{\theta}\right] \exp\left[-i\frac{p}{\hbar k}(\bar{\theta} + 2\pi\nu)\right]$$

We note that the summation over ν does not contain the integration variable $\bar{\theta}$ and we can therefore factor the exponential function $\exp[-2\pi i p/(\hbar k)\nu]$ out of the integral. We therefore arrive at

$$c_n(p) = \frac{1}{\sqrt{\hbar k}} \delta_N^{(1/2)}\left[\frac{p}{\hbar k}\right] \frac{1}{2\pi} \int_0^{2\pi} d\bar{\theta} \, \cos\left[\kappa\sqrt{n}\sin\bar{\theta}\right] \exp\left[-i\frac{p}{\hbar k}\bar{\theta}\right]$$

where we have introduced the function

$$\delta_N^{(1/2)}(\xi) \equiv \frac{1}{\sqrt{N}} \sum_{\nu=0}^{N-1} \exp(-2\pi i\xi\nu).$$

Momentum Quantization

We notice that the function $\delta_N^{(1/2)}(\xi)$ is periodic and has maxima at integer values of ξ. Indeed, at these positions the phase factors are integer multiples of 2π and each term in the sum is unity giving the value $\sqrt{N}$ for the function $\delta_N^{(1/2)}$. Hence, as $N \to \infty$ the maxima of $\delta_N^{(1/2)}$ approach ∞. For non-integer values ξ the individual terms cancel each other.

Hence, this behavior suggests that the function $\delta_N^{(1/2)}$ acts as a comb of δ-functions at integer values of ξ. However, as we show in Appendix P only the square of $\delta_N^{(1/2)}$ acts as a comb of δ-functions. Since the argument of $\delta_N^{(1/2)}$ is $p/(\hbar k)$ we find that the momentum of the atom can take on only multiple integers of the momentum $\hbar k$.

We have therefore found a quantization of the atomic momentum in multiples of the photon momentum. However, the association with the momentum of the light field is slightly misleading. This quantization does not arise from the quantization of the radiation field. It emerges from the periodicity of the potential, namely the mode function of the electromagnetic field.

Since we are interested in momentum distributions and hence probabilities the function $\delta_N^{(1/2)}$ only appears as a square. This guarantees that the quantity $p/\hbar k$ takes on integer values. Therefore, the remaining integral gives the Bessel function $J_p(\kappa\sqrt{n})$ as shown in Appendix O. In this case we find

$$c_n(p) = \frac{1}{\sqrt{\hbar k}} \delta_N^{(1/2)}\left[\frac{p}{\hbar k}\right] \frac{1}{2}\left[1 + (-1)^{p/(\hbar k)}\right] J_{p/(\hbar k)}(\kappa\sqrt{n}) \quad (19.27a)$$

and analogously

$$s_n(p) = \frac{1}{\sqrt{\hbar k}} \delta_N^{(1/2)}\left[\frac{p}{\hbar k}\right] \frac{1}{2i}\left[1 - (-1)^{p/(\hbar k)}\right] J_{p/(\hbar k)}(\kappa\sqrt{n}). \quad (19.27b)$$

We recognize that the probability amplitude $c_n(p)$ is only nonzero for even integer multiples of $\hbar k$. In contrast, the amplitude $s_n(p)$ only takes on nonzero values for odd integer multiples of $\hbar k$. We recall that the amplitudes $c_n(p)$ and $s_n(p)$ are associated with the atom leaving the cavity in the ground or excited state, respectively when it has entered the cavity in the ground state. Therefore, in order to leave it in the ground state it has to undergo an even number of Rabi cycles and thus exchanges an even number of photon momenta. Likewise, an atom leaving in the excited state needs an odd number of momenta exchange in order to make the transition from its initial ground state. This is just another manifestation of the entanglement of the field variables with the momentum of the atom.

Momentum Distribution

We are now in a position to derive explicit expressions for the momentum distributions derived in the preceding section. We start our discussion with the averaged momentum distribution, Eq. (19.26). We can combine the contributions from the atoms leaving the cavity in the ground state or in the excited state corresponding to the probabilities $|c_n(p)|^2$ and $|s_n(p)|^2$ when we note that due to the special form Eq. (19.27) of c_n and s_n the first sum only contains the even multiples of $\hbar k$ whereas the second contribution only contains the odd multiples. However, in both cases the probability is given by the square of the Bessel function. Hence, we arrive at

$$W(p) = \frac{1}{\hbar k} \sum_{\wp=-\infty}^{\infty} \delta(p - \wp \hbar k) \sum_{m=0}^{\infty} W_m J_{p/(\hbar k)}^2(\kappa \sqrt{m})$$

$$= \frac{1}{\hbar k} \sum_{\wp=-\infty}^{\infty} \delta(p - \wp \hbar k) W_\wp$$

where we have introduced the dimensionless and discrete momentum distribution

$$W_\wp [|\psi_{\text{field}}\rangle] \equiv \sum_{m=0}^{\infty} W_m [|\psi_{\text{field}}\rangle] J_\wp^2(\kappa \sqrt{m}) \tag{19.28}$$

of the cavity field in the state $|\psi_{\text{field}}\rangle$ with photon statistics $W_m \equiv |w_m|^2$. Moreover, we have replaced the square of the function $\delta_N^{(1/2)}$ by a comb of δ-functions at integer values of $\hbar k$.

We note that this averaged momentum distribution $W_\wp$ involves only the photon statistics of the cavity field. In particular, it does not bring in the probability amplitudes w_m. In Figs. 19.3, 19.4 and 19.5 we depict the averaged momentum distributions for a number state, a coherent state and a highly squeezed state in the cavity. In order to facilitate a comparison and to bring out most clearly the influence of the photon statistics on the momentum distribution we use the same average number of photons $\bar{m} = 9$ in all three cases. Moreover, we depict the photon statistics scaled in units of $(\wp/\kappa)^2$.

We note that all three momentum distributions are different. For the number state we find oscillations and a dominant maximum at $\wp = \kappa\sqrt{\bar{m}}$. The oscillations are very reminiscent of the Franck-Condon oscillations discussed in Chapter 7. In Sec. 19.5 we show that, indeed, both oscillations have a common origin: Interference in phase space. For the coherent and the squeezed state the oscillations for small momenta have been averaged out but the dominant maximum at $p = \kappa\sqrt{\bar{m}}$ remains. Moreover, for the case of the squeezed state we note that the oscillatory photon statistics manifests itself in the decay of the right side of the maximum.

These phenomena come to light when we use the asymptotic expansion

$$J_\wp(z) \cong \begin{cases} 0 & \text{for } |\wp| > z \\ A_\wp(z) \cos \phi_\wp(z) & \text{for } |\wp| < z \end{cases} \tag{19.29}$$

of the Bessel function derived in Appendix O. Here we have introduced the amplitude

$$A_\wp(z) \equiv \sqrt{\frac{2}{\pi}} \left(z^2 - \wp^2 \right)^{-1/4}$$

and the phase

$$\phi_\wp(z) \equiv \sqrt{z^2 - \wp^2} - \wp \arccos\left(\frac{\wp}{z}\right) - \pi/4.$$

For $|\wp| > z$ the analytical continuation of this expression leads to an exponential decay. Therefore, this region contributes little to the momentum distribution and we have put $J_\wp(z) = 0$ for $\wp > z$.

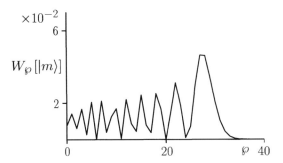

Fig. 19.3: Momentum distribution of atoms scattered off a single mode of a cavity field in a number state $|m\rangle$ with $m = 9$ photons for an interaction parameter $\kappa = 10$. The distribution shows a dominant peak at $\wp \equiv p/(\hbar k) = \kappa\sqrt{m}$ and a strong decay for momenta larger than this critical value. For smaller momenta the distribution is oscillatory. These oscillations result from quantum interference of translational motion. The envelope follows the classical cross section.

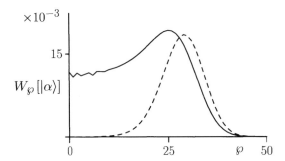

Fig. 19.4: Influence of the photon distribution of the cavity field on the momentum distribution of scattered atoms. The field is in a coherent state $|\alpha\rangle$ of average number of photons $\bar{m} = \alpha^2 = 9$. The Poissonian photon distribution (dashed curve) creates a smooth momentum distribution. The maximum of W_m governs the maximum of $W_\wp$. The right edge of W_m controls the right edge of $W_\wp$.

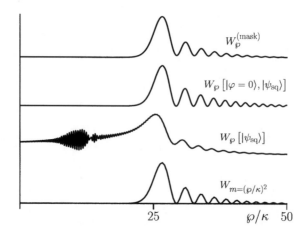

Fig. 19.5: The photon statistics of a squeezed, displaced state and its readout via the momentum distribution of deflected atoms. We scale the photon statistics (lower curve) in units of $(\wp/\kappa)^2$. The curve $W_\wp[|\varphi=0\rangle,|\psi_{sq}\rangle]$ corresponds to a joint measurement of the atomic momentum and the field phase, whereas the distribution $W_\wp[|\psi_{sq}\rangle]$ ignores the field phase. The top curve $W_\wp^{(mask)}$ gives the momentum distribution of atoms filtered by a mask of slit width $d = \lambda/10$ placed at the nodes of the standing wave. The joint measurement strategy gives an adequate readout while ignoring the field phase results in a less effective readout as well as in additional rapid oscillations. Here we have choosen the squeezing parameter $s = 50$ and the displacement parameter $\alpha = 10$.

The expression Eq. (19.29) is our main tool to gain insight into the behavior of the momentum distribution $W_\wp$ determined by the Bessel function sum Eq. (19.28). It brings out two parts of the momentum distribution. For this purpose we substitute Eq. (19.29) into the expression Eq. (19.28) for the momentum distribution. With the help of the trigonometric relation $\cos^2\alpha = (1+\cos(2\alpha))/2$ we indeed find the representation

$$W_\wp \cong W_\wp^{(smooth)} + W_\wp^{(rapid)}$$

where the smooth part

$$W_\wp^{(smooth)} \equiv \frac{1}{\pi\kappa}\sum_{m=m'}^{\infty}\frac{W_m}{\sqrt{m-(\wp/\kappa)^2}}$$

results from the slowly varying prefactor $\mathcal{A}_\wp(\kappa\sqrt{m})$ of the Bessel function expansion and the rapid part

$$W_\wp^{(rapid)} \equiv \frac{1}{\pi\kappa}\sum_{m=m'}^{\infty}\frac{W_m}{\sqrt{m-(\wp/\kappa)^2}}\cos\left[2\phi_\wp(\kappa\sqrt{m})\right]$$

originates from the cosine term in the expansion.

Here we have also used the approximation $J_\wp(z) = 0$ for $\wp > z$. This changes the starting point of the summation from zero to $m' \equiv [(\wp/\kappa)^2]$. The symbol $[z]$ denotes the largest integer smaller than or equal to z.

When we replace the summation in the smooth part by integration we recognize that the momentum distribution

$$W_\wp^{(\text{smooth})} \cong \frac{1}{\pi\kappa} \int_{m'}^{\infty} dm \, \frac{W_m}{\sqrt{m-(\wp/\kappa)^2}}$$

averages the photon statistics with respect to the weight function

$$\tilde{\mathcal{A}}_m = \begin{cases} 0 & \text{for } m < (\wp/\kappa)^2 \\ \left(\pi\kappa\sqrt{m-(\wp/\kappa)^2}\right)^{-1} & \text{for } m > (\wp/\kappa)^2 \end{cases}$$

determined by the amplitude of the Bessel function. This weight has its dominant contribution at $m = (\wp/\kappa)^2$ but due to the square-root dependence it has a broad wing. Therefore the momentum distribution carries information about the photon statistics of the cavity field. However, it is not a perfect read out since the photon statistics is averaged over this broad wing. Especially the features corresponding to small photon numbers get washed out.

We conclude this discussion by noting that the new integration variable $y = \sqrt{m-(\wp/\kappa)^2}$ casts the smooth part into the compact form

$$W_\wp^{(\text{smooth})} \equiv \frac{2}{\pi\kappa} \int_0^{\infty} dy \, W_{m=y^2+(\wp/\kappa)^2}.$$

For a more detailed discussion of the smooth part we refer to Problem 19.1.

Joint Measurements

We now turn to the case of a joint measurement between the momentum of the atom and the field. When we substitute the explicit expressions Eqs. (19.27) for s_n and c_n into the expression for the joint momentum distribution Eq. (19.25) we arrive at

$$W(p, |\tilde{\psi}_{\text{field}}\rangle) = \frac{1}{\hbar k} \sum_{\wp=-\infty}^{\infty} \delta(p - \wp\hbar k) W_\wp[|\tilde{\psi}_{\text{field}}\rangle, |\psi_{\text{field}}\rangle] \quad (19.30)$$

where we have introduced the dimensionless and discrete momentum distribution

$$W_\wp[|\tilde{\psi}_{\text{field}}\rangle, |\psi_{\text{field}}\rangle] \equiv \left| \sum_{m=0}^{\infty} \tilde{\psi}_m^* w_m[|\psi_{\text{field}}\rangle] J_\wp(\kappa\sqrt{m}) \right|^2. \quad (19.31)$$

Nowhere clearer than in the comparison between the averaged and the joint momentum distributions $W_\wp$ and $W(\wp, |\tilde{\psi}_{\text{field}}\rangle)$ do we recognize the power of entanglement. In the averaged distribution we sum the squares of Bessel functions. In the joint distribution we first sum the Bessel functions and then square the result. Since the Bessel functions oscillate between positive and negative values cancelations can occur in the summation over Bessel functions. No such cancelation arises in the averaged distribution since there only the square of the Bessel function appears.

From the asymptotic expansion Eq. (19.29) of the Bessel function we recognize that a dominant maximum occurs when the index is equal to the argument. When we assume that the product $\tilde{\psi}_m^* w_m$ is slowly varying on the scale of the oscillations in the Bessel function the main contribution to the sum arises for $\wp = \kappa\sqrt{m}$, that is $m = (\wp/\kappa)^2$. This yields the approximate expression

$$W(\wp) \cong \mathcal{N} W_{m=(\wp/\kappa)^2}$$

where $\mathcal{N}$ denotes a normalization constant.

This expression clearly shows that in this case the joint momentum distribution follows precisely the photon statistics of the field state in the cavity. This is very different from the case of the averaged momentum distribution where we have to average the photon statistics with respect to the weight function $\tilde{\mathcal{A}}_\wp$.

We illustrate this for the example of a highly squeezed state as the initial state in the cavity and a phase state

$$\left|\tilde{\psi}_{\text{field}}\right\rangle \equiv |\varphi = 0\rangle \equiv \frac{1}{\sqrt{2\pi}} \sum_{n=0}^{\infty} |n\rangle$$

as a reference state. We note, that this reference state satisfies the requirement that the product $\tilde{\psi}_m^* w_m$ is slowly varying since $\tilde{\psi}_m^* = const$. Indeed, we find that the momentum distribution follows precisely the oscillatory photon statistics.

In this context it is interesting to understand why there is such a close connection between the two distributions. We note that this is not true if the reference state is a single photon number state or, likewise, if the initial field state is a photon number state. In this case the summation over the photon number states $|m\rangle$ reduces to a single term and the cancelation due to the oscillatory behavior of the Bessel function does not occur. We obviously need an initial and a reference field state that have broad photon distributions. In the case of the squeezed state and the phase state this condition is satisfied.

There is a simple explanation for this phenomenon of exact read out of photon statistics from momentum statistics. Since we are performing joint measurements, we are selecting very specific atoms from our ensemble. The squeezed state we have chosen has a phase distribution that is centered around 0. Likewise, the phase state corresponds to phase $\varphi = 0$. Hence, the joint measurement selects atoms that have not changed the phase of the field. These are the atoms that have traversed the cavity at the nodes where the electric field vanishes. However, at the nodes the gradient of the field is nonzero. Consequently, the atoms obtain a momentum. The steepness of the gradient depends on the photon number and therefore the momentum transfer depends on the photon number. Since photon numbers are discrete the momentum transfer is discrete. Moreover, the probability for a given deflection angle is determined by the probability to find the corresponding electric field gradient, that is to find the corresponding photon number. Hence, there is a one-to-one correspondence between the momentum distribution and the photon number distribution.

19.4.3 Kapitza-Dirac Scattering with a Mask

In the preceding section we have found that the momentum distribution allows a readout of the photon statistics of the electromagnetic field in the cavity. In particular, we have seen that a joint measurement selecting only those atoms that pass the resonator at the nodes provides a perfect readout. This suggests to replace the joint measurement strategy by a simple mask with narrow slits around the nodes of the field. These slits have to have a period that is half of the period λ of the standing wave and they have to be narrower than $\lambda/2$. In the optical regime it is hard to obtain such mechanical slits. However, a proper choice of the atomic transition together with an extra light field allows to create atomic wave packets on the optical wave length scale. For more details of these experiments we refer to the literature.

Form of the Wave Packet

We therefore assume a de Broglie wave

$$f(x) = \frac{1}{\sqrt{N}} \sum_{\nu=0}^{N-1} g(x - \nu\lambda/2)$$

that is a coherent superposition of N Gaussian wave packets

$$g(x) \equiv \sqrt{\frac{1}{\sqrt{\pi}d}} \exp\left[-\frac{1}{2}\left(\frac{x}{d}\right)^2\right]$$

located at the nodes of the field and with a width $d \ll \lambda/2$. Since the Gaussians are so narrow they have no overlap with each other and therefore the normalization

$$\int_{-\infty}^{\infty} dx\, |f(x)|^2 = 1$$

gives rise to the prefactor $1/\sqrt{N}$ in the definition of $f(x)$.

The Gaussian wave packet $g(x)$ in position space translates itself into a Gaussian wave packet

$$\tilde{g}(p) \equiv \frac{1}{\sqrt{\sqrt{\pi}\Delta p}} \exp\left[-\frac{1}{2}\left(\frac{p}{\Delta p}\right)^2\right]$$

where the width in momentum space is $\Delta p = \hbar/d$.

Momentum Amplitude for Ground State Atoms

We can now substitute this form $f(x)$ of the de Broglie wave into the definition Eqs. (19.23) of $c_n(p)$ and $s_n(p)$ and perform the integration. We start with $c_n(p)$ and after introducing the integration variable $\bar{x} \equiv x - \nu\lambda/2$ find the expression

$$c_n(p) = \frac{1}{\sqrt{N}} \sum_{\nu=0}^{N-1} \exp\left[-i\pi \frac{p}{\hbar k}\nu\right] \times \frac{1}{\sqrt{\sqrt{\pi}\hbar/d}} \times$$

$$\frac{1}{\sqrt{\pi}} \frac{1}{\sqrt{2}d} \int_{-\infty}^{\infty} d\bar{x}\, \exp\left[-\frac{1}{2}\left(\frac{\bar{x}}{d}\right)^2\right] \cos\left[\kappa\sqrt{n}\sin(k\bar{x})(-1)^\nu\right] \exp\left[-i\frac{p\bar{x}}{\hbar}\right]$$

or

$$c_n(p) = \delta_N^{(1/2)}\left[\frac{p}{2\hbar k}\right]\frac{1}{\sqrt{\sqrt{\pi}\Delta p}}\frac{1}{\sqrt{\pi}}\int_{-\infty}^{\infty} dy\, e^{-y^2} \cos\left[\kappa\sqrt{n}\sin(\sqrt{2}dky)\right]\exp\left[-i\frac{p}{\Delta p}\sqrt{2}y\right].$$

We now use the fact that the slit is much narrower than the period, that is $dk = 2\pi d/\lambda \ll 1$. This allows us to linearize the sine function and we can immediately perform the remaining Gauss integrals with the help of the relation

$$\int_0^\infty dy\, e^{-y^2} e^{i\alpha y} = \sqrt{\pi}e^{-(\alpha/2)^2}$$

and arrive at

$$c_n(p) = \delta_N^{(1/2)}\left[\frac{p}{2\hbar k}\right]\frac{1}{2}\frac{1}{\sqrt{\sqrt{\pi}\Delta p}} \times$$

$$\left\{\exp\left[-\frac{1}{2}\left(\frac{p - \kappa\sqrt{n}\hbar k}{\Delta p}\right)^2\right] + \exp\left[-\frac{1}{2}\left(\frac{p + \kappa\sqrt{n}\hbar k}{\Delta p}\right)^2\right]\right\}.$$

When we recall the momentum representation $\tilde{g}(p)$ of the wave packet we find

$$c_n(p) = \delta_N^{(1/2)}\left[\frac{p}{2\hbar k}\right]\frac{1}{2}\left[\tilde{g}(p - \kappa\sqrt{n}\hbar k) + \tilde{g}(p + \kappa\sqrt{n}\hbar k)\right].$$

We first note that the period $\lambda/2$ of the grating instead of λ as in the first example has produced a discreteness of the scattered momenta of integer multiples of $2\hbar k$ rather than $\hbar k$. Moreover, we note that the initial momentum distribution $\tilde{g}$ gets displaced to momenta $\pm\kappa\sqrt{n}\hbar k$. Hence, every number state in the cavity gives rise to a momentum transfer by $\pm\kappa\sqrt{n}\hbar k$. When the initial momentum distribution Δp is smaller than the displacement

$$\delta p \equiv \kappa(\sqrt{n+1} - \sqrt{n})\hbar k \cong \frac{\kappa}{2\sqrt{n}}\hbar k$$

caused by neighboring number states the discreteness of the number states manifests itself in discrete peaks in the momentum distribution of the deflected atoms.

Probability Amplitude for Excited State Atoms

We now turn to the calculation of $s_n(p)$. This is very analogous to the evaluation of $c_n(p)$. However, there is a slight but important difference. That is the reason why we discuss it now. We substitute the de Broglie wave $f(x)$ into the definition of $s_n(p)$ and find

$$s_n(p) = \frac{1}{\sqrt{N}}\sum_{\nu=0}^{N-1}\exp\left[-i\pi\frac{p}{\hbar k}\nu\right] \times \frac{1}{\sqrt{\sqrt{\pi}\hbar/d}} \times$$

$$\frac{1}{\sqrt{\pi}}\frac{1}{\sqrt{2}d}\int_{-\infty}^{\infty} d\bar{x}\, \exp\left[-\frac{1}{2}\left(\frac{\bar{x}}{d}\right)^2\right]\sin\left[\kappa\sqrt{n}\sin(k\bar{x})(-1)^\nu\right]\exp\left[-i\frac{p\bar{x}}{\hbar}\right].$$

When we recall the relation $\sin(-\alpha) = -\sin(\alpha)$ we get

$$s_n(p) = \frac{1}{\sqrt{N}} \sum_{\nu=0}^{N-1} \exp\left[-i\pi\left(\frac{p}{\hbar k}-1\right)\nu\right] \times$$

$$\frac{1}{\sqrt{\sqrt{\pi}\Delta p}} \frac{1}{\sqrt{\pi}} \int_{-\infty}^{\infty} dy\, e^{-y^2} \sin\left[\kappa\sqrt{n}\sin(\sqrt{2}dky)\right] \exp\left[-i\frac{p}{\Delta p}\sqrt{2}y\right].$$

We note that the antisymmetry of the sine function together with the period $\lambda/2$ has created an additional term in the sum over ν. This leads to odd integer multiples of $\hbar k$. Indeed, when we recall the definition of $\delta_N^{(1/2)}$ the above formula reads

$$s_n(p) = \delta_N^{(1/2)}\left[\frac{p}{2\hbar k}-\frac{1}{2}\right] \times$$

$$\frac{1}{\sqrt{\sqrt{\pi}\Delta p}} \frac{1}{\sqrt{\pi}} \int_{-\infty}^{\infty} dy\, e^{-y^2} \sin\left[\kappa\sqrt{n}\sqrt{2}dky\right] \exp\left[-i\frac{p}{\Delta p}\sqrt{2}y\right]$$

where we have also linearized the mode function.

We can perform the remaining Gauss integral and arrive at

$$s_n(p) = \delta_N^{(1/2)}\left[\frac{p}{2\hbar k}-\frac{1}{2}\right]\frac{1}{2i}\left[\tilde{g}(p-\kappa\sqrt{n}\hbar k)-\tilde{g}(p+\kappa\sqrt{n}\hbar k)\right].$$

19.5 Interference in Phase Space

Consider a plane de Broglie wave incident on a single mode cavity field in the nth number state. What is the probability to find the momentum p of an atom scattered from this standing wave? In the Raman-Nath approximation the answer is

$$W(p) = \sum_l \delta(p - l\hbar k) J_l^2(\kappa\sqrt{n}) \tag{19.32}$$

as discussed in the preceding sections. But how can we understand this result from phase space?

A more elementary approach is the concept of interference in phase space. This technique has already proven to provide deep insight into various problems in quantum optics. In the present section we apply it to the problem of the deflection of atoms from a quantized light field and rederive the above probability distribution $W(p)$, Eq. (19.32).

The probability $W(p)$ to find the momentum p is the absolute value squared of the scalar product

$$w(p) \equiv \langle p | \psi \rangle$$

between the momentum state $|p\rangle$ and the quantum state $|\psi\rangle$ of the center-of mass motion of the atom along the standing wave. But what is the quantum state of the motion?

19.5.1 How to Represent the Quantum State?

In the preceding chapter we have derived the state vector $|\Psi\rangle$ of the complete system. However, now we do not need all the details but only focus on the essentials. When the de Broglie wave covering a length L of the standing wave enters the cavity the atomic momentum parallel to the wave vector of the field vanishes. In the Raman-Nath approximation the atom leaving the cavity has acquired the momentum $p = \kappa\sqrt{n}\cos(kx)\hbar k$. Here κ denotes the interaction strength. This suggests to represent the motional state $|\psi\rangle$ of the atoms as the curve $p = \kappa\sqrt{n}\cos(kx)\hbar k$ in phase space. The momentum p and the coordinate x span this phase space. The most elementary version of a phase space distribution for this state is then

$$W_{|\psi\rangle}(x,p) = \frac{1}{L}\delta\left[p - \kappa\sqrt{n}\cos(kx)\hbar k\right].$$

Here we have chosen the normalization such that

$$\int_0^L dx \int_{-\infty}^{\infty} dp\, W_{|\psi\rangle}(x,p) = 1.$$

The momentum eigenstate $|p\rangle$ is obviously a straight line parallel to the position axis. Its phase space representation is then a δ-function and hence,

$$W_{|p\rangle} = \delta(p - p').$$

19.5.2 Area of Overlap

The concept of interference in phase space associates with the probability $W(p)$ the area of overlap

$$\mathcal{A}(p) \equiv \int dx \int dp'\, W_{|p\rangle}(x,p') W_{|\psi\rangle}(x,p')$$

between the momentum eigenstate and the quantum state of the motion represented by the two phase space distribution functions $W_{|p\rangle}$ and $W_{|\psi\rangle}$, respectively.

We distinguish three situations: When $|p| > \kappa\sqrt{n}\hbar k$ there is no overlap. Hence, the resulting probability $W(p)$ vanishes. However, we recognize that the δ-function representation of the motional state is rather primitive. A more complete treatment associates an Airy function behavior as discussed in the context of the uniform asymptotic expansion. Therefore, the probability is exponentially small.

When $p = \kappa\sqrt{n}\hbar k$ the momentum line touches the maxima of the cosine wave tangentially. This gives rise to a large overlap and therefore to a large probability. However, in addition, there is a new effect: Due to the periodicity of the electromagnetic wave there are many such tangential overlaps. All of these overlaps interfere and the relative phase difference becomes important. This leads to the discreteness of the momentum as discussed mathematically in the preceding chapter.

The consequences of interference from the periodic structure come out most clearly in the case of $p < \kappa\sqrt{n}\hbar k$ and one additional feature emerges: In the elementary period $0 < kx < 2\pi$ of the cosine wave we have two different positions

$x_s^{(+)} = k^{-1}\arccos(p/(\kappa\sqrt{n}\hbar k))$ and $x_s^{(-)} = \pi - k^{-1}\arccos(p/(\kappa\sqrt{n}\hbar k))$. The corresponding overlap

$$\mathcal{A}(p) = \int dx \int dp'\,\delta(p'-p)W_{|\psi\rangle}(x,p') = \frac{1}{L}\int dx\,\delta\left[p - \kappa\sqrt{n}\cos(kx)\hbar k\right]$$

at the position $x = x_s^{(+)}$ is

$$\mathcal{A}(p) = \frac{1}{N}\frac{1}{2\pi}\int \frac{dy}{\sqrt{\kappa^2 n(\hbar k)^2 - y^2}}\delta(p-y).$$

Here we have introduced the integration variable $y \equiv \kappa\sqrt{n}\cos(kx)\hbar k$ and have used $kL = 2\pi N$.

After integration we find for the overlap

$$\mathcal{A}(p) = \frac{1}{N}\frac{1}{2\pi}\frac{1}{\sqrt{\kappa^2 n(\hbar k)^2 - p^2}} \equiv \frac{1}{N}A(p).$$

The phase ϕ associated with the overlap is the area enclosed by the cosine wave, the momentum axis and the line $p' = p$ scaled in units of $\hbar$. We therefore find

$$\phi(p) = \frac{1}{\hbar}\left[\int_0^{x_s^{(+)}} dx\,\kappa\sqrt{n}\cos(kx)\hbar k - p x_s^{(+)}\right] - \frac{\pi}{4}.$$

Here we have included a phase shift of $\pi/4$ as discussed in the Chapter 7. We can perform this integral to find

$$\phi(p) = \sqrt{\kappa^2 n - \left(\frac{p}{\hbar k}\right)^2} - \frac{p}{\hbar k}\arccos\left(\frac{p}{\hbar k}\frac{1}{\kappa\sqrt{n}}\right) - \frac{\pi}{4}.$$

The phase associated with the crossing point at $x_s^{(-)} = -\phi(p)$.

Due to the periodicity of the standing electromagnetic field these overlaps repeat themselves. The phase difference between consecutive equivalent phase points is $\lambda p/\hbar = 2\pi p/(\hbar k)$. When the initial de Broglie wave covers N periods of the standing wave we have the interference of N such elementary zones. This is analogous to solid state physics. In particular, it gives rise to the discreteness of the momenta when the number of zones goes to infinity.

19.5.3 Expression for Probability Amplitude

When we combine these ideas we arrive at the following expression for the probability amplitude

$$w(p) = \sum_{\nu=0}^{N-1}\sqrt{\mathcal{A}(p)}\exp\left[i\left(\phi(p) + 2\pi\frac{p}{\hbar k}\nu\right)\right] + \sqrt{\mathcal{A}(p)}\exp\left[-i\left(\phi(p) + 2\pi\frac{p}{\hbar k}\nu\right)\right].$$

We take the phase factors out of the sum and combine it with the factor $1/N$ contained in $\mathcal{A}$. This yields

$$w(p) = \frac{1}{\sqrt{N}}\sum_{\nu=0}^{N-1}\exp\left(-2\pi i\frac{p}{\hbar k}\nu\right)2A(p)\cos\phi(p).$$

When we recall the definition of the function $\delta_N^{(1/2)}$ and the asymptotic expansion

$$J_p(z) \cong \sqrt{\frac{2}{\pi}} \left(z^2 - p^2\right)^{-1/4} \cos\left[\sqrt{z^2 - p^2} - p\arccos(p/z) - \pi/4\right].$$

of the Bessel function discussed in Appendix O we find with $z = \kappa\sqrt{n}$ the momentum probability amplitude

$$w(p) = \delta_N^{(1/2)}\left[\frac{p}{\hbar k}\right] J_{p/(\hbar k)}(\kappa\sqrt{n}),$$

and hence the distribution, Eq. (19.32).

Problems

19.1 Momentum Distribution: Properties of the Smooth Part

In Sec. 19.4 we have derived the momentum distribution of an atom scattered from a quantized light field using the Raman-Nath approximation. We have shown that the exact distribution

$$W_\wp \equiv \sum_{m=0}^{\infty} W_m J_\wp^2(\kappa\sqrt{m})$$

consists of a smooth part

$$W_\wp^{(\text{smooth})} \equiv \frac{2}{\pi\kappa} \int_0^\infty dy\, W_{m=y^2+(\wp/\kappa)^2}$$

and an oscillatory contribution. In this problem we discuss properties of the smooth part and compare the results to the exact expressions.

(a) Show that the smooth momentum distribution is normalized to unity.

(b) Show that the second moment $\langle \wp^2 \rangle$ is connected to the average number of photons $\overline{m}$ by the relation

$$\langle \wp^2 \rangle = \frac{1}{2}\kappa^2\overline{m}.$$

(c) Show that this result is in full agreement with the corresponding expression calculated from the exact momentum distribution.

(d) Derive the relation

$$W_{\wp=0}^{(\text{smooth})} = \frac{1}{\pi\kappa}\frac{1}{\sqrt{\overline{m}}}\left(1 + \frac{3}{8}\frac{\sigma^2}{\overline{m}} + \ldots\right)$$

connecting the contribution at $\wp = 0$ with the normalized variance $\sigma^2 \equiv \overline{m^2}/\overline{m} - \overline{m}$ of the photon distribution.

(e) Prove the scaling law

$$W_\rho^{(\text{smooth})}(\lambda\kappa) = \frac{1}{\lambda} W_{\rho/\lambda}^{(\text{smooth})}(\kappa).$$

What is the physical meaning of this relation?

19.2 Atom Optics in the Dispersive Limit

The interaction of an atomic beam with a standing light field can in the dispersive limit be described by the Hamiltonian

$$\hat{H}_{\text{eff}} = \frac{\hat{P}_x^2}{2M} + \hbar\Omega\left(\hat{a}^\dagger\hat{a} + \frac{1}{2}\right) + \frac{\hbar\omega}{2}\hat{\sigma}_z - \frac{\hbar g_0^2}{\Delta}\sin^2(kx)\left[\hat{\sigma}_z\hat{a}^\dagger\hat{a} + \frac{1}{2}(\hat{\sigma}_z + 1)\right].$$

(a) Show that the state

$$|\Psi_0\rangle = |b\rangle \otimes \sum_{m=0}^{\infty} c_m|m\rangle \otimes \int_{-\infty}^{+\infty} dx\, f(x)\,|x\rangle$$

is transformed into the state

$$|\Psi_w(t)\rangle = |b\rangle \otimes \sum_{m=0}^{\infty} c_m \int_{-\infty}^{+\infty} dx\, e^{-im\frac{g_0^2}{\Delta}\sin^2(kx)t}\, f(x)\,|m, x\rangle$$

due to the interaction (interaction time t) with the light field. The calculation is performed in the interaction picture and the Raman-Nath approximation has been used.

(b) Calculate the position distribution (x-direction) of atoms after they passed the light field.

(c) Calculate the momentum distribution (x-direction) of atoms after they passed the light field for the case that the width of the atomic beam (plane wave) is considerably larger than the wavelength of the light.

(d) What is the position distribution (x-direction) of the atoms after they passed the light field if the field before the interaction was in a coherent state $|re^{i\varphi}\rangle$ and after the interaction it was found to be in a phase state $|\varphi\rangle = 1/\sqrt{2\pi}\sum_n e^{in\varphi}|n\rangle$? Discuss your result.

Hint: For large enough amplitude of the coherent state the Poisson distribution can be approximated by a Gaussian and the sum can be replaced by an integral.

(e) Calculate the field state for the case that the field was in a coherent state before the interaction and after the interaction the atom was found at the position x.

References

Introduction to Atom Optics

For a review of the field of atom optics see

A.P. Kazantsev, G.I. Surdutovich and **V.P. Yakovlev**, *Mechanical Action of Light on Atoms*, World Scientific, Singapore, 1990

S. Stenholm, in: *Laser manipulation of atoms and ions*, edited by E. Arimondo, W.D. Phillips and F. Strumia, Proc. of the International School of Physics "Enrico Fermi", Course CXVIII, Varenna, 1991, North-Holland, Amsterdam, 1992

The experimental situation is summarized by

M. Sigel, C.S. Adams and **J. Mlynek**, *Frontiers in Laser Spectroscopy*, ed. by T.W. Hänsch and M. Inguscio, Proc. of the International School of Physics "Enrico Fermi", Course CXX, Varenna, 1992, North-Holland, Amsterdam, 1993

C.S. Adams, M. Sigel and **J. Mlynek**, *Atom Optics*, Phys. Rep. **240**, 143–210 (**1994**)

and the Special Issues by

J. Mlynek, V. Balykin and **P. Meystre**, *Optics and Interferometry with Atoms*, Appl. Phys. B **54**, (5) (1992)

E. Arimondo and **H.-A. Bachor**, *Atom Optics*, J. Quant. Semicl. Opt. **8**, 495 (1996)

P.R. Berman, *Atom Interferometry*, Adv. At. Mol. Opt. Phys., Academic Press, Boston, 1996

Scattering of Particles from Light

For the original papers on the Stern-Gerlach experiment see

W. Gerlach and **O. Stern**, *Der experimentelle Nachweis der Richtungsquantelung im Magnetfeld*, Z. Physik **9**, 349–352 (1922)

and for the scattering of electrons off an electromagnetic field we refer to

P.L. Kapitza and **P.A.M. Dirac**, *The reflection of electrons from standing light waves*, Proc. Camb. Philos. Soc. **29**, 297–300 (1933)

For the scattering of atoms from a classical light field see

R.J. Cook and **A.F. Bernhardt**, *Deflection of Atoms by a Resonant Standing Electromagnetic Wave*, Phys. Rev. A **18**, 2533–2537 (1987)

A.F. Bernhardt and **B.W. Shore**, *Coherent Atomic Deflection by Resonant Standing Waves*, Phys. Rev. A **23**, 1290–1301 (1981)

For the experiment see

P.E. Moskowitz, P.L. Gould, S.R. Atlas and **D.E. Pritchard**, *Diffraction of an Atomic Beam by Standing-Wave Radiation*, Phys. Rev. Lett. **51**, 370–373 (1983)

For a review of the new field of atom optics as a testing ground for quantum chaos see

M.G. Raizen, in: *Advances in Atomic, Molecular and Optical Physics*, Vol. 41, ed. by B. Bederson and H. Walther, Academic Press, Boston, 1998, pp. 43–81

Atom Optics in Quantized Light Fields

The field of atom optics in quantized light fields has been pioneered by

P. Meystre, E. Schumacher and S. Stenholm, *Atomic Beam Deflection in a Quantum Field*, Opt. Commun. **73**, 443–447 (1989)

P. Meystre, E. Schumacher and E.M. Wright, *Quantum Pendellösung in Atom Diffraction by a Light Grating*, Ann. Physik (Leipzig) **48**, 141–148 (1991)

For an overview of atom optics in a quantized light field see

A.M. Herkommer and W.P. Schleich, *Review of Atom Optics in Quantized Light Fields*, Comments At. Mol. Phys. **33**, 145–157 (1997)

M. Freyberger, A.M. Herkommer, D.S. Krähmer, E. Mayr and W.P. Schleich, in: *Advances in Atomic and Molecular Physics*, Vol. 41, ed. by B. Bederson and H. Walther, Academic Press, Boston, 1999, pp. 143–180

For the readout of the photon statistics by scattering atoms from a quantized light field see

V.M. Akulin, Fam Le Kien and W.P. Schleich, *Deflection of atoms by a quantum field*, Phys. Rev. A **44**, R1462–R1465 (1991)

A. Herkommer, V.M. Akulin and W.P. Schleich, *Quantum Demolition Measurement of Photon Statistics by Atomic Beam Deflection*, Phys. Rev. Lett. **69**, 3298–3301 (1992)

F. Treussart, J. Hare, L. Collot, V. Lefèvre, D.S. Weiss, V. Sandoghdar, J.M. Raimond and S. Haroche, *Quantized Atom-Field Force at the Surface of a Microsphere*, Opt. Lett. **19**, 1651–1653 (1994)

The reconstruction of the field state using the deflection of atoms has been proposed by

M. Freyberger and A.M. Herkommer, *Probing a quantum state via atomic deflection*, Phys. Rev. Lett. **72**, 1952–1955 (1994)

Theory and Experiments on Light Induced Absorption Gratings

D.O. Chudesnikov and V.P. Yakovlev, *Bragg Scattering on Complex Potentials and Formation of Super-Narrow Momentum Distributions of Atoms in Light Fields*, Laser Phys. **1**, 110–119 (1991)

D.S. Krähmer, A.M. Herkommer, E. Mayr, V.M. Akulin, I.Sh. Averbukh, T. van Leeuwen, V.P. Yakovlev and W.P. Schleich, in: *Quantum Optics VI*, ed. by D.F. Walls and J.D. Harvey, Springer, Berlin, 1994

M.K. Oberthaler, R. Abfalterer, S. Bernet, J. Schmiedmayer and A. Zeilinger, *Atom Waves in Crystals of Light*, Phys. Rev. Lett. **77**, 4980–4983 (1996)

20 Wigner Functions in Atom Optics

In the preceding chapter we have given a brief introduction into the field of atom optics in quantized light fields. This treatment has relied on a state vector description. In the present chapter we describe the motion of an atom in a quantized light field using the concept of the Wigner distribution. In particular, we focus on the Stern-Gerlach regime which has not received much attention in Chapter 19. Moreover, we use a nonresonant atom to illustrate the mechanical action of light.

In Sec. 20.1 we briefly summarize our model. We then in Sec. 20.2 start from the Schrödinger equation for the state vector corresponding to the atomic motion and the field and derive the equation of motion for the Wigner function describing the center-of-mass motion of the atom only. The latter turns out to be the sum of the Wigner functions corresponding to the motion of the atom in the individual number states, weighted with the photon statistics. In Sec. 20.3 we give an analytical solution for the equation of motion of the Wigner function for the case when the wavelength of the light is much larger than the de Broglie wavelength of the atom. This regime carries the name Stern-Gerlach regime. From the evolution of the Wigner function we note in Sec. 20.4 that each individual Fock state deflects the atoms in different directions and focuses them at different points. This feature allows us in Sec. 20.5 to read out the photon statistics from the momentum statistics. Finally, in Sec. 20.6 we derive simple expressions for the position and the size of the foci of the individual Fock states based on phase-space considerations.

20.1 Model

We describe the interaction of a nonresonant atom with a quantized electromagnetic field mode shown in Fig. 20.1 by the effective Hamiltonian

$$\hat{H} = \frac{\hat{p}_x^2}{2M} + [\theta(z+L) - \theta(z)]\, g(\hat{x})\, \hat{a}^\dagger \hat{a}. \tag{20.1}$$

The operators $\hat{a}$ and $\hat{a}^\dagger$ are the annihilation and creation operators of the field mode and the coupling constant

$$g(x) \equiv \alpha\, \mathcal{E}_0^2(x) \tag{20.2}$$

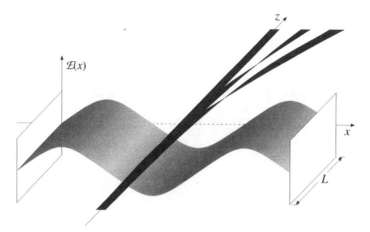

Fig. 20.1: Quantum lens. A beam of nonresonant atoms propagating initially along the z-axis interacts with the light field in the region $-L \leq z \leq 0$. Different Fock states deflect atoms in different directions and focus them at different points.

contains the atomic linear susceptibility

$$\alpha \equiv \frac{\wp^2}{\hbar \Delta}$$

determined by the square of the dipole moment $\wp$ and the detuning Δ, multiplied by the square of the vacuum electric field $\mathcal{E}_0$. For simplicity we assume a rectangular field distribution in z-direction of length L as expressed by the Heaviside step functions $\theta(z)$.

Due to the nonresonant interaction we neglect the internal degrees of freedom of the atom. We treat the transverse motion of the atom quantum mechanically and hence keep the operator nature of the kinetic energy $\hat{p}_x^2/(2M)$. In z-direction we consider the velocities such that the corresponding de Broglie wavelength $\lambda_{\text{dB}} \equiv 2\pi\hbar/(Mv_z)$ is much smaller than the typical dimension of the electromagnetic field estimated by the wavelength λ. Hence we describe the motion along the z-axis classically.

We associate the z-coordinate with time via the relation $z = v_z t - L$ and consider the spatial evolution of the system in this direction as a time evolution given by the Schrödinger equation

$$i\hbar \frac{d|\Psi\rangle}{dt} = \hat{H}|\Psi\rangle. \tag{20.3}$$

For a given L, the velocity v_z defines an interaction time $t_L \equiv L/v_z$.

When the atom enters the field at $t = 0$, the state vector $|\Psi(t = 0)\rangle$ is the direct product of the state $|\psi_{\text{cm}}\rangle$ of the transverse motion with wave function $f(x)$ of the atom and the field state $|\psi_{\text{field}}\rangle$, that is,

$$|\Psi(t=0)\rangle \equiv |\psi_{\text{cm}}\rangle \otimes |\psi_{\text{field}}\rangle \equiv \int_{-\infty}^{\infty} dx\, f(x)|x\rangle \otimes \sum_{n=0}^{\infty} w_n |n\rangle,$$

where the complex coefficients w_n denote the photon probability amplitudes of the field.

The Schrödinger equation Eq. (20.3) with the Hamiltonian Eq. (20.1) couples the degrees of freedom of the field and the motion. As a result of this interaction the states of the field and the atom get entangled which allows us to gain information about one subsystem via the other.

When we substitute the ansatz

$$|\Psi(t)\rangle = \sum_{n=0}^{\infty} w_n \int_{-\infty}^{\infty} dx\ \Psi_n(x,t)\ |x\rangle \otimes |n\rangle$$

for the state vector $|\Psi\rangle$ into the Schrödinger equation Eq. (20.3) with the Hamiltonian Eq. (20.1) we find

$$i\hbar \frac{\partial \Psi_n(x,t)}{\partial t} = \left\{ \frac{\hat{p}_x^2}{2M} + [\theta(t) - \theta(t - t_L)]\, g(x) n \right\} \Psi_n(x,t). \tag{20.4}$$

Each equation of this system is a Schrödinger equation for a particle which moves in the potential

$$U_n(x) \equiv [\theta(z + L) - \theta(z)]\, g(x)\, n$$

created by the individual Fock states. Thus Eq. (20.4) gives the time evolution of the probability amplitude $\Psi_n(x,t)$ of finding the atom at the time t at the position x, and the field in the nth Fock state.

For the time interval $0 \leq t \leq t_L$, that is when the atom is in the standing light field, it feels the potential

$$U_n(x) \equiv g(x)\, n, \tag{20.5}$$

whereas for $t_L < t$, when it is out of the cavity the potential vanishes, that is

$$U_n(x) = 0.$$

Hence, the time evolution of the state $|\Psi\rangle$ of the combined system of the quantum field and the transverse motion is given by the time evolution of the probability amplitudes $\Psi_n(x,t)$ subject to the initial condition $\Psi_n(x, t=0) = f(x)$.

20.2 Equation of Motion for Wigner Functions

The state vector $|\Psi\rangle$ contains the complete information about the field and the atom and our further treatment depends on the specific question we wish to address. For example we might want to consider the properties of the quantum field ignoring the transverse motion of the atom, or vice versa concentrate on the motion while ignoring the field variables. Moreover, we might want to investigate the entanglement between the atomic and field variables via joint measurements as discussed in detail in the preceding chapter.

In the present section, however, we confine ourselves to measurements of the atomic motion only. We are interested in the distribution of atoms in phase space spanned by transverse position x and momentum p. Therefore, our main tool is the Wigner function.

We start from the density matrix

$$\hat{\rho} \equiv |\Psi(t)\rangle\langle\Psi(t)|$$
$$= \sum_{n',n''=0}^{\infty} w_{n'} w_{n''}^* \int_{-\infty}^{\infty} dx' \int_{-\infty}^{\infty} dx'' \, \Psi_{n'}(x',t) \, \Psi_{n''}^*(x'',t) \, |x'\rangle\langle x''| \otimes |n'\rangle\langle n''| \quad (20.6)$$

of the entire system and trace over the field variables to find the reduced atomic density matrix

$$\hat{\rho}_{\text{at}} \equiv \sum_{n=0}^{\infty} \langle n|\hat{\rho}|n\rangle = \sum_{n=0}^{\infty} |w_n|^2 \int_{-\infty}^{\infty} dx' \int_{-\infty}^{\infty} dx'' \, \Psi_n(x',t) \, \Psi_n^*(x'',t) \, |x'\rangle\langle x''|. \quad (20.7)$$

We obtain the Wigner function

$$W(x,p;t) \equiv \frac{1}{2\pi\hbar} \int_{-\infty}^{\infty} d\xi \, \exp\left(-\frac{i}{\hbar}p\xi\right) \langle x + \tfrac{1}{2}\xi|\hat{\rho}_{\text{at}}|x - \tfrac{1}{2}\xi\rangle \quad (20.8)$$

of the transverse atomic motion by substituting the atomic density operator Eq. (20.7) into Eq. (20.8) and arrive at the incoherent superposition

$$W(x,p;t) = \sum_{n=0}^{\infty} |w_n|^2 W_n(x,p;t) \quad (20.9)$$

of the atomic Wigner functions

$$W_n(x,p;t) \equiv \frac{1}{2\pi\hbar} \int_{-\infty}^{\infty} d\xi \, \exp\left(-\frac{i}{\hbar}p\xi\right) \Psi_n^*(x - \tfrac{1}{2}\xi, t) \, \Psi_n(x + \tfrac{1}{2}\xi, t), \quad (20.10)$$

each of which corresponds to the motion of the atom in the potential U_n, Eq.(20.5), given by the nth Fock state. The weight of the nth Wigner function is the photon statistics $|w_n|^2$ of the initial field state.

According to Sec. 3.3 the quantum Liouville equation for $W_n(x,p;t)$ reads

$$\frac{\partial W_n}{\partial t} = -\frac{p}{M}\frac{\partial W_n}{\partial x} + \sum_{l=0}^{\infty} \frac{(-1)^l (\hbar/2)^{2l}}{(2l+1)!} \frac{\partial^{2l+1} U_n(x)}{\partial x^{2l+1}} \frac{\partial^{2l+1} W_n}{\partial p^{2l+1}}. \quad (20.11)$$

Hence we can find the time dependence of the transverse atomic motion by solving either the Schrödinger equation Eq. (20.4) for the probability amplitude Ψ_n or the equation of motion Eq. (20.11) for the corresponding Wigner function W_n.

20.3 Motion in Phase Space

We now consider the phenomenon of deflection and focusing of atoms by a quantized light field in the Stern-Gerlach regime: We take the initial transverse wave packet to be narrow compared to the wavelength of the light field and the interaction time t_L short, so that the wave packet does not change its position in x-direction considerably. In this case the equation of motion for the Wigner function can be solved exactly and immediately provides us via Eq. (20.9) the position or momentum distribution of the atoms.

20.3.1 Harmonic Approximation

We take as the initial condition for Ψ_n a Gaussian wave packet

$$\Psi_n(x, t = 0) \equiv f(x) \equiv (\sqrt{\pi}d)^{-1/2} \exp\left[-\frac{1}{2}\left(\frac{x}{d}\right)^2\right], \quad (20.12)$$

where d denotes the width of the transverse distribution of atoms.

In the case $d \ll \lambda$ we can expand the mode function $g(x)$, Eq. (20.2), around the center $x = 0$ of the wave packet, that is

$$g(x) = g_0 + g_1 x + \frac{1}{2}g_2 x^2 + \ldots \cong g_0 - \frac{1}{2}\frac{g_1^2}{g_2} + \frac{1}{2}g_2\left(x + \frac{g_1}{g_2}\right)^2, \quad (20.13)$$

where $g_1 \equiv dg/dx|_{x=0}$ and $g_2 \equiv d^2g/dx^2|_{x=0}$.

By combining the linear and the quadratic contributions of the potential in a binomial we reduce the problem to the motion in the harmonic potential

$$U_n(x) = U_n(x_f) + \frac{1}{2}M\Omega_n^2 (x - x_f)^2 \quad (20.14)$$

of the displaced harmonic oscillator with minimum $U_n(x_f) \equiv (g_0 - \frac{1}{2}g_1^2/g_2)n$ at $x_f \equiv -g_1/g_2$ and frequency $\Omega_n \equiv \sqrt{ng_2/M}$.

The constant term $U_n(x_f)$ can be omitted from the Schrödinger equation

$$i\hbar \frac{\partial \Psi_n(x,t)}{\partial t} = \left[\frac{\hat{p}_x^2}{2M} + \frac{1}{2}M\Omega_n^2(x - x_f)^2\right]\Psi_n(x,t)$$

for the probability amplitude Ψ_n, since it only results in an irrelevant phase factor.

20.3.2 Motion of the Atom in the Cavity

We can use the well-known Green's function for the harmonic oscillator to solve the equation for the time evolution of Ψ_n. However, in order to gain deeper insight into the dynamics of this wave packet we pursue the more illustrative phase-space approach using the Wigner function.

Wigner Function in Harmonic Approximation

Within the harmonic approximation of Sec. 20.3.1 only the term $l = 0$ in the equation of motion Eq. (20.11) for W_n contributes to the sum, since $\partial^n U/\partial x^n = 0$ for all $n > 2$, and we are left with the classical Liouville equation

$$\frac{\partial W_n}{\partial t} = -\frac{p}{M}\frac{\partial W_n}{\partial x} + \frac{\partial U(x)}{\partial x}\frac{\partial W_n}{\partial p}$$

$$= -\frac{p}{M}\frac{\partial W_n}{\partial x} + M\Omega_n^2(x - x_f)\frac{\partial W_n}{\partial p} \quad (20.15)$$

for the Wigner function.

The harmonic approximation has a big advantage: Since the frequency Ω_n of the harmonic oscillator is independent of the oscillator amplitude all parts of the distribution move in phase space with the same angular velocity, and the Wigner function $W_n(x, p; t)$ at time t follows by a rotation of the initial Wigner function $W_n(x, p; t = 0)$ around the phase space point $(x = x_f; p = 0)$. Hence we find the distribution $W_n(x, p; t)$ at time t from the initial distribution $W_n(x_0, p_0; t = 0)$ by letting each particle of the initial ensemble propagate from the phase space point x_0 and p_0 to x and p along the classical phase space trajectories governed by

$$\dot{\bar{x}} = \frac{\partial H}{\partial \bar{p}} = \frac{\bar{p}}{M} \qquad \dot{\bar{p}} = -\frac{\partial H}{\partial \bar{x}} = -\frac{\partial U_n}{\partial \bar{x}}, \qquad (20.16)$$

where the Hamiltonian

$$H \equiv \frac{\bar{p}^2}{2M} + U_n(\bar{x}) \qquad (20.17)$$

contains the harmonic potential $U_n(\bar{x})$.

Consequently, the Wigner function at time t reads

$$W_n(x, p; t) = \int_{-\infty}^{\infty} dx_0 \int_{-\infty}^{\infty} dp_0 \, \delta[x - \bar{x}(x_0, p_0; t)] \, \delta[p - \bar{p}(x_0, p_0; t)] \, W(x_0, p_0; t = 0)$$

where the two delta functions ensure that the particles travel on the classical trajectories $\bar{x} = \bar{x}(x_0, p_0; t)$ and $\bar{p} = \bar{p}(x_0, p_0; t)$ from the initial phase space point $\bar{x}(x_0, p_0; t = 0) \equiv x_0$ and $\bar{p}(x_0, p_0; t = 0) \equiv p_0$ to x and p.

With the help of the delta functions we can perform the integration and find

$$W_n(x, p; t \leq t_L) = W_n(x_0(x, p; t), p_0(x, p; t); t = 0) \qquad (20.18)$$

where we have expressed the initial phase space points x_0 and p_0 by the final space points x and p via the solutions $\bar{x}$ and $\bar{p}$.

In order to arrive at an explicit expression for the Wigner function Eq. (20.18) for $0 \leq t \leq t_L$ we have to find the solution of the displaced harmonic oscillator

$$M\ddot{\bar{x}} + M\Omega_n^2 \bar{x} = M\Omega^2 x_f \qquad (20.19)$$

resulting from Eq. (20.17) subject to the initial condition $\bar{x}(t = 0) \equiv x_0$ and $\bar{p}(t = 0) \equiv p_0$. It is easy to verify that indeed the functions

$$\bar{x}(x_0, p_0; t) = (x_0 - x_f) \cos(\Omega_n t) + \frac{p_0}{M\Omega_n} \sin(\Omega_n t) + x_f$$
$$\bar{p}(x_0, p_0; t) = p_0 \cos(\Omega_n t) - M\Omega_n (x_0 - x_f) \sin(\Omega_n t) \qquad (20.20)$$

satisfy Eq. (20.19) together with the corresponding initial conditions.

When we express x_0 and p_0 in terms of $x = \bar{x}$ and $p = \bar{p}$ via

$$x_0(x, p; t) = (x - x_f) \cos(\Omega_n t) - \frac{p}{M\Omega_n} \sin(\Omega_n t) + x_f$$
$$p_0(x, p; t) = p \cos(\Omega_n t) + M\Omega_n (x - x_f) \sin(\Omega_n t) \qquad (20.21)$$

the final phase space distribution is given by Eq. (20.18) and Eqs. (20.21).

Example of Gaussian

We now return to the explicit problem of the Gaussian Eq. (20.12). We substitute the initial wave function $\Psi_n(x, t = 0) \equiv f(x)$ of the transverse motion into the definition of the Wigner function Eq. (20.10) and find

$$W_n(x_0, p_0; 0) = \frac{1}{\pi \hbar} \exp\left\{-\left(\frac{x_0}{d}\right)^2 - \left(\frac{d}{\hbar} p_0\right)^2\right\}. \tag{20.22}$$

When we make use of Eq. (20.21) we arrive at

$$W_n(x, p; t \le t_L) = \frac{1}{\pi \hbar} \exp\left[-\frac{1}{d^2}\left((x - x_f)\cos(\Omega_n t) - \frac{p}{M\Omega_n}\sin(\Omega_n t) + x_f\right)^2\right]$$

$$\times \exp\left[-\left(\frac{d}{\hbar}\right)^2 \left(p\cos(\Omega_n t) + (x - x_f)M\Omega_n \sin(\Omega_n t)\right)^2\right]. \tag{20.23}$$

This Wigner function describes the distribution of atoms in phase space when they are still in the standing light wave, that is, for $0 \le t \le t_L$.

20.3.3 Motion of the Atom outside the Cavity

After the interaction, that is for $t > t_L$, there is no potential, that is $U_n \equiv 0$. The free motion of the atoms corresponds to the time evolution of the Wigner function given by the equation

$$\frac{\partial W_n}{\partial t} = -\frac{p}{M}\frac{\partial W_n}{\partial x}. \tag{20.24}$$

Again we can find the solution

$$W_n(x, p; t > t_L) = W_n\left(x - \frac{p}{M}(t - t_L), p; t = t_L\right) \tag{20.25}$$

with the initial condition $W_n(x, p; t = t_L)$ using classical trajectories. Indeed, from Eq. (20.25) we note that the particles move along trajectories of constant momentum, that is, parallel to the x-axis.

With the help of Eq. (20.25) we arrive for the example of the Gaussian initial condition Eq. (20.23) at

$$W_n(x, p; t \ge t_L)$$
$$= \frac{1}{\pi \hbar} \exp\left\{-\frac{1}{d^2}\left[\left(x - \frac{p}{M}(t - t_L) - x_f\right)\cos\varphi_n - \frac{p}{M\Omega_n}\sin\varphi_n + x_f\right]^2\right\}$$
$$\times \exp\left\{-\left(\frac{d}{\hbar}\right)^2 \left[p\cos\varphi_n + \left(x - \frac{p}{M}(t - t_L) - x_f\right)M\Omega_n \sin\varphi_n\right]^2\right\},$$

where $\varphi_n \equiv \Omega_n t_L$.

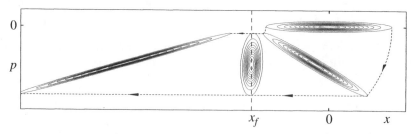

Fig. 20.2: Evolution of the atomic Wigner function inside and outside of the light field for a Fock state $n = 2$. In the light field the initial Gaussian cigar, narrow in momentum but broad in position, rotates due to the evolution in the parabolic potential. Outside the light field the momentum is conserved which results in the shearing of the distribution. The width of the distribution in position reaches a minimum when the cigar stands vertically which corresponds to the focus.

We recall that time translates into the z-coordinate via $z = v_z t - L$. Thus the distribution

$$W_n(x, p; z)$$
$$= \frac{1}{\pi \hbar} \exp\left\{-\frac{1}{d^2}\left[\left(x - \frac{p}{Mv_z}z - x_f\right)\cos\varphi_n - \frac{p}{M\Omega_n}\sin\varphi_n + x_f\right]^2\right\}$$
$$\times \exp\left\{-\left(\frac{d}{\hbar}\right)^2\left[p\cos\varphi_n + \left(x - \frac{p}{Mv_z}z - x_f\right)M\Omega_n\sin\varphi_n\right]^2\right\} \quad (20.26)$$

describes the quantum state of the center-of-mass motion of the atom after it has traversed the cavity in the nth Fock state.

20.3.4 Snap Shots of the Wigner Function

In Fig. 20.2 we show the contour lines of the initial Gaussian Wigner cigar Eq. (20.22) at the origin of phase space aligned along the x-axis. Due to the motion of the atom in the harmonic oscillator potential U_n the Gaussian cigar rotates in phase space around the point $(x = x_f, p = 0)$ by the angle $\varphi_n = \Omega_n t_L$, as given by Eq. (20.23).

The following free evolution given by Eq. (20.26) is depicted in Fig. 20.2 for three typical times, that is, for three typical positions outside of the light field. We note that the width of the Wigner function in x-variable first decreases and then increases. It reaches a minimum when the cigar crosses the phase space line $x = x_f$. This is the physical origin of the focusing of the atoms.

So far we have only considered the motion of the atom in the potential U_n given by the nth Fock state. In the case of a field state consisting of a superposition of Fock states the Wigner function $W(x, p; z)$ of the atomic motion is the incoherent sum Eq. (20.9) of the Wigner functions $W_n(x, p; t)$ weighted with the photon statistics $|w_n|^2$. In Fig. 20.3 we show the Wigner function $W(x, p; t)$ at the exit of the cavity, whereas in Fig. 20.4 we depict its contour lines for various times t, that is, for various positions z.

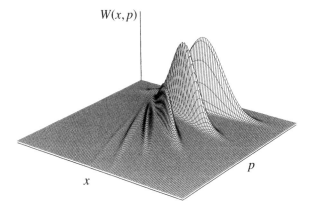

Fig. 20.3: Wigner function of the atom at the exit of the cavity. The field is in a coherent state of average number of photons $\bar{n} = 1$. Due to different angular velocities for different photon numbers the Wigner function of the atom has split into several parts during the interaction with the quantized light field.

The left column of Fig. 20.4 displays the evolution of the initial Gaussian cigar *in* the standing light field. We note that due to the n-dependence of the frequency Ω_n of the oscillator potential, that is, the angular velocity in phase space, the Gaussian cigar splits into many cigars – each of which is weighted with the photon statistics.

Figure 20.4(c) depicts the distribution of atoms at the exit of the cavity and serves as the initial distribution for the free evolution shown on the right column for various positions z *outside* of the light field. Each cigar experiences the shearing effect discussed in Fig. 20.2. Moreover, different cigars move through the focal line $x = x_f$ at different times, which correspond to different positions in z-direction.

20.4 Quantum Lens

So far we have analyzed the distribution of transverse position and momentum in its dependence on the coordinate z. Now we consider the spatial distribution $\mathcal{W}(x, z)$ of atoms in the x-z plane. In particular, we concentrate on the focusing effect and calculate the focal length of the quantum lens.

20.4.1 Distributions of Atoms in Space

We find the spatial distribution $\mathcal{W}(x, z)$ by integrating $W(x, p; z)$ given by Eqs. (20.9) and (20.26) over p and arrive at

$$\mathcal{W}(x, z) = \sum_{n=0}^{\infty} |w_n|^2 \mathcal{W}_n(x, z), \qquad (20.27)$$

where

$$\mathcal{W}_n(x, z) \equiv \int_{-\infty}^{\infty} dp\, W_n(x, p; z)$$

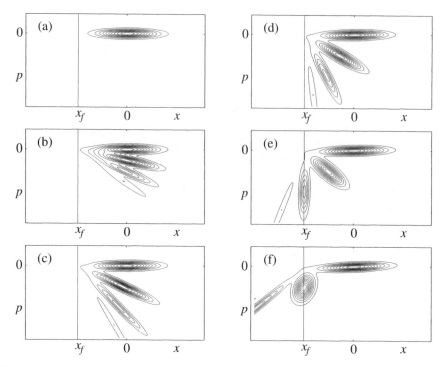

Fig. 20.4: Time evolution of the Wigner function of the atom inside (a–c) and outside of the cavity (d–f). The cavity field is in a coherent state of average photon number $\bar{n} = 1$. Here we display the contour lines of the four dominant Wigner peaks corresponding to the vacuum $n = 0$ and the first three excited Fock states $n = 1$, $n = 2$ and $n = 3$. In (a) we display the initial distribution, and in (b–c) we show the rotation and the splitting due to the interaction with the light field. During the free evolution shown in (d–f) the individual cigars go through $x = x_f$ at different times corresponding to different focal distances.

is the spatial distribution of atoms due to the interaction with the nth Fock state.

After evaluation of the Gaussian integrals we find

$$\mathcal{W}_n(x, z) = \frac{1}{\sqrt{\pi} D_n(z)} \exp\left\{-\frac{1}{D_n^2(z)}\left[x - x_f\left(1 - \cos\varphi_n + \frac{\Omega_n}{v_z} z \sin\varphi_n\right)\right]^2\right\} \quad (20.28)$$

with

$$D_n(z) \equiv \left[\left(\frac{\hbar}{dM\Omega_n}\right)^2 \left(\frac{\Omega_n}{v_z} z \cos\varphi_n + \sin\varphi_n\right)^2 + d^2\left(\cos\varphi_n - \frac{\Omega_n}{v_z} z \sin\varphi_n\right)^2\right]^{1/2}. \quad (20.29)$$

In Fig. 20.5 we show the contour lines of the distribution $\mathcal{W}(x, z)$ of atoms and a cut along the focal line $x = x_f$ for a coherent state of average number of photons $\bar{n} = 1$. The initial atomic beam splits up into a number of partial beams due to the deflection of atoms by the individual Fock states. Moreover, we find that each partial beam corresponding to the nth Fock state focuses at the individual point $(x_f, \mathcal{F}_n)$. The cut through the distribution $\mathcal{W}(x, z)$ along the focal line $x = x_f$ depicted in the

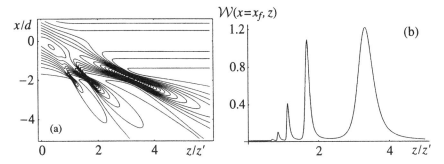

Fig. 20.5: Contour plot (left) of the probability $\mathcal{W}(x,z)$ of finding an atom at the point with coordinates x and z. The Gaussian atomic beam centered at $x=0$ leaves the cavity at $z=0$. The field is in a coherent state of average photon number $\bar{n}=1$. The undeflected and unfocused partial wave associated with the cavity vacuum state represents the profile of the incident beam. The deflected partial waves associated with different photon states of the field focus along the line $x=x_f$. The intensity of atoms along this line is shown on the right. The parameters used here are $\varphi_{\bar{n}} = 0.15$, $\hbar/(d^2 M \omega_{\bar{n}}) = 20$, and $x_f/d = -2$. The values of z are scaled in units of $z' \equiv v_z/\Omega_{\bar{n}}$.

right part of Fig. 20.5 shows the weight of each partial beam reflecting the photon statistics of the initial field state.

20.4.2 Focal Length and Deflection Angle

The distribution $\mathcal{W}$ given by Eqs. (20.27), (20.28) and (20.29) allows us to determine the focal length $\mathcal{F}_n$ and the deflection angle θ_n for each partial wave. Here we assume that we can separate the individual beams as in the example shown in Fig. 20.5.

Exact Expressions

We obtain the focal length $\mathcal{F}_n$ from Eq. (20.29) as the position z where $D_n(z)$ reaches the minimum. For $z > 0$, that is, outside of the light field this corresponds to the condition

$$\cos \varphi_n - \frac{\Omega_n}{v_z} z \sin \varphi_n = 0$$

which yields

$$z_f \equiv \mathcal{F}_n \equiv \frac{v_z}{\Omega_n \tan \varphi_n}. \qquad (20.30)$$

When the atom leaves the light field at $z=0$, the center of the wave packet is located at

$$x_n = x_f (1 - \cos \varphi_n), \qquad (20.31)$$

as predicted by Eq. (20.28).

The packet passes through the focus $x = x_f$ and $z = \mathcal{F}_n$ and hence experiences a deflection by an angle

$$\theta_n \equiv \arctan\left(\frac{x_f - x_n}{\mathcal{F}_n}\right) = \arctan\left(x_f \frac{\Omega_n}{v_z} \sin \varphi_n\right), \qquad (20.32)$$

where we have made use of Eqs. (20.30) and (20.31).

The focal distance $\mathcal{F}_n$ given by Eq. (20.30) depends on the curvature g_2 of the potential U_n, since $\Omega_n = \sqrt{ng_2/M}$ and $\varphi_n = \Omega_n L/v_z$ depend on g_2. In contrast, the deflection angle θ_n depends on the slope g_1 as well as on the curvature g_2.

Small Angle Approximation

However in the limit of $\varphi_n \ll 1$, the angle θ_n becomes independent of g_2, that is,

$$\theta_n \cong \frac{n}{N},$$

where the quantity

$$\frac{1}{N} \equiv \frac{g_1 L}{Mv_z^2}$$

is the deflection due to the first Fock state.

Hence the deflection angle depends linearly on the number n. Since n can take on only discrete values the deflection angles can only take on discrete values. The granular structure of the photon field manifests itself in the deflection angle.

Moreover, in this limit the focal length reads

$$\mathcal{F}_n \cong \frac{1}{n}\mathcal{F}_1,$$

where

$$\mathcal{F}_1 \equiv \frac{Mv_z^2}{g_2 L}$$

is the focal length of the Fock state $|n=1\rangle$.

Thus the focal distance depends on the photon number n and the discreteness of n manifests itself in a discreteness of $\mathcal{F}_n$. However, in contrast to the deflection angle θ_n, which increases linearly with n, the focal length decreases inversely with n.

20.5 Photon and Momentum Statistics

We can obtain the photon statistics by making use of the strong correlation between the field state and the momentum distribution $\widetilde{W}(p)$ of the atoms after they have left the field. We find this distribution by integrating the Wigner function $W(x, p : t)$ over x which yields

$$\widetilde{\mathcal{W}}(p) = \sum_{n=0}^{\infty} |w_n|^2 \, \widetilde{\mathcal{W}}_n(p).$$

Here the momentum distribution

$$\widetilde{\mathcal{W}}_n(p) \equiv \frac{1}{\sqrt{\pi}\widetilde{D}_n} \exp\left[-\left(\frac{d}{\hbar \widetilde{D}_n}\right)^2 (p - M\Omega_n x_f \sin \varphi_n)^2\right] \qquad (20.33)$$

of the atoms due to the interaction with the nth Fock state has a width

$$\widetilde{D}_n \equiv \left[\cos^2 \varphi_n + \left(\frac{d^2 M \Omega_n}{\hbar} \sin \varphi_n\right)^2\right]^{1/2}.$$

20.5 Photon and Momentum Statistics

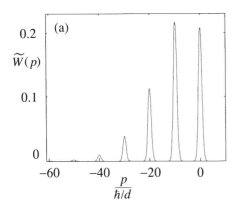
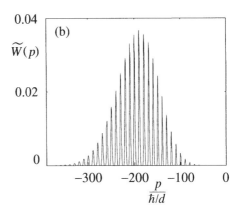

Fig. 20.6: Readout of the photon statistics from the momentum distribution of atoms scattered by a quantized electromagnetic field in a coherent state of average number of photons $\bar{n} = 1$ (a) and $\bar{n} = 20$ (b). We can identify the contributions of the individual Fock states comprising the coherent state. The envelope of the momentum distribution gives the Poissonian photon statistics. Here we have chosen for (a) $\varphi_{\bar{n}} = 1/\sqrt{10}$, $\hbar/(d^2 M \Omega_{\bar{n}}) = \sqrt{10}$, and for (b) $\varphi_{\bar{n}} = \sqrt{2}$, $\hbar/(d^2 M \Omega_{\bar{n}}) = 1/\sqrt{2}$ and in both cases $x_f/d = -100$.

This expression clearly shows that an atom which has entered the light field in the nth Fock state with the average momentum $\langle p \rangle = 0$ leaves the cavity with the average momentum

$$p_n \equiv M \Omega_n x_f \sin \varphi_n. \tag{20.34}$$

The motion in the free field region conserves the momentum and hence the distribution is independent of the position z. It is this momentum transfer which gives rise to the deflection angle θ_n, Eq. (20.32), via the relation

$$\theta_n \equiv \arctan \frac{p_n}{M v_z}.$$

When the difference $\Delta p_n \equiv p_{n+1} - p_n$ in the transferred momentum due to two neighboring Fock states is larger than the momentum uncertainty given by the width $\widetilde{D}_n$ of the two corresponding Gaussians, we can resolve the contribution of each individual Fock state. Since each Gaussian is weighted with the photon statistics, the momentum distribution in this case is a complete readout of the photon statistics as shown in Fig. 20.6.

We conclude this section by noting that in deriving the above results we have expanded the coupling constant $g(x)$ around the center of the wave packet. Hence this solution is only valid provided the wave packet has not moved considerably due to the interaction with the light field, that is, it has not moved into a regime in which the displaced harmonic oscillator potential is not a good approximation to the potential U_n.

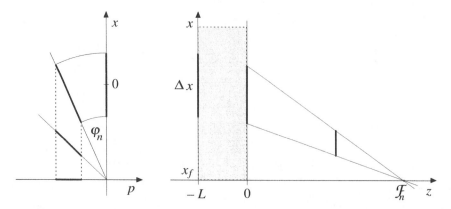

Fig. 20.7: Quantum lens due to the nth Fock state (right) and evolution of the atomic beam in phase space (left). Here we only consider an interval Δx of the x-axis. Due to the interaction with the standing light field this interval is rotated by an angle φ_n. In the limit of small angles $\varphi_n \ll 1$ the displacement in x-direction is small compared to Δx. During the free evolution the considered points in phase space remain on a straight line and cross the focal line $x = x_f$ at the same time t_f. As a result the width Δx of the wave packet decreases until it vanishes at $x = x_f$.

20.6 Heuristic Approach

So far we have discussed the complete phase-space dynamics of the atomic motion in the quantized light field. Now we illustrate this approach in simple geometrical terms. In particular, we rederive the expression Eq. (20.30) for the focal distance $\mathcal{F}_n$ of the nth Fock state quantum lens starting from the pictorial sketch Fig. 20.7 and calculate the size of the focal spot.

20.6.1 Focal Length

Figure 20.7 explains the essential points of the process of focusing. On the right-hand side we summarize the setup, whereas on the left-hand side we show the evolution of the atomic wave packet in phase space.

Geometrical Picture

The atomic wave of transverse width Δx, depicted here by a fat line, enters the light field (shaded area) at $z = -L$ and leaves it slightly displaced and compressed at $z = 0$. Due to the interaction with the light in the nth Fock state, different parts of the wave packet gain different transverse momenta. As a result of the free evolution outside of the cavity the beam focuses at the point $z = \mathcal{F}_n$.

This process is most easily understood from the dynamics in phase space. For the sake of simplicity we take the initial phase space distribution to be a line distribution of length Δx centered at $x = 0$ with vanishing momentum. Since the potential formed by the nth Fock state is harmonic, the fat line representing the initial distribution rotates in phase space by an angle φ_n around $(x = x_f, p = 0)$. This rotation in phase

20.6 Heuristic Approach

space is the origin of the slight shift and compression of the wave packet. Moreover, different parts of the wave packet acquire different momenta which are proportional to their coordinates.

Outside of the cavity the momentum of each part is conserved. As a result all points of the rotated line distribution pass the point $x = x_f$ at the same time t_f as shown by the fat line along the momentum axis. Since time corresponds to the z-coordinate via the relation $z = v_z t - L$, the focal length $\mathcal{F}_n$ of the nth Fock state reads $\mathcal{F}_n = v_z t_f - L$.

Mathematical Treatment

We now rederive the focal distance $\mathcal{F}_n$, Eq. (20.30), using this picture. For this purpose we consider the time evolution of phase space points with vanishing initial momentum, that is, $p(t = 0) = 0$. This set of points is shown on the left hand side of Fig. 20.7 by the fat line centered at $x = 0$.

According to Eq. (20.20) at a later time $t \leq t_L$ these points have moved to the new positions

$$x(t) = (x_0 - x_f) \cos(\Omega_n t) + x_f$$
$$p(t) = -M\Omega_n (x_0 - x_f) \sin(\Omega_n t)$$

and now lie on the line

$$p(x) = -M\Omega_n \tan(\Omega_n t)(x - x_f) \qquad (20.35)$$

in phase space which goes through the point $x = x_f$ at the angle $\varphi_n = \Omega_n t$ with respect to the x-axis.

After the atom has left the light field it moves freely and conserves its momentum

$$p_L \equiv p(t \geq t_L) = -M\Omega_n \tan \varphi_n (x_L - x_f),$$

where $x_L \equiv x(t_L)$. The position x evolves according to

$$x(t) = \frac{p_L}{M}(t - t_L) + x_L = [1 - \Omega_n \tan \varphi_n (t - t_L)](x_L - x_f) + x_f.$$

Therefore, all points on the line Eq. (20.35) reach the focal spot $x = x_f$ at the same time

$$t_f \equiv \frac{1}{\Omega_n \tan \varphi_n} + t_L.$$

The corresponding distance in z-direction yields the desired focal length

$$\mathcal{F}_n \equiv v_z t_f - L = \frac{v_z}{\Omega_n \tan \varphi_n}.$$

This expression is identical to Eq. (20.30) obtained in Sec. 20.4.2 from the spatial distribution $\mathcal{W}(x, z)$ of the atoms.

20.6.2 Focal Size

So far we have considered the extreme of geometrical optics, that is, a line distribution in phase space which is infinitely narrow in momentum. But what happens to the focus in the framework of wave optics, that is, when we admit a finite uncertainty δp in momentum?

This uncertainty δp results in a broadening of the focal point and gives a width to the focus. In order to derive an expression for the focal size we now consider the time evolution of the phase space point $(x = 0, p = \delta p)$. The time t'_f at which this point crosses the line $x = x_f$ provides via the relation $z'_n = v_z t'_f - L$ the width $\delta \mathcal{F}_n \equiv z'_n - \mathcal{F}_n$.

According to Eq. (20.20) at time t_L the point $(x = 0, p = \delta p)$ has arrived at

$$x'_L = -x_f \cos\varphi_n + \frac{\delta p}{M\Omega_n} \sin\varphi_n + x_f \qquad (20.36)$$

and

$$p'_L = \delta p \cos\varphi_n + x_f M\Omega_n \sin\varphi_n. \qquad (20.37)$$

We now find the time t'_f, at which this point passes the focal line $x = x_f$. Since the momentum p'_L is conserved during the free evolution we find

$$\frac{p'_L}{M}(t'_f - t_L) + x'_L = x_f$$

and hence

$$t'_f = t_L + \frac{M}{p'_L}(x_f - x'_L).$$

With the relation $v_z t_L = L$ this time translates into the z-coordinate

$$z'_n \equiv v_z t'_f - L = \frac{M v_z}{p'_L}(x_f - x'_L).$$

We make use of Eqs. (20.36) and (20.37) to express $(x_f - x'_L)$ and p'_l which yields

$$z'_n = \frac{v_z}{\Omega_n \tan\varphi_n}\left[1 - \frac{\delta p}{M\Omega_n x_f \sin\varphi_n}\frac{1}{\cos\varphi_n}\left(1 + \frac{\delta p}{M\Omega_n x_f \sin\varphi_n}\cdot\cos\varphi_n\right)^{-1}\right]$$

and with the help of the formulas Eqs. (20.30) and (20.34) for the focal length $\mathcal{F}_n$ and the transferred momentum p_n this equation takes the from

$$z'_n = \mathcal{F}_n\left(1 - \frac{\delta p}{p_n}\frac{1}{\cos\varphi_n}\frac{1}{1 + \frac{\delta p}{p_n}\cdot\cos\varphi_n}\right).$$

Hence the width $\delta\mathcal{F}_n$ of the nth focus reads

$$\delta\mathcal{F}_n \equiv \mathcal{F}_n - z'_n = \frac{\delta p}{p_n}\cdot\mathcal{F}_n\frac{1}{\cos\varphi_n}\frac{1}{1 + \frac{\delta p}{p_n}\cos\varphi_n} \qquad (20.38)$$

which in the limit $\delta p \cos\varphi_n \ll p_n$ reduces to

$$\delta\mathcal{F}_n \cong \frac{\delta p}{p_n}\mathcal{F}_n.$$

The spread δp in momentum translates into a broadening of the focus. Thus when we want to resolve the individual foci caused by the individual Fock states the focal size has to be smaller than the focal separation. This condition puts severe constraints on the parameters of the lens, as discussed in Problem 20.1.

Problems

20.1 Quantum Lens in Raman-Nath Approximation

Show the following properties of the quantum lens:

(a) The spatial distribution $\mathcal{W}_n^{RN}(x,z)$ of atoms in Raman-Nath approximation reads

$$\mathcal{W}_n^{(RN)}(x,z) = \frac{1}{\sqrt{\pi}D_n^{(RN)}(z)} \exp\left[-\left(\frac{nz+Nx}{ND_n^{(RN)}(z)}\right)^2\right] \quad (20.39)$$

with

$$D_n^{(RN)}(z) = \left[\left(\frac{\hbar}{dMv_z}z\right)^2 + d^2\left(1+\frac{n}{N}\frac{z}{x_f}\right)^2\right]^{1/2}. \quad (20.40)$$

To derive this result start from the Hamiltonian Eq. (20.1) and neglect the kinetic energy operator while the atom is in the cavity but retain it in the free motion.

(b) Show that Eq. (20.28) reduces to Eq. (20.39) in the limit of small rotation angles $\varphi_n = \Omega_n t_L \ll 1$, such that $\cos\varphi_n \cong 1$ and $\sin\varphi_n \cong \varphi_n$. However, the width D_n, Eq. (20.29), reduces to

$$D_n(z) \cong \left\{\left[\frac{\hbar}{dMv_z}(z+L)\right]^2 + d^2\left(1+\frac{nz}{Nx_f}\right)^2\right\}^{1/2}$$

and hence contains an additional term $\hbar L/(dMv_z)$. This contribution reflects the spreading of the wave packet during the motion of the atom in the harmonic oscillator potential. Since the Raman-Nath approximation neglects the kinetic part of the Hamiltonian compared to the electromagnetic interaction energy it cannot account for this term.

(c) Derive the size

$$\delta\mathcal{F}_n \cong \frac{M\hbar v_z^3}{dg_1 g_2 L^2}\frac{1}{n^2},$$

of the focus in the limit of the Raman-Nath approximation.

(d) What is the condition to resolve the individual foci?

Hint: See Averbukh et al. (1994).

References

Quantum Lens

For the quantum lens in atom optics see

I.Sh. **Averbukh**, V.M. **Akulin** and W.P. **Schleich**, *A Quantum Lens for Atomic Waves*, Phys. Rev. Lett. **72**, 437–441 (1994)

E. **Mayr**, D. **Krähmer**, A.M. **Herkommer**, V.M. **Akulin**, W.P. **Schleich** and I.Sh. **Averbukh**, *Phase Space as Arena of Atomic Motion in a Quantized Light Field*, Acta Phys. Pol. A **86**, 81–95 (1994)

P. **Domokos**, P. **Adam**, J. **Janszky** and A. **Zeilinger**, *Atom de Broglie Wave Deflection by a Single Cavity Mode in the Few-Photon Limit: Quantum Prism*, Phys. Rev. Lett. **77**, 1663–1666 (1996)

B. **Rohwedder** and M. **Orszag**, *Quantized light lenses for atoms: The perfect thick lens*, Phys. Rev. A **54**, 5076–5084 (1996)

Discreteness of Photon Field

For the use of the deflection of atoms to demonstrate the granular structure of the photon field we refer to

A. **Herkommer**, V.M. **Akulin** and W.P. **Schleich**, *Quantum Demolition Measurement of Photon Statistics by Atomic Beam Deflection*, Phys. Rev. Lett. **69**, 3298–3301 (1992)

F. **Treussart**, J. **Hare**, L. **Collot**, V. **Lefèvre**, D.S. **Weiss**, V. **Sandoghdar**, J.M. **Raimond** and S. **Haroche**, *Quantized Atom-Field Force at the Surface of a Microsphere*, Opt. Lett. **19**, 1651–1653 (1994)

H. **Mack**, S. **Meneghini** and W.P. **Schleich**, *Atom Optics and the Discreteness of Photons*, in: *Quantum Optics of Small Structures*, edited by B. Lenstra, T.D. Visser and K.A.H. van Leeuwen, Proc. Kon. Ned. Akad. Wetensch., 169–183 (2000)

A Energy Wave Functions of Harmonic Oscillator

In this appendix we discuss the solutions $u(\xi)$ of the time independent Schrödinger equation

$$u''(\xi) + (2\eta - \xi^2)u(\xi) = 0. \qquad (A.1)$$

for the harmonic oscillator. Here ξ is the dimensionless position variable and the two primes denote differentiation with respect to ξ.

For an arbitrary value of η the general solutions of this differential equation are the parabolic cylinder functions. However, these functions do not have the correct asymptotic behavior for large values of ξ. In particular, we have to guarantee the normalizability and hence we must consider solutions which vanish when ξ goes to plus or minus infinity. Only for particular values of η, namely $\eta_m = m + 1/2$ do the parabolic cylinder functions exhibit the right asymptotic behavior. In this case these solutions reduce to polynomials namely Hermite polynomials H_m. The corresponding values η_m are the energy eigenvalues.

In this appendix we first derive these solutions using a polynomial ansatz. We then discuss their asymptotics in the case of large quantum numbers.

A.1 Polynomial Ansatz

To gain some insight into the solution of this differential equation we first consider it in the asymptotic limit $\xi \to \infty$ where we can neglect the parameter η in comparison to ξ. In this case the differential equation (A.1) has the form

$$u''(\xi) - \xi^2 u(\xi) = 0.$$

As a normalizable solution of this equation we try

$$u(\xi) = e^{-\frac{1}{2}\xi^2}.$$

By differentiating this function twice we find

$$u''(\xi) = e^{-\frac{1}{2}\xi^2}(\xi^2 - 1) \cong \xi^2 e^{-\frac{1}{2}\xi^2} = \xi^2 u(\xi).$$

Here we have neglected the term unity compared to ξ^2.

Indeed, this ansatz satisfies the differential equation in the limit of $\xi \to \infty$. This suggests an ansatz for the exact solution of the form

$$u(\xi) = \mathcal{N}\, H(\xi)\, e^{-\frac{1}{2}\xi^2} \tag{A.2}$$

where $H(\xi)$ is a function to be determined.

We substitute the ansatz (A.2) into the differential equation (A.1) and obtain

$$H'' - 2\xi H' + 2(\eta + \tfrac{1}{2}) H = 0. \tag{A.3}$$

We solve this equation using an ansatz

$$H(\xi) = \sum_{k=0}^{\infty} c_k\, \xi^k \tag{A.4}$$

in the form of a power series. For the first and second derivatives we obtain

$$H'(\xi) = \sum_{k=0}^{\infty} k\, c_k\, \xi^{k-1}$$

and

$$H''(\xi) = \sum_{k=0}^{\infty} k(k-1)\, c_k\, \xi^{k-2} = \sum_{k=2}^{\infty} k(k-1)\, c_k\, \xi^{k-2} = \sum_{j=0}^{\infty} (j+2)(j+1)\, c_{j+2}\, \xi^j,$$

where in the last equation we have shifted the summation index according to $j = k-2$.

We insert these power series into Eq. (A.3) and find

$$\sum_{j=0}^{\infty} \left[(j+2)(j+1)\, c_{j+2} - 2(j - \eta - \tfrac{1}{2})\, c_j \right] \xi^j = 0.$$

Since the polynomials ξ^j for $j = 0, 1, 2, \ldots$ are linearly independent functions, their coefficients in square brackets have to vanish. This condition yields the recurrence relation

$$c_{j+2} = \frac{2(j - \eta - \tfrac{1}{2})}{(j+2)(j+1)}\, c_j. \tag{A.5}$$

This recurrence relation is of the type

$$c_{j+2} = d_j\, c_j$$

and couples only terms with odd or even j. We repeatedly apply this formula and obtain

$$c_{j+2} = d_j\, c_j = d_j\, d_{j-2}\, c_{j-2} = \ldots = d_j\, d_{j-2} \cdots d_0\, c_0.$$
$$\phantom{c_{j+2} = d_j\, c_j = d_j\, d_{j-2}\, c_{j-2} = \ldots = d_j\, d_{j-2} \cdots d_}{}_1\, {}_1$$

Depending on whether j is even or odd the series starts with c_0 or c_1 and involves only even or odd powers of ξ.

In order to investigate the asymptotic behavior of H we now consider large powers of ξ which means large values of j. In this limit the recurrence relation (A.5) simplifies to

$$c_{j+2} \cong \frac{2}{j+2}\, c_j$$

which leads to
$$c_\nu \cong \frac{1}{\nu!} \quad \text{for} \quad j = 2\nu \gg 1.$$
Hence, the functions $H(\xi)$ have the asymptotic behavior
$$H(\xi) = \sum_j^\infty c_j \, \xi^j \sim \sum_\nu^\infty \frac{1}{\nu!} \xi^{2\nu} \sim e^{\xi^2},$$
enforcing on the energy eigenfunctions the behavior
$$u(\xi) \propto H(\xi) \, e^{-\frac{1}{2}\xi^2} \sim e^{+\frac{1}{2}\xi^2}.$$

Clearly, these functions are not normalizable.

The only way out of this problem is to require that the series (A.4) does not extend to infinity but stops at a certain value of j, that is $c_j = 0$ for some j. From the recurrence relation (A.5) we see that this occurs when the parameter η takes on half integer values, that is
$$\eta_m \equiv m + \frac{1}{2}.$$
With this choice of η we find that the functions $H(\xi)$ are polynomials of the order m and we denote them by H_m. They are called Hermite polynomials.

Their polynomial character guarantees that the energy eigenfunctions
$$u_m(\xi) = \mathcal{N}_m \, H_m(\xi) \, e^{-\frac{1}{2}\xi^2}$$
with the eigenvalue $\eta_m = m + 1/2$ are normalizable. Indeed, we can calculate the normalization integral
$$1 = \int_{-\infty}^\infty d\xi \, u_m^2(\xi) = \mathcal{N}_m^2 \int_{-\infty}^\infty d\xi \, H_m^2(\xi) \, e^{-\xi^2} = \mathcal{N}_m^2 \, \sqrt{\pi} \, 2^m \, m!.$$

From this result we obtain the normalization constant
$$\mathcal{N}_m = \left(\frac{1}{\pi}\right)^{1/4} \frac{1}{(2^m \, m!)^{1/2}}.$$

We conclude by summarizing the main results. The energy wave functions of a harmonic oscillator are expressed in terms of Hermite polynomials. They involve either only odd or only even powers of the position and therefore have odd or even parity. The mth energy wave function is a polynomial of mth order and displays m zeros.

A.2 Asymptotic Behavior

In this section we derive an asymptotic expression for the energy wave functions of a harmonic oscillator in the limit of large quantum numbers. We use the representation of the Hermite polynomials in terms of a contour integral in complex space and evaluate it using the method of steepest descent discussed in Appendix H. This provides us with the behavior of the energy wave function appropriately away from the turning points.

A.2.1 Energy Wave Function as a Contour Integral

We first derive a representation of the energy wave functions using an integral formula for the Hermite polynomials.

We start from the expression Eq. (4.4)

$$u_m(x) = -\left(\frac{\kappa^2}{\pi}\right)^{1/4} (2^m m!)^{1/2} \frac{1}{2\pi i} \oint dz\, z^{-(m+1)} e^{\kappa x z - z^2/4} e^{-(\kappa x)^2/2}$$

for the energy wave function.

To perform the limit of large quantum numbers in a convenient way it is useful to cast this formula into the form

$$u_m(x) = f_m(x)\, I_m(x). \qquad (A.6)$$

Here we have introduced

$$f_m(x) \equiv \left(\frac{\kappa^2}{(2\pi)^3}\right)^{1/4} \left[2^{m+1/2}\,(m!)\right]^{1/2} e^{-(\kappa x)^2/2} \qquad (A.7)$$

and the contour integral

$$I_m(x) \equiv -(2\pi)^{-1/2} \frac{1}{i} \oint dz\, z^{-1/2} e^{-\Xi_m(x;z)}, \qquad (A.8)$$

which arises from the Hermite polynomial.

In this step we have decomposed the denominator $z^{(m+1)}$ of the integral representation of H_m into a denominator $z^{1/2}$ and have included the remaining part $z^{m+1/2}$ into the phase

$$\Xi_m(x;z) \equiv \tfrac{1}{4} z^2 - \kappa x\, z + (m+\tfrac{1}{2}) \ln z \qquad (A.9)$$

of the integral. We emphasize that this decomposition is not unique but it guarantees that even in the large m-limit the quantum number m always appears in the combination $m+1/2$ familiar from the quantized energy of the harmonic oscillator.

A.2.2 Evaluation of the Integral I_m

We start our analysis by considering the large-m behavior of the integral $I_m(x)$, Eq. (A.8).

Points of Steepest Descent

The method of steepest descent discussed in Appendix H expands the complex-valued function Ξ_m into a Taylor series

$$\Xi_m(x;z) = \Xi_m(x;z_\pm) + \frac{1}{2}\left.\frac{\partial^2 \Xi_m}{\partial z^2}\right|_{z=z_\pm} (z-z_\pm)^2 + \cdots \qquad (A.10)$$

around the points $z_\pm$ governed by the condition

$$\left.\frac{\partial \Xi_m}{\partial z}\right|_{z=z_\pm} = \left[\tfrac{1}{2} z - \kappa x + (m+\tfrac{1}{2})\frac{1}{z}\right]\bigg|_{z=z_\pm} = 0. \qquad (A.11)$$

A.2 Asymptotic Behavior

This yields the quadratic equation

$$z^2 - 2\kappa x\, z + 2(m + \tfrac{1}{2}) = 0 \qquad (A.12)$$

and the two solutions

$$z_\pm = \kappa x \pm i\left[2(m + \tfrac{1}{2}) - \kappa^2 x^2\right]^{1/2} \equiv \kappa x \pm i p_m(x)/(\hbar\kappa). \qquad (A.13)$$

Hence, the two points of steepest descent have the same real part but differ in the sign of their imaginary parts. Moreover, both lie on a circle of radius $\sqrt{2(m+1/2)}$. Indeed, we have

$$|z_\pm|^2 = (\kappa x)^2 + \left(\frac{p_m(x)}{\hbar\kappa}\right)^2 = 2(m + 1/2).$$

This is consistent with a naive phase space picture of an energy eigenstate of energy $\hbar\Omega(m+1/2)$ as a circular phase space trajectory of radius $\sqrt{2(m+1/2)}$.

Evaluation of Second Derivative

The next steps consist of calculating Ξ_m and its second derivative at the points $z_\pm$ and then inserting the expansion (A.10) into the integral (A.8).

The second derivative of Ξ_m follows from Eq. (A.11) and reads

$$\left.\frac{\partial^2 \Xi_m}{\partial z^2}\right|_{z=z_\pm} = \left[\frac{1}{2} - (m + \tfrac{1}{2})\frac{1}{z^2}\right]_{z=z_\pm} = \tfrac{1}{2}\left[z_\pm^2 - 2(m + \tfrac{1}{2})\right]z_\pm^{-2}.$$

When we make use of Eqs. (A.12) and (A.13) we find

$$\frac{1}{2}\left.\frac{\partial^2 \Xi_m}{\partial z^2}\right|_{z=z_\pm} = \tfrac{1}{2}\left[\kappa x z_\pm - 2(m + \tfrac{1}{2})\right]z_\pm^{-2}$$

$$= \tfrac{1}{2}\left\{-\left[2(m + \tfrac{1}{2}) - \kappa^2 x^2\right] \pm i p_m x/\hbar\right\} z_\pm^{-2}$$

$$= \pm\frac{i}{2} p_m/(\hbar\kappa)\left[\kappa x \pm i p_m(x)/(\hbar\kappa)\right] z_\pm^{-2},$$

or

$$\frac{1}{2}\left.\frac{\partial^2 \Xi_m}{\partial z^2}\right|_{z=z_\pm} = \pm\frac{i}{2}\frac{p_m}{\hbar\kappa} z_\pm^{-1}. \qquad (A.14)$$

Hence the second derivative of the phase Ξ_m at the two points $z_\pm$ of steepest descent differ in their signs.

Gauß Approximation for I_m

We substitute the Taylor expansion Eq. (A.10) of the phase Ξ_m with the explicit form Eq. (A.14) for the second derivative into the expression for I_m, Eq. (A.8), and arrive at

$$I_m(x) \cong \frac{1}{i}\left\{-e^{-\Xi(x;z_+)}(2\pi)^{-1/2}\int_{-\infty}^{\infty} dz\, z_+^{-1/2}\exp\left[-i\frac{p_m}{\hbar\kappa}\frac{1}{2z_+}(z-z_+)^2\right]\right.$$

$$\left. + e^{-\Xi(x;z_-)}(2\pi)^{-1/2}\int_{-\infty}^{\infty} dz\, z_-^{-1/2}\exp\left[+i\frac{p_m}{\hbar\kappa}\frac{1}{2z_-}(z-z_-)^2\right]\right\},$$

where we have incorporated in the integrals the clockwise direction of circulation. This expression simplifies to

$$I_m(x) = \left(\frac{\hbar\kappa}{p_m}\right)^{1/2} \frac{1}{i} \left[-e^{-\Xi(x;z_+)} \frac{1}{\sqrt{\pi}} \int_{-\infty}^{\infty} dy\, e^{-iy^2} + e^{-\Xi(x;z_-)} \frac{1}{\sqrt{\pi}} \int_{-\infty}^{\infty} dy\, e^{iy^2} \right].$$

We evaluate the complex Gauß integral with the help of the integral relation

$$\int_{-\infty}^{\infty} dy\, e^{\pm iy^2} = \pi^{1/2} e^{\pm i\pi/4}, \tag{A.15}$$

which is the central tool in the field of semiclassical quantum mechanics. This integral is closely related to the Fresnel integral and the Cornu spiral discussed in Appendix H.

When we make use of Eq. (A.15) we find

$$I_m(x) = \left(\frac{\hbar\kappa}{p_m}\right)^{1/2} \frac{1}{i} \left\{ -e^{-[\Xi(x,z_+)+i\pi/4]} + e^{-[\Xi(x,z_-)-i\pi/4]} \right\}. \tag{A.16}$$

Evaluation of Phase of Wave Functions

We now evaluate the exponents Ξ_m, Eq. (A.9), at $z = z_\pm$. It is advantageous to first rewrite $z_\pm$, Eq. (A.13), in polar coordinates, that is

$$z_\pm \equiv |z_\pm| e^{\pm i\varphi_m} \equiv [2(m+\tfrac{1}{2})]^{1/2}$$
$$\times \exp\left\{ \pm i \arctan\left[\left(2(m+\tfrac{1}{2}) - \kappa^2 x^2\right)^{1/2} (\kappa x)^{-1} \right] \right\},$$

which yields with the help of Eqs. (A.12) and (A.13)

$$\Xi_m(x; z = z_\pm) = -\tfrac{1}{2}\kappa x z_\pm - \tfrac{1}{2}(m+\tfrac{1}{2}) + (m+\tfrac{1}{2})\ln z_\pm$$
$$= -\tfrac{1}{2}\kappa^2 x^2 \mp \frac{i}{2} x p_m/\hbar - \tfrac{1}{2}(m+\tfrac{1}{2})$$
$$+ \tfrac{1}{2}(m+\tfrac{1}{2})\ln[2(m+\tfrac{1}{2})] \pm i(m+\tfrac{1}{2})\varphi_m,$$

or

$$\Xi_m(x; z = z_\pm) = \pm i\left[(m+\tfrac{1}{2})\varphi_m - \tfrac{1}{2}xp_m/\hbar\right]$$
$$- \tfrac{1}{2}(m+\tfrac{1}{2}) + \tfrac{1}{2}(m+\tfrac{1}{2})\ln[2(m+\tfrac{1}{2})] - \kappa^2 x^2/2.$$

This result reduces Eq. (A.16) to

$$I_m(x) = 2\left(\frac{\hbar\kappa}{p_m}\right)^{1/2} \cos[S_m(x) - \pi/4]$$
$$\times \exp\left[\tfrac{1}{2}(m+\tfrac{1}{2}) - \tfrac{1}{2}(m+\tfrac{1}{2})\ln[2(m+\tfrac{1}{2})] + \kappa^2 x^2/2\right], \tag{A.17}$$

where we have defined

$$S_m(x) \equiv (m + \tfrac{1}{2})\varphi_m(x) - \tfrac{1}{2} x p_m(x)/\hbar$$

$$= (m + \tfrac{1}{2}) \arctan\left\{ \left[2(m+\tfrac{1}{2}) - \kappa^2 x^2\right]^{1/2} (\kappa x)^{-1} \right\} - \tfrac{1}{2} \kappa x \left[2(m+\tfrac{1}{2}) - \kappa^2 x^2\right]^{1/2}$$

$$= \frac{1}{\hbar} \int_x^{\xi_m} dx\, p_m(x),$$

and have used the trigonometric relation

$$\sin(\beta + \pi/4) = \sin(\beta - \pi/4 + \pi/2) = \cos(\beta - \pi/4).$$

A.2.3   Asymptotic Limit of f_m

We are now left with the problem of finding the asymptotic limit of the function f_m which involves $m!$. For large values of m we can apply the improved Stirling equation

$$m! \cong (2\pi)^{1/2} (m + \tfrac{1}{2})^{m+1/2} e^{-(m+1/2)} \tag{A.18}$$

which allows us to derive the expression

$$f_m(x) \cong \left(\frac{\kappa}{2\pi}\right)^{1/2} [2(m+\tfrac{1}{2})]^{(m+1/2)/2} \exp\left[-\tfrac{1}{2}(m+\tfrac{1}{2}) - \kappa^2 x^2/2\right],$$

or

$$f_m(x) \cong \left(\frac{\kappa}{2\pi}\right)^{1/2} \exp\left[-\tfrac{1}{2}(m+\tfrac{1}{2}) + \tfrac{1}{2}(m+\tfrac{1}{2}) \ln[2(m+\tfrac{1}{2})] - \kappa^2 x^2/2\right].$$

When we substitute Eq. (A.17) and the above result into Eq. (A.6) we arrive at the large-m limit

$$u_m(x) \cong (2/\pi)^{1/2} \left(\frac{\kappa^2 \hbar}{p_m(x)}\right)^{1/2} \cos\left[S_m(x) - \pi/4\right] \tag{A.19}$$

of the mth energy wave function. This result for u_m is identical with the standard WKB wave function discussed in Chapter 5.

A.2.4   Bohr's Correspondence Principle

In the preceding section we have shown that the main contributions to the integral I_m, and hence to the wave function u_m arise from the two points of steepest descent $z_\pm$. We recall that these points lie on a circle of radius $\sqrt{2(m+1/2)}$.

Note that the factor $1/2$ in the radius is to some degree arbitrary. This comes out most clearly when we include the factor $z^{-1/2}$ in the exponent of Eq. (A.8) as well. In this case we perform the method of steepest decent on the function

$$\tilde{\Xi}_m(x; z) \equiv \tfrac{1}{4} z^2 - \kappa x\, z + (m+1) \ln z.$$

This yields a phase space trajectory of radius $[2(m+1)]^{1/2}$.

Likewise we can confine our steepest descent analysis to the mth power of z only, that is, we consider the function

$$\overline{\Xi}_m(x;z) \equiv \frac{1}{4}z^2 - \kappa x\, z + m \ln z.$$

In this case we find a phase space trajectory of radius $(2m)^{1/2}$.

Moreover, we could have taken any other decomposition of the exponent of the integration variable z in the denominator of the integral. This arbitrariness is also contained in Bohr's correspondence principle. According to Bohr we can replace finite differences of neighboring quantum numbers, m and $m+1$, by the differential, that is,

$$h_{m+1} - h_m \cong \frac{\partial h_m}{\partial m}.$$

This replacement is valid when m is large. Note, however, that the correspondence principle does not tell us at which m value we have to evaluate this derivative. Kramers' detailed Airy function analysis of the Schrödinger equation for a stationary state finally provides the correct WKB quantization rule

$$action = 2\pi\hbar(m + 1/2)$$

and the correct answer to the $1/2$ puzzle: we have to evaluate the derivative $\partial h_m/\partial m$ at $m + 1/2$, that is, half way between the neighboring quantum numbers. This we discuss in detail in Chapter 5.

References

For the properties of Hermite polynomials and their relation to parabolic cylinder functions, see for example

G. Szegö, *Orthogonal Polynomials*, American Mathematical Society, New York, 1939

I.S. Gradsteyn and **I.M. Ryzhik**, *Table of Integrals, Series and Products*, Academic Press, New York, 1965

M. Abramowitz and **I.E. Stegun**, *Handbook of Mathematical Functions*, National Bureau of Standards, Washington D.C., 1964

B Time Dependent Operators

In this appendix we discuss problems associated with the differentiation of an operator

$$\hat{\mathcal{U}}(t) \equiv \exp[\hat{B}(t)]$$

where $\hat{B}(t)$ is a time dependent operator. This problem is central to quantum mechanics since $\hat{B}(t)$ is often the Hamiltonian and therefore $\hat{\mathcal{U}}$ is closely related to the time evolution operator. We show that in general the derivative of an exponential operator $\exp[\hat{B}(t)]$ is not just the same exponential operator times the derivative of $\hat{B}(t)$. This feature is a consequence of the fact that the derivative of $\hat{B}$ does not necessarily commute with $\hat{B}$.

This question of differentiation is closely related to the problem of time ordering of operators. We derive a formula to time order the product of n integrals containing the Hamiltonian.

B.1 Caution when Differentiating Operators

In order to differentiate $\hat{\mathcal{U}}$ we use the Taylor expansion of the exponential function and find

$$\frac{d}{dt}\exp[\hat{B}(t)] = \sum_{n=0}^{\infty} \frac{1}{n!}\frac{d}{dt}\hat{B}^n(t).$$

Since the nth power of $\hat{B}(t)$ is the product of n operators, that is

$$\hat{B}^n(t) = \hat{B}(t) \cdot \hat{B}(t) \cdot \ldots \cdot \hat{B}(t)$$

we obtain the time derivative

$$\begin{aligned}
\frac{d}{dt}\hat{B}^n(t) =\ & \frac{d}{dt}\hat{B}(t) \cdot \hat{B}(t) \cdot \ldots \cdot \hat{B}(t) \\
+\ & \hat{B}(t) \cdot \frac{d}{dt}\hat{B}(t) \cdot \ldots \cdot \hat{B}(t) \\
& \vdots \\
+\ & \hat{B}(t) \cdot \hat{B}(t) \cdot \ldots \cdot \frac{d}{dt}\hat{B}(t).
\end{aligned} \quad (B.1)$$

Note, that here we have kept track of the order of the operators. Only when the time derivative of $\hat{B}$ and $\hat{B}$ commute, that is

$$\left[\frac{d}{dt}\hat{B}(t), \hat{B}(t)\right] = 0,$$

we can combine these terms and arrive at the familiar result

$$\frac{d}{dt}\hat{B}^n(t) = n \cdot \hat{B}^{n-1} \cdot \frac{d}{dt}\hat{B}(t) = n \cdot \frac{d}{dt}\hat{B}(t) \cdot \hat{B}^{n-1}(t)$$

which immediately yields

$$\frac{d}{dt}\exp[\hat{B}(t)] = \frac{d}{dt}\hat{B}(t) \cdot \exp[\hat{B}(t)] = \exp[\hat{B}(t)] \cdot \frac{d}{dt}\hat{B}(t).$$

However, when $\hat{B}$ and its time derivative do not commute we cannot resum the exponential. In this case a much more complicated expression arises.

We now illustrate this property using two examples. The operator

$$\hat{B} \equiv -\frac{i}{\hbar}\hat{H}_0 t$$

commutes with the time derivative

$$\frac{d}{dt}\hat{B}(t) = -\frac{i}{\hbar}\hat{H}_0$$

and therefore yields

$$\frac{d}{dt}\hat{\mathcal{U}}(t, t_0) = -\frac{i}{\hbar}\hat{H}_0\hat{\mathcal{U}}(t, t_0).$$

In contrast the operator

$$\hat{B}(t) \equiv -\frac{i}{\hbar}\int_0^t dt'\, \hat{H}(t')$$

containing the integral over a time dependent Hamiltonian $\hat{H}(t)$ does not necessarily commute with its time derivative

$$\frac{d}{dt}\hat{B}(t) = -\frac{i}{\hbar}\hat{H}(t).$$

Indeed, the two operators only commute provided the time dependent Hamiltonians at different times commute with themselves, that is $[\hat{H}(t), \hat{H}(t')] = 0$ for all times $t' \leq t$.

B.2 Time Ordering

In Sec. 2.4.3 we have introduced the time ordering operator

$$\hat{T}\left[\hat{A}(t_1)\hat{B}(t_2)\right] \equiv \begin{cases} \hat{B}(t_2)\hat{A}(t_1) & \text{for } t_2 > t_1 \\ \hat{A}(t_1)\hat{B}(t_2) & \text{for } t_1 > t_2 \end{cases} \quad (B.2)$$

ensuring that in a product of two operators $\hat{A}(t_1)$ and $\hat{B}(t_2)$ at two different times t_1 and t_2 the one at an earlier time always stands to the right. In this section we illustrate this concept by calculating the time ordered product of the square and the nth power of the integral of a time dependent Hamiltonian.

B.2.1 Product of Two Terms

In order to develop a feeling for time ordering we first evaluate the quantity

$$\hat{\mathcal{H}}_2 \equiv \hat{T}\left[\int_{t_0}^{t} dt'\, \hat{H}(t')\right]^2 = \hat{T}\left[\int_{t_0}^{t} dt_2 \int_{t_0}^{t} dt_1\, \hat{H}(t_2)\hat{H}(t_1)\right] \tag{B.3}$$

consisting of the product of two operators $\hat{H}(t_1)$ and $\hat{H}(t_2)$ at different times. Such a term appears in second order when we expand the time evolution operator

$$\hat{T}\left[\exp\left\{-\frac{i}{\hbar}\int_{t_0}^{t} dt'\, \hat{H}(t')\right\}\right] \cong \mathbb{1} - \frac{i}{\hbar}\int_{t_0}^{t} dt_1\, \hat{H}(t_1) - \frac{1}{2\hbar^2}\hat{T}\left[\int_{t_0}^{t} dt_2 \int_{t_0}^{t} dt_1\, \hat{H}(t_2)\hat{H}(t_1)\right]$$

for a time dependent Hamiltonian $\hat{H}(t)$. Since in the first order term the Hamiltonian $\hat{H}$ appears only linearly, there is no time ordering necessary.

With this motivation in mind we now return to the problem Eq. (B.3) of performing time ordering in the square of the integral of the Hamiltonian. For this purpose we decompose the integration with respect to t_1 in the domains $t_2 > t_1$ and $t_1 > t_2$ as required by the definition Eq. (B.2) of the time ordering operator, that is

$$\int_{t_0}^{t} dt_2 \int_{t_0}^{t} dt_1 = \int_{t_0}^{t} dt_2 \int_{t_0}^{t_2} dt_1 + \int_{t_0}^{t} dt_2 \int_{t_2}^{t} dt_1,$$

and find

$$\hat{\mathcal{H}}_2 = \hat{T}\left[\int_{t_0}^{t} dt_2 \int_{t_0}^{t_2} dt_1\, \hat{H}(t_2)\hat{H}(t_1)\right] + \hat{T}\left[\int_{t_0}^{t} dt_2 \int_{t_2}^{t} dt_1\, \hat{H}(t_2)\hat{H}(t_1)\right].$$

In the first integral we have $t_2 > t_1$ and thus

$$\hat{T}\left[\hat{H}(t_2)\hat{H}(t_1)\right] = \hat{H}(t_2)\hat{H}(t_1).$$

In contrast, the second integral covers the domain $t_1 > t_2$ and with the relation

$$\hat{T}\left[\hat{H}(t_2)\hat{H}(t_1)\right] = \hat{H}(t_1)\hat{H}(t_2)$$

we arrive at

$$\hat{\mathcal{H}}_2 = \int_{t_0}^{t} dt_2 \int_{t_0}^{t_2} dt_1\, \hat{H}(t_2)\hat{H}(t_1) + \int_{t_0}^{t} dt_2 \int_{t_2}^{t} dt_1\, \hat{H}(t_1)\hat{H}(t_2).$$

We now interchange the order of integration in the second double-integral as shown in Fig. B.1 using the formula

$$\int_{t_0}^{t} dt_2 \int_{t_2}^{t} dt_1 = \int_{t_0}^{t} dt_1 \int_{t_0}^{t_1} dt_2$$

B Time Dependent Operators

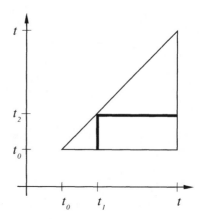

Fig. B.1: Change of order of integration. In order to cover the triangle in (t_1, t_2)-parameter space defined by the diagonal, the point (t_0, t_0) and the vertical line at t we integrate first the variable t_1 from t_2 to t (horizontal bar) and then t_2 from t_0 to t. Equivalently we can first integrate the variable t_2 from t_0 to t_1 (vertical bar) and then integrate t_1 from t_0 to t.

and find

$$\hat{\mathcal{H}}_2 = \int_{t_0}^{t} dt_2 \int_{t_0}^{t_2} dt_1 \, \hat{H}(t_2)\hat{H}(t_1) + \int_{t_0}^{t} dt_1 \int_{t_0}^{t_1} dt_2 \, \hat{H}(t_1)\hat{H}(t_2).$$

When we make the replacement

$$t_1 \to t_2 \quad \text{and} \quad t_2 \to t_1$$

in the second double-integral we recognize that it is identical to the first one and hence

$$\hat{\mathcal{H}}_2 = 2 \cdot \int_{t_0}^{t} dt_2 \int_{t_0}^{t_2} dt_1 \, \hat{H}(t_2)\hat{H}(t_1). \tag{B.4}$$

Consequently we arrive at the representation

$$\hat{T}\left[\exp\left\{-\frac{i}{\hbar}\int_{t_0}^{t} dt' \, \hat{H}(t')\right\}\right] \cong \mathbb{1} - \frac{i}{\hbar}\int_{t_0}^{t} dt_1 \, \hat{H}(t_1) - \frac{1}{\hbar^2}\int_{t_0}^{t} dt_2 \int_{t_0}^{t_2} dt_1 \, \hat{H}(t_2)\hat{H}(t_1)$$

for the lowest orders of the time evolution operator for a time dependent Hamiltonian.

B.2.2 Product of n Terms

We now perform the time ordering in the product of n integrals containing the Hamiltonian. In particular, we prove the relation

$$\hat{\mathcal{H}}_n(t) \equiv \hat{T}\left[\int_{t_0}^{t} dt' \hat{H}(t')\right]^n = n! \int_{t_0}^{t} dt_n \int_{t_0}^{t_n} dt_{n-1} \ldots \int_{t_0}^{t_2} dt_1 \hat{H}(t_n)\hat{H}(t_{n-1})\ldots\hat{H}(t_1). \tag{B.5}$$

At this point we recognize that the result Eq. (B.4) for the square of the integrated Hamiltonian is a special case of this formula. We are therefore tempted to pursue a similar path in order to derive this more general formula. However, here we follow a more convenient approach. We first derive a recurrence-differential equation and then show that the right hand side of Eq. (B.5) satisfies this equation.

When we differentiate the operator

$$\hat{\mathcal{H}}_n(t) \equiv \hat{T}\left[\int_{t_0}^{t} dt_n\, \hat{H}(t_n) \int_{t_0}^{t} dt_{n-1}\, \hat{H}(t_{n-1}) \ldots \int_{t_0}^{t} dt_1\, \hat{H}(t_1)\right]$$

consisting of a product of n identical integrals with respect to time we find

$$\frac{d}{dt}\hat{\mathcal{H}}_n(t) = \hat{T}\left[\hat{H}(t)\int_{t_0}^{t} dt_{n-1}\, \hat{H}(t_{n-1}) \ldots \int_{t_0}^{t} dt_1\, \hat{H}(t_1)\right.$$

$$+ \int_{t_0}^{t} dt_n\, \hat{H}(t_n)\hat{H}(t) \ldots \int_{t_0}^{t} dt_1\, \hat{H}(t_1)$$

$$\vdots$$

$$\left. + \int_{t_0}^{t} dt_n\, \hat{H}(t_n) \int_{t_0}^{t} dt_{n-1}\, \hat{H}(t_{n-1}) \ldots \hat{H}(t)\right]$$

in complete accordance with Eq. (B.1).

In contrast to Eq. (B.1) we can now simplify this expression using the time ordering operator $\hat{T}$, which ensures that the operator at the latest time has to stand furthest to the left. The latest time is t and hence we can move the operator $\hat{H}(t)$ all the way to the left. As a consequence the n terms in the sum become identical. Moreover, this product of $n-1$ integrals of the Hamiltonian has to be still time ordered. Hence we arrive at the recurrence-differential equation

$$\frac{d}{dt}\hat{\mathcal{H}}_n(t) = n\,\hat{H}(t)\hat{T}\left[\int_{t_0}^{t} dt'\, \hat{H}(t')\right]^{n-1} = n\,\hat{H}(t)\hat{\mathcal{H}}_{n-1}(t). \qquad (B.6)$$

When we now differentiate the right hand side of Eq. (B.5) we find indeed

$$\frac{d}{dt}\hat{\mathcal{H}}_n(t) = n!\int_{t_0}^{t} dt_{n-1} \int_{t_0}^{t_{n-1}} dt_{n-2} \ldots \int_{t_0}^{t} dt_1\, \hat{H}(t)\hat{H}(t_{n-1})\ldots\hat{H}(t_1) = n\,\hat{H}(t)\hat{\mathcal{H}}_{n-1}(t),$$

that is the n-dimensional nested integral Eq. (B.5) obeys the two-term recurrence-differential equation Eq. (B.6). Obviously the right hand side of Eq. (B.5) satisfies the initial condition

$$\hat{\mathcal{H}}_n(t_0) = 0$$

as suggested by the definition of $\hat{\mathcal{H}}_n$.

C Süßmann Measure

In this appendix we address the question of an appropriate measure of the width of a normalized probability distribution. We first bring out the problems associated with standard measures such as the variance and then illustrate a proposal by G. Süßmann for a new definition of width: The inverse of the area underneath the square of the distribution. We demonstrate that the so-defined width is meaningful for a Gaussian and a Lorentzian.

C.1 Why Other Measures Fail

How to define the width of a distribution function? Many suggestions offer themselves. However, not all suggestions are meaningful. Some of these definitions overemphasize special features of the distribution. For a detailed discussion of this topic we refer to the literature. However, in the present context we concentrate on a few only.

For the sake of simplicity we assume a distribution $f = f(x)$ with vanishing first moment

$$\langle x \rangle \equiv \int_{-\infty}^{\infty} dx\, x \cdot f(x) = 0.$$

Why not take the variance

$$\delta_V x^2 \equiv \langle x^2 \rangle = \int_{-\infty}^{\infty} dx\, x^2 \cdot f(x)$$

of the distribution f as a measure of the width? Because this measure is only useful for well localized distributions. Indeed, for a Gaussian

$$f(x) = \frac{1}{\sqrt{\pi}} \frac{1}{D} \exp\left[-\left(\frac{x}{D}\right)^2\right]$$

we find immediately

$$\delta_V x = \frac{1}{\sqrt{2}} D.$$

It is instructive to compare and contrast the so-defined width to the decay width $\delta_D = 2D$ governed by the two points $\pm D$ where the Gaussian has decayed to $1/e$. The latter is larger by a factor $2\sqrt{2}$ compared to $\delta_V x$, that is, $\delta_V x < \delta_D x$.

We now turn to the case of a Lorentzian

$$f(x) \equiv \frac{1}{\pi} \frac{D}{D^2 + x^2}.$$

This distribution displays long ranging wings which only decay quadratically. Consequently the second moment does not exist. Clearly due to the quadratic function in the integrand of the second moment the variance over-emphasizes the wings of the distribution. Nevertheless, the Lorentzian has a characteristic width $\delta_D = 2D$ defined by the separation of the two points $\pm D$ where the two terms in the denominator are equal. Here we have the hierarchy $\delta_D x < \delta_V x = \infty$.

C.2 One Way out of the Problem

The example of the Lorentzian illustrates in a striking way the need for a better estimate of the width of a distribution. We now turn to the discussion of the Süßmann measure which does not have the problems discussed above.

Definition

Why not approximate the distribution by a rectangular distribution with the average height $\overline{h}$? But how to define $\overline{h}$?

We can average any function $g(x)$ that depends on the variable x with the help of the distribution function f via the definition

$$\langle g(x) \rangle \equiv \int_{-\infty}^{\infty} dx\, g(x) \cdot f(x).$$

When we are interested in the average height of f we obviously have to average $f(x)$ and hence, $g(x) = f(x)$. Consequently we find the average height

$$\overline{h} \equiv \langle f(x) \rangle \equiv \int_{-\infty}^{\infty} dx\, f(x)\, f(x)$$

of the distribution f.

The Süßmann measure $\delta_S x$ is then determined by the constraint to conserve the normalization of f, that is

$$\delta_S x \cdot \overline{h} = 1$$

and hence

$$\delta_S x = \frac{1}{\overline{h}} = \frac{1}{\langle f(x) \rangle} = \frac{1}{\int_{-\infty}^{\infty} dx\, f^2(x)}. \tag{C.1}$$

According to this definition the width $\delta_S x$ of the distribution is the inverse of the area underneath the square of the distribution.

Examples

We illustrate the Süßmann measure for the two examples of the Gaussian and Lorentzian. When we perform the necessary integrals over the squares of the distributions we find for the width $\delta_S x$ of the Gaussian

$$\delta_S x = \sqrt{2\pi}\, D.$$

Hence, this width is even larger than the width $\delta_D x$ and we have the hierarchy $\delta_V x < \delta_D x < \delta_S x$.

Similarly we arrive for the Lorentzian at

$$\delta_S x = 2\pi D.$$

Hence this width is larger than the width $\delta_D x = 2D$ by a factor π which accounts for the wings beyond $|x| > D$. For the case of the Lorentzian we obtain the hierarchy is $\delta_D x < \delta_S x < \delta_V x = \infty$.

C.3 Generalization to Higher Dimensions

We conclude this appendix by noting that we can extend this notion to distributions with more variables. For the sake of simplicity we confine ourselves to two-dimensional functions. This is of particular importance in the context of phase space distributions. Then the notion of width transforms to area. Indeed, the area A of x-p-phase space where the function $W(x,p)$ takes on considerable values follows as

$$A \equiv \frac{1}{\int\limits_{-\infty}^{\infty} dx \int\limits_{-\infty}^{\infty} dp\, W^2(x,p)}.$$

The Süßmann measure is a special case of a general class of statistical measures summarized in the review article by Wehrl. For more details we refer to the literature. These concepts have become important in the context of quantum state reconstruction.

References

For a detailed study of various definitions of the width of a probability distribution see

I. Bialynicki-Birula, M. Freyberger and **W.P. Schleich**, *Various Measures of Quantum Phase Uncertainty: A Comparative Study*, Physica Scripta **T48**, 113–118 (1993)

G. Süssmann, *Uncertainty Relation: From Inequality to Equality*, Z. Naturforsch. **52a**, 49–52 (1997)

A. Wehrl, *General properties of entropy*, Rev. Mod. Phys. **50**, 221–260 (1978)

D Phase Space Equations

In the process of translating the von Neumann equation into phase space we have encountered rather complicated Fourier transforms of matrix elements. The latter contain combinations of the kinetic and potential energy and the density operator. The goal of the present appendix is to express these quantities in terms of derivatives of the Wigner function.

We encounter similar expressions when we derive the phase space equations determining the Moyal function. Therefore, it is useful to first study more general matrix elements and then apply these results to the problem of the quantum Liouville equation and the Moyal functions.

D.1 Formulation of the Problem

In this appendix we derive the explicit form of the right hand side of the quantum Liouville equation

$$\frac{\partial W}{\partial t} = \mathcal{T} + \mathcal{U}, \qquad (D.1)$$

determining the time evolution of the Wigner function

$$W(x,p;t) \equiv \frac{1}{2\pi\hbar} \int_{-\infty}^{\infty} d\xi \, \exp\left(-\frac{i}{\hbar} p \xi\right) \left\langle x + \tfrac{1}{2}\xi \middle| \hat{\rho}(t) \middle| x - \tfrac{1}{2}\xi \right\rangle.$$

In particular, we express the terms

$$\mathcal{T} = -\frac{i}{\hbar} \frac{1}{2M} \frac{1}{2\pi\hbar} \int_{-\infty}^{\infty} d\xi \, e^{-ip\xi/\hbar} \left\langle x + \tfrac{1}{2}\xi \middle| \hat{p}^2 \hat{\rho} - \hat{\rho} \hat{p}^2 \middle| x - \tfrac{1}{2}\xi \right\rangle$$

and

$$\mathcal{U} \equiv -\frac{i}{\hbar} \frac{1}{2\pi\hbar} \int_{-\infty}^{\infty} d\xi \, e^{-ip\xi/\hbar} \left\langle x + \tfrac{1}{2}\xi \middle| U(\hat{x})\hat{\rho} - \hat{\rho} U(\hat{x}) \middle| x - \tfrac{1}{2}\xi \right\rangle$$

resulting from the kinetic and potential energy in terms of derivatives of the Wigner function.

Moreover, we start from the two equations

$$\frac{1}{2\pi\hbar}\int_{-\infty}^{\infty}d\xi\, e^{-ip\xi/\hbar}\langle x+\tfrac{1}{2}\xi|\frac{1}{2}\{|E''\rangle\langle E'|,\hat{H}\}|x-\tfrac{1}{2}\xi\rangle = \frac{E'+E''}{2}W_{|E''\rangle\langle E'|}$$

and

$$\frac{1}{2\pi\hbar}\int_{-\infty}^{\infty}d\xi\, e^{-ip\xi/\hbar}\langle x+\tfrac{1}{2}\xi|\frac{i}{\hbar}\big[|E''\rangle\langle E'|,\hat{H}\big]|x-\tfrac{1}{2}\xi\rangle = \frac{i}{\hbar}(E'-E'')W_{|E''\rangle\langle E'|}$$

to obtain two partial differential equations in phase space determining the Moyal function

$$W_{|E''\rangle\langle E'|} \equiv \frac{1}{2\pi\hbar}\int_{-\infty}^{\infty}d\xi\, e^{-ip\xi/\hbar}\langle x+\tfrac{1}{2}\xi|E''\rangle\langle E'|x-\tfrac{1}{2}\xi\rangle.$$

The problem of the derivation of the quantum Liouville equation is closely related to the evaluation of the left hand side of the Moyal equation. This stands out most clearly when we substitute the Hamiltonian

$$\hat{H} \equiv \frac{\hat{p}^2}{2M} + U(\hat{x})$$

into the above integrals and find

$$\frac{1}{2}(\mathcal{T}^{(+)}+\mathcal{U}^{(+)}) = \frac{E'+E''}{2}W_{|E''\rangle\langle E'|} \tag{D.2a}$$

and

$$\frac{i}{\hbar}(\mathcal{T}^{(-)}+\mathcal{U}^{(-)}) = \frac{i}{\hbar}(E'-E'')W_{|E''\rangle\langle E'|}. \tag{D.2b}$$

where we have separated the contributions

$$\mathcal{T}^{(\pm)} \equiv \frac{1}{2M}\frac{1}{2\pi\hbar}\int_{-\infty}^{\infty}d\xi\, e^{-ip\xi/\hbar}\langle x+\tfrac{1}{2}\xi|\big[|E''\rangle\langle E'|\hat{p}^2 \pm \hat{p}^2|E''\rangle\langle E'|\big]|x-\tfrac{1}{2}\xi\rangle$$

from the kinetic part and the term

$$\mathcal{U}^{(\pm)} \equiv \frac{1}{2\pi\hbar}\int_{-\infty}^{\infty}d\xi\, e^{-ip\xi/\hbar}\langle x+\tfrac{1}{2}\xi|\big[|E''\rangle\langle E'|U(\hat{x}) \pm U(\hat{x})|E''\rangle\langle E'|\big]|x-\tfrac{1}{2}\xi\rangle$$

corresponding to the potential energy.

D.2 Fourier Transform of Matrix Elements

In order to evaluate the quantities $\mathcal{T}$, $T^{(\pm)}$, U and $\mathcal{U}^{(\pm)}$ it is useful to introduce the Fourier transform

$$t^{(\pm)}(\hat{A}) \equiv \frac{1}{2\pi\hbar}\int_{-\infty}^{\infty}d\xi\, e^{-ip\xi/\hbar}\mathcal{M}^{(\pm)}(x,\xi)$$

of the matrix elements

$$\mathcal{M}^{(\pm)}(x,\xi) \equiv \left\langle x + \tfrac{1}{2}\xi \right| \hat{p}^2 \hat{A} \pm \hat{A}\hat{p}^2 \left| x - \tfrac{1}{2}\xi \right\rangle \tag{D.3}$$

arising from the anti-commutator and the commutator of the kinetic energy with an operator $\hat{A}$. For the special choices $\hat{A} \equiv \hat{\rho}$ and $\hat{A} \equiv |E''\rangle \langle E'|$ we find the relations

$$\mathcal{T} = -\frac{1}{2M} \frac{i}{\hbar} t^{(-)}(\hat{\rho}), \tag{D.4}$$

and

$$\mathcal{T}^{(\pm)} = \frac{1}{2M} t^{(\pm)} \left(|E''\rangle \langle E'| \right) \tag{D.5}$$

connecting the quantities $\mathcal{T}$ and $\mathcal{T}^{(\pm)}$ with $t^{(\pm)}$.

Similarly, we deal with the potential energy $U(x)$ by defining the Fourier transform

$$u^{(\pm)}(\hat{A}) \equiv \frac{1}{2\pi\hbar} \int_{-\infty}^{\infty} d\xi \, e^{-ip\xi/\hbar} \mathcal{V}^{(\pm)}(x,\xi) \tag{D.6}$$

of the matrix elements

$$\mathcal{V}^{(\pm)}(x,\xi) \equiv \left\langle x + \tfrac{1}{2}\xi \right| U(\hat{x})\hat{A} \pm \hat{A}U(\hat{x}) \left| x - \tfrac{1}{2}\xi \right\rangle$$

containing the anti-commutator and the commutator of the operators $U(\hat{x})$ and $\hat{A}$.

In this case we find the relations

$$\mathcal{U} = -\frac{i}{\hbar} u^{(\pm)}(\hat{\rho})$$

and

$$\mathcal{U}^{(\pm)} = u^{(\pm)} \left(|E''\rangle \langle E'| \right),$$

connecting the quantities $\mathcal{U}$ and $\mathcal{U}^{(\pm)}$ with $u^{(\pm)}$.

D.3 Kinetic Energy Terms

We begin our analysis by considering the terms $t^{(\pm)}$ originating from the kinetic energy and first cast the matrix elements $\mathcal{M}^{(\pm)}$ in a different form. For this purpose we recall from Eq. (2.13) the relation

$$\langle x | \hat{p}^2 | \psi \rangle = -\hbar^2 \frac{\partial^2}{\partial x^2} \langle x | \psi \rangle$$

which yields

$$\left\langle x + \tfrac{1}{2}\xi \right| \hat{p}^2 | \psi \rangle = -\hbar^2 \frac{\partial^2}{\partial (x + \tfrac{1}{2}\xi)^2} \left\langle x + \tfrac{1}{2}\xi \middle| \psi \right\rangle$$

and

$$\langle \psi | \hat{p}^2 \left| x - \tfrac{1}{2}\xi \right\rangle = -\hbar^2 \frac{\partial^2}{\partial (x - \tfrac{1}{2}\xi)^2} \left\langle \psi \middle| x - \tfrac{1}{2}\xi \right\rangle.$$

As a consequence we find the formula

$$\mathcal{M}^{(\pm)}(x,\xi) = -\hbar^2 \left[\frac{\partial^2}{\partial(x+\frac{1}{2}\xi)^2} \pm \frac{\partial^2}{\partial(x-\frac{1}{2}\xi)^2} \right] \langle x + \tfrac{1}{2}\xi | \hat{A} | x - \tfrac{1}{2}\xi \rangle$$

for the matrix elements $\mathcal{M}^{(\pm)}$.

Next we express the differentiations with respect to the variables

$$x'' \equiv x + \frac{1}{2}\xi \quad \text{and} \quad x' \equiv x - \frac{1}{2}\xi,$$

that is

$$x = \frac{1}{2}(x' + x'') \quad \text{and} \quad \xi = x'' - x'$$

by differentiations with respect to x and ξ by making use of the chain rule, that is

$$\frac{\partial}{\partial(x + \frac{1}{2}\xi)} \equiv \frac{\partial}{\partial x''} = \frac{\partial}{\partial x}\frac{\partial x}{\partial x''} + \frac{\partial}{\partial \xi}\frac{\partial \xi}{\partial x''} = \frac{1}{2}\frac{\partial}{\partial x} + \frac{\partial}{\partial \xi},$$

and similarly

$$\frac{\partial}{\partial(x - \frac{1}{2}\xi)} \equiv \frac{\partial}{\partial x'} = \frac{1}{2}\frac{\partial}{\partial x} - \frac{\partial}{\partial \xi}.$$

Hence we arrive at the expressions

$$\frac{\partial^2}{\partial(x + \frac{1}{2}\xi)^2} = \frac{1}{4}\frac{\partial^2}{\partial x^2} + \frac{\partial^2}{\partial x \partial \xi} + \frac{\partial^2}{\partial \xi^2}$$

and

$$\frac{\partial^2}{\partial(x - \frac{1}{2}\xi)^2} = \frac{1}{4}\frac{\partial^2}{\partial x^2} - \frac{\partial^2}{\partial x \partial \xi} + \frac{\partial^2}{\partial \xi^2}$$

and the matrix elements $\mathcal{M}^{(\pm)}$ reduce to

$$\mathcal{M}^{(+)} = \left(-\frac{\hbar^2}{2}\frac{\partial^2}{\partial x^2} - 2\hbar^2 \frac{\partial^2}{\partial \xi^2} \right) \langle x + \tfrac{1}{2}\xi | \hat{A} | x - \tfrac{1}{2}\xi \rangle$$

and

$$\mathcal{M}^{(-)} = -2\hbar^2 \frac{\partial^2}{\partial \xi \partial x} \langle x + \tfrac{1}{2}\xi | \hat{A} | x - \tfrac{1}{2}\xi \rangle.$$

We substitute these expressions back into the definition of $t^{(\pm)}$ and replace the differentiations with respect to ξ with the help of partial integration arriving at

$$t^{(+)} = \left(-\frac{\hbar^2}{2}\frac{\partial^2}{\partial x^2} - 2p^2 \right) \frac{1}{2\pi\hbar} \int_{-\infty}^{\infty} d\xi\, e^{-ip\xi/\hbar} \langle x + \tfrac{1}{2}\xi | \hat{A} | x - \tfrac{1}{2}\xi \rangle \quad (D.7)$$

and

$$t^{(-)} = -2i\hbar p \frac{\partial}{\partial x} \frac{1}{2\pi\hbar} \int_{-\infty}^{\infty} d\xi\, e^{-ip\xi/\hbar} \langle x + \tfrac{1}{2}\xi | \hat{A} | x - \tfrac{1}{2}\xi \rangle. \quad (D.8)$$

We are in the position to calculate $\mathcal{T}$ with the help of the explicit expression Eq. (D.8) for $t^{(-)}(\rho)$ and the connection formula Eq. (D.4) which yields

$$\mathcal{T} = -\frac{p}{M}\frac{\partial}{\partial x}W. \tag{D.9}$$

Similarly we find from Eqs. (D.7) and (D.8) together with Eqs. (D.4) and (D.5) the results

$$\mathcal{T}^{(+)} = -\left(\frac{p^2}{M} + \frac{\hbar^2}{4M}\frac{\partial^2}{\partial x^2}\right)W_{|E''\rangle\langle E'|} \tag{D.10a}$$

and

$$\mathcal{T}^{(-)} = -i\hbar\frac{p}{M}\frac{\partial}{\partial x}W_{|E''\rangle\langle E'|} \tag{D.10b}$$

for the Moyal function.

D.4 Potential Energy Terms

We now turn to the potential energy terms $u^{(\pm)}$. When we make use of the eigenvalue equation of the position eigenstates we find with the help of

$$U(\hat{x})\,|x\rangle = U(x)\,|x\rangle$$

the relation

$$\mathcal{V}^{(\pm)} \equiv \left[U(x+\tfrac{1}{2}\xi) \pm U(x-\tfrac{1}{2}\xi)\right]\langle x+\tfrac{1}{2}\xi|\,\hat{A}\,|x-\tfrac{1}{2}\xi\rangle.$$

We expand the potential into a Taylor series

$$U(x\pm\tfrac{1}{2}\xi) = \sum_{m=0}^{\infty}\frac{1}{m!}\frac{d^m U(x)}{dx^m}(\pm\tfrac{1}{2}\xi)^m$$

and find that the sum

$$U(x-\tfrac{1}{2}\xi) + U(x+\tfrac{1}{2}\xi) = 2\sum_{l=0}^{\infty}\frac{(i\hbar/2)^{2l}}{(2l)!}\frac{d^{2l}U}{dx^{2l}}\left(-\frac{i\xi}{\hbar}\right)^{2l} \tag{D.11}$$

of the two shifted potentials depends on the even derivatives of the potential whereas the difference

$$U(x-\tfrac{1}{2}\xi) - U(x+\tfrac{1}{2}\xi) = -i\hbar\sum_{l=0}^{\infty}\frac{(i\hbar/2)^{2l}}{(2l+1)!}\frac{d^{2l+1}U}{dx^{2l+1}}\left(-\frac{i\xi}{\hbar}\right)^{2l+1} \tag{D.12}$$

only involves the odd derivatives.

When we substitute these expressions into the Fourier integral Eq. (D.6) for $u^{(\pm)}$ and write the powers of ξ as derivatives of the exponential we obtain

$$u^{(+)} = 2\sum_{l=0}^{\infty}\frac{(i\hbar/2)^{2l}}{(2l)!}\frac{d^{2l}U}{dx^{2l}}\frac{\partial^{2l}}{\partial p^{2l}}\frac{1}{2\pi\hbar}\int_{-\infty}^{\infty}d\xi\, e^{-ip\xi/\hbar}\langle x+\tfrac{1}{2}\xi|\,\hat{A}\,|x-\tfrac{1}{2}\xi\rangle$$

and

$$\mathcal{U}^{(-)} = -i\hbar \sum_{l=0}^{\infty} \frac{(i\hbar/2)^{2l}}{(2l+1)!} \frac{d^{2l+1}U}{dx^{2l+1}} \frac{\partial^{2l+1}}{\partial p^{2l+1}} \frac{1}{2\pi\hbar} \int_{-\infty}^{\infty} d\xi \, e^{-ip\xi/\hbar} \left\langle x + \tfrac{1}{2}\xi \right| \hat{A} \left| x - \tfrac{1}{2}\xi \right\rangle.$$

Consequently, the potential energy term $\mathcal{U}$ in the quantum Liouville equation reads

$$\mathcal{U} = \sum_{l=0}^{\infty} \frac{(-1)^l (\hbar/2)^{2l}}{(2l+1)!} \frac{d^{2l+1}U}{dx^{2l+1}} \frac{\partial^{2l+1}}{\partial p^{2l+1}} W(x,p). \qquad (D.13)$$

Similarly the potential terms $\mathcal{U}^{(\pm)}$ take the form

$$\mathcal{U}^{(+)} = 2 \sum_{l=0}^{\infty} \frac{(-1)^l (\hbar/2)^{2l}}{(2l)!} \frac{d^{2l}U}{dx^{2l}} \frac{\partial^{2l}}{\partial p^{2l}} W_{|E''\rangle\langle E'|}(x,p) \qquad (D.14a)$$

and

$$\mathcal{U}^{(-)} = -i\hbar \sum_{l=0}^{\infty} \frac{(-1)^l (\hbar/2)^{2l}}{(2l+1)!} \frac{d^{2l+1}U}{dx^{2l+1}} \frac{\partial^{2l+1}}{\partial p^{2l+1}} W_{|E''\rangle\langle E'|}(x,p). \qquad (D.14b)$$

Hence, the contributions due to the potential energy involve derivatives of the Wigner function with respect to the momentum. The number of derivatives depends on the potential.

D.5 Summary

We are now in a position to present the quantum Liouville equation. With the help of Eqs. (D.9) and (D.13) the equation of motion for the Wigner function, Eq. (D.1) reads

$$\left(\frac{\partial}{\partial t} + \frac{p}{M} \frac{\partial}{\partial x} - \frac{dU(x)}{dx} \frac{\partial}{\partial p} \right) W(x,p;t) = \sum_{l=1}^{\infty} \frac{(-1)^l (\hbar/2)^{2l}}{(2l+1)!} \frac{d^{2l+1}U(x)}{dx^{2l+1}} \frac{\partial^{2l+1}}{\partial p^{2l+1}} W(x,p;t).$$

Moreover, with the explicit expressions Eqs. (D.10) and (D.14) for $\mathcal{T}^{(\pm)}$ and $\mathcal{U}^{(\pm)}$ we find that the two phase space equations

$$\left[\frac{p^2}{2M} + U(x) - \frac{\hbar^2}{8M} \frac{\partial^2}{\partial x^2} + \sum_{l=1}^{\infty} \frac{(-1)^l (\hbar/2)^{2l}}{(2l)!} \frac{d^{2l}U}{dx^{2l}} \frac{\partial^{2l}}{\partial p^{2l}} \right] W_{|E''\rangle\langle E'|} = \frac{E' + E''}{2} W_{|E''\rangle\langle E'|}$$

and

$$\left[\frac{p}{M} \frac{\partial}{\partial x} - \frac{\partial U}{\partial x} \frac{\partial}{\partial p} - \sum_{l=1}^{\infty} \frac{(-1)^l (\hbar/2)^{2l}}{(2l+1)!} \frac{d^{2l+1}U}{dx^{2l+1}} \frac{\partial^{2l+1}}{\partial p^{2l+1}} \right] W_{|E''\rangle\langle E'|} = \frac{i}{\hbar} (E'' - E') W_{|E''\rangle\langle E'|}$$

determine the Moyal function.

E Airy Function

The Airy function plays a central role in the construction of WKB wave functions. Here, we briefly review the essential properties of this function. Moreover, we derive asymptotic expansions in various domains, discuss the Stokes and anti-Stokes lines and address the Stokes phenomenon.

E.1 Definition and Differential Equation

In 1838 Airy defined a diffraction integral representing for each color the variation of the electromagnetic wave across a rainbow. For real arguments z this Airy function is defined by the integral

$$\text{Ai}(z) \equiv \frac{1}{2\pi} \int\limits_{-\infty}^{\infty} dt \, \exp\left(\frac{i}{3} t^3 + izt\right). \tag{E.1}$$

For arbitrary complex numbers z it is a similar integral with a modified contour.

There are physical phenomena such as the fractional revivals discussed in Problem 9.4 where Airy functions with complex arguments become important. However, in many situations the functional dependence of $\text{Ai}(z)$ is only of interest along the real axis $\text{Re}(z) \equiv x$.

The definition Eq. (E.1) clearly brings out that the Airy function is a generalization of the δ-function

$$\delta(z) \equiv \frac{1}{2\pi} \int\limits_{-\infty}^{\infty} dt \, \exp\left(izt\right)$$

and involves the integration variable z not only in a linear way but also as the third power. We note that higher order generalizations with the fifth power also exist. For a detailed discussion refer to the literature.

The Airy function satisfies the differential equation

$$\text{Ai}''(z) - z \, \text{Ai}(z) = 0.$$

We can prove this relation by differentiating the definition Eq. (E.1) twice with respect to z which yields

$$\text{Ai}''(z) = \frac{1}{2\pi} \int\limits_{-\infty}^{\infty} dt \, \exp\left(\frac{i}{3} t^3\right) (-t^2) \exp\left(izt\right).$$

When we integrate by parts we arrive at

$$\mathrm{Ai}''(z) = -\frac{1}{2\pi i} \exp\left(\frac{i}{3}t^3\right) \exp(izt)\Big|_{-\infty}^{\infty} + \frac{1}{2\pi} \int_{-\infty}^{\infty} dt \exp\left(\frac{i}{3}t^3\right) \exp(izt).$$

The first term does not converge in an absolute way. However, we could have added an exponential $\exp(-\varepsilon|t|)$ with a small decay. That would have insured the vanishing of the integrand at infinity. We therefore can drop the first term and find the differential equation Eq. (E.1) of the Airy function.

We conclude this section by noting that there is even a rigorous mathematical way of defining the limit of oscillatory functions. This procedure also reappears in the theory of divergent sums. Space does not allow us to go deeper in this fascinating topic where sums such as

$$\Sigma \equiv 1 - 1 + 1 - 1 + \ldots = 1 - (1 - 1 + 1 - 1 + \ldots) = 1 - \Sigma$$

take on the value

$$\Sigma = \frac{1}{2}$$

or, more relevant for our problem, where

$$\sin(\infty) = \cos(\infty) = 0.$$

For more details we refer to the book by Hardy but emphasize that the theory of divergent sums becomes important again in the context of the Airy function when we discuss the Stokes phenomenon in Sec. E.2.3.

E.2 Asymptotic Expansion

In this section we perform an asymptotic expansion of the Airy function, Eq. (5.19). We apply the method of stationary phase to the integral

$$\mathrm{Ai}(z) \equiv \frac{1}{2\pi} \int_{-\infty}^{\infty} dt \, \exp\left[iS(t; z)\right]$$

defining the Airy function where

$$S(t; z) \equiv \frac{1}{3}t^3 + zt \qquad (E.2)$$

denotes the phase.

We expand the phase

$$S(t; z) \cong S(t = t_s; z) + \frac{1}{2} \frac{\partial^2 S}{\partial t^2}\bigg|_{t_s} (t - t_s)^2$$

around the points

$$t_s = \pm\sqrt{-z} \qquad (E.3)$$

of stationary phase following from the condition

$$\left.\frac{\partial S}{\partial t}\right|_{t_s} = t_s^2 + z \stackrel{!}{=} 0.$$

In this appendix we confine ourselves to real values of $z = x$. For the general case of complex arguments of the Airy function we refer to the literature.

According to Eq. (E.3), we find for $x > 0$ that the points

$$t_s = \pm i\sqrt{|x|} \qquad (E.4)$$

of stationary phase are purely imaginary. In contrast for $x < 0$ the points

$$t_s = \pm\sqrt{|x|} \qquad (E.5)$$

are purely real. We also note that for $x = 0$ the two points of stationary phase collapse into a single one. Here our asymptotic expansion breaks down.

We first consider the asymptotic expansion of the Airy function for negative arguments. A brief discussion of the asymptotics for positive arguments concludes this section.

E.2.1 Oscillatory Regime

When we substitute the expression Eq. (E.5) for the stationary phase point for $x < 0$ into the expression Eq. (E.2) for the phase $S(t; x)$ we get

$$S(t_s; x) = \pm \frac{1}{3}|x|^{3/2} - |x|\left(\pm|x|^{1/2}\right) = \mp\frac{2}{3}|x|^{3/2}$$

and

$$\left.\frac{\partial^2 S}{\partial t^2}\right|_{t_s} = 2t|_{t_s} = \pm 2|x|^{1/2}.$$

This allows us to expand the phase $S(t; x)$ around t_s and we arrive at

$$S(t; x) \cong S(t_s; x) + \frac{1}{2}\left.\frac{\partial^2 S}{\partial t^2}\right|_{t_s}(t - t_s)^2 = \mp\frac{2}{3}|x|^{3/2} \pm |x|^{1/2}(t \mp |x|^{1/2})^2.$$

Hence, the integral representation of the Airy function Eq. (5.19) reads

$$\text{Ai}(x) \cong \exp\left[-\frac{2}{3}i|x|^{3/2}\right]\frac{1}{2\pi}\int_{-\infty}^{\infty} dt \exp\left[i|x|^{1/2}\left(t - |x|^{1/2}\right)^2\right]$$

$$+ \exp\left[\frac{2}{3}i|x|^{3/2}\right]\frac{1}{2\pi}\int_{-\infty}^{\infty} dt \exp\left[-i|x|^{1/2}\left(t + |x|^{1/2}\right)^2\right]. \qquad (E.6)$$

With the help of the integral relation

$$\int_{-\infty}^{\infty} dt\, e^{i\alpha t^2} = \sqrt{\frac{\pi}{|\alpha|}}\, e^{i\,\text{sign}\,\alpha\,\pi/4} \qquad (E.7)$$

we find

$$\text{Ai}(x) \cong \frac{1}{\sqrt{\pi}}|x|^{-1/4}\cos\left(\frac{2}{3}|x|^{3/2} - \frac{\pi}{4}\right) \qquad \text{for } x < 0. \qquad (E.8)$$

E.2.2 Decaying Regime

We now turn to the case when $x > 0$. Here the points of stationary phase as well as the phase S become purely imaginary. Indeed, when we substitute the stationary phase point Eq. (E.4) into the expression Eq. (E.2) for the phase we find

$$S(t_s; x) = \mp \frac{i}{3}|x|^{3/2} + |x|\left(\pm i|x|^{1/2}\right) = \pm i \frac{2}{3}|x|^{3/2}.$$

Similarly, the second derivative of the phase reads

$$\left.\frac{\partial^2 S}{\partial t^2}\right|_{t_s} = 2t|_{t_s} = \pm 2i|x|^{1/2}$$

and the Taylor expansion of the phase S takes the form

$$S(t; x) \cong \pm i \frac{2}{3}|x|^{3/2} \pm i|x|^{1/2}\left(t \mp i|x|^{1/2}\right)^2.$$

Since the Airy function involves the phase S in the exponential with a prefactor i we now find

$$\text{Ai}(x) \cong \exp\left[-\frac{2}{3}|x|^{3/2}\right] \frac{1}{2\pi}\int_{-\infty}^{\infty} dt \, \exp\left[-|x|^{1/2}\left(t - i|x|^{1/2}\right)^2\right]$$
$$+ \exp\left[+\frac{2}{3}|x|^{3/2}\right] \frac{1}{2\pi}\int_{-\infty}^{\infty} dt \, \exp\left[+|x|^{1/2}\left(t + i|x|^{1/2}\right)^2\right]. \quad \text{(E.9)}$$

When we compare this expression valid for $x > 0$ with the corresponding formula Eq. (E.6) we find again two contributions from the two points of stationary phase. However, since these points are purely imaginary and there is a prefactor i in the exponential the two contributions have now exponentially decaying and growing terms rather than oscillatory terms. Moreover, in the second integral in Eq. (E.9) a Gaussian with positive sign occurs. Consequently, this integral does not converge. When we analyze the behavior of the Airy function by, for example evaluating the integral numerically, we will find that the Airy function decays exponentially for positive x. Hence, the second term in Eq. (E.9) has to be neglected. When we perform the remaining Gauss integral the appropriate asymptotic expansion of the Airy function for positive x therefore reads

$$\text{Ai}(x) \cong \frac{1}{2\sqrt{\pi}}|x|^{-1/4} \exp\left[-\frac{2}{3}|x|^{3/2}\right].$$

We emphasize that so far we have not given a rigorous mathematical reason for the disappearance of the contribution resulting from the second point of stationary phase. Indeed, this is a rather sophisticated problem which we briefly discuss in the next section.

E.2.3 Stokes Phenomenon

In the discussion of the asymptotics of the Airy function we have found that the Airy function changes its behavior as we go from negative to positive arguments: It goes from an oscillatory function at negative arguments to a decaying function at positive arguments. The oscillations result from the interference of two points of stationary phase giving raise to two complex phases whereas the decaying contribution is due to a single point of stationary phase giving raise to a real-valued exponential. Indeed, in this domain there is a second point of stationary phase but in order to find agreement with the exact behavior we had to discard its contribution. However, we could not give an explanation for this phenomenon of disappearance of a stationary phase point. How is it that on the bright side of the Airy function, that is for $x < 0$, we have two exponentials whereas on the dark side, that is for $x > 0$, we have one exponential? Where does this exponential switch on?

Definition of Stokes and Anti-Stokes Lines

The answer to this question lies in complex space. It is related to the definition of Stokes lines. In the literature there exist essentially two definitions of Stokes and anti-Stokes lines. Here we follow the one used by Heading.

We are considering two independent solutions f_1 and f_2 of a differential equation of second order which have the leading behavior of $\exp[S_1(z)]$ and $\exp[S_2(z)]$, respectively, that is

$$f_1 \sim \exp[S_1(z)] \quad \text{and} \quad f_2 \sim \exp[S_2(z)].$$

Moreover, we assume that in the different domains of complex space the two solutions display quite different behavior: There are domains where one is dominant over the other or where both are about equal. Stokes and anti-Stokes lines define these domains.

Anti-Stokes lines in complex space are defined by the condition

$$\mathrm{Re}\left[S_1(z) - S_2(z)\right] = 0.$$

Hence, on anti-Stokes lines both f_1 and f_2 have the same magnitude.

In contrast, the Stokes lines are defined by the condition

$$\mathrm{Im}\left[S_1(z) - S_2(z)\right] = 0.$$

Hence, on this line the different character of the two asymptotic solutions f_1 and f_2 comes out most clearly since

$$f_1(z) \sim e^{\mathrm{Re}\, S_1(z)} e^{i\varphi} \quad \text{and} \quad f_2(z) \sim e^{\mathrm{Re}\, S_2(z)} e^{i\varphi}$$

where $\varphi = \mathrm{Im}\, S_1(z) = \mathrm{Im}\, S_2(z)$. When we cross a Stokes line we therefore see that the two solutions exchange their character. They go from a dominant (subdominant) to a subdominant (dominant) solution.

E Airy Function

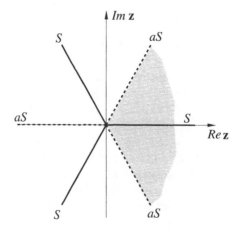

Fig. E.1: Stokes (S) and anti-Stokes (aS) lines of the Airy function in complex space. In the grey domain the Airy function Ai is subdominant. In the remaining part it is dominant.

Application to Airy Function

In the case of the Airy function we have the differential equation

$$f'' - zf = 0$$

with the two independent solutions $f_1(z) \equiv \text{Ai}(z)$ and $f_2(z) \equiv \text{Bi}(z)$. We have already shown that the asymptotic expressions for $\text{Ai}(z)$ read

$$f_1(z) \sim z^{-1/4} \exp\left[-\frac{2}{3}z^{3/2}\right].$$

Following a similar analysis we find

$$f_2(z) \sim z^{-1/4} \exp\left[\frac{2}{3}z^{3/2}\right].$$

Hence, the condition $\text{Re}\, z^{3/2} = 0$ defines the anti-Stokes lines

$$\arg z = \pm\frac{\pi}{3}, \pi,$$

whereas the Stokes lines defined by $\text{Im}\, z^{3/2} = 0$ fulfill the condition

$$\arg z = 0, \pm\frac{2\pi}{3},$$

as shown in Fig. E.1.

Hence, the Airy function is subdominant in the sector $-\pi/3 < \arg z < \pi/3$ of complex space depicted in Fig. E.1 by the grey area. But it is dominant in the complementary domain. Consequently, Bi is dominant for $-\pi/3 < \arg z < \pi/3$ and subdominant in the complementary domain.

Switching Function

In 1847 Stokes devised approximations for Ai(z) for large positive z values and in particular derived the formally exact representation

$$\text{Ai}(z) = \frac{\exp\left[-\frac{2}{3}z^{3/2}\right]}{2z^{1/4}\sqrt{\pi}} \sum_{r=0}^{\infty} (-1)^r T_r$$

with coefficients

$$T_r = \frac{1}{(36z^{3/2})^r} \frac{(3r - \frac{1}{2})!}{r!(r - \frac{1}{2})!}.$$

We can show that for large values of r these coefficients have the form

$$T_r \cong \frac{(r-1)!}{2\pi \left(\frac{4}{3} z^{3/2}\right)^r}.$$

Therefore, when z is large the terms T_r decrease at first and then increase. As a consequence the series diverges. However, as shown by M. Berry one can truncate this sum at an appropriate point and resume the divergent tail using Borel summation. This leads to the representation

$$\text{Ai}(z) = \frac{\exp\left[-\frac{2}{3}z^{3/2}\right]}{2z^{1/4}\sqrt{\pi}} \sum_{r=0}^{r^*} (-1)^r T_r + i\Theta(z) \frac{\exp\left[+\frac{2}{3}z^{3/2}\right]}{2z^{1/4}\sqrt{\pi}} \sum_{r=0}^{r^*} T_r$$

where $r^* \equiv 4z^{3/2}/3$.

Here $\Theta(z)$ switches rapidly but smoothly from 0 to 1 across the Stokes line $z = |z|\exp[i2\pi/3]$ thereby generating the second exponential which is subdominant.

References

G.H. Hardy, *Divergent Series*, Oxford University Press, Oxford, 1963

For the behavior of the Airy function and the Stokes lines see
R.B. Dingle, *Asymptotic Expansions: Their Derivation and Interpretation*, Academic Press, New York, 1973

J. Heading, *An Introduction to Phase Integral Methods*, Methuen, London, 1962

M.V. Berry and **C.J. Howls**, *Infinity interpreted*, Physics World, **6**(6), 35–39, (1993)

M.V. Berry and **C.J. Howls**, *Hyper Asymptotics for Integrals with Saddles*, Proc. R. Soc. London, **A434**, 657–675, (1991)

F Radial Equation

In Sec. 3.5 we have reduced the phase space equations for the Wigner function of an energy eigenstate of a harmonic oscillator to the ordinary differential equation

$$\left(y \frac{d^2}{dy^2} + \frac{d}{dy} - y + 2\eta \right) W_{|\eta\rangle}(y) = 0 \qquad (F.1)$$

for the variable $y \equiv r^2$. In this appendix we show that the normalization condition of the Wigner function enforces the discreteness of the energy eigenvalues and selects the Wigner function. For this purpose we solve the radial equation Eq. (F.1) and find

$$W_{|m\rangle}(y) = \frac{(-1)^m}{\pi\hbar} L_m(2y) \, e^{-y}. \qquad (F.2)$$

Ansatz for Radial Equation

To gain some insight into the solution we first discuss the differential equation in the limit of $y \to \infty$. In this regime Eq. (F.1) reduces to

$$\left(\frac{d^2}{dy^2} - 1 \right) W_{|\eta\rangle}(y) = 0$$

with the solution

$$W_{|\eta\rangle} \sim e^{-y}.$$

This leads us to the ansatz

$$W_{|\eta\rangle}(y) = L(y) \, e^{-y} \qquad (F.3)$$

which when substituted into Eq. (F.1) yields the differential equation

$$y \frac{d^2}{dy^2} L + (1 - 2y) \frac{d}{dy} L + (2\eta - 1) L = 0. \qquad (F.4)$$

We solve this equation using the ansatz

$$L(y) = \sum_{j=0}^{\infty} c_j \, y^j \qquad (F.5)$$

in the form of a power series. The first two derivatives then read

$$\frac{d}{dy}L(y) = \sum_{j=0}^{\infty} j\, c_j\, y^{j-1} = \sum_{j=0}^{\infty}(j+1)\,c_{j+1}\,y^j$$

and

$$\frac{d^2}{dy^2}L(y) = \sum_{j=0}^{\infty} j(j-1)\,c_j\,y^{j-2} = \sum_{j=0}^{\infty}(j+1)j\,c_{j+1}\,y^{j-1}$$

which when substituted into Eq. (F.4) provide

$$\sum_{j=0}^{\infty}\left[(j+1)j\,c_{j+1} + (j+1)\,c_{j+1} - 2j\,c_j + (2\eta-1)\,c_j\right]y^j = 0.$$

We therefore obtain the two-term recurrence relation

$$(j+1)^2\,c_{j+1} = 2(j+\tfrac{1}{2}-\eta)\,c_j$$

or

$$c_{j+1} = \frac{2(j+\tfrac{1}{2}-\eta)}{(j+1)^2}\,c_j. \qquad (F.6)$$

Energy Quantization from Normalization

It is straightforward to solve this recurrence relation. However, this is not even necessary when we only want to understand the quantization condition. For this case we consider the asymptotics of the recurrence relation, that is in the case of large j-values. In this limit the recurrence relation reads

$$c_j \cong \frac{2}{j}\,c_{j-1}$$

which provides the solution

$$c_j \cong \frac{2^j}{j!}\,c_0.$$

Therefore, the asymptotic behavior of $L(y)$ is

$$L(y) = \sum_j c_j\,y^j \sim \sum_j \frac{(2y)^j}{j!} = e^{2y}.$$

When we recall the ansatz Eq. (F.3) for the Wigner function the asymptotic behavior for $y \to \infty$ is

$$W_{|\eta\rangle}(y) \sim e^y.$$

Hence, this solution is not normalizable.

We ensure normalization by the requirement that the series (F.5) truncates at some power of y, say y^m, i.e. $c_{m+1} = 0$. From the recurrence relation (F.6) we then conclude that

$$\eta_m \equiv m + \frac{1}{2} \quad \text{or} \quad E_m = \left(m+\frac{1}{2}\right)\hbar\Omega.$$

Indeed, the requirement of normalization has lead to the quantization of energy.

Explicit Form of L_m

When we truncate the recurrence relation at the mth term we find

$$c_j = -\frac{2(m-j+1)}{j^2} c_{j-1} = \frac{(-2)^j}{j!} \frac{(m-j+1)(m-j+2)\cdots m}{j!} c_0 = \frac{(-2)^j}{j!} \binom{m}{j} c_0$$

and the solution is thus

$$L(y) = c_0 \sum_{j=0}^{m} \binom{m}{j} \frac{(-2y)^j}{j!} \equiv c_0 \, L_m(2y), \tag{F.7}$$

where

$$L_m(x) \equiv \sum_{j=0}^{m} \binom{m}{j} \frac{(-x)^j}{j!}$$

denotes the mth Laguerre polynomials.

We have therefore determined the Wigner function

$$W_{|m\rangle}(y) = c_0 \, L_m(2y) \, e^{-y}$$

of the mth energy eigenstate.

Normalization Constant

Our last task is to find the normalization constant c_0. With the help of this solution and the definition Eq. (F.7) of the Laguerre polynomial the normalization condition

$$\int_{-\infty}^{\infty} dx \int_{-\infty}^{\infty} dp \, W_{|m\rangle}(x,p) = \hbar \int_{0}^{\infty} dr\, r \int_{0}^{2\pi} d\varphi \, W_{|m\rangle}(r) = \pi\hbar \int_{0}^{\infty} dy \, W_{|m\rangle}(y) \stackrel{!}{=} 1$$

translates into

$$\pi\hbar c_0 \sum_{j=0}^{m} \binom{m}{j} \frac{(-2)^j}{j!} \int_{0}^{\infty} dy \, y^j \, e^{-y} = \pi\hbar c_0 \sum_{j=0}^{m} \binom{m}{j} \frac{(-2)^j}{j!} \Gamma(j+1) = \pi\hbar c_0 \sum_{j=0}^{m} \binom{m}{j}(-2)^j.$$

When we evaluate the binomial sum

$$\sum_{j=0}^{m} \binom{m}{j} 1^{m-j}(-2)^j = (1-2)^m = (-1)^m$$

we find

$$c_0 = \frac{(-1)^m}{\pi\hbar}.$$

Therefore, the Wigner function of the mth energy eigenstate takes indeed the form Eq. (F.2).

G Asymptotics of a Poissonian

In Chapter 4 we derive the energy distribution of a coherent state by evaluating the overlap integral between a coherent state and an energy eigenstate. When we approximate the energy wave function by a δ-function located at the turning point of the classical motion we naturally arrive at half integer quantum numbers. Moreover, we derive the same result in Chapter 8 from the area of overlap formalism. However, the standard asymptotics of the Poissonian does not arrive at this result. In the present appendix we use an improved Stirling formula to obtain the half integer expression.

The Poisson distribution

$$W_m = \frac{\lambda^m}{m!} e^{-\lambda} \tag{G.1}$$

shows the familiar Gaussian limit for $\lambda \to \infty$,

$$W_m \cong \frac{1}{\sqrt{2\pi\lambda}} \exp\left\{-\left[\frac{m-\lambda}{\sqrt{2\lambda}}\right]^2\right\}. \tag{G.2}$$

Since for any integer value of λ the exact Poisson distribution, Eq.(G.1), satisfies the relation

$$W_{\lambda-1} = W_\lambda,$$

the maximum of W_m is located at $\lambda - \frac{1}{2}$ in contrast to the naive Gaussian approximation Eq. (G.2).

An improved approximation, valid at least in the neighborhood of $m = \lambda$, is therefore

$$W_m \cong \frac{1}{\sqrt{2\pi\lambda}} \exp\left\{-\left[\frac{m+\frac{1}{2}-\lambda}{\sqrt{2\lambda}}\right]^2\right\}, \tag{G.3}$$

as confirmed by Fig. 4.7.

We now derive the asymptotic result Eq. (G.3) bringing in half-integers based on the improved Stirling formula

$$m! \cong \sqrt{2\pi} \, \frac{(m+\frac{1}{2})^{m+1/2}}{e^{m+1/2}}. \tag{G.4}$$

From the logarithm of Eq. (G.1)

$$\ln W_m = m \ln \lambda - \ln(m!) - \lambda,$$

we find with the help of Eq. (G.4)

$$\ln W_m = \left(m + \frac{1}{2}\right) \ln \left(\frac{\lambda}{m + \frac{1}{2}}\right) + m + \frac{1}{2} - \lambda - \ln \sqrt{2\pi\lambda}.$$

When we introduce $\delta \equiv m + \frac{1}{2} - \lambda$ the above equation reads

$$\ln W_m = -\lambda \left(1 + \frac{\delta}{\lambda}\right) \ln \left(1 + \frac{\delta}{\lambda}\right) + \delta - \ln \sqrt{2\pi\lambda}$$

which for $\delta/\lambda \ll 1$ reduces with the help of $\ln(1 - x) \cong -x - \frac{1}{2}x^2$ to

$$\ln W_m \cong -\frac{1}{2}\frac{\delta^2}{\lambda} - \ln \sqrt{2\pi\lambda},$$

a result equivalent to Eq. (G.3).

References

For a discussion of the improved Stirling formula see
C. Leubner, *Generalized Stirling approximations to N!*, Euro. J. Phys. **6**, 299–301 (1985)

H Toolbox for Integrals

A key tool throughout the book is the method of stationary phase. This technique allows us to evaluate a certain class of integrals in an approximate but analytical way. A crucial ingredient are the Fresnel integrals. In this appendix we first briefly review the method of stationary phase and then analyze the emerging Fresnel integral using the Cornu spiral. This has also important implications on the concept of interference in phase space discussed in Chapter 7.

H.1 Method of Stationary Phase

In order to bring out the main ideas of the method we first concentrate on the approximate evaluation of one-dimensional integrals. We then turn to the case of a multi-dimensional integral.

H.1.1 One-Dimensional Integrals

The method applies to integrals of the form

$$I(x) \equiv \int_{-\infty}^{\infty} dt\, f(t;x) \exp[ig(t;x)].$$

The function g may contain regimes of the integration variable t where it changes slowly. Moreover, it may contain domains where g is a rapidly increasing, decreasing or oscillating function. In this case the function $\exp[ig(t;x)]$ is rapidly varying. When we assume that f is slowly varying compared to these oscillations, this domain of the integration variable does not contribute significantly to the integral. Obviously the main contribution to the integral arises from the slowly varying regime of g, that is where

$$\left. \frac{\partial g(t;x)}{\partial t} \right|_{t=t_s} = 0.$$

We refer to the point t_s where this derivative vanishes as the point of stationary phase.

Single Point of Stationary Phase

We then expand the phase g of the exponential into a Taylor series

$$g(t;x) \cong g(t_s;x) + \frac{1}{2}\left.\frac{\partial^2 g(t;x)}{\partial t^2}\right|_{t=t_s}(t-t_s)^2 = g(t_s;x) + \frac{1}{2}c(t_s;x)(t-t_s)^2$$

around the points t_s where the first derivative of g vanishes. Here we have introduced the curvature

$$c(t_s;x) \equiv \left.\frac{\partial^2 g(t;x)}{\partial t^2}\right|_{t=t_s}$$

at the point of stationary phase.

When we evaluate the slowly varying function f at t_s and factor it out of the integral we arrive at

$$I(x) \cong f(t_s;x)\,\exp\left[ig(t_s;x)\right]\int_{-\infty}^{\infty}dt\,\exp\left[\frac{i}{2}c(t_s;x)(t-t_s)^2\right].$$

Note that we have retained the same limits of integration as in the original integral. This is not quite trivial since we have expanded the phase g into a Taylor series and have factored out the function f evaluated at t_s. Two questions arise: (i) Is the emerging integral convergent? (ii) How large a neighborhood around t_s is contributing to the integral? The first question is easy to answer when we recall the Fresnel integral

$$G(\gamma) \equiv \int_0^{\infty}dt\,e^{i\gamma t^2} = \frac{1}{2}\sqrt{\frac{\pi}{|\gamma|}}\,e^{i\,\mathrm{sign}\gamma\,\pi/4}. \qquad (H.1)$$

The answer to the second question lies in the behavior of the Cornu spiral which we discuss in the next section.

The integral relation Eq. (H.1) finally provides the approximate expression

$$I(x) \cong \sqrt{\frac{2\pi}{|c(t_s;x)|}}\,f(t_s;x)\,\exp\left\{i\left[g(t_s;x) + \mathrm{sign}\,(c(t_s;x))\frac{\pi}{4}\right]\right\}.$$

We note that the curvature c enters in the denominator of the square root. Hence, the approximate expression for I becomes singular when the curvature of g vanishes. In this case we have to include the third derivative of g and the resulting integral is the Airy function discussed in Appendix E. This expansion is usually referred to as the uniform asymptotic expansion.

The problem of a vanishing curvature is relevant when we recall that the point of stationary phase t_s depends on x. Therefore, $g(t_s;x)$ and $c(t_s;x)$ are functions of x. In particular one can imagine x values such that c vanishes.

Many Points of Stationary Phase

So far we have assumed that g enjoys only a single point of stationary phase. However, we can generalize the above result immediately to the case of many points t_j

of stationary phase. In this case we have to take the sum

$$I(x) \cong \sum_j \sqrt{\frac{2\pi}{|c(t_j;x)|}} f(t_j;x) \exp\left\{i\left[g(t_j;x) + \text{sign}\left(c(t_j;x)\right)\frac{\pi}{4}\right]\right\}$$

of contributions arising from all points t_j.

The sign of the curvature at the individual points t_j determines the sign of the constant phase shift.

H.1.2 Multi-Dimensional Integrals

We now apply the method of stationary phase to a multi-dimensional integral

$$I(x) \equiv \int_{-\infty}^{\infty} dt^{(1)} \ldots \int_{-\infty}^{\infty} dt^{(n)}\, f(t^{(1)},\ldots,t^{(n)};x)\, \exp\left[ig\left(t^{(1)},\ldots,t^{(n)};x\right)\right].$$

The parameter x can also take a multi-dimensional form. Since this does not modify our analysis we have suppressed this generalization.

The main contribution to the integral arises from the points

$$(t)_j \equiv \left(t_j^{(1)},\ldots,t_j^{(n)}\right)$$

of stationary phase defined in the n-dimensional space spanned by the n integration variables $t^{(1)},\ldots,t^{(n)}$. We find these points from the condition

$$\left.\frac{\partial}{\partial t^{(k)}} g\left(t^{(1)},\ldots,t^{(n)};x\right)\right|_{(t)_j} = 0.$$

Indeed, the function g depends on n integration-variables. In order to find the point of stationary phase we perform a partial differentiation with respect to all n variables. This yields n equations which can involve in each equation all n variables. The coordinates $t_j^{(1)},\ldots,t_j^{(n)}$ of the point of stationary phase $(t)_j$ satisfy these equations. Here we have also allowed for various different points of stationary phase numbered by the subscript j.

We now expand the function g into a Taylor series around the points $(t)_j$, that is

$$g\left(t^{(1)},\ldots,t^{(n)};x\right) \cong g((t)_j;x) + \frac{1}{2}\sum_{k,l} c_{kl}((t)_j;x)\left(t^{(k)} - t_j^{(k)}\right)\left(t^{(l)} - t_j^{(l)}\right) \quad (H.2)$$

where we have evaluated the curvature matrix

$$c_{kl}(t^{(1)},\ldots,t^{(n)};x) \equiv \frac{\partial^2}{\partial t^{(k)}\, \partial t^{(l)}} g(t^{(1)},\ldots,t^{(n)};x)$$

at the points $(t)_j$ of stationary phase.

When we assume that f is slowly varying we can evaluate it at $(t)_j$ and factor it out of the integral. With the expansion Eq. (H.2) of g we arrive at

$$I(x) \cong \sum_j f((t)_j;x)\, \exp\left[ig((t)_j;x)\right] J_j\left((t)_j;x\right) \quad (H.3)$$

where we have introduced the multi-dimensional integral

$$J_j((t)_j; x) \equiv \int_{-\infty}^{\infty} d\tau^{(1)} \cdots \int_{-\infty}^{\infty} d\tau^{(n)} \exp\left[\frac{i}{2} \sum_{k,l} \tau^{(k)} c_{kl}((t)_j; x) \tau^{(l)}\right]$$

with integration variables

$$\tau^{(k)} \equiv \left(t^{(k)} - t_j^{(k)}\right).$$

This integral is the generalization of the Fresnel integral to higher dimensions. Indeed, our notation brings out most clearly that the phase of the exponential function is a row-vector times a matrix times a column-vector resulting in a number. The matrix is a $n \times n$ matrix since the vectors formed by the integration variables $\tau^{(k)}$ have n components. Moreover, the curvature matrix is real and symmetric provided g is a real-valued function and such that the order of the partial differentiations giving rise to c_{kl} is not important.

Whenever the curvature matrix involves off-diagonal elements the integrations do not separate. However, since the matrix is real and symmetric, we can diagonalize it with the help of an orthogonal matrix U. This allows us to decompose the n-dimensional coupled integrations into n independent integrations. For this purpose we cast the bilinear form

$$\sum_{k,l} \tau^{(k)} c_{kl}((t)_j; x) \tau^{(l)} = \sum_m \lambda_j^{(m)} \xi_m^2$$

into an expression of purely quadratic terms. Here we have introduced the eigenvalues $\lambda_j^{(k)}$ of the curvature matrix and the transformed integration variables

$$\xi_m = \sum_l U_{ml} \tau^{(l)}. \tag{H.4}$$

The index j appears in the eigenvalues since the curvature matrix depends on the point $(t)_j$ of stationary phase.

We now change the integration variables according to the above transformation Eq. (H.4). This brings in the absolute value of the Jacobi determinant. For this linear transformation this quantity is the determinant of U. Since U is orthogonal we have $|\det U| = 1$. Consequently we arrive at

$$d\tau^{(1)} \cdots d\tau^{(n)} = d\xi_1 \cdots d\xi_n.$$

The exponential of the sum over the quadratic terms is the product of exponentials of the individual quadratic terms and hence, the integral J reduces to a product of one-dimensional Fresnel integrals, that is

$$J_j((t)_j; x) = \prod_{m=1}^{n} \int_{-\infty}^{\infty} d\xi_m \exp\left[\frac{i}{2} \lambda_j^{(m)} \xi_m^2\right].$$

With the help of the integral relation Eq. (H.1) we find

$$J_j((t)_j; x) = \sqrt{\frac{(2\pi)^n}{\left|\lambda_j^{(1)} \cdots \lambda_j^{(n)}\right|}} \exp\left[i \sum_m \text{sign}\lambda_j^{(m)} \frac{\pi}{4}\right].$$

When we recall that the product of the eigenvalues $\lambda_j^{(m)}$ is the determinant of the curvature matrix, that is

$$c((t)_j; x) \equiv \det c_{kl}((t)_j; x)$$

and introduce the integer

$$\nu_j(x) \equiv \sum_{m=1}^{n} \operatorname{sign}\lambda_j^{(m)}(x)$$

this result reads

$$J_j((t)_j; x) = \sqrt{\frac{(2\pi)^n}{|c((t)_j; x)|}} \exp\left[i\nu_j(x)\frac{\pi}{4}\right].$$

When we substitute this expression into Eq. (H.3) we arrive at

$$I(x) \cong \sum_j \sqrt{\frac{(2\pi)^n}{|c((t)_j; x)|}} f((t)_j; x) \exp\left[i\left(g((t)_j; x) + \nu_j(x)\frac{\pi}{4}\right)\right].$$

When we compare this expression to the corresponding one-dimensional result we find a great similarity. However, the constant factor 2π is replaced by $(2\pi)^n$, and the curvature is replaced by the determinant of the curvature matrix. The most important difference lies in the phase shift. Here the sign of the curvature is replaced by the sum of the signs of the eigenvalues $\lambda_j^{(m)}$. This quantity is closely related to the Maslov index of semiclassical quantum mechanics.

H.2 Cornu Spiral

An important tool in the method of stationary phase was the integral relation Eq. (H.1)

$$G(\gamma) \equiv \int_0^\infty dt \, e^{i\gamma t^2} = \frac{1}{2}\sqrt{\frac{\pi}{|\gamma|}} e^{i\operatorname{sign}\gamma \, \pi/4}.$$

It appears over and over again in the field of semiclassical quantum mechanics. How can we derive and understand this key formula? A mathematical and a geometrical approach offer themselves.

The mathematical derivation starts from the standard real-valued Gauss integral

$$\widetilde{G}(\sigma) \equiv \int_0^\infty dt \, e^{-\sigma t^2} = \frac{1}{2}\sqrt{\frac{\pi}{\sigma}}$$

valid for $\operatorname{Re}\sigma \geq 0$. When we now substitute into this integral $\sigma = -i\gamma$ and recall the relations

$$-i = e^{-i\pi/2}$$
$$\gamma = |\gamma|e^{-i(\operatorname{sign}\gamma - 1)\pi/2}$$

we find

$$\sigma = -i\gamma = |\gamma|e^{-i\operatorname{sign}\gamma\pi/2}$$

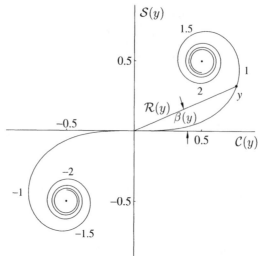

Fig. H.1: The Cornu spiral follows in Cartesian coordinates from the Fresnel integrals $C(y) \equiv \int_0^y dt \cos(\pi t^2/2)$ and $S(y) \equiv \int_0^y dt \sin(\pi t^2/2)$. When we increase the path length y of the spiral the quantities C and S oscillate and converge in the limit of $y \to \infty$ towards steady-state values $C(\infty) = S(\infty) = 1/2$. The convergence towards these values is, however, very slow.

and hence

$$G(\gamma) = \tilde{G}(\sigma = -i\gamma) = \frac{1}{2}\sqrt{\frac{\pi}{|\gamma|}} e^{i\,\mathrm{sign}\gamma\,\pi/4}.$$

This approach provides the correct answer. However, it does not yield any insight into the integral, nor does it provide a clue for the convergence of the integral.

Deeper insight into these questions springs from the geometrical approach where we consider the Cornu spiral

$$F(y) \equiv \int_0^y dt\, e^{i\pi t^2/2}.$$

In this definition we have included for historical reasons the factor $\pi/2$. This integral is familiar from diffraction theory and leads to the Cornu spiral shown in Fig. H.1. Here the horizontal and vertical axis denote the real part

$$\mathcal{C}(y) \equiv \mathrm{Re}\left(\int_0^y dt\, \exp\left(i\frac{\pi}{2}t^2\right)\right) \qquad (H.5a)$$

and the imaginary part

$$\mathcal{S}(y) \equiv \mathrm{Im}\left(\int_0^y dt\, \exp\left(i\frac{\pi}{2}t^2\right)\right) \qquad (H.5b)$$

of F respectively. The spiral is anti-symmetric in the path-length y, that is $F(-y) = -F(y)$, since the integrand is symmetric. A negative path length therefore is measured along the Cornu spiral in the third quadrant.

This picture suggests that an increase in y results in oscillations in $\mathcal{C}$ and $\mathcal{S}$. In the limit of $y \to \infty$, these two functions reach the steady state values

$$\mathcal{C}(y \to \infty) = \mathcal{S}(y \to \infty) = \frac{1}{2}$$

which we can read off of Fig. H.1.

The function

$$\overline{F}(y) \equiv \int_0^y dt\, e^{-i\pi t^2/2} = F^*(y)$$

has a negative imaginary part and is therefore the complex conjugate of the Cornu spiral shown in Fig. H.1.

In order to make contact with the integral G we now express the Fresnel integral in Cartesian coordinates, that is, in the integrals $\mathcal{C}$ and $\mathcal{S}$. We therefore find the relation

$$G(\gamma) = \sqrt{\frac{\pi}{2|\gamma|}}\, \mathcal{C}(y \to \infty) + i\,\mathrm{sign}(\gamma)\, \mathcal{S}(y \to \infty).$$

When we recall the relation

$$e^{\pm i\pi/4} = \frac{1}{\sqrt{2}}(1 \pm i)$$

we find indeed the integral relation Eq. (H.1).

In order to understand the convergence of the Fresnel integral F we recall that the integrand is a circle in complex space that is traversed faster and faster as the integration variable t increases. This speed up is due to the t^2 dependence. As a result the integrand oscillates faster and faster. This guarantees the convergence of the integral.

References

The method of stationary phase is described in many textbooks. See for example
B. Friedmann, *Lectures on Application-oriented Mathematics*, Holden Day, San Francisco, 1969
J. Heading, *An Introduction to Phase Integral Methods*, Methuen, London, 1962
W. Wasow, *Asymptotic Expansions for Ordinary Differential Equations*, Wiley, New York, 1965
N.G. de Bruijn, *Asymptotic Methods in Analysis*, North-Holland, Amsterdam, 1985
The method of steepest descent is summarized in
R. Courant and **D. Hilbert**, *Methoden der mathematischen Physik*, Springer, Heidelberg, 1931
English translation
Methods of Mathematical Physics, Interscience Publishers, New York, 1953
For a nice discussion and representation of the Cornu spiral see
E. Jahnke and **F. Emde**, *Tables of Functions*, Dover, New York, 1945

I Area of Overlap

In this appendix we calculate the diamond shaped area of overlap $\mathcal{A}_{m,n}^{(\diamond)}$ in phase space between the two Planck-Bohr-Sommerfeld bands. For this purpose we use the representation Eq. (7.20)

$$\mathcal{A}_{m,n}^{(\diamond)} = \frac{1}{2} \left. \frac{\partial^2 a(\mathcal{J}^{(1)}, \mathcal{J}^{(2)})}{\partial \mathcal{J}^{(1)} \partial \mathcal{J}^{(2)}} \right|_{\substack{\mathcal{J}^{(1)}=m+\frac{1}{2} \\ \mathcal{J}^{(2)}=n+\frac{1}{2}}}$$

of this area as a double derivative of the area $a(\mathcal{J}^{(1)}, \mathcal{J}^{(2)})$ caught between the orbits corresponding to the reduced actions $\mathcal{J}^{(1)}$ and $\mathcal{J}^{(2)}$. According to Eq. (7.15) this phase space area reads

$$a = a(\mathcal{J}^{(1)}, \mathcal{J}^{(2)}) = 2 \left\{ \int_{\zeta(\mathcal{J}^{(2)})}^{x_c(\mathcal{J}^{(1)}, \mathcal{J}^{(2)})} dx\, p(x; \mathcal{J}^{(2)}) + \int_{x_c(\mathcal{J}^{(1)}, \mathcal{J}^{(2)})}^{\xi(\mathcal{J}^{(1)})} dx\, p(x; \mathcal{J}^{(1)}) \right\}.$$

In this expression we have brought out most clearly the dependence on the two reduced actions $\mathcal{J}^{(1)}$ and $\mathcal{J}^{(2)}$.

I.1 Diamond Transformed into a Rectangle

We start by differentiating with respect to $\mathcal{J}^{(1)}$ and find

$$\frac{1}{2} \frac{\partial a}{\partial \mathcal{J}^{(1)}} = p(x_c; \mathcal{J}^{(2)}) \frac{\partial x_c}{\partial \mathcal{J}^{(1)}} - p(x_c; \mathcal{J}^{(1)}) \frac{\partial x_c}{\partial \mathcal{J}^{(1)}} + p(\xi; \mathcal{J}^{(1)}) \frac{\partial \xi}{\partial \mathcal{J}^{(1)}} + \int_{x_c}^{\xi} dx\, \frac{\partial p(x; \mathcal{J}^{(1)})}{\partial \mathcal{J}^{(1)}}.$$

We recall that at the crossing the momenta of the two orbits are identical, that is

$$p(x_c; \mathcal{J}^{(1)}) = p(x_c; \mathcal{J}^{(2)}) \tag{I.1}$$

and that at the turning point the momentum vanishes

$$p(\xi; \mathcal{J}^{(1)}) = 0.$$

These relations reduce this result to

$$\frac{1}{2}\frac{\partial a}{\partial \mathcal{J}^{(1)}} = \int_{x_c(\mathcal{J}^{(1)},\mathcal{J}^{(2)})}^{\xi(\mathcal{J}^{(1)})} dx\, \frac{\partial p(x;\mathcal{J}^{(1)})}{\partial \mathcal{J}^{(1)}}$$

and the differentiation with respect to $\mathcal{J}^{(2)}$ yields for the area of overlap, Eq. (7.20), the expression

$$\mathcal{A}_{m,n}^{(\diamond)} = -\left[\frac{\partial p(x;\mathcal{J}^{(1)})}{\partial \mathcal{J}^{(1)}}\bigg|_{x=x_c} \frac{\partial x_c}{\partial \mathcal{J}^{(2)}}\right]_{\substack{\mathcal{J}^{(1)}=m+1/2 \\ \mathcal{J}^{(2)}=n+1/2}}. \qquad (I.2)$$

This equation has an interesting interpretation in phase space. Indeed, it suggests to approximate the area of one of the two, black, diamond-shaped, zones of crossover in Fig. 7.3 by a rectangle of height

$$\Delta p^{(\text{PBS})} \equiv p(x_c; m+1) - p(x_c; m) \cong \frac{\partial p(x;\mathcal{J}^{(1)})}{\partial \mathcal{J}^{(1)}}\bigg|_{\substack{x=x_c \\ \mathcal{J}^{(1)}=m+1/2}}$$

and width

$$\Delta x^{(\text{PBS})} \equiv x_c(\mathcal{J}^{(1)}=m+1/2; \mathcal{J}^{(2)}=n+1) - x_c(\mathcal{J}^{(1)}=m+1/2; \mathcal{J}^{(2)}=n)$$
$$= \frac{\partial x_c(\mathcal{J}^{(1)};\mathcal{J}^{(2)})}{\partial \mathcal{J}^{(2)}}\bigg|_{\substack{\mathcal{J}^{(1)}=m+1/2 \\ \mathcal{J}^{(2)}=n+1/2}}$$

as shown in Fig. I.1. Of course neither $\Delta x^{(\text{PBS})}$ nor $\Delta p^{(\text{PBS})}$ individually describes the true extensions of the diamond in position or momentum: The quantity $\Delta x^{(\text{PBS})}$ underestimates the length of the diamond, whereas $\Delta p^{(\text{PBS})}$ overestimates its height. These dimensions are constructed to yield a rectangle whose area in phase space is identical to that of the shaded diamond in Fig. I.1.

I.2 Area of Diamond

The expression Eq. (I.2) allows even another interpretation. To bring this out we rewrite the quantity $\partial x_c/\partial \mathcal{J}^{(2)}$. We start from the two phase space trajectories determined by the energy conservation

$$E^{(1)}(\mathcal{J}^{(1)}) = \frac{[p^{(1)}(x;\mathcal{J}^{(1)})]^2}{2M} + U^{(1)}(x) \qquad (I.3a)$$

and

$$E^{(2)}(\mathcal{J}^{(2)}) = \frac{[p^{(2)}(x;\mathcal{J}^{(2)})]^2}{2M} + U^{(2)}(x). \qquad (I.3b)$$

At the crossing point x_c the two momenta and hence the kinetic energies are equal, and we find

$$E^{(1)}(\mathcal{J}^{(1)}) - E^{(2)}(\mathcal{J}^{(2)}) - U^{(1)}(x_c) + U^{(2)}(x_c) = 0.$$

I.2 Area of Diamond

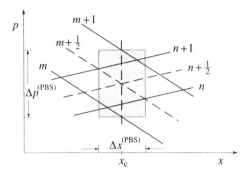

Fig. I.1: The area of the diamond-shaped zone of cross-over between the mth and the nth Planck-Bohr-Sommerfeld band is identical to the area of the rectangle depicted here. Its height $\Delta p^{(\mathrm{PBS})}$ is the difference in *momenta* at the edges of the m band at the position x_c of the crossing Kramers orbits, which are given by the reduced actions $m + \frac{1}{2}$ and $n + \frac{1}{2}$. The edges of the band correspond to the action values m and $m+1$. The width $\Delta x^{(\mathrm{PBS})}$ of the rectangle is the difference in the *position* coordinate of the crossing point between the edges of the n band and the central line of the m band. The quantities $\Delta x^{(\mathrm{PBS})}$ and $\Delta p^{(\mathrm{PBS})}$ do not represent the true extensions of the diamond in these directions, but are constructed as to provide its area.

We differentiate this equation with respect to $\mathcal{J}^{(2)}$ and arrive at

$$-\frac{\partial E^{(2)}(\mathcal{J}^{(2)})}{\partial \mathcal{J}^{(2)}} - \left[\frac{dU^{(1)}}{dx} - \frac{dU^{(2)}}{dx}\right]_{x=x_c} \frac{\partial x_c}{\partial \mathcal{J}^{(2)}} = 0, \qquad (I.4)$$

$$\frac{\partial x_c}{\partial \mathcal{J}^{(2)}} = -\left[\frac{dU^{(1)}}{dx} - \frac{dU^{(2)}}{dx}\right]^{-1}_{x=x_c} \frac{\partial E^{(2)}(\mathcal{J}^{(2)})}{\partial \mathcal{J}^{(2)}}. \qquad (I.5)$$

In order to express the derivatives of the potential we differentiate the expressions Eq. (I.3) with respect to x and find the relations

$$\frac{\partial U^{(j)}}{\partial x} = -\frac{p^{(j)}(x;\mathcal{J}^{(j)})}{M}\frac{dp^{(j)}(x;\mathcal{J}^{(j)})}{dx}$$

which reduce Eq. (I.5) to

$$\frac{\partial x_c}{\partial \mathcal{J}^{(2)}} = \left[\frac{dp^{(1)}(x;\mathcal{J}^{(1)})}{dx} - \frac{dp^{(2)}(x;\mathcal{J}^{(2)})}{dx}\right]^{-1}_{x=x_c} \frac{M}{p^{(2)}(x_c;\mathcal{J}^{(2)})} \frac{\partial E^{(2)}(\mathcal{J}^{(2)})}{\partial \mathcal{J}^{(2)}}.$$

For the interpretation of the expression for the area of overlap it is convenient to re-express the derivative of the energy with respect to the action in terms of the derivative of the momentum with respect to the action. Indeed, with the help of

$$\frac{\partial p}{\partial \mathcal{J}} = \frac{\partial p}{\partial E}\frac{\partial E}{\partial \mathcal{J}} = \frac{M}{p}\frac{\partial E}{\partial \mathcal{J}}$$

following from the phase space trajectory

$$E^{(1)}(\mathcal{J}^{(1)}) = \frac{[p^{(1)}(x; \mathcal{J}^{(1)})]^2}{2M} + U^{(1)}(x)$$

corresponding to the reduced action

$$\mathcal{J} = \frac{1}{2\pi\hbar} \oint dx\, p(x; \mathcal{J})$$

we find

$$\frac{M}{p} \frac{\partial E^{(2)}}{\partial \mathcal{J}^{(2)}} = \frac{\partial p}{\partial \mathcal{J}^{(2)}} \qquad (I.6)$$

which yields

$$\frac{\partial x_c}{\partial \mathcal{J}^{(2)}} = \left[\frac{dp(x; \mathcal{J}^{(1)})}{dx} - \frac{dp(x; \mathcal{J}^{(2)})}{dx} \right]^{-1}\Bigg|_{x=x_c} \frac{\partial p}{\partial \mathcal{J}^{(2)}}.$$

When we substitute this result into Eq. (I.2) we arrive at

$$\mathcal{A}_{m,n}^{(\diamond)} = - \frac{\partial p(x; \mathcal{J}^{(1)})}{\partial \mathcal{J}^{(1)}} \frac{\partial p(x; \mathcal{J}^{(2)})}{\partial \mathcal{J}^{(2)}} \left[\frac{dp(x; \mathcal{J}^{(1)})}{dx} - \frac{dp(x; \mathcal{J}^{(2)})}{dx} \right]^{-1}\Bigg|_{\substack{x=x_c \\ \mathcal{J}^{(1)}=m+1/2 \\ \mathcal{J}^{(2)}=n+1/2}}. \qquad (I.7)$$

Hence, the area of the diamond is the product of three terms. Two contributions arise from the derivatives of the corresponding momenta with respect to the reduced actions. Since in the Bohr correspondence limit we have

$$p(x; m+1) - p(x; m) \cong \frac{\partial p(x; \mathcal{J})}{\partial \mathcal{J}}\Bigg|_{\mathcal{J}=m+\frac{1}{2}}$$

these derivatives denote the heights of the two Planck-Bohr-Sommerfeld bands at their crossing point.

The third term involves the differences of the derivatives with respect to position. This is a measure for the angle under which the two trajectories cross. Hence, the area of the diamond is determined by the heights of the two Planck-Bohr-Sommerfeld bands and their crossing angle. For a more detailed discussion see Problem 7.1.

I.3 Area of Overlap as Probability

In the last step, we now cast the expression Eq. (I.7) for the phase space area of the diamond into a form that brings out most clearly that it determines the probability $\mathcal{A}_{m,n}$. For this purpose we note from the definition of the reduced action, Eq. (7.17), the relation

$$\frac{\partial \mathcal{J}^{(j)}}{\partial E^{(j)}} = M(2\pi\hbar)^{-1} \oint dx\, p^{-1}(x; \mathcal{J}^{(j)}) \equiv (2\pi\hbar)^{-1} T(\mathcal{J}^{(j)}) \qquad (I.8)$$

which together with Eq. (I.6) yields

$$\frac{\partial p^{(1)}}{\partial \mathcal{J}^{(1)}} = \frac{2\pi\hbar M}{T(\mathcal{J}^{(1)})\, p^{(1)}(x;\mathcal{J}^{(1)})}.$$

Analogously we find

$$\frac{\partial p^{(2)}}{\partial \mathcal{J}^{(2)}} = \frac{2\pi\hbar M}{T(\mathcal{J}^{(2)})\, p^{(2)}(x;\mathcal{J}^{(2)})]}.$$

Therefore, Eq. (I.7) now reads

$$\mathcal{A}_{m,n}^{(\diamond)} \cong (2\pi\hbar)^2 M^2 \left[T_m\, p_m^{(1)}(x_c)\, T_n\, p_n^{(2)}(x_c) \right]^{-1} \left[p_n^{(1)\prime}(x_c) - p_m^{(2)\prime}(x_c) \right]^{-1}$$
$$= 2\pi\hbar\, \mathcal{A}_{m,n}.$$

Here we have set $T_m \equiv T(\mathcal{J}^{(1)} = m + 1/2)$, $T_n \equiv T(\mathcal{J}^{(2)} = n + 1/2)$, $p_m(x) \equiv p(x; \mathcal{J}^{(1)} = m + 1/2)$ and $p_n(x_c) \equiv p(x = x_c, \mathcal{J}^{(2)} = n + 1/2)$. Hence, one diamond shaped area of overlap scaled with respect to the area $2\pi\hbar$ represents the probability $\mathcal{A}_{m,n}$.

J P-Distributions

In this appendix we illustrate the formalism developed in Sec. 12.4.3 for converting the Q-function into the P-function. In this way we derive the P-distribution of a thermal, a photon number and a squeezed state. We first briefly summarize the procedure.

According to Eq. (12.43) the P-distribution follows from the Fourier transform

$$\tilde{Q}(\xi) \equiv \tilde{Q}(\xi_r, \xi_i) \equiv \int d\alpha_r \int d\alpha_i \, Q(\alpha_r, \alpha_i) \, e^{-i(\xi_r \alpha_r + \xi_i \alpha_i)} \tag{J.1}$$

of the Q-function of the quantum state of interest with the help of the inversion formula

$$P(\alpha) \equiv P(\alpha_r, \alpha_i) = \frac{1}{(2\pi)^2} \int d\xi_r \int d\xi_i \, e^{(\xi_r^2 + \xi_i^2)/4} \tilde{Q}(\xi_r, \xi_i) e^{i(\xi_r \alpha_r + \xi_i \alpha_i)}. \tag{J.2}$$

Throughout this appendix the limits of integrations are $\pm\infty$, unless specified otherwise.

Hence, the procedure is as follows: We start from the Q-function and find its Fourier transform $\tilde{Q}$. We then substitute this result into the above equation and perform the integration.

J.1 Thermal State

We start from the Q-function, Eq. (12.13),

$$Q(\alpha) = \frac{1}{\pi[\langle \hat{n} \rangle + 1]} \exp\left(-\frac{\alpha_r^2 + \alpha_i^2}{\langle \hat{n} \rangle + 1}\right) \tag{J.3}$$

of a thermal state and first calculate its Fourier transform

$$\tilde{Q}(\xi_r, \xi_i) = \frac{1}{\pi[\langle \hat{n} \rangle + 1]} \int_{-\infty}^{\infty} d\alpha_r \, e^{-\alpha_r^2/[\langle \hat{n} \rangle + 1] - i\xi_r \alpha_r} \int_{-\infty}^{\infty} d\alpha_i \, e^{-\alpha_i^2/[\langle \hat{n} \rangle + 1] - i\xi_i \alpha_i}.$$

With the help of the integral relation

$$\int_{-\infty}^{\infty} dx \, e^{-ax^2 + bx} = \sqrt{\frac{\pi}{a}} \exp\left(\frac{b^2}{4a}\right) \tag{J.4}$$

we find
$$\tilde{Q}(\xi_r, \xi_i) = \exp\left[-\frac{\langle \hat{n}\rangle + 1}{4}(\xi_r^2 + \xi_i^2)\right].$$

We substitute this expression into the inversion formula, Eq. (J.2), of the P-function and find
$$P(\alpha_r, \alpha_i) = \frac{1}{(2\pi)^2}\int_{-\infty}^{\infty} d\xi_r \int_{-\infty}^{\infty} d\xi_i \, \exp\left[-\frac{\langle \hat{n}\rangle}{4}(\xi_r^2 + \xi_i^2)\right] e^{i(\xi_r \alpha_r + \xi_i \alpha_i)} \qquad (J.5)$$

which after another application of the Gauss formula, Eq. (J.4), yields
$$P(\alpha_r, \alpha_i) = \frac{1}{\pi \langle \hat{n}\rangle} \exp\left[\frac{-|\alpha|^2}{\langle \hat{n}\rangle}\right].$$

Hence, the P-distribution of a thermal state is again a Gaussian.

J.2 Photon Number State

In Sec. 12.2 we have derived the Q-function
$$Q(\alpha) = \frac{1}{\pi} \frac{|\alpha|^{2n}}{n!} e^{-|\alpha|^2}$$

of the nth photon number state. In order to find the corresponding P-function we first evaluate the Fourier transform
$$\tilde{Q}(\xi_r, \xi_i) = \frac{1}{\pi}\int d\alpha_r \int d\alpha_i \frac{|\alpha|^{2n}}{n!} e^{-|\alpha|^2} e^{-i(\xi_r \alpha_r + \xi_i \alpha_i)} \qquad (J.6)$$

of this Q-function. With the substitutions
$$\alpha_r \equiv r \cos \varphi_\alpha \qquad \xi_r \equiv \sqrt{\xi_r^2 + \xi_i^2}\cos\varphi_\xi$$
$$\alpha_i \equiv r \sin \varphi_\alpha \qquad \xi_i \equiv \sqrt{\xi_r^2 + \xi_i^2}\sin\varphi_\xi$$

and the definition $\varphi \equiv \varphi_\alpha - \varphi_\xi$ this integral reads
$$\tilde{Q}(\xi_r, \xi_i) = \int_0^\infty dr \, r \frac{r^{2n}}{n!} e^{-r^2} \frac{1}{\pi}\int_{-\pi}^{\pi} d\varphi \, e^{-i\sqrt{\xi_r^2 + \xi_i^2}\, r \cos\varphi}.$$

With the help of the relations
$$\frac{1}{2\pi}\int_{-\pi}^{\pi} d\varphi \, e^{-iz\cos\varphi} = J_0(z)$$

and
$$\int_0^\infty dx \, e^{-x^2} x^{2n+1} J_0(2x\sqrt{z}) = \frac{n!}{2} e^{-z} L_n(z),$$

where $J_0(z)$ and $L_n(z)$ denote the zeroth Bessel function and the nth Laguerre polynomial, respectively we can perform the integrations. We finally arrive at the expression

$$\tilde{Q}(\xi_r, \xi_i) = e^{-(\xi_r^2+\xi_i^2)/4} L_n\left[\frac{1}{4}(\xi_r^2 + \xi_i^2)\right]$$

for the Fourier transform of the Q-function of the nth number state.

We substitute this expression into the inversion formula, Eq. (J.2), and now face the problem to perform the integral

$$P(\alpha_r, \alpha_i) = \frac{1}{(2\pi)^2} \int_{-\infty}^{\infty} d\xi_r \int_{-\infty}^{\infty} d\xi_i \, L_n\left[\frac{1}{4}(\xi_r^2 + \xi_i^2)\right] e^{i(\xi_r \alpha_r + \xi_i \alpha_i)}.$$

There is a serious problem with this integration: Since the function does not vanish for $\xi_r, \xi_i \to \infty$, we cannot perform the inverse transformation and express its solution by an analytic function. However, we can derive a formal expression in terms of derivatives of delta functions.

When we expand the Laguerre polynomial in powers of ξ_r and ξ_i we can cast the P-function in the form

$$P(\alpha_r, \alpha_i) = \frac{1}{(2\pi)^2} \int_{-\infty}^{\infty} d\xi_r \int_{-\infty}^{\infty} d\xi_i \, L_n\left[-\frac{1}{4}\left(\frac{\partial^2}{\partial \alpha_r^2} + \frac{\partial^2}{\partial \alpha_i^2}\right)\right] e^{i(\xi_r \alpha_r + \xi_i \alpha_i)}. \quad (J.7)$$

This allows us to factor the Laguerre polynomial of the Laplace operator

$$\Delta \equiv \frac{\partial^2}{\partial \alpha_r^2} + \frac{\partial^2}{\partial \alpha_i^2}$$

out of the integral, which we now can identify as a delta function. Hence we find

$$P(\alpha_r, \alpha_i) = L_n\left[-\frac{1}{4}\left(\frac{\partial^2}{\partial \alpha_r^2} + \frac{\partial^2}{\partial \alpha_i^2}\right)\right] \delta(\alpha)$$
$$= L_n(-\tfrac{1}{4}\Delta) \, \delta(\alpha).$$

The P-distribution of the nth number state follows from applying the two-dimensional Laplacian n times on a delta function.

J.3 Squeezed State

As a further example we now turn to the discussion of a squeezed state $|\psi_{\text{sq}}\rangle$. We start from the Q-function, Eq. (12.11),

$$Q(\alpha) = \frac{1}{\pi} |\langle \alpha | \psi_{\text{sq}} \rangle|^2 = \frac{2}{\pi} \frac{\sqrt{s}}{s+1} \exp\left[-\frac{2s}{s+1}(\alpha_r - \gamma_r)^2 - \frac{2}{s+1}(\alpha_i - \gamma_i)^2\right] \quad (J.8)$$

of the squeezed state derived in Problem 12.1.

Again we first need the Fourier transform, Eq. (J.1), of the above Q-function. When we substitute the expression Eq. (J.8) into Eq. (J.1) we find

$$\tilde{Q}(\xi_r, \xi_i) = \frac{2}{\pi} \frac{\sqrt{s}}{s+1} \int_{-\infty}^{\infty} d\alpha_r \, \exp\left[-\frac{2s}{s+1}(\alpha_r - \gamma_r)^2 - i\xi_r\alpha_r\right]$$

$$\times \int_{-\infty}^{\infty} d\alpha_i \, \exp\left[-\frac{2}{s+1}(\alpha_i - \gamma_i)^2 - i\xi_i\alpha_i\right].$$

After the substitution of variables

$$\beta_r \equiv \alpha_r - \gamma_r \quad \text{and} \quad \beta_i \equiv \alpha_i - \gamma_i$$

we obtain

$$\tilde{Q}(\xi_r, \xi_i) = \frac{2}{\pi} \frac{\sqrt{s}}{s+1} e^{-i\xi_r\gamma_r} \int_{-\infty}^{\infty} d\beta_r \, \exp\left[-\frac{2s}{s+1}\beta_r^2 - i\xi_r\beta_r\right]$$

$$\times e^{-i\xi_i\gamma_i} \int_{-\infty}^{\infty} d\beta_i \, \exp\left[-\frac{2}{s+1}\beta_i^2 - i\xi_i\beta_i\right].$$

We evaluate the integrals using the Gauss formula, Eq. (J.4), and arrive at

$$\tilde{Q}(\xi_r, \xi_i) = \exp\left[-\frac{s+1}{8s}\xi_r^2 - \frac{s+1}{8}\xi_i^2 - i(\xi_r\gamma_r + \xi_i\gamma_i)\right].$$

According to Eq. (12.41) the Fourier transform of the corresponding P-distribution reads

$$\tilde{P}(\xi_r, \xi_i) = \exp\left[\frac{1}{4}(\xi_r^2 + \xi_i^2)\right] \tilde{Q}(\xi_r, \xi_i)$$

$$= \exp\left[\frac{s-1}{8s}\xi_r^2 - \frac{s-1}{8}\xi_i^2 - i(\xi_r\gamma_r + \xi_i\gamma_i)\right]. \quad (J.9)$$

We emphasize that in general the Fourier transform $\tilde{P}(\xi_r, \xi_i)$ increases exponentially in either the ξ_r- or the ξ_i-coordinate. An exception is the case of a coherent state where $s = 1$. Here, we just have the imaginary term leading to a δ-function. Despite these convergence problems, it is still possible to formally perform the inverse Fourier transform. This is done in a way analogous to the case of the Fock state.

When we substitute the Fourier transform $\tilde{P}$ of the P-distribution, Eq. (J.9), into the inversion formula, Eq. (J.2), we find

$$P(\alpha_r, \alpha_i) = \frac{1}{2\pi} \int_{-\infty}^{\infty} d\xi_r \, \exp\left[\frac{s-1}{8s}\xi_r^2 + i\xi_r(\alpha_r - \gamma_r)\right]$$

$$\times \frac{1}{2\pi} \int_{-\infty}^{\infty} d\xi_i \, \exp\left[-\frac{s-1}{8}\xi_i^2 + i\xi_i(\alpha_i - \gamma_i)\right].$$

We make use of the expansion

$$e^{\lambda x^2} = \sum_{n=0}^{\infty} \frac{\lambda^n}{n!} x^{2n}$$

and arrive at

$$P(\alpha_r, \alpha_i) = \sum_{m=0}^{\infty} \frac{1}{m!} \left(\frac{s-1}{8s}\right)^m \frac{1}{2\pi} \int_{-\infty}^{\infty} d\xi_r \, \xi_r^{2m} \, e^{i\xi_r(\alpha_r - \gamma_r)}$$

$$\times \sum_{n=0}^{\infty} \frac{1}{n!} \left(\frac{1-s}{8}\right)^n \frac{1}{2\pi} \int_{-\infty}^{\infty} d\xi_i \, \xi_i^{2n} \, e^{i\xi_i(\alpha_i - \gamma_i)}.$$

We represent the powers of ξ_r and ξ_i as derivatives of the exponentials which yields

$$P(\alpha_r, \alpha_i) = \sum_{m=0}^{\infty} \frac{1}{m!} \left(\frac{s-1}{8s}\right)^m \left(\frac{\partial}{i\partial\alpha_r}\right)^{2m} \frac{1}{2\pi} \int_{-\infty}^{\infty} d\xi_r \, e^{i\xi_r(\alpha_r - \gamma_r)}$$

$$\times \sum_{n=0}^{\infty} \frac{1}{n!} \left(\frac{1-s}{8}\right)^n \left(\frac{\partial}{i\partial\alpha_i}\right)^{2n} \frac{1}{2\pi} \int_{-\infty}^{\infty} d\xi_i \, e^{i\xi_i(\alpha_i - \gamma_i)}$$

or

$$P(\alpha_r, \alpha_i) = \sum_{m=0}^{\infty} \frac{1}{m!} \left(\frac{1-s}{8s} \frac{\partial^2}{\partial\alpha_r^2}\right)^m \delta(\alpha_r - \gamma_r)$$

$$\times \sum_{n=0}^{\infty} \frac{1}{n!} \left(-\frac{1-s}{8} \frac{\partial^2}{\partial\alpha_i^2}\right)^n \delta(\alpha_i - \gamma_i).$$

Now we are able to perform the summation which provides us with

$$P(\alpha_r, \alpha_i) = \exp\left[\frac{1-s}{8s} \frac{\partial^2}{\partial\alpha_r^2} - \frac{1-s}{8} \frac{\partial^2}{\partial\alpha_i^2}\right] \delta(\alpha - \gamma).$$

This P-distribution is even more singular than the one of a Fock state, which involved only a finite number of derivatives of the delta function.

K Homodyne Kernel

In Sec. 13.2 we have derived an expression for the photon count difference $n_{21} \equiv n_2 - n_1$ of a homodyne detector in terms of a kernel $K_{n_{21}}(\alpha, \beta)$. This kernel involves as the summation index the number of photons n_1 or n_2 on the detectors 1 or 2. In this appendix we express this sum in closed form in terms of Bessel functions and discuss the strong local oscillator limit.

K.1 Explicit Evaluation of Kernel

According to Eq. (13.18) the kernel reads

$$K_{n_{21}}(\alpha, \beta) = \sum_{n_1=0}^{\infty} \left| \left\langle n_1 \left| \frac{\beta - \alpha}{\sqrt{2}} \right\rangle_1 \right|^2 \left| \left\langle n_{21} + n_1 \left| \frac{\beta + \alpha}{\sqrt{2}} \right\rangle_2 \right|^2$$

and represents the probability to count the photon difference n_{21} on the exit ports of a symmetric beam-splitter which has mixed the two coherent states $|\alpha\rangle$ and $|\beta\rangle$.

Since we have to satisfy the condition $0 \leq n_2 = n_{21} + n_1$ we have to consider two cases: (i) When $n_{21} > 0$ the summation is well-defined. (ii) When $n_{21} < 0$ the summation starts at $n_1 = |n_{21}|$. In this case the kernel reads

$$K_{-|n_{21}|}(\alpha, \beta) = \sum_{n_1=|n_{21}|}^{\infty} \left| \left\langle n_1 \left| \frac{\beta - \alpha}{\sqrt{2}} \right\rangle_1 \right|^2 \left| \left\langle -|n_{21}| + n_1 \left| \frac{\beta + \alpha}{\sqrt{2}} \right\rangle_2 \right|^2 .$$

We introduce the summation index $\bar{n}_1 \equiv -|n_{21}| + n_1$ and the kernel takes the form

$$K_{-|n_{21}|}(\alpha, \beta) = \sum_{\bar{n}_1=0}^{\infty} \left| \left\langle \bar{n}_1 + |n_{21}| \left| \frac{\beta - \alpha}{\sqrt{2}} \right\rangle_1 \right|^2 \left| \left\langle \bar{n}_1 \left| \frac{\beta + \alpha}{\sqrt{2}} \right\rangle_2 \right|^2 = K_{|n_{21}|}(-\alpha, \beta).$$

Hence the case of $n_{21} < 0$ follows from the case $n_{21} > 0$ by replacing α by $-\alpha$. Strictly speaking, this situation is not quite identical to the original one since the numbering of the modes is inverted. However, this fact does not matter in the evaluation of the sum.

Since it suffices to consider the case $n_{21} > 0$ we now perform the sum

$$K_{n_{21}} = \sum_{n_1=0}^{\infty} \left| \left\langle n_1 \left| \frac{\beta - \alpha}{\sqrt{2}} \right\rangle_1 \right|^2 \left| \left\langle n_{21} + n_1 \left| \frac{\beta + \alpha}{\sqrt{2}} \right\rangle_2 \right|^2 .$$

With the help of the explicit expression Eq. (13.12) for the Poissonian photon statistics of a coherent state this sum reads

$$K_{n_{21}} = \sum_{n_1=0}^{\infty} \left|\frac{\beta-\alpha}{\sqrt{2}}\right|^{2n_1} \left|\frac{\beta+\alpha}{\sqrt{2}}\right|^{2(n_{21}+n_1)} \frac{1}{n_1!\,(n_{21}+n_1)!} \exp\left[-\frac{1}{2}\left(|\beta-\alpha|^2 + |\beta+\alpha|^2\right)\right].$$

We recall the relation

$$|\beta-\alpha|^2 + |\beta+\alpha|^2 = (\beta-\alpha)(\beta^*-\alpha^*) + (\beta+\alpha)(\beta^*+\alpha^*) = 2\left(|\alpha|^2 + |\beta|^2\right)$$

and find

$$K_{n_{21}} = e^{-|\alpha|^2-|\beta|^2} \left|\frac{\beta+\alpha}{\beta-\alpha}\right|^{n_{21}} \sum_{n_1=0}^{\infty} \frac{\left|\frac{1}{2}(\beta-\alpha)(\beta+\alpha)\right|^{2n_1+n_{21}}}{n_1!\,(n_1+n_{21})!}.$$

The remaining sum is the modified Bessel function

$$I_\nu(z) = \sum_{n=0}^{\infty} \frac{(\frac{1}{2}z)^{2n+\nu}}{n!\,(n+\nu)!} \tag{K.1}$$

and we arrive at

$$K_{n_{21}} = e^{-|\alpha|^2-|\beta|^2} \left|\frac{\beta+\alpha}{\beta-\alpha}\right|^{n_{21}} I_{n_{21}}(|\beta^2-\alpha^2|).$$

For $n_{21} < 0$ or $n_{21} = -|n_{21}|$ we find the expression for $K_{-|n_{21}|}(\alpha, \beta)$ by replacing α by $-\alpha$ in $K_{|n_{21}|}(\alpha, \beta)$, that is

$$K_{-|n_{21}|}(\alpha, \beta) = \left|\frac{\beta-\alpha}{\beta+\alpha}\right|^{|n_{21}|} I_{|n_{21}|}(|\beta^2-\alpha^2|)\, e^{-|\alpha|^2-|\beta|^2}.$$

We can combine both cases to give the formula

$$K_{n_{21}}(\alpha, \beta) = \left|\frac{\beta+\alpha}{\beta-\alpha}\right|^{n_{21}} I_{|n_{21}|}(|\beta^2-\alpha^2|)\, e^{-|\alpha|^2-|\beta|^2}.$$

Indeed, this result satisfies the symmetry relation $K_{-|n_{21}|}(\alpha, \beta) = K_{|n_{21}|}(-\alpha, \beta)$.

K.2 Strong Local Oscillator Limit

In the limit of a coherent state $|\alpha\rangle$ of large amplitude, that is for $|\alpha| \gg 1$, this kernel simplifies considerably. In particular, we consider the limit when $n_{21}/|\alpha|$ remains finite as $|\alpha| \to \infty$.

For this purpose we recall the appropriate asymptotic expansion

$$I_\nu(z) \cong \frac{e^z}{\sqrt{2\pi z}} \exp\left(-\frac{\nu^2}{2z}\right)$$

of the modified Bessel function I_ν, Eq. (K.1) valid for $\nu/z \to 0$ but $\nu/\sqrt{z}$ finite. In this case the kernel $K_{n_{21}}$ reduces to

$$K_{n_{21}}(\alpha, \beta) \cong \left|\frac{\beta+\alpha}{\beta-\alpha}\right|^{n_{21}} \frac{1}{\sqrt{2\pi|\beta^2-\alpha^2|}} \exp\left(-\frac{n_{21}^2}{2|\beta^2-\alpha^2|}\right) e^{-|\alpha|^2-|\beta|^2+|\beta^2-\alpha^2|},$$

or
$$K_{n_{21}}(\alpha, \beta) \cong \left|\frac{\beta/\alpha + 1}{\beta/\alpha - 1}\right|^{n_{21}} \frac{e^{-|\alpha|^2 - |\beta|^2 + |\alpha|^2|1-(\beta/\alpha)^2|}}{\sqrt{2\pi|\alpha|^2|1-(\beta/\alpha)^2|}} \exp\left(-\frac{n_{21}^2}{2|\alpha|^2|1-(\beta/\alpha)^2|}\right). \quad (K.2)$$

When we use the relation
$$|1 - (\beta/\alpha)^2| = \sqrt{1 - (\beta/\alpha)^2 - (\beta^*/\alpha^*)^2 + |\beta/\alpha|^4},$$
neglect the term $|\beta/\alpha|^4$ and expand the square root we find
$$|1 - (\beta/\alpha)^2| \cong \left[1 - \frac{1}{2}(\beta/\alpha)^2 - \frac{1}{2}(\beta^*/\alpha^*)^2\right],$$
which provides us with the result
$$|\alpha|^2 |1 - (\beta/\alpha)^2| \cong |\alpha|^2 - \frac{1}{2}\left(\beta\frac{|\alpha|}{\alpha}\right)^2 - \frac{1}{2}\left(\beta^*\frac{|\alpha|}{\alpha^*}\right)^2 = |\alpha|^2 - \frac{1}{2}\left(\beta e^{-i\theta}\right)^2 - \frac{1}{2}\left(\beta^* e^{i\theta}\right)^2.$$

Here we have used the decomposition $\alpha = |\alpha|e^{i\theta}$ into amplitude $|\alpha|$ and phase θ of the local oscillator.

We substitute this expression back into the exponent in Eq. (K.2) and find
$$K_{n_{21}}(\alpha, \beta) \cong \frac{1}{\sqrt{2\pi|\alpha|^2}} \exp\left(n_{21} \ln\left|\frac{1+\beta/\alpha}{1-\beta/\alpha}\right|\right) \exp\left[-\frac{n_{21}^2}{2|\alpha|^2} - \frac{1}{2}\left(\beta e^{-i\theta} + \beta^* e^{i\theta}\right)^2\right]. \quad (K.3)$$

Here we have also neglected the quadratic terms $(\beta/\alpha)^2$ and $(\beta^*/\alpha^*)^2$ in the denominator of the Gaussian and the square root and have made use of the relation
$$-|\beta|^2 - \frac{1}{2}\left(\beta e^{-i\theta}\right)^2 - \frac{1}{2}\left(\beta^* e^{i\theta}\right)^2 = -\frac{1}{2}\left(\beta e^{-i\theta} + \beta^* e^{i\theta}\right)^2.$$

We are now left with the problem of simplifying the logarithm. With the expansions
$$\left|\frac{1+\beta/\alpha}{1-\beta/\alpha}\right| = \left|\frac{(1+\beta/\alpha)^2}{1-(\beta/\alpha)^2}\right| \cong \left|1 + \frac{\beta}{\alpha}\right|^2 = 1 + \frac{\beta}{\alpha} + \frac{\beta^*}{\alpha^*} + \left|\frac{\beta}{\alpha}\right|^2 \cong 1 + \frac{\beta}{\alpha} + \frac{\beta^*}{\alpha^*}$$
and
$$\ln(1+x) \cong x$$
the logarithm reduces to
$$\ln\left|\frac{1+\beta/\alpha}{1-\beta/\alpha}\right| \cong \frac{\beta}{\alpha} + \frac{\beta^*}{\alpha^*} = \frac{1}{|\alpha|}\left(\beta e^{-i\theta} + \beta^* e^{i\theta}\right).$$

When we substitute this expression back into the asymptotic result Eq. (K.3) for the kernel $K_{n_{21}}$ and combine the terms in the exponent we arrive at
$$K_{n_{21}}(\alpha = |\alpha|e^{i\theta}, \beta) \cong \frac{1}{\sqrt{2\pi|\alpha|^2}} \exp\left\{-\frac{1}{2}\left[\frac{n_{21}}{|\alpha|} - \left(\beta e^{-i\theta} + \beta^* e^{i\theta}\right)\right]^2\right\}. \quad (K.4)$$

Hence, in the strong local oscillator limit the kernel $K_{n_{21}}$ simplifies to a Gaussian distribution in the scaled variable $n_{21}/|\alpha|$ with a maximum at $\beta e^{-i\theta} + \beta^* e^{i\theta}$.

L Beyond the Dipole Approximation

In this appendix we derive the Hamiltonian Eq. (14.37) for an atom moving in an electromagnetic field characterized by $\vec{E}$ and $\vec{B}$. For this purpose we pursue an approach very similar to the one of Sec. 14.6. However, we now take into account the next order in the Taylor expansion Eq. (14.23), that is we include the first derivatives of the vector potential.

We then perform a gauge transformation on this Hamiltonian. First we deal only with a classical Hamiltonian. We therefore find the corresponding Lagrangian, and add a complete time derivative which yields a modified Lagrangian. Here the electromagnetic field enters only through $\vec{E}$ and $\vec{B}$. We then calculate the corresponding Hamiltonian.

In the second approach we concentrate on the quantum case and generalize the gauge transformation discussed in Sec. 14.6.2 to include the center-of-mass motion. This procedure yields a Hamiltonian which is identical to the one obtained from the classical approach.

L.1 First Order Taylor Expansion

In this section we improve the Hamiltonian

$$H^{(0)} \equiv \frac{1}{2M}\vec{P}^2 + \frac{1}{2\mu}\left(\vec{p} - e\vec{A}\right)^2 + V(\vec{r}) \tag{L.1}$$

in zeroth order approximation by including the next order in the Taylor expansion Eq. (14.23).

L.1.1 Expansion of the Hamiltonian

We start from the exact Hamiltonian Eq. (14.21)

$$H = \frac{\vec{P}^2}{2M} + \frac{\vec{p}^2}{2\mu} + V(\vec{r})$$

$$-\frac{e}{\mu}\left\{\frac{\mu}{m_e}\vec{A}\left[\vec{R}+\frac{m_p}{M}\vec{r},t\right]+\frac{\mu}{m_p}\vec{A}\left[\vec{R}-\frac{m_e}{M}\vec{r},t\right]\right\}\vec{p}$$
$$-\frac{e}{M}\left\{\vec{A}\left[\vec{R}+\frac{m_p}{M}\vec{r},t\right]-\vec{A}\left[\vec{R}-\frac{m_e}{M}\vec{r},t\right]\right\}\vec{P}$$
$$+\frac{e^2}{2m_e}\vec{A}^2\left[\vec{R}+\frac{m_p}{M}\vec{r},t\right]+\frac{e^2}{2m_p}\vec{A}^2\left[\vec{R}-\frac{m_e}{M}\vec{r},t\right]$$

obtained from the minimal coupling scheme and use the Taylor expansions

$$\vec{A}\left(\vec{R}+\frac{m_p}{M}\vec{r},t\right)\cong\vec{A}(\vec{R},t)+\frac{m_p}{M}\left(\vec{r}\cdot\vec{\nabla}_{\vec{R}}\right)\vec{A}(\vec{R},t) \quad (L.2a)$$

and

$$\vec{A}\left(\vec{R}-\frac{m_e}{M}\vec{r},t\right)\cong\vec{A}(\vec{R},t)-\frac{m_e}{M}\left(\vec{r}\cdot\vec{\nabla}_{\vec{R}}\right)\vec{A}(\vec{R},t) \quad (L.2b)$$

which yield

$$H^{(1)}\equiv H^{(0)}-\frac{e}{M}\left(\frac{m_p}{m_e}-\frac{m_e}{m_p}\right)\left[\left(\vec{r}\cdot\vec{\nabla}_{\vec{R}}\right)\vec{A}\right]\cdot\vec{p}-\frac{e}{M}\left[\left(\vec{r}\cdot\vec{\nabla}_{\vec{R}}\right)\vec{A}\right]\cdot\vec{P}$$
$$+\frac{e^2}{M}\left(\frac{m_p}{m_e}-\frac{m_e}{m_p}\right)\left[\left(\vec{r}\cdot\vec{\nabla}_{\vec{R}}\right)\vec{A}\right]\cdot\vec{A}+\frac{e^2}{2M^2}\left(\frac{m_p^2}{m_e}+\frac{m_e^2}{m_p}\right)\left[\left(\vec{r}\cdot\vec{\nabla}_{\vec{R}}\right)\vec{A}\right]^2, \quad (L.3)$$

where $H^{(0)}$ is the Hamiltonian, Eq. (L.1) in zeroth approximation.

When we recall the definitions Eqs. (14.13) and (14.17) of the total mass M and the reduced mass μ, respectively, we find with the abbreviation

$$\Delta m \equiv m_p - m_e$$

the relations

$$\left(\frac{m_p}{m_e}-\frac{m_e}{m_p}\right)=(m_p-m_e)\frac{(m_p+m_e)}{m_p m_e}=\frac{\Delta m}{\mu}$$

and

$$\frac{1}{M}+\frac{1}{\mu}\left(\frac{\Delta m}{M}\right)^2=\frac{M\mu+\Delta m^2}{M^2\mu}=\frac{1}{M^2}\frac{(m_p^2-m_p m_e+m_e^2)(m_p+m_e)}{m_p m_e}$$
$$=\frac{1}{M^2}\left(\frac{m_p^2}{m_e}+\frac{m_e^2}{m_p}\right).$$

With these formulas the Hamiltonian Eq. (L.3) reads

$$H^{(1)}\equiv H^{(0)}-\frac{e}{\mu}\frac{\Delta m}{M}\left[\left(\vec{r}\cdot\vec{\nabla}_{\vec{R}}\right)\vec{A}\right]\cdot\vec{p}-\frac{e}{M}\left[\left(\vec{r}\cdot\vec{\nabla}_{\vec{R}}\right)\vec{A}\right]\cdot\vec{P}$$
$$+\frac{e^2}{\mu}\frac{\Delta m}{M}\left[\left(\vec{r}\cdot\vec{\nabla}_{\vec{R}}\right)\vec{A}\right]\cdot\vec{A}+\frac{e^2}{2M}\left[\left(\vec{r}\cdot\vec{\nabla}_{\vec{R}}\right)\vec{A}\right]^2$$
$$+\frac{e^2}{2\mu}\left(\frac{\Delta m}{M}\right)^2\left[\left(\vec{r}\cdot\vec{\nabla}_{\vec{R}}\right)\vec{A}\right]^2.$$

When we now recall the form of the Hamiltonian $H^{(0)}$, Eq. (L.1) in zeroth approximation we can combine the terms to arrive at the rather compact formula

$$H^{(1)}=\frac{1}{2M}\left[\vec{P}-e\left(\vec{r}\cdot\vec{\nabla}_{\vec{R}}\right)\vec{A}\right]^2+\frac{1}{2\mu}\left[\vec{p}-e\vec{A}-e\frac{\Delta m}{M}\left(\vec{r}\cdot\vec{\nabla}_{\vec{R}}\right)\vec{A}\right]^2+V(\vec{r}). \quad (L.4)$$

L.1.2 Extension to Operators

We emphasize that the expression Eq. (L.4) for the Hamiltonian also holds true in the quantum case, that is when we interpret the momenta $\vec{P}$ and $\vec{p}$ as operators according to Jordan's rule

$$\vec{P} \to \hat{\vec{P}} \equiv \frac{\hbar}{i} \vec{\nabla}_{\vec{R}}$$

and

$$\vec{p} \to \hat{\vec{p}} \equiv \frac{\hbar}{i} \vec{\nabla}_{\vec{r}}.$$

Indeed, in Coulomb gauge with $\vec{\nabla} \cdot \vec{A} = 0$, the quantum mechanical cross-term

$$\left\{ \hat{\vec{P}} \cdot \left[\left(\vec{r} \cdot \vec{\nabla}_{\vec{R}} \right) \vec{A} \right] + \left[\left(\vec{r} \cdot \vec{\nabla}_{\vec{R}} \right) \vec{A} \right] \cdot \hat{\vec{P}} \right\} \Phi(\vec{r}, \vec{R}, t)$$

$$= \frac{\hbar}{i} \vec{\nabla}_{\vec{R}} \cdot \left\{ \left[\left(\vec{r} \cdot \vec{\nabla}_{\vec{R}} \right) \vec{A} \right] \Phi \right\} + \left[\left(\vec{r} \cdot \vec{\nabla}_{\vec{R}} \right) \vec{A} \right] \cdot \hat{\vec{P}} \Phi$$

$$= \frac{\hbar}{i} \left\{ \vec{\nabla}_{\vec{R}} \cdot \left[\left(\vec{r} \cdot \vec{\nabla}_{\vec{R}} \right) \vec{A} \right] \right\} \Phi + 2 \left[\left(\vec{r} \cdot \vec{\nabla}_{\vec{R}} \right) \vec{A} \right] \cdot \hat{\vec{P}} \Phi$$

$$= \frac{\hbar}{i} \left[\left(\vec{r} \cdot \vec{\nabla}_{\vec{R}} \right) \left(\vec{\nabla}_{\vec{R}} \cdot \vec{A} \right) \right] \Phi + 2 \left[\left(\vec{r} \cdot \vec{\nabla}_{\vec{R}} \right) \vec{A} \right] \cdot \hat{\vec{P}} \Phi$$

$$= 2 \left[\left(\vec{r} \cdot \vec{\nabla}_{\vec{R}} \right) \vec{A} \right] \cdot \hat{\vec{P}} \Phi$$

is identical to the classical one.

Similarly we find making use of the fact that the vector potential $\vec{A}$ depends only on the center-of-mass coordinate $\vec{R}$ the relation

$$\left\{ \hat{\vec{p}} \cdot \left[\left(\vec{r} \cdot \vec{\nabla}_{\vec{R}} \right) \vec{A} \right] + \left[\left(\vec{r} \cdot \vec{\nabla}_{\vec{R}} \right) \vec{A} \right] \cdot \hat{\vec{p}} \right\} \Phi(\vec{r}, \vec{R}, t)$$

$$= \frac{\hbar}{i} \vec{\nabla}_{\vec{r}} \cdot \left\{ \left[\left(\vec{r} \cdot \vec{\nabla}_{\vec{R}} \right) \vec{A} \right] \Phi \right\} + \left[\left(\vec{r} \cdot \vec{\nabla}_{\vec{R}} \right) \vec{A} \right] \cdot \hat{\vec{p}} \Phi$$

$$= \frac{\hbar}{i} \left\{ \vec{\nabla}_{\vec{r}} \cdot \left[\left(\vec{r} \cdot \vec{\nabla}_{\vec{R}} \right) \vec{A} \right] \right\} \Phi + 2 \left[\left(\vec{r} \cdot \vec{\nabla}_{\vec{R}} \right) \vec{A} \right] \cdot \hat{\vec{p}} \Phi = 2 \left[\left(\vec{r} \cdot \vec{\nabla}_{\vec{R}} \right) \vec{A} \right] \cdot \hat{\vec{p}} \Phi,$$

where in the last step we have applied the relation

$$\vec{\nabla}_{\vec{r}} \cdot \left(\vec{r} \cdot \vec{\nabla}_{\vec{R}} \right) \vec{A} = \frac{\partial}{\partial x_s} \left(x_j \cdot \frac{\partial}{\partial X_j} A_s(X) \right) = \delta_{s,j} \frac{\partial}{\partial X_j} A_s = \frac{\partial A_s}{\partial X_s} = \vec{\nabla} \cdot \vec{A} = 0,$$

and the Coulomb gauge condition.

L.2 Classical Gauge Transformation

The Hamiltonian $H^{(1)}$, Eq. (L.4) is not gauge invariant since it contains the vector potential and its first derivative. In order to express the potential solely in term of the electric and magnetic fields we now first calculate the corresponding Lagrangian $L^{(1)}$ and then add a complete time derivative. In this way we find an equivalent Lagrangian $\tilde{L}^{(1)}$, from which we then construct the corresponding Hamiltonian $\widetilde{H}^{(1)}$. This procedure is completely analogous to the one of Sec. 14.6.1 except that now we include the center-of-mass motion.

L.2.1 Lagrangian with Center-of-Mass Motion

In this section we calculate the Lagrangian

$$L^{(1)} \equiv \dot{\vec{R}} \cdot \vec{P} + \dot{\vec{r}} \cdot \vec{p} - H^{(1)} \qquad (L.5)$$

corresponding to the Hamiltonian $H^{(1)}$. For this purpose we first use the expression Eq. (L.4) for $H^{(1)}$ to evaluate the velocities

$$\dot{\vec{R}} \equiv \frac{\partial H^{(1)}}{\partial \vec{P}} = \frac{1}{M}\left[\vec{P} - e\left(\vec{r} \cdot \vec{\nabla}_{\vec{R}}\right)\vec{A}\right]$$

and

$$\dot{\vec{r}} \equiv \frac{\partial H^{(1)}}{\partial \vec{p}} = \frac{1}{\mu}\left[\vec{p} - e\vec{A} - e\frac{\Delta m}{M}\left(\vec{r} \cdot \vec{\nabla}_{\vec{R}}\right)\vec{A}\right]$$

which yields

$$\vec{P} = M\dot{\vec{R}} + e\left(\vec{r} \cdot \vec{\nabla}_{\vec{R}}\right)\vec{A}$$

and

$$\vec{p} = \mu\dot{\vec{r}} + e\vec{A} + e\frac{\Delta m}{M}\left(\vec{r} \cdot \vec{\nabla}_{\vec{R}}\right)\vec{A}.$$

These relations allow us to express the momenta $\vec{P}$ and $\vec{p}$ in the Lagrangian Eq. (L.5) and in the Hamiltonian $H^{(1)}$ by $\dot{\vec{R}}$ and $\dot{\vec{r}}$ and we arrive at

$$L^{(1)} = M\dot{\vec{R}}^2 + e\dot{\vec{R}} \cdot \left[\left(\vec{r} \cdot \vec{\nabla}_{\vec{R}}\right)\vec{A}\right] + \mu\dot{\vec{r}}^2 + e\dot{\vec{r}} \cdot \vec{A} + e\frac{\Delta m}{M}\dot{\vec{r}} \cdot \left[\left(\vec{r} \cdot \vec{\nabla}_{\vec{R}}\right)\vec{A}\right]$$
$$- \frac{1}{2}M\dot{\vec{R}}^2 - \frac{1}{2}\mu\dot{\vec{r}}^2 - V(\vec{r}).$$

Hence the Lagrangian corresponding to the Hamiltonian Eq. (L.4) reads

$$L^{(1)} = \frac{1}{2}M\dot{\vec{R}}^2 + \frac{1}{2}\mu\dot{\vec{r}}^2 + e\dot{\vec{r}} \cdot \vec{A} - V(\vec{r}) + e\dot{\vec{R}} \cdot \left[\left(\vec{r} \cdot \vec{\nabla}_{\vec{R}}\right)\vec{A}\right] + e\frac{\Delta m}{M}\dot{\vec{r}} \cdot \left[\left(\vec{r} \cdot \vec{\nabla}_{\vec{R}}\right)\vec{A}\right], \quad (L.6)$$

or

$$L^{(1)} = L^{(0)} + e\dot{\vec{R}} \cdot \left[\left(\vec{r} \cdot \vec{\nabla}_{\vec{R}}\right)\vec{A}\right] + e\frac{\Delta m}{M}\dot{\vec{r}} \cdot \left[\left(\vec{r} \cdot \vec{\nabla}_{\vec{R}}\right)\vec{A}\right],$$

where we have recalled the Lagrangian

$$L^{(0)} \equiv \frac{1}{2}M\dot{\vec{R}}^2 + \frac{1}{2}\dot{\vec{\mu}}^2 + e\dot{\vec{r}} \cdot \vec{A} - V(\vec{r})$$

in zeroth order Eq. (L.4). Hence, the center-of-mass motion brings in the two last terms of Eq. (L.6). However, as in the Hamiltonian these terms are not gauge invariant.

L.2.2 Complete Time Derivative

In Sec. 14.6.1 we have added a complete time derivative to the Lagrangian. This addition leaves the equations of motion invariant. Moreover, in this way we can eliminate the vector potential in favor of the electric field.

We now pursue an analogous approach and consider the new Lagrangian

$$\tilde{L}^{(1)} \equiv L^{(1)} - \frac{d}{dt}\left\{e\vec{r}\cdot\vec{A}(\vec{R},t) + \frac{e}{2}\frac{\Delta m}{M}\vec{r}\cdot\left[\left(\vec{r}\cdot\vec{\nabla}_{\vec{R}}\right)\vec{A}\right]\right\}. \tag{L.7}$$

When we perform the differentiation with respect to time and recognize that $\vec{R}$ and $\vec{r}$ are dynamical variables and thus depend on time we find

$$\tilde{L}^{(1)} = L^{(1)} - e\dot{\vec{r}}\vec{A} - e\cdot\vec{r}\cdot\frac{\partial\vec{A}}{\partial t} - e\dot{\vec{R}}\cdot\vec{\nabla}_{\vec{R}}\left(\vec{r}\cdot\vec{A}\right) - \frac{e}{2}\frac{\Delta m}{M}\left[\dot{\vec{r}}\left(\vec{r}\cdot\vec{\nabla}_{\vec{R}}\right)\vec{A}\right.$$

$$\left.+\vec{r}\left[\left(\dot{\vec{r}}\cdot\vec{\nabla}_{\vec{R}}\right)\vec{A}\right] + \left(\vec{r}\cdot\vec{\nabla}_{\vec{R}}\right)\dot{\vec{R}}\cdot\vec{\nabla}_{\vec{R}}\left(\vec{r}\cdot\vec{A}\right) + \vec{r}\cdot\left(\vec{r}\cdot\vec{\nabla}_{\vec{R}}\right)\frac{\partial\vec{A}}{\partial t}\right].$$

We recall that the connection

$$\vec{E} = -\frac{\partial\vec{A}}{\partial t} \tag{L.8}$$

between the electric field $\vec{E}$ and the vector potential in Coulomb gauge and note that we can neglect the term involving the second derivative of the potential since we are only in first order Taylor expansion. When we combine the remaining terms we arrive at

$$\tilde{L}^{(1)} = \frac{1}{2}M\dot{\vec{R}}^2 + \frac{1}{2}\mu\dot{\vec{r}}^2 - V(\vec{r}) + e\vec{r}\cdot\vec{E} + e\dot{\vec{R}}\cdot\left[\left(\vec{r}\cdot\vec{\nabla}_{\vec{R}}\right)\vec{A} - \vec{\nabla}_{\vec{R}}\left(\vec{r}\cdot\vec{A}\right)\right]$$

$$+\frac{e}{2}\frac{\Delta m}{M}\dot{\vec{r}}\cdot\left[\left(\vec{r}\cdot\vec{\nabla}_{\vec{R}}\right)\vec{A} - \vec{\nabla}_{\vec{R}}\left(\vec{r}\cdot\vec{A}\right)\right] + \frac{e}{2}\frac{\Delta m}{M}\left(\vec{r}\cdot\vec{\nabla}_{\vec{R}}\right)\left(\vec{r}\cdot\vec{E}\right),$$

which with the help of the relation

$$\vec{r}\times\vec{B} = \vec{r}\times\vec{\nabla}_{\vec{R}}\times\vec{A} = \vec{\nabla}_{\vec{R}}\left(\vec{r}\cdot\vec{A}\right) - \left(\vec{r}\cdot\vec{\nabla}_{\vec{R}}\right)\vec{A} \tag{L.9}$$

reduces to

$$\tilde{L}^{(1)} = \frac{1}{2}M\dot{\vec{R}}^2 + \frac{1}{2}\mu\dot{\vec{r}}^2 - V(\vec{r}) + e\vec{r}\cdot\vec{E} - e\dot{\vec{R}}\cdot\left(\vec{r}\times\vec{B}\right) - \frac{e}{2}\frac{\Delta m}{M}\dot{\vec{r}}\cdot\left(\vec{r}\times\vec{B}\right)$$

$$+\frac{e}{2}\frac{\Delta m}{M}\left(\vec{r}\cdot\vec{\nabla}_{\vec{R}}\right)\left(\vec{r}\cdot\vec{E}\right). \tag{L.10}$$

Here we have recalled the modified Lagrangian $\tilde{L}^{(0)}$ in zeroth approximation.

L.2.3 Hamiltonian Including Center-of-Mass Motion

We now calculate the Hamiltonian $\tilde{H}^{(1)}$ corresponding to the modified Lagrangian $\tilde{L}^{(1)}$. For this purpose we evaluate the Hamiltonian

$$\tilde{H}^{(1)} = \dot{\vec{R}}\cdot\vec{P} + \dot{\vec{r}}\cdot\vec{p} - \tilde{L}^{(1)} \tag{L.11}$$

by first finding the canonical momenta

$$\vec{P} \equiv \frac{\partial \tilde{L}^{(1)}}{\partial \dot{\vec{R}}} = M\dot{\vec{R}} - e\left(\vec{r} \times \vec{B}\right) \qquad (L.12)$$

and

$$\vec{p} \equiv \frac{\partial \tilde{L}^{(1)}}{\partial \dot{\vec{r}}} = \mu\dot{\vec{r}} - \frac{e}{2}\frac{\Delta m}{M}\left(\vec{r} \times \vec{B}\right) \qquad (L.13)$$

from the Lagrangian $\tilde{L}^{(1)}$, Eq. (L.10).

When we then express $\dot{\vec{R}}$ and $\dot{\vec{r}}$ in the formula Eq. (L.11) for the Hamiltonian and in the Lagrangian L', Eq. (L.10), by $\vec{P}$ and $\vec{p}$, Making use of these relations we arrive at

$$\begin{aligned}\tilde{H}^{(1)} =& \frac{1}{M}\left[\vec{P} + e\left(\vec{r} \times \vec{B}\right)\right]\cdot\vec{P} + \frac{1}{\mu}\left[\vec{p} + \frac{e}{2}\frac{\Delta m}{M}\left(\vec{r} \times \vec{B}\right)\right]\cdot\vec{p} - \frac{1}{2M}\left[\vec{P} + e\left(\vec{r} \times \vec{B}\right)\right]^2 \\ &- \frac{1}{2\mu}\left[\vec{p} + \frac{e}{2}\frac{\Delta m}{M}\left(\vec{r} \times \vec{B}\right)\right]^2 + V(\vec{r}) - e\vec{r}\cdot\vec{E} + \frac{1}{M}\left[\vec{P} + e\left(\vec{r} \times \vec{B}\right)\right]e\left(\vec{r} \times \vec{B}\right) \\ &+ \frac{1}{\mu}\left[\vec{p} + \frac{e}{2}\frac{\Delta m}{M}\left(\vec{r} \times \vec{B}\right)\right]\frac{e}{2}\frac{\Delta m}{M}\left(\vec{r} \times \vec{B}\right) - \frac{e}{2}\frac{\Delta m}{M}\left(\vec{r}\cdot\vec{\nabla}_R\right)\left(\vec{r}\cdot\vec{E}\right),\end{aligned}$$

which after combining appropriate terms yields

$$\begin{aligned}\tilde{H}^{(1)} =& \frac{1}{2M}\left[\vec{P} + e\left(\vec{r} \times \vec{B}\right)\right]^2 + \frac{1}{2\mu}\left[\vec{p} + \frac{e}{2}\frac{\Delta m}{M}\left(\vec{r} \times \vec{B}\right)\right]^2 + V(\vec{r}) - e\vec{r}\cdot\vec{E} \\ &- \frac{e}{2}\frac{\Delta m}{M}\left(\vec{r}\cdot\vec{\nabla}_R\right)\vec{r}\cdot\vec{E}.\end{aligned} \qquad (L.14)$$

Indeed, this Hamiltonian now contains the electromagnetic field through gauge invariant quantities such as $\vec{E}$ and $\vec{B}$.

L.3 Quantum Mechanical Gauge Transformation

In the preceeding section we have derived the gauge invariant Hamiltonian $\tilde{H}^{(1)}$ Eq. (L.14) including the center-of-mass motion. This classical derivation has started from the Hamiltonian Eq. (L.4) in first order approximation and is based on the addition of a complete time derivative to the Lagrangian $L^{(1)}$. In the present section we perform the corresponding quantum mechanical calculation.

L.3.1 Gauge Potential

For this purpose we start from the Schrödinger equation

$$i\hbar \frac{\partial \Phi}{\partial t} = H^{(1)}\Phi \qquad (L.15)$$

L.3 Quantum Mechanical Gauge Transformation

with the Hamiltonian

$$H^{(1)} \equiv H^{(0)} - \frac{e}{M}\left(\frac{m_p}{m_e} - \frac{m_e}{m_p}\right)\left(\vec{r}\cdot\vec{\nabla}_{\vec{R}}\right)\vec{A}\cdot\vec{p} - \frac{e}{M}\left(\vec{r}\cdot\vec{\nabla}_{\vec{R}}\right)\vec{A}\cdot\vec{P}$$
$$+ \frac{e^2}{M}\left(\frac{m_p}{m_e} - \frac{m_e}{m_p}\right)\left[\left(\vec{r}\cdot\vec{\nabla}_{\vec{R}}\right)\vec{A}\right]\cdot\vec{A} + \frac{e^2}{2M^2}\left(\frac{m_p^2}{m_e} - \frac{m_e^2}{m_p}\right)\left[\left(\vec{r}\cdot\vec{\nabla}_{\vec{R}}\right)\vec{A}\right]^2$$

and introduce the wave function $\tilde{\Phi}$ via

$$\Phi(\vec{r}, \vec{R}, t) \equiv \exp\left[\frac{i}{\hbar}e\,\Lambda(\vec{r}, \vec{R}, t)\right]\tilde{\Phi}(\vec{r}, \vec{R}, t) \tag{L.16}$$

where the gauge potential

$$\Lambda(\vec{r}, \vec{R}, t) \equiv \vec{r}\cdot\vec{A}(\vec{R}, t) + \frac{1}{2}\frac{\Delta m}{M}\left(\vec{r}\cdot\vec{\nabla}_{\vec{R}}\right)\left[\vec{r}\cdot\vec{A}(\vec{R}, t)\right] \tag{L.17}$$

is motivated by the complete time derivative Eq. (L.7).

Moreover, we recall that in order to be consistent with the first order Taylor expansion we have to neglect second derivatives of $\vec{A}$ with respect to $\vec{R}$. Derivatives with respect to $\vec{r}$ we have to retain.

We now derive the equation of motion for $\tilde{\Phi}$ by substituting the wave function Φ, Eq. (L.16), into the Schrödinger equation Eq. (L.15), which due to the product of the gauge factor $\exp(ie\Lambda/\hbar)$ and $\tilde{\Phi}$ results in many terms. We investigate each term separately.

Center-of-Mass Motion

We first calculate the term

$$\left[\hat{\vec{P}} - e\left(\vec{r}\cdot\vec{\nabla}_{\vec{R}}\right)\vec{A}\right]\left(\exp\left(\frac{i}{\hbar}e\,\Lambda\right)\tilde{\Phi}\right) = \exp\left(\frac{i}{\hbar}e\,\Lambda\right)\left[\hat{\vec{P}} + e\vec{\nabla}_{\vec{R}}\Lambda - e\left(\vec{r}\cdot\vec{\nabla}_{\vec{R}}\right)\vec{A}\right]\tilde{\Phi}$$

related to the center-of-mass motion, which the help of the expression Eq. (L.17) for Λ reads

$$\left[\hat{\vec{P}} - e\left(\vec{r}\cdot\vec{\nabla}_{\vec{R}}\right)\vec{A}\right]\Phi = \exp\left(\frac{i}{\hbar}e\,\Lambda\right)\left\{\hat{\vec{P}} + e\left[\vec{\nabla}_{\vec{R}}\left(\vec{r}\cdot\vec{A}\right) - \left(\vec{r}\cdot\vec{\nabla}_{\vec{R}}\right)\vec{A}\right]\right\}\tilde{\Phi}.$$

Here we have again neglected the second derivatives of the vector potential.

According to Eq. (L.9) we can combine both terms in the square bracket to give the magnetic field which yields

$$\left[\hat{\vec{P}} - e\left(\vec{r}\cdot\vec{\nabla}_{\vec{R}}\right)\vec{A}\right]\Phi = \exp\left(\frac{i}{\hbar}e\,\Lambda\right)\left[\hat{\vec{P}} + e\left(\vec{r}\times\vec{B}\right)\right]\tilde{\Phi}.$$

We now use this relation to find the action of the center-of-mass of the kinetic energy operator

$$\left[\hat{\vec{P}} - e\left(\vec{r}\cdot\vec{\nabla}_{\vec{R}}\right)\vec{A}\right]^2\Phi = \left[\hat{\vec{P}} - e\left(\vec{r}\cdot\vec{\nabla}_{\vec{R}}\right)\vec{A}\right]\exp\left(\frac{i}{\hbar}e\,\Lambda\right)\left[\hat{\vec{P}} + e\left(\vec{r}\times\vec{B}\right)\right]\tilde{\Phi}$$
$$= \exp\left(\frac{i}{\hbar}e\,\Lambda\right)\left[\hat{\vec{P}} + e\left(\vec{r}\times\vec{B}\right)\right]\left[\hat{\vec{P}} + e\left(\vec{r}\times\vec{B}\right)\right]\tilde{\Phi},$$

that is,

$$\left[\hat{\vec{P}} - e\left(\vec{r}\cdot\vec{\nabla}_{\vec{R}}\right)\vec{A}\right]^2\Phi = \exp\left(\frac{i}{\hbar}e\,\Lambda\right)\left[\hat{\vec{P}} + e\left(\vec{r}\times\vec{B}\right)\right]^2\tilde{\Phi}. \tag{L.18}$$

Relative Motion

We turn to the relative motion and, in particular, consider the expression

$$\left[\hat{\vec{p}} - e\vec{A} - e\frac{\Delta m}{M}\left(\vec{r}\cdot\vec{\nabla}_{\vec{R}}\right)\vec{A}\right]\left(\exp\left(\frac{i}{\hbar}e\Lambda\right)\tilde{\Phi}\right) =$$
$$\exp\left(\frac{i}{\hbar}e\Lambda\right)\left[\hat{\vec{p}} - e\vec{A} - e\frac{\Delta m}{M}\left(\vec{r}\cdot\vec{\nabla}_{\vec{R}}\right)\vec{A} + e\vec{\nabla}_{\vec{r}}\Lambda\right]\tilde{\Phi}.$$

From the definition Eq. (L.17) of the gauge potential Λ we find

$$\vec{\nabla}_{\vec{r}}\Lambda = \vec{A} + \frac{1}{2}\frac{\Delta m}{M}\left[\vec{\nabla}_{\vec{R}}\left(\vec{r}\cdot\vec{A}\right) + \left(\vec{r}\cdot\vec{\nabla}_{\vec{R}}\right)\vec{A}\right]$$

and thus

$$\left[\hat{\vec{p}} - e\vec{A} - e\frac{\Delta m}{M}\left(\vec{r}\cdot\vec{\nabla}_{\vec{R}}\right)\vec{A}\right]\left(\exp\left(\frac{i}{\hbar}e\Lambda\right)\tilde{\Phi}\right) =$$
$$\exp\left(\frac{i}{\hbar}e\Lambda\right)\left\{\hat{\vec{p}} + \frac{e}{2}\frac{\Delta m}{M}\left[\vec{\nabla}_{\vec{R}}\left(\vec{r}\cdot\vec{A}\right) - \left(\vec{r}\cdot\vec{\nabla}_{\vec{R}}\right)\vec{A}\right]\right\}\tilde{\Phi}.$$

Again, we use Eq. (L.9) to express the term in square brackets by the magnetic field and find

$$\left[\hat{\vec{p}} - e\vec{A} - e\frac{\Delta m}{M}\left(\vec{r}\cdot\vec{\nabla}_{\vec{R}}\right)\vec{A}\right]\left(\exp\left(\frac{i}{\hbar}e\Lambda\right)\tilde{\Phi}\right) =$$
$$\exp\left(\frac{i}{\hbar}e\Lambda\right)\left[\hat{\vec{p}} + \frac{e}{2}\frac{\Delta m}{M}\left(\vec{r}\times\vec{B}\right)\right]\tilde{\Phi}.$$

This formula allows us to calculate the kinetic energy operator of the relative motion on the wave function which yields

$$\left[\hat{\vec{p}} - e\vec{A} - e\frac{\Delta m}{M}\left(\vec{r}\cdot\vec{\nabla}_{\vec{R}}\right)\vec{A}\right]^2\left(\exp\left(\frac{i}{\hbar}e\Lambda\right)\tilde{\Phi}\right) =$$
$$\left[\hat{\vec{p}} - e\vec{A} - e\frac{\Delta m}{M}\left(\vec{r}\cdot\vec{\nabla}_{\vec{R}}\right)\vec{A}\right]\exp\left(\frac{i}{\hbar}e\Lambda\right)\left[\hat{\vec{p}} + \frac{e}{2}\frac{\Delta m}{M}\left(\vec{r}\times\vec{B}\right)\right]\tilde{\Phi} =$$
$$\exp\left(\frac{i}{\hbar}e\Lambda\right)\left[\hat{\vec{p}} + \frac{e}{2}\frac{\Delta m}{M}\left(\vec{r}\times\vec{B}\right)\right]^2\tilde{\Phi}. \quad (L.19)$$

Time Derivative

Finally we concentrate on the time derivative

$$i\hbar\frac{\partial\Phi}{\partial t} = i\hbar\frac{\partial}{\partial t}\left[\exp\left(\frac{i}{\hbar}e\Lambda\right)\tilde{\Phi}\right] = \exp\left(\frac{i}{\hbar}e\Lambda\right)\left[-e\frac{\partial\Lambda}{\partial t}\tilde{\Phi} + i\hbar\frac{\partial\tilde{\Phi}}{\partial t}\right]$$

which with the definition Eq. (L.17) of the gauge potential Λ, that is with the formula

$$\frac{\partial\Lambda}{\partial t} = \vec{r}\cdot\frac{\partial\vec{A}}{\partial t} + \frac{1}{2}\frac{\Delta m}{M}\left(\vec{r}\cdot\vec{\nabla}_{\vec{R}}\right)\vec{r}\cdot\frac{\partial\vec{A}}{\partial t} = -\vec{r}\cdot\vec{E} - \frac{1}{2}\frac{\Delta m}{M}\left(\vec{r}\cdot\vec{\nabla}_{\vec{R}}\right)\left(\vec{r}\cdot\vec{E}\right)$$

and the electric field definition Eq. (L.8) reduces to

$$i\hbar\frac{\partial\Phi}{\partial t} = \exp\left(\frac{i}{\hbar}e\Lambda\right)\left\{\left[e\vec{r}\cdot\vec{E} + \frac{e}{2}\frac{\Delta m}{M}\left(\vec{r}\cdot\vec{\nabla}_{\vec{R}}\right)\left(\vec{r}\cdot\vec{E}\right)\right]\tilde{\Phi} + i\hbar\frac{\partial\tilde{\Phi}}{\partial t}\right\}. \quad (L.20)$$

L.3.2 Schrödinger equation for $\widetilde{\Phi}$

We are now in the position to present the Schrödinger equation for $\widetilde{\Phi}$. For this purpose we substitute the results Eqs. (L.18), (L.19) and (L.20) into the Schrödinger equation Eq. (L.15) and find after canceling the gauge factor

$$\left[e\vec{r}\cdot\vec{E} + \frac{e}{2}\frac{\Delta m}{M}\left(\vec{r}\cdot\vec{\nabla}_{\vec{R}}\right)\left(\vec{r}\cdot\vec{E}\right)\right]\widetilde{\Phi} + i\hbar\frac{\partial\widetilde{\Phi}}{\partial t} =$$
$$\frac{1}{2M}\left[\hat{\vec{P}} + e\left(\vec{r}\times\vec{B}\right)\right]^2\widetilde{\Phi} + \frac{1}{2\mu}\left[\hat{\vec{p}} + \frac{e}{2}\frac{\Delta m}{M}\left(\vec{r}\times\vec{B}\right)\right]^2\widetilde{\Phi} + V(\vec{r})\,\widetilde{\Phi},$$

or

$$i\hbar\frac{\partial\widetilde{\Phi}}{\partial t} = \widetilde{H}^{(1)}\widetilde{\Phi}$$

where the Hamiltonian $\widetilde{H}^{(1)}$ is given by Eq. (L.14).

M Effective Hamiltonian

In Sec. 15.4.3 we have shown that the time evolution due to the far off-resonant Jaynes-Cummings-Paul Hamiltonian

$$\hat{H}_{\text{int}}(t) \equiv \hbar g \left(\hat{\sigma} \hat{a}^\dagger e^{i\Delta t} + \hat{\sigma}^\dagger \hat{a} e^{-i\Delta t} \right) \tag{M.1}$$

can be approximated by the one due to the effective Hamiltonian

$$\hat{H}_{\text{eff}} \equiv -\frac{\hbar g^2}{\Delta} \left[\hat{\sigma}_z \hat{a}^\dagger \hat{a} + \frac{1}{2} (\hat{\sigma}_z + \mathbb{1}) \right]. \tag{M.2}$$

We have obtained this result by considering the exact solution of the Jaynes-Cummings-Paul model in the limit of large detuning.

In this Appendix we derive this result using the time evolution operator

$$\hat{\mathcal{U}}(t, t_0 = 0) \cong \mathbb{1} - \frac{i}{\hbar} \int_0^t dt' \, \hat{H}(t') - \frac{1}{\hbar^2} \int_0^t dt' \, \hat{H}(t') \int_0^{t'} dt'' \, \hat{H}(t'') \tag{M.3}$$

discussed in Sec. 2.4 or in Appendix B. Here we only keep terms up to second order in the Hamiltonian.

We now calculate the time evolution operator in the form Eq. (M.3) for the specific example of the nonresonant Jaynes-Cummings-Paul Hamiltonian Eq. (M.1).

We first evaluate the integral

$$\int_0^t dt' \, \hat{H}_{\text{int}}(t') = \frac{\hbar g}{i\Delta} \left[\hat{\sigma} \hat{a}^\dagger \left(e^{i\Delta t} - 1 \right) - \hat{\sigma}^\dagger \hat{a} \left(e^{-i\Delta t} - 1 \right) \right] \tag{M.4}$$

and use this result to obtain the second order contribution

$$\int_0^t dt' \, \hat{H}_{\text{int}}(t') \int_0^{t'} dt'' \, \hat{H}_{\text{int}}(t'')$$

$$= \frac{\hbar^2 g^2}{i\Delta} \int_0^t dt' \left(\hat{\sigma} \hat{a}^\dagger e^{i\Delta t'} + \hat{\sigma}^\dagger \hat{a} e^{-i\Delta t'} \right) \left[\hat{\sigma} \hat{a}^\dagger \left(e^{i\Delta t'} - 1 \right) - \hat{\sigma}^\dagger \hat{a} \left(e^{-i\Delta t'} - 1 \right) \right]$$

$$= \frac{\hbar^2 g^2}{i\Delta} \left\{ \int_0^t dt' \left[\hat{\sigma}^2 \hat{a}^{\dagger 2} \left(e^{2i\Delta t'} - e^{i\Delta t'} \right) + \hat{\sigma} \hat{\sigma}^\dagger \, \hat{a}^\dagger \hat{a} \left(e^{i\Delta t'} - 1 \right) \right. \right.$$

$$\left. \left. - \hat{\sigma}^{\dagger 2} \hat{a}^2 \left(e^{-2i\Delta t'} - e^{-i\Delta t'} \right) - \hat{\sigma}^\dagger \hat{\sigma} \, \hat{a} \hat{a}^\dagger \left(e^{-i\Delta t'} - 1 \right) \right] \right\}. \tag{M.5}$$

This expression simplifies considerably when we recall that

$$\hat{\sigma}^2 = |b\rangle\langle a||b\rangle\langle a| = |b\rangle\langle a|b\rangle\langle a| = 0$$

and thus

$$\hat{\sigma}^{\dagger 2} = 0,$$

which yields

$$\int_0^t dt'\, \hat{H}_{\text{int}}(t') \int_0^{t'} dt''\, \hat{H}_{\text{int}}(t'') = \frac{\hbar^2 g^2}{i\Delta} \int_0^t dt' \left[\hat{\sigma}\hat{\sigma}^\dagger \hat{a}^\dagger \hat{a} \left(e^{i\Delta t'} - 1\right) - \hat{\sigma}^\dagger \hat{\sigma} \hat{a}\hat{a}^\dagger \left(e^{-i\Delta t'} - 1\right) \right].$$

When we perform the remaining integration we get an additional factor of Δ into the denominator and again oscillatory or constant contributions. Moreover, the second and forth term in Eq. (M.5) give rise to terms linear in time. We therefore retain only the linear contribution and arrive at

$$\int_0^t dt'\, \hat{H}_{\text{int}}(t') \int_0^{t'} dt''\, \hat{H}_{\text{int}}(t'') \cong -i\frac{\hbar^2 g^2}{\Delta} \left[\left(\hat{\sigma}^\dagger \hat{\sigma} - \hat{\sigma}\hat{\sigma}^\dagger\right) \hat{a}^\dagger \hat{a} + \hat{\sigma}^\dagger \hat{\sigma} \right] t$$

where we have made use of $[\hat{a}, \hat{a}^\dagger] = 1$.

When we recall that

$$\hat{\sigma}^\dagger \hat{\sigma} - \hat{\sigma}\hat{\sigma}^\dagger = |a\rangle\langle b||b\rangle\langle a| - |b\rangle\langle a||a\rangle\langle b| = |a\rangle\langle a| - |b\rangle\langle b| = \hat{\sigma}_z$$

and

$$\hat{\sigma}^\dagger \hat{\sigma} = |a\rangle\langle b||b\rangle\langle a| = |a\rangle\langle a| = \frac{1}{2}(\hat{\sigma}_z + \mathbb{1})$$

we find

$$\int_0^t dt'\, \hat{H}_{\text{int}}(t') \int_0^{t'} dt''\, \hat{H}_{\text{int}}(t'') \cong -i\frac{\hbar^2 g^2}{\Delta} \left[\hat{\sigma}_z \hat{a}^\dagger \hat{a} + \frac{1}{2}(\hat{\sigma}_z + \mathbb{1}) \right] t. \qquad (M.6)$$

We are now in a position to compare the first and second order contributions Eqs. (M.4) and (M.6). Both contributions involve the detuning linearly in the denominator. Whereas the first order contribution is constant or oscillatory in time, the second order term increases linearly. We can therefore neglect the first order compared to the second order and arrive at

$$\hat{\mathcal{U}}(t, t_0 = 0) \cong \mathbb{1} - \frac{i}{\hbar}\left(-\frac{\hbar g^2}{\Delta}\right) \left[\hat{\sigma}_z \hat{a}^\dagger \hat{a} + \frac{1}{2}(\hat{\sigma}_z + \mathbb{1})\right] t$$

$$\equiv \mathbb{1} - \frac{i}{\hbar}\hat{H}_{\text{eff}}\, t$$

or

$$\hat{\mathcal{U}}(t, t_0 = 0) \cong \exp\left[-\frac{i}{\hbar}\hat{H}_{\text{eff}} \cdot t\right].$$

Here we have recalled the effective Hamiltonian Eq. (M.2).

N Oscillator Reservoir

In this appendix we derive the master equation for a two-level atom interacting with the heat bath of harmonic oscillators. In particular, we evaluate explicitly the double commutator between the interaction Hamiltonians $\hat{V}_2$ and $\hat{V}_1$ at different times and the density operator determining the coarse-grained master equation

$$\dot{\hat{\rho}}_{\text{at}}^{(2)} = -\frac{1}{\hbar^2}\frac{1}{\tau}\int_t^{t+\tau}dt_2\int_t^{t_2}dt_1\,\text{Tr}_{\text{f}}\left\{\left[\hat{V}_2,\left[\hat{V}_1,\hat{\rho}\right]\right]\right\}. \tag{N.1}$$

In order to keep the notation simple we recall the multi-mode annihilation operator

$$\hat{\mathcal{A}}_n \equiv \hat{\mathcal{A}}(t_n) \equiv \frac{1}{\bar{g}}\sum_l g_l \hat{a}_l e^{-i\Delta_l t_n} \tag{N.2}$$

and the corresponding creation operator

$$\hat{\mathcal{A}}_n^\dagger \equiv \hat{\mathcal{A}}^\dagger(t_n) \equiv \frac{1}{\bar{g}}\sum_l g_l^* \hat{a}_l^\dagger e^{i\Delta_l t_n}. \tag{N.3}$$

Here the times $t_n = t_1, t_2$ are the integration variables appearing in the master equation.

In this notation the interaction Hamiltonian takes the form

$$\hat{V}_n \equiv \hat{V}(t_n) \equiv \hbar\bar{g}\left(\hat{\sigma}\hat{\mathcal{A}}_n^\dagger + \hat{\sigma}^\dagger\hat{\mathcal{A}}_n\right)$$

of the Jaynes-Cummings-Paul Hamiltonian.

N.1 Second Order Contribution

We first calculate the double commutator and then turn to the trace over the field oscillators of the reservoir.

N.1.1 Evaluation of Double Commutator

In this notation the double commutator between the interaction Hamiltonians $\hat{V}_1$ and $\hat{V}_2$ at two different times t_1 and t_2 and the density operator $\hat{\rho}(t)$ of the atom and the field at time t reads

$$\left[\hat{\mathcal{V}}_2, [\hat{\mathcal{V}}_1, \hat{\rho}]\right] = (\hbar \bar{g})^2 \left\{ \left[\hat{\sigma}\hat{\mathcal{A}}_2^\dagger, [\hat{\sigma}\hat{\mathcal{A}}_1^\dagger, \hat{\rho}]\right] + \left[\hat{\sigma}\hat{\mathcal{A}}_2^\dagger, [\hat{\sigma}^\dagger\hat{\mathcal{A}}_1, \hat{\rho}]\right] \right. $$
$$\left. + \left[\hat{\sigma}^\dagger\hat{\mathcal{A}}_2, [\hat{\sigma}\hat{\mathcal{A}}_1^\dagger, \hat{\rho}]\right] + \left[\hat{\sigma}^\dagger\hat{\mathcal{A}}_2, [\hat{\sigma}^\dagger\hat{\mathcal{A}}_1, \hat{\rho}]\right] \right\}. \tag{N.4}$$

We now evaluate the individual contributions and start with

$$\left[\hat{\sigma}\hat{\mathcal{A}}_2^\dagger, [\hat{\sigma}\hat{\mathcal{A}}_1^\dagger, \hat{\rho}]\right] = \hat{\sigma}\hat{\mathcal{A}}_2^\dagger \hat{\sigma}\hat{\mathcal{A}}_1^\dagger \hat{\rho} - \hat{\sigma}\hat{\mathcal{A}}_1^\dagger \hat{\rho}\hat{\sigma}\hat{\mathcal{A}}_2^\dagger - \hat{\sigma}\hat{\mathcal{A}}_2^\dagger \hat{\rho}\hat{\sigma}\hat{\mathcal{A}}_1^\dagger + \hat{\rho}\hat{\sigma}\hat{\mathcal{A}}_1^\dagger \hat{\sigma}\hat{\mathcal{A}}_2^\dagger.$$

This expression simplifies when we note that the multi-mode operator of the field commutes with the atomic operator. This feature creates in the first and in the last term of the above equation the squares of $\hat{\sigma}$. We recall that $\hat{\sigma}^2 = \hat{0}$, since a two-level atom can only make a transition to the ground state and cannot fall further. We therefore arrive at

$$\left[\hat{\sigma}\hat{\mathcal{A}}_2^\dagger, [\hat{\sigma}\hat{\mathcal{A}}_1^\dagger, \hat{\rho}]\right] = -\hat{\mathcal{A}}_1^\dagger \hat{\sigma}\hat{\rho}\hat{\sigma}\hat{\mathcal{A}}_2^\dagger - \hat{\mathcal{A}}_2^\dagger \hat{\sigma}\hat{\rho}\hat{\sigma}\hat{\mathcal{A}}_1^\dagger. \tag{N.5}$$

We emphasize that $\hat{\sigma}$ commutes with $\hat{\mathcal{A}}_n$, but does not commute with $\hat{\rho}$. Thus we have to keep track of the order of $\hat{\sigma}$ and $\hat{\rho}$.

Similarly, we find

$$\left[\hat{\sigma}^\dagger\hat{\mathcal{A}}_2, [\hat{\sigma}^\dagger\hat{\mathcal{A}}_1, \hat{\rho}]\right] = \hat{\sigma}^\dagger\hat{\mathcal{A}}_2 \hat{\sigma}^\dagger\hat{\mathcal{A}}_1 \hat{\rho} - \hat{\sigma}^\dagger\hat{\mathcal{A}}_1 \hat{\rho}\hat{\sigma}^\dagger\hat{\mathcal{A}}_2 - \hat{\sigma}^\dagger\hat{\mathcal{A}}_2 \hat{\rho}\hat{\sigma}^\dagger\hat{\mathcal{A}}_1 + \hat{\rho}\hat{\sigma}^\dagger\hat{\mathcal{A}}_1 \hat{\sigma}^\dagger\hat{\mathcal{A}}_2,$$

and note that a two-level atom cannot be excited beyond its upper level, which implies $\left(\hat{\sigma}^\dagger\right)^2 = 0$. We hence arrive at

$$\left[\hat{\sigma}^\dagger\hat{\mathcal{A}}_2, [\hat{\sigma}^\dagger\hat{\mathcal{A}}_1, \hat{\rho}]\right] = -\hat{\mathcal{A}}_1 \hat{\sigma}^\dagger\hat{\rho}\hat{\sigma}^\dagger\hat{\mathcal{A}}_2 - \hat{\mathcal{A}}_2 \hat{\sigma}^\dagger\hat{\rho}\hat{\sigma}^\dagger\hat{\mathcal{A}}_1. \tag{N.6}$$

Unfortunately no further simplification is possible in the two remaining contributions in Eq. (N.4). When we interchange the order of field and atomic operators but still keep track of the order of the individual field operators and the density operator we find

$$\left[\hat{\sigma}\hat{\mathcal{A}}_2^\dagger, [\hat{\sigma}^\dagger\hat{\mathcal{A}}_1, \hat{\rho}]\right] = \hat{\sigma}\hat{\sigma}^\dagger \hat{\mathcal{A}}_2^\dagger\hat{\mathcal{A}}_1 \hat{\rho} - \hat{\mathcal{A}}_1 \hat{\sigma}^\dagger\hat{\rho}\hat{\sigma}\hat{\mathcal{A}}_2^\dagger - \hat{\mathcal{A}}_2^\dagger \hat{\sigma}\hat{\rho}\hat{\sigma}^\dagger\hat{\mathcal{A}}_1 + \hat{\rho}\hat{\sigma}^\dagger\hat{\sigma}\hat{\mathcal{A}}_1\hat{\mathcal{A}}_2^\dagger. \tag{N.7}$$

A similar calculation yields

$$\left[\hat{\sigma}^\dagger\hat{\mathcal{A}}_2, [\hat{\sigma}\hat{\mathcal{A}}_1^\dagger, \hat{\rho}]\right] = \hat{\sigma}^\dagger\hat{\sigma} \hat{\mathcal{A}}_2\hat{\mathcal{A}}_1^\dagger \hat{\rho} - \hat{\mathcal{A}}_1^\dagger \hat{\sigma}\hat{\rho}\hat{\sigma}^\dagger\hat{\mathcal{A}}_2 - \hat{\mathcal{A}}_2 \hat{\sigma}^\dagger\hat{\rho}\hat{\sigma}\hat{\mathcal{A}}_1^\dagger + \hat{\rho}\hat{\sigma}\hat{\sigma}^\dagger\hat{\mathcal{A}}_1^\dagger\hat{\mathcal{A}}_2. \tag{N.8}$$

A quick test allows us to check if Eqs. (N.7) and (N.8) are consistent: We recognize that the left hand sides of the two equations follow from each when we replace the atomic and field operators by their hermitian conjugates. Consequently, the right hand sides are connected by the same transformation. A similar substitution rule connects Eqs. (N.5) and (N.6).

N.1.2 Trace over Reservoir

We are now in a position to take the trace over the field states. For this purpose we assume that before the interaction of the atom with the reservoir the two systems are uncorrelated and the two corresponding density operators are a direct product, that is

$$\hat{\rho} = \hat{\rho}_{\text{at}} \otimes \hat{\rho}_{\text{f}}.$$

When we make use of the fact that the trace is invariant under cyclic permutations, that is,

$$\text{Tr}(\hat{B}\hat{C}) = \text{Tr}(\hat{C}\hat{B})$$

we find from Eqs. (N.5) and (N.6)

$$\text{Tr}_{\text{f}}\left[\hat{\sigma}\hat{\mathcal{A}}_2^\dagger, [\hat{\sigma}\hat{\mathcal{A}}_1^\dagger, \hat{\rho}]\right] = -2\hat{\sigma}\hat{\rho}_{\text{at}}\hat{\sigma} \, \text{Tr}_{\text{f}}\left(\hat{\mathcal{A}}_1^\dagger \hat{\mathcal{A}}_2^\dagger \hat{\rho}_{\text{f}}\right)$$

and

$$\text{Tr}_{\text{f}}\left[\hat{\sigma}^\dagger \hat{\mathcal{A}}_2, [\hat{\sigma}^\dagger \hat{\mathcal{A}}_1, \hat{\rho}]\right] = -2\hat{\sigma}^\dagger \hat{\rho}_{\text{at}} \hat{\sigma}^\dagger \, \text{Tr}_{\text{f}}\left(\hat{\mathcal{A}}_1 \hat{\mathcal{A}}_2 \hat{\rho}_{\text{f}}\right).$$

The two remaining terms resulting from Eqs. (N.7) and (N.8) are more complicated and read

$$\text{Tr}_{\text{f}}\left[\hat{\sigma}\hat{\mathcal{A}}_2^\dagger, [\hat{\sigma}^\dagger \hat{\mathcal{A}}_1, \hat{\rho}]\right] = \hat{\sigma}\hat{\sigma}^\dagger \hat{\rho}_{\text{at}} \, \text{Tr}_{\text{f}}\left(\hat{\mathcal{A}}_2^\dagger \hat{\mathcal{A}}_1 \hat{\rho}_{\text{f}}\right) + \hat{\rho}_{\text{at}} \hat{\sigma}^\dagger \hat{\sigma} \, \text{Tr}_{\text{f}}\left(\hat{\mathcal{A}}_1 \hat{\mathcal{A}}_2^\dagger \hat{\rho}_{\text{f}}\right)$$
$$- \hat{\sigma}^\dagger \hat{\rho}_{\text{at}} \hat{\sigma} \, \text{Tr}_{\text{f}}\left(\hat{\mathcal{A}}_2^\dagger \hat{\mathcal{A}}_1 \hat{\rho}_{\text{f}}\right) - \hat{\sigma} \hat{\rho}_{\text{at}} \hat{\sigma}^\dagger \, \text{Tr}_{\text{f}}\left(\hat{\mathcal{A}}_1 \hat{\mathcal{A}}_2^\dagger \hat{\rho}_{\text{f}}\right).$$

In complete analogy we arrive at

$$\text{Tr}_{\text{f}}\left[\hat{\sigma}^\dagger \hat{\mathcal{A}}_2, [\hat{\sigma}\hat{\mathcal{A}}_1^\dagger, \hat{\rho}]\right] = \hat{\sigma}^\dagger \hat{\sigma} \hat{\rho}_{\text{at}} \, \text{Tr}_{\text{f}}\left(\hat{\mathcal{A}}_2 \hat{\mathcal{A}}_1^\dagger \hat{\rho}_{\text{f}}\right) + \hat{\rho}_{\text{at}} \hat{\sigma}\hat{\sigma}^\dagger \, \text{Tr}_{\text{f}}\left(\hat{\mathcal{A}}_1^\dagger \hat{\mathcal{A}}_2 \hat{\rho}_{\text{f}}\right)$$
$$- \hat{\sigma} \hat{\rho}_{\text{at}} \hat{\sigma}^\dagger \, \text{Tr}_{\text{f}}\left(\hat{\mathcal{A}}_2 \hat{\mathcal{A}}_1^\dagger \hat{\rho}_{\text{f}}\right) - \hat{\sigma}^\dagger \hat{\rho}_{\text{at}} \hat{\sigma} \, \text{Tr}_{\text{f}}\left(\hat{\mathcal{A}}_1^\dagger \hat{\mathcal{A}}_2 \hat{\rho}_{\text{f}}\right).$$

We are now in a position to combine all terms contributing to the master equation which yields

$$\text{Tr}_{\text{f}}\left\{\left[\hat{\mathcal{V}}_2, [\hat{\mathcal{V}}_1, \hat{\rho}]\right]\right\} = (\hbar \bar{g})^2 \Big\{ -2\hat{\sigma}\hat{\rho}_{\text{at}}\hat{\sigma} \, \text{Tr}_{\text{f}}\left(\hat{\mathcal{A}}_1^\dagger \hat{\mathcal{A}}_2^\dagger \hat{\rho}_{\text{f}}\right) - 2\hat{\sigma}^\dagger \hat{\rho}_{\text{at}} \hat{\sigma}^\dagger \, \text{Tr}_{\text{f}}\left(\hat{\mathcal{A}}_1 \hat{\mathcal{A}}_2 \hat{\rho}_{\text{f}}\right)$$
$$+ \left[\left(\hat{\sigma}\hat{\sigma}^\dagger \hat{\rho}_{\text{at}} - \hat{\sigma}^\dagger \hat{\rho}_{\text{at}} \hat{\sigma}\right) \text{Tr}_{\text{f}}\left(\hat{\mathcal{A}}_2^\dagger \hat{\mathcal{A}}_1 \hat{\rho}_{\text{f}}\right) + \left(\hat{\rho}_{\text{at}} \hat{\sigma}^\dagger \hat{\sigma} - \hat{\sigma} \hat{\rho}_{\text{at}} \hat{\sigma}^\dagger\right) \text{Tr}_{\text{f}}\left(\hat{\mathcal{A}}_1 \hat{\mathcal{A}}_2^\dagger \hat{\rho}_{\text{f}}\right)\right]$$
$$+ \left[\left(\hat{\sigma}^\dagger \hat{\sigma} \hat{\rho}_{\text{at}} - \hat{\sigma} \hat{\rho}_{\text{at}} \hat{\sigma}^\dagger\right) \text{Tr}_{\text{f}}\left(\hat{\mathcal{A}}_2 \hat{\mathcal{A}}_1^\dagger \hat{\rho}_{\text{f}}\right) + \left(\hat{\rho}_{\text{at}} \hat{\sigma}\hat{\sigma}^\dagger - \hat{\sigma}^\dagger \hat{\rho}_{\text{at}} \hat{\sigma}\right) \text{Tr}_{\text{f}}\left(\hat{\mathcal{A}}_1^\dagger \hat{\mathcal{A}}_2 \hat{\rho}_{\text{f}}\right)\right] \Big\}.$$
(N.9)

Here we have combined terms with identical traces.

N.2 Symmetry Relations in Trace

Equation (N.9) suggests that next we have to calculate the traces of many combinations of field operators. However, some of them are just the complex conjugate of each other. We now discuss these symmetry relations in more detail.

N.2.1 Complex Conjugates

For this purpose we consider the operator combination

$$\hat{O} \equiv \hat{\mathcal{A}}_1^\dagger \hat{\mathcal{A}}_2 \, \hat{\rho}_f$$

which appears most frequently in the double commutator, and its hermitian conjugate

$$\hat{O}^\dagger = \hat{\rho}_f \left(\hat{\mathcal{A}}_1^\dagger \hat{\mathcal{A}}_2 \right)^\dagger = \hat{\rho}_f \, \hat{\mathcal{A}}_2^\dagger \, \hat{\mathcal{A}}_1.$$

Since the trace is invariant under cyclic permutations we find

$$\mathrm{Tr}_f(\hat{O}^\dagger) = \mathrm{Tr}_f \left(\hat{\rho}_f \, \hat{\mathcal{A}}_2^\dagger \, \hat{\mathcal{A}}_1 \right) = \mathrm{Tr}_f \left(\hat{\mathcal{A}}_1 \, \hat{\rho}_f \, \hat{\mathcal{A}}_2^\dagger \right) = \mathrm{Tr}_f \left(\hat{\mathcal{A}}_2^\dagger \, \hat{\mathcal{A}}_1 \, \hat{\rho}_f \right),$$

which with the help of the relation

$$\mathrm{Tr}(\hat{O}^\dagger) = \mathrm{Tr}(\hat{O})^*$$

yields

$$\mathrm{Tr}_f \left(\hat{\mathcal{A}}_1^\dagger \, \hat{\mathcal{A}}_2 \, \hat{\rho}_f \right)^* = \mathrm{Tr}_f \left(\hat{\mathcal{A}}_2^\dagger \, \hat{\mathcal{A}}_1 \, \hat{\rho}_f \right). \tag{N.10}$$

Similarly, we derive

$$\mathrm{Tr}_f \left[\left(\hat{\mathcal{A}}_1^\dagger \, \hat{\mathcal{A}}_2^\dagger \, \hat{\rho}_f \right)^\dagger \right] = \mathrm{Tr}_f \left(\hat{\rho}_f \, \hat{\mathcal{A}}_2 \, \hat{\mathcal{A}}_1 \right) = \mathrm{Tr}_f \left(\hat{\mathcal{A}}_2 \, \hat{\mathcal{A}}_1 \, \hat{\rho}_f \right) = \mathrm{Tr}_f \left(\hat{\mathcal{A}}_1 \, \hat{\mathcal{A}}_2 \, \hat{\rho}_f \right)$$

and hence arrive at the formula

$$\mathrm{Tr}_f \left(\hat{\mathcal{A}}_1^\dagger \, \hat{\mathcal{A}}_2^\dagger \, \hat{\rho}_f \right)^* = \mathrm{Tr}_f \left(\hat{\mathcal{A}}_2 \, \hat{\mathcal{A}}_1 \, \hat{\rho}_f \right). \tag{N.11}$$

N.2.2 Commutator Between Field Operators

In order to combine various traces in Eq. (N.9) it is useful to calculate the commutator between $\hat{\mathcal{A}}_1$ and $\hat{\mathcal{A}}_2^\dagger$. With the help of the definitions Eqs. (N.2) and (N.3) of the multi-mode operators we find

$$\left[\hat{\mathcal{A}}_1, \hat{\mathcal{A}}_2^\dagger \right] = \frac{1}{\bar{g}^2} \sum_{l,m} g_l \, g_m^* \, e^{-i(\Delta_l t_1 - \Delta_m t_2)} \left[\hat{a}_l, \hat{a}_m^\dagger \right]$$

which due to the familiar commutation relation

$$\left[\hat{a}_l, \hat{a}_m^\dagger \right] = \delta_{l,m}$$

reduces to

$$\left[\hat{\mathcal{A}}_1, \hat{\mathcal{A}}_2^\dagger \right] = \frac{1}{\bar{g}^2} \sum_{l,m} |g_l|^2 \, e^{-i \Delta_l (t_1 - t_2)} \equiv G(t_1 - t_2). \tag{N.12}$$

Hence we have established the formula

$$\mathrm{Tr}_f \left(\hat{\mathcal{A}}_1 \hat{\mathcal{A}}_2^\dagger \hat{\rho}_f \right) = \mathrm{Tr}_f \left\{ \left(\hat{\mathcal{A}}_2^\dagger \hat{\mathcal{A}}_1 + \left[\hat{\mathcal{A}}_1, \hat{\mathcal{A}}_2^\dagger \right] \right) \hat{\rho}_f \right\}$$
$$= \mathrm{Tr}_f \left(\hat{\mathcal{A}}_2^\dagger \hat{\mathcal{A}}_1 \hat{\rho}_f \right) + G(t_1 - t_2), \tag{N.13}$$

where in the last step we have made use of $\mathrm{Tr} \, \hat{\rho}_f = 1$.

N.3 Master Equation

The master equation takes a more compact form when we introduce the abbreviations

$$\beta \equiv \beta_r + i\,\beta_i \equiv \frac{\bar{g}^2}{\tau} \int_t^{t+\tau} dt_2 \int_t^{t_2} dt_1 \; \mathrm{Tr}_\mathrm{f}\left(\hat{\mathcal{A}}_2 \hat{\mathcal{A}}_1 \hat{\rho}_\mathrm{f}\right) \qquad (N.14)$$

and

$$\Gamma \equiv \Gamma_r + i\,\Gamma_i \equiv \frac{\bar{g}^2}{\tau} \int_t^{t+\tau} dt_2 \int_t^{t_2} dt_1 \; \mathrm{Tr}_\mathrm{f}\left(\hat{\mathcal{A}}_2^\dagger \hat{\mathcal{A}}_1 \hat{\rho}_\mathrm{f}\right) \qquad (N.15)$$

and

$$\tilde{G} \equiv \tilde{G}_r + i\,\tilde{G}_i \equiv \frac{\bar{g}^2}{\tau} \int_t^{t+\tau} dt_2 \int_t^{t_2} dt_1 \; G(t_1 - t_2). \qquad (N.16)$$

When we now make use of the symmetry relations Eqs. (N.10), (N.11) and (N.13) in the expression Eq. (N.9) and substitute the result into the formula Eq. (N.1) for the contribution $\dot{\hat{\rho}}_\mathrm{at}^{(2)}$ in second order perturbation theory we arrive at the master equation

$$\begin{aligned}\dot{\hat{\rho}}_\mathrm{at}^{(2)} =\;& 2\,\beta^*\,\hat{\sigma}\,\hat{\rho}_\mathrm{at}\,\hat{\sigma} + 2\,\beta\,\hat{\sigma}^\dagger\,\hat{\rho}_\mathrm{at}\,\hat{\sigma}^\dagger \\ & - \Gamma\left\{\hat{\sigma}\,\hat{\sigma}^\dagger\,\hat{\rho}_\mathrm{at} + \hat{\rho}_\mathrm{at}\,\hat{\sigma}^\dagger\,\hat{\sigma} - \hat{\sigma}^\dagger\,\hat{\rho}_\mathrm{at}\,\hat{\sigma} - \hat{\sigma}\,\hat{\rho}_\mathrm{at}\,\hat{\sigma}^\dagger\right\} \\ & - \Gamma^*\left\{\hat{\sigma}^\dagger\,\hat{\sigma}\,\hat{\rho}_\mathrm{at} + \hat{\rho}_\mathrm{at}\,\hat{\sigma}\,\hat{\sigma}^\dagger - \hat{\sigma}\,\hat{\rho}_\mathrm{at}\,\hat{\sigma}^\dagger - \hat{\sigma}^\dagger\,\hat{\rho}_\mathrm{at}\,\hat{\sigma}\right\} \\ & - \tilde{G}\left\{\hat{\rho}_\mathrm{at}\,\hat{\sigma}^\dagger\hat{\sigma} - \hat{\sigma}\,\hat{\rho}_\mathrm{at}\,\hat{\sigma}^\dagger\right\} - \tilde{G}^*\left\{\hat{\sigma}^\dagger\,\hat{\sigma}\,\hat{\rho}_\mathrm{at} - \hat{\sigma}\,\hat{\rho}_\mathrm{at}\,\hat{\sigma}^\dagger\right\}.\end{aligned}$$

Here we have made use of the abbreviations Eqs. (N.14), (N.15) and (N.16) for β, Γ and $\tilde{G}$.

We can simplify this formula slightly by combining various terms

$$\begin{aligned}\dot{\hat{\rho}}_\mathrm{at}^{(2)} =\;& 2\,\beta^*\,\hat{\sigma}\,\hat{\rho}_\mathrm{at}\,\hat{\sigma} + 2\,\beta\,\hat{\sigma}^\dagger\,\hat{\rho}_\mathrm{at}\,\hat{\sigma}^\dagger \\ & - \left[\left(\Gamma^* + \tilde{G}^*\right)\hat{\sigma}^\dagger\,\hat{\sigma}\,\hat{\rho}_\mathrm{at} + \left(\Gamma + \tilde{G}\right)\hat{\rho}_\mathrm{at}\,\hat{\sigma}^\dagger\,\hat{\sigma} - \left(\Gamma + \Gamma^* + \tilde{G} + \tilde{G}^*\right)\hat{\sigma}\,\hat{\rho}_\mathrm{at}\,\hat{\sigma}^\dagger\right] \\ & - \left[\Gamma\,\hat{\sigma}\,\hat{\sigma}^\dagger\,\hat{\rho}_\mathrm{at} + \Gamma^*\,\hat{\rho}_\mathrm{at}\,\hat{\sigma}\,\hat{\sigma}^\dagger - \left(\Gamma + \Gamma^*\right)\hat{\sigma}^\dagger\,\hat{\rho}_\mathrm{at}\,\hat{\sigma}\right]\end{aligned}$$

and identifying the real and imaginary parts of Γ and $\tilde{G}$ which yields

$$\begin{aligned}\dot{\hat{\rho}}_\mathrm{at}^{(2)} =\;& 2\,\beta^*\,\hat{\sigma}\,\hat{\rho}_\mathrm{at}\,\hat{\sigma} + 2\,\beta\,\hat{\sigma}^\dagger\,\hat{\rho}_\mathrm{at}\,\hat{\sigma}^\dagger \\ & - \left(\Gamma_r + \tilde{G}_r\right)\left[\hat{\sigma}^\dagger\,\hat{\sigma}\,\hat{\rho}_\mathrm{at} + \hat{\rho}_\mathrm{at}\,\hat{\sigma}^\dagger\,\hat{\sigma} - 2\,\hat{\sigma}\,\hat{\rho}_\mathrm{at}\hat{\sigma}^\dagger\right] - \Gamma_r\left[\hat{\sigma}\,\hat{\sigma}^\dagger\,\hat{\rho}_\mathrm{at} + \hat{\rho}_\mathrm{at}\,\hat{\sigma}\,\hat{\sigma}^\dagger - 2\,\hat{\sigma}^\dagger\,\hat{\rho}_\mathrm{at}\,\hat{\sigma}\right] \\ & + i\left(\Gamma_i + \tilde{G}_i\right)\left[\hat{\sigma}^\dagger\,\hat{\sigma},\hat{\rho}_\mathrm{at}\right] - i\,\Gamma_i\left[\hat{\sigma}\,\hat{\sigma}^\dagger,\hat{\rho}_\mathrm{at}\right].\end{aligned}$$

In the last step we have expressed the terms due to the imaginary parts by two commutators.

We recall the formulas

$$\hat{\sigma}^\dagger\,\hat{\sigma} = |a\rangle\langle b|\,|b\rangle\langle a| = |a\rangle\langle a| = \tfrac{1}{2}\left(\mathbb{1} + \hat{\sigma}_z\right)$$

and
$$\hat{\sigma}\,\hat{\sigma}^\dagger = |b\rangle\langle a||a\rangle\langle b| = |b\rangle\langle b| = \tfrac{1}{2}(\mathbb{1}-\hat{\sigma}_z)$$
and thus arrive at the final form
$$\dot{\hat{\rho}}_{\text{at}}^{(2)} = i\left(\Gamma_i + \tfrac{1}{2}\widetilde{G}_i\right)[\hat{\sigma}_z,\hat{\rho}_{\text{at}}]$$
$$-\left(\Gamma_r+\widetilde{G}_r\right)\left[\hat{\sigma}^\dagger\hat{\sigma}\,\hat{\rho}_{\text{at}} + \hat{\rho}_{\text{at}}\,\hat{\sigma}^\dagger\hat{\sigma} - 2\hat{\sigma}\,\hat{\rho}_{\text{at}}\,\hat{\sigma}^\dagger\right] - \Gamma_r\left[\hat{\sigma}\,\hat{\sigma}^\dagger\hat{\rho}_{\text{at}} + \hat{\rho}_{\text{at}}\,\hat{\sigma}\,\hat{\sigma}^\dagger - 2\hat{\sigma}^\dagger\hat{\rho}_{\text{at}}\,\hat{\sigma}\right]$$
$$+2\beta^*\,\hat{\sigma}\hat{\rho}_{\text{at}}\,\hat{\sigma} + 2\beta\,\hat{\sigma}^\dagger\,\hat{\rho}_{\text{at}}\,\hat{\sigma}^\dagger$$
of the master equation in second order perturbation theory.

N.4 Explicit Expressions for Γ, β and $\widetilde{G}$

Next we analyze the parameters Γ, β and $\widetilde{G}$. The corresponding expressions are more complicated than the effective Rabi frequency $g(t)$ of first order theory since they involve the product of two field operators. The operators $\hat{\mathcal{A}}_n$ and $\hat{\mathcal{A}}_n^\dagger$ are multi-mode operators, that is they involve sums over modes. Hence it is convenient to separate in this double sum the products with the same mode operators from the ones with different mode indices. Indeed, when we make use of the definitions Eqs. (N.2) and (N.3) of the multi-mode operators we arrive at

$$\text{Tr}_{\text{f}}\left[\hat{\mathcal{A}}_2^\dagger\hat{\mathcal{A}}_1\hat{\rho}_{\text{f}}\right] = \frac{1}{\bar{g}^2}\sum_{l,m} g_l^*\,g_m\,e^{i(\Delta_l t_2 - \Delta_m t_1)}\,\text{Tr}_{\text{f}}\left(\hat{a}_l^\dagger\hat{a}_m\hat{\rho}_{\text{f}}\right)$$
$$= \frac{1}{\bar{g}^2}\sum_l |g_l|^2\,e^{i\Delta_l(t_2-t_1)}\,\text{Tr}_{\text{f}}\left(\hat{a}_l^\dagger\hat{a}_l\hat{\rho}_{\text{f}}\right)$$
$$+ \frac{1}{\bar{g}^2}\sum_{l\ne m} g_l^*\,g_m\,e^{i(\Delta_l t_2 - \Delta_m t_1)}\,\text{Tr}_{\text{f}}\left(\hat{a}_l^\dagger\hat{a}_m\hat{\rho}_{\text{f}}\right).$$

When the modes of the reservoir are not entangled with each other we can replace the density operator of the field by the corresponding density operator $\hat{\rho}_l$ of this mode, that is
$$\text{Tr}_{\text{f}}\left(\hat{a}_l^\dagger\hat{a}_l\hat{\rho}_{\text{f}}\right) = \text{Tr}_{\text{f}}\left(\hat{a}_l^\dagger\hat{a}_l\hat{\rho}_l\right) = \bar{n}_l$$
and
$$\text{Tr}_{\text{f}}\left(\hat{a}_l^\dagger\hat{a}_m\hat{\rho}_{\text{f}}\right) = \text{Tr}_{\text{f}}\left(\hat{a}_l^\dagger\hat{\rho}_l\right)\text{Tr}_{\text{f}}\left(\hat{a}_m\hat{\rho}_m\right) \equiv \langle\hat{a}_l^\dagger\rangle\langle\hat{a}_m\rangle.$$

Here we have recalled the average number $\bar{n}_l$ of photons in the lth mode.

Hence we find
$$\text{Tr}_{\text{f}}\left[\hat{\mathcal{A}}_2^\dagger\hat{\mathcal{A}}_1\hat{\rho}_{\text{f}}\right] = \frac{1}{\bar{g}^2}\sum_l |g_l|^2\,\bar{n}_l\,e^{i\Delta_l(t_2-t_1)}$$
$$+ \frac{1}{\bar{g}^2}\sum_{l\ne m} g_l^*\,g_m\,\langle\hat{a}_l^\dagger\rangle\langle\hat{a}_m\rangle\,e^{i(\Delta_l t_2 - \Delta_m t_1)},$$

which upon substitution into the definition Eq. (N.15) of Γ results in
$$\Gamma = \sum_l |g_l|^2\,\bar{n}_l\,I(\Delta_l,\Delta_l) + \sum_{l\ne m} g_l^*\,g_m\,\langle\hat{a}_l^\dagger\rangle\langle\hat{a}_m\rangle\,I(\Delta_l,\Delta_m).$$

Here we have introduced the double time integral

$$I(\Delta_l, \Delta_m) \equiv \frac{1}{\tau} \int_t^{t+\tau} dt_2 \int_t^{t_2} dt_1\, e^{i(\Delta_l t_2 - \Delta_m t_1)}. \tag{N.17}$$

Similarly, we find for the parameter β the expression

$$\beta = \sum_l g_l^2 \langle \hat{a}_l^2 \rangle I(-\Delta_l, \Delta_l) + \sum_{l \neq m} g_l g_m \langle \hat{a}_l \rangle \langle \hat{a}_m \rangle I(-\Delta_l, \Delta_m).$$

We finally turn to the quantity $\tilde{G}$. When we substitute the expression Eq. (N.13) for G into the definition Eq. (N.16) for $\tilde{G}$ we arrive at

$$\tilde{G} = \sum_l |g_l|^2 I(\Delta_l, \Delta_l),$$

where we have recalled the definition Eq. (N.17) for the double integral I.

N.5 Integration over Time

We now perform the double time integral

$$I(\Delta_l, \Delta_m) = \frac{1}{\tau} \int_t^{t+\tau} dt_2\, e^{i \Delta_l t_2} \int_t^{t_2} dt_1\, e^{-i \Delta_m t_1}$$

which yields

$$I(\Delta_l, \Delta_m) = \int_t^{t+\tau} dt_2\, e^{i \Delta_l t_2} \frac{e^{-i \Delta_m t_2} - e^{-i \Delta_m t}}{-i \Delta_m \tau},$$

that is

$$I = \frac{1}{(-i \Delta_m \tau)} \int_t^{t+\tau} dt_2\, e^{i(\Delta_l - \Delta_m) t_2} + \frac{e^{-i \Delta_m t}}{i \Delta_m \tau} \int_t^{t+\tau} dt_2\, e^{i \Delta_l t_2},$$

and find

$$I(\Delta_l, \Delta_m) = \frac{e^{i(\Delta_l - \Delta_m)t}}{(-i \Delta_m)} \frac{e^{i(\Delta_l - \Delta_m)\tau} - 1}{i(\Delta_l - \Delta_m)\tau} + \frac{e^{-i(\Delta_m - \Delta_l)t}}{i \Delta_m} \frac{e^{i \Delta_l \tau} - 1}{i \Delta_l \tau}.$$

For $\Delta_l = \Delta_m$ this expression reduces to

$$I(\Delta_l, \Delta_l) = \frac{i}{\Delta_l} - \frac{e^{i \Delta_l \tau} - 1}{\Delta_l^2 \tau}$$

which with the help of the relation

$$e^{i\theta} - 1 = \cos\theta - 1 + i \sin\theta = -2\sin^2(\theta/2) + i \sin\theta$$

reduces to

$$I(\Delta_l, \Delta_l) = 2\frac{\sin^2(\Delta_l \tau/2)}{\Delta_l^2 \tau} + i\frac{1}{\Delta_l}\left[1 - \frac{\sin(\Delta_l \tau)}{\Delta_l \tau}\right]. \tag{N.18}$$

O Bessel Functions

Bessel functions appear in many problems of theoretical physics. In the field of quantum optics they manifest themselves in the momentum distribution of atoms scattered off a standing light field. We therefore derive in this appendix various properties of them and then apply the method of stationary phase to their integral representation.

O.1 Definition

Bessel functions $J_p(z)$ of the first kind arise as the Fourier coefficients in the expansion

$$e^{i\kappa \sin \theta} = \sum_{p=-\infty}^{\infty} J_p(\kappa) e^{ip\theta}.$$

This immediately yields the integral representations

$$J_p(z) = \frac{1}{2\pi} \int_{-\pi}^{\pi} d\theta \, \exp[-i(p\theta - z \sin \theta)] = \frac{1}{2\pi} \int_0^{2\pi} d\theta \, \exp[-i(p\theta - z \sin \theta)] \quad (O.1)$$

of the Bessel function. Both representations are equivalent since we are integrating over a 2π-periodic function.

In the problem of the deflection of atoms the two integrals

$$c(p) \equiv \frac{1}{2\pi} \int_0^{2\pi} d\theta \, \cos(\kappa \sin \theta) e^{-ip\theta}$$

and

$$s(p) \equiv \frac{1}{2\pi} \int_0^{2\pi} d\theta \, \sin(\kappa \sin \theta) e^{-ip\theta}$$

occur. We can express them in terms of Bessel functions when we represent the trigonometric functions cosine and sine as the sum and difference of exponentials and shift the integration in one of the integrals by π. After minor calculations we then find

$$c(p) = \frac{1}{2}[1 + (-1)^p] \, J_p(\kappa)$$

and

$$s(p) = -\frac{i}{2}[1 - (-1)^p] \, J_p(\kappa).$$

Hence, $c(p)$ vanishes for odd values of p whereas $s(p)$ vanishes for even values.

O.2 Asymptotic Expansion

The Bessel function $J_p(z)$ displays characteristic features: It decays rapidly for $|p| > z$ and oscillates for $|p| < z$. In the present section we perform an asymptotic expansion of $J_p(z)$ by applying the method of stationary phase to the integral representation of the Bessel function.

For this purpose we expand the phase

$$S(\theta) \equiv p\theta - z \sin \theta \qquad (O.2)$$

of the Bessel function

$$J_p(z) = \frac{1}{2\pi} \int_{-\pi}^{\pi} d\theta \, \exp[-iS(\theta)]$$

around its point of stationary phase θ_s, governed by

$$\left.\frac{dS}{d\theta}\right|_{\theta=\theta_s} = p - z \cos \theta|_{\theta=\theta_s} = 0, \qquad (O.3)$$

or

$$\frac{p}{z} = \cos \theta_s. \qquad (O.4)$$

Hence for $|p/z| < 1$ we find within the integration region the two real-valued points of stationary phase

$$\theta_s = \pm \arccos\left(\frac{p}{z}\right).$$

However, for $|p/z| > 1$ we find only purely imaginary points of stationary phase. When we substitute them into the phase $S(\theta)$ the phase in the exponent becomes purely imaginary and together with the prefactor i leads to exponentially small terms. This explains the rapid decay of the Bessel function for $|p| > z$.

We now turn to the case of the two real-valued points of stationary phase for $|p| < z$. When we substitute them into the definition Eq. (O.2) of the phase we find

$$S(\theta_s) = \pm \left[p \arccos\left(\frac{p}{z}\right) - \sqrt{z^2 - p^2} \right].$$

The second derivative of $S(\theta)$ follows from Eq. (O.3) together with Eq. (O.4)

$$\left.\frac{\partial^2 S}{\partial \theta^2}\right|_{\theta=\theta_s} = z \sin \theta|_{\theta=\theta_s} = \pm\sqrt{z^2 - p^2}.$$

The Bessel function integral Eq. (O.1) then reads

$$J_p(z) \cong \frac{1}{\sqrt{2\pi}} \exp\left\{i\left[\sqrt{z^2 - p^2} - p \arccos\left(\frac{p}{z}\right)\right]\right\}$$
$$\times \frac{1}{\sqrt{2\pi}} \int_{-\infty}^{\infty} d\theta \, \exp\left[-\frac{i}{2}\sqrt{z^2 - p^2}\,\theta^2\right] + \text{c.c.},$$

which with the help of the formula

$$\int_{-\infty}^{\infty} dy \, e^{-i\alpha y^2} = \sqrt{\frac{\pi}{|\alpha|}} e^{-i\frac{\pi}{4} \operatorname{sign} \alpha}$$

for real α reduces to

$$J_p(z) \cong \sqrt{\frac{2}{\pi}} \left(z^2 - p^2\right)^{-1/4} \cos\left[\sqrt{z^2 - p^2} - p \arccos\left(\frac{p}{z}\right) - \pi/4\right].$$

We note that the interference of the two real-valued points of stationary phase $\pm \theta_s$ determined by

$$p = z \cos \theta_s$$

leads to an oscillatory behavior of the Bessel function $J_p(z)$ for $p < z$.

References

The asymptotic expansion of the Bessel function presented in this appendix is the most elementary example of a Debye expansion. For more details we refer to
P. Debye, *Näherungsformeln für die Zylinderfunktionen für große Werte des Arguments und unbeschränkt veränderliche Werte des Index*, Math. Ann. **67**, 535–558 (1909)

P Square Root of δ

In this Appendix we discuss the properties of the function

$$\delta_N^{(1/2)}(\xi) \equiv \frac{1}{\sqrt{N}} \sum_{\nu=0}^{N-1} \exp(-i\,\xi\,2\pi\nu).$$

In particular, we show that in the limit of $N \to \infty$ the square $\left|\delta_N^{(1/2)}\right|^2$ is an infinite sum of delta functions located at integers j, that is,

$$\lim_{N \to \infty} \left|\delta_N^{(1/2)}(\xi)\right|^2 = \sum_{j=-\infty}^{\infty} \delta(j - \xi).$$

In this sense we can interpret $\delta_N^{(1/2)}(\xi)$ as a model for the square root of a δ-function.

When we recall the relation

$$\sum_{\nu=0}^{N-1} x^\nu = \frac{1 - x^N}{1 - x}$$

we can perform the sum in the definition of $\delta_N^{(1/2)}(\xi)$ and find

$$\delta_N^{(1/2)}(\xi) = \frac{1}{\sqrt{N}} \frac{1 - \exp(-i\,\xi\,2\pi N)}{1 - \exp(-i\,\xi\,2\pi)}.$$

This yields for the square of this function the expression

$$\left|\delta_N^{(1/2)}(\xi)\right|^2 = \frac{1}{N} \left(\frac{\sin(\xi N \pi)}{\sin(\xi \pi)}\right)^2.$$

We note that for any integer value of ξ, numerator and denominator vanish providing us with

$$\lim_{N \to \infty} \left|\delta_N^{(1/2)}(j)\right|^2 = \lim_{N \to \infty} N \to \infty.$$

For any other value of ξ the ratio of the two sine functions is bounded and hence

$$\lim_{N \to \infty} \left|\delta_N^{(1/2)}(\xi)\right|^2 < \lim_{N \to \infty} \frac{c}{N} \to 0,$$

where c is a constant. This behavior suggests that this function is indeed a comb of δ-functions located at integers j. However, we still have to test the appropriate normalization.

For this purpose we now calculate the area underneath the peak at an integer j by introducing the integration variable $\bar{\xi} \equiv \xi \pi N$, that is,

$$\int_{j-\epsilon}^{j+\epsilon} d\xi \left|\delta_N^{(1/2)}(\xi)\right|^2 = \frac{1}{N} \int_{-\epsilon}^{\epsilon} d\xi \, \frac{\sin^2(\xi N \pi)}{\sin^2(\xi \pi)} = \frac{1}{\pi} \int_{-\epsilon N \pi}^{\epsilon N \pi} d\bar{\xi} \, \frac{\sin^2(\bar{\xi})}{N^2 \sin^2(\bar{\xi}/N)}$$

or

$$\lim_{N \to \infty} \int_{j-\epsilon}^{j+\epsilon} d\xi \left|\delta_N^{(1/2)}(\xi)\right|^2 = \frac{1}{\pi} \int_{-\infty}^{\infty} d\bar{\xi} \, \frac{\sin^2 \bar{\xi}}{\bar{\xi}^2} = 1.$$

Hence the area underneath a single peak is normalized to unity.

References

For the notion of the square of a delta function see
M.V. Berry, *Semiclassical theory of spectral rigidity*, Proc. R. Soc. Lond. A **400**, 229–251 (1985)

Q Further Reading

Textbooks on Classical Optics, Laser Physics and Nonlinear Optics

M. Born and **E. Wolf**, *Principles of Optics* (7th Ed.), Cambridge University Press, Cambridge, 1999
J.R. Klauder and **E.C.G. Sudarshan**, *Fundamentals of Quantum Optics*, Benjamin, New York, 1968
H. Haken, *Laser Theory*, Springer Verlag, Heidelberg and Berlin, 1984.
Appeared originally in: *Encyclopedia of Physics*, edited by S. Flügge, Springer, Heidelberg, 1970
H.M. Nussenzveig, *Introduction to Quantum Optics*, Gordon and Breach, London, 1973
W.H. Louisell, *Quantum Statistical Properties of Radiation*, Wiley, New York, 1973
M. Sargent III, **M.O. Scully** and **W.E. Lamb Jr.**, *Laser Theory*, Addison-Wesley, Reading, Mass., 1974
L. Allen and **J.H. Eberly**, *Optical Resonance and Two-Level Atoms*, Wiley, New York, 1975. This book was reprinted by Dover, 1987
R. Loudon, *The Quantum Theory of Light* (3rd Ed.), Oxford Science Publishing, Oxford, 2000
J. Perina, *Coherence of Light*, Reidel, Dordrecht, 1985
M. Schubert and **B. Wilhelmi**, *Nonlinear Optics and Quantum Electronics*, Wiley, New York, 1986
P.W. Milonni and **J.H. Eberly**, *Lasers*, Wiley, New York, 1988
M.D. Levenson and **S.S. Kano**, *Introduction to Nonlinear Laser Spectroscopy*, Academic Press, San Diego, 1988

Textbooks on Quantum Optics

M. Weissbluth, *Photon-Atom Interactions*, Academic Press, Boston, 1989
C. Cohen Tannoudji, **J. Dupont-Roc** and **G. Grynberg**, *Photons and Atoms, Introduction to Quantum Electrodynamics*, Wiley, New York, 1989
B.W. Shore, *The Theory of Coherent Atomic Excitations*, Wiley, New York, 1990
P. Meystre and **M. Sargent III**, *Elements of Quantum Optics*, Springer Verlag, Berlin, 1990
C.W. Gardiner, *Quantum Noise*, Springer Verlag, Berlin, 1991

H. Carmichael, *An Open Systems Approach to Quantum Optics*, Springer Verlag, Heidelberg and Berlin, 1993

D.F. Walls and **G.J. Milburn**, *Quantum Optics*, Springer Verlag, Heidelberg and Berlin, 1994

L. Mandel and **E. Wolf**, *Optical Coherence and Quantum Optics*, Cambridge U.P., New York, 1995

E.R. Pike and **S. Sarkar**, *The Quantum Theory of Radiation*, Clarendon, Oxford, 1995

W. Vogel and **D.G. Welsch**, *Lectures on Quantum Optics*, Akademie Verlag, Berlin, 1996

M.O. Scully and **M.S. Zubairy**, *Quantum Optics*, Cambridge U.P. New York, 1996

S.M. Barnett and **P.M. Radmore**, *Methods in Theoretical Quantum Optics*, Oxford University Press, Oxford, 1997

H.J. Carmichael, *Statistical Methods in Quantum Optics 1, Master Equations and Fokker-Planck-Equations*, Springer, New-York, 1999

Reprint Collections

Selected papers were reprinted in various collections, see for example

L. Mandel and **E. Wolf**, *Selected Papers on Coherence and Fluctuations of Light*, Dover, New York, 1970

P.L. Knight and **L. Allen**, *Concepts of Quantum Optics*, Pergamon, Oxford, 1983

P. Meystre and **D.F. Walls**, *Nonclassical Effects in Quantum Optics*, American Institute of Physics, New York, 1991

J. Perina, *Selected Papers on Photon Statistics and Optical Bistability*, SPIE, Bellingham, 1991

G.S. Agarwal, *Selected Papers on Fundamentals of Quantum Optics*, SPIE, Bellingham, 1995

G.S. Agarwal, *Selected Papers on Resonant and Collective Phenomena in Quantum Optics*, SPIE, Bellingham, 1995

Summer Schools and Conferences

A summary of the state of the art of quantum optics is presented in the proceedings of the following summer schools and conferences

C. DeWitt, **A. Blandin** and **C. Cohen-Tannoudji**, *Quantum Optics and Electronics*, Gordon and Breach, New York, 1965

R.J. Glauber, *Quantum Optics: Proceedings of the International School of Physics "Enrico Fermi"*, Academic, New York, 1969

P. Meystre and **M.O. Scully**, *Quantum Optics, Experimental Gravitation and Measurement Theory*, Plenum, New York, 1981

G. Grynberg and **R. Stora**, *New Trends in Atomic Physics: Les Houches 1982, Session XXX-VIII*, North-Holland, Amsterdam, 1984

J.Dalibard, J.-M. Raimond and **J. Zinn-Justin**, *Fundamental Systems in Quantum Optics: Les Houches Lecture Notes, Session LIII*, North-Holland, Amsterdam, 1992

P. Mandel, *Quantum Optics: Proceedings of the XXth Solvay Conference on Physics*, Physics Reports **219**, 78–348 (1991)

E. Arimondo, W.D. Phillips and **F. Strumia**, *Laser Manipulation of Atoms and Ions: Proceedings of the International School of Physics "Enrico Fermi"*, North-Holland, Amsterdam, 1992

Index

Aharonov-Anandan phase, *see* non-adiabatic phase
Airy function
    asymptotic expansion, 622
    complex argument, 249
    differential equation, 621
    integral representation, 83, 621
    points of stationary phase of Airy integrand, 622
    Stokes and anti-Stokes lines, 625
amplification, *see* damping
anti-photon, 282
area-of-overlap formalism, 192
    Bohr's correspondence rule, 200
    energy distribution, 205
        coherent state, 211
        squeezed state, 213
    evaluation of scalar product, 196
    Franck-Condon transitions, 200
    Kramers trajectory, 190
    phase states, 221
    Planck-Bohr-Sommerfeld bands, 190
    points of stationary phase, 195
    transition probability, 191
asymptotology, 29, 95, 99
atom interferometer, 25
atom optics, 23
atom-reservoir interaction, 532
atomic inversion, 233, 444, 465, 508, 509, 518, 530, 536
autocorrelation function, 233–236
averages using phase space functions, 330, 333–335

Baker-Hausdorff theorem, 297
Barnett-Pegg phase states, 226
beam splitter
    classical transformation, 350–353
    conditions on reflectivity and transmittivity, 352
    count statistics, 356, 357
    symmetric, 352, 353
    transformation of density operators, 355, 356
    transformation of mode operators, 353
    transformation of states, 353, 354, 374
Bernoulli numbers, 275, 284
Berry phase, 172
    adiabatic theorem, 172
    and WKB, 171
    dynamical phase, 175
    flux in Hilbert space, 175
    geometrical phase, 175
    harmonic oscillator, 167
    non-adiabatic, 185
    phase jump at turning point, 182
    topological phase, 175
Bessel function
    modified
        asymptotic expansion, 377, 656
        integral representation, 377
        Taylor expansion, 377, 656
    ordinary
        asymptotic expansion, 680, 681
        generating function, 679
        integral representation, 679
Bloch equations, 535, 536, 548
Bloch sphere, 537
Bogoliubov transformation, *see* squeezed state of electromagnetic field
Bohr's correspondence rule, *see* area-of-overlap formalism
Boltzmann distribution, 509, 517
Bopp operators, 91

boundary conditions for electromagnetic field, 261, 264, 284

carpet, 253
Casimir effect
  classical, 289
  experiment, 277, 278
  parallel plates, 273, 276
  two spheres, 277
cavity quantum electrodynamics, 16, 467
center-of-mass coordinates, 387
classical limit, 75, 80, 93, 158, 331, 332, 339
classical-quantum transition, 94
coarse graining, 510, 513, 520, 671
coherent state of electromagnetic field
  displaced vacuum, 297
  eigenvalue equation, 295
  electric field distribution, 299
  electric field expectation values, 305
  expansion into coherent states, 303
  nonorthogonality, 302
  number state expansion, 296
  over–completeness, 301
  photon statistics, 298
coherent state of mechanical oscillator, 108
  definition, 109
  displacing the ground state, 109
  energy distribution, 110
    asymptotic treatment, 111
    exact, 110
    overlap integral, 113
  time evolution, 113
    state vector, 113
    Wigner function, 118
commutator
  creation and annihilation operators, 271, 292, 332–334, 674
  electric field, 281, 286
  position and momentum operators, 38, 67, 88, 271, 366
  vector potential, 281, 286, 674
  connection between position and momentum wave function, 39, 40
Cornu spiral, see stationary phase method

Coulomb gauge, 259–261, 263, 265, 268, 386
curlicues, 248, 249, 254

damping and amplification of cavity mode
  approximate master equation, 514
  exact master equation, 519, 520
  model, 508, 509
  probability flow, 516
de Broglie optics, 22
decoherence, 546
deflection of atoms, 559
density operator, 44
  definition, 48
  field+atom: Jaynes-Cummings-Paul model, 542
  formal solutions, 61, 62
  interwoven commutators, 510
  quantum state as operator, 46
  thermal phase state, 53
  thermal state, 52
  time evolution, 61
  trace of operator, 49
  von Neumann equation, 61
diamagnetic Hamiltonian, 395
diffraction of atoms, 23
dipole approximation, 389
  for two particles, 410
  in general, 389
Dirac-Heisenberg gauge phase, 401, 410
displacement operator, see coherent state
dressed states, 552

effective potential, 552
eigenvalue problem for the creation operator, 314
eight-port interferometer
  kernels, 363, 364
  photon count statistics, 363, 364, 369
  Q-function measurement, 367, 369
  relation to EPR, 365, 367, 379
  setup, 361
  strong local oscillator, 365, 369
  transformation of states, 361, 363
electromagnetic energy in resonator, 267

electron optics, 22
energy eigenstate: harmonic oscillator
  arbitrary representation, 41
  contour integral, 101
  large-m limit, 103
  large-m limit, 101
  points of stationary phase, 102
  position representation, 42–44, 105
  position wave function, 100, 599
  simple phase space representation, 100
  time evolution, 108
  Wigner function, 106
  WKB wave function, 164

field commutation relations, 286
field quantization with running waves, 285
Floquet solution, 490, 492, 493
Floquet theorem, 476, 490
Fock state, see number state
Fock state preparation, 456
focussing of atoms, see quantum lens
forced harmonic oscillator, 314
Franck-Condon transitions, 200

gauge invariance of electrodynamics, 257, 384
gauge transformation
  global, 385
  local, 385
Gauss sums, 246, 248
Gaussian cigar, 124, 214, 216, 218, 219, 586, 587
geometrical phase, see Berry phase
giant oscillations, 133–135
Glauber-Sudarshan function, see P-distribution

harmonic oscillator, 40, 84
  asymptotic expansion, 102, 599
  energy quantization from phase space, 86
  energy quantization from wave function, 44
  energy wave function, 43, 597
  two-dimensional, 85
  Wigner function, 87, 105

Helmholtz equation
  general case, 261
  rectangular box, 262
Hermite polynomials
  definition, 64, 599
  generating function, 64, 101, 115, 300
  integral representation, 64, 101
Hill determinant, 477
homodyne detector
  difference count statistics, 358–360
  kernel, 359, 360, 655–657
  photon count statistics, 358
  quadrature operator measurement, 443, 444
  strong local oscillator limit, 359, 360, 656
Husimi-Kano function, see Q-function

interaction Hamiltonian
  atom in electromagnetic field, 386
  charged particle in electromagnetic field, 386
  electric field-dipole interaction, 393
  equivalence of $\vec{A} \cdot \vec{p}$ and $\vec{r} \cdot \vec{E}$, 396
  higher order corrections, 392
  in dipole approximation, 389
  Röntgen Hamiltonian, 392, 393
interaction picture
  general definition, 56
  Jaynes-Cummings-Paul model, 413, 426
  Paul trap, 495, 496, 499
interference in phase space, 13, 189
  photon statistics of squeezed states as a result, 13
  scalar product as interfering areas, 13
  Young's double-slit, 14

Jaynes-Cummings-Paul model, 402, 413
  collapse and revival, 233, 444
  definition, 402
  dispersive interaction, 451
  dynamics represented in state space, 418, 419, 421

effective Hamiltonian for large detuning, 430
  for Paul-trap, 494
  Laplace transform, 424
  multi-phonon Hamiltonian, 498
  Raman-coupled model, 433
  role of detuning, 420
  semiclassical version, 431
  state preparation, 451
  state vector for combined system, 422, 426
  time evolution operator using operator algebra, 414
  with time dependent coupling, 433
joint measurements on entangled systems, 435–437, 453, 560, 567

Kapitza-Dirac regime, 561–563
Kapitza-Dirac scattering, 568
Kennard's wave packets, 119
kinetic energy
  matrix elements, 63, 551

Lagrangian, 397
  beyond dipole approximation, 663
  time derivative, 397, 663
Laguerre polynomials
  definition, 631
  generating function, 346
  of Laplacian, 342
Lamb shift, 539, 548
Lamb-Dicke
  parameter, 495
  regime, 496
Laplace transformation, 424, 425
Laplacian
  eigenvalues, 284
  Laguerre polynomials of Laplacian, 342
laser cooling, 479, 503
London phase state, see phase states
Lorentz gauge, 258

Mathieu equation, 475, 476, 482, 503
Maxwell's equations, 256
measurement of
  atomic dipole, 441

  atomic population, 437
  field, 443
  Rabi oscillations
    cavity field, 450
    one-atom maser, 448, 449
    Paul trap, 451
micromaser, see one-atom maser
minimal coupling, 382, 383, 400
modes, 261
  definition, 261
  dimensionless mode function, 267
  orthonormality, 265
  volume, 267
Mollow spectrum, see resonance fluorescence
momentum eigenstate
  completeness relation, 37, 40
  eigenvalue equation, 36
  Wigner function, 138, 139
momentum transfer, 559
  averaged measurement, 561
  joint measurement, 560
  quantization, 563
  smooth part, 566, 574
Moyal functions
  definition, 76, 77, 616
  for harmonic oscillator, 93
  phase space equations, 77, 78, 620
Mulliken principle, 202
multiports, 379

negative probability, 97
neutron optics, 22
nonlinear quantum mechanics, 473, 505

one-atom maser
  average photon number, 19, 524
  collapse and revivals, 448
  experimental set-up, 17
  linewidth, 532
  master equation, 523
  phase diffusion, 529
  photon statistics, 524
  resonance line, 18
  trapping states, 527
  variance of photon statistics, 526

operator ordering
    antinormal, 332–336, 344
    normal, 332, 334, 335, 343
    symmetric, 90
    Weyl-Wigner, 89, 90
operators for
    electric field, 280
    magnetic field, 281
    vector potential, 278
optical parametric oscillator, see squeezed states
optical Stern-Gerlach effect, 24

P-distribution
    averages using P-distribution, 335, 336
    definition, 336, 337
    from Q-function, 339, 340
    number state, 342, 650, 651
    squeezed state, 343, 651, 653
    thermal state, 341, 649, 650
    time evolution, 544
parallel transport, 176
paramagnetic Hamiltonian, 395
particle in a box, 250
Paul trap
    analogy to cavity QED, 26, 504
    end-cap trap, 4, 474, 479
    Laplace equation, 474
    linear Paul trap, 473, 503
    micromotion, 480, 500
    motion in phase space, 486–488
    no trapping in 3-D by static electric fields, 474
    quantum treatment, 451
    stability diagram, 478
Pauli spin matrices, 404, 405
Penning trap, 475
phase space functions, 321, 330, 333–335
phase states, 221, 370
    action and phase variables, 223
    from interference in phase space, 221
    London phase operator, 529
    measured phase operators, 370

phase amplitude of energy eigenstate, 224
    phase distribution, 227
    position wave function, 229
    time evolution, 226
    two-mode phase operators, 374
    Wigner function, 229
photon
    origin of name, 255, 282
    wave function, 289
photon number state, 281, 292
    average electric field, 292
    electromagnetic field eigenstates, 293
    multi-mode, 282
    single mode, 292
Poisson summation formula, 228, 242, 465
poissonian distribution
    asymptotic limit, 111, 113, 466, 633
    sub-poissonian, 6, 18, 127, 310, 312, 317
    super-poissonian, 310, 312
polarization vectors, 263, 285
position eigenstate
    completeness relation, 37
    eigenvalue equation, 36
    Wigner function, 138
propagator of free particle, 63

Q-function
    averages using Q-function, 334, 335
    coherent state, 324, 325
    damped harmonic oscillator, 542
    definition, 321, 324
    from P-distribution, 338, 339
    number state, 325, 326
    squeezed state, 326, 327, 343
    thermal phase state, 329, 330, 343
    thermal state, 328
    vacuum state, 326
quality factor of cavity, 15, 21
quantization conditions
    Bohr-Sommerfeld-Kramers, 161, 163, 164, 190, 193
    EBK, 169
    from phase space, 86, 630

from wave function, 43, 44, 597, 599
WKB, 159, 163
quantization of the radiation field, 269
quantum computer, 27, 33
quantum electronics, 2
quantum information processing, 26
quantum jump method, 544
quantum jumps, 473
quantum lens, 587
    deflection angle, 589
    focal length, 589, 592
    focal size, 594
    harmonic approximation, 583
    model, 579
    motion in phase space, 582
    photon and momentum statistics, 590
    Raman-Nath approximation, 595
quantum Liouville equation, 75, 582, 615, 620
quantum optics, 1
quantum prism, 596
quantum state reconstruction, 143, 150

Rabi equations, 422
    generalized, 550
    Jaynes-Cummings-Paul model, 423
    semiclassical, 431
race track trap, 503
Radon transformation, 144
Raman-Nath approximation, 557, 558, 574
Ramsey method, 452, 468
reduced mass, 388
relative coordinates, 387
resonance fluorescence
    anti-bunching, 5–7
    elastic peak, 2–4
    heterodyne detection, 3, 6
    incoherent contribution, 3
    Mollow spectrum, 3, 4
    second-order correlation function, 5–7
    squeezing, 8
    three-peak spectrum, 3, 4

revivals
    electronic wave packets, 234, 235, 251
    fractional, 234, 236, 239, 240, 246, 247, 249, 252, 253, 446, 447
    Jaynes-Cummings-Paul model, 444–447, 465, 468
    molecular wave packets, 234, 236, 252
    one-atom maser, 448, 449
rotated quadrature states
    definition of quadrature operator, 136, 357
    wave function, 140, 142
    Wigner function, 142
rotating wave approximation, 406, 407

s-parameterized phase space distributions, 345
Schrödinger cat state
    amplitude cat, 316, 317
    damping, 543
    definition of state, 307
    experimental realization, 452, 453
    original cat paradox, 306
    phase cat, 317
    photon statistics, 310–313
    squeezing, 309
    Wigner function, 307, 308
Schrödinger equation, 36
    in phase space, 91
    Jaynes-Cummings-Paul model, 413
    one-dimensional harmonic oscillator, 43, 597
    two-dimensional harmonic oscillator, 85
second harmonic generation, *see* squeezed states
semiclassical limit, 80, 153, 192, 208
size of quantum state, *see* Wigner function, support
spontaneous emission, 432, 516, 532, 541, 548
square root of δ-function, 683

squeezed state of electromagnetic field
  Bogoliubov transformation, 315
  detection of squeezing, 11
  experiment, 10
  optical parametric oscillator, 9
  oscillatory photon statistics, 12
  second harmonic generation, 9
squeezed state of mechanical oscillator, 119
  aligned, 125
  definition, 121
  distributions of fluctuations, 122
  energy distribution, 125
    asymptotic treatment, 128
    exact, 125
    overlap integral, 129
  generalized, 124
  phase distribution, 229
  rotated squeezed state, 127, 132, 147
  squeeze operator, 122
  squeezed vacuum, 132, 146
  time evolution, 135, 136
  Wigner function, 135, 136
squeezed vacuum
  energy distribution
    asymptotic treatment, 13, 14, 132
    exact treatment, 146
  experiment, 13
  oscillatory photon statistics, 14, 132
  time evolution, 120
Stark effect, 440, 441
state
  engineering, 454
  measurement, 451
  of subsystem after measurement, 439
  preparation, 451
stationary phase method, 192, 635
  Cornu spiral, 196, 202, 639
  in many dimensions, 637
  in one dimension, 635
Stern-Gerlach experiment, 23
Stern-Gerlach regime, 561, 579
Stirling formula, 377, 603, 633
sub-fluctuant, 123
super-fluctuant, 123

swallow tail integral, 83

Talbot effect, 250, 253
time evolution operator, 58, 605
  as infinite product, 58
  general Hamiltonians, 60
  Jaynes-Cummings-Paul model, 416
  sum of integrals, 60
  time ordered exponential, 60
time ordering operator, 60, 606
  definition, 60, 606
  evolution operator, 60
  product of two integrals, 607
  product of n integrals, 608
time scales in cavity QED, 15
tomographic cut of Wigner function, see Wigner function
topological phase, see Berry phase
total mass, 387
transverse delta function, 286
trapping states, see one-atom maser
trigonometry
  classical, 370–372
  quantum, 372
truncated Hilbert space, 226

vacuum electric field, 280, 293
Volterra-Schlesinger product integral, 59, 65
von Neumann equation
  derivation, 61
  translation into phase space, 74

wave packet
  electronic, 234, 235, 251
  Kennard, 119
  molecular, 234, 236, 252
  photonic, 233
Wehrl entropy, 613
Weisskopf-Wigner decay, 540
Weyl-Wigner ordering, see operator ordering
whispering gallery modes, 21
width of function
  decay width, 612
  Süßmann measure, 611–613
  variance, 611

Wigner function
- asymptotology, 95
- coherent state, 118, 119, 212
- determined from phase space, 76
- integral representation, 68, 69
- marginals, 69, 70
- negative values, 73
- number state, 79, 85, 87, 105–108, 629, 631
- of tunneling, 93
- phase space equations for energy eigenstate, 78
- photon statistics, 228
- product rule, 71
- represented as displacement plus parity operator, 347
- scalar products as overlap of Wigner functions, see Wigner function, product rule
- Schrödinger cat state, 308, 309, 316
- squeezed state, 123, 124, 135, 136, 213
- thermal state, 92
- time evolution, see quantum Liouville equation
- tomographic reconstruction, 120, 143, 144, 146, 151
- upper bound, 73

WKB method
- applied to harmonic oscillator, 167
- applied to Morse potential, 166
- Berry phase, see Berry phase
- Bohr-Sommerfeld-Kramers quantization condition, 161
- classical phase space trajectory, 153
- classical probability, 154
- energy quantization, 159
- Langer transformation, 166, 168
- matching of solutions, 160
- particle ansatz, 156
- primitive wave function, 163
- uniform asymptotic expansion, 164
- validity, 158
- wave ansatz, 156